ÆDES CHRISTI
in Academia Oxoniensi

# PRINCIPLES OF POLYMER PROCESSING

SPE MONOGRAPHS

*Injection Molding*

**Irvin I. Rubin**

*Nylon Plastics*

**Melvin I. Kohan**

*Introduction to Polymer Science and Technology: An SPE Textbook*

**Herman S. Kaufman and Joseph J. Falcetta**

*Principles of Polymer Processing*

**Zehev Tadmor and Costas G. Gogos**

# PRINCIPLES OF POLYMER PROCESSING

**ZEHEV TADMOR**

*Professor of Chemical Engineering*
*Department of Chemical Engineering*
*Technion-Israel Institute of Technology*
*Haifa, Israel*

**COSTAS G. GOGOS**

*Professor of Chemical Engineering*
*Department of Chemistry and Chemical Engineering*
*Stevens Institute of Technology*
*Hoboken, New Jersey*

A WILEY-INTERSCIENCE PUBLICATION

JOHN WILEY & SONS

NEW YORK · BRISBANE · CHICHESTER · TORONTO

**Library of Congress Cataloging in Publication Data:**

Tadmor, Zehev, 1937–
Principles of polymer processing.

"A Wiley-Interscience publication."
Includes index.
1. Polymers and polymerization. I. Gogos, Costas G., joint author. II. Title.

TP1087. T32 668 78–17859
ISBN 0-471-84320-2

Printed in the United States of America

10 9 8 7 6 5 4 3 2

# Series Preface

The Society of Plastics Engineers is dedicated to the promotion of scientific and engineering knowledge of plastics and to the initiation and continuation of educational activities for the plastics industry.

An example of this dedication is the sponsorship of this and other technical books about plastics. These books are commissioned, directed, and reviewed by the Society's Technical Volumes Committee. Members of this committee are selected for their outstanding technical competence; among them are prominent authors, educators, and scientists in the plastics field.

In addition, the Society publishes *Plastics Engineering*, *Polymer Engineering and Science* (*PE & S*), proceedings of its Annual, National and Regional Technical Conferences (*ANTEC, NATEC, RETEC*) and other selected publications. Additional information can be obtained by writing to the Society of Plastics Engineers, Inc., 656 West Putnam Avenue, Greenwich, Connecticut 06830.

William Frizelle,

*Chairman, Technical Volumes Committee*
*Society of Plastics Engineers*
*St. Louis, Missouri*

# Preface

This book deals with polymer processing, which is the manufacturing activity of converting raw polymeric materials into finished products of desirable shape and properties.

Our goal is to define and formulate a coherent, comprehensive, and functionally useful engineering analysis of polymer processing, one that examines the field in an integral, not a fragmented fashion. Traditionally, polymer processing has been analyzed in terms of specific processing methods such as extrusion, injection molding, calendering, and so on. Our approach is to claim that what is happening to the polymer in a certain type of machine is *not unique*: polymers go through similar experiences in other processing machines, and these experiences can be described by a set of *elementary processing steps* that prepare the polymer for any of the *shaping* methods available to these materials. On the other hand, we emphasize the *unique* features of particular polymer processing methods or machines, which consist of the particular elementary step and shaping *mechanisms* and *geometrical solutions* utilized.

Because with the approach just described we attempt to answer questions not only of "how" a particular machine works but also "why" a particular design solution is the "best" among those conceptually available, we hope that besides being useful for students and practicing polymer engineers and scientists, this book can also serve as a tool in the process of creative design.

The introductory chapter highlights the technological aspects of the important polymer processing methods as well as the essential features of our analysis of the subject. Parts I and II deal with the fundamentals of polymer science and engineering that are necessary for the engineering analysis of polymer processing. Special emphasis is given to the "structuring" effects of processing on polymer morphology and properties, which constitute the "meeting ground" between polymer engineering and polymer science. In all the chapters of these two parts, the presentation is utilitarian; that is, it is limited to what is necessary to understand the material that follows.

Part III deals with the elementary processing steps. These "steps" taken together make up the total thermomechanical experience that a polymer may have in any polymer processing machine prior to shaping. Examining these steps separately, free from any particular processing method, enables us to discuss and understand the range of the mechanisms and geometries (design solutions) that are available. Part III concludes with a chapter on the modeling of the single screw extruder, demonstrating the *analysis* of a complete processor in terms of the elementary steps. We also deal with a new polymer processing device to demonstrate that *synthesis* (invention) is also facilitated by the elementary step approach.

We conclude the text with the discussion of the classes of shaping methods available to polymers. Again, each of these shaping methods is essentially treated independently of any particular processing method. In addition to classifying the shaping methods in a logical fashion, we discuss the "structuring" effects of processing that arise because the macromolecular orientation occurring during shaping is fixed by rapid solidification.

The last chapter, a guide to the reader for the analysis of any of the major processing methods in terms of the elementary steps, is necessary because of the unconventional approach we adopt in this book.

For engineering and polymer science students, the book should be useful as a text in either one-semester or two-semester courses in polymer processing. The selection and sequence of material would of course be very much up to the instructor, but the following syllabi are suggested: *For a one-semester course*: Chapter 1; Sections 5.2, 4, and 5; Chapter 6; Sections 7.1, 2, 7, 9, and 10; Sections 9.1, 2, 3, 7, and 8; Chapter 10; Section 12.1; Sections 13.1, 2, 4, and 5; Section 14.1; Section 15.2; and Chapter 17—students should be asked to review Chapters 2, 3, and 4, and for polymer science students the course content would need to be modified by expanding the discussion on transport phenomena, solving the transport methodology problems, and deleting Sections 7.7, 9, and 10. *For a two-semester course: in the first semester,* Chapters 1, 5, and 6; Sections 7.1, 2, and 7 to 13; Sections 8.1 to 4, and 7 to 13; Chapters 9 and 10; and Sections 11.1 to 4, 6, 8, and 10—students should be asked to review Chapters 2, 3, and 4; and *in the second semester,* Chapters 12 and 13; Section 14.1, and Chapters 15, 16, and 17.

The problems included at the end of Chapters 5 to 16 provide exercises for the material discussed in the text and demonstrate the applicability of the concepts presented in solving problems not discussed in the book.

The symbols used follow the recent recommendations of the Society of Rheology; SI units are used. We follow the stress tensor convention used by Bird, et al.,* namely, $\boldsymbol{\pi} = P\boldsymbol{\delta} + \boldsymbol{\tau}$, where $\boldsymbol{\pi}$ is the total stress tensor, $P$ is the pressure, and $\boldsymbol{\tau}$ is that part of the stress tensor that vanishes when no flow occurs; both $P$ and $\tau_{ii}$ are positive under compression.

We acknowledge with pleasure the colleagues who helped us in our efforts. Foremost, we thank Professor J. L. White of the University of Tennessee, who reviewed the entire manuscript and provided invaluable help and advice on both the content and the structure of the book. We further acknowledge the constructive discussions and suggestions offered by Professors R. B. Bird and A. S. Lodge (University of Wisconsin), J. Vlachopoulos (McMaster University), A. Rudin (University of Waterloo), W. W. Graessley (Northwestern University), C. W. Macosko (University of Minnesota), R. Shinnar (CUNY), R. D. Andrews and J. A. Biesenberger (Stevens Institute), W. Resnick, A. Nir, A. Ram, and M. Narkis (Technion), Mr. S. J. Jakopin (Werner-Pfleiderer Co.) and Mr. W. L. Krueger (3M Co.). Special thanks go to Dr. P. Hold (Farrel Co.), for the numerous constructive discussions and the many valuable comments and suggestions. We also thank Mr. W. Rahim (Stevens), who measured the rheological and thermophysical properties that appear in Appendix A, and Dr. K. F. Wissbrun (Celanese Co.), who helped us with

* R. B. Bird, W. E. Stewart, and E. N. Lightfoot, *Transport Phenomena*, Wiley, New York, 1960; and R. B. Bird, R. C. Armstrong, and O. Hassager, *Dynamics of Polymeric Liquids*, Wiley, New York, 1977.

the rheological data and measured $\eta_0$. Our graduate students of the Technion and Stevens Chemical Engineering Departments deserve special mention because their response and comments affected the form of the book in many ways.

We express our thanks to Ms. D. Higgins and Ms. L. Sasso (Stevens) and Ms. N. Jacobs (Technion) for typing and retyping the lengthy manuscript, as well as to Ms. R. Prizgintas who prepared many of the figures. We also thank Brenda B. Griffing for her thorough editing of the manuscript, which contributed greatly to the final quality of the book.

This book would not have been possible without the help and support of Professor J. A. Biesenberger and Provost L. Z. Pollara (Stevens) and Professors W. Resnick, S. Sideman, and A. Ram (Technion).

Finally, we thank our families, whose understanding, support, and patience helped us throughout this work.

ZEHEV TADMOR
COSTAS G. GOGOS

*Haifa, Israel*
*Hoboken, New Jersey*
*March 1978*

# Contents

## APPENDICES

# 1

# Introduction to Polymer Processing



## 1.1 Survey of Polymer Processing Methods and Machinery

Polymeric materials possess a surprisingly useful combination of chemical, physical, and electrical properties, rendering them probably the most versatile raw materials available to mankind. Moreover, the "plastic" deformable state achieved by thermoplastic polymers at elevated temperatures, and possessed by thermosetting polymers before being chemically "set," allows them to be shaped into a myriad of finished products, some of them of great geometrical complexity. The shaping operations used are relatively quick and easy, and are superbly fitted for mass production without necessarily sacrificing quality and aesthetics. Without a doubt, the usefulness of polymers and plastics, and their profound impact on modern technology and way of life, are due just as much to the versatility and flexibility of the available shaping methods as to their inherent properties.

*Polymer processing is defined as the "engineering specialty concerned with operations carried out on polymeric materials or systems to increase their utility"* (1). Primarily it deals with the conversion of polymeric raw materials into finished products. This chapter briefly surveys the main processing methods practiced today and introduces the reader to what we believe to be a logical framework for analyzing them.

### *Historical Notes*

Polymer processing machinery and methods used today in the plastics industry evolved from the rubber industry. Firsthand accounts of the early developments of the rubber and plastics industries can be found in the writings of Hancock (2), Goodyear (3), Hyatt (4), and Du Cros (5), all of whom made significant contributions to the field. Interesting historical reviews of various aspects of these early days appear in the literature (6–12), including a recent careful and detailed account by White (13) of the historical development of elastomer processing.

Our interest here is primarily the evolution of polymer processing machinery. The earliest documented example of such a machine is a rubber "masticator" consisting of a toothed rotor turned by a winch inside a toothed cylindrical cavity called the "pickle." It was developed in 1820, by Thomas Hancock, in England, to reclaim rubber scraps left from the manual manufacture of elastic straps. A few years later, in 1835, Edwin Chaffee in the United States developed the two roll, steam heated mill for mixing additives into rubber, as well as the calender, which consisted of a series of heated rolls for the continuous coating of cloth and leather by rubber. Chaffee's inventions were outstanding contributions to natural rubber processing and are still used in the elastomer and thermoplastics industries.

The first ram extruders were apparently developed by Henry Bewley and Richard Brooman in 1845 in England (14). A combination of their machines was used for wire covering. The first submarine cable, laid between Dover and Calais in 1851, was manufactured by such an extruder.

But ram extrusion was an intermittent operation. The great need for a continuous extrusion, in particular in the wire and cable field, brought about the single most important development in the processing field—the screw extruder. Circumstantial evidence indicates that A. G. DeWolfe, in the United States, may have developed the first screw extruder in the early 1860s (15). The Phoenix Gummiwerke has published a drawing of a screw dated 1873 (16), and William Kiel and John Prior, in the United States, both claimed the development of such a machine in 1876 (13). But the birth of the extruder, which plays such a dominant role in polymer processing, is linked to the patent of Mathew Gray, in England, in 1879 (17); this patent is the first clear exposition of this type of machine. The Gray machine also included a pair of heated feeding rolls. Independent of Gray, Francis Shaw, in England, developed a screw extruder in 1879, as did John Royle in the United States in 1880.

Another solution to continuous extrusion was proposed by Willoughby Smith in 1887 in England. The raw material was fed from a hopper through hot feed rolls into a chamber that was fitted with a gear pump (14).

The injection molding process was also developed during the same period by John C. Smith and Jesse A. Locke, who in 1870 invented a machine for "Manufacturing of Castings Under Pressure." Although intended for the metal die casting industry, it formed the basis of the plunger injection molding machine. Two years later John Wesley Hyatt obtained a patent on a piston driven injection molding machine for polymers (12). Hyatt was a pioneering figure in the field of polymers: he invented Celluloid, he contributed to many innovations in the processing machinery, which helped in the quick use of the phenol-formaldehyde phenolic thermosetting polymer invented by Dr. Leo Baekeland in 1906, and he developed the first blow molding process. In the ensuing years, the reciprocating plunger injection molding machine underwent many developments, the most important of which is the incorporation of the "torpedo," which contributed to the efficient heat transfer to the polymer until it was replaced by the superior reciprocating screw injection molding machine developed by William H. Willert in 1952 in the United States.

The need for mixing into rubber large amounts of fine carbon black particles and poisonous organic accelerators brought about the development of internal mixers of the type utilized by Hancock in his "pickle." Notably, Frenley H. Banbury developed the still widely used "Banbury" mixer in 1916 (18). Finally, various mixings needs appear to provide also the main motivation for the parallel development of the various multiple screw extruders, used for years in other industries. Detailed documented history of these machines, including the intermeshing twin screw extruders and the various continuous "mixers," is given by Herrmann (19).

These were the most important machines available for polymer processing at the start of the explosive development of polymers and the plastics industry after World War II. In the years since, many improvements and new developments have led to today's diverse arsenal of polymer processing machines and methods, some of which we briefly describe in the remainder of this section.

### *Screw Extrusion*

The word "extrusion" is derived from the Latin words *ex* and *trudere* meaning, respectively, "out" and "to thrust" or "to push." These words describe literally the process of extrusion, where a polymer melt is pushed across a metal die that continuously *shapes* the melt into the desired form. Polymer products that are "infinite" in one direction are manufactured by the extrusion process. These include wires, cables, rods, tubes, pipes, and a variety of profiles (Fig. 1.1), which include filaments, films, and sheets that are products of great volume and importance; with ingenious engineering, even nets and corrugated tubes can be continuously extruded. With a few exceptions, all polymers can be extruded, and

**Fig. 1.1** Samples of die formed products manufactured by the extrusion process. [Reprinted with permission from R. F. Westover, *Encyclopaedia of Polymer Science and Technology*, **8**, 533 (1970) John Wiley & Sons.]

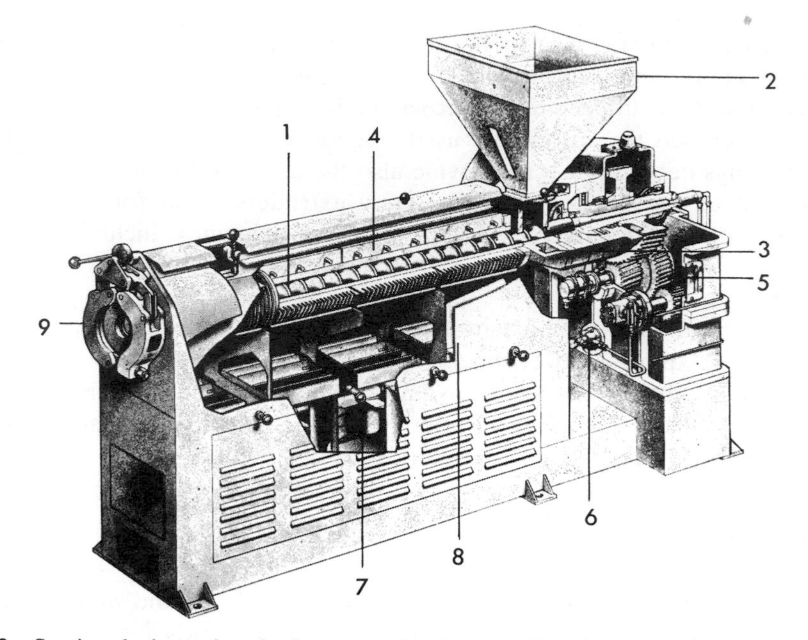


**Fig. 1.2** Sectional view of a single screw plastics extruder. 1, screw; 2, hopper; 3, feed section; 4, barrel heaters; 5, gear box; 6, lubrication system; 7, air blowers to control barrel heating and cooling temperatures; 8, double walled hood for balanced air flow; 9, die clamp assembly. (Courtesy of Francis Shaw & Co., Ltd., Manchester, England.)

many may pass a screw extruder not once but twice during their journey from the reactor to the finished product—first a pelletizing extruder after the reactor, then the shaping extruder.

The heart of the screw extrusion process is an Archimedean screw rotating in a heated barrel (Fig. 1.2). The raw polymer in the form of particulate solids (pellets, powder, etc.) is gravitationally fed onto the screw through a hopper. The solids are conveyed forward, plasticated, homogenized, and pressurized along the screw. Thus a uniformly molten polymer is pumped or pushed across the die attached to the extruder "head." The screw is rotated by electric motors through a gear reducer. The barrel is heated electrically or by a fluid heat exchanger system. Thermocouples placed in the metal barrel wall record and help to control barrel temperature settings. Sections of barrel, however, are often cooled to remove the excessive heat generated by viscous dissipation. The main operating variables are the frequency of screw rotation and the barrel temperature profile. The main design variables are screw diameter and length—usually expressed as length-to-diameter ratio $L/D$. These determine to a large extent extruder throughput, polymer residence time in the extruder, and available barrel surface for heat transfer. Details of screw design, such as number of parallel flights, helix angle, channel depth profile (radial distance between the tip of the flight and the root of the screw), and various geometrical modifications contribute to the quality of plastication and mixing. Some screw designs are illustrated in Fig. 1.3. The screw design is the most important single factor in determining the success of an extrusion operation; therefore a great deal of attention is given to it. Most screws are designed with a lead (or pitch) equal to the diameter (referred to as square pitched screws), which

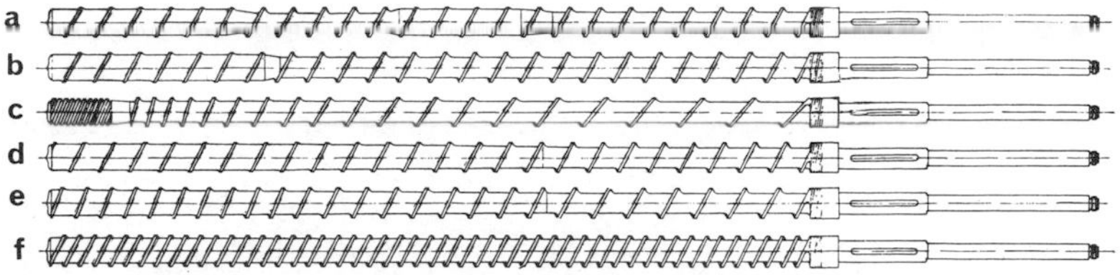

a b c d e f

**Fig. 1.3** Six screw designs. (Courtesy of L. Bandera Costruzioni Mechaniche s.p.a., Busto Arsizio, Italy.)

results in a helix angle of 17.6°. They have a deep feed section, to accommodate and properly convey the low density particulate solid feed, and a final shallow channel for achieving thorough mixing and pressure buildup, reducing sensitivity to the process parameters, and providing a high flow resistance section for "washing out" flow rate fluctuations introduced earlier. The two sections are connected by an intermediate transition section of constant taper.

Extruders are manufactured in a very broad range of sizes, starting from diameters of 2 cm, used for laboratory purposes, up to diameters of 50 cm and above, delivering polymers at a rate of 10 tons/hr. Typical length-to-diameter ratios are 24:1, which gradually evolved from the short (8–10 $L/D$) rubber extruders. The trend is still toward larger $L/D$ ratios, in particular for two stage

**Fig. 1.4** A Davis Standard, Thermatic III single screw extruder. (Courtesy of Davis Standard Co., Pawcatuck, Conn.)

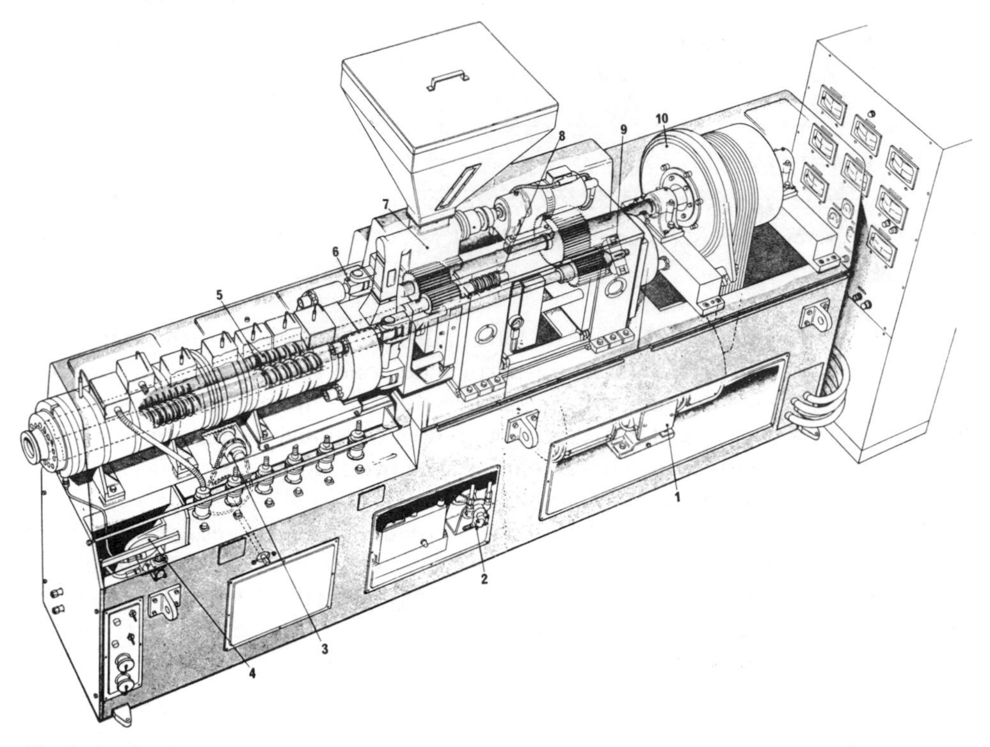


**Fig. 1.5** GKN Windsor TS.250 twin screw extruder: 1, main motor; 2, oil circulation pump; 3, manually operated withdrawal mechanism for trolley mounted barrel; 4, independent oil cooling for front barrel zone; 5, twin screws with interchangeable sections; 6, quick release barrel clamp; 7, controlled material feed; 8, multiple element screw thrust bearing; 9, large diameter standard screw thrust bearing; 10, epicyclic primary gearbox. (Courtesy of GKN Windsor, Surrey, England.)

vented extruders and for special purpose extruders, which may reach an $L/D$ ratio of 36–40. Figure 1.4 shows a typical extruder.

In addition to single screw extruders, there are twin and multiscrew extruders performing essentially the same functions. Among these, the intermeshing twin screw extruders are the most important ones. They partly compete with single screw extruders and partly expand screw extrusion potential. For example, polyvinylchloride (PVC) powder, regrind scrap, and other difficult-to-feed materials are easier to process with twin extruders than with single screw extruders. There are notable differences in the pumping, melting, and mixing mechanisms of single screw and twin screw extruders, as well as among the various types of twin screw extruders. A typical twin screw extruder appears in Fig. 1.5.

The finished product of the extrusion process is determined by the die, which shapes the product, and by the "sizing" equipment, the cooling system, and the cutting equipment, which in turn set the final size and surface quality of the product. The postextrusion (downstream) devices are unique to each process and product (16, 20–23).

Wire coating and cable jacketing were among the first applications of the extrusion process. A schematic view of a wire coating line appears in Fig. 1.6. The polymer is coated over the conducting wire, providing primary electric insulation.

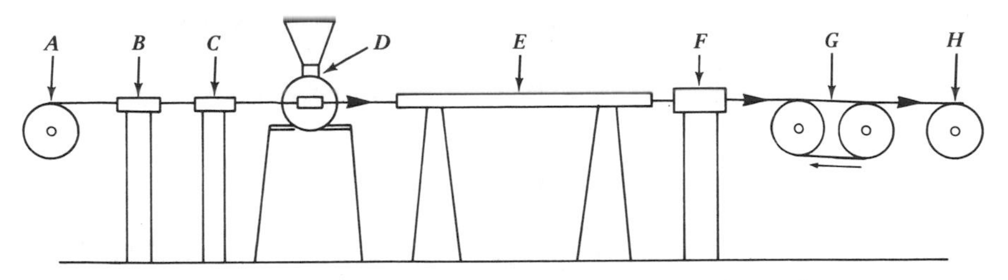


**Fig. 1.6** Schematic view of a wire coating line: *A*, unwind of wire (payoff); *B*, straightener; *C*, preheater; *D*, extruder; *E*, water trough; *F*, tester; *G*, capstan (puller); *H*, windup of coated wire. (Reprinted with permission from P. N. Richardson, *Introduction to Extrusion*, Society of Plastics Engineers, Inc., Greenwich, Conn., 1974.)

The metal wire is fed to the die from a "payoff' through a straightener and a preheater to a cross-head die (Fig. 1.7). The coated wire leaves the die and enters a cooling water trough. It passes through an electric tester (for insulation quality) through a capstan (puller) onto the windup equipment. The speed of a wire coating line depends on the polymer and wire gauge, but for primary insulation with low density polyethylene (LDPE) or PVC, line speeds of 1000–1500 m/min are common. Cable jacketing lines are similar in principle but operate at much lower speeds.

Another application of great industrial importance is film blowing. The vast majority of polymer films are manufactured by this process. Film is extruded into a thin walled tube vertically, either in the upward or downward direction. The film is cooled by an air ring over the die as in Fig. 1.8. Through an air duct inside the die, air is also blown into the tube and is trapped inside because the pair of pull rolls at the upper end of the line, following the collapsing frame, seal the tube outlet. Thus a big bubble is formed. The blow-up ratio, coupled with the rate of haul-off, permits both the control of film thickness and the degree of uni- and biaxial orientation introduced into the film by this process. Hence the blow-up ratio (i.e.,

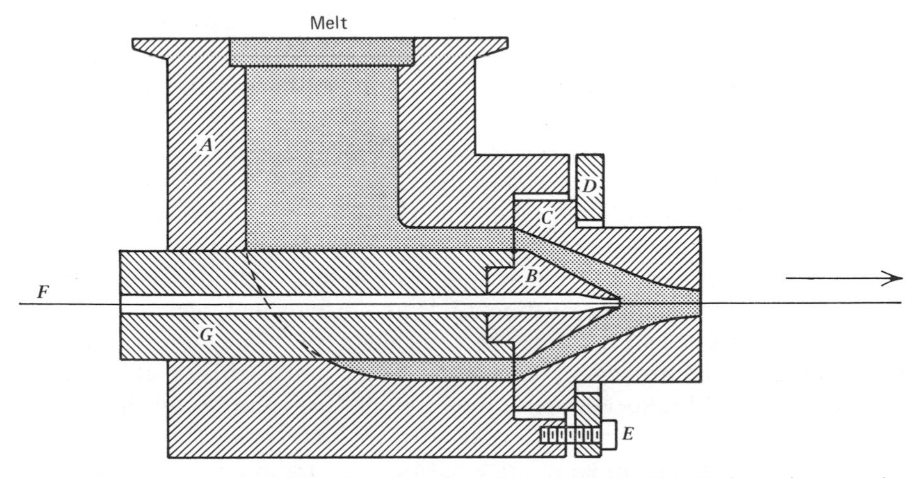


**Fig. 1.7** Schematic cross section of a "cross-head" wire coating die: *A*, die body, cross-head; *B*, guider tip; *C*, die; *D*, die retaining ring; *E*, die retaining bolt; *F*, wire; G, core tube. (Reprinted with permission from P. N. Richardson, *Introduction to Extrusion*, Society of Plastics Engineers, Inc., Greenwich, Conn., 1974.)

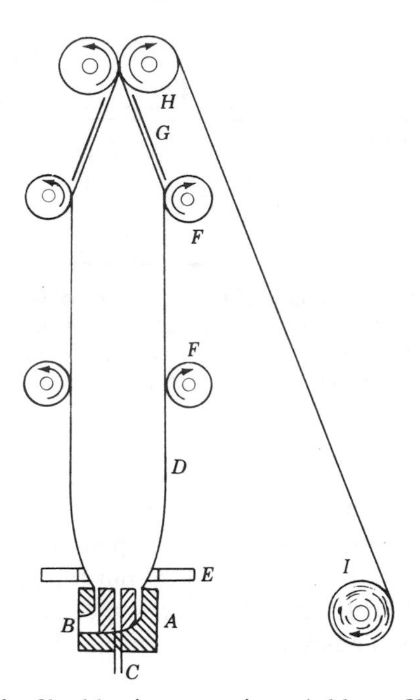


**Fig. 1.8** Schematic view of a film blowing operation: *A*, blown film die; *B*, die inlet; *C*, air hole and valve; *D*, plastic tube (bubble); *E*, air ring for cooling; *F*, guide rolls; *G*, collapsing frame; *H*, pull rolls; *I*, windup roll. (Reprinted with permission from P. N. Richardson, *Introduction to Extrusion*, Society of Plastics Engineers, Inc., Greenwich, Conn., 1974.)

the ratio of the diameter of bubble to that of the die) is of great importance. Common blow-up ratios are in the range of 1.5–4. They determine the orientation in the transverse direction. The haul-off speed determines the drawdown in the machine direction and, consequently, the orientation in that direction. Common die openings are about 0.05 cm, and film thicknesses range from 0.0005 to 0.025 cm. Die diameters range from less than 10 cm to over 120 cm. The stringent quality and economic requirements on film thickness have led to feedback gauge control systems that continuously monitor the thickness by backscatter $\beta$ gauges and control it by varying the pulling rate.

Films can also be manufactured by "roll-casting." The melt is extruded into a thin web, straight onto a chilled, polished, rotating roll (Fig. 1.9). This kind of film is called "chill-roll" film, cast film, or flat film. The product is glossy and generally of high optical quality, but usually is more expensive than blown film.

Sheets are produced in a similar way. The melt is extruded generally via a "coat hanger" type of die (see Section 13.4) into a flat sheet, which is fed onto a stack of three highly polished chrome coated chilled rolls. The temperature of each roll is individually controlled by circulating water and is driven at uniform but controllable speed. Line speeds are of the order of 20 m/min for thin sheets and as low as 2 m/min for heavy sheets. The gap between the first pair of rolls is somewhat smaller than die lip opening, so that a thickness reduction of about 10% is obtained.

Pipe, tube, and profile extrusion is another extrusion operation of outstanding practical importance. We usually distinguish between pipes and tubes by size—

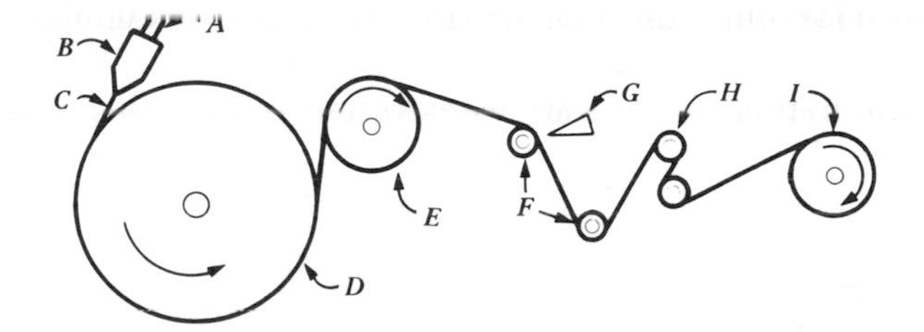


**Fig. 1.9** Schematic view of an extrusion roll-cast film operation: *A*, die inlet; *B*, cast film die; C, air gap with molten web; *D*, casting roll; *E*, stripping roll; *F*, idler roll; *G*, edge trim slitter; H, pull rolls; *I*, windup roll. (Reprinted with permission from P. N. Richardson, *Introduction to Extrusion*, Society of Plastics Engineers, Inc., Greenwich, Conn., 1974.)

below 1.25 cm (0.5 in.) diameter is called a tube; more than this, a pipe. Die design for all these products, in particular profiles, is not simple and is mostly empirical, as discussed in Chapter 13. The most widely used technique for sizing pipes and tubing features vacuum calibrators. The molten tube is fed into a vacuum tank that contains water. The atmospheric pressure inside the tube causes it to expand against a series of sizing plates immersed in the tank across which the tube is pulled. The solidification process soon makes the tube strong enough to resist the pressure difference. The tubes and pipes are pulled downstream in the sizing tank by "caterpillars." Sizing of profiles is more difficult because of their tendency to warp and twist while cooling. Adjustable metal "fingers," which may be cooled, are used to steady and hold the moving soft profiles. Air jets are used to cool and, occasionally, to heat them. Sometimes profiles run through a chilled sizing unit, where they are drawn to the sizing walls by vacuum applied through the perforated walls.

By using more than one extruder and the appropriate die design, it is possible to coextrude multicomponent, layered products. Multilayered films for special applications (in particular, in food packaging) are very common. Bicomponent fibers are coextruded as well as sheets, providing them with special properties.

Thermoplastic foams are produced by introducing in the polymer a blowing agent. There are chemical blowing agents, coated on the pellets or incorporated into them, and liquid blowing agents, injected into the polymer melt. The high pressure in the extruder prevents foaming in the machine; but as soon as the melt leaves the die, the foaming process starts. The expanding bubbles introduce local orientation into the polymer. In addition, by pulling, additional orientation can be introduced in the machine direction. Depending on the polymer, the density of the finished product, and the blowing agent, the process uses one single screw extruder, two single screw extruders in tandem, or twin extruders. Applications include sheets that are later thermoformed into packaging products (polystyrene, polyolefins), shapes, and profiles made of polystyrene (PS), polyolefins, PVC, and acrylonitrile-butadiene-styrene (ABS) for wood replacement, slabs for insulation, and foamed wire coatings.

Finally, other important extrusion processes are the extrusion coating of a number of substrates and monofilament extrusion (spinning). Paper milk cartons coated on both sides with polyethylene (PE) are a good example for the former. The coating provides a water barrier and also permits heat sealing.

Further information on the various extrusion processes can be found in the literature (16, 20–27).

## *Calendering*

Calendering is a process for manufacturing films and sheets by pressing the molten polymer between rotating rolls (cylinders) (21, 22, 24, 25, 28). As we noted earlier, it is one of the earliest processing methods. The calender usually consists of four rolls, which may be arranged in many different ways, the most common configuration used at present being the inverted "L" shape (Fig. 1.10). The polymer mass is fed between the nips of the first two rolls. It emerges as a sheet below this pair and passes over and between the remaining rolls. The first nip controls the feed rate; the second and the third set the final product thickness. Transfer from one roll to the next is accomplished by a combination of temperature, speed, and surface finish differences between rolls. The sheet is taken off the last roll by a higher speed "strip" roll. This can also be used to stretch the sheet. An embosser pair of rolls providing certain surface finishes may follow. Subsequently, the sheet passes a series of chilled rolls for cooling and solidification. The process terminates with the windup process. The feed to the first pair of rolls is usually molten. Preliminary steps may include a Banbury-type mixer and a roll-mill to "sheet" the fluxed product. A more or less continuous feed from the roll is fed into the calender nip. Between the roll mill and the calender, either a metal detector or an extruder is placed to screen solid contaminants. Alternatively, a preblend may be fed directly to the calender. Other alternatives are to feed the calender directly by a plasticating extruder equipped with a simple spreader die.

The outstanding feature of the calender is high production rates. Throughputs on a single line may reach 4000 kg/hr. Figure 1.11 shows a 32 in. calender. Sheets made from polypropylene (PP), PS, and PE are generally made by extrusion, but

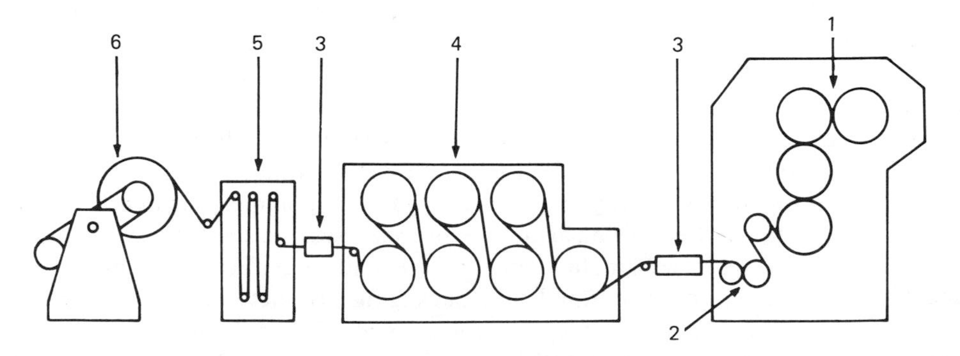


**Fig. 1.10** Schematic view of an inverted "L" calender plant for the production of plastics sheeting: 1, calender; 2, embossing calender; 3, thickness gauges; 4, water cooled cooling train; 5, windup accumulator; 6, windup station. (Reprinted with permission from W. A. Holmes-Walker, *Polymer Conversion*, Halsted Press, London, 1975.)

**Fig. 1.11** A 32×96 in. four roll inverted "L" calender. (Courtesy of Farrell Co., Ansonia, Conn.)

PVC, both plasticized and rigid films and sheets, as well as rubber, are almost always made by calendering, since this process is less likely to cause degradation.

### Injection Molding

Injection molding is the process for producing identical articles from a hollow mold. It is an offshoot of the ancient casting process, and it evolved from metal die casting. Because of their high viscosity, polymer melts cannot be *poured* into the mold; that is, gravitational forces are inadequate for effecting appreciable flow rates. Thus the melt must be *injected* into the mold cavity by applying large forces on it with a plunger. Moreover, once the mold is filled with melt and solidification starts, an additional amount of melt must be *packed* into the mold to offset polymer shrinkage during solidification and to achieve an accurate reproduction of the mold.

The number of articles manufactured by injection molding is immense, covering a size range from tiny gears to articles as large as automobile bumpers and bathtubs. Most polymers can be injection molded, including fiber reinforced, tough "engineering" thermoplastic, as well as thermosetting polymers.

Injection molding equipment (Fig. 1.12) consists of two parts: the injection unit and the clamping unit. The function of the former is to melt the polymer and inject it into the mold, whereas the clamping unit holds the mold, opens and closes it automatically, and ejects the finished product. Present injection units are almost exclusively of the in-line reciprocating screw type. The screw both rotates and undergoes axial reciprocal motion. When it rotates, it acts like a screw extruder, melting and pumping the polymer. When it moves axially, it acts like an injection plunger. The screw is, in general, rotated by a hydraulic motor and its axial motion is activated and controlled by a hydraulic system.

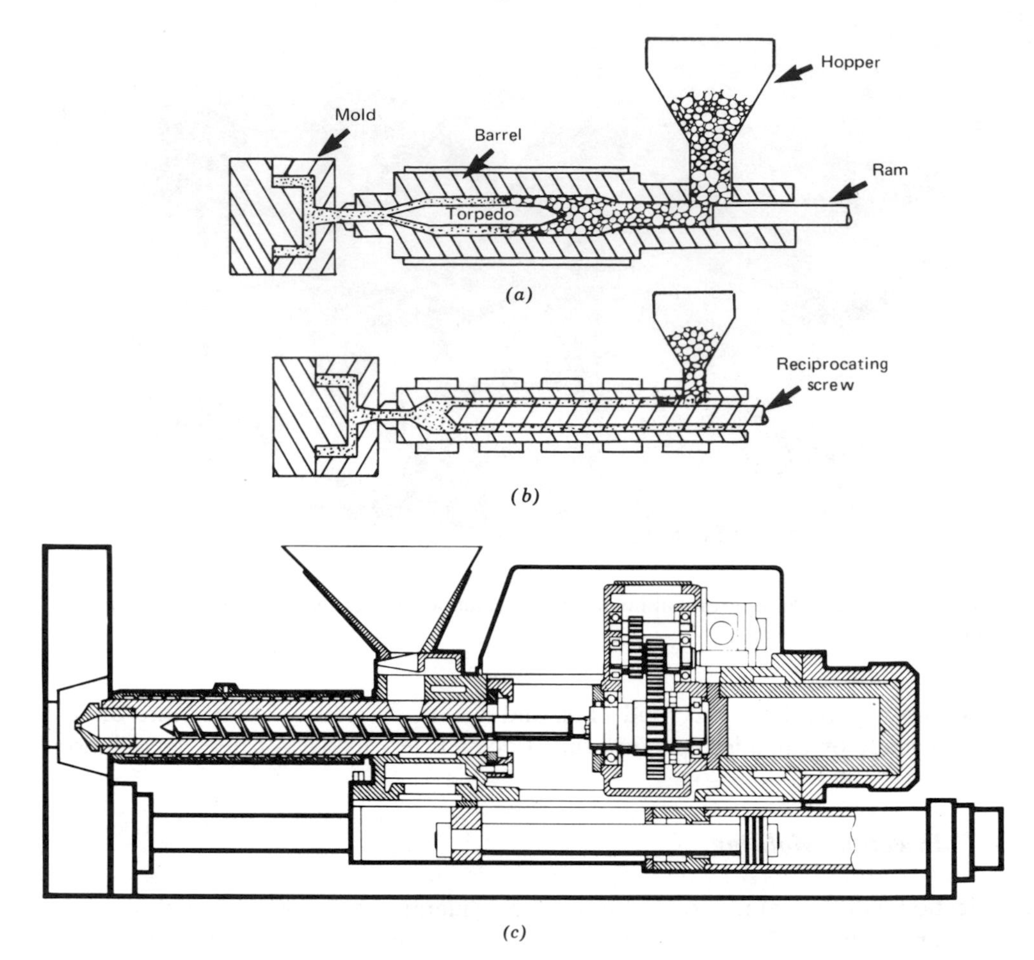


**Fig. 1.12** Schematic view of conventional injection molding machines. (*a*) Ram-fed injection molding machine. (*b*) Screw-fed injection molding machine. (*c*) Layout of a typical screw injection molding machine. (*a* and *b*, reprinted with permission from W. A. Holmes-Walker, *Polymer Conversion*, Halsted Press, London, 1975; (*c*) courtesy of Werner & Pfleiderer, Waldwick, N.J.)

Figure 1.13 depicts the injection molding cycle. The screw moves forward and fills the mold with melt and maintains the injected melt under high pressure, during what is called the "hold" time. (A no-return valve at the end of the screw prevents the polymer from flowing back onto the screw.) During hold time additional melt is injected, offsetting contraction due to cooling and solidification. Later, the "gate," which is a narrow entrance to the mold, freezes, thus isolating the mold from the injection unit. The melt within the mold is still at high pressure. As the melt cools and solidifies, pressure drops to a level that is high enough to ensure the absence of sinkmarks, but not so high that it becomes difficult to remove parts. After gate freezing, screw rotation commences. The "extruded" melt is accommodated in the increasing cylindrical space in front of the screw created by its backward axial motion. Flow rate is maintained by controlling the back pressure (i.e., the hydraulic pressure exerted on the screw), which in turn determines the pressure in the melt generated in front of the screw. After sufficient melt generation for the next "shot," screw rotation ceases. The polymer on the stationary screw continues to melt by heat conduction from the hot barrel. Hence this period is also called "soak" time. Meanwhile, the solidified part is ejected from the mold and closes to accept the next shot. All these operations are automated and are generally performed without any manual intervention. Moreover, the trend is toward installing computer control systems to improve quality and reduce variation in properties from part to part.

The mold filling process itself reflects all the interesting and complicating facets of polymer processing—nonisothermal, transient flow of non-Newtonian fluids in complex geometries with simultaneous structuring and solidification. Chapter 14 discusses all these matters in some detail. In view of these complexities, and in spite of recent intensive theoretical efforts to analyze the mold filling process, mold design remains by and large an empirical process. Injection molding machines, like extruders, vary over a very broad size range. Their size is being specified by injection capacity and by clamping force. Present sizes range from a few grams to a few kilograms, with clamping forces of up to 5000 tons. Figure 1.14 shows an injection molding machine.

In addition to thermoplastic polymers, thermosetting polymers and elastomers are successfully injection molded. Structural injection molding, which produces parts composed of a cellular core and an integral solid "skin," is another important development in injection molding. The "skin" and core may be of different

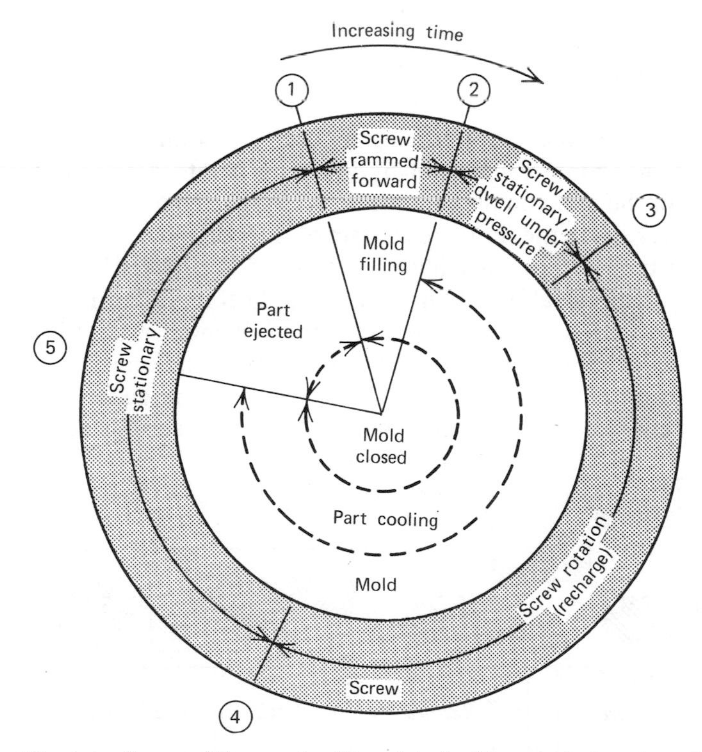


**Fig. 1.13** The injection molding cycle. [Reprinted with permission from R. C. Donovan, *Polym. Eng. Sci.*, **11**, 353 (1971).]

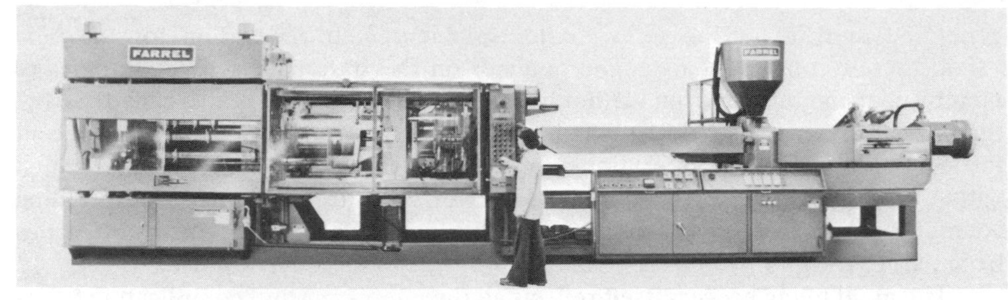

**Fig. 1.14** A typical injection molding machine. (Courtesy of Farrel Co. Ansonia, Conn.)

polymers, whereby two injection units are needed shooting one melt after the other, or the same polymer using one injection unit. In the latter case, the "skin" is formed by the quickly solidifying layer on the cold mold wall surfaces. The quick solidification prevents foaming in the skin. Reaction injection molding (RIM) and liquid injection molding (LIM), differing in the manner of mixing ingredients, currently involve the injection molding of liquid polyurethane systems that polymerize in the mold, with the possibility, at the same time, of generating a cellular structure. The inherent capability to control polymer rigidity and cellular structure makes the process flexible in manufacturing a variety of products, in particular large ones, where the low viscosity of the injected fluids is fully utilized.

Further technological details on the injection molding process can be found in the literature (21, 22, 24, 25, 29).

### *Compression and Transfer Molding*

Compression molding consists of forcing a certain amount of polymer into the desired shape of the mold cavity, not by injecting it into a closed mold, but by closing one half of the mold on the other (Fig. 1.15). The compression, effected by a hydraulic press, results in the intimate contact of the polymer charge with the mold and a squeezing type of flow that fills the cavity. The process is widely used for thermosetting polymers, although it is also applicable, in principle, for thermoplastic polymers. Heat is conducted from the hot mold walls to the polymer, inducing the chemical process of polymerization and crosslinking. The feed can be a preweighted mixture of molding powder, a prepelletized charge, or a preplasticized charge by a screw extruder. Compression molding has certain advantages over

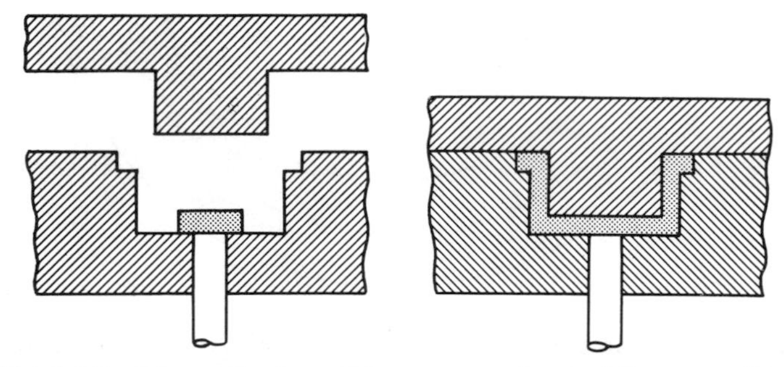

**Fig. 1.15** Schematic view of the compression molding operation.

injection molding: the molds are simpler than those used in injection molding, fillers remain relatively undamaged, and little material is wasted. But the process is slower, and there are geometrical limitations with respect to the parts that can be molded.

To overcome some of these limitations, *transfer molding* was developed. This logical extension of compression molding is an intermediate stage to injection molding. The charge is melted in a separate "pot," which is part of the heated mold. The mold is closed and the molten polymer is transferred from the "pot" by a ram, through runners and gates into the closed cavity. Figure 1.16 depicts this process.

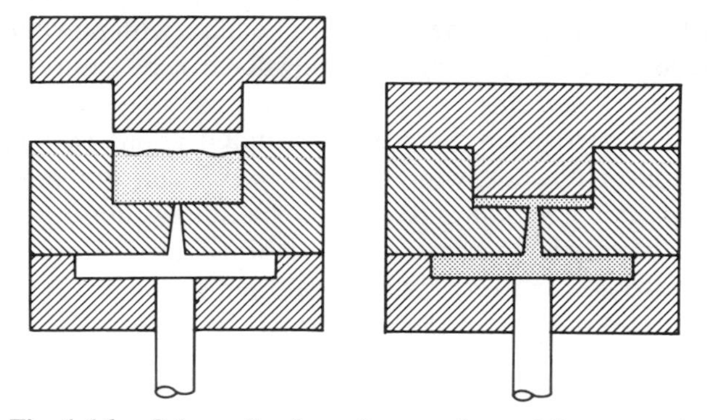

**Fig. 1.16.** Schematic view of a transfer molding operation.

Transfer molding cycles tend to be shorter than those of compression molding because mold temperatures can be maintained at higher levels. The premelted polymer charge flows more easily into the mold; thus it is possible to mold larger, more intricate parts, exhibiting less warping. An alternative to the integral transfer molding is the more flexible plunger transfer mold, which has the "pot," the plunger, and the separate hydraulic cylinder outside the stationary platen, with the "pot" connected to the runner system through the stationary platen very much as in injection molding.

### *Casting, Slush Molding, Dip Coating, Powder Coating, and Rotational Molding*

High viscosity fluids cannot be *cast* into a mold but must be *injected, compressed,* or *transferred* into it at high pressures. But we can process by casting a prepolymer, a monomer, or a polymer dispersed in plasticizer. All these are low viscosity systems, convenient for casting operations. Thus polyesters and epoxy resins are frequently cast into heated molds where chemical crosslinking and setting take place. Styrene and acrylic monomers or polymer-monomer solutions are also frequently processed into finished products by casting. A very common method of processing highly plasticized, flexible PVC is casting "plastisols," which are low viscosity systems of dispersed PVC in a plasticizer. Upon heating, the PVC swells and gels into a rubbery product.

An important variation of straight casting of plastisols is *slush molding*. The material is poured into the hot mold, and after a thick casting has formed on the mold cavity walls, the excess material is poured out. The molds are often rotated to ensure uniform coating. This shaping process is also used in the metal forming and ceramic industries. By inverting the process, hot metal parts can be dipped into plastisol, a polymer power bed, or a fluidized bed of polymer powder, and coated by it. *Powder coating* is used with PVC, PE, PP, polyesters, and acrylics. An improved version of powder coating is *electrostatic coating* and *spraying*, where the particles of powder are given an electric charge of one sign and the article to be coated is given the opposite. The results are uniform coating and a more efficient process.

Fine polymer powder can be coated on the inside of a mold in a number of variations of slush molding (30). The cavity can be filled with powder and the excess poured out, just as in slush molding. The powder can be poured into a rotating mold resulting in *centrifugal casting*, or the powder can be poured into a mold that then moves into a heated oven, where it is rotated about two axes, ensuring a uniform distribution of powder, which sinters into a uniform coating. The latter process is called *rotational molding*. Rotational molding is finding increasing applications in the area of very large items, such as storage tanks. Rotational molding of multilayered "sandwich" structures, structural foams, and similar structures have greatly expanded the versatility of this process (31).

### Blow Molding

Blow molding is a very important polymer processing method for manufacturing hollow articles such as bottles. It is a method borrowed from the glass industry. The process involves first the forming of a molten *parison*, which is a preshaped sleeve, usually made by extrusion. The parison is engaged between two mold halves, and air is blown into it (like a balloon), causing the parison to take the shape of the mold. The polymer quickly solidifies upon contacting the cold mold, and the finished hollow article is ejected. Figure 1.17 represents the process schematically.

There are a number of types of blow molding process, such as *extrusion*, *injection*, *stretch* and *dip blow molding*; some are briefly discussed below. There are two types of extrusion blow molding process, *continuous* and *intermittent*. The former, employed commonly for parts less than 1 gal, has a continuously rotating screw extruder, extruding parisons through one or more dies. The latter consists of three types: reciprocating screw, ram-accumulator, and accumulator head. It is employed usually for heavier articles. Figure 1.18 illustrates the extrusion, reciprocating screw, and ram-accumulator blow molding methods. The intermittent parison production methods are more suited to the nature of the blow molding process; greater flexibility and control are possible in the forming of the parison itself, which is the heart of the process. The blowing step is fast, and little flexibility is allowed in the control of the "bubble" thickness. Thus by controlling the rate of extrusion during parison forming (which results in different degrees of swelling of the extrudate), the thickness of the parison can be programmed to result in a product of more or less uniform thickness. Figure 1.19 shows a parison of programmed thickness. The same result can be obtained by varying the annular die gap and extruding the polymer at a constant rate. Parison forming is very sensitive to rheological properties, hence to temperature. Chapter 15 discusses these aspects of

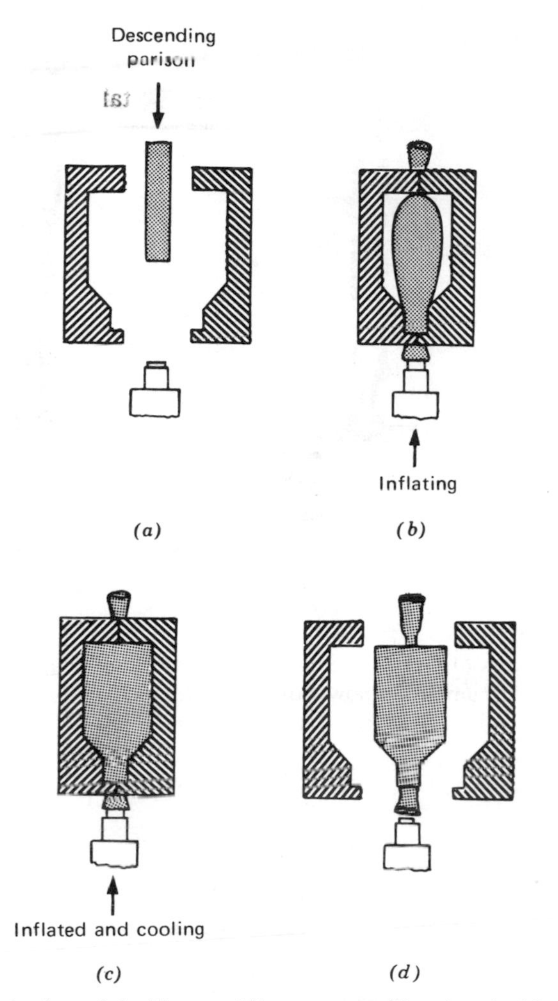


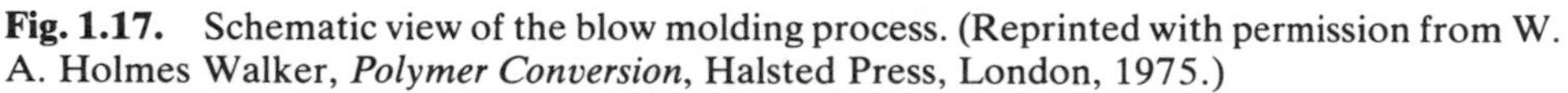

**Fig. 1.17.** Schematic view of the blow molding process. (Reprinted with permission from W. A. Holmes Walker, *Polymer Conversion*, Halsted Press, London, 1975.)

the process. Although the blowing process is very quick the cooling is relatively slower, hence various means have been devised to increase the rate of cooling by injecting liquid carbon dioxide after blowing, or by blowing with high pressure moist air (21). Free hanging parison formation is limited in size because of sagging and parison deformation (curtaining). In the injection blow molding process, the parison is formed by injection of the polymer on a steel rod (Fig. 1.20). The rod with the molded thread already completed is moved to a blowing station, where the article is blown free of scrap. The parison thickness distribution is determined in the injection mold without additional need for control. Some axial orientation is introduced during injection, resulting in an article of partial biaxial orientation. A process that greatly improves product properties by introducing biaxial orientation is the stretch blow molding process (Fig. 1.21). The parison or preform is mechanically stretched axially to introduce orientation. The stretching step is followed by blowing, which introduces the tangential orientation. Mechanical and

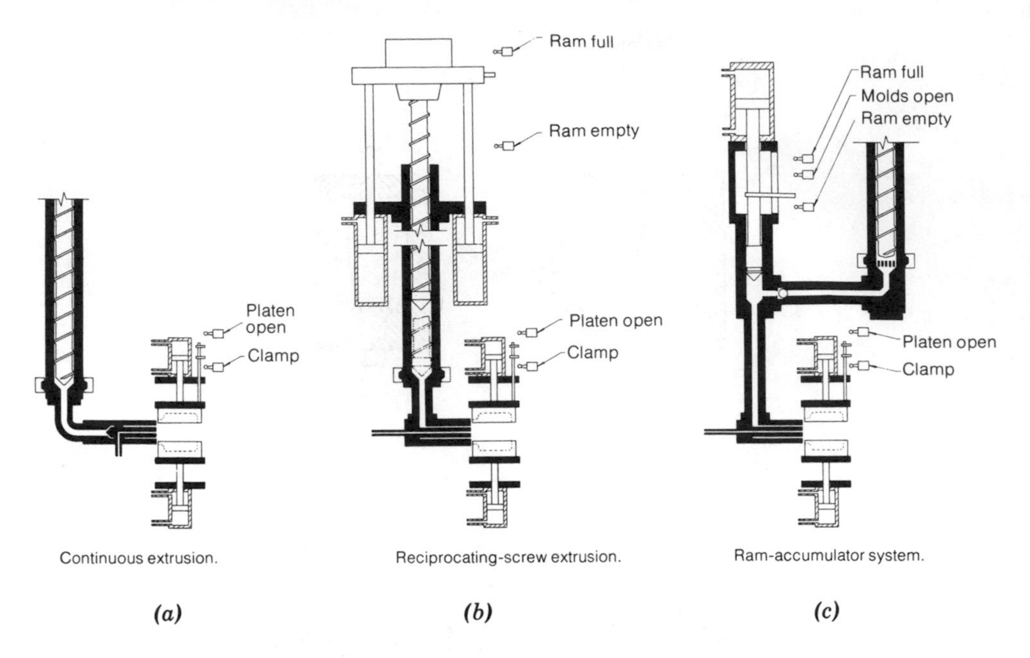


**Fig. 1.18** Schematic view of (*a*) continuous extrusion blow molding, (*b*) reciprocating screw blow molding, (*c*) ram-accumulator blow molding. (Reprinted with permission from *Modern Plastics Encyclopedia*, Vol. 53, McGraw-Hill, New York, 1976–1977.)

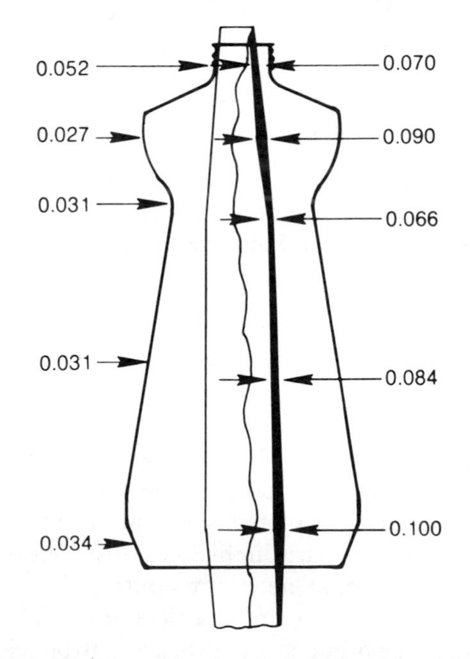


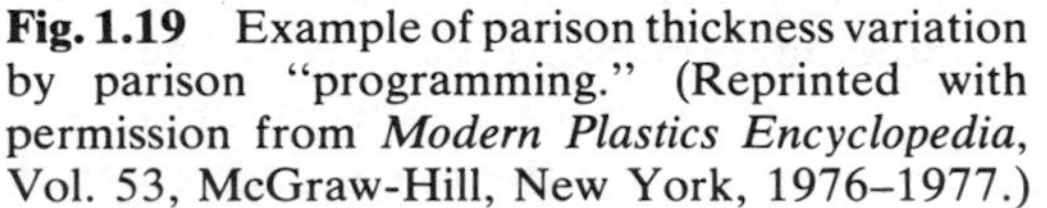
**Fig. 1.19** Example of parison thickness variation by parison "programming." (Reprinted with permission from *Modern Plastics Encyclopedia*, Vol. 53, McGraw-Hill, New York, 1976–1977.)

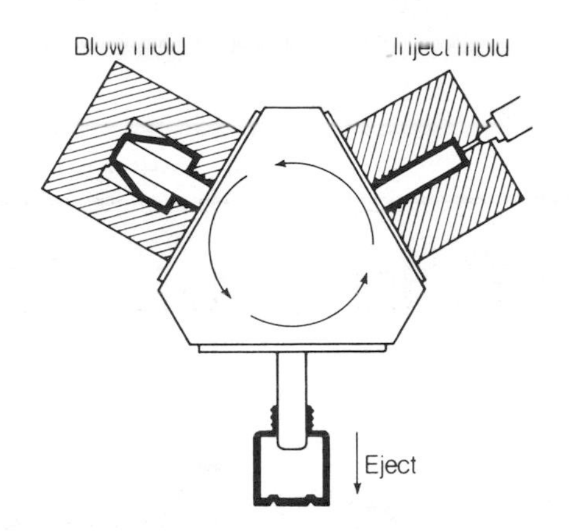


**Fig. 1.20** Three station injection blow molding rotary machine. Third position is easily accessible for removing container or unblown parison. (Reprinted with permission from *Modern Plastics Encyclopedia*, Vol. 53, McGraw-Hill, New York, 1976–1977.)

optical properties are greatly improved, permeability is reduced, and the weight of the product can be substantially reduced. Melt temperature homogeneity and stretch temperature level are the critical parameters of this process.

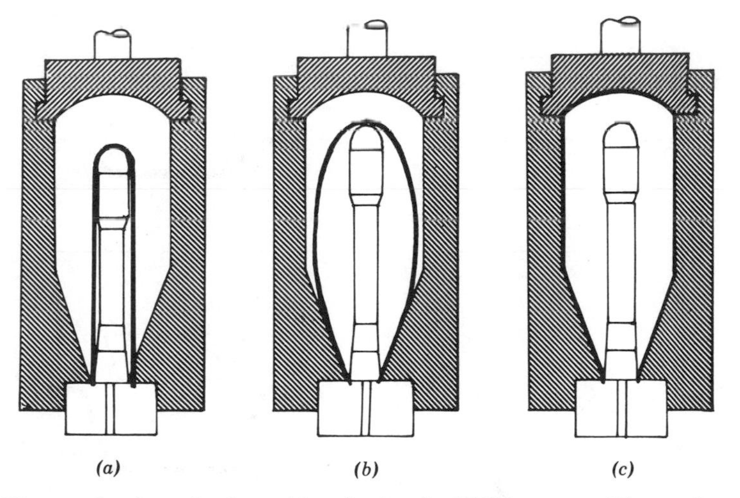


**Fig. 1.21** The production of oriented bottles by the VPM process. The preform has been closed and heated and carried into the mold. (*a*) The preform is in the mold. (*b*) The mandrel has moved up to bring the preform into contact with the tip of the mold cavity, and the first part of the blowing has started. (*c*) The blowing is completed. (Reprinted with permission from W. A. Holmes-Walker, *Polymer Conversion*, Halsted Press, London, 1975.)

### *Thermoforming*

Flat polymer sheets or films can be shaped into fairly deep-drawn container forms by a variety of specific forming methods, which are known as thermoforming. In all

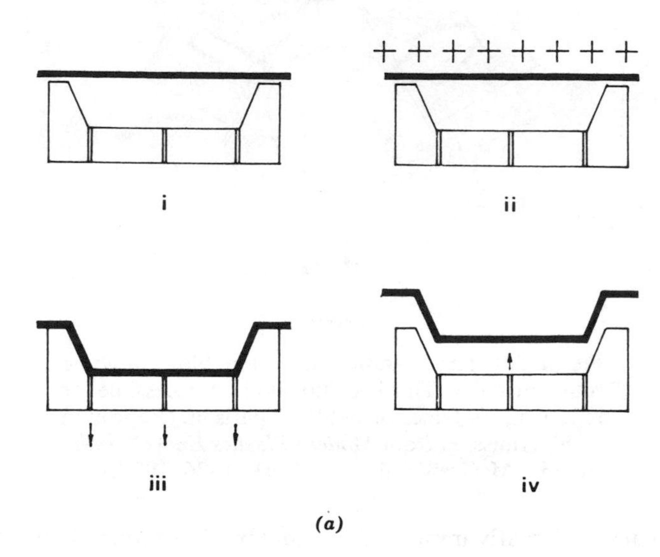


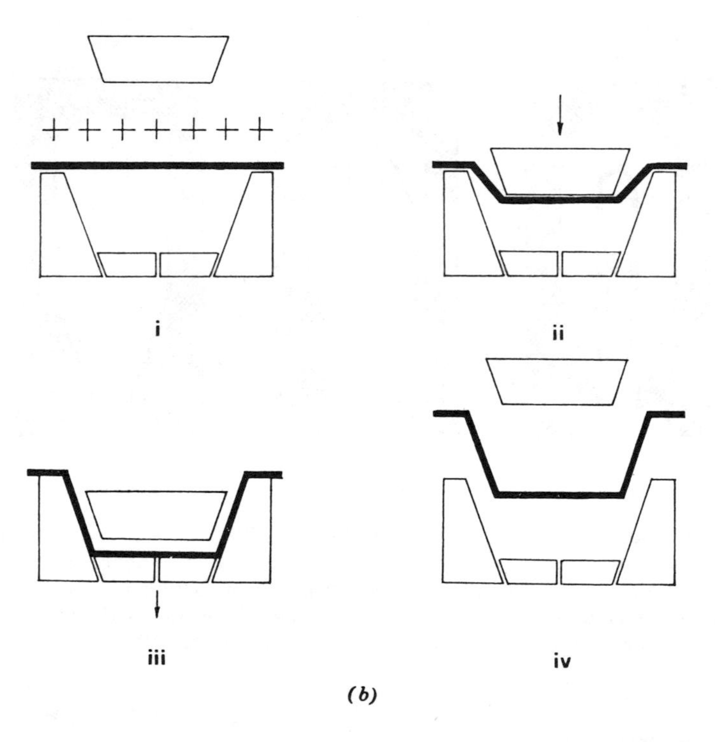


**Fig. 1.22** Schematic view of two thermoforming processes. (*a*) Vacuum forming. (*b*) Plug-assist vacuum forming. (Reprinted with permission from *Modern Plastics Encyclopedia*, Vol. 53, McGraw-Hill, New York, 1976–1977.)

these processes the flat stock is heated slightly above its melting ($T_m$) or glass transition ($T_g$) temperature and is placed in a clamp frame that clamps it along the part perimeter. Because during heating the sheet is unsupported, the polymers used in thermoforming must not be susceptible to creep, so that they do not sag. This is notably the case with ABS and high impact polystyrene (HIPS), which are commonly used in thermoforming. Once this is achieved, the deformable sheet or film is forced to conform to the shape of the mold by using any of the following mechanical means: (*a*) vacuum, which is applied through holes or channels in the mold—*vacuum forming*; (*b*) air or gas pressure in the range of $5 \times 10^4$–$2 \times 10^6$ N/m$^2$, which does not draw but forces the sheet onto the mold—because of the higher forces exerted on the flat stock, *pressure forming* is used with sheets of heavier and tougher polymers; (*c*) a force applied by a plunger (much as in metal forging and stamping), which has a shape that conforms to that of the "female" part of the mold. This third method is known as *matched mold forming*. At temperatures below $T_m$ or $T_g$ this method becomes forging, or *cold forming*. With the applied forces on, cooling is allowed to take place by conductive heat transfer to the cold mold wall(s). This is not a problem with thin film ("blisters," cups) thermoforming where cycles of 1–2 s are common. It can, however, be controlling with heavier sheet thermoforming, in the range of 0.25–1.25 cm, especially with polycrystalline polymers, which are characterized by low crystallization rates. In sheet forming, the forming cycles are long (because of the heating and cooling steps), but large objects (up to 4 m in diameter) can be formed. Figure 1.22 shows schematically the vacuum and a type of the matched mold forming processes.

The main advantage of the thermoforming processes is the low cost of the shaping equipment (tooling). Prototypes and low volume production can be achieved with wood or epoxy molds, and even production molds are made out of aluminum, since low pressures are used and high heat transfer rates are required. Hardened steel molds may be required for the cold forming process. Another advantage is that it can be carried out in the plant where the product to be packed is produced (as in blow molding). The main disadvantage of thermoforming is that it is limited to simple shapes with slight "undercuts"; otherwise, expensive segmented cavities must be used.

Thermoforming operations are referred to as secondary operations because they represent a shaping step that is applied after the primary film or sheet extrusion shaping. A typical thermoforming line, though, may or may not include the extrusion process. Unless the product volume is very high, thin gauge thermoforming utilizes rolls of film manufactured elsewhere. Thermoforming is further discussed in Chapter 15.

An ingenious process that utilizes thermoforming in a much more integral way is the Topformer (22) (or in-line preform forming), illustrated in Fig. 1.23. Melt is generated in a reciprocating screw machine, injected into a parallel plate channel, compression molded to form the stock to be thermoformed, clamped onto the thermoforming unit, and forced onto the mold by any of the three methods discussed earlier. There are many advantages associated with this process: little material is wasted, since the preform can be of any shape; the thickness profile of the preform can be varied; and, very important, the flat stock is not first cooled and heated up again, but remains hot until it touches the thermoforming mold.

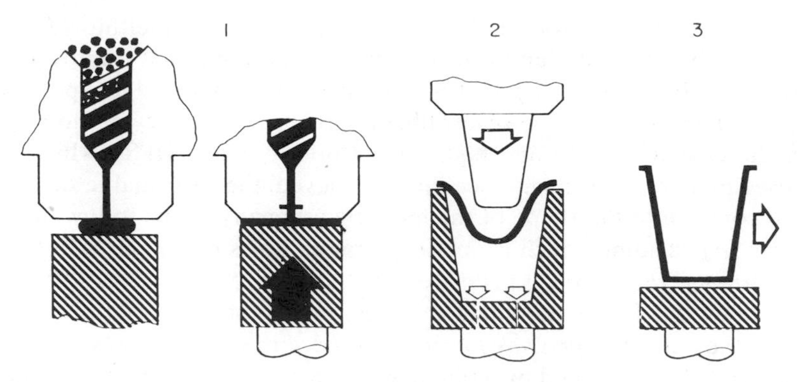


**Fig. 1.23** Topforming, introduced by Daniels Hamilton. (Reprinted with permission from W. A. Holmes-Walker, *Polymer Conversion*, Halsted Press, London, 1975.)

### *Other Processes, Compounding, and Finishing*

In addition to the principal processing methods already discussed, there are numerous others of lesser importance. Moreover, each of the principal processing methods can be broken down into a large number of highly specific processes. Discussion of these, however, is beyond the scope of this book, which stresses underlying principles common to all processing methods. The technical details of the individual processes, which are of great importance to polymer engineering, are generally accessible in the open literature.

Finally, the compounding and finishing operations, preceding and following the main processes bear heavily on the success of any processing method. The former is carried out in a wide variety of mixing equipment, analyzed in some detail in Chapter 11. The finishing operations consist of numerous processes such as mechanical fastening, bonding (adhesive, electromagnetic, etc.), sealing (heat, dielectric, ultrasonic), welding (contact, hot gas, friction, spin, groove, etc.), dyeing, electroplating, vacuum metalizing, printing, decorating, painting, and flock coating. Each one involves specialized technology and machinery. Once again, discussing their detailed intricacies or even their operating principles fall beyond the scope of this book, and the reader is referred to the abundant literature.

## 1.2 Analysis of Polymer Processing in Terms of Elementary Steps and Shaping Methods

The field of polymer processing has been traditionally and consistently analyzed (24) in terms of the prevailing processing methods, that is, extrusion, injection molding, blow molding, calendering, mixing and dispersion, rotational molding, and so on. In analogy to chemical engineering,* these processes have been viewed

* Systematic engineering analysis of chemical processes led to the definition of a series of "unit operations" such as distillation, absorbtion, and filtration, which are common to different chemical processes (e.g. see W. L. McCabe and J. C. Smith, *Unit Operations in Chemical Engineering*, 2nd ed., McGraw-Hill, New York, 1967).

as the "*unit operations*" of polymer processing. As polymer processing matures into a well-defined and well-studied engineering discipline, it is necessary to reexamine this classical way of analyzing the field, because the manner in which a field is broken down into its component elements has profound educational implications. A carefully worked out analysis should evolve into an abstract structure of the field that accomplishes the following objectives:

1. Focuses attention on underlying engineering and scientific principles, which are also the basis of the unifying elements to all processes.
2. Helps develop creative engineering thinking, leading to new improved design.
3. Provides an overall view of the field, facilitating quick and easy assimilation of new information.

The first step of such an analysis of polymer processing is to clearly define its objective(s). In this case the objective is undoubtedly *shaping polymer products*. The shaping operation may be preceded and followed by many manipulations of the polymer to prepare the polymer for shaping and to modify its properties and improve its appearance. Nevertheless, the essence of polymer processing remains the shaping operation.

The selection of the shaping method is dictated by product geometries and sometimes, when alternative shaping methods are available, by economic considerations. Reviewing the various shaping methods practiced by industry, we can classify them in the following groups.

1. Calendering and coating.
2. Die forming.
3. Mold coating.
4. Molding and casting.
5. Secondary shaping.

The first shaping method is a steady continuous process. It is among the oldest methods used extensively in the rubber and plastics industry. It includes the classical calendering as well as various continuous coating operations, such as knife and roll coating. Die forming, which is perhaps industrially the most important shaping operation, includes all possible shaping operations that consist of forcing a melt through a die. Among these are fiber spinning, film and sheet forming, pipe, tube and profile forming, and wire and cable coating. This is also a steady continuous process in contrast to the last three shaping methods, which are cyclic.

The term "mold coating" is assigned to shaping methods such as dip coating, slush molding, powder coating, and rotational molding. All these involve the formation of a relatively thick coating on either the inner or the outer surfaces of molds. The next shaping method is molding and casting, which comprises all the different ways for "stuffing" molds with thermoplastics or thermosetting polymers. These include the common injection molding, transfer molding, and compression molding approaches, as well as the ordinary casting of monomers or low molecular weight polymers, and *in situ* polymerization. Finally, secondary shaping, as implied by the name, involves shaping of preformed polymers. Thermoforming, blow molding, and cold forming can be classified as secondary shaping operations.

The complex rheological properties of polymeric melts play a dominant role in the shaping operations. Applying the large body of knowledge accumulated in melt rheology to the shaping operations is a difficult, yet also challenging task.

During the shaping operation or subsequent to it, a great deal of *structuring* (e.g., molecular orientation and morphological changes; see Section 3.1) is also being imposed on the polymer, to modify and improve physical and mechanical properties. This intimate interaction between shaping and structuring is of great practical significance. It is often the clue to a successful process, and is currently receiving growing attention by polymer engineers.

The polymer is usually supplied to the processor in a particulate form. Shaping of the polymer takes place only subsequent to a series of preparatory operations. The nature of these operations determines to a large extent the shape, size, complexity, and cost of the processing machinery. Hence the significance of a thorough understanding of these operations cannot be overemphasized. One or more of the operations can be found in all existing machinery, and we refer to them as *"elementary steps" of polymer processing.* There are five clearly defined elementary steps:

1. Handling of particulate solids.
2. Melting.
3. Pressurization and pumping.
4. Mixing.
5. Devolatilization and stripping.

Defining "handling of particulate solids" as an elementary step is justified, considering the unique properties exhibited by particulate solids systems. Subjects such as particle packing, agglomeration, stress distribution in hoppers, gravitational flow, arching, compaction, and mechanically induced flow, must be well understood to ensure sound engineering design of processing machines and processing plants.

Subsequent to some operation involving solids handling, the polymer must be melted or heat softened prior to shaping. Often this is the slowest and hence the *rate determining* step in polymer processing. Severe limitations are imposed on attainable rates by the thermal and physical properties of the polymers—in particular, the low thermal conductivity and thermal degradation. The former limits the rate of heat transfer, and the latter places rather low bounds on the temperature and time the polymer can be exposed. The difficulty is further compounded by the very high viscosity of the molten polymer. All these factors stress the need to find the best geometrical configuration for obtaining the highest possible rates of melting.

The molten polymer must be pumped and pressure must usually be generated to bring about shaping—for example, flow through dies or into molds. This elementary step, called "pressurization and pumping," is completely dominated by the rheological properties of polymeric melts and profoundly affects the physical design of processing machinery. Pressurization and melting may be simultaneous, and the processes may interact with each other. Moreover, at the same time, the polymer melt is also mixed. Mixing the melt to obtain a uniform melt temperature or a uniform composition (when the feed consists of a mixture rather than straight polymer), "working" the polymer for improving properties, and a broad range of mixing operations involving dispersions of noncompatible polymers, breakup of agglomerates and fillers—all these belong to the elementary step of "mixing."

The last elementary step of devolatilization and stripping is not discussed in this text. It is of particular importance to post-reactor processing, although it also occurs in commonly used processes, e.g., devolatilizing in vented extruders. This elementary step involves mass transfer phenomena, the detailed mechanisms of which have not yet been fully investigated.

The theoretical analysis of processing in terms of elementary steps, which leads to considerations of the basic physical principles and mechanisms involved in each elementary step, should be helpful in gaining a better insight into the presently used processing methods, encouraging further work on their mathematical formulations, and perhaps also stimulating creative engineering thinking on improved processing methods. We hope that it will help provide answers not only to "how" a certain product is made and "how" a certain machine works, but to "why" a product is made a certain way and, foremost, "why" a particular machine configuration is the "best" or the appropriate one to use. The latter question is indeed the essence of engineering.

The elementary steps as well as the shaping operations are firmly based on the fundamentals of *transport phenomena*, in particular, fluid mechanics and heat transfer, *polymer melt rheology*, *solid mechanics*, and general *principles of mixing*. These provide the basic tools for quantitatively analyzing polymer processing. Another fundamental imput necessary to understand polymer processing is the physics and chemistry of polymers. As was noted above, product properties can be immensely improved by "*structuring*." Clearly, to be able to fully utilize this added degree of freedom for design provided by the inherent polymer properties, a good understanding of polymer structure and properties is needed.

Figure 1.24 summarizes schematically the proposed breakdown of the study of polymer processing. Raw material is prepared for shaping through the elementary steps. The elementary steps may precede shaping, or they may be simultaneous with it. Throughout these processes, and subsequent to them, "structuring" takes place. Finally, postshaping operations other than structuring (printing, decorating, etc.) may follow. The main processing operations are based on the engineering fundamentals mentioned above, in particular, transport phenomena, melt rheology, and mixing, and on polymer physics and chemistry.

Chapters 2, 3, and 4 briefly review those aspects of the nature of polymers, which are most relevant to polymer processing. A similar brief review is presented next (Chapters 5–7) on the main engineering fundamentals, and we introduce many of the basic equations used later in the book. This is followed by the discussion of the elementary steps (Chapters 8–11) and shaping operations (Chapters 13–16). By breaking down the field of polymer processing into elementary steps and shaping operations, we have dismantled it into its "basic components." Chapter 12 demonstrates how the important screw extrusion process is analyzed in terms of the "elementary steps" and how a new processor can be synthesized from them. Chapter 17 gives a short outline of the "dismantling" and subsequent synthesis of complete common polymer processing operations.

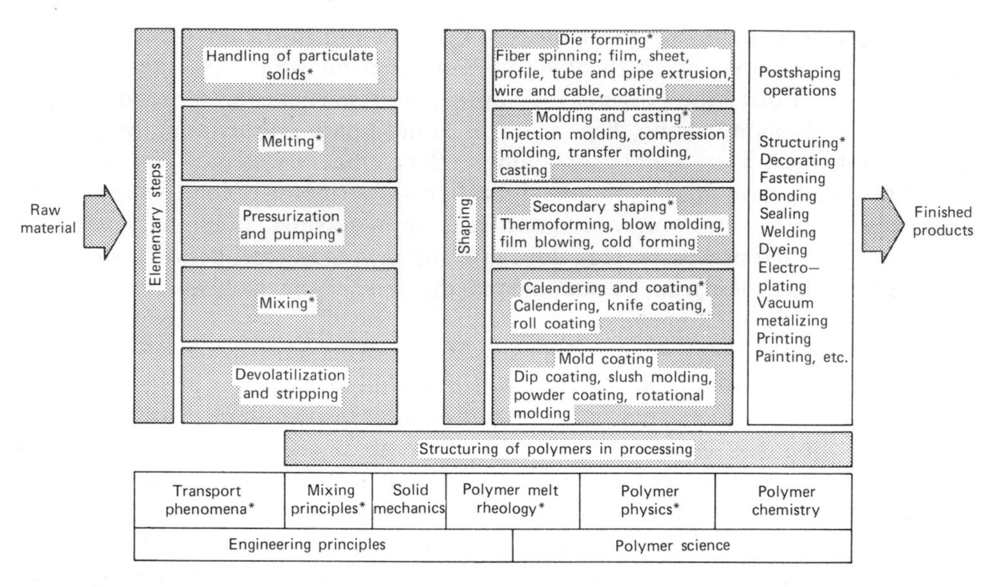


**Fig. 1.24** Conceptual breakdown of polymer processing into elements; asterisk indicates topic discussed in this text.

## *REFERENCES*


1. E. C. Bernhardt and J. M. McKelvey, *Mod. Plast.*, p. 154, July 1958.
2. T. Hancock, *Personal Narrative of the Origin and Progress of Caoutchouc or India Rubber Manufacture in England*, Longmans, Brown, Green, Longmans and Roberts, London, 1857.
3. C. Goodyear, *Gum Elastic* (private printing), New Haven, Conn., 1855.
4. J. W. Hyatt, *Ind. Eng. Chem.*, **6**, 158 (1914).
5. Sir Arthur Du Cros, *Wheels of Fortune: A Salute to Pioneers*, Chapman and Hall, London, 1938.
6. M. Kaufman, *The First Century of Plastics, Celluloid and Its Sequel*, Plastics Institute, London, 1964.
7. H. R. Jacobi, "The Historical Development of Plastics Processing Techniques," *Kunststoffe*, **55**, March 1955.
8. W. Haynes, *Cellulose, the Chemical that Grows*, Doubleday, Garden City, N.Y., 1959.
9. W. Woodruff, *The Rise of the British Rubber Industry During the Nineteenth Century*, Liverpool University Press, Liverpool, 1958.
10. P. Schidrowitz and T. R. Dawson, eds., *History of the Rubber Industry*, Heffer, Cambridge, 1952.
11. E. C. Worden, *Nitrocellulose Industry*, Van Nostrand, New York, 1911.
12. J. H. DuBois, *Plastics History U.S.A.*, Cahners Books, Boston, 1975.
13. J. L. White, "Elastomer Rheology and Processing," *Rubber Chem. Technol.*, **42**, 257–338 (1969).
14. M. Kaufman, "The Birth of the Plastics Extruder," *Polym. Plast.*, (June), 243–251 (1969).
15. V. M. Hovey, "History of Extrusion Equipment for Rubber and Plastics," *Wire and Wire Prod.*, **36**, 192 (1961).
16. G. Schenkel, *Plastics Extrusion Technology and Theory*, Iliffe Books, Ltd., London, 1966.

17. M. Gray, British Patent 5056 (1879).
18. D. H. Killeffer, *Banbury the Master Mixer*, Palmerton, N.Y., 1962.
19. H. Herrmann, *Schneken Maschinen in der Versahrenstechnik*, Springer-Verlag, Berlin, 1972.
20. P. N. Richardson, *Introduction to Extrusion*, Society of Plastics Engineers, Inc., Greenwich, Conn., 1974.
21. *Modern Plastics Encyclopedia*, Vol. 53, McGraw-Hill, New York, 1976–1977.
22. W. A. Holmes-Walker, *Polymer Conversion*, Halsted Press, London, 1975.
23. E. G. Fisher, *Extrusion of Plastics*, Iliffe Books, Ltd., London, 1954.
24. E. C. Bernhardt, *Processing of Thermoplastic Materials*, Reinhold, New York, 1959.
25. J. M. McKelvey, *Polymer Processing*, Wiley, New York, 1962.
26. Z. Tadmor and I. Klein, *Engineering Principles of Plasticating Extrusion*, Van Nostrand Reinhold, New York, 1970.
27. R. M. Ogorkiewicz, ed., *Thermoplastics*, Iliffe Books, Ltd., London, 1969.
28. R. A. Eldon and A. D. Swan, *Calendering of Plastics*, Butterworths, London, 1971.
29. I. Rubin, *Injection Molding*, Wiley, New York, 1973.
30. A. B. Zimmerman, "Processing Powdered Polyethylene," *J. Macromol. Sci.*; *Rev., Poly. Technol.*, **D1** (1), 51–96 (1971).
31. M. A. Rao and J. L. Throne, "Principles of Rotational Molding," *Polym. Eng. Sci.*, **12**, 237–250 (1972).

*PART ONE*

# STRUCTURE AND PROPERTIES OF POLYMERS

# 2

# The Nature of Polymers and Plastics



Polymers are giant molecules of molecular weight in the range of $10^4$–$10^7$; some occur naturally (cellulose, silk, natural rubber, DNA, etc.), whereas polyethylene, polyester, nylon, and others are synthetic. The basis for the formation of such macromolecules lies in the ability of certain monomers to link with each other repeatedly through primary (covalent) chemical bonds. This chemical process is called polymerization, and the resulting chainlike molecules (each "chain loop" is called a mer*) can be either linear, branched, or three-dimensional networks.

Commercially, polymers are almost always mixed together with various *additives*, which can be either monomeric or polymeric and are in the solid, liquid, or gaseous state. Their presence is intentional; it is aimed at achieving specific desired properties of the end product or ease of processing. Examples of additives used in plastics are fillers, reinforcing agents, foaming agents, plasticizers, stabilizers (thermal as well as environmental), lubricants, and pigments. A later section of this chapter briefly discusses the role and utility of such additives.

From the point of view of properties and end use, we can categorize pure and compounded polymers as *plastics*, *elastomers*, and *fibers*. We deal more specifically with each category later in this chapter. On the other hand, throughout this book the term "polymer" is used for all the foregoing without distinction, since our treatment of processing applies to all three.

## 2.1 Polymers

Starting with a given monomer, a great variety of polymers can be obtained *chemically*. The first set of structural variables that can be altered by changing polymerization conditions consists of molecular weight, degree of branching, and crosslink density. Furthermore, since polymerizations are subject to a multitude of random events, it is quite improbable to find that all the chain molecules in a polymer sample have the same length, number of branches, and so on. Rather, there exists a *distribution* of such structural variables, which may be narrow or quite

* The term "mer" derives from the Greek word *meros*, meaning "a part."

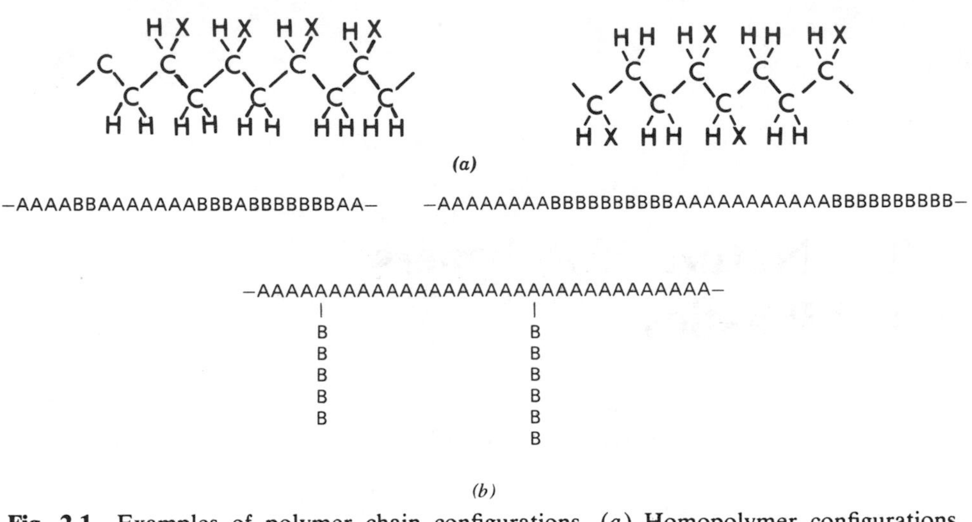


**Fig. 2.1** Examples of polymer chain configurations. (*a*) Homopolymer configurations. Left—head-to-tail, right—head-to-head. (*b*) Copolymer configurations.

wide. It is necessary, therefore, to report the values of molecular weight, branching, and crosslink density in terms of *averages*. Two average molecular weights are commonly used, depending on how the contributions of sample chains having the same length are taken into account statistically. If it is with respect to their number fraction in the sample, the number average molecular weight $\bar{M}_n$ or average chain length $\bar{x}_n$ is obtained; if their weight fraction is considered, the weight average $\bar{M}_w$, or chain length $\bar{x}_w$ is obtained:

$$\bar{M}_n = M_0\bar{x}_n = M_0 \sum_{x=1}^{\infty} xy_x \tag{2.1-1}$$

$$\bar{M}_w = M_0\bar{x}_w = M_0 \sum_{x=1}^{\infty} xw_x \tag{2.1-2}$$

where $M_0$ is the molecular weight of the mer and $y_x$ and $w_x$ are the mole and weight fractions, respectively, of chains containing $x$ mers. Average molecular weights can be determined from measurements of properties of dilute polymer solutions including colligative properties and distributions of molecular weights—as well as averages—with the use of various methods, including gel permeation chromatography.* The relationship $\bar{M}_w/\bar{M}_n \geq 1$ holds for all polymers, with the equality representing a monodispersed sample, where all chains have the same molecular weight.

The second structural variable that is determined chemically is *structural regularity*, which can be broken down into recurrence regularity and stereoregularity. The former refers to the regularity with which the mer recurs along the chain. An example of recurrence regularity can be found in the predominant head-to-tail configurations of substituted vinyl monomers upon polymerization (Fig. 2.1*a*), although head-to-head or tail-to-tail configurations are also conceivable. Polymeric chains possessing spatial regularity are called stereoregular. This is

* J. C. Moore [*J. Polym. Sci.*, 2 (4), 835 (1964)] first described this fractionation method for polymer molecules.

particularly important in considering poly($\alpha$ olefins) such as polypropylene. Isotactic polypropylene, for example, is a strong semicrystalline polymer with a melting point of 165°C, whereas its atactic counterpart is amorphous, soft, and tacky at room temperature.

Polymer molecules can be formed by involving in the polymerization process two or more monomers. If two monomers A and B are involved, the resulting polymer is called a *copolymer*; a *terpolymer* is the result of the simultaneous polymerization of three monomers. Copolymerization is used to incorporate into the resulting molecule a mixture of desirable properties of each of the homopolymers—pure A or B. Examples of copolymers are the random, alternating block, and graft (Fig. 2.1*b*). Copolymerization has greatly increased the utility of some homopolymers; such is the case with the "impact modified" polystyrene, a styrene-butediene copolymer that is not brittle but has high impact resistance. It is worth emphasizing that copolymers and terpolymers are different from mechanical mixtures of two or three polymers, which can be found in compounded plastics.

Being large chain-like molecules containing many single covalent bonds, polymers are permitted, to degrees that depend in the quiescent state on temperature and steric factors, to assume a multitude of *conformations.** Conformations are shapes of the chain molecules brought about—unlike configurations—by the mere *rotation* about single primary bonds. They may be viewed locally (random, zigzag, or helical) or as a property describing the shape of the entire chain (folded, random coil, or extended chain conformations). Long-range conformations, such as extended chain or long period folding, can be induced by imposing on a polymer sample shear or tensile *deformation*, or a specific thermal treatment, that is, annealing. Thus polymer processing, which involves stresses and annealing, does affect the state of macroscopic conformations of polymers and, consequently, their properties. Figure 2.2 illustrates local and long-range conformations. The folded chain conformation is characteristic of polymer crystalline regions; random coil conformations characterize amorphous polymers in either the liquid and rubbery (flexible) or solid (rigid coils) states.

Each mer unit in a polymer molecule is a site for *intermolecular bonds.* Since there is a large number of mer units in any polymer molecule, the level of intermolecular bonding between chains is very high. The structural implications of this property are that polymeric substances are either very viscous liquids or solids. Polymeric substances satisfy the following relation†

$$\sum_{1}^{\bar{x}_n} \begin{bmatrix} \text{strengths of inter-} \\ \text{molecular forces in} \\ \text{each mer unit} \end{bmatrix} > \begin{bmatrix} \text{strength of any} \\ \text{one type of primary} \\ \text{bond present} \end{bmatrix}$$

Polymer molecules with strong intermolecular forces (hydrogen bonding, dipoles) satisfy the relation above at $\bar{x}_n \sim 200$, and those having weak secondary forces (dispersion forces) at $\bar{x}_n \sim 500$. Of course the higher the molecular weight, the stronger polymeric substances become mechanically. Besides intermolecular

* Not all polymer chains are flexible. Rigid chain systems exist (poly-*p*-phenylene terephthalamide, "double ladder" polymers, etc.), and they are interesting both because of their structure and because of their high temperature properties.

† This relation also explains why, at elevated temperatures, thermal degradation or chain scission occurs rather than vaporization.

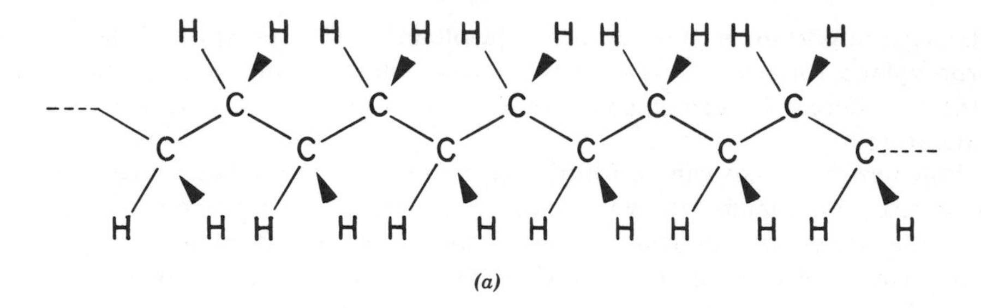


*(a)*

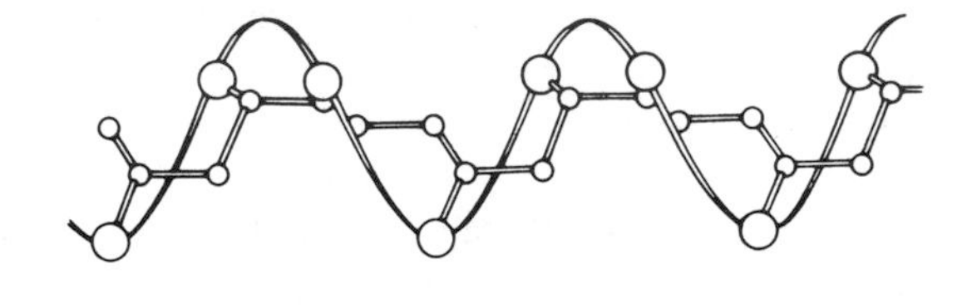

*(b)*

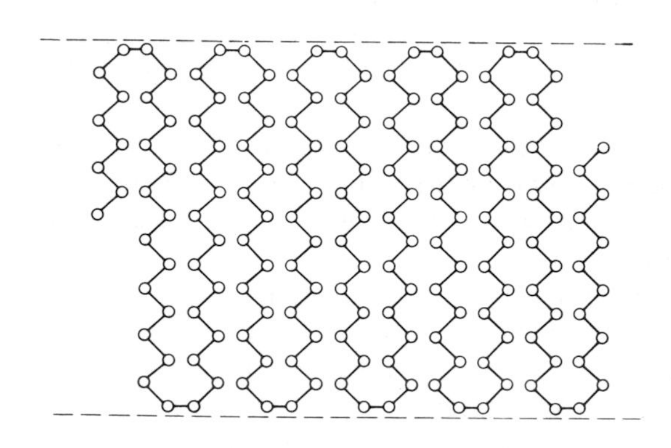

*(c)*

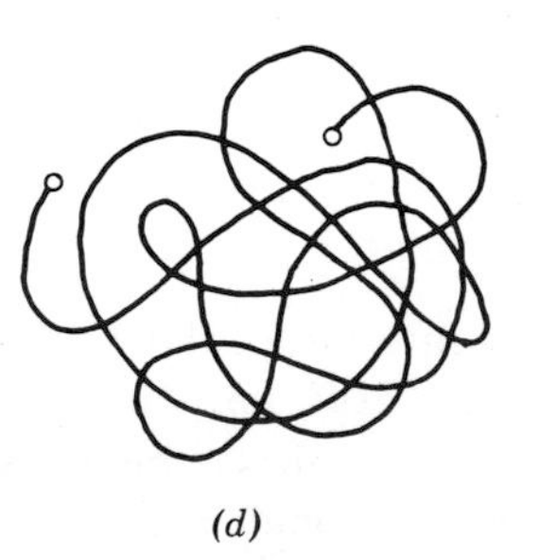

*(d)*

**Fig. 2.2** Local and long-range polymer chain conformations. (*a*) Linear polyethylene in planar zigzag local conformation. (*b*) Isotactic polypropylene in a helical conformation, O=$CH_3$. Helix is imaginary. (*c*) Chain folding long-range conformation of a single polymer molecule. (*d*) Random coil long-range conformation.

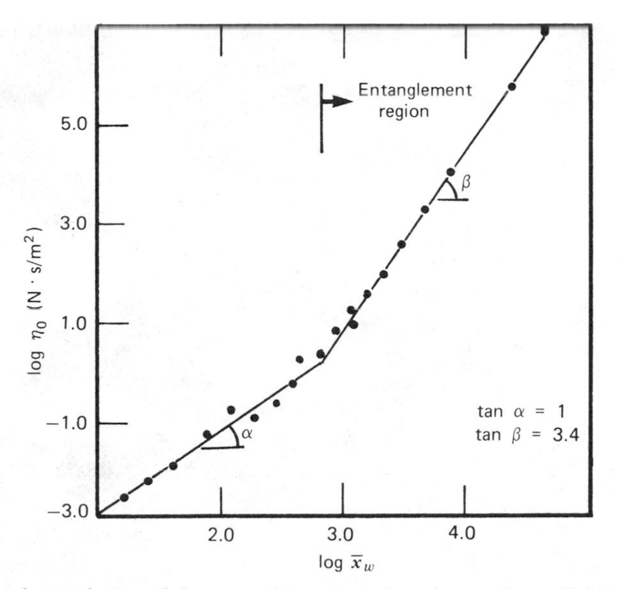


**Fig. 2.3** The dependence of the zero shear rate viscosity on the weight average chain length for polyisobutylene at 217°C. [Reprinted with permission from T. G. Fox, S. Gratch, and S. Loshaek, in *Rheology*, Vol. 1, F. R. Eirich, ed., Academic Press, New York, 1956, Chapter 12.]

bonding, *chain entanglements* may contribute to keeping macromolecular aggregates together and may affect their properties. Thus the break in the zero shear viscosity versus molecular weight curve in Fig. 2.3 is attributed to the onset of the entanglement phenomenon.

The degree to which intermolecular forces are utilized depends on the distance between the chemical entities partaking in the bonds, since to a first approximation these forces decrease with the inverse seventh power of this distance. We should, therefore, discuss the states in which polymers find themselves that are characterized by different values of specific volume and macromolecular order (conformations). There are two classes of polymer: those that are almost totally *amorphous* and those that are *semicrystalline*. Amorphous polymers are composed of chains that are randomly packed and are characterized by a major second order transition, the glass transition temperature, where they change from brittle, glassy solids to rubberlike substances. Below $T_g$ the random coil chains are rigid, whereas above it they are flexible. Semicrystalline polymers, below their melting temperature, form both amorphous and crystalline regions. The amorphous regions respond to temperature as described earlier. The crystalline regions are crystallite (folded chain) aggregates, usually in the morphology of *spherulites*. Figure 2.4 gives a micrograph of spherulitic morphology. Polymer molecules have different polarizabilities along and across the chain. Since the chains are perpendicular to the spherulite radii, such aggregates are *birefringent* and scatter light if they are of a size comparable to the wavelength of visible light (cf. amorphous polymers, e.g., PS, which are optically clear). The size of spherulites affects not only the optical properties of polymers, but also their mechanical response and diffusional properties. The percentage of crystallinity, the number and size of spherulites, and the

**Fig. 2.4** Spherulitic morphology of isotactic polypropylene viewed through a polarizing microscope, ×200.

rate of crystallization depend strongly on the crystallization (annealing) temperature as well as the degree of macromolecular orientation during crystallization, brought about by the application of a stress field. Thus again polymer processing affects polymer properties by affecting the morphology of a processed sample. Chapter 3 deals with the effects of processing on the properties and macrostructure of polymers. Crystalline morphologies other than spherulitic, which is the most common in commercial polymers and plastics, are polymer single crystals formed by folded chains and nonspherulitic morphologies obtained in polymers that contain nucleating agents or are crystallized at high degrees of supercooling.

Figure 2.5 gives the specific volume versus temperature curves for both amorphous and semicrystalline polymers, and Table 2.1 lists the glass transition temperature and melting point of some common polymers, together with their uses. The values of $T_g$ and $T_m$ relative to room temperature can be used together with structural properties to classify polymers as indicated below:

1. *Elastomers* are polymers with crosslinks, or intense entanglements, or microcrystalline regions, whose $T_g$ is such that $T_g + 75°C \leq T_{room}$.
2. *Glassy* polymers are amorphous substances whose $T_g$ is such that $T_g \geq T_{room} + 75°C$.
3. *Semicrystalline* polymers are substances of 50–90% crystallinity whose $T_g$ is much below and $T_m$ much above $T_{room}$.
4. *Fibers* are highly crystalline polymers that can be easily oriented with a $T_m$ such that $T_m > T_{room} + 150°C$.

Glassy and semicrystalline polymers as categorized above form the class of materials usually referred to as *plastics*. Some polymers (e.g., nylon 6, PP, PET) are used, either as plastics or as fibers.

As mentioned previously, polymer molecules in all the amorphous states—glassy, rubbery, molten, and solution—can be ideally treated by assuming that they

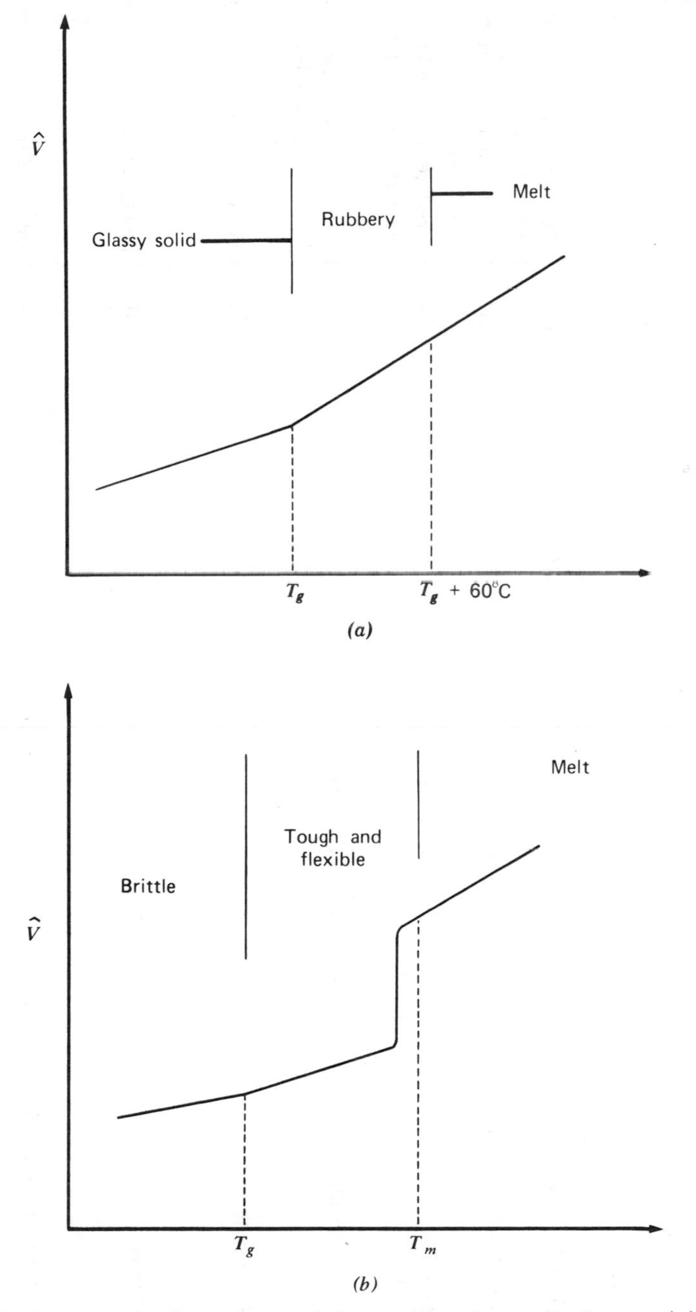


**Fig. 2.5** The temperature dependence of the specific volume of polymers. (*a*) Amorphous. (*b*) Semicrystalline.

**Table 2.1 The Glass Transition and Melting Temperatures of Some Common Polymers, and Their Main Uses**

| Polymer | Repeat Unit | Abbreviation | $T_g$ (°C) | $T_m$ (°C) | Main Uses |
|---|---|---|---|---|---|
| Linear polyethylene | $(-CH_2-CH_2-)$ | HDPE | −110 | 134 | Extruded and injection molded articles, bottles, and containers |
| Branched polyethylene | $(-CH_2-CH_2-)$ | LDPE | −110 | 115 | Flexible packaging film; flexible extruded and molded articles |
| Polystyrene | $(-CH_2-CH(C_6H_5)-)$ | PS | 90–100 | — | Extruded and molded articles that are transparent; foamed articles |
| Polyvinylchloride | $(-CH_2-CH(Cl)-)$ | PVC | 87 | — | Extruded rigid or plasticized articles, tubes, sheets, profiles |
| Polypropylene | $(-CH_2-CH(CH_3)-)$ | PP | −10 | 165 | Extruded and molded articles that are rigid |
| Nylon 6–6 | $(-NH-(CH_2)_6-NH-C(=O)-(CH_2)_4-C(=O)-)$ | — | 50 | 240 | Fibers; molded and extruded rigid articles |
| Polyethylene terephthalate | $(-O-(CH_2)_2-O-C(=O)-C_6H_4-C(=O)-)$ | PET | 70 | 260 | Fibers and transparent strong films |

| | | | | | |
|---|---|---|---|---|---|
| Polyoximethylene | $(-CH_2-O-)$ | Acetal | −50 | 180 | Molded "engineering" structural components; tough |
| Polycarbonate | $(-O-C_6H_4-C(CH_3)_2-C_6H_4-O-C(=O)-)$ | PPO | 150 | — | Molded tough and transparent articles; structural components |
| Polymethylmethacrylate | $(-CH_2-C(CH_3)(COOCH_3)-)$ | PMMA | 90–100 | — | Cast transparent sheet; molded articles |
| Polytetrafluoroethylene | $(-CF_2-CF_2-)$ | Teflon, PTFE | 125 | 327 | "Extruded' tubes; sintered blocks for machining; tapes; solvent resistant |
| Polyacrylonitrile | $(-CH_2-CH(CN)-)$ | PAN | 105 | >250 | High strength fibers |
| Polyisobutylene | $(-CH_2-C(CH_3)_2-)$ | PIB | −70 | — | Adhesives, paper coatings; together with isoprene, butyl rubber |
| Polybutadiene | $(-CH_2-CH=CH-CH_2-)$ | — | −88 | — | Styrene-butadiene rubber (SBR); nitrile rubbers (with PAN); oil resistant |

exist in equilibrium in random coil long-range conformations. In the glassy state there is no chain mobility, but in all the rest there is, except in rigid chain systems, thus segmental chain motions brought about by thermal energy permit polymer molecules to move about much like an entangled collection of long earthworms.* When such chain systems are deformed, the overall degree of randomness is decreased, and there is a tendency of the system to return to a state of maximum conformational entropy. It is precisely this tendency that gives rise to the retractive force $f$ in ideal rubbers (rubber elasticity theory)

$$f = -T\left(\frac{\partial S}{\partial \varepsilon}\right)_{T,P} \tag{2.1-3}$$

where $T$ is the temperature, $\varepsilon$ is the total strain, and $S$ is the conformational entropy

$$S = k \ln \Omega \tag{2.1-4}$$

where $k$ is the Boltzmann constant and $\Omega$ is the total number of microscopic states (long-range conformations) available to the system at a given strain level. Ideal rubbers are entropy elastic; that is, as solids they exhibit a resistance to deformation because their conformational entropy is reduced. This behavior is to be contrasted to the energy elasticity, where the resistance to deformation is due to straining of the intermolecular force field (crystals, glasses). Rubbers are solids with a cross-linked chain network that prohibits flow, and therefore exhibit a certain constant value of retractive force for a given deformation; polymer solutions, melts, and the amorphous regions in semicrystalline polymers above $T_g$, however will show a retractive force that decreases with time. That is, they will exhibit a force (or stress) *relaxation* following the imposition of a sudden strain $\varepsilon_0$. The reason for this behavior is evident in Eq. 2.1-3. The absolute value of $\Delta S$ decreases with time because the flexible chain molecules can return to random conformations, since they are not permanently tied in the structure. Chain coiling, though, is not instantaneous because the chain segments must overcome secondary forces and entanglements, which render the viscosity of such systems very high. Figure 2.6 illustrates these conditions. At the end of the relaxation process such a material will remain permanently in the deformed state, since the retractive force has decayed to zero. That is, as $t$ goes to infinity, such a material will behave like a liquid that is "contented" to be in any shape. On the other hand, the earlier the strain is removed, the higher the value of the internal retractive force, and the closer the substance will return to its initial shape, behaving more like an elastic solid.

The rate of relaxation of the stress can be conveniently measured by the time constant $\lambda$, called the *relaxation time*, assuming a simple exponential decay of the stress (Fig. 2.6*b*). The relaxation time is characteristic of the motions of the flexible coils, through which maximum entropy is attained. A simple dimensionless number can be defined which is helpful in understanding the response of flexible chain substances; it is the Deborah number† De, defined as

$$\text{De} = \frac{\lambda}{t_{\text{exp}}} \tag{2.1-5}$$

* We apologize for this rather graphic but unappealing analogy.

† "הָרִים נָזְלוּ מִפְּנֵי יהוה" From Deborah's song (Judges, 5:5). This line inspired Professor M. Reiner to define the Deborah number (see *Physics Today*, p. 62, January 1964).

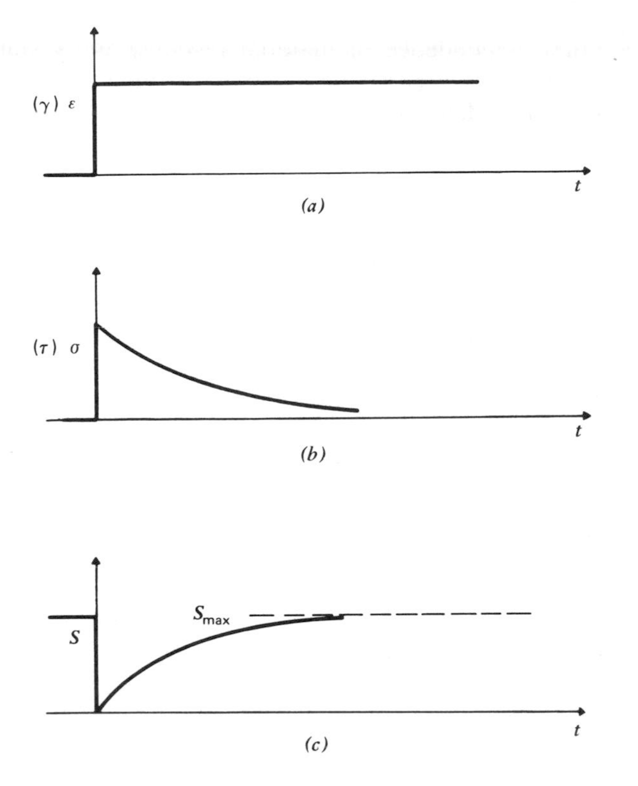


**Fig. 2.6.** Schematic representation of the stress relaxation process in amorphous flexible polymer systems. (*a*) The imposed strain. (*b*) The resulting stress (retractive force) relaxation. (*c*) The time dependent conformational entropy of the system.

where $t_{exp}$ is the time of duration of the experiment. The Deborah number represents the ratio of two characteristic times: the time needed for flexible chain motions to occur to the time allowed for such motions by the experimenter. Thus we have three conditions:

1. $De \to \infty$: flexible uncrosslinked substances behave as solids.
2. $De \to 0$: flexible uncrosslinked substances behave as liquids.
3. For any values of the Deborah number other than the two extremes above, such materials are *viscoelastic*; that is, they behave partly as elastic solids and partly as viscous liquids.

The same arguments can be used to explain the dependence of polymer melt behavior on the applied shear rate, $\dot{\gamma}(s^{-1})$. As $\dot{\gamma}$ is increased, $1/\dot{\gamma}$, which is the experimental time allowed for chain flow (interactions), is decreased, making polymer melts less viscous and more elastic. Since the response of viscoelastic substances depends on the value of the Deborah number, a dimensionless time, such materials are *time* (*or rate*) *dependent*, as far as the properties that are influenced by flexible chain motions are concerned. Another equivalent way of expressing viscoelasticity or time dependence of material response is by stating that such materials have a *fading memory*; the larger the experimental time, the less they remember their natural, undeformed state.

## 2.2 Polymer Additives

As mentioned earlier, most commercial polymers are compounds of macromolecules mixed with a number of additives selected to impart the desired properties to the end product and to facilitate its fabrication. For example, a typical rubber formulation for a tire or a lithographic blanket may contain as many as 15 additives.

Although arriving at a specific complex formulation may be the result of an engineering art, several general statements can be made concerning the role or actions of certain classes of additives on polymer matrices.

### *Stabilizers*

Most polymers must be brought to the molten state during processing at temperatures much above those of their melting or glass transition. This is done to lower their viscosity and to extend the upper limit of possible processing rates without melt fracture (see Chapter 13). Consequently there is the real danger of thermal degradation during processing. For this reason heat stabilizers, such as free radical scavengers, are used, especially for thermally sensitive polymers like PVC.

Burning can be broadly considered to be a process that starts by rapid thermal degradation of polymer chains into volatile and flammable products. Oxygen diffuses into the layer of these volatiles, and ignition occurs. Some of the heat generated by the flame is conducted and radiated to the condensed polymer layer; thus further rapid degradation and volatilization occur, feeding the flame. It is easy to understand why the degradation products of most common polymers are flammable, since being rich in carbon and, most often hydrogen, they are high in fuel content. Certain halogen-containing compounds act as fire retardants in that they interfere with the oxidation (burning) process of the chain fragments; others, such as $Sb_2O_3$ in PVC, act by possibly interfering with the heat transfer to the condensed state or by increasing the heat loss from it by radiation. Fire retarding and thermal stabilizing additives are mixed with polymers before processing and must be stable and compatible with the resin. Furthermore, they must not be toxic, either when ingested orally or when the vapor products generated by the limited burning are inhaled.

Polymer chains are also sensitive to forms of energy other than thermal. In particular, polymers that are intended for outdoor applications—films, tires, building materials—must be able to withstand ultraviolet (UV) radiation, for which purpose UV stabilizers are added. Finally, almost all polymers are subject to oxidative degradation, both short term at elevated processing temperatures, and long term during storage and use. Even in saturated polymers—the principal example being polypropylene—oxygen is absorbed and produces free radicals that react with the chains, usually autocatalytically, and degrade them. Most of the antioxidants combine with the oxygen-generated free radicals and inactivate them.

### *Colorants*

Polymers absorb very little light in the visible range and are, therefore, almost colorless. Furthermore, semicrystalline polymers may scatter light, depending on their crystalline morphology, thus appearing turbid or "milky." Thus for decorative reasons, colorants such as pigments and dyes that absorb light at specific wavelengths are added to some polymers. The colorant must be mixed very well with the rest of the polymer components (otherwise, aesthetically adverse results will be obtained), and it must be compatible with them.

### *External Plasticizers*

External plasticizers are usually monomeric molecules that when mixed with polar or hydrogen bonded polymers, position themselves between these intermolecular bonds and increase the spacing between adjacent bonds. Of course they must also either be polar or be able to form hydrogen bonds. The result of this action is to lower the level of the strength of intermolecular forces, thus decreasing the mechanical strength and increasing the flexibility of the rigid structure. PVC, which is polar, is plasticized by substances such as dioctylphthalate (DOP), and nylon, which is hydrogen bonded, is plasticized by water. Since the softening effect of external plasticizers is the same as that of an increase in temperature, processing of plasticized compounds requires lower temperatures for comparable viscosities.* Thus there is a smaller danger of thermal degradation. In this respect external plasticizers are, indirectly, thermal stabilizers.

If a polymer is rigid because of steric factors that hinder chain mobility, external plasticizers do not help. The *chain* must be made more flexible, and this can sometimes be achieved through copolymerization. In this context copolymerization is sometimes referred to as *internal* plasticization.

### *Reinforcing Agents*

This category of additives is very broad and yet very important in that such additives improve the mechanical properties of the base polymers, chiefly their strength and stiffness. Plastics that are used as structural engineering materials contain reinforcing agents. Asbestos fiber, short and long glass fiber, and, more recently, graphite fiber are common additives in applications calling for improved mechanical properties, including the absence of creep (dimensional stability). Solid reinforcing agents also extend the upper temperature limit of the use of the base polymer.

* Indeed, the term "plasticizer" stems from this process of making the polymer more susceptible to plastic flow.

### *Fillers*

The main function of fillers is to reduce the cost of the end product. A very inexpensive filler, occupying a fraction of the volume of a plastic article, will have such an economic benefit. Nevertheless, fillers are also often specialty additives; they may be present to reduce the thermal expansion coefficient of the base polymer, to improve its dielectric properties, or to "soften" polymers such as PVC (e.g., calcium carbonate).

### *Lubricants*

Lubricants are very low concentration additives that are mixed with polymers to facilitate their flow behavior during processing. There are two categories of lubricant, external and internal. External lubricants are incompatible at all temperatures with the polymer they are used with; therefore during processing they migrate to the melt-metal interface, promoting some effective slippage of the melt by reducing interfacial layer viscosity. Internal lubricants, on the other hand, are polymer compatible at processing temperatures, but incompatible at the use temperature. Therefore, during processing they reduce chain-to-chain intermolecular forces, thus melt viscosity. As the processed plastic products cool, they become incompatible (phase separation) and can eventually migrate to the surface; thus product properties are not permanently affected.

This chapter has very briefly reviewed some elementary properties of polymers. We have noted that polymer properties can be controlled and improved either by imparting chemically desired structural features to the polymeric chains or by incorporating additives into the polymer. But polymer engineers are becoming increasingly aware that in addition to these methods, polymer processing can, per se, impart specific and improved properties to polymers. The following chapter summarizes some of the available scientific and engineering information that suggests ways in which polymer processing can result in the *structuring* of finished products to achieve this goal.

### *REFERENCES*

A. X. Schmidt and C. A. Marlies, *Principles of High-Polymer Theory and Practice*, McGraw-Hill, New York, 1948.

P. J. Flory, *Principles of Polymer Chemistry*, Cornell University Press, Ithaca, N.Y., 1953.

C. E. Schildknecht, *Polymer Processes*, Wiley, Interscience, New York, 1956.

L. R. G. Treloar, *The Physics of Rubber Elasticity*, 2nd ed., Clarendon Press, Oxford, 1958.

A. V. Tobolsky, *Properties and Structure of Polymers*, Wiley, New York, 1960.

C. Tanford, *Physical Chemistry of Macromolecules*, Wiley, New York, 1961.

F. Bueche, *Physical Properties of Polymers*, Wiley-Interscience, New York, 1962.

M. L. Miller, *The Structure of Polymers*, Van Nostrand Reinhold, New York, 1966.

R. W. Lenz, *Organic Chemistry of Synthetic High Polymers*, Wiley-Interscience, New York, 1967.

F. A. Bovey, *Polymer Conformation and Configuration*, Academic Press, New York, 1969.
J. D. Ferry, *Viscoelastic Properties of Polymers*, 2nd ed., Wiley, New York, 1970.
F. Rodriguez, *Principles of Polymer Systems*, McGraw-Hill, New York, 1970.
F. W. Billmeyer, *Textbook of Polymer Science*, 2nd ed., Wiley-Interscience, New York, 1971.
D. J. Williams, *Polymer Science and Engineering*, Prentice-Hall, Englewood Cliffs, N.J., 1971.

# 3

# Morphology and Structuring of Polymers



## 3.1 Definition of Structuring

As we saw in the preceding chapter, polymers, because of their macromolecular nature, have a wide spectrum of conformations available to them and, commonly, also possess the ability to change from one to another in times that are quite short at processing temperatures but very long at use temperatures. Thus the following opportunity presents itself to processing engineers: to design the processing steps in such a way that the desired conformations (order) are achieved at high temperatures in the molten state and, subsequently, "locked in" the structure as the processed articles assume the use temperature. We define *structuring* to be precisely this kind of engineering effort.*

One can find many examples of structuring in commercial polymer processing. As a matter of fact, it is very difficult to avoid structuring of polymers, except in relatively slow processes such as casting and compression molding. Often, however, structuring during polymer processing is accidental, ill understood, and thought to be unavoidable—even a necessary evil in some cases (especially if it results in loss of dimensional stability). Classical examples of purposeful structuring in polymer processing can be found, on the other hand, in the production (spinning and subsequent drawing) of fibers and the extrusion of films that are uni- or biaxially oriented or cast to achieve structures that give to the film the desired mechanical and optical properties.

* In a standard dictionary definition, structuring is (the process of) constructing "something" of interdependent parts in a definite pattern of organization.

Structuring, of course, is not unique to polymer processing. It has been practiced in metallurgy for a long time. As an example, consider the range of properties that steel can acquire by virtue of heat treatment and cold working.

The ability of polymers to be structured during processing is inherent to their anisotropic macromolecular structure, which is held together with strong chemical bonds along the long chain axis and by weak, but many, bonds in the transverse direction from chain to chain. Prepared under quiescent and careful conditions, a system of such chains might attain a unique equilibrium structure and properties. But as Maxwell et al. (1) point out, commercial processing and fabrication operations must be performed at rates that do not permit the formation of equilibrium structures. It is when these nonequilibrium structures (conformations) are radically different from the equilibrium that the effects of structuring are sizable and worth investigating. That the effects can be sizable indeed is demonstrated in Table 3.1.

**Table 3.1 Tensile Moduli of Polymers and Other Engineering Materials: Approximate Ambient Values**

| Material | Modulus $E$ ($N/m^2$ $\times 10^{-9}$) | Density, $\rho$ ($kg/m^3$ $\times 10^{-3}$) | Specific Modulus, $E/\rho$ ($\times 10^{-6}$) |
|---|---|---|---|
| **Polymers** | | | |
| 1. Commonly processed HDPE | 1–7 | 1 | 1–7 |
| 2. "Extrusion drawn" HDPE fibers[a] | ~70 | 1 | ~70 |
| 3. Specially cold drawn HDPE fibers[b] | 68 | 1 | 68 |
| 4. DuPont "Kevlar" fibers | 132 | 1.45 | 92 |
| 5. Theoretical limit of polymers other than HDPE and PVA—fully extended[d] | <140 | ~1 | <140 |
| 6. Theoretical limit of HDPE and PVA—fully extended[d] | 240–250 | 1 | 240–250 |
| **Other Materials** | | | |
| 1. Aluminum alloys | <70 | | |
| 2. "E" glass fiber[c] | 63 | 2.54 | 35 |
| 3. Steels | ~200 | | |
| 4. RAE carbon filaments[c] | 420 | 2.0 | 210 |

[a] J. H. Southern and R. S. Porter, *J. Appl. Polym. Sci.*, **14**, 2305 (1970).
[b] G. Capaccio and I. M. Ward, *Polym. Eng. Sci.*, **15**, 219 (1975).
[c] W. Watt, N. Phillips, and W. Johnson, *The Engineer*, May 27, 1960.
[d] I. Sakurada, T. Ito, and K. Nakamae, *J. Polym. Sci.*, **C15**, 75 (1966).

The tensile modulus of commonly processed HDPE is 1–2 orders of magnitude lower than those of steel, aluminum, glass, and carbon filaments. But the modulus of the HDPE fibers, resulting from the work of Southern and Porter and Ward and his co-workers, is a decade higher than that of bulk HDPE and is comparable to aluminum and glass fibers. Moreover, the theoretically calculated (2) tensile modulus of a fully extended HDPE chain is comparable to steel and carbon filaments. Furthermore, the specific tensile modulus of this ideal structure is the highest known. Thus one can think of the ultimate HDPE "composite," which consists of fully extended HDPE molecules acting as reinforcing whiskers in the matrix of the same polymer. This type of hypothetical structure has been named "homogeneous composite" by Lindenmeyer (3). The engineering desirability of achieving this type of structure is evident.

Not only the tensile modulus of polymers can be radically increased by structuring, but also its optical clarity and thermal conductivity. Figure 3.1 shows the

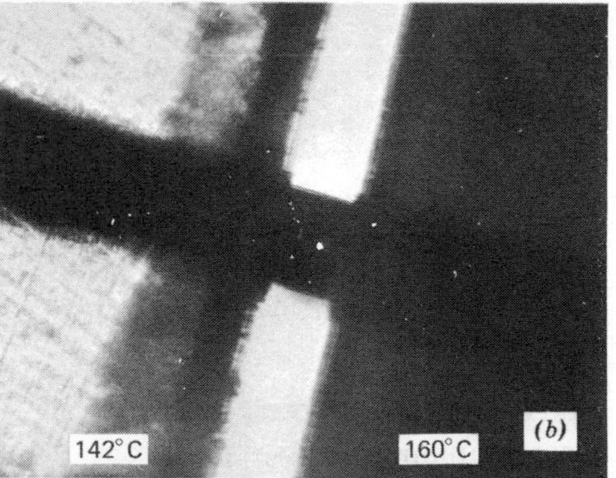


**Fig. 3.1** (*a*) Two strands of the same HDPE. The opaque strand is conventionally extruded. The clear strand prepared under special conditions (extrusion drawn). [Reprinted with permission from R. S. Porter, J. H. Southern, and N. Weeks, *Polym. Eng. Sci.*, **15**, 213 (1975).] (*b*) Two HDPE samples over an india ink line. *Left*—sample flow-crystallized at 142°C; *right*—compression molded sample cooled from 160°C. [Reprinted with permission from V. Tan and C. G. Gogos, *Polym. Eng. Sci.*, **16**, 512 (1976).]

transparency of HDPE samples that were extrusion drawn and crystallized above the normal melting point while flowing in simple shear. Although the morphological reasons for the increased transparency are not entirely clear, the application implications are again evident. Hansen and Bernier (4) have obtained thermal conductivities in the orientation direction in drawn HDPE that are 10 times that of the bulk (same as the modulus, as in the work of Ward).

The experimental evidence and discussion of the potential of structuring just presented centered around HDPE, which is one of the most flexible, "slimmest," and thus most readily oriented polymer chains; similar effects, however, have been found and are expected with other commonly used polymers. The key to understanding structuring lies in the effects on the detailed and overall morphology of the deformation and thermal histories that are imposed on a polymer during processing. Only recently has this vital area received due attention. Of course investigations have been conducted in the last 30 years aimed at unraveling the order, structure, and morphology of crystallizable and amorphous polymers under quiescent (equilibrium) conditions. Understanding the nature of quiescent morphology is a basic prerequisite for appreciating the potential of structuring. We proceed with such a discussion in the next section.

## 3.2 Polymer Single Crystal Morphology

All crystallizable materials used in engineering applications are polycrystalline—that is, made up of a multitude of small crystalline regions, each one bordering with other crystalline or amorphous regions. Thus the overall morphology is quite complex. For this reason, the basic crystalline features and properties of such materials are studied with *single crystals*. Polymers are not an exception. Single polymer crystals are grown from dilute solutions. At the crystallization temperature used, the crystallizable polymer precipitates, forming very tiny platelets (lamellae) that have all the characteristic features of a crystal, such as regular facets (when examined with an electron microscope) and single crystal diffraction patterns. The use of an electron microscope, or a very high magnification optical microscope, is necessary because polymer single crystals are very small. Single HDPE crystals appear in Fig. 3.2*a*; the large dimensions are of the order of a few microns, and the thickness has been found to be very small, of the order of 100 Å. Other polymer single crystals exhibit not flatness but a hollow pyramidal shape with frequent occurrence of spiral growth, reminiscent of screw dislocations. Detailed discussions of single crystal features are given by Geil (5), Keller (6), and Schultz (7). The single most important and unsuspected feature of these single crystals is that the polymer chains are perpendicular, or at a steep angle, to the flat surface, as shown by electron diffraction. This feature can be explained only by *chain folding*, since typical polymers are about $10^4$ Å long and the lamellar thickness is about 100 Å. Figure 3.2*b* represents chain folding schematically. This feature is characteristic of all crystallizable polymers. At a given crystallization temperature, the fold lengths are uniform and, as the crystallization temperature is increased, the fold length increases (Fig. 3.3). The reasons behind the occurrence of chain folding are discussed by Lindenmeyer (8). Essentially, crystallization in polymers is kinetically controlled, and chain folding minimizes the free energy barrier for this process.

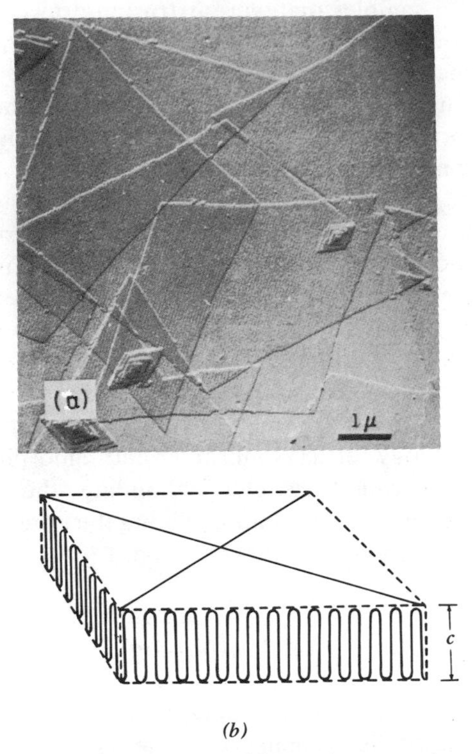


*(b)*

**Fig. 3.2** (*a*) HDPE single crystals as seen with an electron microscope. [Reprinted with permission from V. F. Holland and P. H. Lindenmeyer, *J. Polymer Sci.*, **57**, 589 (1962).] (*b*) Diagrammatic representation of chain folding in polymer crystals (the folds are drawn sharp and regular). [Reprinted with permission from A. Keller, *MTP International Review of Science*, Physical Chemistry Series, **8** (1972) **105** (Butterworths, London).]

Single crystals are composed of *unit cells*, the simplest unique volume regions, which when repeated spatially, form the single crystal. Thus the unit cell describes the way in which the molecules pack in the crystal. The unit cell of polyethylene

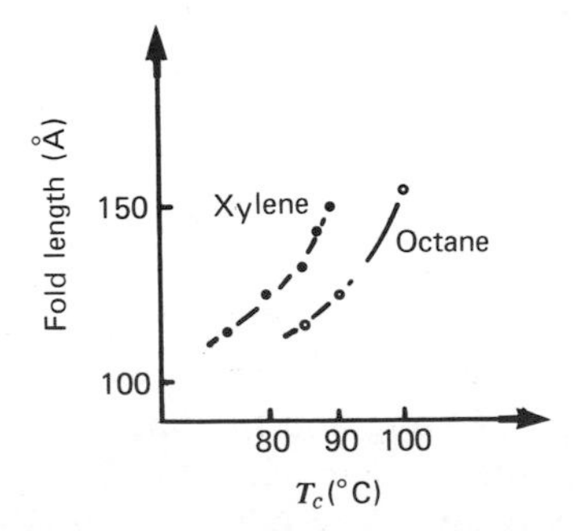


**Fig. 3.3** Long spacing (denoting fold length) in HDPE single crystals as a function of crystallization temperature for two solvents, xylene and octane. [Reprinted with permission from T. Kawai and A. Keller, *Phil. Mag.*, **11**, 1165 (1965).]

(Fig. 3.4) is orthorhombic, that is, it is a cell that is specified by three mutually perpendicular axes of unequal lengths, *a*, *b*, and *c*. The *c*-axis coincides with the direction of the axes of the folded chains of polyethylene in the single crystal. Thus a measure of molecular orientation in uniaxial stretching can be obtained by determining the angle formed by the crystallographic axis *c* with the direction of stretching. In polycrystalline structures, one would need to determine the *average* value of this angle for all the crystallites (single crystals) present. We return to this point in Section 3.9.

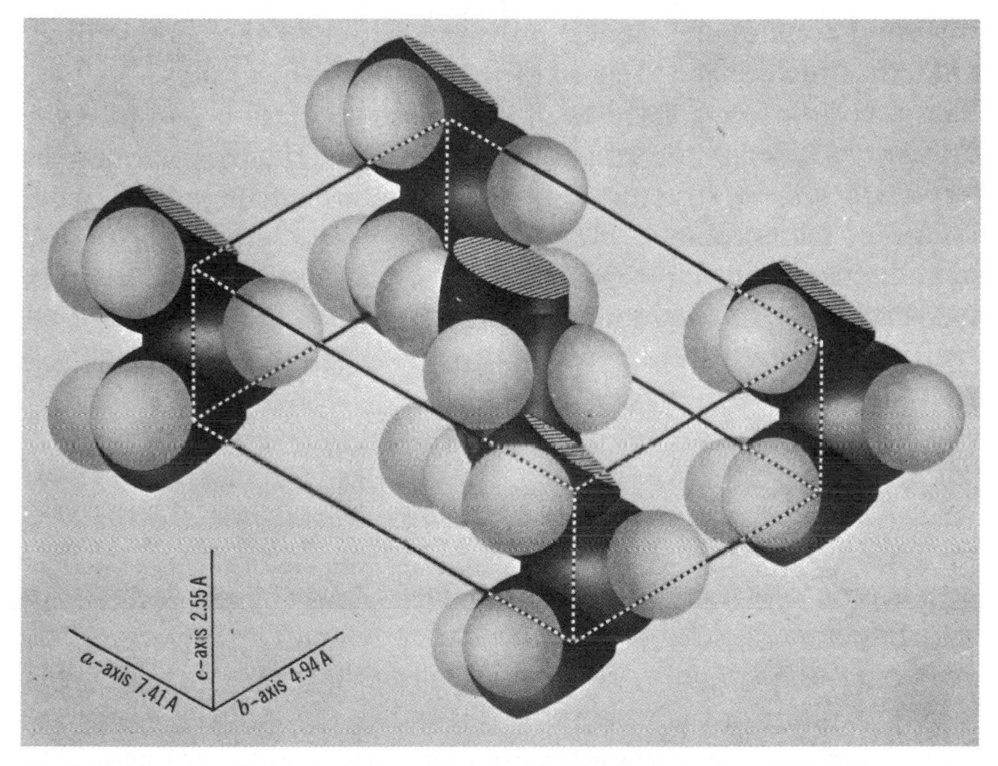


**Fig. 3.4** The unit cell of polyethylene. [Reprinted with permission from F. W. Billmeyer, Jr., *Textbook of Polymer Science*, Wiley, New York, 1965.]

Some polymers are polymorphic, which means that they can crystallize in different crystal forms characterized by different molecular packing or different unit cells (5). The properties of these structures are different. It turns out that one can induce specific crystal forms by varying the thermal or deformation history. In this way polymorphism offers another opportunity of "structuring" through polymer processing. Some examples are of interest. Isotactic polypropylene ordinarily crystallizes in a monoclinic form, but upon quick quenching it forms a predominant smectic phase of spherical agglomerates composed of imperfect hexagonal single crystals (9, 10). White et al. (11) support these findings in spinning isotactic PP in air and into cold water. Deformation of the polymer melt in polybutene-1, followed by isothermal crystallization, induces the formation of the stable crystal form I rather than form II, which is usually obtained when crystallizing the melt (12). Form I is

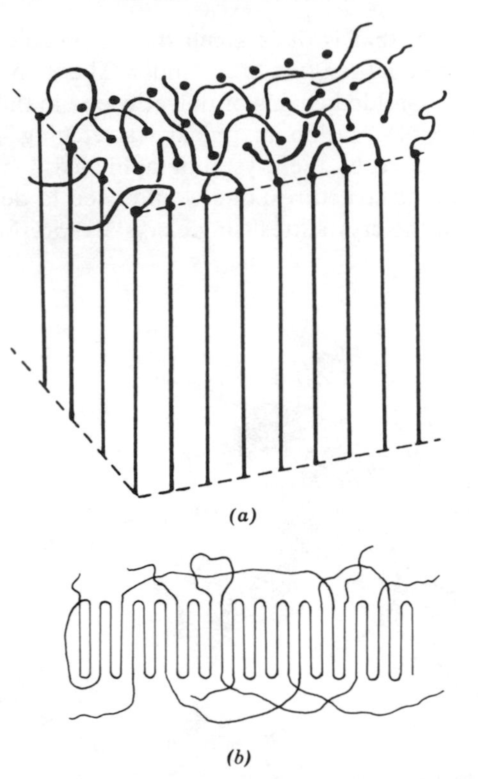


**Fig. 3.5** Disorder in single polymer crystals. (*a*) Model of an intrinsically disordered fold surface in the representation of E. W. Rischer [*Kolloid Z., Z. Polym.*, **231**, 458 (1969), reprinted with permission from P. J. Flory, *J. Am. Chem. Soc.*, **84**, 2857 (1962).] (*b*) Schematic representation of the origin of cilia formation and of the resulting surface looseness. [Reprinted with permission from A. Keller, *Kolloid Z., Z. Polym.*, **231**, 386 (1969).]

denser ($\rho_I = 930$, $\rho_{II} = 877$ kg/m$^3$). Furthermore, there exists a crystal-crystal transformation from II to I having a maximum rate at room temperature (13). Thus any product made out of polybutene-1 is expected to shrink with "shelf time." The degree of this shrinkage diminishes with increasing melt deformation. Thus polymer processing, through melt orientation, offers the possibility of eliminating an undesirable feature of an otherwise very useful polymer.

The chain-folded morphology is not free of defects, and most of them are concentrated along the fold surface, as in Fig. 3.5*a*. Thus single crystals may be visualized as being "sandwiches" of crystalline chain-folded regions between "amorphous" folds. Of particular importance are the loose cilia (Fig. 3.5*b*) and chains, which even at very dilute solutions partake in more than one single crystal. Since single crystals are multilayered—not just a single pyramid—the cilia can act as "ties" between layers. Furthermore, the "amorphous" layers would impart some ductility to multilayered crystals and would facilitate the occurrence of the observed phenomena during annealing and the application of a flow field on crystallizing dilute solutions.

Annealing has profound effects on the structure and properties of single crystals. Chain folding increases with increasing annealing temperature,

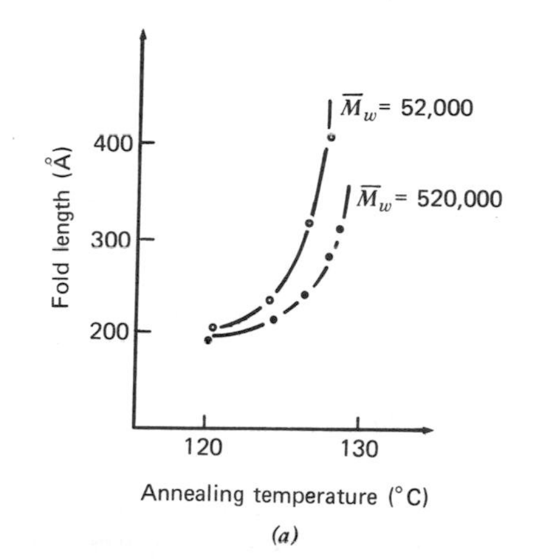


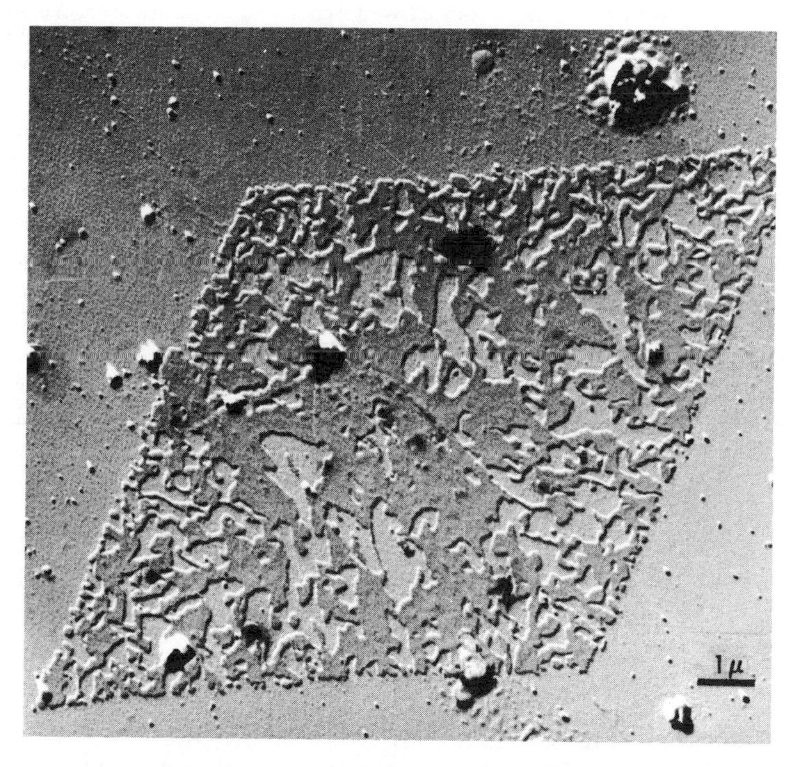


(b)

**Fig. 3.6** (a) Effect of annealing on HDPE single crystals of two different molecular weights. [Reprinted with permission from M. Takayanagi and F. Nagatashi, *Mem. Fac. Eng. Kyushu Univ.*, **4**, 33 (1965).] (b) A single crystal of HDPE crystallized from perchloroethylene solution, then annealed for 30 min. at 125°C, 10° below $T_m$. The fold period increased from about 100 Å to almost 200 Å during annealing. [Reprinted with permission from W. O. Statton and P. H. Geil, *J. Appl. Polym. Sci.*, **3**, 357 (1960).]

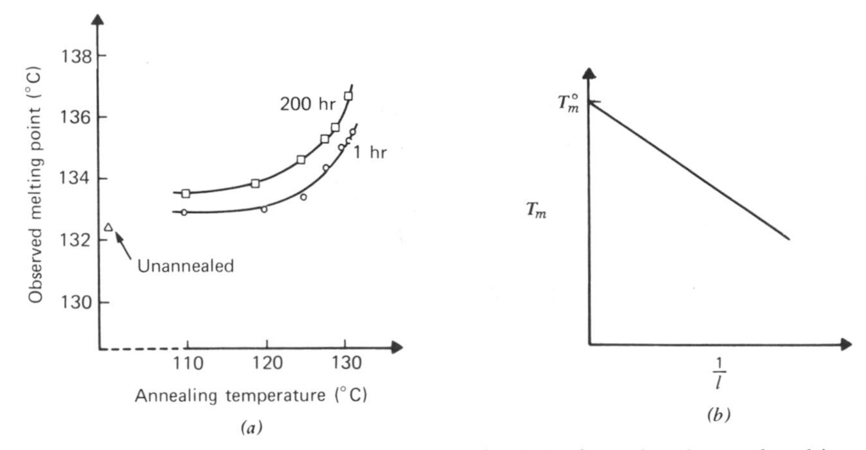


**Fig. 3.7** (*a*) Effect of annealing temperature (and time) on the observed melting point of HDPE. [Reprinted with permission from E. W. Fischer and G. Hinrichsen, *Kolloid-Z., Z. Polym.*, **213**, 93 (1968).] (*b*) Theoretical dependence of the melting point on fold length in HDPE: slope $= 2\sigma_e T_m^\circ/\Delta H_f$ where $\sigma_e$ is the surface free energy, $T_m^\circ$ is the equilibrium melting point at infinite fold length, and $\Delta H_f$ is the heat of fusion. (Ref. 7, p. 80.)

resulting morphologically in a "pitting" effect on the single crystal, as illustrated for HDPE in Figs. 3.6*a* and *b*. The refolding process implies mobility, and it is believed that motion is possible because of the layer defects. Lamellar thickening also results in an increase of the observed melting point. A theoretical dependence of the melting point on fold length can also be derived (14), and an *equilibrium melting point* at "infinite" chain length be estimated, $T_m^0 = 141°C$. Both the theoretical and experimental results appear in Fig. 3.7.

Conducting the dilute solution polymer crystallization, while imposing a flow field, *can* result in single crystal morphologies that differ from the simple folded lamellar. These structures can be termed *flow induced single crystal morphologies.* They were obtained while stirring dilute polymer solutions in a beaker (Couette flow) above a critical speed of the stirrer (15). These morphologies (Fig. 3.8) are termed, very appropriately, "shish kebabs." The "shish kebab" is sometimes covered with a "veil," believed to be formed by long chains that partake in the crystallization of two adjacent "kebabs," as would be expected if disorder were to be present at the folds as discussed earlier (16).

The gross "shish kebab" morphology suggested that the "shish" (sword) is made up entirely of fully extended chains, whereas the "kebabs" (meat pieces) consist of folded chain regions epitaxially grown on the core. The "kebab" structure has been confirmed, but there is strong evidence that the "shish" contains a large number of chain folds and defects (as well as extended chains). Upon annealing they can all become extended. Rijke and Mandelkern (17) annealed stirred crystallized polyethylene at 142°C with a "melting tail" of 152°C, indicative of a perfect, fully extended chain system. The critical stirrer speed needed to create these morphologies is thought to be associated with the appearance of Taylor vortices (18), which are ramifications of secondary flows. At the boundaries of such vortices, the flow field is extensional and not shear. The foregoing explanation indicates that extensional flow is a *necessary* condition for flow induced crystallization of the "shish kebab" type in dilute solutions. As we shall see later, this conclusion does not seem to hold for polymer melt flow induced crystallization.

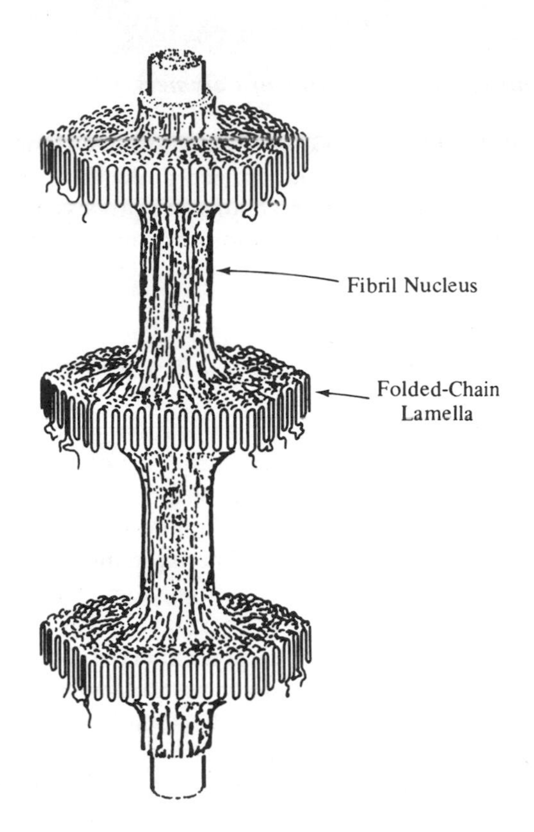


*(a)*

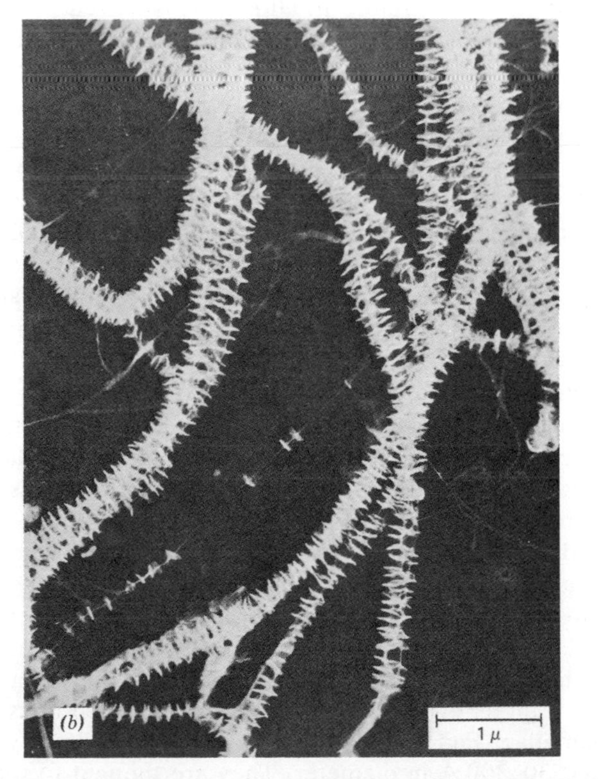


**Fig. 3.8** (a) Model of the "shish kebab" morphology. (b) Fibrillar polyethylene ("shish kebabs") crystallized at 101.8°C during Couette stirring of a 5% solution in *p*-xylene. [Reprinted with permission from A. J. Pennings, J. M. A. A. van der Mark, and A. M. Kiel, *Kolloid-Z., Z. Polym.*, **237**, 336 (1970).]

Pennings and his co-workers (19), who have done most of the work in this field, have also shown that the longer molecules in stirrer crystallization get involved first. Thus an effective fractionation process is at work. This is understood in terms of the increased chances of long chains to stretch and nucleate in the flow field.

## 3.3 Polymer Melt Crystallized Morphology

When considering dilute solution crystallization, we can think of each chain attaching itself to the single crystal(s) as an entity, free of entanglements with other chains and free of competition with them. In concentrated polymer solutions and polymer melts, which are characterized by the random coils sharing the same volume in space, this picture is no longer true. The basic morphological blocks may again be folded chain lamellae, but interchain entanglements and competition during crystallization are expected to result in morphologically more imperfect and complex structures. Another structural feature expected with crystallizing concentrated polymer systems that are not *entirely* monodispersed is *dendrites.* Dendrites are three-dimensional treelike structures radiating outward through branching. Branching is an expected instability feature of crystallizing polydispersed systems (20). The instability is due to concentration gradients created during the crystallization process because of the preferential crystallization of the long chains, which have a higher $T_m^0$, thus are under a large degree of supercooling. Dendrites do give rise to spherical symmetry. Therefore in all melt crystallized polymers we expect spherical polycrystalline regions, made up of imperfect, but distinct, folded chain lamellae.

Experimentally, the expected spherical polycrystalline regions, the *spherulites,* are easy to detect because of their relatively large size (50–1000 $\mu$). Viewed with a polarizing microscope, they appear circular, with a Maltese cross extinction pattern, characteristic of structures that are made of birefringent units arranged in spherical symmetry. Polymer chains *are* birefringent; in most cases, their polarizability along the chain axis is greater than that in the transverse direction. Therefore the Maltese cross pattern suggests that the chains are placed in a spherically symmetrical fashion in the spherulites. Birefringence studies have shown that in all spherulites, polymer chains are aligned *tangentially.* Furthermore, detailed examination of spherulites with electron microscopy, using fractured surfaces or various surface replication techniques (21), has revealed that the tangential arrangement of chains in spherulites is because of their folding, normal to the axis of lamellar ribbons, which radiate outward. These chain-folded ribbons form the basis of the dendrite spherulitic structure. In some cases, notably in HDPE, they twist regularly, giving rise to "radial banding." Figure 3.9 shows spherulites with radial banding, as well as an example of detailed spherulitic microstructure.

In addition to the existence of elongated chain-folded ribbons in spherulites versus flat platelets and pyramids in single crystals, another significant structural difference is the high degree of molecular connection between the spherulitic crystallites. There is direct evidence of the existence of such links, in the form of fibrils between lamellae (Fig. 3.10). These connecting fibrils may measure up to 15,000 Å long and 30–300 Å in diameter. They are thought to form, initially, by

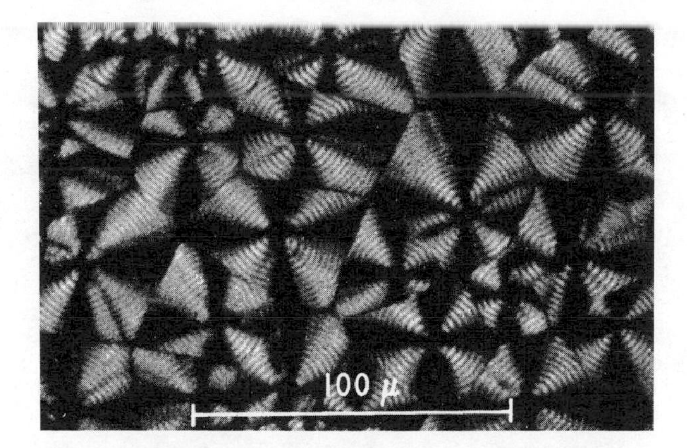


(a)

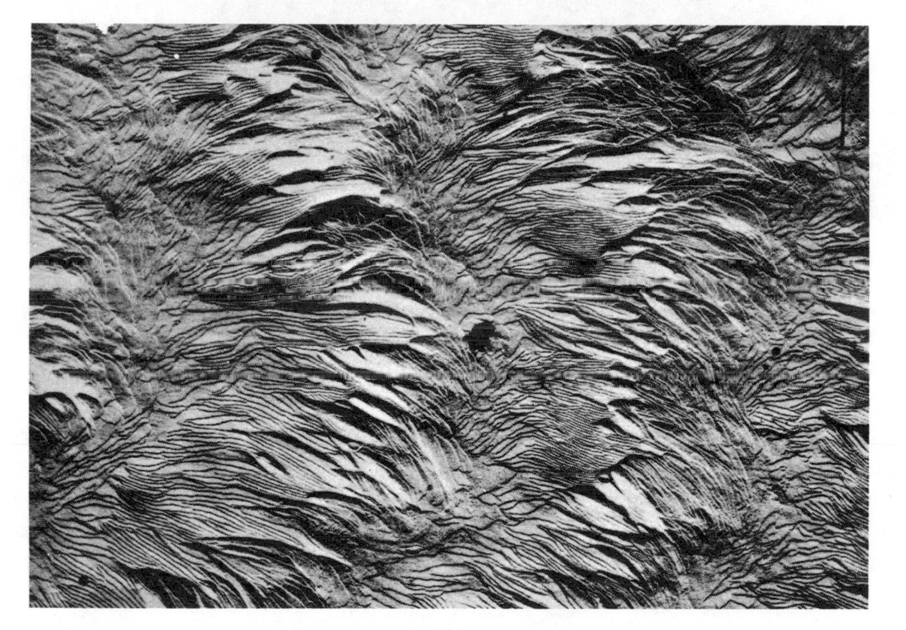

(b)

**Fig. 3.9** (*a*) Optical micrographs of HDPE cast on glass at 120°C using polarizing microscopy. [F. P. Price, *J. Polym. Sci.*, **37**, 71 (1959).] (*b*) Replica of the free (top) surface of an HDPE film cast on a hot liquid and crystallized isothermally, showing ribbonlike lamellae. [R. Eppe, E. W. Fischer, and H. A. Stuart, *J. Polym. Sci.*, **34**, 721 (1959).]

chains involved simultaneously in the crystallization of adjacent lamellae. As such, they may be composed to a large extent of extended chains. The fibrils play an important role in both the intra- and interspherulitic strengths. Another factor affecting the mechanical properties of melt crystallized structures is the "exclusion" from the crystallization process of the lower molecular weight fraction, with a lower degree of supercooling; this fraction ends up at the periphery of the spherulites and in interlamellar amorphous regions (22).

The spherulites and the detailed spherulitic morphology will vary in size, size distribution, type, and degrees of perfection and connectedness depending on (*a*)

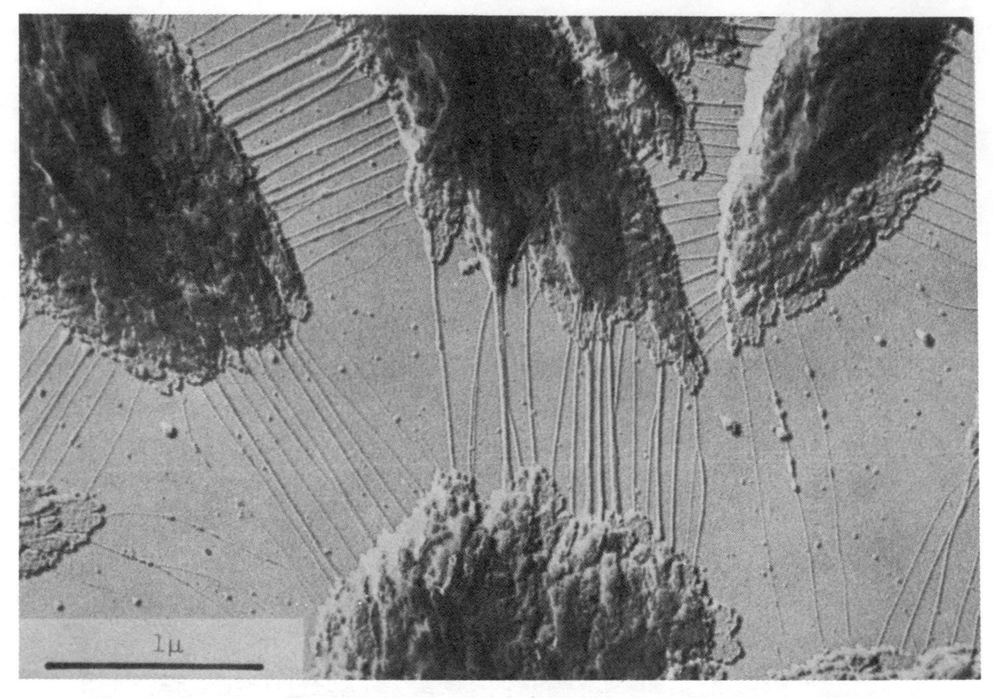


**Fig. 3.10** Interlamellar links in a system of 50% HDPE, $M_w = 726{,}000$, and 50% $nC_{32}H_{66}$ crystallized at 95°C. Paraffin was subsequently removed. The extended chain length of the HDPE fraction used is about 7.5 $\mu$, the observed fibril lengths are of the order of 1–1.5 $\mu$. [Reprinted with permission from H. D. Keith, F. J. Padden and R. G. Vadimsky, *J. Poly. Sci.*, **4**, **A-2**, 267 (1966).]

the conditions under which they were formed, and (*b*) the postcrystallization treatment imposed on them. Examples of variables of the first type are crystallization temperature history, pressure, presence of additives, and imposition of a flow field; examples of postcrystallization treatments are annealing and deformation.

## 3.4 Effects of Crystallization Temperature

On the lamellar scale, as was found in single polymer crystals, increasing the temperature in isothermal crystallization increases the lamellar thickness. The effect, which can be appreciable, is illustrated for HDPE in Fig. 3.11. There is also experimental evidence that the perfection of the crystals in the ribbon lamellae increases with increasing crystallization temperature.

Crystallization temperature has a very pronounced effect on the overall rate of crystallization and on spherulitic morphology.

### *Kinetics of Crystallization*

For polymers to crystallize under quiescent conditions, they must be brought to the supercooled state. The rate of crystallization in pure polymers depends on the

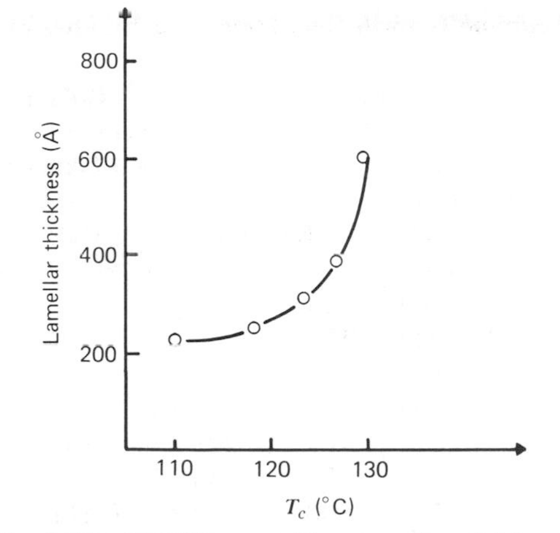


**Fig. 3.11** HDPE crystalline lamellar thickness, in melt crystallized samples, as a function of crystallization temperature. [S. Kavesh and J. M. Schultz, *J. Polym. Sci.*, **9**, **A-2**, 85 (1971).]

product of the rates of two processes: nucleation and crystal growth. Nucleation rates are high at low crystallization temperatures, where chains are characterized by low energy levels. On the other hand, high crystallization temperatures favor rapid crystal growth rates; since the chains must be pulled in from the melt to the crystallizing surface, high temperatures decrease melt viscosity, thus increasing chain mobility and the rate of crystal growth. It is expected then that the rate of crystallization will have a maximum at some crystallization temperature, as indicated in Fig. 3.12, where reciprocal half-times (i.e., the time to attain half the final crystallinity as measured by dilatometry) of a family of polyethylene succinates are plotted against crystallization temperature. Reciprocal half-times are proportional to the overall crystallization rate $G$. The crystallization rate becomes nucleation controlled at high temperatures and growth controlled at low

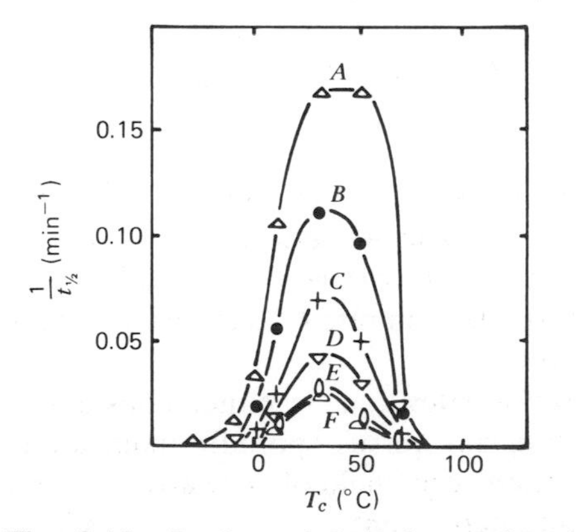


**Fig. 3.12** Reciprocal half-times of polyethylene succinates, $1/t_{\frac{1}{2}}$, as a function of the crystallization temperature. Polymer labeled $A$ has the lowest molecular weight, $F$, the highest. [Reprinted with permission from K. Steiner et al., *Kolloid-Z., Z. Polym.*, **214**, 23 (1966).]

**Table 3.2 Maximum Growth Rate Data for a Number of Polymers**[a]

| Polymer | Average Molecular Weight | $T_m^0$(°K) | $T_g$(°K) | $T_R$(°K) Temperature of Maximum Growth Rate | $G_{max}(\mu/\text{min})$ | $\frac{T_R}{T_m^\circ}$ |
|---|---|---|---|---|---|---|
| Isotactic polystyrene | 190,000 | 523 | | 347 | 6.0 | 0.66 |
| | 1,380,000 | 523 | | 347 | 1.5 | 0.66 |
| | Not given | 523 | | 341 | 0.03 | 0.65 |
| | 185,000 | 523 | | 343 | 0.2 | 0.66 |
| Polyoxymethylene | 40,000 | 453 | 203 | 361 | 400 | 0.80 |
| Nylon 66 | 17,200 | 538 | 333 | 413 | | 0.77 |
| Nylon 6 | 24,700 | 500 | 300 | 411 | 120 | 0.82 |
| Poly(tetramethyl-*p*-silphenylene siloxane) | 8,700 | 418 | 223 | 338 | 101 | 0.81 |
| | 1,400,000 | 423 | 258 | 338 | 21 | 0.80 |
| Nylon 56 | Not given | 539 | 318 | 458 | 200 | 0.85 |
| Nylon 96 | Not given | 519 | 318 | 452 | 130 | 0.87 |
| 1-Poly(propylene oxide) | 10,300 | 348 | | 285 | 50 | 0.82 |
| Poly(ethylene succinate) | 1,500 | 370 | | 328 | 41 | 0.89 |
| | 2,700 | 376 | | 328 | 23 | 0.87 |

[a] From J. M. Schultz, *Polymer Materials Science*, Prentice-Hall, Englewood Cliffs, N.J., 1974, p. 187.

temperatures. It also decreases markedly with increasing molecular weight, since the melt mobility needed for crystal growth is decreased with increasing melt viscosity. Table 3.2 gives the temperatures at which maximum crystallization rates are obtained, as well as their values and molecular weight dependence of the latter, for a number of polymers.

The crystallization process can be described in terms of the Avrami equation (23)

$$\frac{V_\infty - V_{(t)}}{V_\infty - V_0} = \exp(-Gt^m) \tag{3.4-1}$$

where $V$ is the sample volume and the subscripts 0, $\infty$, and $t$ denote the initial, infinite, and crystallization times. The Avrami exponent $m$, ranges from 1 to 4 and depends on the type of nucleation, growth geometry (rod, disk, sphere), and growth control mechanism. The crystallization rate $G$ is temperature and molecular weight dependent, as discussed earlier.

The number of nucleation sites can be either constant of increasing as the crystallization proceeds—constant versus sporadic nucleation. Spherulitic nuclei "appear" in the polarizing microscope field of view after a certain induction time $t_i$, which is strongly temperature dependent

$$t_i \propto \Delta T^{-\varepsilon} \tag{3.4-2}$$

where $\Delta T$ is the degree of supercooling and the values of the parameter $\varepsilon$ range between 2.5 and 9.0.

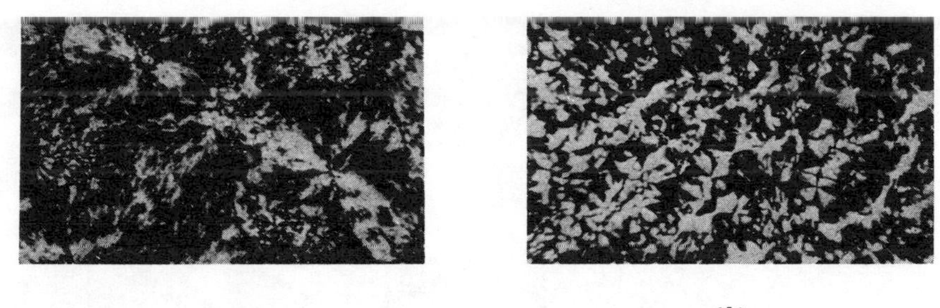

(a) (b)

**Fig. 3.13** Spherulitic morphology of polyoximethylene grown isothermally at (*a*) 136.4°C and (*b*) 170°C. [F. Rybnikař, Collection Czech. Commun., **31**, 4080 (1966).]

## *Spherulitic Morphology*

It is evident from the preceding section that spherulitic morphology varies with crystallization temperature. The extent of variation of spherulitic size can be seen in Fig. 3.13. The low temperature morphology is "grainy," with many very small spherulites because of the high nucleation rate (many sites). Such structures are mechanically ductile because of the great number of molecular ties and amorphous regions between the small spherulites. For the same reasons they have a lower modulus, while they are optically uniform, though translucent in the case of HDPE. On the other hand, spherulites can grow to be quite large at the high crystallization temperature, since there are few nuclei and the growth rates are high. Such spherulitic morphology, containing also more perfect crystals, results in high moduli, brittle, and optically nonuniform substances. Maxwell et al. (1) have reported the interphase region between spherulites of this morphology to be areas along which fracture can take place.

During the solidification steps common in polymer processing operations, cooling starts from the surfaces of the processed articles and proceeds toward the center. Based on the discussion above, if the product is crystallizable, we would expect it to have a grainy morphology near the surfaces and coarser spherulites in the central region, which stays hotter for a longer period (polymers are good thermal insulators). This has been experimentally verified.

## *Annealing Effects*

The effects of annealing of melt crystallized structures are very similar to those in single crystals, if the polymers crystallize to a high degree of crystallinity, as in HDPE. That is:

1. Lamellar thickening is observed which is quite pronounced at annealing temperatures near the normal melting point. Lamellar thickening increases linearly with log $t$. This morphological change may again result in "pitting," but in this case there is void formation. Such an effect is indicated for HDPE in Fig. 3.14.

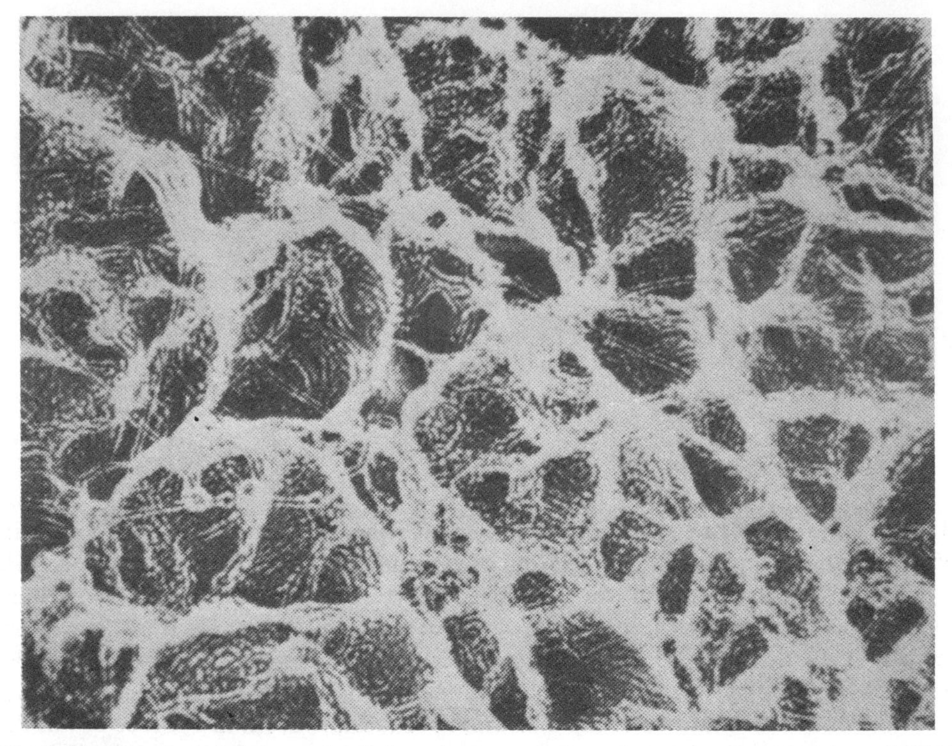

**Fig. 3.14** Replica of linear polyethylene fracture surface after annealing for 3 hr at 126°C. Specimen initially melt crystallized at 110°C. [Reprinted with permission from A. J. McHugh and J. M. Schultz, *Phil. Mag.*, **24**, 155 (1971).]

2. Annealing increases the melting point of the polymer. The effect is due to lamellar thickening and again is larger and faster, the higher the annealing temperature used.
3. A unique feature of annealing in melt crystallized structures is the partial exclusion of low molecular fractions from the new crystalline structure (24).

Quenching and annealing of materials of low crystallinity, such as polyethylene terephthalate, is poorly understood. Annealing greatly improves crystallinity and changes mechanical properties to those associated with stronger, if slightly more brittle, materials (25). Crystal-crystal transformations are also known to occur, and partial transformation from folded to extended chain crystallization is suspected. Figure 3.15 represents the effect of annealing on the tensile modulus of isotactic polypropylene. The tensile modulus doubles in value, with large improvements occurring at high temperatures. The ultimate elongation decreases, as expected.

### *Nucleating Agents*

Nucleating agents are small crystalline particles that are dispersed in the crystallizing melt and remain solid at the crystallization temperature. Examples of nucleating agents are other high melting point polymers and inorganic or organic crystal-

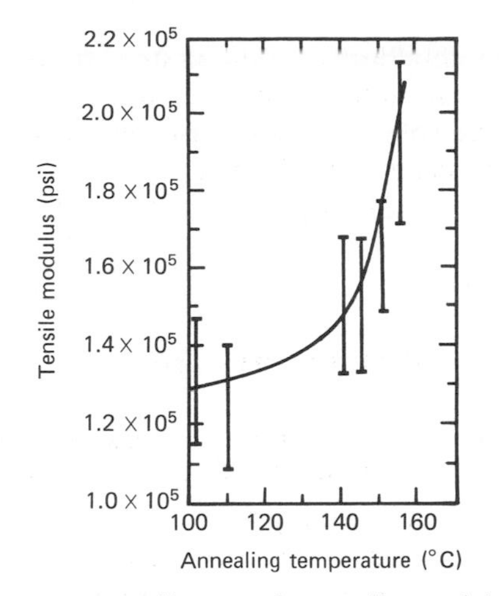


**Fig. 3.15** Effect of long-term annealing on the tensile modulus of *PP*. [Reprinted with permission from P. S. Schotland, *Polym. Eng. Sci.*, **6**, 244 (1966).]

line matter, all in finely dispersed form. As such, they act as *heterogeneous nuclei*. Therefore they favor the nucleation step, and thus shift to a lower region the temperature at which the maximum crystallization rate occurs. Consequently, at quench conditions, the resulting degree of crystallinity is higher, and at crystallization conditions, characterized by higher crystallization temperatures, the resulting spherulitic morphology is more or less grainy, like the one ordinarily obtained at low crystallization temperatures (26). Figure 3.16 illustrates the latter effect. The mechanical properties of externally nucleated structures are those of grainy morphologies discussed previously.

In polymer processing there is an additional reason for the use of nucleating agents; namely, they control the spherulitic size distribution in processed articles. As already discussed, in all processed polymers except very thin films, cooling

**Fig. 3.16** Effect of nucleating agents on the spherulitic morphology of isotactic polypropylene (*a*) without nucleating agents, (*b*) with 1% finely dispersed indigo particles. [Reprinted with permission from *Mechanical Properties of Polymers*, N. M. Bikales, ed., Wiley, New York; p. 62 (1970).]

results in lower surface temperatures, thus grainy morphology, and higher core temperatures with larger spherulites. Nucleating agents make the core morphology more grainy, rendering the spherulitic size distribution less nonuniform. It is not clear that this effect is always desirable from a mechanical properties point of view. It is certainly beneficial to optical properties, if the nucleating agents have compatible refractive indices.

## 3.5 Effects of Crystallization Pressure

The crystallizing of polymer melts under high hydrostatic pressure is quite common in processes such as injection molding. It turns out that pressure is a variable that affects all aspects of the crystallization process and the properties of the resulting structures. In the first place, according to the Clausius–Clapeyron equation, the equilibrium melting point at any pressure is related to the atmospheric pressure melting point as follows:

$$(T_m)_P = (T_m)_{P_{atm}} \exp\left[\frac{(\hat{V}_a - \hat{V}_c)(P-1)}{\Delta H_f}\right] \tag{3.5-1}$$

where $\hat{V}_a$ and $\hat{V}_c$ are the amorphous and crystalline specific volumes, $P$ is the hydrostatic pressure in atmospheres, and $\Delta H_f$ is the atmospheric heat of fusion of the polymer. The equation tells us that there is an appreciable melting point elevation with pressure. It means that a pressurized melt at a given crystallization temperature is under a higher degree of supercooling. The spherulitic morphology and the rate of crystallization will be affected along the lines discussed in Section 3.4. If the crystallization temperature is to the right of the temperature that results in a maximum crystallization rate ($T_R$), the rate of crystallization in the pressurized melt will increase. The opposite will occur at $T_c < T_R$. The rate of nucleation is also expected to increase at the effectively increased degrees of supercooling.

Following the results of Fig. 3.7b, lamellar thickening might be expected in pressurized melts. This has indeed been observed (27). This work indicates that at very high pressures, about 500 $MN/m^2$ (75,000 psi), even complete extended chain crystallization is possible.

Maxwell et al. (1) point to some very interesting considerations of the effects of pressure during processing. Consider Fig. 3.17, which describes the compressibility of HDPE at various pressures and very low rates of increase of pressure. Although at low pressures the obtained response is that expected of liquids, at higher pressures, specific for each temperature but in the range of 3–20,000 psi, large "compressibilities" are observed. Of course this result, due to the phenomenon of pressure induced crystallization (28), is a consequence of the melting point elevation discussed earlier. In a similar fashion, and because molecular mobility is hindered at high pressures, the value of $T_g$ is also increased. Consider now the pressure-temperature history of a polymer melt in an injection mold cavity. Initially it is hot and at a high pressure, but as its "skin" cools and the mold gate freezes, or the valve is closed, the core hydrostatic pressure decreases. In Figure 3.17, the polymer melt would be going toward the origin in a *nonisothermal* fashion. In doing so, and if the pressure were high, pressure induced crystallization would be operative and a spectrum of structures and morphologies would be obtained; that is, near

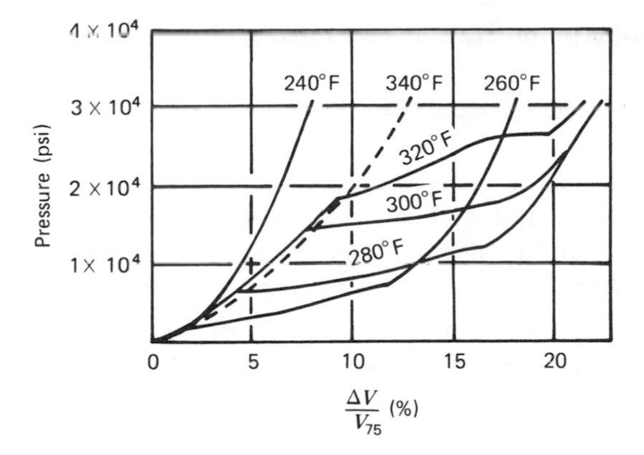


**Fig. 3.17** Bulk compressibility of HDPE at various pressures showing the phenomenon of pressure induced crystallization. [Reprinted with permission from S. Matsuoka and B. Maxwell, *J. Polym. Sci.*, **62**, 174 (1962).]

the surface (high pressure) structures with a high $T_m$ and lamellar thickening are obtained, whereas near the center, structures that were associated with atmospheric pressure crystallization would result. This effect occurs simultaneously with the effects of conductive cooling, which cools the part surface first. As a matter of fact, the pressure reinforces the effects of conductive cooling, since it increases the degree of supercooling in the surface region.

In our discussion of the possible effects of the temperature and pressure history of a polymer melt in an injection mold cavity on the crystallization process, we did not consider molecular orientation effects resulting from flow during the mold filling stage. Such effects are considered next.

## 3.6 Strain and Flow Induced Crystallization

Strain induced crystallization was first reported for natural rubber (29). Under large uniaxial (and later biaxial) elongations, this material crystallizes at room temperature. The resulting degree of crystallinity is small, about 10%, and it is lost reversibly upon removal of the load.

The phenomenological and thermodynamic explanations for this phenomenon are as follows. The externally applied strain results in internal chain stretching and alignment in the same direction as the deformation. The presence of the crosslinks does not allow the chains to relax and become random coils. Thus the major resistance to crystallization, the attainment of local parallel chain alignment, is greatly reduced, and the process of crystallization is favored.

Thermodynamically during any change of phase

$$\Delta G = \Delta H - T\Delta S \tag{3.6-1}$$

where $G$ is the Gibbs free energy, $H$ is the enthalpy, and $S$ is the conformational entropy; $\Delta$'s represent the change in these variables during the phase change. For an equilibrium process, $\Delta G = 0$. Such would be the case for crystallization or melting at the melting point $T_m$. Thus we have

$$T_m = \frac{\Delta H}{\Delta S} \tag{3.6-2}$$

Chain alignment will greatly decrease $\Delta S$ from the amorphous rubber to the crystalline, since the oriented chains have lower conformational entropy. On the other hand, the orientation will not affect the enthalpy of the amorphous rubber, and the value of $\Delta H$ in Eq. 3.6-2, as dictated by the theory of rubber elasticity. It follows from Eq. 3.6-2, then, that there can be an appreciable melting point elevation for the stretched rubber, which will increase the degree of supercooling, the driving force of crystallization.

Although extensive studies of strain induced crystallization in natural rubber gave clues about the resulting morphology, it was Keller who started elucidating the subject (30, 31). Working with HDPE, which he crystallized while drawing it from the melt, he found that primary nucleation occurs along a line, or *row*, rather than at a point, which is the usual quiescent observation leading to a spherulitic morphology. This phenomenon is known as *row nucleation*, where the rows are parallel to the draw direction. Crystalline growth occurs in the perpendicular plane, and cylindrical crystallites or highly anisotropic spherulites result. The morphological picture is one that involves a core of extended chain nuclei on which folded chains can grow into lamellae. The picture is very similar to the solution stirred single crystals, except that there is a much higher degree of connectedness in the drawn melt or stretched rubber morphologies. Incidentally, the nature and extent of these tie molecule connections play a major role in the resulting mechanical properties. Morphological details of row nucleated structures in stretched natural rubber, appear schematically in Fig. 3.18.

The cases of the stretched natural rubber and the HDPE drawn from the melt represent examples of crystallization induced by strain and by flow. Nevertheless,

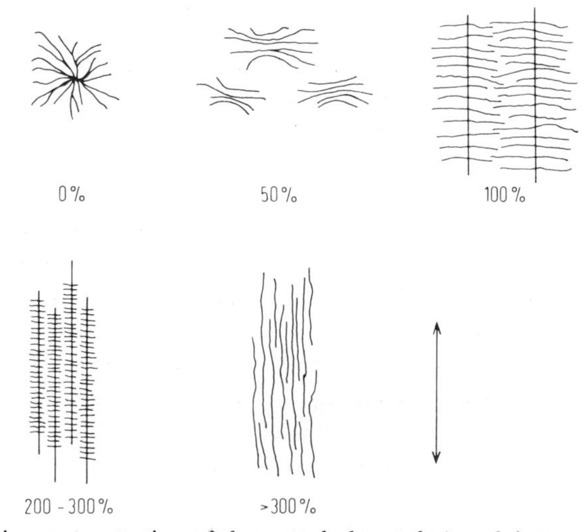


**Fig. 3.18** Schematic representation of the morphology of *cis*-polyisoprene at various levels of stretching. [Reprinted with permission from E. H. Andrews, *Angew. Chem. Int. Ed.*, **13**, 113 (1974).]

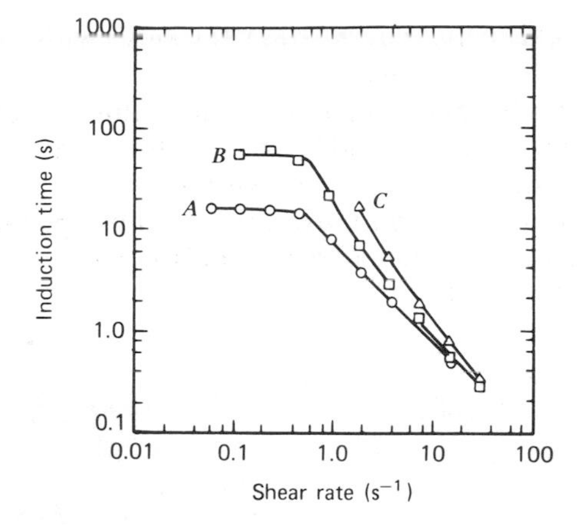


**Fig. 3.19** Induction time for flow-induced crystallization as a function of shear rate of HDPE with $M_w = 1.89 \times 10^4$, $\bar{M}_n = 1.38 \times 10^4$, $[\eta] = 2.60$. Curve *A*, $T = 125°C$; curve *B*, $T = 126°C$; curve *C*, $T = 129°C$. [Reprinted with permission from R. R. Lagasse and B. Maxwell, *Polym. Eng. Sci.*, **16**, 189 (1976).]

the two phenomena are really one and the same in that they are both caused by *local chain alignment.* In the first case, the chains stay aligned because of the entanglements, whereas in the HDPE melt case the chains have to crystallize *before* relaxing to random coils in order to give rise to row nucleated morphologies. Figure 3.19 indicates the role of the chain alignment present in flowing melts, causing nucleation. The nucleation induction time in HDPE is plotted versus the applied shear rate (parallel plate flow) at three crystallization temperatures. First, the induction time decreases dramatically with increasing shear rate, while crystallization rate increases dramatically. This is because chain alignment favors the formation of nuclei. Second, at low shear rates (below $0.5\ s^{-1}$ in these cases), the induction time is not affected. This is because the Deborah number is small; thus the melt has a small relaxation time compared to the experimental time $1/\dot{\gamma}$, and the chains relax during the flow process.

A very useful and interesting application of strain and flow induced crystallization, and an excellent example of structuring, is the discovery and development of the so-called hard elastic films and fibers of polypropylene, polyoxymethylene-acetal copolymer, and other polymers (32). These structures are obtained by melt extrusion and subsequent crystallization under high stress. The temperature, temperature gradients, and stress level are characteristic of each polymer. Extensive morphological investigation of the hard elastic structures has revealed a superstructure of stacked lamellar aggregates. This morphology is a consequence of row nucleation and had been observed before (33). Upon annealing, the stacked lamellar morphology (the lamellae are almost perpendicular to the draw direction) is perfected, and lamellar thickening also takes place. Figure 3.20 is an electron micrograph showing lamellar stacking. Specifically, the following set of properties is observed:

1. A rubbery, elasticlike stress-strain response with 50–95% recovery from high strains; slow recovery.
2. Energy elastic behavior; that is, the modulus decreases with temperature rather than increase as would be the case for normal entropy elastic rubbers.
3. The bulk density decreases with extension.
4. The porosity increases with extension and disappears upon recovery. The increase is large, and the pores are surface connected, making such materials suitable for membranes.

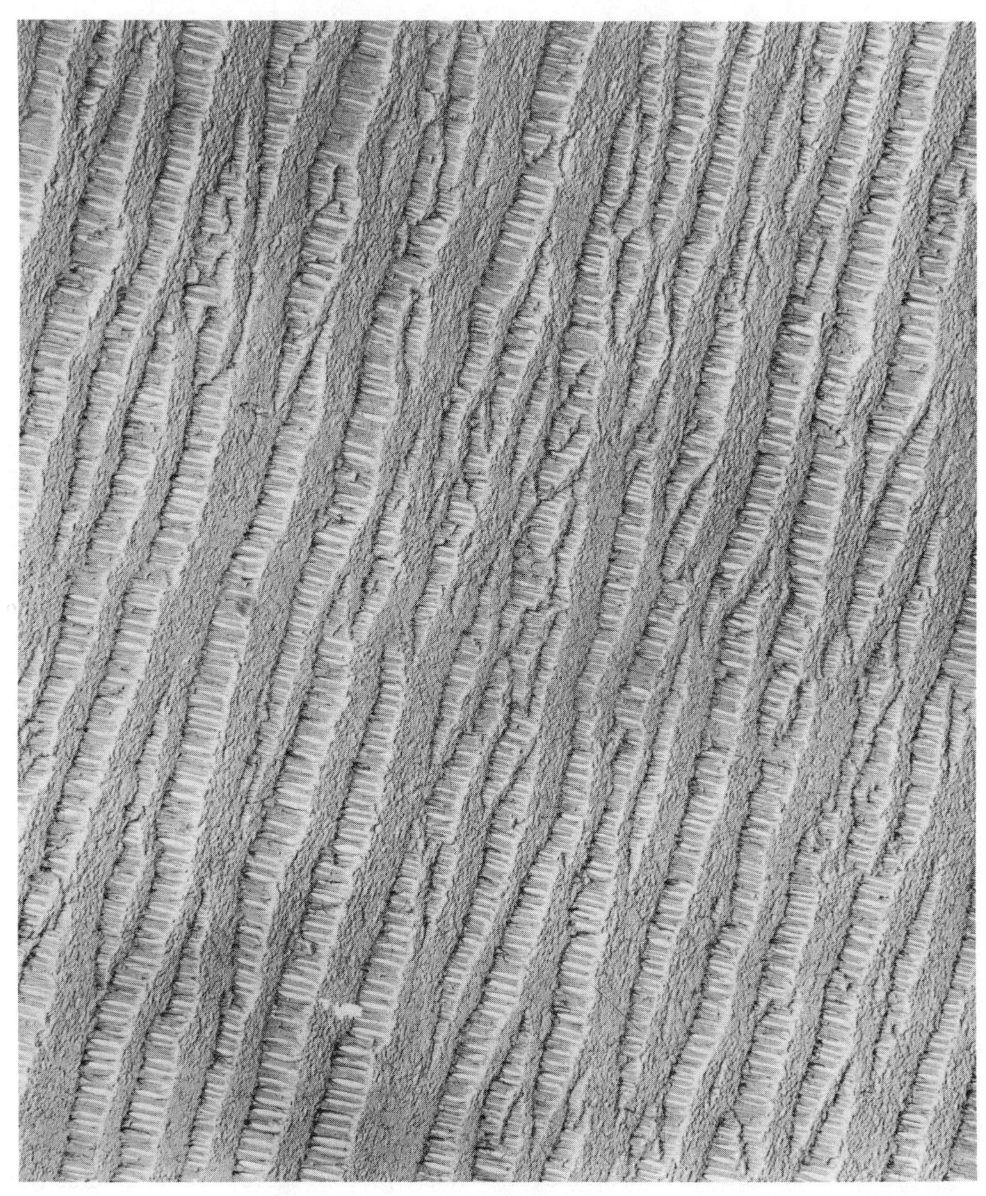

**Fig. 3.20** Electron micrograph of "Celgard" hard elastic film, 15,000×. (Courtesy of I. L. Hay, Celanese Res. Co., Summit N.J.)

The explanation of the above-described behavior in terms of the structure is as follows. The stacked lamellae are pulled apart locally, under tension, while staying "glued" together at other spots. These local spots of connectedness, together with the initially formed extended chains and tie molecules, supply the overall mechanical strength. The separation occurs at spots that contain interlamellar amorphous material or after plastic deformation of the folded chains. This explains the porosity that is generated upon stretching and the rubberlike (hard elastic) behavior. The energy elastic behavior is explained in terms of the bent lamellae, which store elastic energy during deformation.

Porter and his co-workers (34) have applied a combination of high pressure and flow orientation in structuring HDPE. The polymer is extruded at 134°C through a conical die that gives an effective draw ratio of 46. Since this temperature is below the prevailing melting point, because of the orientation, the polymer crystallizes in the die. This increases the pressure necessary for extrusion to very high levels (2000–2500 atm). Thus the process is in effect solid state extrusion. As mentioned earlier in this chapter, the resulting fibers are very strong, having a modulus of $7 \times 10^{10}\ N/m^2$, and are quite transparent. The high modulus, a little over one-third the theoretical (–C–C–) extended chain value, is due to the chain extension and alignment in the extrusion direction, and to the presence of long crystallites that are also axially oriented. The possibility of the presence of extended chains in extrusion drawn HDPE is indicated in the recent characterization work of Porter (35). The high effective draw ratios (the process is called "extrusion drawing") that can take place because of the high hydrostatic pressure make possible the foregoing reinforcing features in these fibers. Their transparency can be explained in terms of the same features and the complete absence of spherulites.

Recently, transparent structures of HDPE that also exhibit increased strength and melting point, have been reported. The first structuring method (36) involves continuous shearing of HDPE in a double cone and plate closed cavity apparatus in the temperature range of 136–146°C. The shear rates used were in the range of $1\text{–}10\ s^{-1}$, and the shearing times necessary to complete the process were of a few minutes duration, decreasing with increasing shear rates and decreasing temperature. The kinetic expressions that have been obtained (shear rate and temperature dependent nucleation and growth rates) point to the potential of simple shear flow in creating the necessary conditions for chain extension, under circumstances very close to those found in common polymer processing.

The second structuring method of HDPE (37) is a potential improvement on the Porter approach. An ordinary flat entrance capillary is used, and care is taken not to have crystallization occur in the entrance region, which would "plug up" the system and require the use of high pressures. Crystallization is induced *in* the capillary by applying steep temperature gradients, with temperature decreasing with increasing capillary length. Again, the chains are induced to crystallize while they are fairly well aligned. The resulting process, which is still experimental, does avoid the need to use high pressure, and again indicates that flow alone (in this case probably a combined tensile-shear field, due to entance elongations) is sufficient for chain stretching, which gives rise to structuring.

Recently, experimental studies have been carried out on the crystallization kinetics and morphology development during melt spinning (38, 39). It was found with HDPE that the effect of the applied tensile stress is to increase the crystallization rates by several orders of magnitude and decrease the crystallization induction times at least a hundred fold. The morphology changes from spherulitic at low spinline stresses to a row (or cylindritic) type.

Finally, HDPE films which are transparent, but mechanically inferior to the structured HDPE obtained by the extrusion drawing process, have been obtained in a rolling operation (40). The rolling is very fast and it takes place under rapid quenching. Thus the extensional flow field probably creates row nucleation and the rapid quenching, a great many tie molecules.

From the above it is evident that structuring can be induced in pure drag, shear—elongational, and pressure flows, with or without the assistance of externally applied pressures and temperature gradients. An excellent example in polymer processing, where all of the above effects exist, is the injection mold filling process (see Chapter 14). Morphological studies by Clark (41) and others have shown that the surface layer consists of row nuclei formed by chains aligned in the direction of flow. Lamellae grow on them and they are normal to both the surface and the direction of flow. This morphology persists over a very small thickness. It is followed by a layer where the lamellae are still normal to the surface but randomly oriented with respect to the filling direction. Thermal gradients play an important role in the formation of this transcrystalline region. A central spherulitic core essentially free of orientation effects exists in thick articles. This layered morphology, the result of flow, temperature, and pressure levels and gradients, can be varied by carefully choosing molding conditions. Controlled variations, though, are not yet possible.

## 3.7 Effects of Cold Drawing

Chain rearrangement below the melting point can be obtained with annealing as well as chain rearrangement and alignment in specific directions with a number of "cold forming" operations such as cold rolling and cold drawing. We shall examine the effects of the latter on the morphology of initially spherulitic samples. The cold drawing process is carried out at temperatures between $T_g$ and $T_m$.

Looking at the conventional polymer morphology first, we see that crystals are chain folded lamellae in a spherulitic overall arrangement. Amorphous components reside at the lamellar surfaces in the form of loose cilia or tie molecules. Tie molecules and other amorphous material connect neighboring spherulites. Thus the structure be viewed as a composite of the amorphous component, which is rubbery, and solid crystals containing defects. When we deform such a sample, we expect to produce changes in the crystal lattice as in any other crystalline material. Additionally, we can force lamellae to slip past one another, and we can break them up by pulling and unfolding of chains. During all these crystalline state changes, the amorphous material provides the major resistance to deformation.

The events above result in changing not only the morphology but also the overall shape of the sample. Sample extension is not uniform, but it occurs through the propagation of a "neck" (a reduction in the cross-sectional area) occurring at some axial point of the stretched sample. Figure 3.21 shows the necking and drawing configurations, as well as the stress-strain response. The upper yield point is associated with the neck formation and the low slope region with the cold

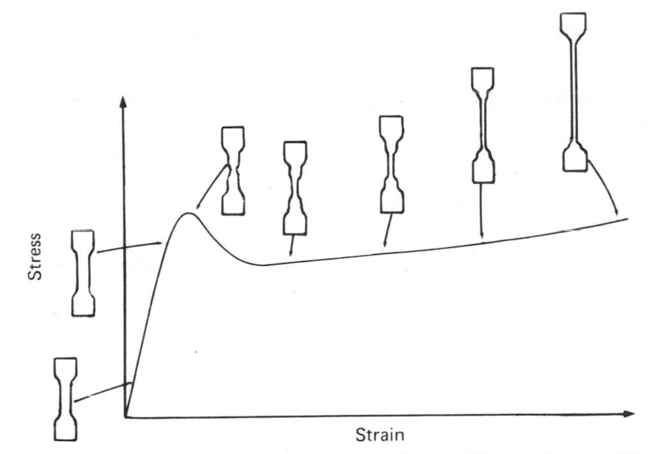


**Fig. 3.21** Schematic stress-strain curves for a semicrystalline polymer. The shape of tensile specimens at several stages is indicated. [Reprinted with permission from J. M. Schultz, *Polymer Materials Science*, Prentice-Hall, Englewood Cliffs, N.J. 1974.]

drawing or neck propagation process. Finally, the last high slope region is one that indicates strain hardening resulting in brittle fracture.

Necking and cold drawing also occur during the uniaxial stretching of fibers and films. Commercial fibers are always drawn after spinning, to increase the modulus of elasticity. Films are uniaxially oriented occasionally to make them anisotropic; they are also stretched to large strains to induce the phenomenon of fibrillation, in which the film separates in the transverse direction into loosely connected fibers that can be spun or twisted into yarns and ropes. Because of the large industrial interest, morphological changes of spun fibers during cold drawing have been studied in great detail (42). The results of these studies indicate that the onset of necking is not brought about by local temperature rises, which would promote melting and flow, thus structured rearrangements. Furthermore, overall sample geometry softening of the specimen is insufficient to explain yielding. The conclusion is that yielding is the result of the ability of polycrystalline "composites" to accommodate stress induced destruction of the crystalline units. In this process both the amorphous and the crystalline phases are involved. A "molecular" model of the morphological changes initiated with necking and propagated by cold drawing is the following (7), also indicated in Fig. 3.22:

1. The lamellae slip rigidly past one another. Lamellae parallel to the direction of draw cannot slip; thus spherulites become anisotropic. At this stage, at which necking begins, the strain is accommodated almost entirely by the interlamellar amorphous component.
2. Since the amorphous "ties" are almost completely extended, slip-tilting of the lamellae is induced.
3. Lamellar breakup occurs through chain pulling and unfolding; the chains pulled still connect the fragments of the lamellae.
4. The lamellar fragments slip further in the direction of draw and get aligned. They now form fibrils of alternating crystal blocks and stretched amorphous regions, which may also contain free chain ends, and some chain folds. Thus the lamellae break into fragments that end up stacked in the axial direction.

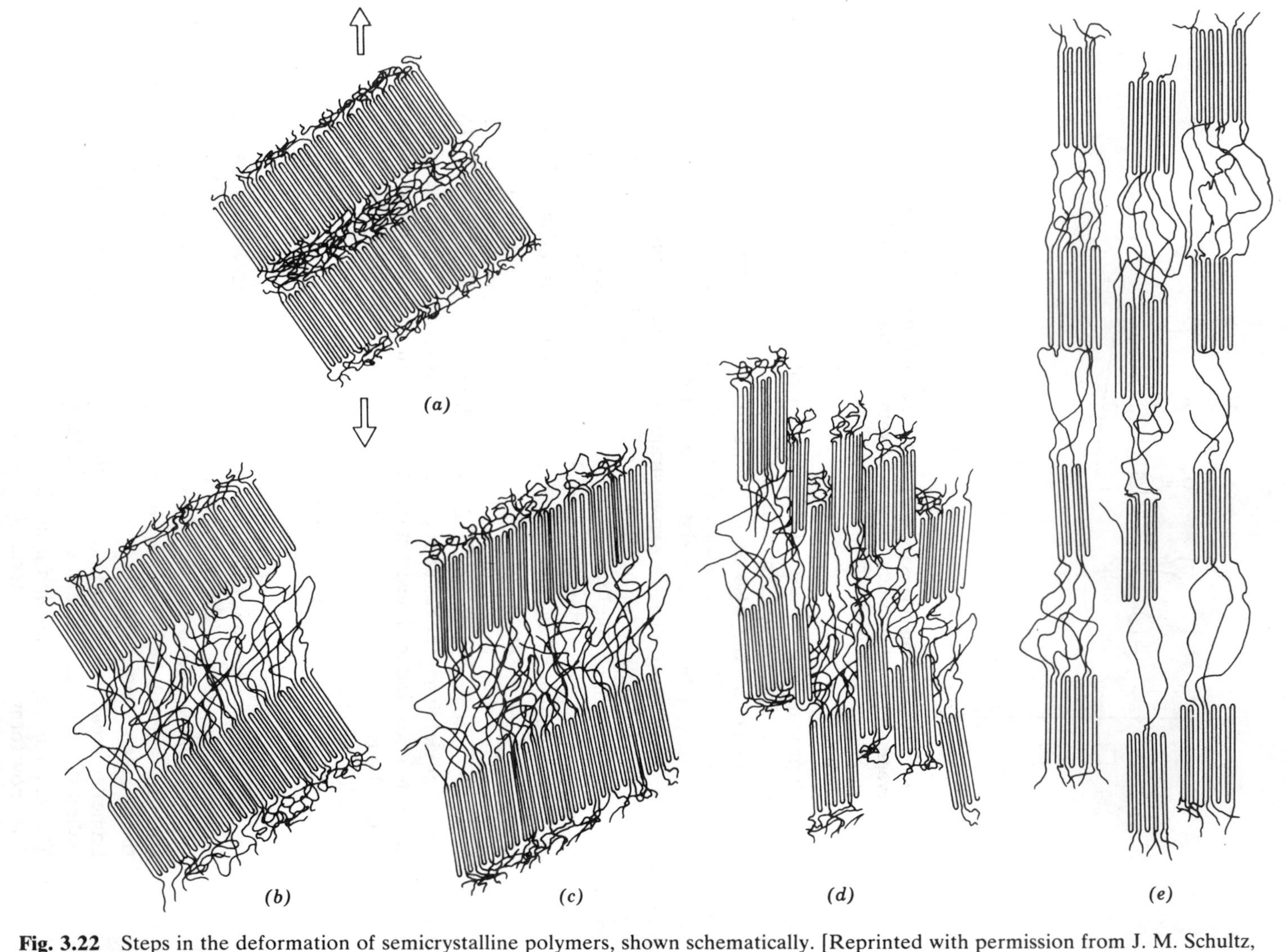


**Fig. 3.22** Steps in the deformation of semicrystalline polymers, shown schematically. [Reprinted with permission from J. M. Schultz, *Polymer Materials Science*, Prentice-Hall, Englewood Cliffs, N.J. 1974.]

Tie molecules that connect these fragments in the draw direction provide the strength of the microfibrils in the fiber. Thus the goal in a fiber structuring operation is to employ the values of the parameters of spinning and drawing processes, which increase the fraction of tie molecules.

The effect of annealing highly drawn fibers is to more or less destroy the stacked microfibrillar structure and to reconstitute the initial structure. Similarly, if the drawing process is very slow and occurs at high enough temperatures, realignment of the torn off fragments tends to reconstruct the lamellar morphology. It follows, then, that the cold drawing processing variables (rate of extension and temperature), as well as initial morphology and molecular weight and distribution, play a decisive role in fiber drawing. The work of Ward cited in the beginning of this chapter, resulting in very high modulus HDPE drawn fibers, is an example of structuring by properly selecting the starting material and the values of the fiber drawing process variables.

## 3.8 Amorphous Polymers

Morphologically, the generally accepted view of amorphous polymers is that they are *orderless* structures composed of coiled and entangled molecules. At low temperatures relative to their glass transition, these molecules are motionless. Only their atoms vibrate, with amplitudes that increase with temperature. Near the glass transition, atomic vibrations of neighboring atoms become more cooperative, resulting in motions of segments of the chains at $T_g$. The binding segmental (secondary force) energy becomes equal to the thermal energy locally at this temperature. The segmental motion frequency is high enough to impart ductility to amorphous polymers (semicrystalline too, since they contain amorphous regions), but too small to make flow possible at practical processing rates with levels of viscosities that are manageable from the processing point of view. Thus only at 40–50°C above $T_g$ can typical amorphous polymers be considered as processable melts, where interchain slippage does not require very high stresses.

The lack of order in amorphous polymers has as a consequence the existence of structural "voids," which are stationary below $T_g$ and mobile above $T_g$. Above $T_g$ the role of voids is pivotal, since unoccupied space is needed for segmental diffusion (which results in flow) to occur. That is, polymer segments "jump into" voids (leaving new ones behind) in diffusional and flow processes. The rate of these segmental jumps increases with increasing temperature and decreases with increasing segmental binding energy, that is, intermolecular forces, which are usually expressed in terms of the flow activation energy. The kinetic theory of liquids of Eyring (43) is based on this "structural" picture. It was first developed for monomers, where the voids are taken to be of molecular, not segmental, size.

Stresses applied to amorphous polymers below the glass transition temperatures result in instantaneous deformations of the order of 5%—which are larger than those obtained with inorganic glasses—with plastic deformation occurring. Both shear and normal stress yielding (crazing) is possible. In the latter case, studies of crazes in polystyrene films subjected to uniaxial load have shown the craze

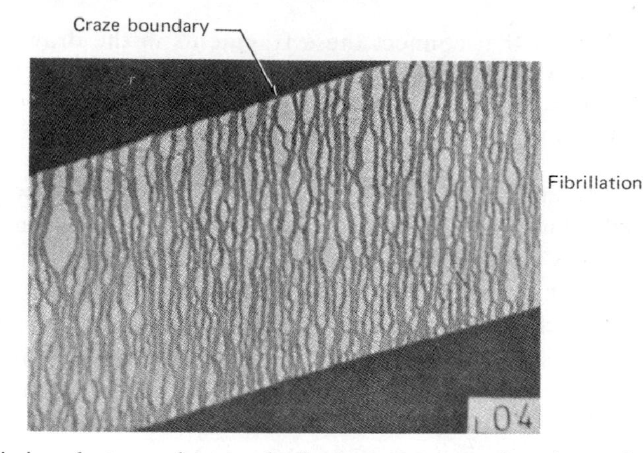


**Fig. 3.23** Transmission electron micrograph of polystyrene craze interiors, taken from a thin film deformed in the electron microscope. [Reprinted with permission from P. Beahan, M. Bevis, and D. Hull, *Phil. Mag.*, **24**, 1267 (1971).]

region to be occupied by drawn fibrils and voids (Fig. 3.23). This structure is reminiscent of the hard elastic morphology discussed in Section 3.6, but it is of small structuring interest because in glassy polymers crazes initiate cracks that are responsible for the overall brittle fracture at the 5% level elongations. The morphology of Fig. 3.23 serves to demonstrate that there is local flow below the $T_g$.

Evidence of motion below $T_g$ has also been reported in the absence of externally applied forces. Geil and his co-workers have reported crystallization of crystallizable polymers, that is, annealing, below $T_g$ (44, 45). This phenomenon, together with other experimental results and thermodynamic considerations indicating that chain folding is possible in amorphous polymers (46, 47), has led to the view that amorphous polymers are not totally orderless but are characterized by ordered nodular regions in both the melt and glassy states. Although evidence obtained from electron microscopy on the nodular structure may not be free of experimental errors, the idea of small ordered regions in amorphous polymers cannot be excluded.

The foregoing discussion serves to point out that there is very little structuring possible while polymers are in the glassy state, since the chains are "practically" motionless. It also indicates that any chain orientation present in the glassy state will be more or less permanent, unless the polymer is brought near or above $T_g$. Such "frozen in" strains, which impart anisotropic properties to polymers in the glassy state, are brought about by chain orientation in flow and deformation processes at $T > T_g$. As such, they can be the result of deliberate structuring, analogous to obtaining special crystalline morphologies by deforming crystallizable polymer melts.

When deformation, or flow, is imposed on polymer melts, polymer chains acquire a preferred orientation and their mean end-to-end distance increases. Experimentally, the degree of orientation can be measured by streaming birefringence (see Section 3.9), making use of the optical anisotropy of the polymer chain. If the sample is thin and quickly quenched, anisotropic shrinkage will yield similar results upon annealing. To be able to predict a priori the degree of orientation from the imposed flow field, the duration of flow, and the polymer properties, however, molecular theories are needed that accurately describe the behavior of polymer chains. Such theories are still limited, although a great deal of work is conducted in this field (48). Molecular theories for dilute solutions are in a more advanced state and can provide information about the behavior of macromolecular chains in flow fields. These theories are based primarily on a detailed analysis of the forces at work. Such forces include hydrodynamic drag, which tends to orient the chain in the direction of flow, the Brownian motion forces, which tend to randomize (coil) the chain, and the intramolecular (intersegmental) forces. One particular theory developed by Rouse (49) predicts the following relationship between the mean square end-to-end distance $\langle (r_N - r_1)^2 \rangle$ and shear rate $\dot{\gamma}$ in parallel plate flow

$$\langle (r_N - r_1)^2 \rangle = \langle (r_N - r_1)^2 \rangle_{\text{equil}} + \frac{2kT}{G}\left( \sum_{1}^{N-1} \lambda_n \right)^2 \dot{\gamma}^2 \tag{3.8-1}$$

where $G$ is the chain "modulus" (dynes/molecule · cm) and $\lambda_n$ is relaxation time. As expected, an increase in shear rate results in an elongation of the chain, the elongation being more effective, the larger are the relaxation times $\lambda_n$.

Dilute solution molecular theories also provide information on the effect of the rate of orientation relaxation on cessation of flow. The driving force for such relaxation is Brownian motion, and the retarding force is that representing the hydrodynamic drag, which depends on the viscosity of the solution. Thus temperature, as expected, has a large effect on the rate of orientation loss. It is worth noting that the "frozen in" strains that are associated with long relaxation times will persist the longest.

Orientation and structuring in polymer processing, however, involves melts, rather than dilute solutions, which have experienced complex deformation and temperature histories. In the absence of adequate constitutive equations to yield information on the degree of orientation resulting from such an experience, we rely heavily on experimental evidence, obtained from flow birefringence experiments (50, 51). The relationship between deformation history of vitrified polymers and orientation measured by birefringence was recently studied by White et al. (52). They concluded that orientation, as measured by birefringence, may be predictable from the knowledge of the stress field at the time of vitrification.

Of practical interest is also the work of Cleereman et al. (53) on polystyrene monofilament orientation. Their results support the expectation that a constant stress results in elastic, delayed elastic, and viscous (irrecoverable) deformations. It is the delayed elastic deformation that appears as "frozen in" strain upon quenching. The elastic strain relaxes immediately, and the viscous has no tendency to recover. Furthermore, they found that orientation in the glassy state increases linearly as the amount of deformation above $T_g$ increases as in Fig. 3.24. The effect of temperature is very pronounced.

Although orientation, as measured by birefringence, showed the above-described dependence on filament temperature and stretch rate, other physical properties did not. This was explained by Cleereman in the following way. At low filament temperatures, stretching results in the deformation of small relaxation time structural units; the large relaxation time deformations (rearrangement of

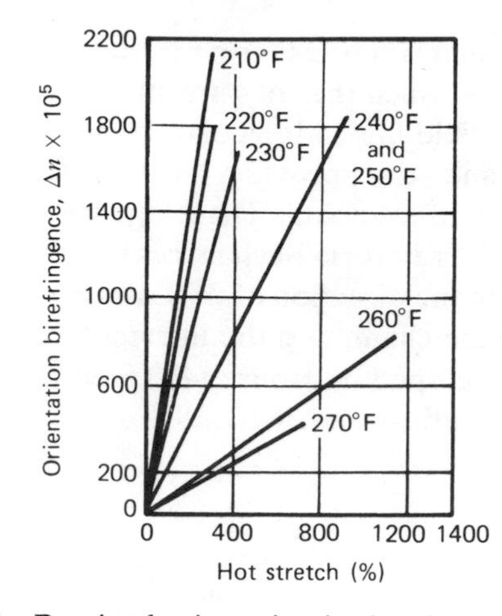


**Fig. 3.24** Retained orientation in the glassy state as a function of hot filament temperature; samples were drawn at a constant rate and quenched to 49°C. [Reprinted with permission from K. J. Cleereman et al., Modern Plastics, May 1956, p. 19.]

entire chains) take too long. Thus the quenched sample from a low temperature stretch process may contain many oriented chain segments, giving rise to optical anisotropy, but these segments would randomize quickly upon annealing. This is exactly what their shrinkage experiments at $T > T_g$ show (Fig. 3.25). The time derivative of the shrinkage is plotted versus time, for samples stretched at constant stretch rate but different temperatures. The orientation resulting from low temperature stretching is large but shorter lived than that achieved at high temperatures. The orientation in the latter case would be longer lasting in the glassy state and would affect physical properties more, since long molecular segments, or entire chains, are involved. This is what is observed.

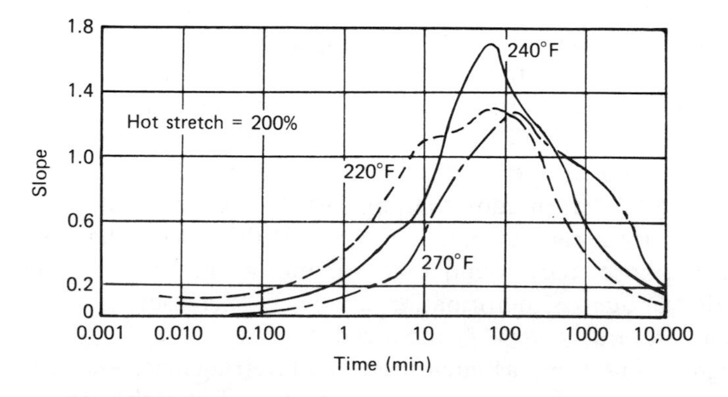


**Fig. 3.25** Slopes of the time dependent shrinkage of hot filament drawn fibers indicating the type of structural units involved in hot melt stretching. The temperatures indicate the temperature during stretching. All shrinkage experiments were conducted at 90°C. [Reprinted with permission from K. J. Cleereman et al., Modern Plastics, May 1956, p. 19.]

## 3.9 Experimental Determination of Polymer Molecular Orientation

We have discussed how orientation of polymer molecules is brought about during polymer processing and how this orientation affects the kinetics and morphology of the crystalline phase and the properties of polymer products. Before closing this chapter, we consider briefly the experimental methods that are available for measuring the macromolecular orientation of amorphous and polycrystalline polymers undergoing simple types of deformation. Several investigators have pioneered in this field, notably Stein (54–56), Hermans et al. (57), Wilchinsky (58), Bunn (59), and Samuels (60) and the University of Tennessee investigators (38, 39, 52) have made more recent contributions.

Following Samuels (60), if $P$ is an observed property of a polycrystalline polymer, $P^\circ_{cr}$ and $P^\circ_{am}$, the associated *intrinsic* properties of the crystalline and amorphous portions, and $\beta$ the volume fraction occupied by the crystalline portion, then the following relation can be stated

$$P = \beta P^\circ_{\text{cr}} + (1-\beta)P^\circ_{\text{am}} \tag{3.9-1}$$

Such an expression (a mixing rule) holds, no matter what the orientation of the polymer molecules, for properties that are not affected by the anisotropicity of the polymer chain. Density is an example of such a property, which is observed isotropically. As a matter of fact, the crystalline volume fraction can be calculated from Eq. 3.9-1 when applied to the density of a polycrystalline sample

$$\beta = \frac{\rho - \rho_{\text{am}}}{\rho_{\text{cr}} - \rho_{\text{am}}} \tag{3.9-2}$$

where $\rho$ is the measured sample density, $\rho_{\text{cr}}$ can be calculated from the unit cell dimensions and unit cell monomer molecular weight, and $\rho_{\text{am}}$ can be measured directly from the pure amorphous structure, when available, or extrapolated from the molten state.

But some material properties are affected by the anisotropic nature of the polymer chains, thus their state of orientation, becoming *anisotropic* (different in different directions) when the polymer molecules are oriented. Thus although Eq. 3.9-1 is adequate for such properties when the sample is *unoriented*, a knowledge of orientation is needed in general. This is expressed as follows:

$$\Delta P_{\text{oriented}} = \beta P^\circ_{\text{cr}} f_{\text{cr}} + (1-\beta)P^\circ_{\text{am}} f_{\text{am}} \tag{3.9-3}$$

where $\Delta P_{\text{oriented}}$ is the observed anisotropic property and $f_{\text{cr}}$ and $f_{\text{am}}$ are *orientation functions* characterizing the average orientation of the polymer chains in the crystalline and amorphous phases, respectively (60). That is, in defining $f_{\text{cr}}$ and $f_{\text{am}}$ in Eq. 3.9-3, averaging over the other two crystallographic axes has taken place. A specific example of this equation can be found with the birefringence of uniaxially stretched polymers (e.g., fibers). If the difference in the refractive index of the stretched sample in the directions parallel and perpendicular to stretching, $n_{\parallel} - n_{\perp}$, is defined as the birefringence $\Delta n$, then

$$\Delta n = \beta \Delta n^\circ_{\text{cr}} f_{\text{cr}} + (1-\beta)\Delta n^\circ_{\text{am}} f_{\text{am}} \tag{3.9-4}$$

where $\Delta n^\circ_{\text{cr}}$ and $\Delta n^\circ_{\text{am}}$ are the birefringence values of the *perfectly oriented* crystalline and amorphouse phases, respectively (i.e., the refractive index difference in the directions parallel and perpendicular to the polymer chains).

Thus $f_{cr}$ and $f_{am}$ are of interest from a structuring point of view, since they denote the average chain orientation in the crystalline and amorphous phases of processed or simply deformed polymers. Their values can be obtained experimentally through the study of wide angle X-ray scattering (WAXS), birefringence, sonic modulus, and shrinkage upon annealing above $T_g$. We briefly discuss each method, giving literature references for more detailed treatment.

The crystalline content can be determined from density measurements (Eq. 3.9-2); common instruments are the dilatometer and the density gradient column (61). Comparison between the WAXS behavior of the polycrystalline and amorphous (molten) states also leads to good estimates of $\beta$ (62).

WAXS measurements of oriented polycrystalline samples can yield not only the orientation function of the main polymer chain, but also those of the other two crystallographic axes. Briefly, having a single crystal composed of unit cells of crystallographic axes $a$, $b$, and $c$ (let us assume for the moment that they are mutually perpendicular, as in orthorhombic PE), we can consider planes composed of lattice points that act as mirrors to incident radiation. Whether the reflections from successive planes of spacing $d$ are reinforced (so that there is a *measurable* result) depends on the angle $\theta$ between the incident ray and the reflecting plane, as expressed by Bragg's law

$$n\lambda = 2d \sin \theta \tag{3.9-5}$$

where $\lambda$ is the radiation wavelength and $n = 1, 2, 3, \ldots$ As the crystal is rotated, there will be angles that result in reinforced reflections, thus in definite "spots" on a

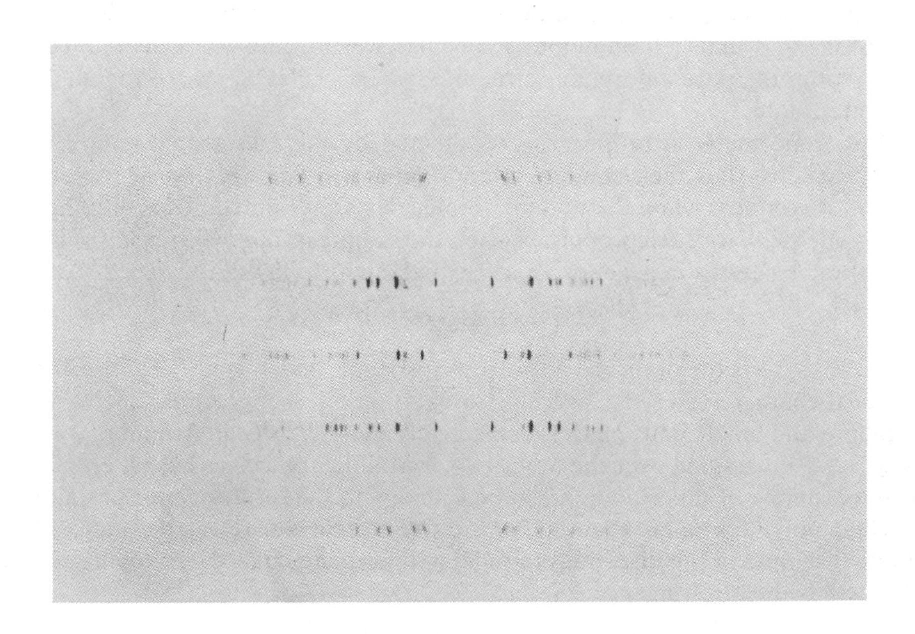

**Fig. 3.26** Single crystal rotation X-ray diffraction pattern of salicylic acid. [Reprinted with permission from R. J. Samuels, *Structured Polymers*, Wiley, New York, 1974.]

photographic plate, as in Fig. 3.26. With a polycrystalline sample, however, where the crystallites are randomly oriented, the resulting reflections are cones having the incident beam as their axis. Thus rings are observed on the photographic plate (the Debye-Scherrer rings) as in Fig. 3.27. Since the crystallites are oriented in a given direction, the corresponding circular rings become arcs and, under "complete" orientation, dots. Specifically, the orientation function for each crystallographic axis

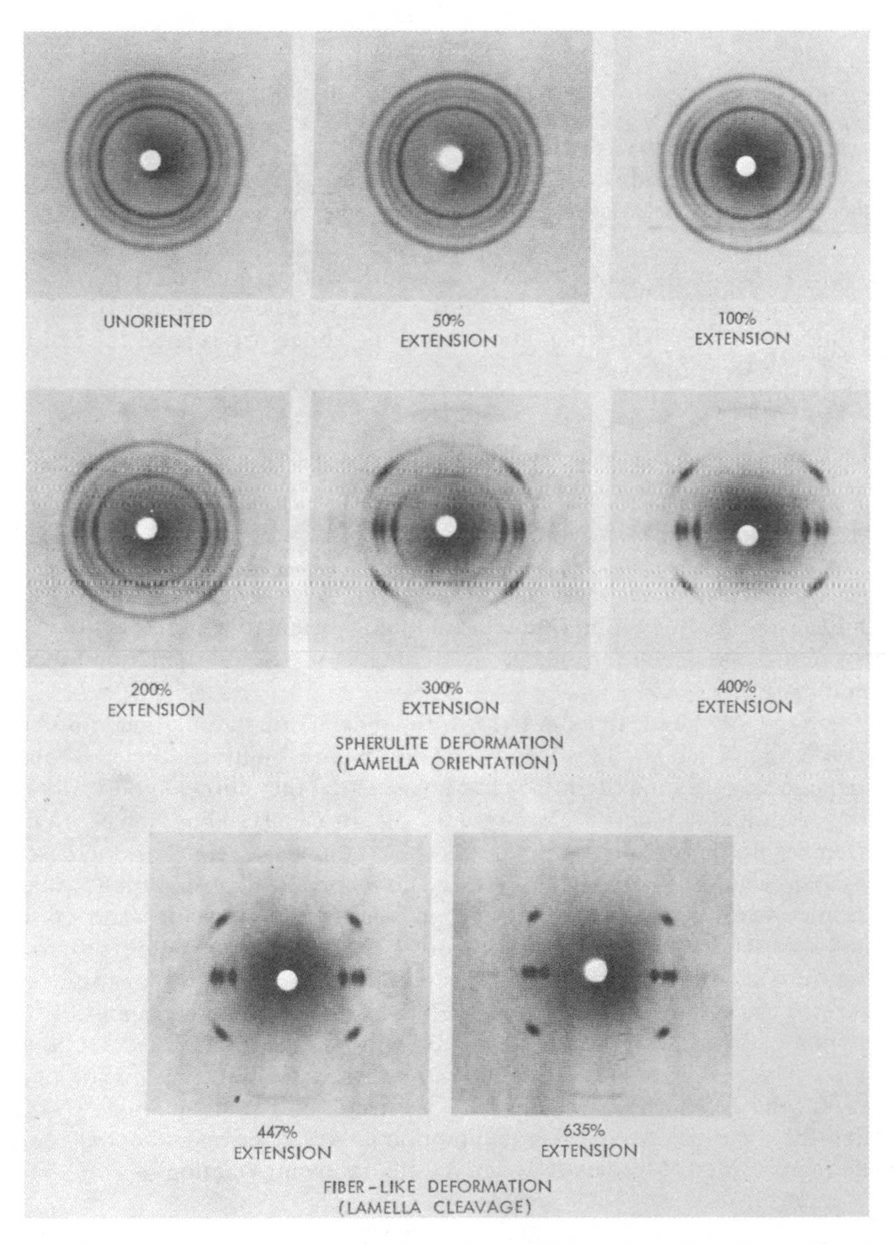


**Fig. 3.27** Effect of isotactic polypropylene film extension on WAXS patterns. [Reprinted with permission from R. J. Samuels, *Structured Polymers*, Wiley, New York, 1974.]

$a$, $b$, $c$ can be obtained through the relations

$$f^a = \frac{3\,\overline{\cos^2 \Phi_{a,z}} - 1}{2}$$
$$f^b = \frac{3\overline{\cos^2 \Phi_{b,z}} - 1}{2} \qquad (3.9\text{-}6)$$
$$f^c = \frac{3\overline{\cos^2 \Phi_{c,z}} - 1}{2}$$

The values of $\overline{\cos^2 \Phi_{j,z}}$, $j = a, b, c$, can be evaluated as follows:

$$\overline{\cos^2 \Phi_{j,z}} = \frac{\int_0^{\pi/2} I_{hkl}(\Phi_{j,z}) \cos^2 \Phi_{j,z} \sin \Phi_{j,z}\, d\Phi_{j,z}}{\int_0^{\pi/2} I_{hkl}(\Phi_{j,z}) \sin \Phi_{j,z}\, d\Phi_{j,z}} \qquad (3.9\text{-}7)$$

where $z$ is the stretch axis, which is also an axis of symmetry (e.g., fiber) and $I_{hkl}(\Phi_{j,z})$ is the intensity diffracted from the $(hkl)$ planes, which are normal to the $j$ crystallographic axis. The orientation functions above are related for orthogonal unit cell crystals as follows:

$$f^a + f^b + f^c = 0 \qquad (3.9\text{-}8)$$

$$\sum_{j=1}^{3} \overline{\cos^2 \Phi_{j,z}} = 1 \qquad (3.9\text{-}9)$$

When a particular crystallographic axis, say $c$, is perpendicular to $z$, $f^c = -0.5$; when it is parallel, $f^c = 1$, and when it is randomly oriented with respect to $z$, $f^c = 0$. In Eqs. 3.9-3 and 3.9-4, for uniaxial stretching, $f_{cr} = f^c$, the orientation function of the *main polymer chain*.

Stein (54–56) has dealt extensively with uniaxial orientation functions obtained from WAXS; Wilchinsky (58) and Sack (63) have analyzed the problem with nonorthogonal cells, and Stein (55) has investigated the more complex problem of biaxial orientation. Figure 3.28 presents the three crystallographic orientation functions of PE during melt spinning (39). As the takeup velocity is increased, the chains align with the fiber axis, and axes $a$ and $b$ become almost perpendicular to it.

Sonic waves propagate with different speeds in amorphous and crystalline domains of polycrystalline materials. Furthermore, with polymeric systems their propagation is different in the directions parallel and perpendicular to the chains. This is because in the chain direction, stretching of C–C bonds is responsible for the phenomenon, whereas in the transverse direction weaker intermolecular bonds are involved. The *sonic modulus E* is a measure of the speed of transmission of sound waves. It follows from the foregoing that the transverse sonic modulus $E_t^0$ is much smaller than the $E_{ax}^{\circ}$. With this assumption and a continuum model for polycrystalline polymers, Moseley (64) derived the following relation

$$\frac{1}{E} = \frac{1 - \overline{\cos^2 \theta}}{E_t^{\circ}} \qquad (3.9\text{-}10)$$

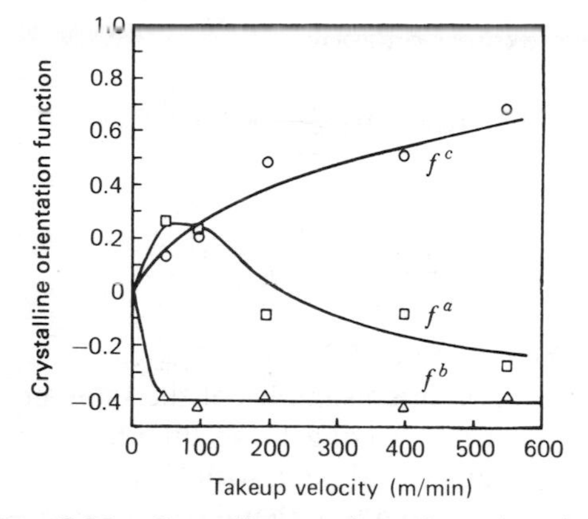


**Fig. 3.28** Crystalline orientation functions versus takeup velocity developed during melt spinning of HDPE: polymer flow rate, $1.93 \pm 0.02$ g/min; extrusion temperature, $207 \pm 3$°C. [Reprinted with permission from J. E. Spruiell and J. L. White, *Polym. Eng. Sci.*, **15**, 660 (1975).]

Where $\theta$ is the angle between the polymer chains and the direction of transmission. The sonic modulus is the Young modulus (65), and as such it is related to the compressibility $K$ and the bulk modulus $B$, as follows:

$$K = \frac{1}{B} = \frac{3(1-2\nu)}{E} \tag{3.9-11}$$

where $\nu$ is Poisson's ratio. Furthermore, the compressibility can be expressed by the mixing rule

$$K = \beta K_{\text{cr}} + (1-\beta) K_{\text{am}} \tag{3.9-12}$$

where $K_{\text{cr}}$ and $K_{\text{am}}$ are the compressibilities of the crystalline and amorphous phases, respectively.

Combining the three equations above and using $\nu = \frac{1}{3}$ (for this case, $K = 1/E$), we obtain for an unoriented sample

$$\frac{3}{2E_u} = \frac{\beta}{E^0_{t,\text{cr}}} + \frac{1-\beta}{E^0_{t,\text{am}}} \tag{3.9-13}$$

and for oriented samples

$$\frac{1}{E_{\text{or}}} = \frac{\beta}{E^\circ_{t,\text{cr}}}(1 - \overline{\cos^2 \theta_{\text{cr}}}) + \frac{1-\beta}{E^\circ_{t,\text{am}}}(1 - \overline{\cos^2 \theta_{\text{am}}}) \tag{3.9-14}$$

Defining as before an orientation function for the chain axis with respect to the direction of stretching $z$, we write

$$f^c = \frac{3\,\overline{\cos^2 \Phi_{c,z}} - 1}{2}$$

We obtain from the two expression above

$$\tfrac{3}{2}(\Delta E^{-1})=\frac{\beta f_{\rm cr}}{E^{\circ}_{t,\rm cr}}+\frac{1-\beta}{E^{\circ}_{t,\rm am}}f_{\rm am} \tag{3.9-15}$$

where

$$(\Delta E^{-1})\equiv(E_{\rm u}^{-1}-E_{\rm or}^{-1}) \tag{3.9-16}$$

From Eq. 3.9-13, the quantities $E^{\circ}_{t,\rm cr}$ and $E^{\circ}_{t,\rm am}$ can be evaluated by measuring the sonic modulus of two unoriented samples of the same polymer but of different degrees of crystallinity $\beta$. Then, having these quantities as well as $f_{\rm cr}=f^{c}$ from WAXS measurements, $f_{\rm am}$ can be evaluated with Eqs. 3.9-15 and 3.9-16, that is, by conducting sonic modulus experiments with uniaxially oriented fibers.

Thus it is possible to determine $f_{\rm cr}$ and $f_{\rm am}$ independently from WAXS and sonic modulus measurements. Still, *birefringence* measurements are useful because of the ease of the experimental method. Referring to Eq. 3.9-4, if $\Delta n^{\circ}_{\rm cr}$ and $\Delta n^{\circ}_{\rm am}$ are known, then in conjunction with WAXS measurements yielding $f_{\rm cr}$, $f_{\rm am}$ can be obtained. But the intrinsic refractive indices are not easy to measure (they require perfectly oriented chains), and they are usually estimated; $\Delta n^{\circ}_{\rm cr}$ from $C_{36}$ paraffin single crystals and $\Delta n^{\circ}_{\rm am}$ from bond polarizability values have been calculated by Bunn (59). Samuels (60) describes an experimental method of evaluating these quantities. In both cases birefringence offers another way of determining $f_{\rm am}$.

The value of $\Delta n$ can be related to the stress field $\sigma$ through the stress optical coefficient $C$. Specifically, the principal birefringences are proportional to the differences in principal stresses (52). Thus for uniaxial stretching the uniform $\Delta n$ is proportional to the spinline stress $\sigma$

$$\Delta n = C\sigma \tag{3.9-17}$$

and since the melt is amorphous, we obtain from Eq. 3.9-4 that

$$\Delta n = \Delta n^{\circ}_{\rm am} f_{\rm am} \tag{3.9-18}$$

Combining the two equations above we obtain that

$$f_{\rm am}=\frac{C}{\Delta n^{\circ}_{\rm am}}\sigma \tag{3.9-19}$$

The quantity $(C/\Delta n^{\circ})$ turns out to be constant for all polymers. Thus $f_{\rm am}$ is proportional to $\sigma$, during flow and during melt stress relaxation. Similar expressions relating orientation to the stress field can be obtained for shear flows (52).

*Shrinkage* upon annealing is certainly the most widely used method for determining macromolecular orientation in processed articles. It is quite simple to carry out and is based on the intuitive concept that there is a direct relation between the observed dimensional changes of the processed specimen and the randomization of oriented polymer chains. The method can be used on the entire specimen to obtain an average orientation, or on microtomed sections to obtain an orientation distribution, as we shall see in Section 14.1.

For amorphous polymers annealed above $T_g$, it seems reasonable that most of the molecular randomization would result in dimensional changes of the specimen.

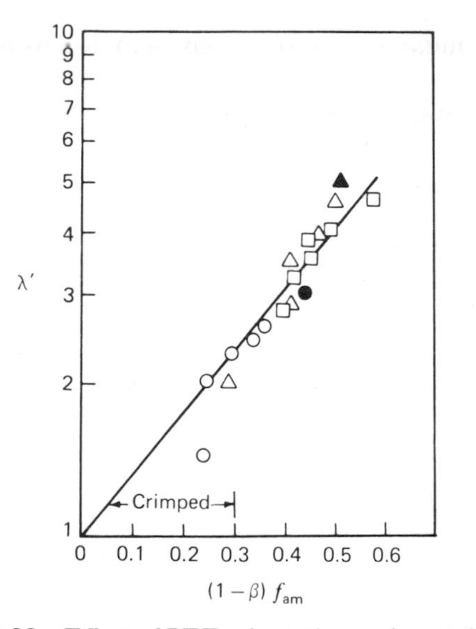


**Fig. 3.29** Effect of PET orientation and crystallinity on the residual extension ratio after shrinkage following annealing: △ 5× draw ratio, oil annealed fibers; □, 5× draw ratio, air annealed fibers; ○ 3× draw ratio, oil annealed fibers; ▲, ●, unannealed fibers. [Reprinted with permission from R. J. Samuels, *Structured Polymers*, Wiley, New York, 1974.]

This is because of the presence of intra-and intermolecular entanglements. Since the randomization process is a strain recovery phenomenon, it is characterized by relaxation times that increase with decreasing annealing temperature (see Figs. 3.24 and 3.25). Upon complete recovery, the amount of shrinkage should be directly related to both the amorphous orientation function $f_{am}$ and the frozen in stress present in the unannealed system (Eq. 3.9-19).

It is somewhat surprising that a similar relationship exists between the amorphous orientation function and annealing shrinkage in low crystallinity polymers such as PET. Samuels (66) describes results with annealing PET fibers that were drawn at 80°C to different extensions after having been spun in an essentially unoriented and amorphous form. Upon drawing ($T_{g_{unoriented}} = 67°C$), chains become oriented and crystallization occurs (about 30%). Upon annealing [$T_{g_{oriented}} \simeq 100$–160°C (66)], amorphous orientation decreases and crystallites thicken. Despite the latter phenomenon, a direct relationship between shrinkage and $f_{am}$ exists, as demonstrated in Fig. 3.29, indicating that the amorphous portion controls the results of thermal annealing. In this figure $\lambda' = L_a/L_0$, the ratio of the sample length after annealing to the original length. A high value of $\lambda'$ denotes that annealing was not allowed to occur. It is worth noting in Fig. 3.29 that the initial draw ratio and the annealing environment and temperature are not important to the relationship between $\lambda'$ and $f_{am}$. The factor $(1-\beta)$ corrects for crystallinity content changes during the process.

It is evident from the foregoing discussion that much theoretical and experimental work is needed to elucidate the questions of how much orientation (involving what chain units) results from a given melt flow and temperature history. This question is central to efforts of structuring both amorphous and crystallizable polymers through polymer processing, since orientation affects the bulk mechanical, optical, and dielectric properties of solid polymers. It is beyond the scope of this book to discuss the solid properties of polymers. This subject has been treated in a number of texts such as those by Alfrey (68), Leaderman (69), Treloar (70), Tobolsky (71), Ferry (72), Bueche (73), Nielsen (74), and more recently by Vincent (75), McCrum, Read, and Williams (76), Stein (77), Ward (78), and Samuels (60).

In this chapter we have attempted to highlight the molecular reasons that make structuring possible and to point out its great potential. We discuss specific structuring questions and effects in connection with the various shaping operations during which they take place.

## *REFERENCES*


1. B. Maxwell, C. G. Gogos, L. L. Blyler, Jr., and R. M. Mineo, *Soc. Plast. Eng. Trans.*, **4**, 165 (1964).
2. L. R. G. Treloar, *Polymer*, **1**, 95 (1960); F. C. Frank, *Proc. Roy. Soc.*, **A319**, 127 (1970).
3. P. H. Lindenmeyer, address at the Plastics Institute of America Annual Conference, 1974.
4. D. Hansen and G. A. Bernier, *Polym. Eng. Sci.*, **12**, 204 (1972).
5. P. H. Geil, *Polymer Single Crystals*, Wiley-Interscience, New York, 1963.
6. A. Keller, *Rep. Prog. Phys.*, **31**, Pt. 2, 623 (1968).
7. J. M. Schultz, *Polymer Materials Science*, Prentice-Hall, Englewood Cliffs, N.J., 1974.
8. P. H. Lindenmeyer, *Polym. Eng. Sci.*, **14**, 456 (1974).
9. D. M. Gezovich and P. H. Geil, *Polym. Eng. Sci.*, **8**, 202 (1968).
10. S. Kapur and C. E. Rogers, *J. Polymer Sci.*, **10**, 2107 (1972).
11. H. P. Nadella, H. M. Henson, J. E. Spruiell and J. L. White, *J. Appl. Polym. Sci.*, **21**, 3003 (1977).
12. A. Wereta, Jr., and C. G. Gogos, *Polym. Eng. Sci.*, **11**, 19 (1971).
13. J. Powers, J. D. Hoffman, J. J. Weeks, and F. A. Quinn, Jr., *J. Res. NBS—A, Phys. Chem.*, **69A**, 335 (1965).
14. K. H. Illers and H. Hendus, *Macromol. Chem.*, **113**, 1 (1968).
15. A. J. Pennings and A. M. Kiel, *Kolloid-Z., Z. Polym.*, **205**, 160 (1965).
16. A. J. Pennings and A. M. Kiel, *Kolloid-Z, Z. Polym.*, **237**, 336 (1970).
17. A. M. Rijke and L. Mandelkern, *J. Polym. Sci.*, **8A-Z**, 225 (1970).
18. G. I. Taylor, *Phil. Trans. Roy. Soc.*, **A223**, 289 (1923).
19. A. J. Pennings and M. F. J. Pijpers, *Macromolecules*, **4**, 261 (1970).
20. J. W. Rutter and B. Chalmers, *Can. J. Phys.*, **31**, 96 (1953).
21. A. Keller, *J. Polymer Sci.*, **17**, 351 (1955).
22. H. D. Keith, F. J. Padden, and R. G. Vadimsky, *J. Polym. Sci.*, **4A-2**, 267 (1966).
23. M. Avrami, *J. Chem. Phys.*, **7**, 1103 (1939); **8**, 212 (1940); **9**, 177 (1941).
24. H. D. Keith and F. J. Padden, *J. Appl. Phys.*, **35**, 1220; 1286 (1964).
25. R. L. Miller, *Polymer*, **1**, 135 (1960).
26. S. Inoue, *J. Polym. Sci.*, **A1**, 2013 (1963).
27. P. H. Geil, R. R. Anderson, B. Wunderlich, and T. Arakawa, *J. Polym. Sci.*, **2A**, 370 (1964).
28. S. Matsuoka and B. Maxwell, *J. Polym. Sci.*, **62**, 174 (1962).
29. M. Katz, *Nature*, **13**, 410 (1925).
30. A. Keller, *J. Polym. Sci.*, **15**, 31 (1955).

31. A. Keller and M. J. Machin, *J. Macromol. Sci.-Phys.*, **B1**, 41 (1967).
32. B. J. Sprague, *J. Macromol. Sci.-Phys.*, **B8**, 157 (1973).
33. M. J. Hill and A. Keller, *J. Macromol. Sci.-Phys.*, **B3**, 153 (1969).
34. J. H. Southern and R. J. Porter, *J. Appl. Polym. Sci.*, **14**, 2305 (1970).
35. N. E. Weeks, S. Mori, and R. S. Porter, *J. Polym. Sci.*, **13**, 2031 (1975); N. E. Weeks and R. S. Porter, *ibid.*, 2049.
36. V. Tan and C. G. Gogos, *Polym. Eng. Sci.*, **16**, 512 (1976).
37. J. R. Collier, T. Y. T. Tam, J. Newcome, and N. Dinos, *Polym. Eng. Sci.*, **16**, 204 (1976).
38. J. R. Dees and J. E. Spruiell, *J. Appl. Polym. Sci.*, **18**, 1053 (1974).
39. J. E. Spruiell and J. L. White, *Polym. Eng. Sci.*, **15**, 660 (1975).
40. T. K. Kwei, T. T. Wang, and H. E. Bain, *J. Polym. Sci.*, **C31**, 87 (1970).
41. E. S. Clark, *Appl. Polym. Symp.* **20**, 325 (1973).
42. (a) J. L. White, K. C. Dharod, and E. S. Clark, *J. Appl. Polym. Sci.*, **18**, 2539 (1974); (b) A. Peterlin, *Polym. Eng. Sci.*, **9**, 172 (1969).
43. S. Glasstone, K. J. Laidler, and H. Eyring, *Theory of Rate Processes*, McGraw-Hill, New York, 1941, Chapter 9.
44. A. Siegmann and P. H. Geil, *J. Macromol. Sci.-Phys.*, **B4**, 239 (1970).
45. K. Neki and P. H. Geil, *J. Macromol. Sci.-Phys.*, **B8**, 295 (1973).
46. G. S. Y. Yeh and P. H. Geil, *J. Macromol. Sci.-Phys.*, **B1**, 235 (1967).
47. G. S. Y. Yeh, *J. Macromol. Sci.-Phys.*, **B6**, 465 (1972).
48. R. B. Bird, O. Hassager, R. C. Armstrong, and C. F. Curtis, *Dynamics of Polymeric Liquids*, Vol. II, *Kinetic Theory*, Wiley, New York, 1977.
49. P. E. Rouse, Jr., *J. Chem. Phys.*, **21**, 1272 (1953).
50. W. Philippoff, *Nature*, **178**, 811 (1956).
51. J. L. S. Wales, *The Application of Flow Birefringence to Rheological Studies in Polymer Melts*, Monograph, Delft University Press (1976).
52. K. Oda, J. L. White and E. S. Clark, *Polym. Eng. Sci.*, **18**, 53 (1978).
53. K. J. Cleereman, H. J. Karam and J. L. Williams, *Modern Plastics*, May 1953, p. 19.
54. R. S. Stein and F. H. Norris, *J. Polym. Sci.*, **21**, 381 (1956).
55. R. S. Stein, *J. Polym. Sci.*, **31**, 327 (1958), and **24**, 383 (1957).
56. C. R. Desper and R. S. Stein, *J. Appl. Phys.*, **37**, 3990 (1976).
57. J. J. Hermans, P. H. Hermans, D. Vermaas, and A. Weidinger, *Rec. Trav. Chim.*, **65**, 427 (1946).
58. Z. W. Wilchinsky, *J. Appl. Phys.*, **30**, 792 (1959).
59. C. W. Bunn and R. de P. Danbeny, *Trans. Faraday Soc.*, **50**, 1173 (1954).
60. R. J. Samuels, *Structured Polymers*, Wiley, New York, 1974.
61. G. M. Brauer and E. Horowitz, in *Analytical Chemistry of Polymers*, Vol. III, G. M. Kline, ed., Wiley-Interscience, New York, 1962, Chapter 1.
62. W. Roland, *Polymer*, **5**, 89 (1964).
63. R. A. Sack, *J. Polym. Sci.*, **54**, 543 (1961).
64. W. W. Moseley, Jr., *J. Appl. Polym. Sci.*, **3**, 266 (1960).
65. J. W. Ballos and S. Silverman, *Text. Res. J.*, **14**, 282 (1944).
66. R. J. Samuels, *J. Polym. Sci.*, **A2**, 781 (1972).
67. J. H. Dumbleton, T. Murayama, and J. P. Bell, *Kolloid Z., Z. Polym.*, **228**, 54 (1968).
68. T. Alfrey, Jr., *Mechanical Behavior of High Polymers*, Wiley-Interscience, New York, 1948.
69. H. Leaderman, "Viscoelasticity Phenomena in Amorphous High Polymeric Systems," in *Rheology—Theory and Applications*. F. R. Eirich, ed., Academic Press, New York, 1958, Chapter 1.
70. L. R. G. Treloar, *The Physics of Rubber Elasticity*, 2nd ed., Clarendon Press, Oxford, 1958.
71. A. V. Tobolsky, *Properties and Structure of Polymers*, Wiley, New York, 1960.
72. J. D. Ferry, *Viscoelastic Properties of Polymers*, 2nd ed., Wiley, New York, 1970.

73. F. Bueche, *Physical Properties of Polymers*, Wiley-Interscience, New York, 1962.
74. L. E. Nielsen, *Mechanical Properties of Polymers*, Reinhold, New York, 1962.
75. P. I. Vincent, "Mechanical Properties of Polymers: Deformation," in *Physics of Plastics*, P. D. Ritchie, ed., Van Nostrand, Princeton, N.J., 1965, Chapter 2.
76. N. G. McCrum, B. Read, and G. Williams, *Anelastic and Dielectric Effects in Polymeric Solids*, Wiley, New York, 1967.
77. R. S. Stein, "Studies of the Deformation of Crystalline Polymers," in *Rheology—Theory and Applications*, Vol. 5, F. R. Eirich, ed., Academic Press, New York, 1969, Chapter 6.
78. I. M. Ward, *Mechanical Properties of Solid Polymers*, Wiley, New York, 1971.

# 4

# Surface Properties



This chapter deals with surface phenomena and surface properties that are pertinent to polymer processing. We first briefly examine surface free energy, cohesive and adhesive energies, and other basic concepts associated with surfaces that have a concrete theoretical base; then we use these concepts in the description of surface phenomena, such as adhesion, dry friction, wear, and lubrication, which are not completely understood and depend heavily on experimental and material conditions.

## 4.1 Surface Tension

The concept of *surface tension* is necessary to explain the tendency of a free liquid surface to assume a shape that has a minimum area. Surface tension, though, is not to be associated with a modulus of elasticity of the surface; surfaces are not elastic, and films of viscous liquids will stretch at constant rates (i.e., will flow) before they break.

Consider a segment of surface one meter wide. Let the surface tension be $\Gamma$ in units of newtons per meter, and allow the surface to be stretched such that 1 $m^2$ new surface area is created. This process requires $\Gamma$ joules of work. If this surface then contracts reversibly to its original size, it will perform $\Gamma$ joules of work. It can be shown that surface tension is equal numerically, as well as in units, to the *surface free energy* per unit area.*

When a molecule of a liquid migrates to the surface, it does so by moving against unbalanced intermolecular forces, since the surface is not surrounded symmetrically by other liquid molecules. Thus molecules in a liquid surface layer possess additional energy because work has been done to bring them to the surface. The same argument holds for solids.

* The definition $\Gamma = dG/dA$ holds for single component phases in contact, neglecting adsorption at the interface. For multiple component phase surface tension expressions see W. J. Moore, *Physical Chemistry*, 3rd ed., Prentice-Hall, Englewood Cliffs, 1962, Chapter 18.

Because of the equivalence of surface tension and surface free energy per unit area, the equilibrium thermodynamic relation (1)

$$G = H + T\left(\frac{\partial G}{\partial T}\right)_p \tag{4.1-1}$$

when applied to the process of surface formation, becomes

$$G_s = H_s + T\left(\frac{\partial G_s}{\partial T}\right)_p \quad \text{or} \quad \Gamma = H_s + T\frac{\partial \Gamma}{\partial T} \tag{4.1-2}$$

The term $\partial\Gamma/\partial T$ is negative for all liquids; thus $H_s$, which is the energy involved in the creation of one unit area new surface, is greater than $\Gamma$, the work expended in forming the surface. The difference $(H_s - \Gamma)$ is supplied by the bulk through a cooling of the neighboring layers during stretching.

The surface tension is the quantity that controls the spreading of a liquid droplet on a planar liquid or solid surface; this phenomenon is usually called *wetting*. For spreading to be possible, the process must result in a negative or no change in free energy.

### *Spreading on Liquids*

A drop of liquid A resting on liquid B, as in Fig. 4.1*a*, spreads to cover an additional unit area of B. This process results in an increase of the upper surface of liquid A and the interface AB, and a corresponding loss of the same amount of area of liquid B. Thus since surface tension is surface free energy per unit area, we see that the process results in the following change* of free energy per unit of surface area changed:

$$\Gamma_A + \Gamma_{AB} - \Gamma_B$$

Therefore, for spreading to occur

$$\Gamma_A + \Gamma_{AB} - \Gamma_B \leq 0 \tag{4.1-3}$$

From the relation above we can conclude that if liquid A spreads on B, then liquid B does not spread on A.

The free energy change per unit area involved in the process of liquid A *adhering* to liquid B is

$$\Gamma_{AB} - \Gamma_A - \Gamma_B \tag{4.1-4}$$

On the other hand, in bringing together two surfaces of unit area of liquid A, the free energy of *cohesion* would be $-2\Gamma_A$. Thus for the process of adhesion of A on B to be favored over cohesion of A, the free energy change for the former has to be smaller than the latter

$$\Gamma_{AB} - \Gamma_A - \Gamma_B \leq -2\Gamma_A$$

* $\Gamma_A$, $\Gamma_B$ are usually written as $\Gamma_{AV}$, $\Gamma_{BV}$ to denote that the surface tensions are measured in equilibrium with the respective vapors.

or

$$\Gamma_A + \Gamma_{AB} - \Gamma_B \leq 0$$

This expression, which is identical to Eq. 4.1-3, implies the conclusion that the criteria of liquid adhesion and spreading are identical.

### Spreading on Solids

Referring to Fig. 4.1*b*, for the liquid drop to rest in equilibrium on the plane solid surface, the three surface tensions $\Gamma_{LV}$, $\Gamma_{SV}$, and $\Gamma_{SL}$ must form an equilibrium system of forces for a given contact angle $\theta$, that is (2),

$$\Gamma_{SV} - \Gamma_{SL} = \Gamma_{LV} \cos\theta \tag{4.1-5}$$

Using the relationships between adhesion and cohesion developed above, adhesion or wetting would occur when

$$\Gamma_{SL} - \Gamma_{SV} - \Gamma_{LV} < -2\Gamma_{LV} \tag{4.1-6}$$

From Eq. 4.1-5, the left-hand side of Eq. 4.1-6 becomes equal to $-\Gamma_{LV}\ (1 + \cos\theta)$. Thus the condition of Eq. 4.1-6 is never satisfied as an inequality; only $2\Gamma_{LV} = \Gamma_{LV}(1 + \cos\theta)$, at $\theta = 0$, i.e., when adhesion is energetically equivalent to cohesion. In summary then, adhesion between two phases, liquid-solid or liquid-liquid, is equivalent to wetting or spreading of one liquid phase.

Zisman and his co-workers (3, 4) studied the equilibrium contact angles of various pure liquids, of known surface tensions, on both high and low energy solid surfaces. Hard solids of high melting point have high surface free energy per unit area (0.5–5.0 $J/m^2$), whereas soft organic solids of low melting point have a low surface free energy (generally $< 0.1\ J/m^2$). Their studies indicate that for a given solid, a linear relationship exists between the surface tension of the liquid and the cosine of the contact angle $\theta$. When this relationship is extrapolated to $\cos\theta = 1$, the condition for wetting, one obtains the *critical surface tension* $\Gamma_c$ useful in evaluating adhesive liquids for given solids.

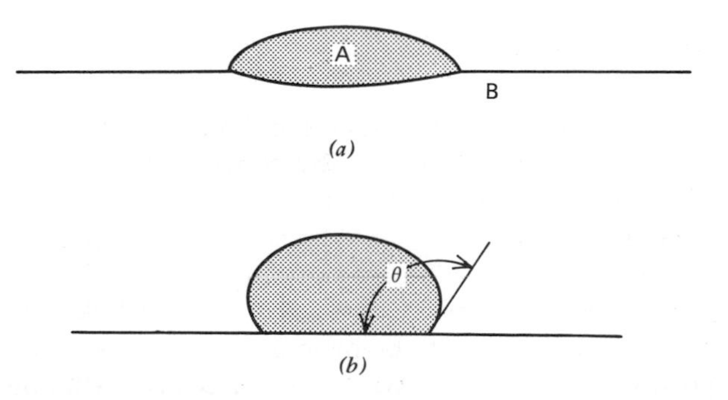


**Fig. 4.1** Interfacial force equilibrium. (*a*) A liquid droplet resting on a liquid. (*b*) A liquid droplet resting on a solid surface.

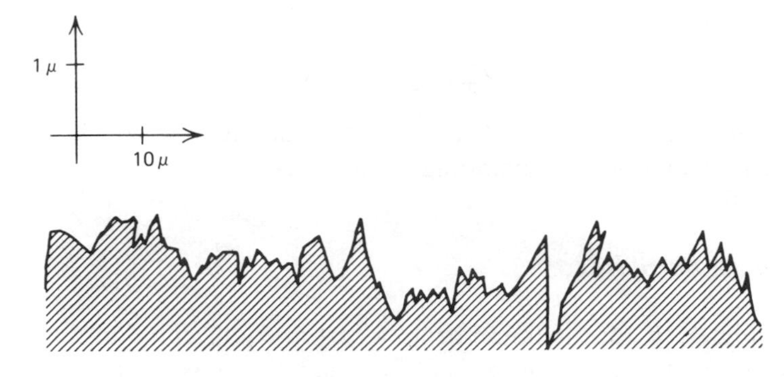


**Fig. 4.2** Oblique section of a finely abraded steel surface. The section gives a vertical magnification of 10,000 and a horizontal magnification of 1000. (Reprinted with permission from *Encyclopaedia Britannica*, p. 931.)

## 4.2 Adhesion

### *Ahesion Between Solids*

The relation between adhesion, surface free energy, and surface forces was discussed in the previous section. For two solids to adhere, there must be intimate interfacial contact because van der Waals forces are negligible for distances greater than a few Ångstroms. Because solid surfaces are not perfectly smooth but are characterized by asperities (Fig. 4.2), the real area of contact is only a very small fraction of the nominal contact area, as Bowden and Tabor (5) have shown. In solid interfacial contact, not only the real area of contact is important to solid-solid adhesion, but also the presence of surface contaminants, organic or oxide, which tend to decrease adhesion. Bowden and Tabor have also pointed out the role of the local elastic recovery at the points of contact. Unless the interface is annealed, while the normal load is on, there will be an elastic recovery at the contact points after the normal load is removed and loss of real contact area. Because of these shortcomings, liquids are often used to promote adhesion between two solids. The liquid adhesive most often solidifies and is in that state at use conditions.

### *Liquid-Solid Adhesion*

From the discussion of liquid-solid adhesion and wetting in Section 4.1, the role of the critical surface tension $\Gamma_c$ as well as the contact angle is evident. Two points need to be further discussed. The first is that in the presence of asperities on the solid surface, incomplete spreading ($\theta > 0°$) results in the entrapment of air pockets. The second point is that in real adhesive bonding, the equilibrium contact angle is not always achieved. For polymeric melts, the ability of the liquid to attain the *equilibrium* contact angle is found to be proportional to its surface tension $\Gamma_{LV}$ and inversely proportional to its zero shear rate viscosity. Although $\partial\Gamma/\partial T < 0$, as mentioned earlier, $\partial\mu/\partial T$ is much more negative; thus increasing the temperature level enables polymer melts to attain the equilibrium contact angle much faster.

Extensive interfacial contact is a necessary, but not sufficient, condition for strong adhesion. This is certainly true when a mechanically weak surface layer is present in one of the adherents. The weak layer would preclude a strong adhesive joint. Crystallizable polymers, in the process of spherulitic growth, reject lower molecular weight chains and push them to the periphery of spherulites. This phenomenon would create a weak surface layer unless extensive surface nucleation occurred during the crystallization process. Another way to eliminate the weak surface layer in the solid state is to induce surface crosslinking. The problem has been treated by Schonhorn and Sharpe (6). Melting a crystallizable polymer on a high energy substrate, such a polyethylene on aluminum, creates a strong adhesive joint; a weak joint is created between a polymer film crystallized in contact with the atmosphere and then joined to aluminum. There is evidence that profuse nucleation occurs at the high energy interface, resulting in the expulsion from the surface layer of low molecular weight species. Additionally, because of the profuse nucleation, many intercrystalline entanglements are created in the surface layer. Both effects would create a strong surface layer and would improve the strength of the adhesive bonding. Surface activation, by corona discharge or similar methods, is achieved by the creation of polar surface groups at certain reactive sites through collisions between these sites and the charged particles (7). The activated polar surfaces are more wettable, thus improve adhesion, if spreading is the limiting factor. On the other hand, if wettability is not the problem to start with, surface activation methods can still be used for the purpose of surface crosslinking. In all these methods the excited and metastable rare gas particles remove a hydrogen atom by impingement on the surface and create polymer radicals. Some of these radicals form double bonds and others crosslink. The latter effect eliminates the weak surface layer and improves the strength of the adhesive joint.

## 4.3 Tribology

Tribology is the science of friction, wear, and lubrication (8). Solid-solid friction is of great importance in polymer processing. Most processing operations commence with the polymer in a particulate form. Handling such a material and processing it are functions that depend on the frictional properties of the polymer. The wear of processing equipment due to the flow of polymers at high temperatures and pressures is certainly an important practical problem, and processing engineers must understand the mechanisms involved. Finally, we deal with lubrication. Although the classical subject of the behavior of lubricating oils is of no direct concern to us in this book, the broader problem of the effect of additives that greatly influence the solid-solid interface properties as a result of material accumulation at the interface, is of importance in polymer processing.

### *Solid-Solid (Dry) Friction*

Friction is the tangential resistance offered to the sliding of one solid over another. Bowden and Tabor (5), held that two main factors are responsible for dry friction:

> The first, and usually the more important factor, is the adhesion which occurs at the regions of real contact: these adhesions, welds, or junctions have to be sheared if sliding is to occur. . . . The second factor arises from the ploughing, grooving, of one surface by the asperities of the other.

In the case of *static* friction only adhesion at the contacts sites is important, whereas in either *sliding* or *rolling* friction the second factor enters the picture. By neglecting the second factor relative to the first, we can explain two important experimental observations in dry friction which, incidentally, were first recorded by Leonardo da Vinci around the year 1500.

1. The frictional force ($F_T$) is independent of the nominal contact area.
2. The frictional force is proportional to the normal force ($F_N$) between the two solid surfaces, that is, $F_T \propto F_N$. The proportionality constant for any pair of solid surfaces is called the *coefficient of friction f.*

As mentioned earlier, the real contact area between two metallic surfaces may be only a small fraction of the nominal contact area, of the order of $10^{-4}$. This implies that even if the normal load is small, the pressure at the contact points is sufficiently high to reach the value of the yield stress of the softer metal $\sigma_y$. Assuming that this is indeed the case—that is, that plastic flow occurs—we can argue that the area at a point of contact $A_i$ is

$$A_i = \frac{F_{N_i}}{\sigma_y} \tag{4.3-1}$$

where $F_{N_i}$ is the load supported by the contact point. An adhering contact point forms a joint that can be broken only when the value of the applied tangential force $F_{T_i}$ reaches the level

$$F_{T_i} = \tau_y A_i \tag{4.3-2}$$

where $\tau_y$ is the shear strength of the softer metal. If we assume that the total tangential frictional force is simply the sum of all $F_{T_i}$, we obtain that

$$F_T = \sum F_{T_i} = \frac{\tau_y}{\sigma_y} \sum F_{N_i} = \left(\frac{\tau_y}{\sigma_y}\right) F_N \tag{4.3-3}$$

Equation 4.3-3 suggests that the *static* coefficient of friction is a material property characteristic of the pair of metallic surfaces and, specifically, of the softer metal, since

$$f' = \frac{\tau_y}{\sigma_y} \tag{4.3-4}$$

By extension, Eqs. 4.3-3 and 4.3-4, known as Amonton's law, are assumed to hold for kinematic friction too, assuming that adhesion predominates. The statement that *f* is a material property, independent of the geometric nature of the surface and frictional process conditions, is only a *rough approximation*. If two surfaces are highly polished, the real contact area will be increased and the two parts will appear to stick together. If the surfaces of two identical metals are cleaned of all impurities, all contact points will apparently form real welds, since the atoms cannot distinguish the surface to which they belong and the coefficient of friction will be increased severalfold. It is therefore hard to speak about the coefficient of friction as a true material property because it greatly depends on the nature of the surface.

Furey (8) described some of the basic problems in studying friction. There is as yet no completely satisfactory theory describing the structure of the solid surfaces. This is true at the atomic level as well as on a larger scale. Moreover, sliding of one solid on another introduces a new set of circumstances and unknowns. It may lead to high and unknown local temperatures and pressures, generating fresh and chemically different surfaces, and mostly altering the topography of the surface as a result of deformation and wear. For these reasons the coefficients of static and sliding friction are different. The static coefficient is larger than the sliding (kinematic) coefficient. This difference is probably responsible for the "slip-stick" motion that is usually observed during dry sliding. Following Nielsen (9), the real area of contact increases gradually during the "stick" stage as the surfaces are pulled into more intimate contact by the increasing tangential force. "Slip" occurs when the forces become sufficient to shear and plough the material. During "slip," the area of real contact and the friction rapidly decreases.

In view of these complexities, it is remarkable that Eq. 4.3-3 represents numerous metal-metal dry frictional data rather well, for both the static and sliding cases. Polymers, on the other hand, exhibit a more complex frictional behavior on metal. This is perhaps not surprising since the physical situation involves a relatively soft, viscoelastic, and temperature dependent material in contact with a hard, elastic, and much less temperature and rate dependent material. Empirical evidence of these complexities is the nonlinear relationship between the frictional force and the normal load

$$F_T = CF_N^{\alpha} \tag{4.3-5}$$

from which a load dependent coefficient of friction can be deduced

$$f = CF_N^{\alpha-1} \tag{4.3-6}$$

where $C$ is a constant and the exponent $\alpha$ is found to vary between the values of 1 and 0.666. It has been suggested (10) that $\alpha = \frac{2}{3}$ corresponds to the case of pure elastic deformation at the contact points, whereas $\alpha = 1$, according to Eq. 4.3-3, to purely plastic (yielding) deformation. Hence values in between appear to reflect viscoelastic deformation at the contact points. If this is the case, the total contact area would be expected to depend on the normal load, the time of contacts, the temperature, and the speed of sliding. As we shall see below, these effects are generally observed. At this point it is worth noting that the expression for the dry coefficient of friction (Eq. 4.3-6) has the same form as that of the viscosity ("internal friction") of a power law fluid that describes the non-Newtonian behavior of polymer melts (see Eq. 6.5-2).

From the foregoing it follows that except for $\alpha = 1$, the coefficient of friction decreases with increasing load $F_N$. This is supported experimentally over limited load ranges (11, 12) as shown in Fig. 4.3.

The effect of temperature on the coefficient of friction can be quantitatively deduced from Eq. 4.3-2. An increase in temperature results in a reduction of the shear strength and an increase of the contact area. Since these two effects are competing, the coefficient of friction may either increase or decrease with

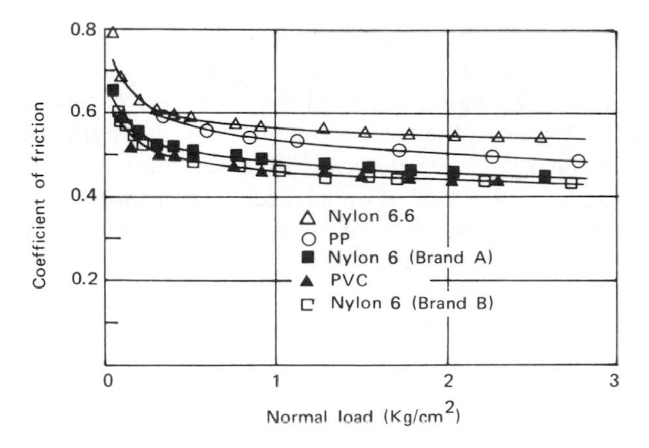


**Fig. 4.3** Coefficients of friction of various polymers in pellet form or steel versus normal load at 30°C and 1 cm/s sliding speed. The reported coefficients of frictions are "rubbed-in" values. (Reprinted with permission from K. Schneider, *Kunststoffe*, **59**, 97 (1969).)

increasing temperature. Several investigators (11–15) have observed minima at temperatures much below the melting or softening point (Fig. 4.4). The observed dramatic increase of the coefficient of friction near the melting or glass transition temperature (15) reflects the effect of a thin molten film, formed at the interface, undergoing viscous drag flow; it is not indicative of dry friction.

The sliding speed has only a moderate effect on the polymer-metal coefficient of friction (11, 13). An increase in speed of one order of magnitude brings about a friction coefficient increase of the order of 20%. This effect is expected from Eq. 4.3-4, since the shear strength of viscoelastic materials increases with increasing rate of deformation. However the expected results may be diminished by the generation of high local temperatures and vibrations at elevated sliding speeds.

The effect of surface roughness on the coefficient of friction is rather complicated. We have noted for metals that highly polished surfaces give rise to an

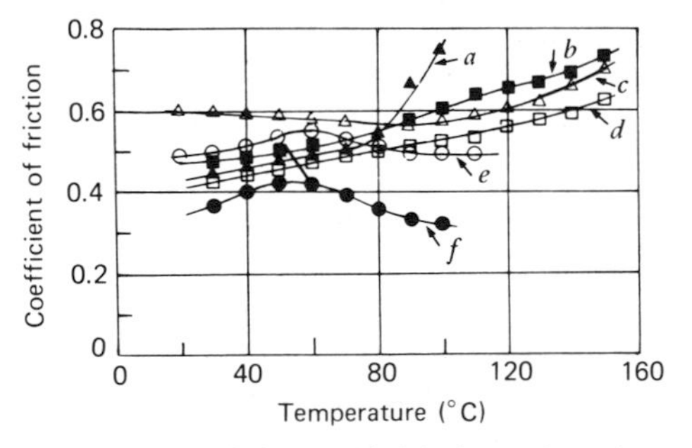


**Fig. 4.4** Coefficients of friction of various polymers in pellet form on steel versus temperature with normal load of 1.7 $Kg/cm^2$ and 1 cm/s sliding speed. The reported coefficients of friction are "rubbed-in" values. (*a*) PVC, (*b*) Nylon-6 (Brand A), (*c*) Nylon 66, (*d*) Nylon-6 (Brand B), (*e*) PP, (*f*) polyethylene. (Reprinted with permission from K. Schneider, *Kunststoffe*, **59**, 97 (1969).)

increase in contact area resulting in an increase in the static coefficient of friction, however increased surface roughness has also been observed to increase the kinematic coefficient of friction, an effect that can perhaps be attributed to increased "ploughing, grooving, and cracking." In polymer-metal dry friction, an increase in metal surface roughness also brings about an increase in the coefficient of friction (11, 13, 15). Working with PVC powder, Chang and Daane (13) observed an increase in the coefficient of friction with increasing metal surface roughness over a wide temperature range. The magnitude of the increase, however, varied with temperature, the maximum occurring at about 150°F. Their results may not be valid, strictly speaking, because the metal surfaces used differed chemically: "Xaloy," stainless steel, and chromium plated steel.

The effect of rubbing time between a metal and a polymer surface was studied by Schneider (11, 14). The coefficient of friction measured on unused, clean metal surfaces is low, whereas the one measured after "rubbing in" is much higher. The limiting values are reproducible. For example, the coefficient of friction of nylon-6 on a clean metal surface is 0.05 and increases to 0.42 after extensive rubbing in a reciprocating friction tester (11). Similar observations were reported by Schneider (11) for other polymers. This effect he attributed to the transfer of polymeric material to the metal mating surface (see adhesive wear). Thus, in reality, after complete "rubbing in" the coefficient of friction approaches that of polymer to polymer. Nevertheless, one does not expect this value to be reached, since the physical situation is one of a "soft" polymer solid in contact with a thin layer of the same material coating a hard metallic surface. It is interesting to note that if after complete "rubbing in" a new polymer in pellet form is used on the coated metal surface, the coefficient of friction is lower. This may be due to the partial tearing of the deposited polymer layer by the sharp edges of the new pellets. Furthermore, if after every cycle the polymer pellets are renewed, a new steady state value of the coefficient of friction is reached, the value of which is intermediate to the two limiting values mentioned previously. For nylon-6 this value was 0.28 (11). Such a value may be representative of the friction coefficient in the solid feed sections of processing equipment where a new polymer is continuously sliding on the same metal surface. At any rate, in discussing sliding friction phenomena between metals and polymers, the wear characteristics of polymers must be taken into consideration.

Tabor and his co-workers have done considerable work that points out the interesting and complex frictional behavior of polymers (12). First they argue that in smooth surface polymer-metal friction the "ploughing" contribution is negligible; adhesion at the interface is the dominant factor. This is because polymers are too soft, thus are easily ploughed. If the adhesive forces between the metal and the polymer are greater than the cohesive polymer forces, sliding occurs at a plane within the polymer, giving rise to kinematic friction coefficients $f > 0.2$. On the other hand, if the adhesive forces are weaker than the cohesive, sliding occurs *at* the interface, and $f < 0.1$. In the first case, appreciable transfer of polymer on the metal surface takes place, a film of thickness of the order of 1 $\mu$ is formed, and in the second, negligible polymer transfer occurs. Most polymers fall into the first category, as can be seen by the values of $f$ in Table 4.1. Outstanding are the very low values of $f$ for PTFE and HDPE. These very low values (which are of great practical importance) are characteristic of the steady response of a sample sliding on a clean, smooth surface at low sliding speeds of the order of 0.1–1.0 cm/s. When the sliding speed is increased to 10 cm/s, $f \cong 0.30$, typical of the majority of polymers. The same rise in $f$ results when the metal surface is rough, about 1000 Å.

**Table 4.1 Static and Kinematic Friction Coefficients of Some Polymers on Smooth Glass Surfaces (12)**

| Polymer | $f'$ | $f$ |
|---|---|---|
| FEP | 0.22 | 0.20 |
| KelF | 0.28 | 0.28 |
| PTFE | 0.20 | 0.06 |
| HDPE | 0.20 | 0.08 |
| LDPE | 0.30 | 0.30 |
| PVC | 0.40 | 0.40 |

The low value of $f$ is preceded by a large value of $f \cong 0.2$. At the peak point a thick "lump" of PTPE is transferred to the glass surface. Its thickness is greater than 10,000 Å. Following the peak, in the region of $f \cong 0.06$, a film 30–100 Å thick is deposited on the glass surface and can be easily detached from it. If the experiment is repeated with the same sample sliding in the same direction on a new portion of the smooth surface, only the film is formed and $f = f' \cong 0.06$. Finally, if the sample is rotated 90° about its axis, the behavior reverts to that of the virgin sample, with a high $f'$ and the transfer of a "lump" of polymer. Optical microscopy, electron microscopy, and electron diffraction studies have contributed to the conclusion by Tabor that once the first lump pulling has occurred with the virgin sample, the PTFE chains are pulled at the surface and deposit themselves in a highly oriented fashion. The force to slide PTFE is then equal to the force to *draw* the chains out of the bulk. This process is not destroyed by stopping and restarting sliding in the same sample direction, but, obviously, is disrupted (and must be reinitiated) by rotation of the sample to an orientation 90° from the original.

## Wear

Wear, together with breakage and obsolescence, is one of the main reasons for the cessation of usefulness of inanimate objects. Wear has been defined as "the removal of material from solid surfaces as a result of mechanical action" (16). The amount of material lost by the surfaces involved is usually very small and difficult to detect. Nevertheless, especially in applications involving high pressures and speed, it is sufficient to seriously interfere with the proper mechanical function of the system.

There are four main types of wear processes and often they occur concurrently.

*Adhesive wear* arises from strong forces at actual contact points between two surfaces. Thus when the surfaces are forced to slide over each other, fragments of one are pulled off and adhere to the other; these fragments may be transferred back to the original surface or may form loose debris at later stages of the sliding process. From our discussion of adhesion and dry friction, it follows that adhesive wear is influenced by the sliding friction coefficient and the actual contact area. Rabinowicz and Shooter (17) have obtained data indicating that if two solids have a hardness ratio $R$, their adhesive wear rates vary with $1/R^2$. This is why in metal-polymer dry sliding friction, a polymer coat is formed on the metal surface, increasing the value of the friction coefficient as discussed earlier. It is interesting to note, though, that even low adhesion soft polymers, having a low $\Gamma_c$, will remove quite sizable amounts of matter from metals as strong as low carbon steel (17). The implication of this observation in the construction and surface treatment of polymer processing machinery is obvious; the wear of the extruder barrel and screw flights is not due solely to metal-to-metal wear.

*Abrasive wear* occurs when a rough hard surface slides on a softer surface and "ploughs" grooves onto it. Hard surface particles between two softer surfaces can have the same wear effect. The material removed from the softer surface usually forms loose debris. Abrasive wear is then the phenomenon that probably occurs when a thoroughly rubbed-in metal surface with polymer is placed in sliding contact with fresh polymer pellets, as is the case in the feed section of extruders discussed above. Abrasive wear can be eliminated, initially at least, if the surfaces are smooth and free of hard particles. Adhesive wear, with the creation of rough surfaces and loose debris, can promote abrasive wear. This is especially true if the adhesive wear debris is made harder by oxidation.

*Corrosive wear* occurs when two surfaces slide in a corrosive environment. The adhesive and abrasive wear creates continuously new surfaces; thus the corrosive action of the environment continues deeper into each surface. Fortunately, polymers are rather inert substances that resist corrosion, and since their adhesive or abrasive wear on hard metals is not sizable, they also do not promote corrosive wear on metals. A notable exception is PVC, which degrades when improperly stabilized, producing hydrogen chloride, a corrosive substance.

*Surface fatigue wear* is observed after repeated cycles of loading and unloading of surfaces, such as in a rolling of a cylinder or a ball or the cyclic application of pressure. In principle, ideal elastic solids with no surface defects should show no fatigue wear. But with real materials, surface or subsurface microcracks are formed which eventually grow into large cracks of the surface. Polymers are known to fatigue under the cyclic application of stress.

### Lubrication

Lubricants are materials that when placed at the interface of two solids in contact, change the nature and degree of their surface interaction. That is, lubricants are added to sliding systems to reduce the dry friction coefficient, the amount of wear, or the degree of surface adhesion. Occasionally lubricants are introduced to reduce the interfacial temperature, especially in situations of abrasive wear, such as in metal cutting.

When a thin film of liquid or gas, completely separating two moving solid surfaces, is present at the interface, we have the case of *fluid lubrication*. The flow behavior of the fluid—that is, its viscosity—will govern the motion, as in the case of fluid bearings. The phenomena that take place in fluid lubricated bearings have been under intensive study since the classical work of Osborn Reynolds in 1886

(18), which laid the foundation of hydrodynamic lubrication. The interest stems, of course, from the great practical importance of fluid (oil) lubrication. When two machine elements (e.g., a journal and a bearing) separated by a small fluid-filled clearance are in relative motion, they cannot come in contact with each other because of the extremely large pressure that this process would create. As a consequence, the revolving journal is separated from the bearing, and solid-solid contact leading to wear is thus avoided. The essential feature of this phenomenon is that an eccentricity between the rotating journal and the stationary bearing results in a motion of the axis of the journal.

Such problems can be solved approximately by making the so-called *lubrication approximation*, which we discuss in Section 5.4 and deal with in many other instances in the pressurization and flow and shaping chapters.

The lubrication approximation is extensively used in polymer processing, because although clearances and tapers (clearance variations) may be appreciable, the viscosity of melts, thus the viscous forces, are several orders of magnitude higher than those of lubricating oils. It is worth noting that in polymer processing hydrodynamic lubrication is often achieved by the processed polymer itself. In screw extrusion, for example, the polymer melt that flows in the small flight clearance between the screw tip and the barrel serves as a lubricant, prevents excessive metal wear, and indeed makes screw extrusion practically feasible.

*Boundary lubrication*, on the other hand, is characterized by the presence of a very thin film of lubricant—a monolayer—covering the entire actual interface contact area, or part of it. *Solid film lubrication*, which is closely related to boundary lubrication, is characterized by a soft, solid film of substantial thickness, covering, again, either the entire actual surface contact area or part of it.

The calculation of the reduction of the frictional forces or wear by the action of boundary or solid film lubrication depends, to a large extent, on the "modeling" of the actual physical situation and the assumptions involved, especially relative to the sizes of lubricated patches, the total lubricated area, and the wear particles (19). The lubricating action of 1-octadecanone (stearyl alcohol), $CH_3–(CH_2)_{17}–OH$, on two copper surfaces is illustrated in Fig. 4.5. The melting point of stearyl alcohol is 59.6°C and the boiling point is 210.5°C. The first increase of the friction coefficient and wear occurs near the melting point of the lubricant. That is, solid lubrication changes over to liquid boundary lubrication and the metal-to-metal contact area increases from zero to a value characteristic of the system. The second dramatic increase occurs at a temperature region much below the lubricant's boiling point; the new level of the friction coefficient and wear is that of the unlubricated surfaces. Therefore it appears that the lubricant becomes desorbed. That is, at these high temperatures the lubricant is not held tightly by the surface and is pushed aside, giving rise to solid contact only. This second transition occurs at a temperature region that depends on the metal surface hardness and the lubricant-metal surface tension. Decreasing the first and increasing the second shifts the transition to higher temperatures. A long liquid boundary lubrication plateau is achieved with high energy surfaces such as metals, when the lubricants used are polar, such as fatty acids and alcohols, or contain many dipoles and have flexible chains. The lubricants used must also have a very high boiling point and resist thermal and oxidative degradation. Lubricants such as stearic acid are often incorporated in the pro-

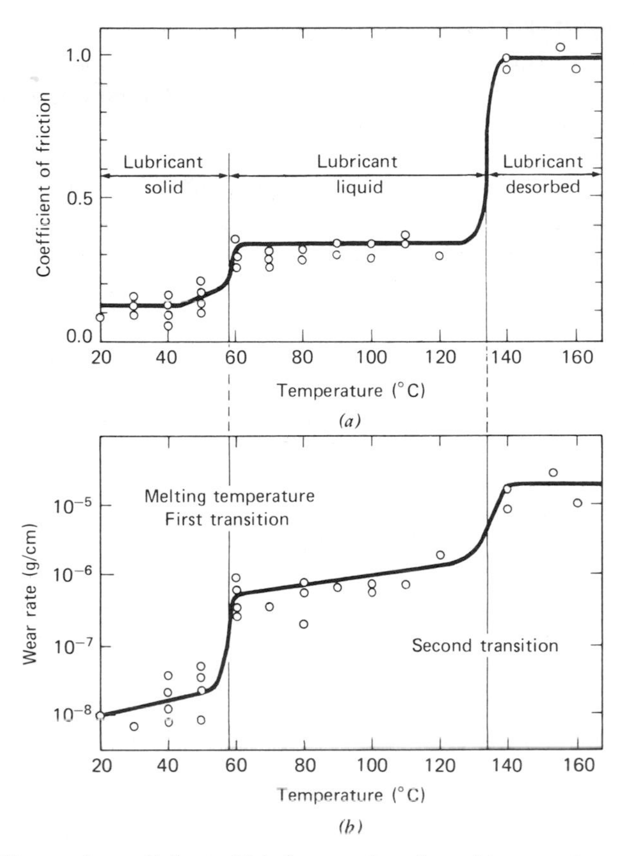


**Fig. 4.5** (*a*) Kinematic coefficient of friction as a function of temperature; copper on copper, 2 kg, 0.01 cm/s, lubricated with octadecyl (stearyl) alcohol. (*b*) Metal transfer as a function of temperature. (Reprinted with permission from E. Rubinowicz, *Friction and Wear of Materials*, Wiley, New York, 1965.)

cessing of polymers having high viscosity and friction coefficients, as in the case of PVC.

## 4.4 Triboelectricity

Triboelectric phenomena refer to the electric charge transfer when two dissimilar materials are brought into contact, the extent of which increases greatly by rubbing them together. Thus contact charging is equivalent to the triboelectric effect. In polymer processing, triboelectric problems arise whenever the polymer is handled (20)—dust particles are attracted to molded parts, foreign particles become trapped in coating operations, plastic chips adhere to "deflashed" molded parts, films wrap around rolls and stick to plates and belts. Fibers in the spinning operation may accumulate charge, making the material difficult to handle in subsequent drawing and weaving operations. When the accumulated charge is large, it may arc to nearby objects, causing fires or giving "shocks" upon touch.

The significance of contact charge forces becomes evident by considering that a 1 $\mu$ polymer particle would be electrostatically held to a surface by about 30 electron charges.

Turning to the basic contact charging phenomenon, we consider first two metals, A and B, coming in contact with each other while being electrically insulated from their environment. Upon contact, a potential difference is created simply because conduction electrons find it easier to move, say, from metal A to metal B than from B to A. If the two metals are carefully separated and kept isolated from other contact, they will remain charged. This is then contact charge transfer in conducting metals, the charge usually being between tenths of a volt to 2 V (21). More quantitatively, during the metal contact there will be electron charge transfer from one metal to the other until the new potentials in the metals align the Fermi levels. The conduction electrons occupy this level. An amount of energy $\phi$ is necessary to allow them to escape from the metal. The work function $\phi$ varies from metal to metal and is a function of temperature. From the foregoing discussion it follows that contact potential generated between metals A and B will be equal to $(\phi_A - \phi_B)$. Metal A, having the smaller work function, will lose electrons and will be positively charged, whereas B will be negatively charged. For example, metal work function values are 4.5 V for tungsten and 1.9 V for cesium. The resulting contact potential is 2.6 V, with tungsten being negative.

The phenomena that take place during the contact between one or two *nonconductors* are less well understood. It is believed that initially only electrons of the surface are involved in the charging process. Deeper layers may become partially charged after a characteristic time known as the "charge relaxation time." There are very few conduction electrons, and the extent and quality of the contact are very important to the phenomenon of contact charge transfer.

In both metal and nonconductor contact though, the following statement holds: what are termed "triboelectric effects" are not the result of rubbing; that is, no (or negligible) charge is formed during the rubbing frictional phenomena. What is accomplished with rubbing is a large contact area, by more frequent contact of old surfaces or by the generation of new surfaces (see preceding sections). In metals, there is no need for a plethora of contacts; however in nonconductors there is such a requirement, and this is where the illusion originated. Rubbing is the *vehicle* for, not the cause of, triboelectric effects.

The work function of nonconductors has been measured by electrification techniques (22). These data are useful and needed, but it is emphasized that they are not reproducible. Apparently surface cleanliness, treatment, and humidity affect the work function values in a complex manner. Having this in mind, and only to give an indication of which materials are likely to become negative and which positive upon contact, we present Table 4.2. The term "triboelectric series" denotes a list, on which materials at the top would be positively charged coming in contact with materials at the bottom of the series. Triboelectric series are also not exactly reproducible. An extensive table reviewing contact charging in many materials over the period 1937–1972 is presented by Hendricks (21).

In practice we are usually interested in *eliminating* triboelectric effects—the effects of contact charge transfer. There are a number of practical solutions.

One common method is making both substances "conductive." This can be achieved by incorporating into the polymer conductive additives such as carbon black and ethylated amines. This method is used in items such as fuel lines, fuel tanks, conveyor belts, and conductive shoe soles. It provides a permanent solution. Coating products with a thin layer of conductive antistatic finish is often practiced

**Table 4.2 Triboelectric Series**[a]

| **Negative End of Series** |
| --- |
| Wool |
| Nylon |
| Viscose |
| Cotton |
| Natural silk |
| Acetate rayon |
| Lucite of Perspex |
| Polyvinyl alcohol |
| Dacron |
| Orlon |
| PVC |
| Dynel (copolymer of acrylonitrile and vinyl chloride) |
| Velon (copolymer of vinylidene chloride and vinyl chloride) |
| Polyethylene |
| Teflon |
| **Positive End of Series** |

[a] From D. J. Montgomery, *Solid State Phys.*, **9**, 139 (1959).

in the synthetic fiber field. This is not a truly permanent solution because the coating layer continuity might be broken in the process or during washing of the fabric.

Antistatic surface additives should not be confused with "antiblocking" agents, which are used to keep flat polymer surfaces (films) from *intimate* contact. Such a Coating products with a thin layer of conductive antistatic finish is often practiced in be required in the process of unrolling. Antiblocking agents are fine, small particles, mostly inorganic, present in a concentration of about 0.1%. Examples of such particles, some of which would be found at the surfaces to prevent blocking, are silica and calcium carbonate.

During processing, triboelectric problems are handled by taking measures of either eliminating charge accumulation or removing it. Working in a high humidity atmosphere belongs to the former approach, since the surface becomes more conductive. Accumulated charge can be removed by neutralizing it. This is done by creating an ionic atmosphere by splitting air molecules into positive and negative ions, either by radiation or by creating a high voltage electric field across an air gap, that is, corona discharge. Many forms of "static bars" are manufactured on this principle as well as ionizated air guns that blow ionized air on any type of surface to discharge static electricity.

## *REFERENCES*

1. J. W. Gibbs, "Equilibrium of Heterogeneous Substances—Influence of Surfaces of Discontinuity Upon the Equilibrium of Heterogeneous Masses—Theory of Capillarity," *The Collected Works of J. W. Gibbs*, Vol. 1, Longmans Green, New York, 1931, p. 219.
2. T. Young, *Phil. Trans. Roy. Soc.*, **95**, 65 (1805).
3. E. G. Shafrin and W. A. Zisman, *J. Phys. Chem.*, **64**, 519 (1960).

4. E. F. Hare and W. A. Zisman, *J. Phys. Chem.*, **59**, 335 (1955).
5. F. P. Bowden and D. Tabor, *Friction and Lubrication of Solids*, Oxford University Press, London, 1950.
6. H. Schonhorn and L. H. Sharpe, *J. Polym. Sci.* (*A*), **3**, 3087 (1965).
7. J. M. McKelvey, *Polymer Processing*, Wiley, New York, 1962, p. 162.
8. M. J. Furey, *Ind. Eng. Chem.*, **61**, 12–29 (1969).
9. L. E. Nielsen, *Mechanical Properties of Polymers*, Reinhold, New York, 1962, p. 225.
10. A. S. Lodge and H. G. Howell, *Proc. Phys. Soc. B Band*, **67** (1954).
11. K. Schneider, *Kunststoffe*, **59**, 97–102 (1969).
12. B. J. Briscoe, C. M. Pooley, and D. Tabor, in *Advances in Polymer Friction and Wear*, Vol. 5A, L. H. Lee, ed., Plenum Press, New York, 1975.
13. H. Chang and R. A. Daane, Society of Plastics Engineers, *32nd Annual Technical Conference*, San Francisco, May 1974, p. 335.
14. K. Schneider, "Conveying Mechanism in Feed Sections of an Extruder," Doctoral dissertation, Technischen Hochschule, Aachen, 1968.
15. R. B. Gregory, *Soc. Plast. Eng. J.*, **25**, 55–59 (1969).
16. E. Rabinowicz, *Friction and Wear of Materials*, Wiley, New York, 1965, p. 109.
17. E. Rabinowicz and K. V. Shooter, *Proc. Phys. Soc.*, **65B**, 671 (1952).
18. O. Reynolds, *Phil. Trans. Roy. Soc.*, Pt. I, **177**, 157 (1886).
19. E. Rabinowicz and D. Tabor, *Proc. Roy. Soc.*, **A208**, 455 (1951).
20. W. W. Levy, *Soc. Plast. Eng. J.*, **18**, 10 (1962).
21. C. D. Hendricks, "Charging Macroscopic Particles," in *Electrostatics and Its Applications*, A. D. Moore, ed., Wiley, New York, 1973.
22. D. K. Davies, *Brit. J. Appl. Phys.* (*J. Phys. D.*), **2**, 1533 (1969).

*PART TWO*

# ENGINEERING FUNDAMENTALS

# 5

# Transport Phenomena



The engineering science of transport phenomena deals with the transfer of momentum, energy, and mass. It is the theoretical tool with which engineers handle problems involving fluid flow, heat transfer, and diffusion in multicomponent systems. Since much of the engineering design and analysis in polymer processing involves these transport operations, we devote this chapter to a brief review of this material.*

We follow in this book the unified approach to the subject to stress that in polymer processing, transport phenomena usually occur simultaneously. For example, as pointed out in Chapter 1, melting (heat transfer) and pressurization (fluid flow) frequently occur simultaneously. Moreover, because of the high viscosity of polymer melts, their flow in processing equipment is nonisothermal, necessitating the simultaneous solution of flow and heat transfer. Finally, in certain processing operations that involve devolatilization or the incorporation of certain additives, all three transfer operations (mass, momentum, and heat) take place in the same conduit.

The following sections describe the basic equations of transport phenomena—the equations of conservation of mass, momentum, and energy—the relevant constitutive equations, and the relevant transport and thermodynamic properties. Furthermore, we discuss the stress and rate of deformation tensors and devote a section to the "lubrication approximation," which is very important in polymer processing.

## 5.1 The Balance Equations

Since in transport processes mass, momentum, and energy are transported from one part of the medium to another, it is essential that the proper "bookkeeping" be

* For a detailed treatment of transport phenomena, consult Refs. 1–4.

applied to keep track of these quantities. This can be done with the use of *balance equations*, which simply state mathematically the physical laws of conservation of mass, momentum, and energy. That is, balance equations are *universal physical laws* that apply equally well to all media: solids or fluids, stationary or flowing. These equations can be formulated over a specified macroscopic volume, such as an extruder, or they can have the form of a differential (field) equation that holds at every point of the medium. In the former case the balance holds over the extensive quantities of mass, momentum, and energy, whereas in the latter their intensive counterparts (i.e., density, specific momentum, and specific energy) are affected.

The molecular nature of matter is ignored in the balance equations; matter is viewed as a *continuum*. Specifically, the assumption is made that the mathematical points over which the balance field equations hold are "big enough" to be characterized by property values that have been averaged over a large number of molecules, so that from point to point there are no discontinuities. Furthermore, *local equilibrium* is assumed. That is, even though transport processes can be fast and irreversible (dissipative), thus far from thermodynamic equilibrium, the assumption is made that locally the molecules establish equilibrium very fast (5).

### *The Equation of Continuity*

It is instructive to derive the simplest of the balance equations, the *equation of continuity*, which deals with the principle of the conservation of mass. If we consider a region that can be described in Cartesian coordinates (i.e., $x$, $y$, $z$ or $x_1$, $x_2$, $x_3$, or in short, $x_i$, $i = 1, 2, 3$) in which a pure fluid of density $\rho(x_i, t)$ is flowing with a velocity $\boldsymbol{v}(x_i, t)$, the mass conservation principle over a stationary volume element $\Delta V = \Delta x \, \Delta y \, \Delta z$ (Fig. 5.1) can be expressed as follows

$$\left\{\begin{matrix}\text{Rate of mass}\\ \text{accumulation}\\ \text{in } \Delta V\end{matrix}\right\} = \left\{\begin{matrix}\text{Rate of mass}\\ \text{transferred}\\ \text{in } \Delta V\end{matrix}\right\} - \left\{\begin{matrix}\text{Rate of mass}\\ \text{transferred}\\ \text{out of } \Delta V\end{matrix}\right\} \tag{5.1-1}$$

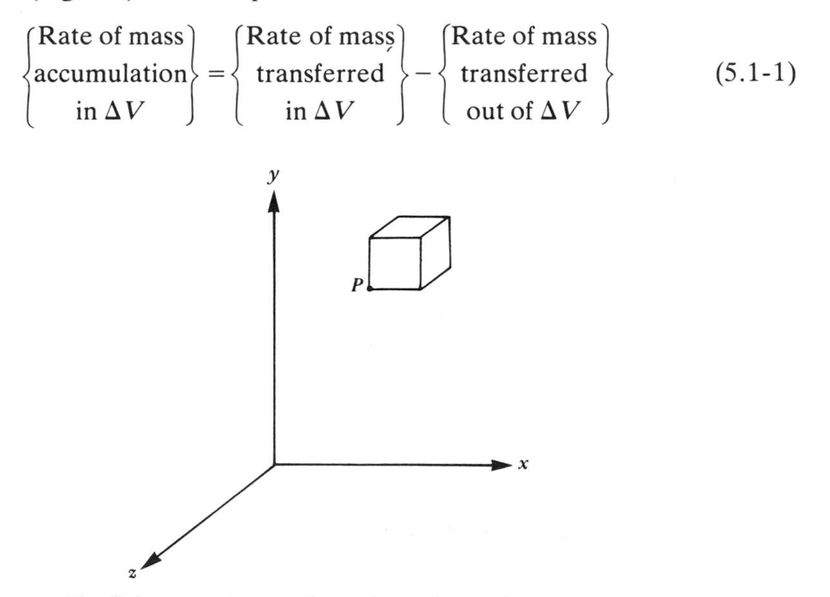


**Fig. 5.1** A stationary Cartesian volume element over which the mass conservation principle is applied.

Assuming a density value $\rho$ for the entire $\Delta V$, the rate of mass accumulation is equal to $\Delta V\ \partial\rho/\partial t$. This term is then equal to the algebraic sum of all the mass fluxes $(\rho v_i)$ entering or leaving the six faces of the cube, as in Fig. 5.1. The resulting expression is

$$\Delta x\ \Delta y\ \Delta z \frac{\partial \rho}{\partial t} = \Delta y\ \Delta z[(\rho v_x)|_x - (\rho v_x)|_{x+\Delta x}] + \Delta x\ \Delta z[(\rho v_y)|_y - (\rho v_y)|_{y+\Delta y}]$$

$$+ \Delta x\ \Delta y[(\rho v_z)|_z - (\rho v_z)|_{z+\Delta z}] \tag{5.1-2}$$

Each bracket on the right-hand side of Eq. 5.1-2 expresses the *net flux* of mass crossing the three principal planes of the cube.* Dividing this expression by $\Delta V$ and taking the limit as the cube dimensions approach zero, we obtain

$$\frac{\partial \rho}{\partial t} = -\left[\frac{\partial}{\partial x}(\rho v_x) + \frac{\partial}{\partial y}(\rho v_y) + \frac{\partial}{\partial z}(\rho v_z)\right] \tag{5.1-3}$$

Expanding the right-hand side, we obtain

$$\frac{\partial \rho}{\partial t} + \left(v_x \frac{\partial \rho}{\partial x} + v_y \frac{\partial \rho}{\partial y} + v_z \frac{\partial \rho}{\partial z}\right) = -\rho\left(\frac{\partial v_x}{\partial x} + \frac{\partial v_y}{\partial y} + \frac{\partial v_z}{\partial z}\right) \tag{5.1-4}$$

Both these equations are expressions of the equation of continuity. They can be represented in a coordinate free form, utilizing vector notation,† respectively, as follows:

$$\frac{\partial \rho}{\partial t} = -(\nabla \cdot \rho \boldsymbol{v}) = -\text{div}\ (\rho \boldsymbol{v}) \tag{5.1-5}$$

►

$$\frac{D\rho}{Dt} = -\rho(\nabla \cdot \boldsymbol{v}) = -\rho\ \text{div}\ \boldsymbol{v} \tag{5.1-6}$$

where $\nabla$, known as "del" or "nabla" is a differential operator,‡ which in rectangular coordinates§ is defined as

$$\nabla = \boldsymbol{\delta}_1 \frac{\partial}{\partial x_1} + \boldsymbol{\delta}_2 \frac{\partial}{\partial x_2} + \boldsymbol{\delta}_3 \frac{\partial}{\partial x_3} \tag{5.1-7}$$

where $\boldsymbol{\delta}_i$ are the unit vectors. The "substantial derivative" $D/Dt$ is defined as

$$\frac{D}{Dt} = \frac{\partial}{\partial t} + \boldsymbol{v} \cdot \nabla \tag{5.1-8}$$

* Since we are dealing with a pure fluid, there are no diffusional-conductive fluxes across the surface.

† We assume the reader is familiar with vector notation, and except for brief explanatory comments, no summary on vector operation is presented. This is available in many texts (e.g., Refs. 1a, 6, 7a). The tabulated components of the balance equations in various coordinate system presented in this chapter should enable the reader to apply them without any detailed knowledge of vector operations.

‡ Recall that the operation of $\nabla$ on a scalar is the "gradient" e.g., $\nabla P$ (which is a vector); whereas its operation on a vector field can either be the "divergence" or the "curl" of the vector field. The former is obtained by a dot product (also called scalar product) as $\nabla \cdot \boldsymbol{v}$ or div $\boldsymbol{v}$ in Eq. 5.1-6, where the result is a scalar; the latter is obtained by the cross product (also called vector product) $\nabla \times \boldsymbol{v}$ or curl $\boldsymbol{v}$ and the result is a vector.

§ For the use of the $\nabla$ operator in curvilinear coordinates, see Problem 5.1.

and it is the time change of the variables as measured by a viewer translating with the "substance." Thus Eq. 5.1-5 states the mass conservation principle as measured by a *stationary* observer. The derivative $(\partial/\partial t)$ is evaluated at *any fixed position* in space (Eulerian point of view). On the other hand, Eq. 5.1-6 is the same conservation principle as measured (reported) by an observer who is following the path of the fluid, moving with the fluid velocity (Lagrangean point of view). Table 5.1 expresses Eq. 5.1-5 in various coordinate systems.

**Table 5.1 The Equation of Continuity in Several Coordinate Systems**[a]

**Rectangular coordinates** $(x, y, z)$

$$\frac{\partial \rho}{\partial t}+\frac{\partial}{\partial x}(\rho v_x)+\frac{\partial}{\partial y}(\rho v_y)+\frac{\partial}{\partial z}(\rho v_z)=0$$

**Cylindrical Coordinates** $(r, \theta, z)$

$$\frac{\partial \rho}{\partial t}+\frac{1}{r}\frac{\partial}{\partial r}(\rho r v_r)+\frac{1}{r}\frac{\partial}{\partial \theta}(\rho v_\theta)+\frac{\partial}{\partial z}(\rho v_z)=0$$

**Spherical Coordinates** $(r, \theta, \phi)$

$$\frac{\partial \rho}{\partial t}+\frac{1}{r^2}\frac{\partial}{\partial r}(\rho r^2 v_r)+\frac{1}{r \sin\theta}\frac{\partial}{\partial \theta}(\rho v_\theta \sin\theta)+\frac{1}{r\sin\theta}\frac{\partial}{\partial \phi}(\rho v_\phi)=0$$

[a] Reprinted with permission from R. B. Bird, W. E. Stewart, and E. N. Lightfoot, *Transport Phenomena*, Wiley, New York, 1960.

For an incompressible fluid, $\rho \neq \rho(x_i, t)$, thus $D\rho/Dt=0$, which results in

$$\nabla \cdot \boldsymbol{v}=0 \tag{5.1-9}$$

If the density changes of the flowing fluid cannot be neglected, then an appropriate equation of state of the form $\rho=\rho(T, P)$ must be used in conjunction with the balance equations, including the continuity equation. Section 5.5 covers the density of polymers.

An alternative, more elegant, and coordinate free derivation of the balance equations is obtained by considering a spatially fixed and closed region in the medium, of volume $V$ and surface $S$, as in Fig. 5.2. The enclosed volume of

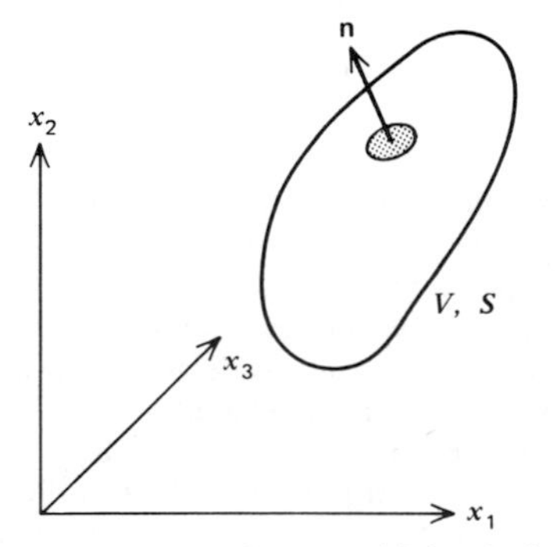


**Fig. 5.2** A "control volume" over which a balance expression is developed.

arbitrary shape is called the *control volume*. The orientation for any surface area element $dS$ is denoted by an outward pointing normal unit vector $\mathbf{n}$. The balance equations are set up simply by equating the total net flux of the conserved variable over the "enclosing" surfaces to the rate of change of the variable within the controlled volume. Thus if the conserved variable is mass

$$\frac{d}{dt}\int_V \rho\, dV = -\int_S (\mathbf{n} \cdot \rho \boldsymbol{v})\, dS \tag{5.1-10}$$

Note that $\rho\boldsymbol{v}$ is the mass flux and that $\mathbf{n} \cdot \rho\boldsymbol{v}$ is the mass flux component normal to the surface element. The negative sign appears because an increase in density is obtained as a result of an influx of mass ($(\mathbf{n} \cdot \rho\boldsymbol{v})$ is positive for outward flux). We can convert the surface integral in Eq. 5.1-10 to a volume integral using the Gauss divergence theorem*

$$\frac{d}{dt}\int_V \rho\, dV = -\int_V (\boldsymbol{\nabla} \cdot \rho \boldsymbol{v})\, dV \tag{5.1-11}$$

Since the control volume is fixed in space, the ordinary derivative is changed into partial derivatives, and since by definition we have selected a fixed volume, $\partial[\int_V \rho\, dV]/\partial t = \int_V (\partial\rho/\partial t) dV$ and Eq. 5.1-11 can be written as

$$\int_V \left[\frac{\partial \rho}{\partial t} + \left(\boldsymbol{\nabla} \cdot \rho \boldsymbol{v}\right)\right] dV = 0 \tag{5.1-12}$$

This equation holds for an arbitrary volume $V$; thus the integral must vanish, resulting in the equation of continuity, 5.1-5

$$\frac{\partial \rho}{\partial t} + (\boldsymbol{\nabla} \cdot \rho \boldsymbol{v}) = 0$$

### *The Equation of Motion*

Turning to the equation of conservation of momentum, we follow the control volume approach.† First we recall that momentum is a vector, and it is *independently* conserved in the three spatial directions. Hence the equation of motion is a *vectorial* equation having three components. Momentum can be transported across the surface of the control volume by two mechanisms—*convection* and *conduction*. The former is simply due to the bulk flow of the fluid across the surface. Associated to it is a momentum flux (i.e., momentum per unit area per unit time) given by $\rho\boldsymbol{v}\boldsymbol{v}$. The second mechanism by which momentum is transported in and out the volume element is due to intermolecular forces on each side of the

* The divergence theorem of Gauss states that if $V$ is a volume bounded by a closed surface $S$ and $\mathbf{A}$ is a continuous vector field, then

$$\int_V (\boldsymbol{\nabla} \cdot \mathbf{A})\, dV = \int_S (\mathbf{n} \cdot \mathbf{A})\, dS$$

† For a detailed derivation of the equation of motion using a differential volume element in rectangular coordinates, see Ref. 1. It is instructive to carry out such a detailed derivation at least once.

surface $S$. The momentum flux associated with this mechanism is $\boldsymbol{\pi}$, the *stress tensor.* In interpreting the stress tensor as a momentum flux, the component $\pi_{ij}$ represents the flux of positive $j$ momentum in the positive $i$-direction. This definition of $\boldsymbol{\pi}$ leads to a convention that is consistent with the other transport phenomena as shown in Example 5.1 below.

The rate of accumulation of momentum in the control volume $V$ equals the sum of the net influx of momentum by both mechanisms and the body forces acting on the fluid volume as a whole, such as gravity $\mathbf{g}$

$$\frac{d}{dt}\int_V \rho \boldsymbol{v}\, dV = -\int_S [\mathbf{n}\cdot \rho \boldsymbol{v}\boldsymbol{v}]dS - \int_S [\mathbf{n}\cdot \boldsymbol{\pi}]dS + \int_V \rho \mathbf{g}\, dV \qquad (5.1\text{-}13)$$

Following the same arguments as before, utilizing the divergence theorem of Gauss, we obtain the equation of motion also called Cauchy's equation (3)

$$\blacktriangleright \qquad \frac{\partial}{\partial t}\rho \boldsymbol{v} = -[\nabla \cdot \rho \boldsymbol{v}\boldsymbol{v}] - [\nabla \cdot \boldsymbol{\pi}] + \rho \mathbf{g} \qquad (5.1\text{-}14)$$

or in terms of the substantial derivative, with the aid of the equation of continuity, Eq. 5.1-14 can be written as

$$\blacktriangleright \qquad \rho \frac{D\boldsymbol{v}}{Dt} = -[\nabla \cdot \boldsymbol{\pi}] + \rho \mathbf{g} \qquad (5.1\text{-}15)$$

It is convenient to divide the stress tensor $\boldsymbol{\pi}$ into two parts

$$\boldsymbol{\pi} = P\boldsymbol{\delta} + \boldsymbol{\tau} \qquad (5.1\text{-}16)$$

where $\boldsymbol{\tau}$ is the *dynamic* or extra stress associated with flow, $P$ is a scalar called the "pressure," and $\boldsymbol{\delta}$ is the unit (identity) tensor

$$\boldsymbol{\delta} = \begin{pmatrix} 1 & 0 & 0 \\ 0 & 1 & 0 \\ 0 & 0 & 1 \end{pmatrix} \qquad (5.1\text{-}17)$$

Equation 5.1-16 in component form is $\pi_{ij} = P\delta_{ij} + \tau_{ij}$, where $\delta_{ij} = 1$ for $i = j$, and $\delta_{ij} = 0$ for $i \neq j$. For convenience, the tensor $\boldsymbol{\pi}$ is called the *total* stress tensor and $\boldsymbol{\tau}$ simply the stress tensor. Clearly $\pi_{ij} = \tau_{ij} (i \neq j)$ and $\pi_{ii} = P + \tau_{ii}$. Thus the total normal stress incorporates the contribution of the "pressure" $P$, which is isotropic. In the absence of flow, at equilibrium, the "pressure" $P$ becomes identical to the thermodynamic pressure, which for pure fluids is related to density and temperature, $P(\rho, T)$. Two difficulties are associated with $P$. First, flow implies nonequilibrium conditions, and it is not obvious that $P$ appearing during flow is the same pressure as defined by thermodynamics. Second, when the incompressibility assumption is made (generally used in solving polymer processing problems) the meaning of $P$ is not clear, and $P$ is regarded as an arbitrary variable. No difficulty, however, arises in solving problems, because we need to know only the pressure gradient.

By substituting Eq. 5.1-16 into Eqs. 5.1-14 and 5.1-15, we get, respectively,

$$\blacktriangleright \qquad \frac{\partial}{\partial t}(\rho \boldsymbol{v}) = -[\nabla \cdot \rho \boldsymbol{v}\boldsymbol{v}] - \nabla P - [\nabla \cdot \boldsymbol{\tau}] + \rho \mathbf{g} \qquad (5.1\text{-}18)$$

and

$$\rho \frac{Dv}{Dt} = -\nabla P - [\nabla \cdot \tau] + \rho g \tag{5.1-19}$$

Table 5.2 lists the components of the equation of motion in various coordinate systems.

Equation 5.1-15 has the familiar form of Newton's second law, stating that the rate of change of momentum of a fluid particle equals the sum of the forces acting

**Table 5.2 The Equation of Motion in Terms of $\tau$ in Several Coordinate Systems[a]**

**Rectangular Coordinates** $(x, y, z)$

$$\rho\left(\frac{\partial v_x}{\partial t} + v_x \frac{\partial v_x}{\partial x} + v_y \frac{\partial v_x}{\partial y} + v_z \frac{\partial v_x}{\partial z}\right) = -\frac{\partial P}{\partial x} - \left(\frac{\partial \tau_{xx}}{\partial x} + \frac{\partial \tau_{yx}}{\partial y} + \frac{\partial \tau_{zx}}{\partial z}\right) + \rho g_x$$

$$\rho\left(\frac{\partial v_y}{\partial t} + v_x \frac{\partial v_y}{\partial x} + v_y \frac{\partial v_y}{\partial y} + v_z \frac{\partial v_y}{\partial z}\right) = -\frac{\partial P}{\partial y} - \left(\frac{\partial \tau_{xy}}{\partial x} + \frac{\partial \tau_{yy}}{\partial y} + \frac{\partial \tau_{zy}}{\partial z}\right) + \rho g_y$$

$$\rho\left(\frac{\partial v_z}{\partial t} + v_x \frac{\partial v_z}{\partial x} + v_y \frac{\partial v_z}{\partial y} + v_z \frac{\partial v_z}{\partial z}\right) = -\frac{\partial P}{\partial z} - \left(\frac{\partial \tau_{xz}}{\partial x} + \frac{\partial \tau_{yz}}{\partial y} + \frac{\partial \tau_{zz}}{\partial z}\right) + \rho g_z$$

**Cylindrical Coordinates** $(r, \theta, z)$

$$\rho\left(\frac{\partial v_r}{\partial t} + v_r \frac{\partial v_r}{\partial r} + \frac{v_\theta}{r} \frac{\partial v_r}{\partial \theta} - \frac{v_\theta^2}{r} + v_z \frac{\partial v_r}{\partial z}\right) = -\frac{\partial P}{\partial r} - \left(\frac{1}{r}\frac{\partial}{\partial r}(r\tau_{rr}) + \frac{1}{r}\frac{\partial \tau_{r\theta}}{\partial \theta} - \frac{\tau_{\theta\theta}}{r} + \frac{\partial \tau_{rz}}{\partial z}\right) + \rho g_r$$

$$\rho\left(\frac{\partial v_\theta}{\partial t} + v_r \frac{\partial v_\theta}{\partial r} + \frac{v_\theta}{r} \frac{\partial v_\theta}{\partial \theta} + \frac{v_r v_\theta}{r} + v_z \frac{\partial v_\theta}{\partial z}\right) = -\frac{1}{r}\frac{\partial P}{\partial \theta} - \left(\frac{1}{r^2}\frac{\partial}{\partial r}(r^2\tau_{r\theta}) + \frac{1}{r}\frac{\partial \tau_{\theta\theta}}{\partial \theta} + \frac{\partial \tau_{\theta z}}{\partial z}\right) + \rho g_\theta$$

$$\rho\left(\frac{\partial v_z}{\partial t} + v_r \frac{\partial v_z}{\partial r} + \frac{v_\theta}{r} \frac{\partial v_z}{\partial \theta} + v_z \frac{\partial v_z}{\partial z}\right) = -\frac{\partial P}{\partial z} - \left(\frac{1}{r}\frac{\partial}{\partial r}(r\tau_{rz}) + \frac{1}{r}\frac{\partial \tau_{\theta z}}{\partial \theta} + \frac{\partial \tau_{zz}}{\partial z}\right) + \rho g_z$$

**Spherical Coordinates** $(r, \theta, \phi)$

$$\rho\left(\frac{\partial v_r}{\partial t} + v_r \frac{\partial v_r}{\partial r} + \frac{v_\theta}{r} \frac{\partial v_r}{\partial \theta} + \frac{v_\phi}{r \sin\theta} \frac{\partial v_r}{\partial \phi} - \frac{v_\theta^2 + v_\phi^2}{r}\right)$$

$$= -\frac{\partial P}{\partial r} - \left(\frac{1}{r^2}\frac{\partial}{\partial r}(r^2\tau_{rr}) + \frac{1}{r\sin\theta}\frac{\partial}{\partial \theta}(\tau_{r\theta}\sin\theta) + \frac{1}{r\sin\theta}\frac{\partial \tau_{r\phi}}{\partial \phi} - \frac{\tau_{\theta\theta} + \tau_{\phi\phi}}{r}\right) + \rho g_r$$

$$\rho\left(\frac{\partial v_\theta}{\partial t} + v_r \frac{\partial v_\theta}{\partial r} + \frac{v_\theta}{r} \frac{\partial v_\theta}{\partial \theta} + \frac{v_\phi}{r \sin\theta} \frac{\partial v_\theta}{\partial \phi} + \frac{v_r v_\theta}{r} - \frac{v_\phi^2 \cot\theta}{r}\right)$$

$$= -\frac{1}{r}\frac{\partial P}{\partial \theta} - \left(\frac{1}{r^2}\frac{\partial}{\partial r}(r^2\tau_{r\theta}) + \frac{1}{r\sin\theta}\frac{\partial}{\partial \theta}(\tau_{\theta\theta}\sin\theta) + \frac{1}{r\sin\theta}\frac{\partial \tau_{\theta\phi}}{\partial \phi} + \frac{\tau_{r\theta}}{r} - \frac{\cot\theta}{r}\tau_{\phi\phi}\right) + \rho g_\theta$$

$$\rho\left(\frac{\partial v_\phi}{\partial t} + v_r \frac{\partial v_\phi}{\partial r} + \frac{v_\theta}{r} \frac{\partial v_\phi}{\partial \theta} + \frac{v_\phi}{r \sin\theta} \frac{\partial v_\phi}{\partial \phi} + \frac{v_\phi v_r}{r} + \frac{v_\theta v_\phi}{r}\cot\theta\right)$$

$$= -\frac{1}{r\sin\theta}\frac{\partial P}{\partial \phi} - \left(\frac{1}{r^2}\frac{\partial}{\partial r}(r^2\tau_{r\phi}) + \frac{1}{r}\frac{\partial \tau_{\theta\phi}}{\partial \theta} + \frac{1}{r\sin\theta}\frac{\partial \tau_{\phi\phi}}{\partial \phi} + \frac{\tau_{r\phi}}{r} + \frac{2\cot\theta}{r}\tau_{\theta\phi}\right) + \rho g_\phi$$

[a] Reprinted with permission from R. B. Bird, W. E. Stewart, and E. N. Lightfoot, *Transport Phenomena*, Wiley, New York, 1960.

on it. Thus $-[\nabla \cdot \boldsymbol{\pi}]$ is the net force exerted by the surrounding fluid on the fluid element.

### *The Stress Tensor*

Consider a point $P$ in a continuum on an arbitrary surface element $\Delta S$ (defined by the normal $\mathbf{n}$), as in Fig. 5.3. Let $\Delta \mathbf{f}_i$ be the resultant force exerted by the material

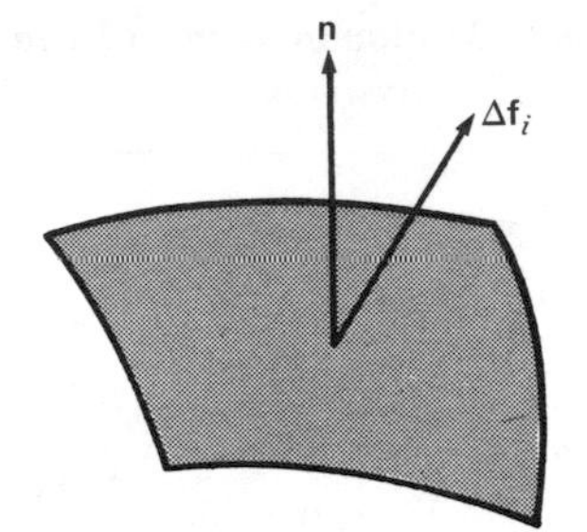


**Fig. 5.3** A generalized surface used in the definition of the stress vector.

on the positive side of the surface on that on the negative side across $\Delta S$. The average force per unit area is $\Delta \mathbf{f}_i/\Delta S$. This quantity attains limiting nonzero value as $\Delta S$ approaches zero, at point $P$ (Cauchy's stress principle). This limiting quantity is called the *stress vector*, or traction vector $\mathbf{T}'$. But $\mathbf{T}'$ depends on the orientation of the area element $\Delta S$, that is, the direction of the surface normal $\mathbf{n}$. Thus it would appear that there are an infinite number of unrelated ways of expressing the state of stress at point $P$. It turns out, however, that the state of stress at $P$ can be completely specified by giving the stress vector components of *any three mutually perpendicular planes* passing through the point. That is, a total of nine components, three for each vector, is needed. Each component can be described by two indices $ij$, the first denoting the orientation of the surface and the second the direction of the force. Figure 5.4 gives these components for three Cartesian planes. The nine stress vector components form a second order Cartesian tensor, the stress* tensor $\boldsymbol{\pi}'$. Furthermore, certain argumentation based on principles of mechanics, experimental observations as well as molecular theories, leads to the conclusion that the stress tensor is symmetric

$$\pi'_{ij} = \pi'_{ji} \tag{5.1-20}$$

Hence only six independent components of the stress tensor are needed to fully define the state of the stress at point $P$. The $\pi'_{ii}$ are called the *normal stress* components and the $\pi'_{ij}(i \neq j)$ the *shear stress* components.

By considering the forces that the material on the positive side of the surface (i.e., material on the side of the surface in which the outer normal vector points) exerts on the material on the negative side, a stress component is *positive* when it acts in the positive direction of the coordinate axes and on a plane whose outer

* Note that we differentiate the stress tensor $\boldsymbol{\pi}'$ discussed in this section, from the previously discussed stress tensor $\boldsymbol{\pi}$, because they are defined on the basis of different sign convention, as discussed below.

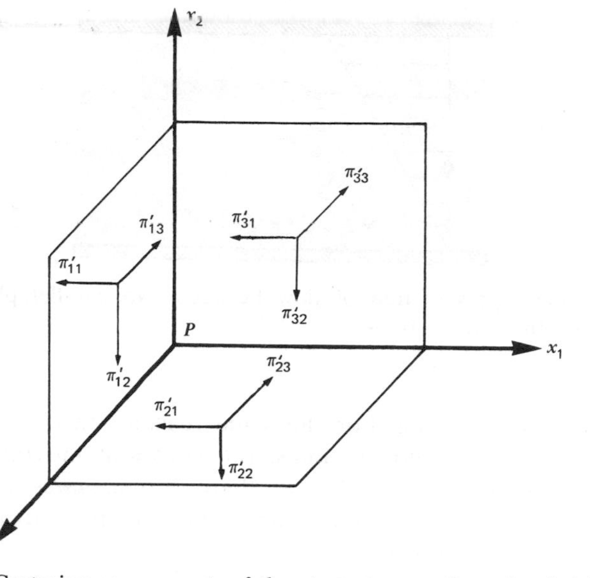


**Fig. 5.4** The nine Cartesian components of the stress tensor; the cube shrinks to a point $P$ in the limit.

normal points in one of the positive coordinate directions. The $\pi'_{ij}$ are also positive if both the directions above are negative; they are negative if any *one* direction is negative. Hence by this sign convention, generally used in mechanics and mechanical engineering, tensile stresses are positive and compressive stresses are negative. In this sign convention all the stresses shown in Fig. 5.4 are positive. Unfortunately this sign convention is *opposite* to that resulting from momentum transport considerations, thus $\boldsymbol{\pi}'^{\dagger} = -\boldsymbol{\pi}$ (where † stands for "transpose"). In the latter sign convention, as pointed out by Bird et al. (7b), if we consider the stress vector $\boldsymbol{\pi}_n = \mathbf{n} \cdot \boldsymbol{\pi}$ acting on surface $dS$ of orientation $\mathbf{n}$, the force $\boldsymbol{\pi}_n\, dS$ is that exerted by the material on the negative side on that on the positive side. (By Newton's third law, this force is equal and opposite to that exerted by the material of the positive side to the material of the negative side.) It follows then that in this convention tensile stresses are negative.

In this book we follow the latter sign convention; that is, that of $\boldsymbol{\pi}$ introduced during the momentum transfer considerations, used by Bird et al. (1). As we pointed out in the introductory remarks, polymer processing *is* the simultaneous occurrence of momentum, heat, and occasionally mass transfer. This sign convention satisfies a physical symmetry among the various transport phenomena as we shall see below. Nevertheless, it is worth emphasizing that the sign convention used in no way affects the solution of flow problems. Once constitutive equations are inserted into the equation of motion, and stress components are replaced by velocity gradients, the two sign conventions lead to identical expressions.

### Example 5.1 *The Similarity Between the Three Transport Phenomena*

Let us examine a viscous fluid between two parallel plates, the top of which is moving (Fig. 5.5). Because of intermolecular forces, the fluid layer next to the

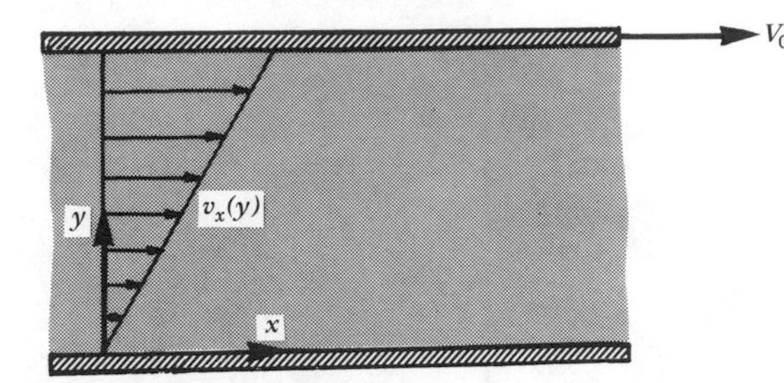


**Fig. 5.5** Schematic representation of flow between two parallel plates, one of them stationary, showing the velocity profile.

top plate will start moving. For the same reasons, this layer *will transmit momentum* to the layer below it, and so on. Thus $x$–momentum is transferred from the top to the bottom; that is, there is a *negative* momentum flux according to the coordinate system used. But momentum flux is nothing more than a shear stress in this case.

The negative momentum flux creates a *positive* velocity gradient as shown in Fig. 5.5. A *linear* relationship between these two variables defines an important class of fluids called "Newtonian fluids." Newton's law of viscosity in simple shear flow is written as

$$\tau_{yx} = -\mu \frac{dv_x}{dy} \tag{5.1-21}$$

where $\mu$ is the *viscosity*. The negative sign in this equation is introduced to obtain the correct direction of the momentum flux. Similarly, in one-dimensional conductive heat transfer we relate the positive heat flux to a negative temperature gradient by Fourier's law

$$q_y = -k \frac{dT}{dy} \tag{5.1-22}$$

where $k$ is the thermal conductivity; and in one-dimensional diffusion we relate the positive mass flux of species $A$ to a negative concentration gradient by Fick's law (constant density and low concentrations of $A$)

$$j_{\mathrm{A}y} = -\mathscr{D}_{\mathrm{AB}} \frac{dc_{\mathrm{A}}}{dy} \tag{5.1-23}$$

where $\mathscr{D}_{\mathrm{AB}}$ is the binary diffusivity. Thus all three transport phenomena have the same sign convention.

### *The Rate of Strain Tensor*

We noted earlier that the (deviatoric) stress tensor $\boldsymbol{\tau}$ is related to fluid flow. The kinematic quantity by which fluid flow is expressed is the velocity gradient $dv_i/dx_j$. In Example 5.1 Newton's law of viscosity indicates that in simple shear flow between parallel plates, the single nonvanishing stress component $\tau_{yx}$ is proportional to the single nonvanishing velocity gradient $dv_x/dy$. In a general flow field, however, there is more than one possible velocity gradient. Each of the three velocity components may vary in each of the three spatial coordinates, giving rise to *nine* possible components. We can therefore define a *velocity gradient tensor* denoted by $\nabla \boldsymbol{v}$ (dyadic product of $\nabla$ and $\boldsymbol{v}$) which in Cartesian coordinates can be written as

$$\nabla \boldsymbol{v} = \begin{pmatrix} \frac{\partial v_1}{\partial x_1} & \frac{\partial v_2}{\partial x_1} & \frac{\partial v_3}{\partial x_1} \\ \frac{\partial v_1}{\partial x_2} & \frac{\partial v_2}{\partial x_2} & \frac{\partial v_3}{\partial x_2} \\ \frac{\partial v_1}{\partial x_3} & \frac{\partial v_2}{\partial x_3} & \frac{\partial v_3}{\partial x_3} \end{pmatrix} \tag{5.1-24}$$

A fluid in motion may simultaneously flow and rotate. These motions can be separated by decomposing the velocity gradient tensor into two parts

$$\nabla \boldsymbol{v} = \tfrac{1}{2}(\dot{\boldsymbol{\gamma}} + \boldsymbol{\omega}) \tag{5.1-25}$$

where $\dot{\boldsymbol{\gamma}}$ and $\boldsymbol{\omega}$ are the *rate of strain tensor* and the *vorticity tensor*, respectively, defined as

$$\dot{\boldsymbol{\gamma}} = \nabla \boldsymbol{v} + (\nabla \boldsymbol{v})^{\dagger} \tag{5.1-26}$$

$$\boldsymbol{\omega} = \nabla \boldsymbol{v} - (\nabla \boldsymbol{v})^{\dagger} \tag{5.1-27}$$

where $(\nabla \boldsymbol{v})^{\dagger}$ is the *transpose** of $\nabla \boldsymbol{v}$. Table 5.3 lists the components of $\dot{\gamma}$ in various coordinates. For the simple parallel plate flow just discussed, $\dot{\boldsymbol{\gamma}}$ reduces to

$$\dot{\boldsymbol{\gamma}} = \begin{pmatrix} 0 & 1 & 0 \\ 1 & 0 & 0 \\ 0 & 0 & 0 \end{pmatrix} \dot{\gamma} \tag{5.1-28}$$

where $\dot{\gamma}$ is the shear rate, which is a scalar and is related to the second invariant† of the tensor $\dot{\boldsymbol{\gamma}}$

$$\dot{\gamma} = \sqrt{\tfrac{1}{2}(\dot{\boldsymbol{\gamma}} : \dot{\boldsymbol{\gamma}})} \tag{5.1-29}$$

In simple shear flow the shear rate equals, of course, the single nonvanishing velocity gradient, which can be written as follows:

$$\dot{\gamma} = \frac{dv_x}{dy} = \frac{d}{dy}\left(\frac{dx}{dt}\right) = \frac{d}{dt}\left(\frac{dx}{dy}\right) = \frac{d\gamma}{dt} \tag{5.1-30}$$

where $\gamma$ is the total shear strain.

* The $(\nabla \boldsymbol{v})^{\dagger}$ has the same components as $\nabla \boldsymbol{v}$ but with the indices transposed (interchanging its rows and columns).

† The second invariant of $\dot{\boldsymbol{\gamma}}$ is $II_{\dot{\gamma}} = \dot{\boldsymbol{\gamma}} : \dot{\boldsymbol{\gamma}} = \Sigma_i \Sigma_j \dot{\gamma}_{ij} \dot{\gamma}_{ji}$.

**Table 5.3 The Components of $\boldsymbol{\tau} = -\mu[\dot{\boldsymbol{\gamma}} - \frac{2}{3}(\nabla \cdot \boldsymbol{v})\boldsymbol{\delta}]$ in Several Coordinate Systems**[a]

**Rectangular Coordinates** $(x, y, z)$

$$\tau_{xx} = -\mu\left[2\frac{\partial v_x}{\partial x} - \frac{2}{3}(\nabla \cdot \boldsymbol{v})\right]$$

$$\tau_{yy} = -\mu\left[2\frac{\partial v_y}{\partial y} - \frac{2}{3}(\nabla \cdot \boldsymbol{v})\right]$$

$$\tau_{zz} = -\mu\left[2\frac{\partial v_z}{\partial z} - \frac{2}{3}(\nabla \cdot \boldsymbol{v})\right]$$

$$\tau_{xy} = \tau_{yx} = -\mu\left[\frac{\partial v_x}{\partial y} + \frac{\partial v_y}{\partial x}\right]$$

$$\tau_{yz} = \tau_{zy} = -\mu\left[\frac{\partial v_y}{\partial z} + \frac{\partial v_z}{\partial y}\right]$$

$$\tau_{zx} = \tau_{xz} = -\mu\left[\frac{\partial v_z}{\partial x} + \frac{\partial v_x}{\partial z}\right]$$

$$(\nabla \cdot \boldsymbol{v}) = \frac{\partial v_x}{\partial x} + \frac{\partial v_y}{\partial y} + \frac{\partial v_z}{\partial z}$$

**Cylindrical Coordinates** $(r, \theta, z)$

$$\tau_{rr} = -\mu\left[2\frac{\partial v_r}{\partial r} - \frac{2}{3}(\nabla \cdot \boldsymbol{v})\right]$$

$$\tau_{\theta\theta} = -\mu\left[2\left(\frac{1}{r}\frac{\partial v_\theta}{\partial \theta} + \frac{v_r}{r}\right) - \frac{2}{3}(\nabla \cdot \boldsymbol{v})\right]$$

$$\tau_{zz} = -\mu\left[2\frac{\partial v_z}{\partial z} - \frac{2}{3}(\nabla \cdot \boldsymbol{v})\right]$$

$$\tau_{r\theta} = \tau_{\theta r} = -\mu\left[r\frac{\partial}{\partial r}\left(\frac{v_\theta}{r}\right) + \frac{1}{r}\frac{\partial v_r}{\partial \theta}\right]$$

$$\tau_{\theta z} = \tau_{z\theta} = -\mu\left[\frac{\partial v_\theta}{\partial z} + \frac{1}{r}\frac{\partial v_z}{\partial \theta}\right]$$

$$\tau_{zr} = \tau_{rz} = -\mu\left[\frac{\partial v_z}{\partial r} + \frac{\partial v_r}{\partial z}\right]$$

$$(\nabla \cdot \boldsymbol{v}) = \frac{1}{r}\frac{\partial}{\partial r}(rv_r) + \frac{1}{r}\frac{\partial v_\theta}{\partial \theta} + \frac{\partial v_z}{\partial z}$$

**Spherical Coordinates** $(r, \theta, \phi)$

$$\tau_{rr} = -\mu\left[2\frac{\partial v_r}{\partial r} - \frac{2}{3}(\nabla \cdot \boldsymbol{v})\right]$$

$$\tau_{\theta\theta} = -\mu\left[2\left(\frac{1}{r}\frac{\partial v_\theta}{\partial \theta} + \frac{v_r}{r}\right) - \frac{2}{3}(\nabla \cdot \boldsymbol{v})\right]$$

$$\tau_{\phi\phi} = -\mu\left[2\left(\frac{1}{r\sin\theta}\frac{\partial v_\phi}{\partial \phi} + \frac{v_r}{r} + \frac{v_\theta \cot\theta}{r}\right) - \frac{2}{3}(\nabla \cdot \boldsymbol{v})\right]$$

$$\tau_{r\theta} = \tau_{\theta r} = -\mu\left[r\frac{\partial}{\partial r}\left(\frac{v_\theta}{r}\right) + \frac{1}{r}\frac{\partial v_r}{\partial \theta}\right]$$

$$\tau_{\theta\phi} = \tau_{\phi\theta} = -\mu\left[\frac{\sin\theta}{r}\frac{\partial}{\partial\theta}\left(\frac{v_\phi}{\sin\theta}\right) + \frac{1}{r\sin\theta}\frac{\partial v_\theta}{\partial\phi}\right]$$

$$\tau_{\phi r} = \tau_{r\phi} = -\mu\left[\frac{1}{r\sin\theta}\frac{\partial v_r}{\partial\phi} + r\frac{\partial}{\partial r}\left(\frac{v_\phi}{r}\right)\right]$$

$$(\nabla\cdot\boldsymbol{v}) = \frac{1}{r^2}\frac{\partial}{\partial r}(r^2 v_r) + \frac{1}{r\sin\theta}\frac{\partial}{\partial\theta}(v_\theta\sin\theta) + \frac{1}{r\sin\theta}\frac{\partial v_\phi}{\partial\phi}$$

[a] Reprinted with permission from R. B. Bird, W. E. Stewart, and E. N. Lightfoot, *Transport Phenomena*, Wiley, New York, 1960.

### The Newtonian Fluid

Constitutive equations relate the stress tensor $\boldsymbol{\tau}$ to the rate of deformation tensor $\dot{\boldsymbol{\gamma}}$. For a Newtonian fluid in an arbitrary flow, Newton's law of viscosity can be generalized to

► $$\boldsymbol{\tau} = -\mu\dot{\boldsymbol{\gamma}} + (\tfrac{2}{3}\mu - \kappa)(\nabla\cdot\boldsymbol{v})\boldsymbol{\delta} \tag{5.1-31}$$

where $\kappa$ is the dilatational viscosity. For incompressible fluids (and polymers are generally treated as such), we recall that $\nabla\cdot\boldsymbol{v} = 0$, and Eq. 5.1-31 reduces to

► $$\boldsymbol{\tau} = -\mu\dot{\boldsymbol{\gamma}} \tag{5.1-32}$$

Although polymers are non-Newtonian fluids (i.e., they do not obey the relationship above), many problems in polymer processing are initially solved using the Newtonian assumption because (*a*) such solutions provide simple results that give an insight into the nature of problem, (*b*) they provide quick quantitative estimates, and (*c*) frequently it is too difficult to solve the problem using a more representative constitutive equation. Yet polymer processing can be understood in depth only by considering the non-Newtonian character of polymer melts. The study of non-Newtonian behavior forms the active branch of the science of rheology, which Chapter 6 discusses in detail.

Inserting Eq. 5.1-32 into Eq. 5.1-19—that is, assuming Newtonian flow behavior, constant viscosity, and constant density— we have*

► $$\rho\frac{D\boldsymbol{v}}{Dt} = -\nabla P + \mu\nabla^2\boldsymbol{v} + \rho\boldsymbol{g} \tag{5.1-33}$$

which is the well-known Navier–Stokes equation. The symbol $\nabla^2$ is called the "Laplacian," defined as $\nabla^2 = \nabla\cdot\nabla$. Table 5.4 lists the components of the Navier–Stokes equation in various coordinates. Equation 5.1-33, together with the equation of continuity, and assuming the appropriate initial and boundary conditions, determines the velocity and pressure fields in incompressible Newtonian fluids flowing in isothermal flow.

* $-\nabla\cdot\boldsymbol{\tau} = +\mu\nabla\cdot\dot{\boldsymbol{\gamma}} = \mu\nabla\cdot[\nabla\boldsymbol{v} + (\nabla\boldsymbol{v})^\dagger] = \mu[\nabla^2\boldsymbol{v} + \nabla\cdot(\nabla\boldsymbol{v})^\dagger]$
$= \mu[\nabla^2\boldsymbol{v} + \nabla(\nabla\cdot\boldsymbol{v})] = \mu\nabla^2\boldsymbol{v}$

**Table 5.4 The Navier–Stokes Equation in Several Coordinate Systems**[a]

**Rectangular Coordinates** $(x, y, z)$

$$\rho\left(\frac{\partial v_x}{\partial t}+v_x\frac{\partial v_x}{\partial x}+v_y\frac{\partial v_x}{\partial y}+v_z\frac{\partial v_x}{\partial z}\right)=-\frac{\partial P}{\partial x}+\mu\left(\frac{\partial^2 v_x}{\partial x^2}+\frac{\partial^2 v_x}{\partial y^2}+\frac{\partial^2 v_x}{\partial z^2}\right)+\rho g_x$$

$$\rho\left(\frac{\partial v_y}{\partial t}+v_x\frac{\partial v_y}{\partial x}+v_y\frac{\partial v_y}{\partial y}+v_z\frac{\partial v_y}{\partial z}\right)=-\frac{\partial P}{\partial y}+\mu\left(\frac{\partial^2 v_y}{\partial x^2}+\frac{\partial^2 v_y}{\partial y^2}+\frac{\partial^2 v_y}{\partial z^2}\right)+\rho g_y$$

$$\rho\left(\frac{\partial v_z}{\partial t}+v_x\frac{\partial v_z}{\partial x}+v_y\frac{\partial v_z}{\partial y}+v_z\frac{\partial v_z}{\partial z}\right)=-\frac{\partial P}{\partial z}+\mu\left(\frac{\partial^2 v_z}{\partial x^2}+\frac{\partial^2 v_z}{\partial y^2}+\frac{\partial^2 v_z}{\partial z^2}\right)+\rho g_z$$

**Cylindrical Coordinates** $(r, \theta, z)$

$$\rho\left(\frac{\partial v_r}{\partial t}+v_r\frac{\partial v_r}{\partial r}+\frac{v_\theta}{r}\frac{\partial v_r}{\partial \theta}-\frac{v_\theta^2}{r}+v_z\frac{\partial v_r}{\partial z}\right)$$

$$=-\frac{\partial P}{\partial r}+\mu\left[\frac{\partial}{\partial r}\left(\frac{1}{r}\frac{\partial}{\partial r}(rv_r)\right)+\frac{1}{r^2}\frac{\partial^2 v_r}{\partial \theta^2}-\frac{2}{r^2}\frac{\partial v_\theta}{\partial \theta}+\frac{\partial^2 v_r}{\partial z^2}\right]+\rho g_r$$

$$\rho\left(\frac{\partial v_\theta}{\partial t}+v_r\frac{\partial v_\theta}{\partial r}+\frac{v_\theta}{r}\frac{\partial v_\theta}{\partial \theta}+\frac{v_r v_\theta}{r}+v_z\frac{\partial v_\theta}{\partial z}\right)$$

$$=-\frac{1}{r}\frac{\partial P}{\partial \theta}+\mu\left[\frac{\partial}{\partial r}\left(\frac{1}{r}\frac{\partial}{\partial r}(rv_\theta)\right)+\frac{1}{r^2}\frac{\partial^2 v_\theta}{\partial \theta^2}+\frac{2}{r^2}\frac{\partial v_r}{\partial \theta}+\frac{\partial^2 v_\theta}{\partial z^2}\right]+\rho g_\theta$$

$$\rho\left(\frac{\partial v_z}{\partial t}+v_r\frac{\partial v_z}{\partial r}+\frac{v_\theta}{r}\frac{\partial v_z}{\partial \theta}+v_z\frac{\partial v_z}{\partial z}\right)$$

$$=-\frac{\partial P}{\partial z}+\mu\left[\frac{1}{r}\frac{\partial}{\partial r}\left(r\frac{\partial v_z}{\partial r}\right)+\frac{1}{r^2}\frac{\partial^2 v_z}{\partial \theta^2}+\frac{\partial^2 v_z}{\partial z^2}\right]+\rho g_z$$

**Spherical Coordinates** $(r, \theta, \phi)$[b]

$$\rho\left(\frac{\partial v_r}{\partial t}+v_r\frac{\partial v_r}{\partial r}+\frac{v_\theta}{r}\frac{\partial v_r}{\partial \theta}+\frac{v_\phi}{r\sin\theta}\frac{\partial v_r}{\partial \phi}-\frac{v_\theta^2+v_\phi^2}{r}\right)$$

$$=-\frac{\partial P}{\partial r}+\mu\left(\nabla^2 v_r-\frac{2}{r^2}v_r-\frac{2}{r^2}\frac{\partial v_\theta}{\partial \theta}-\frac{2}{r^2}v_\theta\cot\theta-\frac{2}{r^2\sin\theta}\frac{\partial v_\phi}{\partial \phi}\right)+\rho g_r$$

$$\rho\left(\frac{\partial v_\theta}{\partial t}+v_r\frac{\partial v_\theta}{\partial r}+\frac{v_\theta}{r}\frac{\partial v_\theta}{\partial \theta}+\frac{v_\phi}{r\sin\theta}\frac{\partial v_\theta}{\partial \phi}+\frac{v_r v_\theta}{r}-\frac{v_\phi^2\cot\theta}{r}\right)$$

$$=-\frac{1}{r}\frac{\partial P}{\partial \theta}+\mu\left(\nabla^2 v_\theta+\frac{2}{r^2}\frac{\partial v_r}{\partial \theta}-\frac{v_\theta}{r^2\sin^2\theta}-\frac{2\cos\theta}{r^2\sin^2\theta}\frac{\partial v_\phi}{\partial \phi}\right)+\rho g_\theta$$

$$\rho\left(\frac{\partial v_\phi}{\partial t}+v_r\frac{\partial v_\phi}{\partial r}+\frac{v_\theta}{r}\frac{\partial v_\phi}{\partial \theta}+\frac{v_\phi}{r\sin\theta}\frac{\partial v_\phi}{\partial \phi}+\frac{v_\phi v_r}{r}+\frac{v_\theta v_\phi}{r}\cot\theta\right)$$

$$=-\frac{1}{r\sin\theta}\frac{\partial P}{\partial \phi}+\mu\left(\nabla^2 v_\phi-\frac{v_\phi}{r^2\sin^2\theta}+\frac{2}{r^2\sin\theta}\frac{\partial v_r}{\partial \phi}+\frac{2\cos\theta}{r^2\sin^2\theta}\frac{\partial v_\theta}{\partial \phi}\right)+\rho g_\phi$$

[a] Reprinted with permission from R. B. Bird, W. E. Stewart, and E. N. Lightfoot, *Transport Phenomena*, Wiley, New York, 1960.

[b] In these equations

$$\nabla^2=\frac{1}{r^2}\frac{\partial}{\partial r}\left(r^2\frac{\partial}{\partial r}\right)+\frac{1}{r^2\sin\theta}\frac{\partial}{\partial \theta}\left(\sin\theta\frac{\partial}{\partial \theta}\right)+\frac{1}{r^2\sin^2\theta}\left(\frac{\partial^2}{\partial \phi^2}\right)$$

Polymer processing flows are generally creeping flows. A creeping flow is one in which viscous forces predominate over inertial forces. Classic examples of such flows include those treated by the hydrodynamic theory of lubrication, Hele–Shaw flows, and the flow of very viscous fluids past immersed bodies. The equation of motion reduces to the following form

$$\rho \frac{\partial \boldsymbol{v}}{\partial t} = -\nabla P - [\nabla \cdot \boldsymbol{\tau}] + \rho \mathbf{g} \tag{5.1-34}$$

Creeping flows are treated in detail by Happel and Brenner (8).

### *The Equation of Energy*

Following the same procedure as before for the conservation of mass and monentum, we can obtain a balance equation for the rate of change of kinetic and internal energy. A separate balance equation for the kinetic energy can be obtained from the equation of motion by forming the dot product of the fluid velocity $\mathbf{v}$ with the equation of motion. Subtracting the latter equation from the former results in the *thermal energy balance*, which can be written as

$$\rho \frac{DU}{Dt} = -(\nabla \cdot \mathbf{q}) - P(\nabla \cdot \boldsymbol{v}) - (\boldsymbol{\tau} : \nabla \boldsymbol{v}) + \dot{S} \tag{5.1-35}$$

where $U$ is the specific internal energy (per unit mass), $\mathbf{q}$ the heat flux vector, and $\dot{S}$ is a thermal energy source term. The term $P(\nabla \cdot \boldsymbol{v})$ is the *reversible* rate of internal energy increase per unit volume by compression, and the term $(\boldsymbol{\tau} : \nabla \boldsymbol{v})$ is the *irreversible* rate of internal energy increase per unit volume by viscous dissipation. It is this latter term, together with the temperature dependence of viscosity, which couples the equations of motion and thermal energy and necessitates their simultaneous solution. It can be significant and occasionally dominant in viscous polymer flow. The heat flux can be expressed in terms of the temperature gradient using a generalized form of Fourier's equation

► $$\mathbf{q} = -k\nabla T \tag{5.1-36}$$

The internal energy, assuming equilibrium, can be expressed in terms of the temperature and specific volume, $U = U(T, \hat{V})$. If we further assume that the fluid is incompressible (which also implies that $C_p = C_v$), we obtain the following useful form of the equation of thermal energy:

► $$\rho C_v \frac{DT}{Dt} = (\nabla \cdot k\nabla T) - (\boldsymbol{\tau} : \nabla \boldsymbol{v}) + \dot{S} \tag{5.1-37}$$

which is listed in various coordinate systems in terms of heat and momentum fluxes, in Table 5.5*a*. For incompressible Newtonian fluids with constant thermal conductivity, Eq. 5.1-37 reduces to

► $$\rho C_v \frac{DT}{Dt} = k\nabla^2 T + \tfrac{1}{2}\mu(\dot{\boldsymbol{\gamma}} : \dot{\boldsymbol{\gamma}}) + \dot{S} \tag{5.1-38}$$

Equation 5.1-38 in various coordinates is also listed in Table 5.5*b*.

**Table 5.5a The Equation of Energy in Terms of Energy and Momentum Fluxes in Several Coordinate Systems**[a]

**Rectangular Coordinates** $(x, y, z)$

$$\rho C_v\left(\frac{\partial T}{\partial t}+v_x\frac{\partial T}{\partial x}+v_y\frac{\partial T}{\partial y}+v_z\frac{\partial T}{\partial z}\right)$$

$$=-\left[\frac{\partial q_x}{\partial x}+\frac{\partial q_y}{\partial y}+\frac{\partial q_z}{\partial z}\right]-T\left(\frac{\partial P}{\partial T}\right)_\rho\left(\frac{\partial v_x}{\partial x}+\frac{\partial v_y}{\partial y}+\frac{\partial v_z}{\partial z}\right)-\left\{\tau_{xx}\frac{\partial v_x}{\partial x}+\tau_{yy}\frac{\partial v_y}{\partial y}+\tau_{zz}\frac{\partial v_z}{\partial z}\right\}$$

$$-\left\{\tau_{xy}\left(\frac{\partial v_x}{\partial y}+\frac{\partial v_y}{\partial x}\right)+\tau_{xz}\left(\frac{\partial v_x}{\partial z}+\frac{\partial v_z}{\partial x}\right)+\tau_{yz}\left(\frac{\partial v_y}{\partial z}+\frac{\partial v_z}{\partial y}\right)\right\}$$

**Cylindrical Coordinates** $(r, \theta, z)$

$$\rho C_v\left(\frac{\partial T}{\partial t}+v_r\frac{\partial T}{\partial r}+\frac{v_\theta}{r}\frac{\partial T}{\partial \theta}+v_z\frac{\partial T}{\partial z}\right)$$

$$=-\left[\frac{1}{r}\frac{\partial}{\partial r}(rq_r)+\frac{1}{r}\frac{\partial q_\theta}{\partial \theta}+\frac{\partial q_z}{\partial z}\right]-T\left(\frac{\partial P}{\partial T}\right)_\rho\left(\frac{1}{r}\frac{\partial}{\partial r}(rv_r)+\frac{1}{r}\frac{\partial v_\theta}{\partial \theta}+\frac{\partial v_z}{\partial z}\right)$$

$$-\left\{\tau_{rr}\frac{\partial v_r}{\partial r}+\tau_{\theta\theta}\frac{1}{r}\left(\frac{\partial v_\theta}{\partial \theta}+v_r\right)\right.$$

$$\left.+\tau_{zz}\frac{\partial v_z}{\partial z}\right\}-\left\{\tau_{r\theta}\left[r\frac{\partial}{\partial r}\left(\frac{v_\theta}{r}\right)+\frac{1}{r}\frac{\partial v_r}{\partial \theta}\right]+\tau_{rz}\left(\frac{\partial v_z}{\partial r}+\frac{\partial v_r}{\partial z}\right)+\tau_{\theta z}\left(\frac{1}{r}\frac{\partial v_z}{\partial \theta}+\frac{\partial v_\theta}{\partial z}\right)\right\}$$

**Spherical Coordinates** $(r, \theta, \phi)$

$$\rho C_v\left(\frac{\partial T}{\partial t}+v_r\frac{\partial T}{\partial r}+\frac{v_\theta}{r}\frac{\partial T}{\partial \theta}+\frac{v_\phi}{r\sin\theta}\frac{\partial T}{\partial \phi}\right)$$

$$=-\left[\frac{1}{r^2}\frac{\partial}{\partial r}(r^2q_r)+\frac{1}{r\sin\theta}\frac{\partial}{\partial \theta}(q_\theta\sin\theta)+\frac{1}{r\sin\theta}\frac{\partial q_\phi}{\partial \phi}\right]$$

$$-T\left(\frac{\partial P}{\partial T}\right)_\rho\left(\frac{1}{r^2}\frac{\partial}{\partial r}(r^2v_r)+\frac{1}{r\sin\theta}\frac{\partial}{\partial \theta}(v_\theta\sin\theta)+\frac{1}{r\sin\theta}\frac{\partial v_\phi}{\partial \phi}\right)$$

$$-\left\{\tau_{rr}\frac{\partial v_r}{\partial r}+\tau_{\theta\theta}\left(\frac{1}{r}\frac{\partial v_\theta}{\partial \theta}+\frac{v_r}{r}\right)+\tau_{\phi\phi}\left(\frac{1}{r\sin\theta}\frac{\partial v_\phi}{\partial \phi}+\frac{v_r}{r}+\frac{v_\theta\cot\theta}{r}\right)\right\}$$

$$-\left\{\tau_{r\theta}\left(\frac{\partial v_\theta}{\partial r}+\frac{1}{r}\frac{\partial v_r}{\partial \theta}-\frac{v_\theta}{r}\right)+\tau_{r\phi}\left(\frac{\partial v_\phi}{\partial r}+\frac{1}{r\sin\theta}\frac{\partial v_r}{\partial \phi}-\frac{v_\phi}{r}\right)\right.$$

$$\left.+\tau_{\theta\phi}\left(\frac{1}{r}\frac{\partial v_\phi}{\partial \theta}+\frac{1}{r\sin\theta}\frac{\partial v_\theta}{\partial \phi}-\frac{\cot\theta}{r}v_\phi\right)\right\}$$

[a] Reprinted with permission from R. B. Bird, W. E. Stewart, and E. N. Lightfoot, *Transport Phenomena*, Wiley, New York, 1960.

**Table 5.5b The Equation of Energy in Terms of the Transport Properties in Several Coordinate Systems**[a]

*(for Newtonian fluids of constant $\rho$, $\mu$ and $k$ note that the constancy of $\rho$ implies that $C_v = C_p$)*

**Rectangular Coordinates** $(x, y, z)$

$$\rho C_v\left(\frac{\partial T}{\partial t}+v_x\frac{\partial T}{\partial x}+v_y\frac{\partial T}{\partial y}+v_z\frac{\partial T}{\partial z}\right)$$

$$=k\left[\frac{\partial^2 T}{\partial x^2}+\frac{\partial^2 T}{\partial y^2}+\frac{\partial^2 T}{\partial z^2}\right]+2\mu\left\{\left(\frac{\partial v_x}{\partial x}\right)^2+\left(\frac{\partial v_y}{\partial y}\right)^2+\left(\frac{\partial v_z}{\partial z}\right)^2\right\}$$

$$+\mu\left\{\left(\frac{\partial v_x}{\partial y}+\frac{\partial v_y}{\partial x}\right)^2+\left(\frac{\partial v_x}{\partial z}+\frac{\partial v_z}{\partial x}\right)^2+\left(\frac{\partial v_y}{\partial z}+\frac{\partial v_z}{\partial y}\right)^2\right\}$$

**Cylindrical Coordinates** $(r, \theta, z)$

$$\rho C_v\left(\frac{\partial T}{\partial t}+v_r\frac{\partial T}{\partial r}+\frac{v_\theta}{r}\frac{\partial T}{\partial \theta}+v_z\frac{\partial T}{\partial z}\right)$$

$$=k\left[\frac{1}{r}\frac{\partial}{\partial r}\left(r\frac{\partial T}{\partial r}\right)+\frac{1}{r^2}\frac{\partial^2 T}{\partial \theta^2}+\frac{\partial^2 T}{\partial z^2}\right]+2\mu\left\{\left(\frac{\partial v_r}{\partial r}\right)^2+\left[\frac{1}{r}\left(\frac{\partial v_\theta}{\partial \theta}+v_r\right)\right]^2+\left(\frac{\partial v_z}{\partial z}\right)^2\right\}$$

$$+\mu\left\{\left(\frac{\partial v_\theta}{\partial z}+\frac{1}{r}\frac{\partial v_z}{\partial \theta}\right)^2+\left(\frac{\partial v_z}{\partial r}+\frac{\partial v_r}{\partial z}\right)^2+\left[\frac{1}{r}\frac{\partial v_r}{\partial \theta}+r\frac{\partial}{\partial r}\left(\frac{v_\theta}{r}\right)\right]^2\right\}$$

**Spherical Coordinates** $(r, \theta, \phi)$

$$\rho C_v\left(\frac{\partial T}{\partial t}+v_r\frac{\partial T}{\partial r}+\frac{v_\theta}{r}\frac{\partial T}{\partial \theta}+\frac{v_\phi}{r\sin\theta}\frac{\partial T}{\partial \phi}\right)$$

$$=k\left[\frac{1}{r^2}\frac{\partial}{\partial r}\left(r^2\frac{\partial T}{\partial r}\right)+\frac{1}{r^2\sin\theta}\frac{\partial}{\partial \theta}\left(\sin\theta\frac{\partial T}{\partial \theta}\right)+\frac{1}{r^2\sin^2\theta}\frac{\partial^2 T}{\partial \phi^2}\right]$$

$$+2\mu\left\{\left(\frac{\partial v_r}{\partial r}\right)^2+\left(\frac{1}{r}\frac{\partial v_\theta}{\partial \theta}+\frac{v_r}{r}\right)^2+\left(\frac{1}{r\sin\theta}\frac{\partial v_\phi}{\partial \phi}+\frac{v_r}{r}+\frac{v_\theta\cot\theta}{r}\right)^2\right\}$$

$$+\mu\left\{\left[r\frac{\partial}{\partial r}\left(\frac{v_\theta}{r}\right)+\frac{1}{r}\frac{\partial v_r}{\partial \theta}\right]^2+\left[\frac{1}{r\sin\theta}\frac{\partial v_r}{\partial \phi}+r\frac{\partial}{\partial r}\left(\frac{v_\phi}{r}\right)\right]^2\right.$$

$$\left.+\left[\frac{\sin\theta}{r}\frac{\partial}{\partial \theta}\left(\frac{v_\phi}{\sin\theta}\right)+\frac{1}{r\sin\theta}\frac{\partial v_\theta}{\partial \phi}\right]^2\right\}$$

[a] Reprinted with permission from R. B. Bird, W. E. Stewart, and E. N. Lightfoot, *Transport Phenomena*, Wiley, New York, 1960.

### *The Equation of Continuity for Binary Mixtures*

Following polymer manufacture it is often necessary to remove dissolved volatile solvents, unreacted monomer, moisture, and impurities from the product. Moreover, devolatilization is also needed sometimes just prior to the shaping operation, to reduce the volatile content to very low levels, ensuring the manufacture of high quality and safe products. In extruders, for example, devolatilization is accomplished in a vented section, where the polymer melt fills the screw channel only partially, exposing to vacuum a film of melt repeatedly deposited on the barrel surface. The volatile components diffuse to the polymer-vapor interface, evaporate into the vapor phase, and are removed through the vent. The devolatilization process is generally *diffusion* controlled, with negligible interfacial and vapor phase "resistances" (9). The surface concentration is the equilibrium concentration between vapors and the solution. Its value is a function of the a partial vapor pressure, temperature, and solvent—polymer interaction (10).

For a binary system of constant density, where a low concentration component A is diffusing through the other component, the equation of continuity for component A is

$$\frac{Dc_A}{Dt} = \mathscr{D}_{AB}\nabla^2 c_A + \dot{R}_A \tag{5.1-39}$$

where the diffusivity $\mathscr{D}_{AB}$ was assumed constant, $c_A$ is the molar concentration of species A, and $\dot{R}_A$ is the molar rate of production of A per unit volume (e.g., by chemical reaction). This equation is identical *in form* to Eq. 5.1-38 (in the absence of viscous dissipation), hence the components of Eq. 5.1-39 in various coordinate systems can also be obtained from Table 5.5.

## 5.2 Modeling of Engineering Processes

Engineering design, analysis, control, optimization, trouble shooting, and any other similar engineering activity relating to specific industrial *processes* can best be done by a quantitative study of effects of design and process variables and material parameters on the process. This endeavor brings about the need for *mathematical modeling* of the specific industrial process or system. Hence engineering mathematical modeling, as the name implies, refers to the attempt to *mimic* (describe) the real engineering system or process through mathematical equations, which will always contain simplifications about the nature of the substances involved, the relative magnitudes of the various physical effects, and the geometry of the space in which the phenomena take place. A mathematical model, therefore, will always remain an *approximation* of the real system. The better the model, the closer it will approximate the real system. In the field of polymer processing, the *process* to be modeled is the *processing method* (or part of it), which consists of a series of intricate, mostly transport-based, physical phenomena occurring in complex geometrical configurations.

It is worthwhile to note at this point that the various *scientific theories* that quantitatively and mathematically formulate natural phenomena are in fact *mathematical models of nature.* Such, for example, are the kinetic theory of gases, the kinetic theory of rubber elasticity, Bohr's atomic model, molecular theories of polymer solutions, and even the equations of transport phenomena cited earlier in this chapter. They, just like the engineering mathematical models, contain simplifying assumptions. For example, the transport equations involve the assumption that matter can be viewed as a continuum and that even in fast, irreversible processes, local equilibrium can be achieved. The paramount difference between a mathematical model of a natural process and that of an engineering system is the required level of accuracy and, of course, the generality of the phenomena involved.

An engineering mathematical model may consist of a single algebraic equation, sets of partial differential equations, or any possible combinations of various kinds of equations and mathematical operations, often in the form of large computer programs. Indeed, the availability of high speed digital computers has increased the modeling possibilities immensely, bringing all engineering models closer to the real process.

The quantitative study of the *process*, which as we stressed at the outset is the reason for modeling, is called *simulation*. But modeling and simulation have useful functions beyond the quantitative study of the process. An attempt to build a model for a complex process requires first of all a clear definition of objectives, which often by itself is both useful and educational. In addition, by repeated simulations, a better understanding of the process is achieved, greatly improving our insight and developing our "engineering intuition." With a model, extrapolation or scale-up problems can be studied, the effect of individual variables isolated, and sensitivity and stability problems explored. All these are often hard, costly, or occasionally impossible to carry out in the real process.

Model building consists of assembling sets of various mathematical equations, originating from engineering fundamentals such as the balance equations which, together with appropriately selected boundary conditions, bear the interrelations between variables and parameters corresponding to those in the real process. Modeling a complex process, such as a polymer processing operation, is done by breaking it down into clearly defined *subsystems*. Then by introducing appropriate sets of simplifying assumptions and making use of well-founded engineering fundamentals, a mathematical model is built for each subsystem. These are then assembled into the complete model. The latter is tested for experimental verification. A mathematical model, no matter how sophisticated and complicated, is of little use if it does not reflect reality to a satisfactory degree (engineering approximation).

There are various ways to classify mathematical models (11). First, according to the nature of the process, they can be classified as *deterministic* or *stochastic*. The former refers to a process in which each variable or parameter acquires a certain specific value or sets of values according to the operating conditions. In the latter, an element of uncertainty enters; we cannot specify a certain value to a variable, but only a most probable value. Transport based models (excluding random errors) are deterministic; residence time distribution models in well-stirred tanks are, for example, stochastic. Mathematical models can be also classified according to the mathematical basis on which the model is built. Thus we have *transport phenomena* based models (most of this book deals with such models), *empirical* models (based on experimental correlations), and *population balance* models such as the previously quoted residence time distribution models. Models can be further classified as *steady* or *unsteady*, *lumped parameter* or *distributed parameter* (implying, respectively, no variation or variation with spatial coordinates), and *linear* or *nonlinear*.

In polymer processing, in summary, the mathematical models are deterministic (as are the processes), generally transport based, either steady (continuous processes, except when dynamic models for control purposes are needed) or unsteady (cyclic processes), linear generally only to a first approximation, and distributed parameter (although when the process is broken into small finite elements, locally lumped parameter models are used).

## 5.3 Common Boundary Conditions and Simplifying Assumptions

### *The No-Slip Condition*

At a solid–liquid interface the velocity of the liquid is that of the solid surface. This is commonly known as the *no-slip* assumption, because the liquid is assumed to adhere, thus to have no velocity relative to the wall.

This assumption holds very well for all viscous liquids. It also seems to be valid for most polymer melts, which are viscoelastic, under nearly all flow conditions. Experimental support for the no-slip condition for slow flows of polymer melts was given by den Otter (12), who utilized the indigenous gel particles in polyethylene melts to observe the flow conditions at the wall. Previous experiments using "large" tracer particles indicated possible slip at the wall (13, 14). Some of these findings were interpreted by den Otter to be artifacts of the experimental system and the large tracer particles. Slip at the wall was also invoked at high flow rates, in the melt fracture region (see Chapter 13). This is the case, typically for example, for HDPE melts (15). The phenomenon that may be taking place at higher flow rates is that of "slip-stick," whereby the melt snaps away from the surface (adhesion is overcome) under the influence of elastic straining; it sticks again after the strain is recovered (14). In any case, especially below the region of melt fracture, the no-slip boundary condition is used universally.

At this point it is appropriate and necessary to comment on the precise action of the "slip agents" mentioned in Chapter 2, which are used with very viscous and heat sensitive polymers during processing. The "slip agents," being incompatible with the polymer at processing temperatures, migrate to the metal surface of the processing equipment and *replace* the polymer at the melt–metal interface. Since the viscosity of the slip agent is much lower than that of the melt, and the shear stress level is very high, large velocity gradients are developed within them. Thus even if the thickness of the "slip agent" is minimal, the melt moves with an appreciable velocity relative to the metal surface, appearing to be slipping; in reality, neither the slip agent nor the melt slips. As an example, a 100 Å thick layer of a slip agent of viscosity of 0.1 $N{\cdot}s/m^2$ at a surface where the shear stress level is $5 \times 10^4\ N/m^2$ (not uncommon in polymer processing) would have a velocity at the "slip agent"–melt interface of 0.5 cm/s.

### *Liquid–Liquid Interface*

At the interface between two immiscible liquids, the boundary conditions that must be satisfied are (*a*) a continuity of both the tangential and the normal velocities (this implies a no-slip condition at the interface), (*b*) a continuity of the shear stress, and (*c*) the balance of the difference in normal stress across the interface by the interfacial (surface) forces. Thus the normal stresses are not continuous at the interface, but differ by an amount given by the following expression

$$P_1 - P_0 = \Gamma\left(\frac{1}{R_1} - \frac{1}{R_2}\right) \tag{5.3-1}$$

where $P_1 - P_0$ is the pressure difference due to the surface tension $\Gamma$ acting on a curved surface of radii of curvature $R_1$ and $R_2$.

### The Steady State Assumption

A physical process has reached a *steady state* when a stationary observer, located at *any* point of the space where the process is taking place, observes no changes in the response of the system with time. Mathematically this statement reduces to the need that in the field equations describing the process, all the $(\partial/\partial t)$ terms be zero. In reality, very rarely are processes truly steady. Boundary conditions, forcing functions, system "resistances," and composition or constitution of the substances involved change periodically, randomly or monotonically by small amounts. These changes bring about process response fluctuations. In such cases the process can still be treated as if it were steady using the *pseudo-steady state approximation.* To illustrate this approximation, let us consider a pressure flow in which the driving force pressure drop varies with time. The flow is isothermal; thus it is described by Eqs. 5.1-5 and 5.1-18. We set $\partial\rho/\partial t$ and $\partial(\rho\boldsymbol{v})/\partial t$ equal to zero and proceed to solve the problem, as if it were a steady state one; that is, we solve assuming $\Delta P$ to be constant and not a function of time. The solution is of the form

$$\boldsymbol{v} = \boldsymbol{v}(x_i, \Delta P, \text{geometric, and material variables})$$

Because $\Delta P$ was taken to be a constant, $\boldsymbol{v}$ is also a constant with time. The pseudo-steady state approximation "pretends" that the foregoing solution holds for any level of $\Delta P$ and that the functional dependence of $\boldsymbol{v}$ on time is simply

$$\boldsymbol{v}(x_i, t) = \boldsymbol{v}(x_i, \Delta P(t), \text{geometric, and material variables})$$

The pseudo-steady state approximation is not valid if the values of $\Delta(\rho\boldsymbol{v})/\Delta t$ ($\Delta t$ being the characteristic time of fluctuation of $\Delta P$) obtained using this approximation contribute to an appreciable fraction of the mean value of the applied $\Delta P$.

### The Constant Thermophysical Properties Assumption

The last commonly used set of assumptions in liquid flow (isothermal as well as nonisothermal) and in conductive heat transfer is *to treat $k$, $C_p$, and $\rho$ as constant* quantities, independent of $T$ and $P$. Section 5.5 discusses the pressure and temperature dependence of these quantities in polymer solids and melts. In polymer processing, where both heat transfer and flow take place, typical temperature variations may attain 200°C and pressure variations 50 $MN/m^2$. Under such variations the density of a typical polymer would change by 10 or 20% depending on whether it is amorphous or crystalline, while $k$ and $C_p$ would undergo variations of 30–40%.

Assuming that polymer melts are incompressible fluids does not introduce large errors in the momentum and energy equations, although we should carefully evaluate the melt density at the prevailing pressures and temperatures. Also, assuming constant $C_p$ and $k$ may affect the results of heat transfer or coupled heat transfer and flow in polymer processing. Chapter 9 presents examples of results obtained with constant or variable $k$ and $C_p$.

## 5.4 The Lubrication Approximation: Reynolds' Equation

In polymer processing we frequently encounter creeping viscous flows in slowly varying, relatively narrow, gaps. These flows are usually solved by the well-known *lubrication approximation*, which originated with the famous work by Osborne Reynolds (16) in which he laid the theoretical foundations of hydrodynamic lubrication.* The theoretical analysis of lubrication deals with the hydrodynamic behavior of thin films from a fraction of a mil to a few mils thick. High pressures of the order of thousands of psi (millions of newtons per square meter) may develop in these films as a result of the relative motion of the confining walls. In polymer processing we are dealing generally with "films" that are several orders of magnitude thicker, but since the viscosity of polymeric melts is also several orders of magnitude higher than the viscosity of lubricating oils, the assumptions leading to the lubrication theory are still valid. It is, therefore, instructive to review the principles of hydrodynamic lubrication (17).

The assumptions on which the lubrication theory rests are as follows: (*a*) the flow is laminar, (*b*) the flow is steady in time, (*c*) the flow is isothermal, (*d*) the fluid is incompressible, (*e*) the fluid is Newtonian, (*f*) there is no slip at the walls, (*g*) inertial forces resulting from acceleration of the fluid are negligible as compared with viscous shear forces, (*h*) any motion of the fluid in a direction normal to the surfaces can be neglected in comparison with motion parallel to them, and (*i*) there is no transverse flow.

According to these assumptions, for a film extending in the $x$ and $z$ directions, the only nonvanishing velocity components are $v_x$ and $v_z$, and the equations of continuity and motion in Cartesian coordinates reduce, respectively, to (see Tables 5.1 and 5.2)

$$\frac{\partial v_x}{\partial x}+\frac{\partial v_z}{\partial z}=0 \tag{5.4-1}$$

$$\frac{\partial P}{\partial x}=\mu\frac{\partial^2 v_x}{\partial y^2} \tag{5.4-2}$$

$$\frac{\partial P}{\partial y}=0 \tag{5.4-3}$$

$$\frac{\partial P}{\partial z}=\mu\frac{\partial^2 v_z}{\partial y^2} \tag{5.4-4}$$

* Honoring his contribution to this field, the unit commonly used in engineering viscosity $lb_f{\cdot}s/in^2$ is called "reyn" after Reynolds, as the "poise" is called after Poiseuille.

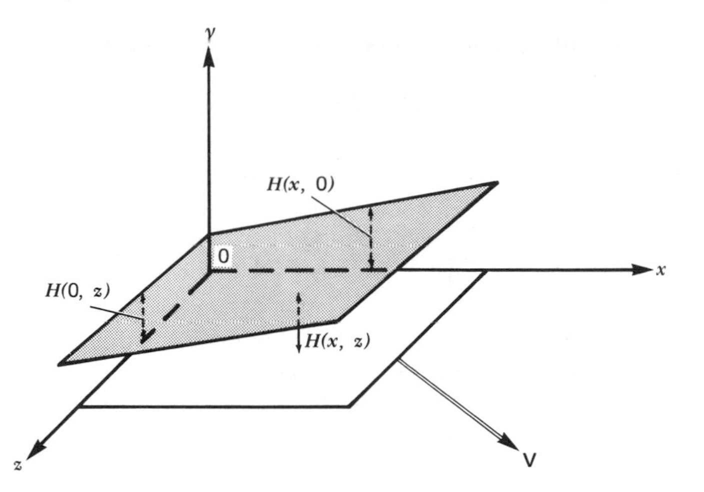


**Fig. 5.6** A flow region formed by two "almost parallel" plates of spacing $H(x, z)$; one of them is stationary, while the other is moving in the $xz$ plane.

Equation 5.4-3 implies that there is no transverse pressure gradient. The geometrical configuration is a two-dimensional gap with slowly varying thickness (Fig. 5.6). We allow one of the confining plates (the lower one in Fig. 5.6) to move at constant velocity $V$ with components $V_x$ and $V_z$.

The boundary conditions are $v_x(H) = v_z(H) = 0$, $v_z(0) = V_z$, $v_x(0) = V_x$, where $H(x, z)$ is the local separation between the plates.

The foregoing set of assumptions and equations has led to the *lubrication approximation*, which in physical terms is tantamount to stating that locally a fully developed flow between parallel plates with a gap equal to the *local* gap is capable of describing the actual flow.

Equations 5.4-2 and 5.4-4 can be directly integrated, and with the boundary conditions used, they lead to the following velocity profiles:

$$v_x(y) = V_x\left(1 - \frac{y}{H}\right) + \frac{yH}{2\mu}\left(\frac{\partial P}{\partial x}\right)\left(\frac{y}{H} - 1\right) \tag{5.4-5}$$

$$v_z(y) = V_z\left(1 - \frac{y}{H}\right) + \frac{yH}{2\mu}\left(\frac{\partial P}{\partial z}\right)\left(\frac{y}{H} - 1\right) \tag{5.4-6}$$

which upon integration over $y$ lead to the respective volumetric flow rates per unit width $q_x$ and $q_z$

$$q_x = \frac{V_x H}{2} + \frac{H^3}{12\mu}\left(-\frac{\partial P}{\partial x}\right) \tag{5.4-7}$$

$$q_z = \frac{V_z H}{2} + \frac{H^3}{12\mu}\left(-\frac{\partial P}{\partial z}\right) \tag{5.4-8}$$

The equation of continuity is now integrated over $y$:

$$\int_0^H \left(\frac{\partial v_x}{\partial x} + \frac{\partial v_z}{\partial z}\right) dy = 0 \tag{5.4-9}$$

to give the following differential equation:

$$\frac{\partial q_x}{\partial x}+\frac{\partial q_z}{\partial z}=0 \tag{5.4-10}$$

Finally, by substituting Eqs. 5.4-7 and 5.4-8 into Eq. 5.4-10, the following equation is obtained

$$\blacktriangleright \quad \frac{\partial}{\partial x}\left(H^3\frac{\partial P}{\partial x}\right)+\frac{\partial}{\partial z}\left(H^3\frac{\partial P}{\partial z}\right)=6\mu\frac{\partial H}{\partial x}V_x+6\mu\frac{\partial H}{\partial z}V_z \tag{5.4-11}$$

which is known as the *Reynolds equation* for incompressible fluids.

By solving Reynolds'equation for any $H(x, z)$ the two-dimensional pressure distribution $P(x, z)$ is obtained, from which the local pressure gradients can be evaluated and, via Eqs. 5.4-5 to 5.4-8, the local velocity profiles and flow rates can be calculated.

The lubrication approximation facilitates solutions to flow problems in complex geometries, where analytical solutions either cannot be obtained (necessitating tedious numerical methods) or are lengthy and difficult. The utility of this approximation can be well appreciated by comparing the *almost exact* solution of pressure flow in a slightly tapered channel to that obtained using the lubrication approximation (see Problem 5.16).

The lubrication approximation as derived earlier is valid for purely viscous and Newtonian fluids. But polymer melts are viscoelastic and exhibit normal stresses in shearing flows. Because of this rheological characteristic, there is another aspect to this approximation with these fluids, which Chapters 6 and 16 mention and discuss.

## 5.5 Transport and Thermodynamic Properties of Polymers

The general balance equations are associated with certain categories of materials via constitutive equations. These equations contain material parameters such as the viscosity, the thermal conductivity, and the diffusivity. Chapter 6 covers rheological constitutive equations for polymers. This section discusses the other two transport properties, to provide some useful relationships for solving problems in polymer processing. Additionally, we shall discuss the density and the specific heat of polymers and their dependence on temperature and pressure, as well as structural polymer variables.

### *Thermal Conductivity*

There are no useful theoretical expressions for the thermal conductivity of polymers. This is because there are but a few approximate theories, even for spherical and nonpolar liquids, and none for solids. Physically it is known that in metals the heat transfer is carried out by the conduction electrons, whereas the phenomenon in "nonconductors" occurs by atomic and molecular motions. The same is true in nonconducting liquids.

The thermal conductivity of the polymer chain is thought to be anisotropic, as is polarizability, discussed in Section 3.9. That is, heat can be transmitted along the backbone primary bonds with much less scattering than is possible in transmitting it from chain to chain along secondary bonds. In glassy polymers heat transfer may take the backbone path, but it is random; the same is true in polymer melts. Thus we should observe neither a microscopic nor a macroscopic anisotropy. We should expect, though, an increase in $k$ with increasing molecular weight. This has been reported by a number of investigators and is shown for polyethylene melt in Fig. 5.7. For similar reasons we should expect anisotropic $k$ in oriented amorphous polymers in the glassy state (Fig. 5.8). Such a behavior should have an effect on processes such as thermoforming. It is worth noting that both the abovenamed effects, molecular weight and orientation, result in appreciable changes of $k$.

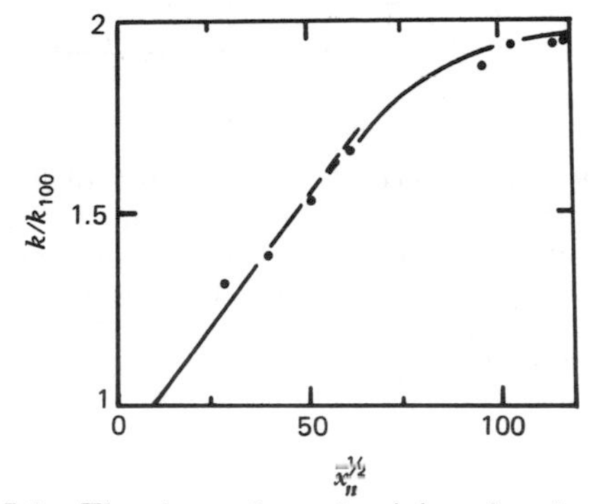


**Fig. 5.7** The thermal conductivity of molten polyethylene (140°C) versus the square root of the degree of polymerization. $K_{100}$ is the conductivity of a polyethylene sample of $N = 100$. [Reprinted with permission from D. Hansen and C. C. Ho, *J. Polym. Sci.*, **43**, 659 (1965).]

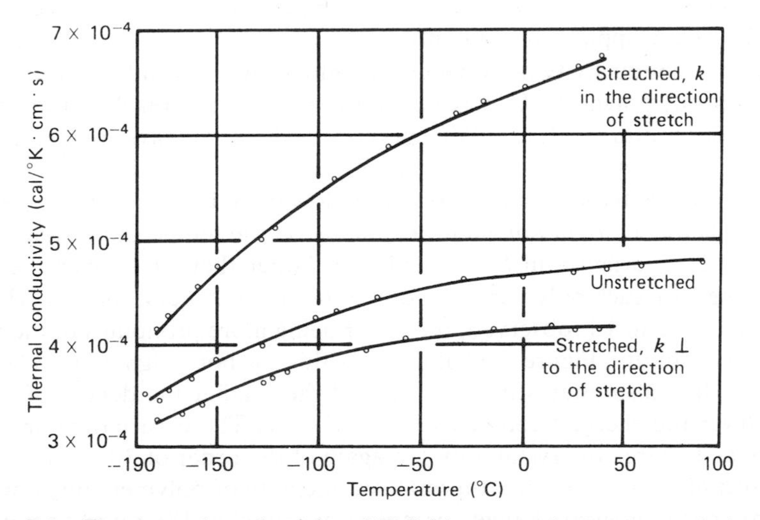


**Fig. 5.8** The thermal conductivity of PMMA, unstretched and stretched to 375%, based on K. Eirmann and K. H. Hellwege [*J. Polym. Sci.*, **57**, 99 (1962)], showing some of their experimental points; reprinted with permission.

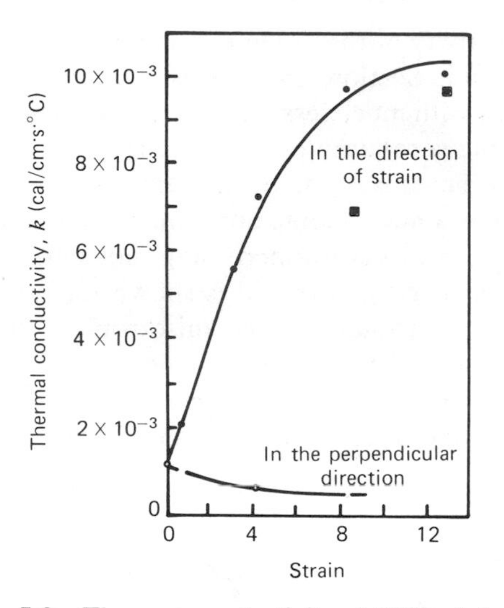


**Fig. 5.9** Thermal conductivity at 50°C of sheared HDPE (Marlex 5003) as a function of residual strain after shearing (●) at 65°C, and residual strain after tension (■) at 100°C. [Reprinted with permission from D. Hansen and G. A. Bernier, *Polym. Eng. Sci.*, **12**, 204 (1972).]

The influence of orientation on conductivity is very significant in flexible crystallizable polymers such as HDPE. Whereas in the usual folded chain spherulitic morphology there is order but no overall anisotropy, under structuring conditions similar to those discussed in Section 3.6, the effects of chain orientation are significant. Hansen and Bernier (18) observed a twentyfold difference in $k$ parallel and perpendicular to the orientation direction (Fig. 5.9). This structuring effect is large enough to have applications potential.

The thermal conductivity of polymers is temperature dependent. In the glassy state of amorphous polymers, $k$ increases with temperature, reaches a maximum, then either drops (natural rubber, PVC, polyisobutylene) or stays constant. Figure 5.10 gives the temperature dependent $k$ for PVC, rigid and plasticized. Because of their effect on $T_g$, external plasticizers either lower or increase the value of $k$, compared to that of the rigid polymer, depending on the temperature range of the measurement. The variation in $k$ is usually not larger than 30% over the entire processing range for such polymers. In polycrystalline polymers, such as HDPE, a continuous decrease in $k$ is observed, the effect being more pronounced, the higher the level of crystallinity. Figure 5.11 illustrates this for both high and low density polyethylenes. It is also interesting to note that the smaller the degree of crystallinity, the lower the thermal conductivity, at $T < T_m$. These variations in $k$ with temperature and degree of crystallinity are again of the order of 30–40%.

The effect of pressure on the thermal conductivity of polymer solids, without inducing any morphological changes, is not large, roughly 1% increase per 1000 psi. As such, it might be neglected in processing. The effect of pressure on the conductivity of melts is also not expected to be large.

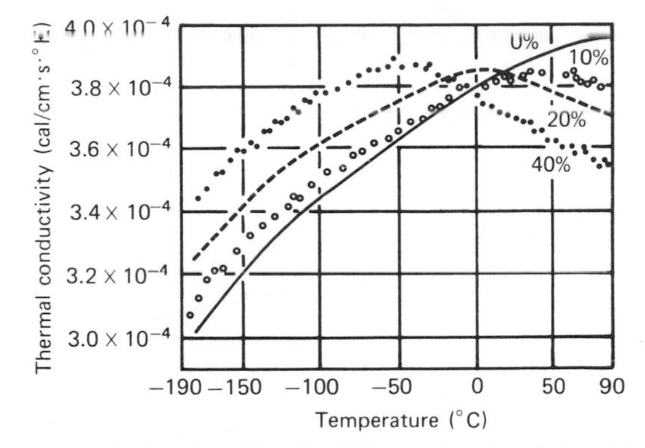


**Fig. 5.10** Thermal conductivity of PVC with different levels of plasticizer. (Reference cited in Fig. 5.8.) Reprinted with permission.

It is noteworthy that the values of $k$ of most common polymers not only does not vary with temperature in excess of 30–40% but is in the range of $3\text{–}12 \times 10^{-4}$ cal/cm·s·°C (0.12–0.50 W/m·°C) for all of them. For more information on the thermal conductivity coefficient of polymers, including a fair amount of data, the reader is referred to the work of Kline and Hansen (19).

Finally, a useful quantity in solving heat conduction problems is the *thermal diffusivity* $\alpha = k/\rho C_p$. Its value can be calculated from the values of the components of the product, but it is usually the other way around; thermal diffusivity measurements are more accurate than those of thermal conductivity. Figure 5.12 gives polycarbonate thermal diffusivity values, together with those of $\rho$, $k$, and $C_p$, as a function of temperature.

It is worth dealing briefly with the thermal conductivity of heterogeneous and particulate systems, since such systems are encountered in polymer processing. Examples of the former are blends and polymers with fillers, the second category includes loosely packed powders as in rotational molding, or compressed pellet or powder "beds" as in extrusion and injection molding. The heat conduction problem in these systems can be generally treated as in homogeneous solids using "*effective*"

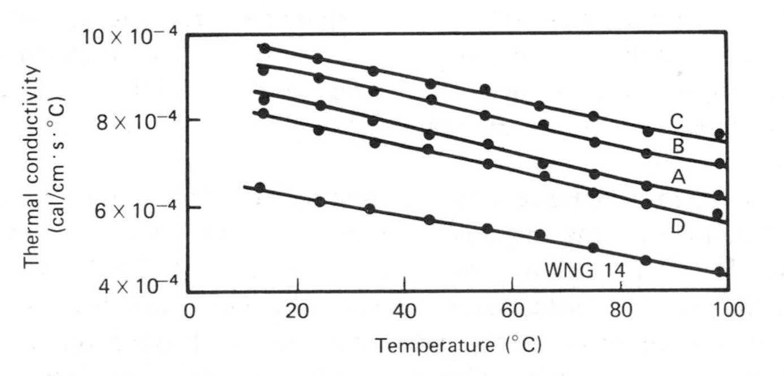


**Fig. 5.11** Thermal conductivity of various polyethylenes at $T < T_m$. The samples have different crystalline content, as indicated by their densities: C, 0.951; B, 0.948; A, 0.940; D, 0.935; WNG14, 0.918. [Reprinted with permission from R. P. Sheldon and S. K. Lane, *Polymer*, **6**, 205 (1965).]

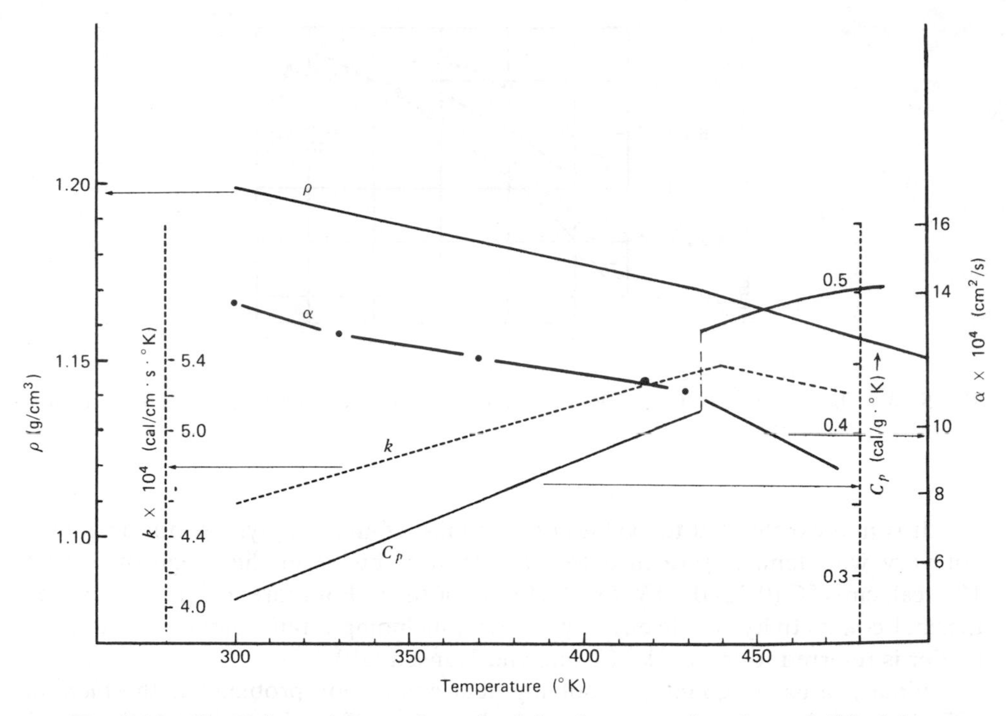


**Fig. 5.12** Coefficient of thermal conductivity, heat capacity, thermal diffusivity, and density for polycarbonate.

*thermophysical properties*. Thus the thermal conductivity of a filled system consisting of a low concentration, randomly distributed, uniformly sized, and spherical minor components in a continuous polymer matrix can be shown to be (20)

$$k_e = k_p \left[ \frac{2k_p + k_f - 2x_v(k_p - k_f)}{2k_p + k_f + x_v(k_p - k_f)} \right] \tag{5.5-1}$$

where $k_e$, $k_p$, and $k_f$ are, respectively, the effective, the polymer, and the "filler" thermal conductivities, and $x_v$ is the volume fraction of the minor component (filler). Equation 5.5-1 is reportedly good for $x_v < 0.1$ as long as the filler particles do not agglomerate and are wetted by the polymer matrix, and provided $k_p$ and $k_f$ are not very different in value. A number of variations of this equation for higher $x_v$ values, nonuniform and nonspherical particles, and very different $k_f$ and $k_p$ (e.g., metal powder fillers in a polymeric matrix) have been developed and are reviewed by Orr (20).

Particulate solid polymeric systems conduct heat less readily than do homogeneous polymeric systems, because the thermal conductivity of most gases is considerably less than those of polymers ($k_{\text{air}} = 0.026\ \text{W/m}\cdot{}^\circ\text{K}$; $k_{\text{LDPE}} = 0.182\ \text{W/m}\cdot{}^\circ\text{K}$) and the contact area between the solid particles is small. Heat is transmitted by a number of mechanisms: conduction through the solid particles, conduction across contact surfaces between solid particles, conduction through the gas film near contact points, conduction in the gas phase, radiation between solid surfaces, and radiation between neighboring voids. Clearly, compaction will affect most of these modes of heat transfer; it is not surprising, therefore, to find that the

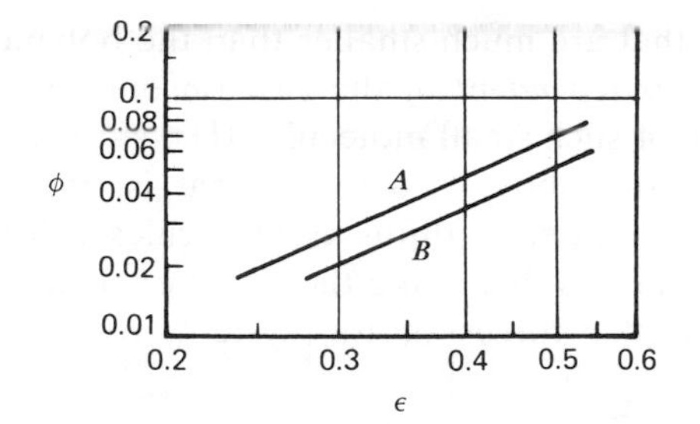


**Fig. 5.13** Correlation factor $\phi$ for particle beds with air filled interstices in terms of the void fraction $\varepsilon$. Curves based on experimental data points (deleted from the figure) of solids packings. Curve A, carborundum and quartz beds; curve B, beds of steel and lead spheres and cylinders. [Reprinted with permission from S. Yagi and D. Kunii, *A.J.Ch.E.J.*, **3**, 373 (1957).]

effective heat conduction is sensitive to compaction. Guided by experimental data, Yagi and Kunii (21) suggested a mathematical model for the thermal conductivity of the bed, which for fine particles and low temperatures simplifies to

$$k_e = \frac{k_p(1-\varepsilon)}{1+(k_p\phi/k_g)} \tag{5.5-2}$$

where $k_g$ is the thermal conductivity of the gas occupying the voids, $\varepsilon$ is the "bed" porosity, and $\phi$ is the function plotted in Fig. 5.13.

### *Mass Diffusivity*

Theories for the prediction of the mass diffusivity in binary systems $\mathscr{D}_{AB}$ exist only for low density gases (1*b*). In liquid binary systems two approximate theories have been used. The first is a hydrodynamic theory, that starts with the Nernst–Einstein equation describing the motion of a single particle of the "diffusant" in a stationary, continuous medium. This results in the Stokes–Einstein equation (1*b*)

$$\mathscr{D}_{AB}\mu = \frac{kT}{6\pi R_A} \tag{5.5-3}$$

where $\mu$ is the viscosity of the medium, $k$ is Boltzmann's constant, and $R_A$ is the "radius" of the diffusing molecule. The second theory is that of Eyring (1*c*), which considers liquid diffusion to be an activated rate process (like fluid flow) in which a single molecule is involved. Both theories give results that are only in qualitative agreement with experimental data.

In polymer processing, as discussed in Section 5.2, we are interested in the diffusion of either small gas molecules (originating from entrapped air or formed during foaming processes) or dissolved monomer or solvent in polymer melts. We do not deal with the much more complex subject of diffusion of small or large molecules in amorphous or semicrystalline solids.

For gas molecules that are much smaller than the polymer unit, it is observed that the diffusivity $\mathscr{D}$ is *independent* of the gas concentration. This observation is attributed to the failure of such small molecules ($H_2$, $N_2$, $O_2$, $CO_2$, etc.) to interact with the polymer melt or rubber structures; that is, they *do not plasticize* it. Therefore an increased number of diffusing molecules will not affect $\mathscr{D}$. On the other hand, all the parameters that make the polymer structure more "porous" or decrease the intermolecular forces of the system, result in an increase of $\mathscr{D}$. Specifically

1. Decreasing the molecular diameter of the diffusant increases $\mathscr{D}$, since smaller "holes" are required.
2. Increasing temperature increases $\mathscr{D}$, since the average "hole" size increases. The dependence is of the form

$$\mathscr{D}(T)=\mathscr{D}_0 \exp \frac{-E_d}{RT} \tag{5.5-4}$$

3. Decreasing the number of polar groups in the polymer increases $\mathscr{D}$. As in item 2, the cohesive energy density is decreased.
4. Increasing the chain flexibility also results in an increase $\mathscr{D}$. The effects above are discussed in detail by Stannett (22a), who also presents reliable experimental data.

For the case of organic molecules of a size commensurate to that of the mer, diffusivity is concentration dependent. The diffusant size is such that it interacts with the structure of the polymer, forcing segmental motions in the polymer chain, thus resulting in chain conformational changes. It follows then that increasing the concentration of such molecules results in a "loosening" of the polymer melt structure, that is, it results in plasticization. This explains the observation that $\mathscr{D}$ increases with increasing concentration of the diffusant. The degree to which this occurs should depend on the molecular nature of both the polymer and the diffusant. Not only $\mathscr{D}$ is concentration dependent but also $E_d$. Thus the simple equation 5.5-4 does not hold. Instead, the diffusivity can be expressed as

$$\mathscr{D}=\mathscr{D}^{\circ}(T)F(c,T) \tag{5.5-5}$$

where $\mathscr{D}^{\circ}$ is the diffusivity at zero diffusant concentration [corresponding to $\mathscr{D}(T)$ in Eq. 5.5-4]. The function $F(c, T)$ is experimentally obtained. The diffusivity increases exponentially with concentration, the slope of the straight line relationship being related to the free volume fraction of this polymer. This observation has led to a number of theories based on the concept of the Doolittle-type free volume, which has been successfully used in the viscosity of polymers in the region $T_g < T < T_g + 100$ (i.e., the WLF equation). These theories, as well as experimental data, are presented by Fujita (22b) and Kumins and Kwei (22c).

During the diffusion of organic vapors in polymer solids, $\mathscr{D}$ is time dependent, for the following reason: organic molecules have again (as above $T_g$) the tendency to "plasticize" the structure of polymers, but this effect is not instantaneous, since the relaxation times are large. The long-term molecular motions that are brought about by such molecules often result in morphological changes in semicrystalline polymers (23).

### *Density*

Polymers, as we have seen in earlier chapters, can exist in the molten liquid and semicrystalline or amorphous solid states. Polymer chains tend toward random conformations in the melt state, and they arrive at them from oriented conformations after times that are characteristic of the polymer, the temperature, and the pressure. Thus there is a thermodynamic state for polymer melts ($T > T_g$ for amorphous, and $T > T_m$ for semicrystalline) at which an equation of state is assumed to hold under a hydrostatic stress field and when the time dependent effects either have disappeared or are neglected, that is, $P = P(\hat{V}, T)$.

Two approaches can be used in the determination of the equation of state. The first is experimental, yielding empirical equations that are the result of curve fitting of the obtained response. It is time-consuming and difficult at high pressures and for prolonged times of exposure of the sample to high temperatures. The second approach is essentially to arrive at the equation of state from a knowledge of the intermolecular force field. These forces are almost always assumed to follow the Lennard–Jones potential. Then statistical mechanics averages are taken over molecular variables to obtain macroscopically observed quantities. This is achieved through the calculation of the partition function (24). The partition function is very difficult to calculate, necessitating many assumptions about molecular structure and forces, which must be considered before applying the resulting equation of state.

Table 5.6 presents a number of essentially empirical equations of state for polymers, together with limited data on some materials.

Order of magnitude values for melt compressibility and thermal expansion coefficients are $1.5 \times 10^{-9}$ $(\mathrm{N/m^2})^{-1}$ and $5 \times 10^{-4}$ $(°\mathrm{K})^{-1}$, respectively. Therefore the errors involved in assuming that the polymer melt density is constant and independent of temperature and pressure, are small. We note the phenomenon of pressure induced crystallization in crystallizable polymer melts that can occur not very far above the normal melting point, and at processing pressures. This phenomenon results in a time dependent density because of the kinetics of crystallization, as discussed in Chapter 3.

Although polymer melts attain their equilibrium densities very soon after thermal and pressure equilibrium (i.e., $\rho = \rho(P, T)$), near and below either $T_g$ or $T_m$, polymer density is no longer uniquely determined by temperature and pressure alone. Annealing temperature and time and rate of cooling, or in general the entire thermal history, play an important role in the value of the density at any time (25); that is, $\rho = \rho(T, P, t)$.

### *Specific Heat*

Specific heat or heat capacity is the quantity of heat needed to increase the temperature of a unit mass of a body by a unit of temperature. It is defined, in the limit, as

$$C = \frac{dQ}{dT} \tag{5.5-6}$$

where $C$ is the heat capacity (J/kg·°K) and $Q$ is the quantity of heat ($J$). It can be measured either at constant volume $C_v$ or at constant pressure $C_p$. Heat capacity at constant pressure is usually larger than that at constant volume, since at constant pressure part of the heat added is used in the work of expansion of the substance. This can be seen by the term $P(\partial V/\partial T)_p$, which is caused by the volume change of the substance against the external pressure $P$, in the thermodynamic relationship

$$C_p - C_v = \left[P + \left(\frac{\partial U}{\partial V}\right)_T\right]\left(\frac{\partial V}{\partial T}\right)_P \tag{5.5-7}$$

The other term $(\partial U/\partial V)_T \cdot (\partial V/\partial T)_P$ is the expansion work needed to overcome the cohesive intermolecular forces $(\partial U/\partial V)_T$ which are negligible in gases but appreciable in liquids and solids. Since $(\partial V/\partial T)_P$ is small for polymers we can assume that $C_p = C_v$.

The heat capacity of ideal gases and crystals can be calculated using statistical thermodynamics and quantum mechanics. This is not true for liquids, especially those that are polar and nonspherical. Thus no theoretical expressions exist for the heat capacity of amorphous or polycrystalline polymers and their melts. Experimentally it is found that the heat capacity at constant pressure of amorphous solids increases with temperature, increases discontinuously at $T_g$ (onset of segmental

**Table 5.6a Empirical Equations for $\rho(T, P)$ or Specific Volume $\hat{V}(T, P)$ in Polymers**

|  |  |
|---|---|
| **Spencer and Gilmore**[a] | |
| $(P + \pi)(\hat{V} - \omega) = RT$ | $\pi$ = "internal pressure"<br>$\omega$ = specific volume at 0°K<br>$R$ = material constant |
| **Breuer and Rehage**[b] | |
| $\hat{V}(T, P) = \hat{V}_0 + \Phi_0 T - \frac{K_0}{a}(1 + bT)\ln(1 + aP)$ | $a, b$ = material parameters<br>$V_0$ = the volume at 0°C, and zero pressure<br>$\Phi_0 = (d\hat{V}/dT)_P$ at 0°C, and zero pressure<br>$K_0 = (d\hat{V}/dP)_T$ at 0°C, and zero pressure |
| **Kamal and Levan**[c] | |
| $\rho(T, P) = \rho_{00} + \left(\frac{\partial \rho}{\partial P}\right)_{P=0} \cdot T + (a + bT)P + \frac{1}{2}(c + dT)P^2$ | $\rho_{00}$ = density at 0°K and zero pressure<br>$a, b, c, d$ = adjustable material parameters |
| **Simha and Olabisi**[d] | |
| $1 - \frac{\hat{V}(T, P)}{\hat{V}(T, 0)} = 0.0894 \ln\left[1 + \frac{P}{B(T)}\right]$ | (modified Tait equation) |
| $B(T) = B_0 \exp(-B_1 T)$ | $B_0, B_1$ = material parameters |
| **For polyethylenes** | |
| $\hat{V}(T, 0) = a_0 + a_1 T + a_2 T^2$ for $T < T_m$ | $a_i$ = material parameters |
| $\hat{V}(T, 0) = \hat{V}_0 \exp(\alpha_1 T)$ for $T > T_m$ | $\alpha_1$ = material parameter |
| **For methacrylates** | |
| $\hat{V}(T, 0) = a_0 + a_1 T + a_2 T^2 + a_3 T^3$ | $a_i$ = material parameters |

[a] R. S. Spencer and G. D. Gilmore, *J. Appl. Phys.*, **20**, 502 (1949); **21**, 523 (1950).
[b] H. Breuer and G. Rehage, *Kolloid Z. Z. Polym.*, **216**, 166 (1967).
[c] M. R. Kamal and N. T. Levan, *Polym. Eng. Sci.*, **13**, 131 (1973).
[d] O. Olabisi and R. Simha, *Macromolecules*, **8**, 206 (1975); 211 (1975).

**Table 5.6*b* Experimental Data for Some of the Equations of Table 5.6*a***

**Spencer and Gilmore Equations**

| Polymer[e] | $R$ (psi·cm$^3$/g°K) | $\omega$ (cm$^3$/g) | $\pi$ (psi) |
|---|---|---|---|
| PS (atactic) | 11.6 | 0.822 | 27.000 |
| PMMA | 12.05 | 0.734[f] | 31.300 |
| PE | 43.0 | 0.875[f] | 47.600 |

**Simha and Olabisi Equations**

| Polymer[g] | $T_m$ (°K) | $a_0$ (cm$^3$/g) | $a_1 \times 10^3$ (cm$^3$/g°K) | $a_2 \times 10^6$ (cm$^3$/g°K$^2$) | $V_0$ (cm$^3$/g) | $\alpha_1 \times 10^4$ (°K$^{-1}$) | Temperature Range (°K) |
|---|---|---|---|---|---|---|---|
| HDPE | 403 | 1.2556 | −1.7743 | 3.3368 | 0.9172 | $7.80_6$ | 293–388<br>415–473 |
| LDPE | 386 | 1.8778 | −5.7855 | 10.3720 | 0.9399 | $7.34_1$ | 292–371<br>398–473 |

| Polymer[g] | Temperature Range (°C) | $B_0$ (bar) | $B_1 \times 10^3$ (°C$^{-1}$) |
|---|---|---|---|
| HDPE | 142–200 | 1767 | 4.661 |
| LDPE | 130–200 | 1771 | 4.699 |
| PMMA | | | |
| Glass | 17.2–91 | 3564 | 3.229 |
| Liquid | 113.5–160 | 2875 | 4.146 |

| Polymer | Temperature Range (°C) | $a_0$ (cm$^3$/g) | $a_1 \times 10^4$ (cm$^3$/g°C) | $a_2 \times 10^6$ (cm$^3$/g°C$^2$) | $a_3 \times 10^{10}$ (cm$^3$/g°C$^3$) |
|---|---|---|---|---|---|
| PMMA | 17.2–56.8 | 0.8417 | 1.3711 | 0.5765 | 0.0 |
| PMMA | 67.7–100 | 0.8394 | 1.8365 | 0.4049 | 0.0 |
| PMMA | 110–160 | 0.8254 | 2.8383 | 0.7792 | 0.0 |

[e] The PMMA is Lucite HM 140; the PE is duPont P1000 PM1.
[f] Values uncertain owing to lack of reliable values of equilibrium densities.
[g] Linear (HDPE) and branched (LDPE) polyethylene standards SRM 1475 and SRM 1476 respectively of the National Bureau of Standards.

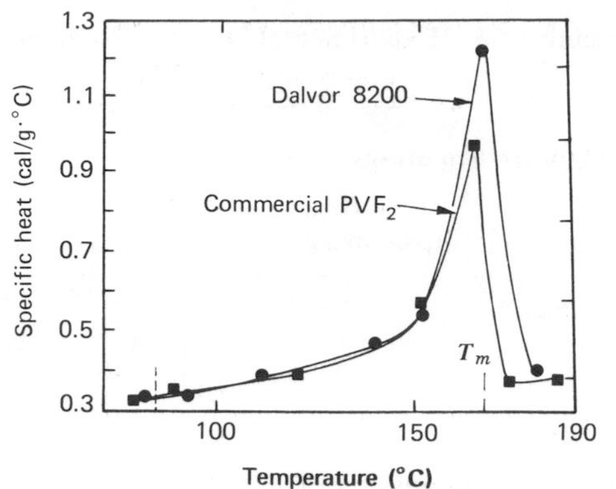


**Fig. 5.14** Specific heat of two commercial polyvinylidene fluoride ($PVF_2$) samples. [Reprinted with permission from J. P. Stallings and S. G. Howell, *Polym. Eng. Sci.*, **11**, 507 (1971).]

motions), and increases usually at a slower rate in the melt region (Fig. 5.12). For polycrystalline polymers, no discontinuity is observed around $T_g$ because the amorphous portion of the structure is usually low. The value of $C_p$ increases up to the region of the melting point. At the melting point, $C_p$ is theoretically infinite. In practice, where there is a melting temperature range, $C_p$ goes through a sharp maximum and decreases to values that are lower than those just below the melting range. As mentioned previously, $C_p$ of the melt increases at a slight rate with temperature (Fig. 5.14). The area under each curve of Fig. 5.14, in the neighborhood of $T_m$, is equal to the product of the crystalline volume fraction and the heat of fusion $\lambda$. Both quantities are affected by the flow and the thermal history

**Table 5.7 Heat of Fusion of Some Polymers**

| Polymer | $\lambda\,(J/kg) \times 10^{-4}$ | Reference |
|---|---|---|
| Polyoxymethylene | 24.9 | *c* |
| Polybutene-1 | 24.7 | *b* |
| HDPE ("Super Dylan") | 24.5 | *a* |
| PP | 23.4 | *b* |
| HDPE ("Marlex 50") | 21.8 | *a* |
| Nylon 66 | 20.5 | *b* |
| LDPE | 13.8 | *b* |
| PET | 13.7 | *b* |
| Natural rubber (*cis*-polyisoprene) | 6.4 | *b, d* |
| PTFE | 5.7 | *c* |

[a] B. Ke, *J. Polym. Sci.*, **42**, 15 (1960).
[b] L. Nielsen, *Mechanical Properties of Polymers*, Reinhold, New York, 1962.
[c] H. W. Starkweather, Jr., and R. H. Boyd, *J. Phys. Chem.*, **64**, 410 (1960).
[d] R. E. Roberts and L. Mandelkern, *J. Am. Chem. Soc.*, **77**, 781 (1955).

imposed on the polymer melt, as discussed in Chapter 3; $\lambda$ values for a number of polymers appear in Table 5.7.

Anderson and his co-workers (26), who have worked extensively on the effects of pressure on the thermal properties of polymers, report that the specific heat of amorphous polymers in the glassy state decreases only very slightly with increasing pressure. The same holds true for polymer melts. Of course, there is a significant $C_p$ decrease if the applied pressure induces a glass transition, and drastic changes (an increase followed by a decrease) if crystallization is induced. Thus in polymer processing we expect the effects of pressure on $C_p$ to be significant at temperature levels *just* above the atmospheric $T_g$ and $T_m$ but not in the temperature region just below them. For simplification purposes, we can consider $C_p$ to be pressure independent, mildly increasing in the region below $T_g$ or $T_m$ and in the melt state (15–30% per 100°C), while becoming large at the melting region (five to tenfold increase) and increasing discontinuously at $T_g$ by about 10%. Table 5.8 gives room temperature $C_p$ values for a number of common polymers, together with values for the density, thermal expansion coefficients, and thermal conductivity.

**Table 5.8 Density, Thermal Expansion Coefficient, Thermal Conductivity, and Heat Capacity of Some Polymers at Room Temperature**[a]

| Polymer | Density[b] $(kg/m^3 \times 10^{-3})$ | Thermal Conductivity[c] $(J/m \cdot s \cdot {}^\circ K)$ | Thermal Expansion Coefficient[d] $[(m^3/m^3 \cdot {}^\circ K) \times 10^4]$ | Heat Capacity[e] $[(J/kg \cdot {}^\circ K) \times 10^{-3}]$ |
|---|---|---|---|---|
| ABS | 1.01–1.04 | 0.188–0.335 | 2.85–3.90 | 1.25–1.67 |
| Nylon 66 | 1.13–1.15 | 0.243 | 2.40 | 0.46 |
| Polycarbonate | 1.2 | 0.192 | 2.00 | 1.25 |
| Polyester | 1.37 | 0.289 | 1.80 | 1.25 |
| LDPE | 0.910–0.925 | 0.335 | 3.00–6.00 | 2.30 |
| HDPE | 0.940–0.965 | 0.460–0.519 | 3.30–3.90 | 2.30 |
| PMMA | 1.17–1.20 | 0.167–0.251 | 1.50–2.70 | 1.46 |
| Polyoxymethylene | 1.42 | 0.230 | 2.43 | 1.46 |
| PS | 1.04–1.09 | 0.100–0.138 | 1.80–2.40 | 1.34 |
| PTFE | 2.0–2.14 | 0.250 | 3.00 | 1.05 |
| Polyurethane | 1.05–1.25 | 0.070–0.310 | 3.00–6.00 | 1.67–1.88 |
| PVC (rigid) | 1.30–1.45 | 0.125–0.293 | 1.50–5.55 | 0.84–1.25 |

[a] From *Modern Plastics Encyclopedia*, McGraw-Hill, New York, 1976.
[b] Measured with ASTM Method D-792.
[c] Measured with ASTM Method C-177.
[d] Three times the value of the linear thermal expansion coefficient measured with ASTM Method D-696.
[e] Method not reported (possibly differential scanning colorimeter).

## REFERENCES

1. R. B. Bird, W. E. Stewart, and E. N. Lightfoot, *Transport Phenomena*, Wiley, New York, 1960: (a) Appendix A; (b) Chapter 15; (c) Chapter 1.

2. J. R. Welty, C. E. Wicks, and R. E. Wilson, *Fundamentals of Momentum, Heat, and Mass Transport*, Wiley, New York, 1969.
3. C. Truesdell and R. A. Toupin, "The Classical Field Theories," in *Handbuch der Physik*, Vol. III, Springer, Berlin, 1960.
4. W. J. Beek and K. M. Muttzall, *Transport Phenomena*, Wiley, New York, 1975.
5. J. G. Kirkwood and B. L. Crawford, Jr., *Phys. Chem.*, **56**, 1048 (1952).
6. P. M. Morse and H. Feshbach, *Methods of Theoretical Physics*, McGraw-Hill, New York, 1953, Chapter 1.
7. R. B. Bird, R. C. Armstrong, and O. Hassager, *Dynamics of Polymeric Liquids*, Vol. I, *Fluid Mechanics*, Wiley, New York, 1977: (a) Appendix A; (b) p. 4.
8. J. Happel and H. Brenner, *Low Reynolds Number Hydrodynamics with Special Applications to Particulate Media*, Prentice-Hall, Englewood Cliffs, N.J., 1965.
9. G. A. Latinen, "Devolatilization of Viscous Polymer Systems," *Advances in Chemistry Series*, **34**, 235 (1962).
10. D. P. Maloney and J. M. Prausnitz, *A.I.Ch.E.J.*, **22**, 74 (1976).
11. D. M. Himmelblau and K. B. Bischoff, *Process Analysis and Simulation*; *Deterministic Systems*, Wiley, New York, 1968.
12. J. L. den Otter, *Rheol. Acta*, **10**, 200 (1971).
13. J. J. Benbow, R. V. Charley, and P. Lamb, *Nature*, **192**, 223 (1961).
14. B. Maxwell and J. C. Galt, *J. Polym. Sci.*, **62**, 850 (1962).
15. L. L. Blyler, Jr., and A. C. Hart, Jr., *Polym. Eng. Sci.*, **10**, 193 (1970).
16. Osborne Reynolds, "On the Theory of Lubrication and Its Application to Mr. Beauchamps Tower's Experiments," *Phil. Trans. Royal Soc.*, **177**, 157–234 (1886).
17. D. D. Fuller, "Lubrication Mechanics," in *Handbook of Fluid Dynamics*, V. L. Streeter, ed., McGraw-Hill, New York, 1961, Section 22.
18. D. Hansen and G. A. Bernier, *Polym. Eng. Sci.*, **12**, 204 (1972).
19. D. E. Kline and D. Hansen, in *Techniques and Methods of Polymer Evaluation*, Vol. IV, P. E. Slade Jr., and L. T. Jenkins, eds., Dekker, New York, 1970, Chapter 5.
20. C. Orr, Jr., *Particulate Technology*, Macmillan, New York, 1966.
21. S. Yagi and D. Kunii, *A.I.Ch.E.J.*, **3**, 373 (1957).
22. *Diffusion in Polymers*, J. Crank and G. J. Park, eds., Academic Press, London, 1968: (a) V. Stannett, "Simple Gases," Chapter 2; (b) H. Fujita, "Organic Vapors Above the Transition Temperature," Chapter 3; (c) C. A. Kumins and T. K. Kwei, "Free Volume and Other Theories," Chapter 4.
23. C. E. Rogers, J. R. Semancik, and S. Kapur, "Transport Processes in Polymers," in *Structure of Properties of Polymer Films*, R. W. Lenz and R. S. Stein, eds., Plenum Press, New York, 1973.
24. W. Kauzmann, *Thermodynamics and Statistics*, Benjamin, New York, 1967.
25. A. J. Kovacs, *Adv. Polym. Sci.*, **3**, 394 (1963).
26. P. Anderson and B. Sundgvist, *J. Polym. Sci.*, **13**, 243 (1975).

## PROBLEMS

**5.1** ***Coordinate Transformations****. (*a*) Verify the following relationships for the conversion of any function in rectangular coordinates $\phi(x, y, z)$, into one in cylindrical coordinates $\psi(r, \theta, z)$

$$x = r\cos\theta, \qquad y = r\sin\theta, \qquad z = z$$
$$r = \sqrt{x^2 + y^2}, \qquad \theta = \arctan\frac{y}{x}, \qquad z = z$$

* For a detailed discussion of vector and tensor algebra in curvilinear coordinates see Appendix A in R. B. Bird, R. C. Armstrong, and O. Hassager, *Dynamics of Polymeric Liquids*, Vol. I, Wiley, New York, 1977.

(*b*) Show that the derivatives of any scalar function (including components of vectors and tensors) in rectangular coordinates can be obtained from the derivatives of the scalar function in cylindrical coordinates via the following operators:

$$\frac{\partial}{\partial x} = \cos\theta \frac{\partial}{\partial r} + \left(-\frac{\sin\theta}{r}\right)\frac{\partial}{\partial \theta}$$

$$\frac{\partial}{\partial y} = \sin\theta \frac{\partial}{\partial r} + \left(\frac{\cos\theta}{r}\right)\frac{\partial}{\partial \theta}$$

$$\frac{\partial}{\partial z} = \frac{\partial}{\partial z}$$

***Hint***: Use the "chain rule" of partial differentiation.

(*c*) The unit vectors in rectangular coordinates are $\boldsymbol{\delta}_x$, $\boldsymbol{\delta}_y$, $\boldsymbol{\delta}_z$, those in cylindrical coordinates are $\boldsymbol{\delta}_r$, $\boldsymbol{\delta}_\theta$, and $\boldsymbol{\delta}_z$.

Show that the following relationship between the unit vectors exists

$$\boldsymbol{\delta}_r = \cos\theta\, \boldsymbol{\delta}_x + \sin\theta\, \boldsymbol{\delta}_y$$

$$\boldsymbol{\delta}_\theta = -\sin\theta\, \boldsymbol{\delta}_x + \cos\theta\, \boldsymbol{\delta}_y$$

$$\boldsymbol{\delta}_z = \boldsymbol{\delta}_z$$

and

$$\boldsymbol{\delta}_x = \cos\theta\, \boldsymbol{\delta}_r - \sin\theta\, \boldsymbol{\delta}_\theta$$

$$\boldsymbol{\delta}_y = \sin\theta\, \boldsymbol{\delta}_r + \cos\theta\, \boldsymbol{\delta}_\theta$$

(*d*) From the results of (*c*), prove that

$$\frac{\partial}{\partial r}\boldsymbol{\delta}_r = 0 \qquad \frac{\partial}{\partial r}\boldsymbol{\delta}_\theta = 0 \qquad \frac{\partial}{\partial r}\boldsymbol{\delta}_z = 0$$

$$\frac{\partial}{\partial \theta}\boldsymbol{\delta}_r = \boldsymbol{\delta}_\theta \qquad \frac{\partial}{\partial \theta}\boldsymbol{\delta}_\theta = -\boldsymbol{\delta}_r \qquad \frac{\partial}{\partial \theta}\boldsymbol{\delta}_z = 0$$

$$\frac{\partial}{\partial z}\boldsymbol{\delta}_r = 0 \qquad \frac{\partial}{\partial z}\boldsymbol{\delta}_\theta = 0 \qquad \frac{\partial}{\partial z}\boldsymbol{\delta}_z = 0$$

(*e*) The operator $\boldsymbol{\nabla}$ in rectangular coordinates is

$$\boldsymbol{\nabla} = \boldsymbol{\delta}_x \frac{\partial}{\partial x} + \boldsymbol{\delta}_y \frac{\partial}{\partial y} + \boldsymbol{\delta}_z \frac{\partial}{\partial z}$$

Using the results of (*b*) and (*d*), derive the expression for $\boldsymbol{\nabla}$ in cylindrical coordinates.

(*f*) Evaluate $\boldsymbol{\nabla} \cdot \boldsymbol{v}$ in cylindrical coordinates.

**5.2** ***Interpretation of the Equation of Continuity.*** (*a*) Show that the equation of continuity can be written as

$$\frac{D\rho}{Dt} = -\rho(\boldsymbol{\nabla} \cdot \boldsymbol{v})$$

where $D/Dt$ is the *substantial derivative* defined as

$$\frac{D}{Dt} = \frac{\partial}{\partial t} + \boldsymbol{v} \cdot \boldsymbol{\nabla}$$

(*b*) Show that for incompressible, fully developed isothermal flow in a tube, the equation of continuity indicates the $v_r = 0$.

***NOTE*:** Problems 5.3 to 5.11, dealing with laminar and isothermal flow of Newtonian and incompressible fluids, give readers who have not dealt with transport phenomena an opportunity to develop the ability to solve such problems. We suggest the following solution methodology: (*a*) pick the appropriate coordinate system, draw the flow channel, and visualize the flow on physical grounds, that is, "say something" about the velocity components; (*b*) reduce the continuity equation to the form appropriate to the flow problem; (*c*) reduce the equation of motion or the Navier–Stokes equation into the forms appropriate to the problem; (*d*) state the boundary and, if any, initial conditions; (*e*) solve for the velocity profile and, if appropriate, the volumetric flow rate; (*f*) solve for the interfacial forces by the fluid on the channel wall; (*g*) sketch out the velocity and velocity gradient fields (profiles).

**5.3** ***Parallel Plate Flow.*** Consider the flow between two horizontal parallel plates a distance $H$ apart. One of the plates is moving with a velocity $v_x = V$ and the other is stationary, as in Fig. 5.5.

**5.4** ***Couette Flow.*** "Couette" flow is the flow in the annular space between two long concentric cylinders of radii $R_o$ and $R_i$, created by the rotation of one of them. Consider Couette flow by neglecting the effect of the base of the instrument; an annular ring of width $(R_o - R_i)$ used to retain the fluid. Treat the problems where (*a*) the outer cylinder is rotating with angular velocity $\Omega$ ($s^{-1}$), (*b*) the inner cylinder is rotating with angular velocity $-\Omega$ ($s^{-1}$).

Obtain the result also by making a *torque balance* over a thin fluid shell formed by two imaginary fluid cylinders of radii $r$ and $r + \Delta r$ and length $L$. ($R_i < r < R_o$.)

**5.5** ***Axial Drag Flow between Concentric Cylinders.*** Consider the flow created in the space formed by two concentric cylinders of radii $R_o$ and $R_i$, by the inner cylinder moving with an axial velocity $V$. Neither cylinder rotates, and their length $L$ is much larger than $\Delta R$. The system is open to the atmosphere at both ends.

Obtain the result also by making a *force balance* on a thin fluid shell discussed above. (This problem is related to wire coating; see Chapter 13.)

**5.6** ***Capillary Pressure Flow.*** Solve the problem of flow in a capillary of radius $R$ and length $L$, where $L \gg R$. The fluid is fed from a reservoir under the influence of an applied pressure $P_0$. The exit end of the capillary is at atmospheric pressure. Consider three physical situations: (*a*) a horizontal capillary, (*b*) a downward vertical capillary flow, and (*c*) an upward vertical capillary flow.

**5.7 *Axial Pressure Flow between Concentric Cylinders.*** Solve the problem of flow in the horizontal concentric annular space formed by two long cylinders of length $L$ and radii $R_i$ and $R_o$, caused by an entrance pressure $P_0$, which is higher than the exit (atmospheric) pressure. Consider the limit as $(R_o - R_i/R_o + R_i)$ approaches zero.

**5.8 *Helical Flow between Concentric Cylinders.*** Consider the helical flow in an annular space created by a pressure drop $(P_0 - P_i)$ and the rotation of the inner cylinder with an angular velocity $\Omega(\mathrm{s}^{-1})$.

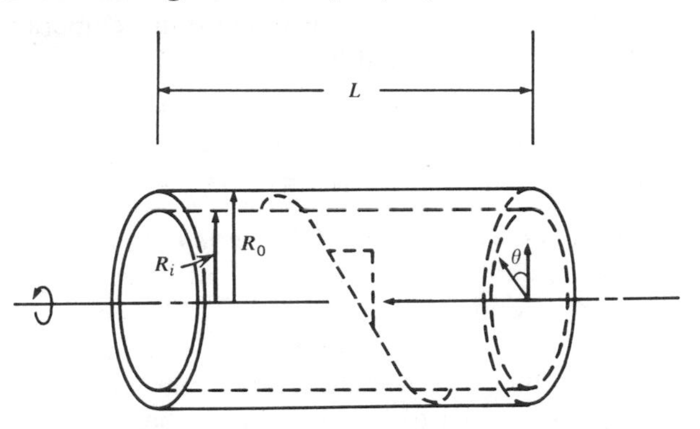


**5.9 *Torsional Drag Flow between Parallel Disks.*** Solve the torsional drag flow problem between two parallel disks, one of which is stationary while the other is rotating with an angular velocity $\Omega(\mathrm{s}^{-1})$. (***Note:*** $v_\theta/r = \text{constant}$.)

**5.10 *Radial Pressure Flow between Parallel Disks.*** Solve the problem of radial pressure flow between two parallel disks. The flow is created by a pressure drop $(P|_{r=0} - P|_{r=R})$. Disregard the "entrance" region, where the fluid enters from a small hole at the center of the top disk.

**5.11 *Flow near a Wall Suddenly Set in Motion.*** Set up the parallel plate drag flow problem (Problem 5.3) during its start-up period $t \leq t_{tr}$, when $v_x = f(t)$ in the entire flow region, and show that the resulting velocity profile, after solving the differential equation (see Ref. 1, Example 4.1-1), is $v_x/V = 1 - \mathrm{erf}(y/\sqrt{4\mu t/\rho})$ if $H$ is very large.

***NOTE:*** Problems 5.12 to 5.14 deal with steady state conductive heat transfer in solids of constant density and thermal conductivity. They are included so that readers who have not dealt with transport phenomena may develop an ability for solving such heat transfer problems common to polymer processing. We suggest the solution methodology: (*a*) after picking an appropriate coordinate system, visualize the heat conduction problem, stating your assumptions; (*b*) state the appropriate form of the thermal energy equations; (*c*) state the boundary conditions; (*d*) solve for the temperature profile and heat losses at the surfaces; (*e*) plot the temperature profile.

**5.12 *Heat Conduction across Flat Solid Slab.*** Solve the problem of heat transfer across an "infinitely" large flat plate of thickness $H$, for the following three physical situations: (*a*) the two surfaces are kept at $T_1$ and $T_2$, respectively; (b) one surface is kept at $T_1$ while the other is exposed to a fluid of temperature $T_b$, which causes a heat flux $q_y|_{y=H} = h_2(T_2 - T_b)$, $h_2$ being the heat transfer coefficient (W/m$^2$·°K); (*c*) both surfaces are exposed to two different fluids of temperatures $T_a$ and $T_b$ with heat transfer coefficients $h_1$ and $h_2$, respectively.

**5.13** ***Heat Transfer in Pipes.*** Solve the problem of conductive heat transfer across an "infinitely long" tube of inside and outside radii of $R_i$ and $R_o$. Consider the following two physical situations: (*a*) the surface temperatures at $R_i$ and $R_o$ are maintained at $T_i$ and $T_o$; (*b*) both the inside and outside tube surfaces are exposed to heat transfer fluids of constant temperatures $T_a$ and $T_b$ and heat transfer coefficients $h_i$ and $h_o$.

**5.14** ***Heat Transfer in Insulated Pipes.*** Solve case *b* above for a composite tube made of material of thermal conductivity $k_i$ for $R_i \le r \le R_m$ and of material of thermal conductivity $k_o$ for $R_m \le r \le R_o$.

**5.15** ***Parallel Plate Flow with Viscous Dissipation.*** Consider the nonisothermal flow of a Newtonian fluid whose $\rho$, $C_p$, and $k$ are constant, while its viscosity varies with temperature as $\mu = Ae^{\Delta E/RT}$. The flow is between two "infinite" parallel plates, one of which is stationary while the other is moving with a velocity $V$. The fluid has a considerably high viscosity so that the energy dissipated ($\frac{1}{2}\mu(\dot{\gamma}:\dot{\gamma})$ in Eq. 5.1-38) cannot be neglected. State the equations of continuity, momentum, and energy for the following two physical situations and suggest a solution scheme:

(*a*) $T(0) = T_0$, $T(H) = T_1$
(b) $q_y|_{y=0} = q_y|_{y=H} = 0$

**5.16** ***Flow between Tapered Plates* The Lubrication Approximation.*** Consider the steady isothermal pressure flow of a Newtonian and incompressible fluid flowing in a channel formed by two *slightly* tapered plates of "infinite" width. Using the cylindrical coordinate

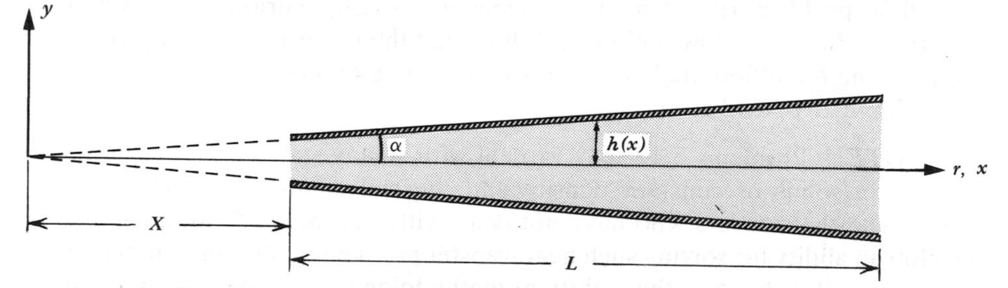


system in Fig. P5.16 and assuming that $v_r(r, \theta)$, $v_\theta = v_z = 0$, show that the continuity and momentum equations reduce to

$$\frac{1}{r}\frac{\partial}{\partial r}(rv_r) = 0 \qquad \text{or} \qquad v_r = \frac{F(\theta)}{r} \tag{a}$$

$$\frac{\partial P}{\partial r} = \frac{\mu}{r^2}\left(\frac{\partial^2 v_r}{\partial \theta^2}\right) \tag{b}$$

* W. E. Langlois, *Slow Viscous Flows*, Ch. VIII, McMillan, London, 1964.

$$\frac{\partial P}{\partial \theta}=\frac{2\mu}{r}\left(\frac{\partial v_r}{\partial \theta}\right) \tag{c}$$

Differentiate Eq. $b$ with respect to $\theta$ and Eq. $c$ with respect to $r$ and equate. Solve the resulting equations using the boundary condition

$$v_r(r, \pm\alpha)=0, \; Q=\int_{-\alpha}^{\alpha} v_r r\, d\theta$$

to obtain the velocity and pressure fields:

$$v_r(r, \theta)=\frac{Q}{r}\frac{\sin^2\alpha-\sin^2\theta}{\sin\alpha\cos\alpha-\alpha+2\alpha\sin^2\alpha} \tag{d}$$

$$P(r, \theta)=P_0+\frac{\mu Q\,(\cos^2\theta-\sin^2\theta)(X^2/r^2-1)}{X^2\sin\alpha\cos\alpha-\alpha+2\alpha\sin^2\alpha} \tag{e}$$

where $P(X, 0)=P_0$.

Show that the two nonvanishing pressure gradients in Cartesian coordinates are

$$\frac{\partial P}{\partial x}=-\frac{2\mu Q(1+D^2)D^3 h}{E}\frac{h^2-3D^2y^2}{(h^2+D^2y^2)^3} \tag{f}$$

$$\frac{\partial P}{\partial y}=\frac{2\mu Q(1+D^2)D^4 y}{E}\frac{3h^2-D^2y^2}{(h^2+D^2y^2)^3} \tag{g}$$

where $D=\tan\alpha$, $h=D(x-X)$ and $E=D-(1-D^2)\arctan D$.

From the Reynolds equation 5.4-11, show that for the tapered channel pressure flow,

$$\frac{\partial P}{\partial x}=-\frac{3Q\mu}{2h^3} \tag{h}$$

Plot the ratio of pressure drops obtained by Eqs. $h$ and $f$ to show that for $\alpha<10°$, the error involved using the lubrication approximation is very small.

# 6

# Polymer Melt Rheology



A major portion of all the polymer processing shaping operations and elementary steps involves either the isothermal or, most often, the nonisothermal flow of polymer melts in geometrically complex conduits. Before dealing with the realistic polymer processing flow problems, therefore, it is appropriate to examine separately the rheological (flow) behavior of polymer melts in simple flow situations and in the absence of temperature gradients. Our aims are to clarify the physical meaning of terms such as "non-Newtonian" or "viscoelastic behavior," "primary normal stress coefficient," and "viscosity functions," to discuss from a mathematical viewpoint briefly the constitutive equations that either quantitatively or semiquantitatively describe the observed behavior of polymer melts, and to examine the experimental methods that yield the rheological information needed to characterize polymer melt flow behavior in simple flows.

The science of rheology, and in particular polymer melt and solution rheology, is undergoing intensive study. This chapter is not a comprehensive review of the field: this was recently accomplished by Bird, et al. (1), whose work we frequently quote. We simply present a summary of the subject and introduce the rheological elements that are of common use in polymer processing. For more detailed discussions on the theoretical and experimental aspects of rheology, the reader is referred to a number of excellent texts (1–6).

## 6.1 Constitutive Equations

Constitutive equations describe the response of a substance when it is forced out of its equilibrium state. This response varies from substance to substance, and for a given material it may also vary in both degree and kind with the level of the applied stimulus. The relationship between the applied stimulus and the resulting response is characteristic of, and unique to, the *constitution* of the substance, hence the name "*constitutive equation.*" The nature and the exact level of the response are determined by the interatomic and intermolecular forces. But generally speaking, the state of our detailed knowledge of these forces is incomplete; thus it is not possible to accurately predict the macroscopic response of a substance from microscopic molecular information. Therefore constitutive equations are most often *empirical.* On the other hand, guided by experimental results one can develop approximate molecular models for many classes of substances and form "molecular theories" that result in constitutive equations.

Examples of *linear* empirical constitutive equations are Newton's law of viscosity, Fourier's law for heat conduction, and Fick's law for mass diffusion, presented in Section 5.1.

Newton's law of viscosity describes the rheological behavior of an important class of fluids—called Newtonian fluids—which have a constant viscosity that does not depend on the level of the applied stress or the resulting velocity gradient; it depends only on the temperature and pressure. This dependence is approximately given by

$$\mu(T, P) = \mu_0 \exp \frac{\Delta E}{R}\left[\frac{T_0 - T}{T_0 T}\right] \exp \beta(P - P_0) \qquad (6.1\text{-}1)$$

where $\mu_0$ is the viscosity at $T_0$ and $P_0$ (the reference temperature and pressure respectively), $\Delta E$ is the activation energy for flow, $R$ is the gas constant, and $\beta$ is a material property ($m^2/N$).

However there exist important classes of substances whose rheological properties are dependent on the stress (stimulus) or the rate of strain (response). The resultant constitutive equations are thus *nonlinear* or non-Newtonian. Outstanding among these classes of materials are polymer melts and solutions. But this is not the only difference between the rheological responses of Newtonian fluids and polymer melts and solutions. The next section describes those non-Newtonian effects, which are important in polymer processing. Section 6.3 deals with constitutive equations that attempt to describe either some or all rheological properties of these materials.

## 6.2 Non-Newtonian Behavior of Polymer Melts

In the examples that follow, the flow behavior of a typical polymer melt is contrasted to that of Newtonian liquids (1). Both classes of fluids are considered to be incompressible (see Chapter 5). In addition to demonstrating the non-Newtonian effects, these examples indicate that it is not possible to describe the observed flow behavior of polymer melts with the Newtonian constitutive equation for incompressible fluids

$$\boldsymbol{\tau} = -\mu \dot{\boldsymbol{\gamma}} \qquad (6.2\text{-}1)$$

### *Shear Rate Dependent Viscosity*

Consider two identical capillaries like the one in Fig. 6.1. One contains a Newtonian fluid and the other a polymer melt. Experimentally it is found that when various levels of pressure drop $\Delta P$ across the capillary are applied, both fluids flow with increasing flow rate $Q$, but ($a$) the Newtonian fluid exhibits a constant $Q/\Delta P$ ratio, indicating a constant resistance to the applied $\Delta P$; and ($b$) the polymer melt exhibits a $Q/\Delta P$ ratio that for very small values of $\Delta P$ is constant, although it increases by as much as 2 orders of magnitude with increasing $\Delta P$. Thus the resistance of the fluids to the applied $\Delta P$ is decreasing, as if these materials are "yielding." Hence the name *pseudoplastic* or *shear thinning*.

The rheological implication of this experiment, which is characteristic of most polymer melts, is that with increasing rates of deformation, the response of polymer melts changes from Newtonian to non-Newtonian shear thinning behavior. The latter response is generally dominant in the practical processing rates of deformation. *Indeed, the most important non-Newtonian property in polymer processing is shear thinning.* From a practical point of view, it is worth noting that this rheological property facilitates fast flow and eases the problem of excessive viscous heat generation. The Newtonian constitutive equation 6.2-1 cannot, of course, account for shear thinning behaviour.

The opposite rheological effect to shear thinning is *shear thickening*, commonly called *dilatant* behavior. Although no polymer melts have been found to be dilatant under any conditions, certain polymer solutions have been reported to exhibit dilatant behavior (7, 11).

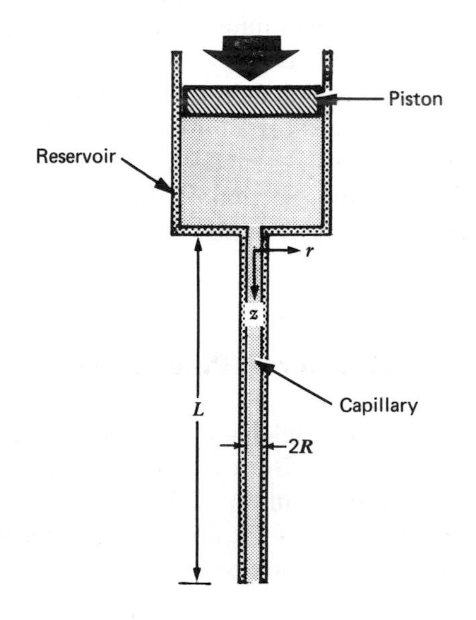


**Fig. 6.1** Schematic representation of a capillary flow apparatus.

### *Normal Stress Differences in Shear Flows*

Consider two identical stationary cup-rotating rod arrangements, one containing a Newtonian fluid (Fig. 6.2*a*), and the other a polymer melt (Fig. 6.2*b*). Upon rotation of the rod, a vortex is formed in the Newtonian fluid in the vicinity of the shaft. This can be explained in terms of the centrifugal forces that "throw" the fluid radially out; that is, $P(r_2) > P(r_1)$, where $r_2 > r_1$. On the other hand, the surface profile in the polymer melt is exactly the opposite; the fluid "mushrooms" on the rotating rod. This inward motion is against the centrifugal forces. Moreover, this phenomenon, which is called the Weissenberg effect,* is observed even at low shaft rotational rates. This phenomenon is often attributed to the appearance of so-called strangulation stresses. The polymer molecules become oriented during the

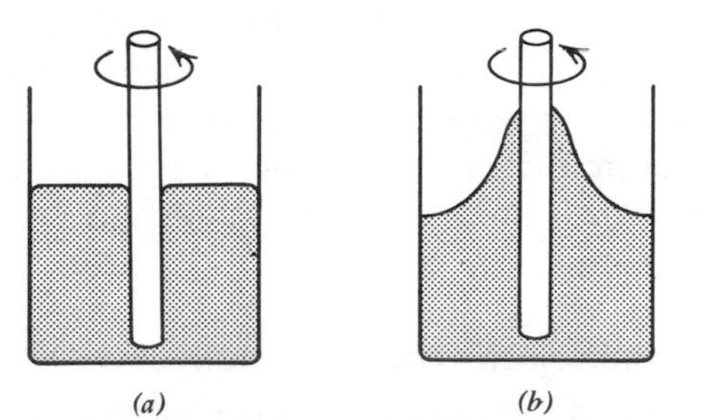

**Fig. 6.2** Tangential annular (Couette) flow. Schematic representations of the situation with (*a*) a Newtonian fluid, (*b*) a non-Newtonian fluid. (*c*) Sample of 18% by weight polymethylmethacrylate in *n*-butyl acetate, clearly showing the Weissenberg effect. (Reprinted with permission from *Modern Chemical Engineering*, Vol. 1, A. Acrivos, ed., Reinhold Publishing Co., New York, p. 201.)

* After K. Weissenberg who reported the phenomenon [*Nature*, **159**, 310 (1947)]. F. H. Garner and A. H. Nissan reported such effects earlier but gave the wrong interpretation [*Nature*, **158**, 634 (1946)].

annular flow, and in their tendency to return to the random coil state, they exert a hoop stress component on the layer of fluid next to them toward the rod.

The foregoing "molecular explanation" of the Weissenberg phenomenon has the following continuum mechanical counterpart. The annular tangential flow causes an extra tension in the (tangential) direction of flow compared to the perpendicular radial direction. That is, $\tau_{\theta\theta} - \tau_{rr} < 0$. From the mechanics point of view then, the *normal stress difference* $(\tau_{\theta\theta} - \tau_{rr})$ is responsible for the tendency of the fluid to climb the rotating rod and for the increase in pressure with decreasing radius. Nonzero normal stress differences for simple shear flows such as the foregoing cannot be predicted by Eq. 6.2-1, which predicts that all normal stresses vanish. Such normal stress effects, however, are observed with polymer melts and solutions. Their values can be of the order of the shear stress.

In addition to the Weissenberg effect, many rheological phenomena are attributed to the generation of normal stress differences in simple shear flows by polymer melts and solutions. Since these phenomena involve a variety of shear flows and geometries, it is useful and necessary to establish a convention for labeling them. Considering the shear flows that have only one nonvanishing velocity component that varies along only one direction,* we state the following generally accepted convention:

| | |
|---|---|
| Direction "1" | the velocity component direction |
| Direction "2" | the direction along which the velocity changes |
| Direction "3" | neutral direction |

The labeling of the normal stress differences follow the foregoing convention

| | |
|---|---|
| $\tau_{11} - \tau_{22}$ | the first (primary) normal stress difference |
| $\tau_{22} - \tau_{33}$ | the second (secondary) normal stress difference |

In Couette flow, then, the rod climbing effect was attributed to the existence of the primary normal stress difference. Other flow phenomena attributed primarily to the first normal stress difference are as follows:

1. In the torsional flow between two parallel disks, a pressure is generated that increases with decreasing radius. This geometry has indeed been utilized in creating a melt pump called the normal stress extruder, discussed in Section 10.6.
2. In the torsional flow between a flat disk and a cone (Fig. 6.11), the pressure generated again increases with decreasing radius. This flow geometry is utilized in viscometry for evaluating the primary $\tau_{\phi\phi} - \tau_{\theta\theta}$ as well as the secondary $\tau_{\theta\theta} - \tau_{rr}$ normal stress differences.
3. Polymer melts upon exiting from long capillaries, under the force of a pressure drop, "swell." That is, their diameter $D$ is larger than that of the capillary $D_0$. The melt while flowing in the capillary is under an extra axial tension that forces the free jet to contract axially and "swell" radially. Thus

* Such flows are often called viscometric, because they are utilized in viscometry, or rheometry in general. For a broader definition of viscometric flows, see Ref. 1, p. 130.

the magnitude of the primary normal stress difference $\tau_{11}-\tau_{22}$ reflects this phenomenon.

The secondary normal stress difference is probably an order of magnitude smaller than the primary, as Section 6.7 demonstrates. Its precise measurement has not yet been achieved. Nevertheless, it is responsible for a small number of rheological phenomena such as:

1. The observation that the pressure in axial annular flow is larger at the inner cylinder. The pressure difference is not very large.
2. The observation that in wire coating, if the wire is eccentric to the die, there exist forces that tend to diminish this eccentricity. It has been shown that such forces may be partly attributed to the second normal stress difference (8).

### *Stress Relaxation, Creep Recovery, and "Stress Overshoot" in Polymer Melts (Viscoelastic Behavior)*

Consider two identical parallel plate flow experimental setups such as the one in Fig. 6.3*a*. The first contains a Newtonian fluid and the second a polymer melt. The bottom plates are fixed, and the top ones are displaced suddenly by a distance $\Delta x$, so that $\gamma_{yx} = \Delta x/\Delta y$ is the instantaneously applied strain. The shear stresses $\tau_{yx} = F_x(t)/A_y$, observed with the two fluids, are illustrated in Fig. 6.3*b* and 6.3*c*. The immediate stress relaxation of the Newtonian fluid is consistent with Eq. 6.2-1,

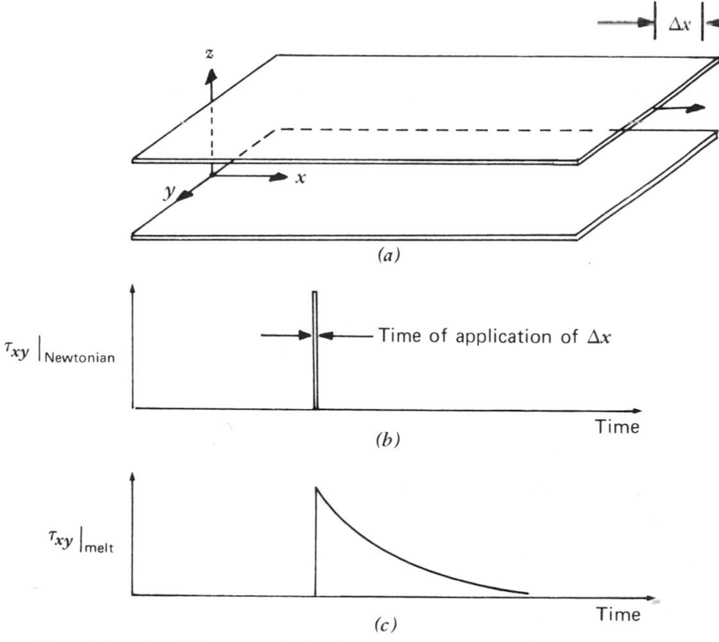


**Fig. 6.3** (*a*) The parallel plate flow used in the stress relaxation experiment. (*b*) The resulting Newtonian shear stress. (*c*) The resulting polymer melt shear stress.

since except for the infinitesimally small interval of time when $\Delta x$ is applied, $d(\Delta x/\Delta y)/dt = dv_x/dy \equiv \dot{\gamma}$ is zero. Consequently, the slow stress relaxation of the polymer melt, when $\dot{\gamma} = 0$, cannot be explained in terms of the Newtonian constitutive equation. It can be explained, though, in terms of the theory of viscoelasticity (see Sections 2.1 and 6.4).

Another ramification of the viscoelasticity of polymer melts is their ability to recover all, some, or none of the strain applied on them, depending on the value of the Deborah number. Creep recovery was mentioned earlier in connection with extrudate swelling. A clearer demonstration of recoil is offered by the experiments of Kapoor (9). Consider again two capillaries, such as the one in Fig. 6.1, one containing a Newtonian fluid and the other a polymer melt. Let a tracer be injected diametrically while both fluids are at rest. Then apply a constant pressure for a short fixed interval, removing it after that. Figure 6.4 depicts the resulting shape of the tracers for a viscoelastic solution. The behavior of the Newtonian fluid would be consistent with Eq. 6.2-1; after frame 5 there is no externally applied stress, therefore no flow would be observed. The behavior of the polymer melt that shows creep recovery is then at least partly elastic. The qualifying term "partly" refers to the *incomplete* and *delayed* strain recovery, which is expected from flexible uncrosslinked polymer systems (see Section 2.1).

Finally, polymer melts and solutions differ from Newtonian fluids in their response during the startup of simple shearing experiments. Turning to Fig. 6.5, we note that the startup stress to a cone and plate flow of a PS melt exhibits an "overshoot"; it does not increase monotonically and asymptotically to the steady state stress value, as observed with Newtonian fluids or with polymer melts at very low shear rates ($De \rightarrow 0$). The phenomenon can be attributed to the entanglement density changes brought about during moderate and fast flows. For such Deborah numbers the flowing molecules are forced to disentangle faster than their natural response, resulting in excessive structural straining, thus stress overshoot. The equilibrium entanglement density can be obtained after long times of cessation of flow. That is, the above-named structural changes are *reversible.* Stress overshoot is a nonlinear viscoelastic phenomenon, since it cannot be predicted by linear viscoelastic theories.

From a polymer processing point of view, we note that the "overshoot" region persists for periods that are commensurate to filling times of molds in injection molding, or times to produce a parison in intermittent blow molding. Thus the transient stress overshoot behavior is expected to contribute during such operations. Furthermore, the viscoelastic behavior of polymer melts, and in particular their ability to stress-relax and strain-recover, is of importance to polymer processing in general and to structuring and shaping in particular. As we know from Chapter 3, the stresses and resulting strains remaining in the melt after processing will to a large extent determine the final product morphology and properties, both short and long range.

### *Spinnability of Polymer Melts*

A very important engineering property of polymers is their ability to be spun (free jet stretching) rather easily and stably. In the spinning process, which Section 15.1

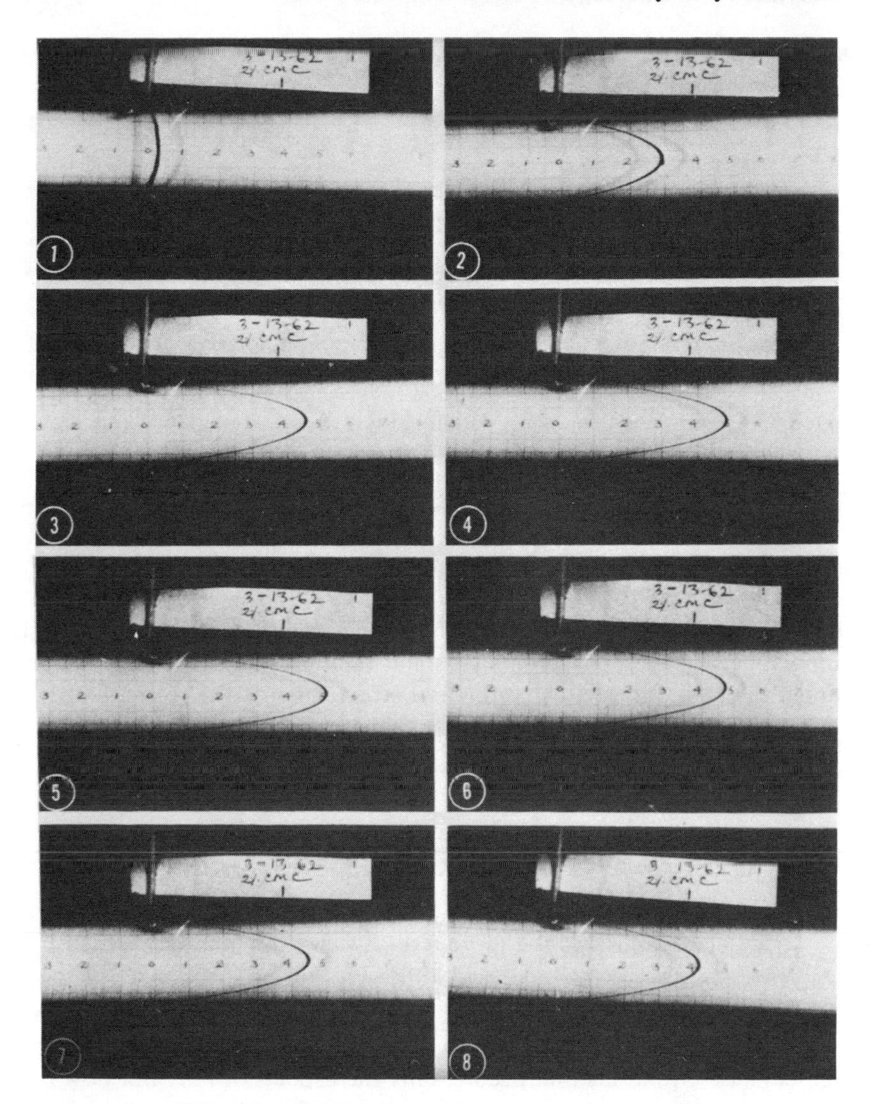

**Fig. 6.4** Recoil of 2% by weight carboxymethyl cellulose 70H in water after the cessation of capillary flow, after frame 5. A pressure gradient of 1.54 $kN/m^3$ was applied at frame 1 and removed just before frame 5. (Reprinted with permission from N. N. Kapoor, M.S. thesis, Department of Chemical Engineering, University of Minnesota, 1963.)

treats in some detail, the filament-shaped extrudates can be stretched a number of times their original length without breaking. The underlying reasons for melt spinnability may be macromolecular entanglements and very large extensional viscosity of such systems (or the large "melt strength"), which in turn is due to the macromolecular nature of the fluid, and the resulting orientation during the extensional flow occurring during the filament stretching. The forces resulting from the high extensional viscosity may predominate over the surface tension forces and stabilize the filament drawing flow. The phenomenon of the tubeless siphon (10), where the macromolecular fluid can be siphoned out of a container with the *bare*

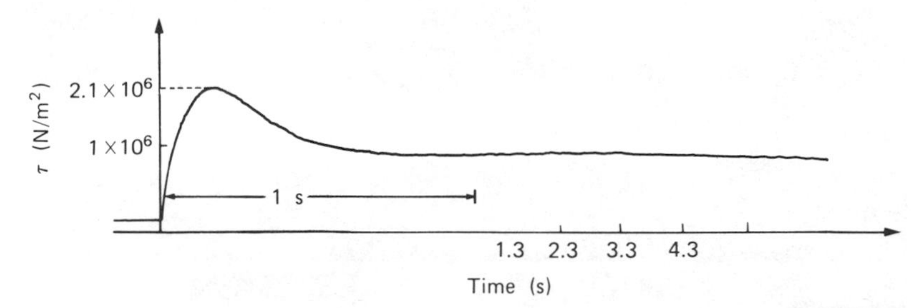


**Fig. 6.5** Stress overshoot during the flow of a polystyrene melt (PS 160,000) in a biconical rheometer: 160°C, 15.15 $s^{-1}$. The magnitude of overshoot depends on the time of rest between successive startups. (Reprinted with permission from K. J. Madonia, Ph.D. dissertation, Department of Chemical Engineering, Stevens Institute of Technology, Hoboken, N.J., 1970.)

fluid stream forming the siphon, is attributable to the same factors that enable melts to be spun.

Of the rheological characteristics of polymer melts described previously, the most important to polymer processing is shear thinning. This statement is made not to diminish the importance to shaping and structuring of the normal stresses and viscoelastic response, but to emphasize that shear thinning effects in polymer melts are very large.

## 6.3 Polymer Melt Constitutive Equations

From our discussion in the previous section, it is obvious that the rheological behavior of polymer melts and solutions is complex. It is not surprising to find, therefore, that despite 30 years of concentrated effort by rheologists, there is no *usable* constitutive equation that describes quantitatively *all* the flow phenomena involving polymer melts. In the absence of such a constitutive equation, scientists and engineers use equations that predict only the aspects of polymer flow behavior that are of main interest to them or are important to the particular problem at hand. There is a multitude of constitutive equations proposed for polymer melts. Only a few have been used to solve polymer processing problems. Nevertheless, we feel that it is instructive to trace their origin and to indicate the interrelationship among them. We shall do this qualitatively, without dealing in detail with the mathematical complexities of the subject.

We follow here the systematic and clear classification and description of the constitutive equations of Bird et al. (1), and we refer the reader who is interested in the detailed development of the subject to that source. There is a general agreement that all the constitutive equations for polymer melts and solutions are special cases of a very general constitutive relation according to which the stress at any point in a flowing fluid and at any time depends on the entire flow history of the fluid element occupying that point. Because it does not depend on the flow history of adjacent elements, the dependence is "simple," and the general relation is called the *simple fluid* constitutive equation (12).

One physical restriction, translated into a mathematical requirement, must be satisfied: it is that the simple fluid relation must be "objective," which means that its predictions should not depend on whether the fluid rotates as a rigid body. This can be achieved by casting the constitutive equation (expressing its terms) in special frames. One is the *corotational* frame, which follows (translates with) each particle and rotates with it. The other is the *codeformational* frame, which translates, rotates, and deforms with the flowing particles. In either frame, the "observer" is oblivious to rigid body rotation. Thus a constitutive equation cast in either frame is "objective" or, as it is commonly expressed, "obeys the principle of material objectivity."* From these coordinate frames, two classes of rheological equations have arisen: the *codeformational* and the *corotational.* Both codeformational and corotational constitutive equations can be transformed in the fixed (laboratory) frame in which the balance equations appear and where experimental results are obtained. The transformations are similar to, but more complex than, that from the substantial frame to the fixed (see Section 5.1). Finally, a corotational constitutive equation can be transformed to a codeformational one.†

Goddard (13) expressed the notion of the simple fluid constitutive equation in a corotational integral series. The integral series expansion had been used in the codeformational frame by Green and Rivlin (14) and Coleman and Noll (15). The corotational expansion takes the form (1a):

$$\boldsymbol{\tau}(x, t) = -\int_{-\infty}^{t} G_1(t-t')\dot{\boldsymbol{\Gamma}}'\,dt' - \frac{1}{2}\int_{-\infty}^{t}\int_{-\infty}^{t} G_{11}(t-t', t-t'')[\dot{\boldsymbol{\Gamma}}'\cdot\dot{\boldsymbol{\Gamma}}'' + \dot{\boldsymbol{\Gamma}}''\cdot\dot{\boldsymbol{\Gamma}}']\,dt''\,dt' - \cdots \tag{6.3-1}$$

where $G_1$, $G_{11}$, . . . , are characteristic material functions, $\dot{\boldsymbol{\Gamma}}$ is the corotating rate of strain (velocity gradient) tensor, $t'$, $t''$, $t'''$ are integration variables, and $t$ is the "present" time. Equation 6.3-1 is in an unusable form. There exist two routes through which useful constitutive equations can be obtained (see Ref. 1, Fig. 8.3-1):

1. Expand $\dot{\boldsymbol{\Gamma}}$ in a Taylor series about $t' = t$ (1e)

$$\dot{\boldsymbol{\Gamma}}(t, t') = \dot{\boldsymbol{\gamma}}(t) - (t-t')\frac{\mathscr{D}\dot{\boldsymbol{\gamma}}}{\mathscr{D}t} + \cdots \tag{6.3-2}$$

where

► $$\frac{\mathscr{D}\dot{\boldsymbol{\gamma}}}{\mathscr{D}t} = \frac{\partial\dot{\boldsymbol{\gamma}}}{\partial t} + \{\boldsymbol{v}\cdot\nabla\dot{\boldsymbol{\gamma}}\} + \tfrac{1}{2}(\{\boldsymbol{\omega}\cdot\dot{\boldsymbol{\gamma}}\} - \{\dot{\boldsymbol{\gamma}}\cdot\boldsymbol{\omega}\}) \tag{6.3-3}$$

is the *corotational* derivative or *Jaumann* derivative measuring the time rate of change of $\dot{\boldsymbol{\gamma}}$ as measured by an "observer" that is translating and rotating with the local fluid velocity and vorticity. Keeping only the first two terms of the Taylor series (which means that the flow under consideration is

* The concept of the corotational frame applied to viscoelastic fluids is due to S. Zaremba [*Bull. Int. Acad. Sci., Cracov.*, 594 (1903)]. H. Fromm [*Z. Angew. Math. Mech.*, **25/27**, 146 (1947); **28**, 43 (1948)] apparently rediscovered it. Later Oldroyd, Noll, and De Witt made use of both convected coordinate systems and the principle of material objectivity and made very significant contributions to the development of the theories discussed below [J. G. Oldroyd, *Proc. Roy. Soc.*, **A200**, 523 (1950); W. Noll, *J. Rat. Mech. Anal.*, **4**, 3 (1955); T. W. DeWitt, *J. Appl. Phys.*, **26**, 889 (1955)].

† All the above-mentioned transformations are treated systematically and in detail by Bird et al. (1).

*almost* steady), one can obtain the *second order fluid* constitutive equation

$$\boldsymbol{\tau} = -\alpha_1 \dot{\boldsymbol{\gamma}} + \alpha_2 \frac{\mathscr{D}\dot{\boldsymbol{\gamma}}}{\mathscr{D}t} - \alpha_{11}\{\dot{\boldsymbol{\gamma}} \cdot \dot{\boldsymbol{\gamma}}\} - \cdots \tag{6.3-4}$$

where $\alpha_i$ are constants related to $G_1, G_{11}, \ldots$. For *steady shear flows*, the Criminale–Ericksen–Filbey (CEF) constitutive equation can be obtained (16)

► $$\boldsymbol{\tau} = -\eta \dot{\boldsymbol{\gamma}} - (\tfrac{1}{2}\Psi_1 + \Psi_2)\{\dot{\boldsymbol{\gamma}} \cdot \dot{\boldsymbol{\gamma}}\} + \frac{1}{2}\Psi_1 \frac{\mathscr{D}\dot{\boldsymbol{\gamma}}}{\mathscr{D}t} \tag{6.3-5}$$

where $\eta, \Psi_1$, and $\Psi_2$ are the viscosity, first normal stress difference coefficient, and second normal stress difference coefficient functions, respectively. They are all functions of the magnitude of the rate of strain tensor $\dot{\gamma} = \sqrt{\frac{1}{2}(\dot{\boldsymbol{\gamma}} : \dot{\boldsymbol{\gamma}})}$. Because many polymer processing flows are steady shear flows and because of the physical significance of the material functions $\eta, \Psi_1$, and $\Psi_2$, the CEF equation is considered in detail in the following section.

If the normal stress coefficient functions $\Psi_1$ and $\Psi_2$ are ignored, the CEF equation reduces to the *generalized Newtonian fluid* (*GNF*) equation

► $$\boldsymbol{\tau} = -\eta \dot{\boldsymbol{\gamma}} \tag{6.3-6}$$

From the GNF equation, the constitutive equation for an incompressible Newtonian fluid is obtained if the viscosity is taken to be constant (Eq. 6.2-1)

► $$\boldsymbol{\tau} = -\mu \dot{\boldsymbol{\gamma}}$$

2. If in Eq. 6.3-1 a single integral term is retained the *Goddard-Miller* (G-M) constitutive equation is obtained (17):

► $$\boldsymbol{\tau} = -\int_{-\infty}^{t} G(t-t')\dot{\boldsymbol{\Gamma}}\, dt' \tag{6.3-7}$$

For *small deformation* flows it is evident from Eqs. 6.3-2 and 6.3-3 that $\dot{\boldsymbol{\Gamma}}$ equals $\dot{\boldsymbol{\gamma}}$; thus the G–M equation yields the *general linear viscoelastic* (LVE) fluid (1a, 14, 15)

► $$\boldsymbol{\tau} = -\int_{-\infty}^{t} G(t-t')\dot{\boldsymbol{\gamma}}(t')\, dt' \tag{6.3-8}$$

where $G(t-t')$ is the relaxation modulus, which can take specific forms depending on the LVE "mechanical model" used to simulate the real LVE behavior. For example, if a single "Maxwell element" consisting of a spring $G$ and a dashpot $\mu$ in series is used (see Section 6.4), the *Maxwell* constitutive equation is obtained

$$\boldsymbol{\tau} + \lambda_0\, d\boldsymbol{\tau}/dt = -\eta_0 \dot{\boldsymbol{\gamma}} \tag{6.3-9}$$

where $\lambda_0 = \eta_0/G$. When $\lambda_0 = 0 (G \to \infty)$, the Newtonian constitutive equation for an incompressible fluid is obtained, Eq. 6.2-1.

From the G–M equation, while still in the corotational frame, we can choose a specific form of the relaxation modulus. Thus for a single Maxwell element we can obtain

$$\boldsymbol{\tau}+\lambda_0 \frac{\mathscr{D}\boldsymbol{\tau}}{\mathscr{D}t} = -\eta_0 \dot{\boldsymbol{\gamma}} \tag{6.3-10}$$

This is called the *Zaremba–Fromm–DeWitt* (*ZFD*) equation (see footnote, p. 155).

As stated earlier, the simple fluid concept can be expressed in a series of codeformational integrals (1b, 14, 15)

$$\boldsymbol{\tau} = -\int_{-\infty}^{t} G_1(t-t')\boldsymbol{\gamma}^{[1]'}\,dt' - \frac{1}{2}\int_{\infty}^{t}\int_{\infty}^{t} G_2(t-t', t-t'')[\boldsymbol{\gamma}^{[1]'}\cdot\boldsymbol{\gamma}^{[1]''}+\boldsymbol{\gamma}^{[1]''}\cdot\boldsymbol{\gamma}^{[1]'}]\,dt''\,dt' - \cdots \tag{6.3-11}$$

where $G_1, G_2, \ldots,$ are material functions and $\boldsymbol{\gamma}^{[1]}$ is the codeforming rate of strain tensor using covariant differentiation. If contravariant derivatives are used (1b)

$$\boldsymbol{\tau} = -\int G^1(t-t')\boldsymbol{\gamma}'_{[1]}\,dt' - \frac{1}{2}\int_{-\infty}^{t}\int_{-\infty}^{t} G^2(t-t', t-t'')[\boldsymbol{\gamma}'_{[1]}\cdot\boldsymbol{\gamma}''_{[1]}+\boldsymbol{\gamma}''_{[1]}\cdot\boldsymbol{\gamma}'_{[1]}]\,dt''\,dt' - \cdots \tag{6.3-12}$$

where $G^1, G^2, \ldots,$ are material functions and $\boldsymbol{\gamma}_{[1]}$ is the codeforming rate of strain tensor using contravariant differentiation.

As was the case with Eq. 6.3-1, Eqs. 6.3-11 and 6.3-12 are not usable. But the same means of making them usable are available (see Ref. 1, Fig. 9.5-3, and Table 9.4-1). They are two specific steps,

1. For almost steady flows one can expand $\boldsymbol{\gamma}^{[1]}$ or $\boldsymbol{\gamma}_{[1]}$ about $t=t'$ and obtain second order fluid constitution equations in the codeforming frame. When steady shear flows are considered, the CEF equation is obtained which, in turn, reduces to the GNF equation for $\Psi_1=\Psi_2=0$ and to a Newtonian equation if, additionally, the viscosity is constant.
2. Setting $G_2, G_3, \ldots,$ or $G^2, G^3, \ldots,$ equal to zero, Eqs. 6.3-11 and 6.3-12 reduce to G–M type equations. For example,

$$\boldsymbol{\tau} = -\int_{-\infty}^{t} G(t-t')\boldsymbol{\gamma}'_{[1]}\,dt' \tag{6.3-13}$$

is the so-called *Oldroyd* (18)-*Walters* (19)-*Fredrickson* (20) equation. This equation when integrated by parts yields the *Lodge rubberlike liquid* equation (3)

$$\boldsymbol{\tau} = \int_{-\infty}^{t} M(t-t')\boldsymbol{\gamma}'_{[0]}\,dt' \tag{6.3-14}$$

where $M(t-t') = dG(t-t')/dt'$ and $\boldsymbol{\gamma}_{[0]}$ is the strain tensor in a codeforming frame using contravariant differentiation.

For small deformations, Eq. 6.3-13 is reduced to the LVE equations, 6.3-8 and 6.3-9 ($\boldsymbol{\gamma}_{[1]}=\dot{\boldsymbol{\gamma}}$). On the other hand, for large deformations, while still in the codeforming frame, one can use a particular linear viscoelastic model to represent $G(t-t')$ in Eq. 6.3–13. If, as before, a single Maxwell element is used, one can obtain the following analog to Eq. 6.3-10

$$\blacktriangleright \qquad \boldsymbol{\tau} + \lambda_0 \boldsymbol{\tau}_{(1)} = -\eta_0 \dot{\boldsymbol{\gamma}} \qquad (6.3\text{-}15)$$

where $\tau_{(1)}$ is a codeforming time derivative (1b) equal to

$$\boldsymbol{\tau}_{(1)} = \frac{D}{Dt}\boldsymbol{\tau} - \{(\nabla \boldsymbol{v})^{\dagger} \cdot \boldsymbol{\tau} + \boldsymbol{\tau} \cdot (\nabla \boldsymbol{v})\} \qquad (6.3\text{-}16)$$

Together with Eq. 6.3-16, Eq. 6.3-15 is the *White–Metzner* constitutive equation, which has been frequently used as a nonlinear viscoelastic model. Of course for small deformations $\boldsymbol{\tau}_{(1)} = d\boldsymbol{\tau}/dt$, and the single Maxwell fluid equation 6.3-9 is obtained.

Finally, a number of commonly used constitutive equations are derived from Eq. 6.3-12 by specifying $G^1, G^2, \ldots,$ (or $M_1, M_2, \ldots,$), instead of specifying only $G^1$ and setting $G^2, \ldots,$ equal to zero. Moreover, in these equations $M_i$ are allowed to be functions of the invariants of the strain or rate of strain tensors, since there is experimental evidence supporting this dependence (21). Examples of such usable integral codeformational constitutive equations are

$$\blacktriangleright \quad \boldsymbol{\tau} = +\int_{-\infty}^{t} [M_1(t-t', \mathrm{I}_{\gamma'_{[0]}}, \mathrm{II}_{\gamma'_{[0]}})\boldsymbol{\gamma}'_{[0]} + M_2(t-t', \mathrm{I}_{\gamma'_{[0]}}, \mathrm{II}_{\gamma'_{[0]}})\{\boldsymbol{\gamma}'_{[0]} \cdot \boldsymbol{\gamma}'_{[0]}\}]\, dt' \qquad (6.3\text{-}17)$$

This is the *Bernstein–Kearsley–Zappas* (*BKZ*) (22) constitutive equation.

$$\blacktriangleright \qquad \boldsymbol{\tau} = +\int_{-\infty}^{t} M(t-t', \mathrm{II}_{\dot{\gamma}}(t'))\left[\left(1+\frac{\varepsilon}{2}\right)\boldsymbol{\gamma}'_{[0]} - \frac{\varepsilon}{2}\boldsymbol{\gamma}^{[0]'}\right] dt' \qquad (6.3\text{-}18)$$

This is the *Bogue* (23) or Bird–Carreau (24) or Chen–Bogue (25) constitutive equation, depending on the representation of the dependence of $M$ on $\mathrm{II}\dot{\boldsymbol{\gamma}}$; $\varepsilon$ is a constant.

We have tried to give a quick glimpse of the interrelationships among some commonly used constitutive equations for polymer melts and solutions. None predicts quantitatively the entire spectrum of rheological behavior of these materials. Some are better than others, though more complex to use in connection with the equation of motion. Table 6.1 briefly summarizes the predictive abilities of some of the foregoing, as well as other constitutive equations.

In the next three sections we shall discuss three of the above-given rheological equations, the LVE, the GNF, and the CEF: the first because it reveals the viscoelastic nature of polymer melts; the second because in its various specific forms, it is widely used in polymer processing, the third because of its ability to predict normal stress differences in steady shearing flows, a property that is useful per se, but also very helpful in Section 6.7 on viscometry.

## 6.4 Linear Viscoelasticity

Section 6.3 presented the linear viscoelastic fluid constitutive equation, Eq. 6.3-8, and indicated its origins and the assumptions used. Furthermore, we stated that the

**Table 6.1 Selected Constitutive Equations that Have Been Used for Polymer Melts, and Comments on Their Predictive Abilities[a]**

| Equation | $\eta(\gamma)$ | $\Psi_1(\gamma)$, $\Psi_2(\gamma)$ | Stress Overshoot | VE Response |
|---|---|---|---|---|
| Newtonian fluid (5.1-32) | Constant | Zero | No | No |
| All GNF fluids (Section 6.5) | $\eta(\dot{\gamma})$ fit depends on model | Zero | No | No |
| LVE fluids (6.3-8) | Constant | Zero | No | Predicts small deformation *linear* response |
| G-M (6.3-7) | Good fit can be obtained | $\Psi_2 = -0.5\Psi_1$; $\Psi_1 = f(\eta)$ (6.7-23) | Yes; followed by spurious oscillations | Predicts nonlinear response in terms of $G(t-t')$ determined from LVE. |
| ZFD (6.3-10) | Abrupt drop for single element; better fit for several $\lambda_{0i}$, $\eta_{0i}$ | $\Psi_2 = -0.5\Psi_1$; $\Psi_i = f(\dot{\gamma})$ | Yes, followed by spurious oscillations | Yes the pairs $\eta(\dot{\gamma})$, $\eta'(\omega)$ and $\frac{1}{2}\Psi_1(\dot{\gamma})$, $\eta''(\omega)$ are identical; semiquantitatively correct |
| Second order fluids (6.3-4) | Constant | $\Psi_i$ are constant and related to each other | No | No |
| CEF fluids (6.3-5) | $\eta(\dot{\gamma})$ unspecified | $\Psi_i(\dot{\gamma})$ unspecified | No | No |
| Lodge rubberlike liquid (6.3-14) | Constant | $\Psi_1 =$ constant; $\Psi_2 = 0$ | No; predicts elongational stress growth $\eta^+(t, \dot{\varepsilon})$ | Yes |
| White-Metzner (6.3-15); (6.3-16) | Constant | $\Psi_1 =$ constant; $\Psi_2 = 0$ | No | Yes |
| BKZ (6.3-17) | Predicts $\eta(\dot{\gamma})$ | $\Psi_1 = f(\dot{\gamma})$ and related to $\eta(\dot{\gamma})$; this relationship tests out semiquantitatively | Yes | Yes |
| Bogue (Bird-Carreau) (6.3-18) | Good fit; depends on model for $M$ | Good fit; depends on model for $M$ | Yes | Yes |

[a] For more details on the predictive abilities of the constitutive equations listed see Ref. 1.

relaxation modulus $G(t)$ depends on the "mechanical model" used to represent the linear viscoelastic response. Let us elaborate on this point.

Linear viscoelastic response, quite logically, can be obtained, at least qualitatively, if the material is thought to have the "dual nature" of a Newtonian viscous fluid and a Hookean elastic solid with both aspects of this dual nature acting concurrently. This idea can be expressed in terms of the simple mechanical models in Fig. 6.6.

If the Maxwell element, for example, is subjected to a shear stress relaxation experiment, $\gamma=0$ for $t<0$, $\gamma=\gamma_0$ for $t>0$, then the resulting time dependent stress is (see Problem 6.1)

$$\tau(t)=\gamma_0 G e^{-t/\lambda} \tag{6.4-1}$$

where the relaxation time $\lambda$ is the ratio of the "viscosity" and the "modulus" of the elements of the Maxwell model. Equation 6.4-1 can be used to define the *time dependent (relaxation) modulus*

$$G(t)\equiv\frac{\tau(t)}{\gamma_0}=Ge^{-t/\lambda} \tag{6.4-2}$$

where $G(t)$ is *independent* of the stress relaxation experiment, that is, the level of applied strain, which in this experiment is the stimulus. This follows because the Maxwell element is a *linear system.*

Similarly, if a Voigt element is subjected to a creep experiment, $\tau=0$ for $t<0$, $\tau=\tau_0$ for $t\geq 0$, the resulting time dependent strain will be (see Problem 6.3)

$$\gamma(t)=\tau_0\frac{1}{G}[1-e^{-t/\lambda}] \tag{6.4-3}$$

We can again define an experiment-independent quantity, the *time dependent compliance*

$$J(t)=\frac{\gamma(t)}{\tau_0}=\frac{1}{G}[1-e^{-t/\lambda}] \tag{6.4-4}$$

The *actual* stress relaxation and creep behavior of flexible uncrosslinked polymers is similar *only qualitatively* to the responses predicted by Maxwell and Voigt models, Eqs. 6.4-2 and 6.4-4, even at small strain or stress levels where such materials are linear viscoelastic. Figure 6.7 indicates the similarities and differences. The main difference is that the response predicted by the simple viscoelastic bodies is sharp and abrupt, characteristic of the simple exponential time dependence of both $G(t)$ and $J(t)$. On the other hand, Fig. 6.7 suggests that the experimentally obtained $G(t)$ and $J(t)$ could be *curve fitted* by polynomials of exponential terms such as those in Eqs. 6.4-2 and 6.4-4. Specifically,

$$G(t)_{\exp}\cong\sum_{i=1}^{N}G_i e^{-t/\lambda_i} \tag{6.4-5}$$

and

$$J(t)_{\exp}\cong\sum_{i=1}^{N}\left(\frac{1}{G_i}\right)[1-e^{-t/\lambda_i}] \tag{6.4-6}$$

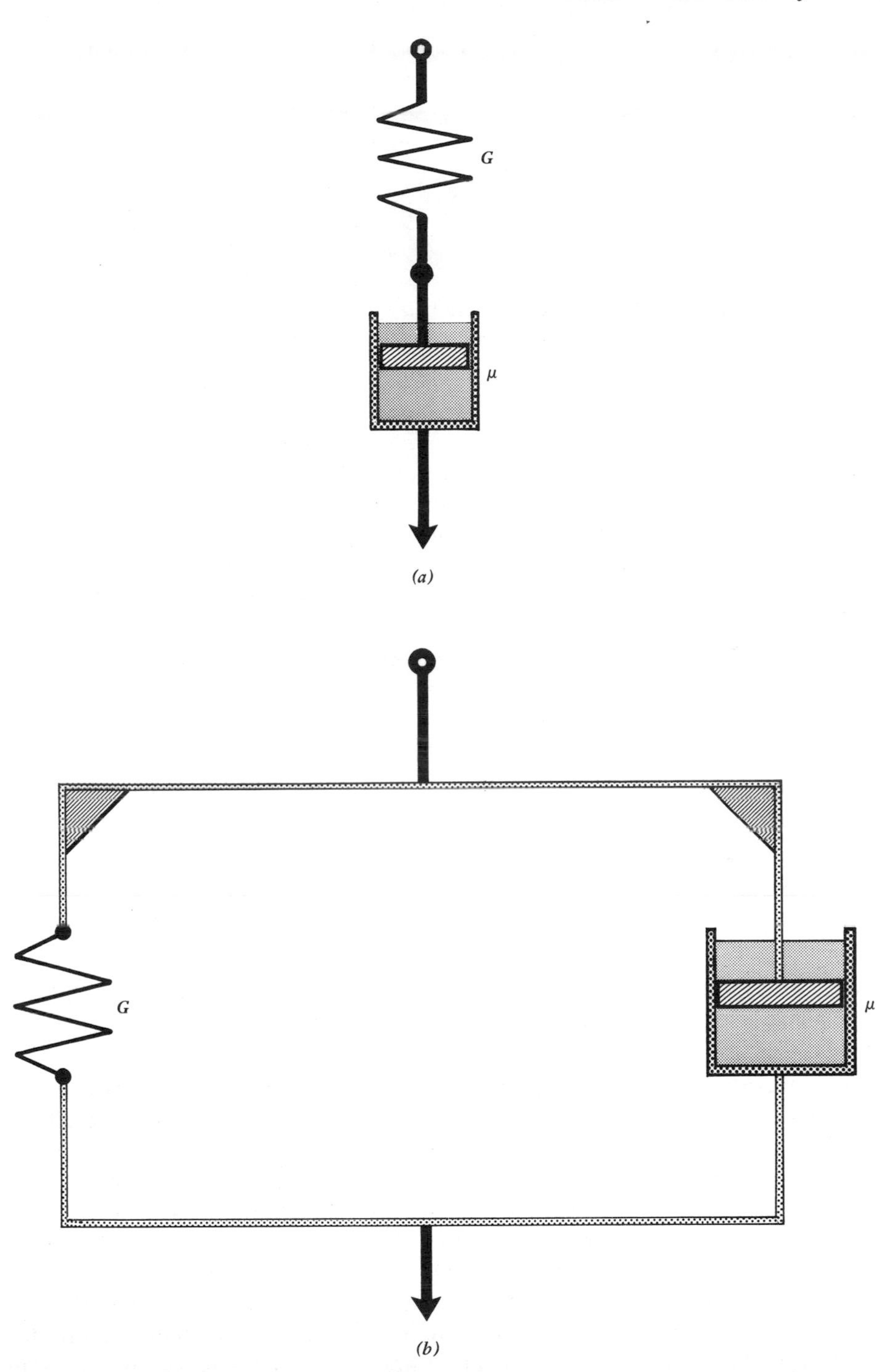


**Fig. 6.6** The Maxwell ($a$) and Voigt ($b$) mechanical models for simple linear viscoelastic response.

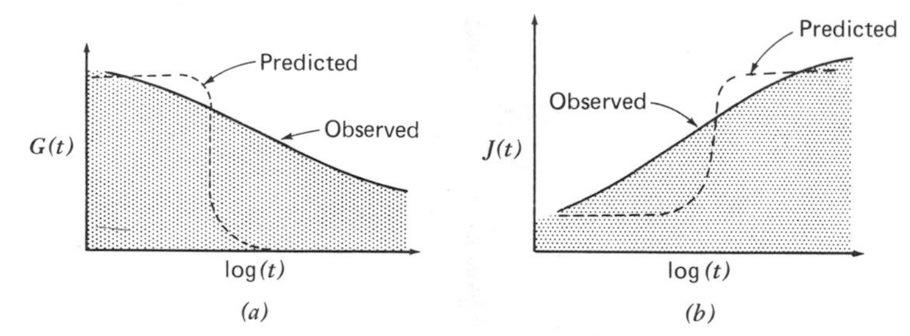


**Fig. 6.7** Actual and predicted ($a$) stress relaxation (Maxwell) and ($b$) creep (Voigt) responses of flexible uncrosslinked polymers.

The two equations above suggest a *discrete spectrum* of relaxation and retardation times. Physically the concept of discrete spectral response is quite reasonable and denotes that a system of strained flexible polymer chains returns to the state of maximum conformational entropy through a large number ($N$) of types of molecular motions, some of which occur fast (short $\lambda$), while others occur slowly. The "longest" relaxation time is probably the characteristic time for the reorganization of the entire chain or the *entire system of chains.* Guided by this reasoning and the view of continuum mechanics that *approximates* discrete molecular systems with continuous media for mathematical convenience, the discrete spectra can be extended to continuous ones as follows:

$$G(t)_{\text{exp}} = \int_0^{\infty} G(\lambda)\, e^{-t/\lambda}\, d\lambda \equiv \int_{-\infty}^{+\infty} H(\ln \lambda)\, e^{-t/\lambda}\, d(\ln \lambda) \qquad (6.4\text{-}7)$$

and

$$J(t)_{\text{exp}} = \int_0^{\infty} J(\lambda)[1 - e^{-t/\lambda}]\, d\lambda \equiv \int_{-\infty}^{+\infty} Y(\ln \lambda)[1 - e^{-t/\lambda}]\, d \ln \lambda \qquad (6.4\text{-}8)$$

where $G(\lambda)$, $H(\ln \lambda)$ are continuous relaxation spectral functions, and
$J(\lambda)$, $Y(\ln \lambda)$ are continuous retardation spectral functions.

Stress relaxation and creep experiments at small strains and stresses are used in conjunction with Eqs. 6.4-5 and 6.4-6 or 6.4-7 and 6.4-8 to determine either the discrete or continuous spectra of relaxation times of flexible uncrosslinked polymers. The techniques used are graphical and approximate (26).

The mere fact that the relaxation or creep of linear viscoelastic behavior depends on $e^{-t/\lambda}$, irrespectively whether $\lambda$ is single, multiple, or continuously varying, brings about the realization that time and temperature have equivalent effects. The reason for this equivalence rests on the dependence of the relaxation—or retardation—time on temperature. By using an Arrhenius temperature dependence for the viscosity and a linear temperature dependence, which is much weaker, as suggested by rubber elasticity theory, we obtain

$$\lambda = \frac{\mu}{G} = \frac{Ce^{\Delta E/RT}}{G_0 \rho T} \cong C_0\, e^{\Delta E/RT} \qquad (6.4\text{-}9)$$

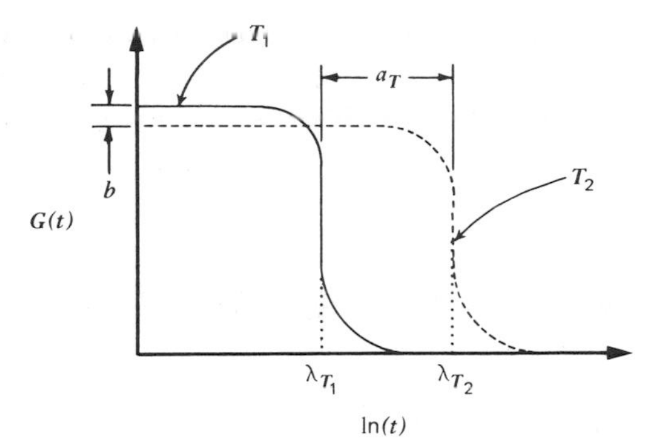


**Fig. 6.8** Illustration of the time-temperature shift for a Maxwell element in a stress relaxation experiment.

Therefore the term $t/\lambda = 1/De$ will have values depending on the absolute temperature; that is, constancy of $De$ can be obtained at long times if the temperature is low, thus $\lambda$ is large, or at short times when the temperature is high. This time-temperature equivalence can be simply stated: the lower the temperature of a flexible polymer, the longer it takes to relax or creep, and vice versa. Figure 6.8 illustrates this graphically for the relaxation of a Maxwell body. Time-temperature equivalence also holds for linear viscoelastic bodies with discrete or continuous spectra of relaxation times, Eqs. 6.4-5 to 6.4-8, if the assumption of a common $\Delta E$ for all $\lambda$ is made.* Experimentally small strain stress relaxation, and small stress creep behavior of flexible polymers at various temperatures can be shifted to obtain "master" curves (27).

The shift factor $a_T$, which is simply the ratio of the relaxation times at two different temperatures (Fig. 6.8), has been found empirically to be expressible for all amorphous materials in the range $T_g < T < T_g + 100°\text{K}$ as

► $$\log a_T = \frac{-17.44(T - T_g)}{51.6 + (T - T_g)} \tag{6.4-10}$$

In this expression, known as the Williams–Landel–Ferry (WLF) equation (28), the reference temperature is the $T_g$ of each material. The two constants are related to the free volume fraction of the amorphous materials at $T_g$. Since the shift factor is the ratio of relaxation times at two temperatures, using Eq. 6.4-9 we obtain the following information on low shear rate viscosity:

$$\log a_T = \log\left(\frac{\lambda}{\lambda_{T_g}}\right) \cong \log\left(\frac{\mu}{\mu_{T_g}}\right) \tag{6.4-11}$$

The WLF equation is useful in predicting viscosity values only in the range $T_g < T < T_g + 100°\text{C}$. Most polymers are processed at temperatures above this range. A

* The assumption that the activation energy is the same for all viscoelastic motions rests on experimental evidence, especially on the paraffin homologous series, as discussed in S. Glasstone, K. J. Laidler, and H. Eyring, *Theory of Rate Processes*, McGraw-Hill, New York, 1941, Chapter 9.

notable exception is unplasticized ("rigid") PVC with a $T_g = 87°C$. This material, because of its susceptibility to thermal degradation, is processed below the range of $T_g + 100°C$. In applying the WLF equation, be cautious to use the limiting, zero shear rate, viscosity values and the proper value of $T_g$.

**Example 6.1** ***Predicting the Temperature Dependence of Viscosity with the WLF Equation***

Collins and Metzger [*Polym. Eng. Sci.*, **10**, 57 (1970)] report the following viscosity data for a relatively low molecular weight PVC resin:

| | Viscosity ($N \cdot s/m^2$) | |
|---|---|---|
| Temperature (°C) | At "Zero" Shear Rate | At $10\ s^{-1}$ |
| 190 | $4.2 \times 10^3$ | $3.0 \times 10^3$ |
| 160 | $1.0 \times 10^5$ | $3.5 \times 10^4$ |

How do the ratios of the tabulated viscosities at the two temperatures compare with the prediction of the WLF equation?

Uplasticized PVC has a $T_g = 87°C$, and according to Eq. 6.4-10

$$a_{T=160} = \log\left(\frac{\eta_{160}}{\eta_{87}}\right) = \frac{(-17.44)(73)}{51.6+73} = -10.2$$

$$a_{T=190} = \log\left(\frac{\eta_{190}}{\eta_{87}}\right) = \frac{(-17.44)(103)}{51.6+103} = -11.6$$

Subtracting the second from the first expression above, we obtain the WLF prediction

$$\frac{\eta_{160}}{\eta_{190}} = 25.3$$

The ratios of viscosities from the reported data are at zero shear rate 23.8, whereas at a shear rate of $10\ s^{-1}$ 11.7. Thus the prediction of the WLF equation is good for very low shear rates, but *poor* for processing rates, where large changes of viscosity are predicted.

We now turn to the LVE Eq. 6.3-8 and ask the question: What stresses are necessary to support a viscometric flow $v_1 = \dot{\gamma} x_2$, $v_2 = 0$, $v_3 = 0$. For simplicity, let us assume a single Maxwell element (i.e., $G(t-t') = Ge^{-(t-t')/\lambda}$. For this flow (Table

5.3) the rate of strain tensor $\boldsymbol{\gamma}$ is

$$\dot{\boldsymbol{\gamma}} = \begin{pmatrix} 0 & \dot{\gamma} & 0 \\ \dot{\gamma} & 0 & 0 \\ 0 & 0 & 0 \end{pmatrix}$$

$$\boldsymbol{\tau} = -\int_0^t G e^{-t/\lambda} e^{+t'/\lambda} \begin{pmatrix} 0 & \dot{\gamma} & 0 \\ \dot{\gamma} & 0 & 0 \\ 0 & 0 & 0 \end{pmatrix} dt'^{*} \tag{6.4-12}$$

$$\boldsymbol{\tau} = -G e^{-t/\lambda} \begin{pmatrix} 0 & \dot{\gamma} & 0 \\ \dot{\gamma} & 0 & 0 \\ 0 & 0 & 0 \end{pmatrix} \int_0^t e^{+t'/\lambda} \, dt' \tag{6.4-13}$$

$$\boldsymbol{\tau} = -\mu \begin{pmatrix} 0 & \dot{\gamma} & 0 \\ \dot{\gamma} & 0 & 0 \\ 0 & 0 & 0 \end{pmatrix} e^{-t/\lambda} e^{+t'/\lambda} \Big|_{t'=0}^{t'=t} = -\mu \begin{pmatrix} 0 & \dot{\gamma} & 0 \\ \dot{\gamma} & 0 & 0 \\ 0 & 0 & 0 \end{pmatrix} (1 - e^{-t/\lambda}) \tag{6.4-14}$$

Thus the LVE equation predicts only shear stresses $\tau_{12} = \tau_{21}$ which build up asymptotically to the value $-\mu\dot{\gamma}$. The steady state response of the LVE fluid is Newtonian, since $-\tau_{12}/\dot{\gamma} = \mu$. Newtonian behavior in polymer melts is observed at very small shear rates. Furthermore, as Section 6.3 stated, the LVE equation is limited to very small strains. For large strains and strain rates, nonlinear viscoelastic constitutive equations are used such as the ZFD, the White–Metzner, the G–M, the BKZ, the Lodge elastic liquid, or the Bogue, which were presented in Section 6.3. Only the more complex of these equations predict semiquantitatively the polymer melt behavior; the rest are qualitative in their predictions. The linear viscoelasticity theory, however, is useful for the following reasons:

1. It gives a molecular insight into the viscoelastic nature of polymers as well as qualitative trends for their time-dependent mechanical properties.
2. It explains the observed time-temperature equivalence in the mechanical behavior of polymers.
3. It can be used to interpret the stress relaxation, creep, and dynamic testing response of polymers. Dynamic testing, in which a small sinusoidally varying strain $\gamma(t) = \gamma_0 \sin \omega t$ is applied on a polymeric sample, is a particularly useful tool in separating the elastic and viscous response of materials; additionally, the ratio of the viscous-to-elastic stress is a measure of the energy dissipated per unit time. Example 6.2 illustrates the linear viscoelastic response to dynamic testing, as well as its utility.

**Example 6.2** ***Small Amplitude Oscillatory Motion of a Linear Viscoelastic Body***

Derive the steady state response of a linear viscoelastic body to an externally applied sinusoidal shear strain (dynamic testing), using the constitutive equation 6.3-8

$$\tau(t) = -\int_{-\infty}^{t} G(t-t') \frac{d\gamma}{dt'} \, dt$$

* The lower limit of the integral is not $-\infty$ but zero because the flow commenced at $t' = 0$.

and

$$\frac{d\gamma}{dt'} = \gamma_0 \omega \cos \omega t'$$

Let the linear viscoelastic body be represented by a continuous spectrum of relaxation times, that is,

$$G(t-t') = \int_{-\infty}^{+\infty} H(\ln \lambda)\, e^{-(t-t')/\lambda}\, d \ln \lambda$$

Substituting in the constitutive equation and integrating, we have

$$\tau(t) = -\int_{-\infty}^{t} \left[ \int_{-\infty}^{\infty} H(\ln \lambda)\, e^{-t/\lambda}\, e^{+t'/\lambda}\, d \ln \lambda \right] \gamma_0 \omega \cos \omega t'\, dt'$$

$$= -\omega \gamma_0 \int_{-\infty}^{\infty} H(\ln \lambda)\, e^{-t/\lambda} \left[ \int_{-\infty}^{t} e^{t'/\lambda} \cos \omega t'\, dt' \right] d \ln \lambda$$

$$= -\gamma_0 \int_{-\infty}^{\infty} \frac{H(\ln \lambda)}{1+\omega^2 \lambda^2} [\omega\lambda \cos \omega t + \omega^2 \lambda^2 \sin \omega t]\, d(\ln \lambda)$$

$$= -\gamma_0 \left[ \int_{-\infty}^{\infty} \frac{H(\ln \lambda) \omega^2 \lambda^2}{1+\omega^2 \lambda^2}\, d(\ln \lambda) \right] \sin \omega t - \gamma_0 \left[ \int_{-\infty}^{\infty} \frac{H(\ln \lambda) \omega \lambda}{1+\omega^2 \lambda^2}\, d(\ln \lambda) \right] \cos \omega t$$

Thus according to the result just given, the response of a linear viscoelastic body to a sinusoidal strain (*a*) lags in time the applied strain, and (*b*) is composed of purely elastic and purely viscous parts.

Figure 6.9 illustrates these features. Furthermore it is useful to define the following quantities* associated with dynamic mechanical testing:

► $G'(\omega) = \int_{-\infty}^{+\infty} \frac{H(\ln \lambda) \omega^2 \lambda^2}{1+\omega^2 \lambda^2}\, d(\ln \lambda)$ — the in-phase or elastic dynamic modulus

► $G''(\omega) = \int_{-\infty}^{+\infty} \frac{H(\ln \lambda) \omega \lambda}{1+\omega^2 \lambda^2}\, d(\ln \lambda)$ — the out-of-phase or loss dynamic modulus

► $\tan \delta = \frac{G''}{G'}$ — the loss tangent or dissipation factor; the ratio of the mechanical energy dissipated to that stored per cycle

Dynamic mechanical testing is a useful structural characterization tool for polymers, because the onset of structural motions for such materials at certain frequency or temperature ranges can be detected by changes in $G'$, $G''$, and tan $\delta$.

* Note that $\omega$ and $\lambda$ always appear as a product; this is nothing but a ramification of the time-temperature equivalence of polymer mechanical response.

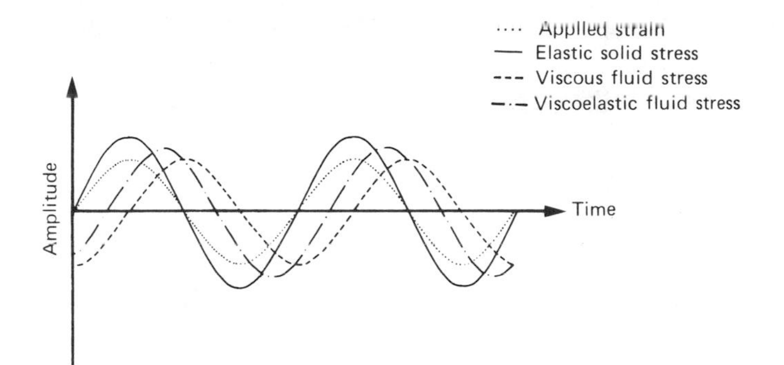


**Fig. 6.9** The schematic stress response of an elastic, a viscous, and a viscoelastic body to a sinusoidally applied strain.

## 6.5 The Generalized Newtonian Fluid (GNF) Equations

GNF is the generic expression for a family of empirical, semiempirical, or molecular origin equations that were developed to account for the non-Newtonian (shear dependent) behavior of liquids. These constitutive equations comprise the different ways of expressing the dependence of the viscosity function $\eta$ on the applied (or resulting) rate of strain. There is only one general prerequisite on $\eta$. Since the viscosity is a scalar, it must be a function of only the three (scalar) invariants of $\dot{\boldsymbol{\gamma}}$.

For an incompressible fluid $\mathrm{I}_{\dot{\gamma}} = 2(\nabla \cdot \boldsymbol{v}) = \sum_i \dot{\gamma}_{ii} = 0$. For shearing flows $\mathrm{III}_{\dot{\gamma}} = \det \dot{\boldsymbol{\gamma}} = \sum_i \sum_j \sum_k \dot{\gamma}_{ij}\dot{\gamma}_{jk}\dot{\gamma}_{ki} = 0$, and even for close to shearing flows the dependence of $\eta$ on $\mathrm{III}_{\dot{\gamma}}$ can be neglected. The non-Newtonian viscosity $\eta$ will, therefore, be taken to be dependent only on the second invariant $\mathrm{II}_{\dot{\gamma}} = (\dot{\boldsymbol{\gamma}} : \dot{\boldsymbol{\gamma}}) = \sum_i \sum_j \dot{\gamma}_{ij}\dot{\gamma}_{ji}$. Actually, instead of $\mathrm{II}_{\dot{\gamma}}$ the *magnitude* of $\dot{\boldsymbol{\gamma}}$ defined as

$$\blacktriangleright \qquad \dot{\gamma} = \sqrt{\tfrac{1}{2}\mathrm{II}_{\dot{\gamma}}} \tag{6.5-1}$$

is preferred. In simple shear flows $v_1 = f(x_2)$, $v_2 = 0$, $v_3 = 0$, $\mathrm{II}_{\dot{\gamma}} = 2\dot{\gamma}_{21}^2$ and the magnitude of $\dot{\boldsymbol{\gamma}}$ is simply the shear rate ($\dot{\gamma} = |\dot{\gamma}_{21}|$).

With these preliminaries we list and discuss a number of commonly used empirical expressions $\eta = \eta(\dot{\gamma})$ and $\eta = \eta(\tau)$ where $\tau$ is the magnitude of $\boldsymbol{\tau}$.

### *The Power Law Model*

The power law model, proposed by Ostwald and de Waele (29), is empirical, and its origin can be found in the logarithmic plot of $\eta(\dot{\gamma})$(Fig. 6.10). The straight line in the shear rate region $10 < \dot{\gamma} < 10^3\ s^{-1}$ in the figure suggests the functional form

$$\blacktriangleright \qquad \eta(\dot{\gamma}) = m\dot{\gamma}^{n-1} \tag{6.5-2}$$

where $m\,(N \cdot s^n/m^2)$ and the dimensionless $n$ are parameters, commonly called the consistency and power law index, respectively. Thus a power law constitutive

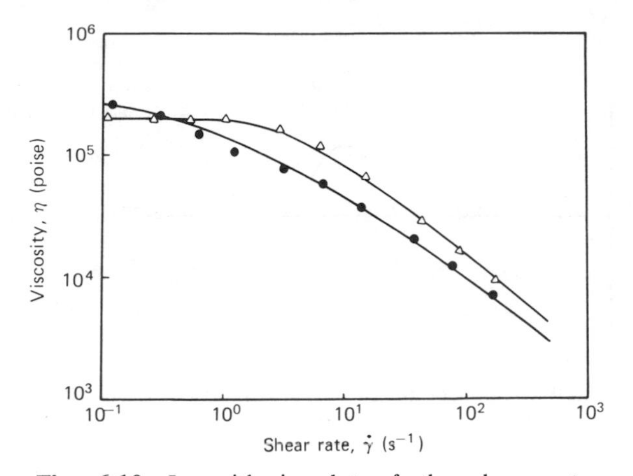


**Fig. 6.10** Logarithmic plot of the shear rate dependent viscosity of a narrow molecular weight distribution PS(Δ) at 180°C, showing the "Newtonian plateau" and the "power law" regions and a broad distribution PS (●). [Reprinted with permission from W. W. Graessley et al., *Trans. Soc. Rheol.*, **14**, 519 (1970).]

equation can be arrived at:

$$\blacktriangleright \qquad \boldsymbol{\tau} = -m\dot{\gamma}^{n-1}\dot{\boldsymbol{\gamma}} = -m\left[\sqrt{\tfrac{1}{2}(\dot{\boldsymbol{\gamma}}:\dot{\boldsymbol{\gamma}})}\right]^{n-1}\dot{\boldsymbol{\gamma}} \qquad (6.5\text{-}3)$$

The parameter $m$ is a sensitive function of temperature, obeying an Arrhenius-type relationship, see Eq. 6.1-1

$$\blacktriangleright \qquad m = m_0 \exp\left[\frac{\Delta E}{R}(1/T - 1/T_0)\right] \qquad (6.5\text{-}4)$$

where $m_0$ is the value of $m$ at $T_0$, and $\Delta E$ is the flow activation energy. For mathematical convenience, frequently a simpler relationship is used

$$\blacktriangleright \qquad m = m_0\, e^{-a(T-T_o)} \qquad (6.5\text{-}5)$$

where $a$ is an empirical parameter. Equation 6.5-5 holds well over relatively narrow temperature ranges.

The dependence of $m$ on hydrostatic pressure is exponential as stated in Eq. 6.1-1. Goldblatt and Porter (30) have reviewed the subject recently.

It is important to note once again that in the relationship above, as well as with other non-Newtonian constitutive equations, the viscosity depends on the magnitude of the entire tensor $\dot{\boldsymbol{\gamma}}$, through $(\dot{\boldsymbol{\gamma}}:\dot{\boldsymbol{\gamma}})$. This means that the viscosity is a function of *all* the velocity gradients, not just the one that we may be interested in a given momentum balance.

Strictly speaking, Eq. 6.5-3 is an adequate empirical constitutive equation for predicting steady viscometric flow behavior. It predicts neither normal stress differences nor viscoelastic response such as stress relaxation. Furthermore, the flow index $n$ is not constant, as Fig. 6.10 indicates; it approaches unity for $\dot{\gamma} \to 0$ ("Newtonian plateau").

The following comments can be made about the power law equation and the viscosity or "flow" curve in Fig. 6.10:

1. The upper limit of the Newtonian plateau is dependent on $\bar{M}_w$ and the melt temperature. It is roughly in the region $\dot{\gamma} = 10^{-2}\ s^{-1}$. Nylon and PET are important exceptions. They persist as Newtonian fluids at high shear rates.
2. This upper limit decreases with increasing $\bar{M}_w$, MWD at constant $\bar{M}_w$, and decreasing melt temperature. It is believed to terminate roughly where the Deborah number reaches unity.
3. If the power law equation is used in pressure flows, where $0 \le \dot{\gamma} \le \dot{\gamma}_{max}$, an error is introduced in the very low shear rate region by not assuming that $n = 1$ (31) (see Problem 6.6).
4. The transition from the Newtonian plateau to the power law region is sharp for monodispersed polymer melts and broad for polydispersed melts.
5. The slope of the viscosity curve in the power law region is not exactly constant. The flow index $n$ decreases with increasing shear rate. Thus the power law equation holds exactly only for limited ranges of shear rate, for a given value of $n$.

Despite its limitations, the power law model is one of the most widely used empirical relations in polymer fluid dynamics, and it gives surprisingly good results even for nonviscometric and slightly transient flows.

### The Ellis Model

This equation expresses the viscosity in terms of the magnitude of the shear stress (33)

$$\frac{\eta_0}{\eta(\tau)} = 1 + \left(\frac{\tau}{\tau_{1/2}}\right)^{\alpha-1} \tag{6.5-6}$$

The resulting constitutive equation is

$$\boldsymbol{\tau} = \eta(\tau)\dot{\boldsymbol{\gamma}} \tag{6.5-7}$$

where $\eta(\tau)$ is defined by Eq. 6.5-6 and $\tau = \sqrt{\frac{1}{2}\mathrm{II}_\tau}$ is the magnitude of the stress tensor $\boldsymbol{\tau}$, which, in simple shear flow is the shear stress ($\tau = |\tau_{21}|$).

The $\eta_0$ in Eq. 6.5-6 is the zero shear viscosity (limit of $\eta$ as $\dot{\gamma} \to 0$), $\tau_{1/2}$ is the shear stress value, where $\eta = \eta_0/2$, and $\alpha - 1$ is the slope of the curve $\log(\eta_0/\eta - 1)$ versus $\log(\tau/\tau_{1/2})$. This constitutive equation is not as easy to use as the power law but is superior to it, since it predicts a Newtonian plateau at very low shear rates.

### The Carreau Model

This equation, also quite successful in its non-Newtonian viscosity predictions and easy to use, was proposed by Carreau (34)

$$\frac{\eta(\dot{\gamma}) - \eta_\infty}{\eta_0 - \eta_\infty} = [1 + (\lambda\dot{\gamma})^2]^{(n-1)/2} \tag{6.5-8}$$

Again the constitutive equation is that of the GNF (Eq. 6.3-6), with $\eta(\dot{\gamma})$ given by Eq. 6.5-8; $\eta_0$ is the zero shear rate viscosity, and $\eta_\infty$ is the infinite-shear-rate viscosity that can be usually taken to be either the solvent viscosity (for polymer solutions) or zero for polymer melts.

### The Bingham Fluid

This empirical constitutive equation describes semiquantitatively the rheological response of lattices such as the polymer emulsion systems used in "no drip" paints, and of pastes and slurries (36) that are common rheological materials in the food processing industry (e.g., ketchup).

The Bingham "fluid" is a deforming solid below a characteristic yield stress $\tau_y$ and a flowing fluid above it

$$\begin{aligned} \eta &= \infty \qquad \tau \leq \tau_y \\ \eta(\dot{\gamma}) &= \mu_0 + \frac{\tau_y}{\dot{\gamma}} \qquad \tau \geq \tau_y \end{aligned} \tag{6.5-9}$$

Appendix A gives the constants of the power law, Ellis, and Carreau models for a number of commercial polymers. The constants were evaluated by fitting the experimental $\log \eta$ versus $\log \dot{\gamma}$ curves obtained with the Instron Capillary Viscometer using capillaries of $L/D = 40.0$ and 8. Low shear rate data were obtained with dynamic experiments by making use of the Cox–Merz rule (35) as discussed in the appendix.

**Example 6.3** ***Flow of a Power Law Model Fluid in Tubes***

Consider a steady, isothermal, laminar, fully developed pressure flow of an incompressible power law model fluid in a horizontal tube. Derive expressions for ($a$) velocity profile and ($b$) flow rate.

*Solution*: ($a$) For tubular flow we use a cylindrical coordinate system. Since flow is isothermal and the fluid is incompressible, the equations of motion and continuity together with the constitutive equation fully describe the flow. On the basis of symmetry, we assume that there is no $\theta$ dependence and that $v_\theta = 0$. Fully developed flow implies that $\partial v_z/\partial z = 0$. The equation of continuity reduces to

$$\frac{\partial}{\partial r}(r v_r) = 0 \tag{6.5-10}$$

which can be integrated to give $r v_r = C$, where $C$ is a constant. Since $v_r = 0$ at the tube radius, we conclude that $C = 0$ and therefore $v_r = 0$. Hence the only nonvanishing velocity component is $v_z$, which is a function of only $r$. Turning to the equation of motion in Table 5.2, the three components of the equation reduce to

$$\frac{\partial P}{\partial r} = 0$$

$$\frac{\partial P}{\partial \theta} = 0 \tag{6.5-11}$$

$$\frac{\partial P}{\partial z} = -\frac{1}{r}\frac{\partial}{\partial r} r(\tau_{rz})$$

Clearly $P$ is a function of only $z$, whereas the right-hand side of the last equation is a function of only $r$; therefore partial differentials can be replaced by ordinary differentials and, following integration, we get

$$\tau_{rz} = -\left(\frac{r}{2}\right)\frac{dP}{dz} + C_1 \tag{6.5-12}$$

where $C_1$ is an integration constant.

The only nonvanishing velocity gradient in this flow is $dv_z/dr$. The rate of strain tensor $\dot{\boldsymbol{\gamma}}$ from Table 5.3 reduces to

$$\dot{\boldsymbol{\gamma}} = \begin{pmatrix} 0 & 0 & \dfrac{dv_z}{dr} \\ 0 & 0 & 0 \\ \dfrac{dv_z}{dr} & 0 & 0 \end{pmatrix} \tag{6.5-13}$$

and the constitutive equation reduces to

$$\tau_{rz} = -m\dot{\gamma}^{n-1}\dot{\gamma}_{rz} \tag{6.5-14}$$

but

$$\dot{\gamma} = \sqrt{\tfrac{1}{2}(\dot{\boldsymbol{\gamma}}:\dot{\boldsymbol{\gamma}})} = \sqrt{\left(\frac{dv_z}{dr}\right)^2} = \left|\frac{dv_z}{dr}\right| \tag{6.5-15}$$

which on substitution in the constitutive equation gives

$$\tau_{rz} = -m\left[\left(\frac{dv_z}{dr}\right)^2\right]^{(n-1)/2}\frac{dv_z}{dr} = -m\left|\frac{dv_z}{dr}\right|^{n-1}\frac{dv_z}{dr} \tag{6.5-16}$$

Note that $\dot{\gamma}$, the magnitude of $\dot{\boldsymbol{\gamma}}$, is always positive, therefore, we maintain an absolute value sign over the term that reflects the shear rate dependence of the viscosity. Equation 6.5-16 indicates the $\tau_{rz} = 0$ at $r = 0$, where $dv_z/dr = 0$. Hence the constant $C_1$ in 6.5-12 is zero. Combining Eqs. 6.5-16 and 6.5-12 gives

$$m\left|\frac{dv_z}{dr}\right|^{n-1}\frac{dv_z}{dr} = \frac{r}{2}\left(\frac{dP}{dz}\right) \tag{6.5-17}$$

In tubular flow* for all $r$, $dv_z/dr < 0$; therefore

$$\left|\frac{dv_z}{dr}\right| = -\frac{dv_z}{dr}$$

and Eq. 6.5-17 can be written as

$$\left(-\frac{dv_z}{dr}\right) = \left(-\frac{r}{2m}\frac{dP}{dz}\right)^s \tag{6.5-18}$$

where $s = 1/n$. Note that $dP/dz < 0$, and the term in parentheses on the right-hand side of Eq. 6.5-18 is positive. This equation can be integrated with the boundary condition $v_r(R) = 0$ to give

► $$v_z(r) = \left(\frac{R}{s+1}\right)\left[-\frac{R}{2m}\frac{dP}{dz}\right]^s\left[1 - \left(\frac{r}{R}\right)^{s+1}\right] \tag{6.5-19}$$

(*b*) The volumetric flow rate is obtained from Eq. 6.5-19

► $$Q = \int_0^R 2\pi r v_z\, dr = \frac{\pi R^3}{s+3}\left(-\frac{R}{2m}\frac{dP}{dz}\right)^s \tag{6.5-20}$$

Since $dP/dz$ is constant, Eq. 6.5-20 can be written as

► $$Q = \frac{\pi R^3}{s+3}\left[-\frac{R}{2m}\frac{\Delta P}{L}\right]^s \tag{6.5-21}$$

where $\Delta P = P_L - P_0$, $P_0$ is the pressure at $z = 0$ and $P_L$ at $z = L$. Equation 6.5-21 for $s = 1$ reduces to the familiar Hagen-Poiseuille "law"

$$Q = \frac{\pi R^4}{8\mu L}(P_0 - P_L) \tag{6.5-22}$$

## 6.6 The Criminale–Ericksen–Filbey (CEF) Equation

Section 6.3 traced the origin and development of the CEF equation, which holds the steady shear flows. The material functions $\eta$, $\Psi_1$, and $\Psi_2$ are functions of $\dot{\gamma}$, the magnitude of $\dot{\boldsymbol{\gamma}}$. Thus

$$\eta = \eta(\dot{\gamma}), \qquad \Psi_1 = \Psi_1(\dot{\gamma}), \qquad \Psi_2 = \Psi_2(\dot{\gamma}) \tag{6.6-1}$$

Since the viscosity functions in the both generalized Newtonian and the CEF equations are expected to be the same, it is assumed that a CEF fluid has the velocity field, in a steady viscometric flow, of a purely viscous fluid. Then in connection with Eq. 6.3-5, rheologists ask: given a viscometric flow field, calculate the stress field (components) necessary to maintain it, if the flowing fluid is incompressible and obeys the CEF equation. The example that follows illustrates both the question and the method of obtaining the answer.

* Note that in more complex flow problems where the velocity gradient changes sign over the flow region, different solutions result for different regions of flow (see Chapters 10 and 13).

**Example 6.4** ***The CEF Equation in Steady, Fully Developed Flow in Tubes***

Calculate the stresses generated during the steady flow in a tube of a fluid whose rheological behavior is described by the CEF equation.

*Solution:* We have the following results available from Example 6.3, since the flow fields of a GNF and a CEF fluid are of the same form

$$\dot{\boldsymbol{\gamma}} = \begin{pmatrix} 0 & 0 & \dot{\gamma}_{rz} \\ 0 & 0 & 0 \\ \dot{\gamma}_{rz} & 0 & 0 \end{pmatrix}$$

To calculate the stresses generated by a CEF fluid, from Eq. 6.3-5, we need to calculate the quantities $\dot{\boldsymbol{\gamma}} \cdot \dot{\boldsymbol{\gamma}}$ and $\mathscr{D}\dot{\boldsymbol{\gamma}}/\mathscr{D}t$ appearing in the constitutive equation. First

$$\{\dot{\boldsymbol{\gamma}} \cdot \dot{\boldsymbol{\gamma}}\} = \begin{pmatrix} \dot{\gamma}_{rz}^2 & 0 & 0 \\ 0 & 0 & 0 \\ 0 & 0 & \dot{\gamma}_{rz}^2 \end{pmatrix}$$

Next we calculate

$$\frac{\mathscr{D}}{\mathscr{D}t}\dot{\boldsymbol{\gamma}} = \frac{\partial}{\partial t}\dot{\boldsymbol{\gamma}} + \{\boldsymbol{v} \cdot \boldsymbol{\nabla}\dot{\boldsymbol{\gamma}}\} + \frac{1}{2}(\{\boldsymbol{\omega} \cdot \dot{\boldsymbol{\gamma}}\} - \{\dot{\boldsymbol{\gamma}} \cdot \boldsymbol{\omega}\})$$

$\partial\dot{\boldsymbol{\gamma}}/\partial t = 0$ because the flow is steady. For the components of the tensor $\boldsymbol{v} \cdot \boldsymbol{\nabla}\dot{\boldsymbol{\gamma}}$ we make use of Table 6.2. We will work out just one component.

$$(\boldsymbol{v} \cdot \boldsymbol{\nabla}\dot{\boldsymbol{\gamma}})_{rz} = (\boldsymbol{v} \cdot \boldsymbol{\nabla})\dot{\gamma}_{rz} - \frac{v_\theta}{r}\dot{\gamma}_{\theta z}$$

$$= \left(v_r \frac{\partial}{\partial r} + \frac{v_\theta}{r}\frac{\partial}{\partial \theta} + v_z \frac{\partial}{\partial z}\right)\dot{\gamma}_{rz} - \frac{v_\theta}{r}\dot{\gamma}_{\theta z}$$

Since $v_r = 0$, $v_\theta = 0$, and $\partial v_z/\partial z = 0$ for a developed capillary flow, the term $(\boldsymbol{v} \cdot \boldsymbol{\nabla}\dot{\boldsymbol{\gamma}})_{rz} = 0$. Similarly we evaluate all other components and conclude that $(\boldsymbol{v} \cdot \boldsymbol{\nabla}\dot{\boldsymbol{\gamma}}) = 0$. The vorticity tensor $\boldsymbol{\omega}$ can be obtained for this flow from Table 6.3.

$$\boldsymbol{\omega} = \boldsymbol{\nabla}\boldsymbol{v} - (\boldsymbol{\nabla}\boldsymbol{v})^\dagger = \begin{pmatrix} 0 & 0 & \dot{\gamma}_{rz} \\ 0 & 0 & 0 \\ -\dot{\gamma}_{rz} & 0 & 0 \end{pmatrix}$$

$$\{\boldsymbol{\omega} \cdot \dot{\boldsymbol{\gamma}}\} = \begin{pmatrix} 0 & 0 & \dot{\gamma}_{rz} \\ 0 & 0 & 0 \\ -\dot{\gamma}_{rz} & 0 & 0 \end{pmatrix}\begin{pmatrix} 0 & 0 & \dot{\gamma}_{rz} \\ 0 & 0 & 0 \\ \dot{\gamma}_{rz} & 0 & 0 \end{pmatrix} = \begin{pmatrix} \dot{\gamma}_{rz}^2 & 0 & 0 \\ 0 & 0 & 0 \\ 0 & 0 & -\dot{\gamma}_{rz}^2 \end{pmatrix}$$

Similarly

$$\{\dot{\boldsymbol{\gamma}} \cdot \boldsymbol{\omega}\} = \begin{pmatrix} -\dot{\gamma}_{rz}^2 & 0 & 0 \\ 0 & 0 & 0 \\ 0 & 0 & \dot{\gamma}_{rz}^2 \end{pmatrix}$$

Thus we get

$$\frac{1}{2}(\{\boldsymbol{\omega}\cdot\dot{\boldsymbol{\gamma}}\}-\{\dot{\boldsymbol{\gamma}}\cdot\boldsymbol{\omega}\})=\begin{pmatrix}\dot{\gamma}_{rz}^2 & 0 & 0\\ 0 & 0 & 0\\ 0 & 0 & -\dot{\gamma}_{rz}^2\end{pmatrix}$$

The stresses generated by a CEF fluid in a fully developed capillary flow are according to Eq. 6.3-5

$$\begin{pmatrix}\tau_{rr} & \tau_{r\theta} & \tau_{rz}\\ \tau_{\theta r} & \tau_{rr} & \tau_{\theta z}\\ \tau_{zr} & \tau_{z\theta} & \tau_{zz}\end{pmatrix}=-\eta(\dot{\gamma})\begin{pmatrix}0 & 0 & \dot{\gamma}_{rz}\\ 0 & 0 & 0\\ \dot{\gamma}_{rz} & 0 & 0\end{pmatrix}-[\tfrac{1}{2}\Psi_1(\dot{\gamma})+\Psi_2(\dot{\gamma})]\begin{pmatrix}\dot{\gamma}_{rz}^2 & 0 & 0\\ 0 & 0 & 0\\ 0 & 0 & \dot{\gamma}_{rz}^2\end{pmatrix}$$

$$+\tfrac{1}{2}\Psi_1(\dot{\gamma})\begin{pmatrix}\dot{\gamma}_{rz}^2 & 0 & 0\\ 0 & 0 & 0\\ 0 & 0 & -\dot{\gamma}_{rz}^2\end{pmatrix}$$

or

$$\tau_{rz}=\tau_{zr}=-\eta\dot{\gamma}_{rz}$$

$$\tau_{rr}=-(\tfrac{1}{2}\Psi_1+\Psi_2)\dot{\gamma}_{rz}^2+\tfrac{1}{2}\Psi_1\dot{\gamma}_{rz}^2=-\Psi_2\dot{\gamma}_{rz}^2$$

$$\tau_{\theta\theta}=0$$

$$\tau_{zz}=-(\tfrac{1}{2}\Psi_1+\Psi_2)\dot{\gamma}_{rz}^2-\tfrac{1}{2}\Psi_1\dot{\gamma}_{rz}^2=-(\Psi_1+\Psi_2)\dot{\gamma}_{rz}^2$$

All other $\tau_{ij}$ terms for $i\neq j$ are zero.

Recognizing that the flow is viscometric, let us adopt the notation convention mentioned in the preceding section. For capillary flow $z$ is "1," $r$ is "2," and $\theta$ is "3." Thus the shear stress $\tau_{12}$ and first and second normal stress differences $\tau_{11}-\tau_{22}$ and $\tau_{22}-\tau_{33}$ are

► $$\tau_{12}=\tau_{21}=-\eta\dot{\gamma}_{21} \tag{6.6-2}$$

► $$\tau_{11}-\tau_{22}=\tau_{zz}-\tau_{rr}=-\Psi_1\dot{\gamma}_{21}^2 \tag{6.6-3}$$

► $$\tau_{22}-\tau_{33}=\tau_{rr}-\tau_{\theta\theta}=-\Psi_2\dot{\gamma}_{21}^2 \tag{6.6-4}$$

Thus the three material functions of the CEF equation are identified as follows: $\eta(\dot{\gamma})$ is the viscosity function, $\Psi_1(\dot{\gamma})$ is the first normal stress difference coefficient, and $\Psi_2(\dot{\gamma})$ is the second normal stress difference coefficient.

The same relations (6.6-2 to 6.6-4) are obtained for the flow of a CEF fluid in *all viscometric flows*, if coordinate "1" is taken to be in the direction of flow, coordinate "2" in the direction of changing velocity, and coordinate "3" is the "neutral" coordinate. Experimentally the viscosity function is the same in all viscometric flows.

**Table 6.2 The Components of $(\boldsymbol{v} \cdot \nabla \dot{\gamma})$ in Three Coordinate Systems**[a]

**Rectangular Coordinates**[b] $(x, y, z)$

$$(\boldsymbol{v} \cdot \nabla \dot{\gamma})_{xx} = (\boldsymbol{v} \cdot \nabla)\dot{\gamma}_{xx}$$

$$(\boldsymbol{v} \cdot \nabla \dot{\gamma})_{yy} = (\boldsymbol{v} \cdot \nabla)\dot{\gamma}_{yy}$$

$$(\boldsymbol{v} \cdot \nabla \dot{\gamma})_{zz} = (\boldsymbol{v} \cdot \nabla)\dot{\gamma}_{zz}$$

$$(\boldsymbol{v} \cdot \nabla \dot{\gamma})_{xy} = (\boldsymbol{v} \cdot \nabla \dot{\gamma})_{yx} = (\boldsymbol{v} \cdot \nabla)\dot{\gamma}_{xy}$$

$$(\boldsymbol{v} \cdot \nabla \dot{\gamma})_{yz} = (\boldsymbol{v} \cdot \nabla \dot{\gamma})_{zy} = (\boldsymbol{v} \cdot \nabla)\dot{\gamma}_{zy}$$

$$(\boldsymbol{v} \cdot \nabla \dot{\gamma})_{zx} = (\boldsymbol{v} \cdot \nabla \dot{\gamma})_{xz} = (\boldsymbol{v} \cdot \nabla)\dot{\gamma}_{xz}$$

**Cylindrical Coordinates**[c] $(r, \theta, z)$

$$(\boldsymbol{v} \cdot \nabla \dot{\gamma})_{rr} = (\boldsymbol{v} \cdot \nabla)\dot{\gamma}_{rr} - \frac{v_\theta}{r}(\dot{\gamma}_{r\theta} + \dot{\gamma}_{\theta r})$$

$$(\boldsymbol{v} \cdot \nabla \dot{\gamma})_{\theta\theta} = (\boldsymbol{v} \cdot \nabla)\dot{\gamma}_{\theta\theta} + \frac{v_\theta}{r}(\dot{\gamma}_{r\theta} + \dot{\gamma}_{\theta r})$$

$$(\boldsymbol{v} \cdot \nabla \dot{\gamma})_{zz} = (\boldsymbol{v} \cdot \nabla)\dot{\gamma}_{zz}$$

$$(\boldsymbol{v} \cdot \nabla \dot{\gamma})_{r\theta} = (\boldsymbol{v} \cdot \nabla \dot{\gamma})_{\theta r} = (\boldsymbol{v} \cdot \nabla)\dot{\gamma}_{\theta r} + \frac{v_\theta}{r}(\dot{\gamma}_{rr} - \dot{\gamma}_{\theta\theta})$$

$$(\boldsymbol{v} \cdot \nabla \dot{\gamma})_{\theta z} = (\boldsymbol{v} \cdot \nabla \dot{\gamma})_{z\theta} = (\boldsymbol{v} \cdot \nabla)\dot{\gamma}_{\theta z} + \frac{v_\theta}{r}\dot{\gamma}_{rz}$$

$$(\boldsymbol{v} \cdot \nabla \dot{\gamma})_{rz} = (\boldsymbol{v} \cdot \nabla \dot{\gamma})_{zr} = (\boldsymbol{v} \cdot \nabla)\dot{\gamma}_{rz} - \frac{v_\theta}{r}\dot{\gamma}_{\theta z}$$

**Spherical Coordinates**[d] $(r, \theta, \phi)$

$$(\boldsymbol{v} \cdot \nabla \dot{\gamma})_{rr} = (\boldsymbol{v} \cdot \nabla)\dot{\gamma}_{rr} - \frac{2v_\theta}{r}\dot{\gamma}_{r\theta} - \frac{2v_\phi}{r}\dot{\gamma}_{r\phi}$$

$$(\boldsymbol{v} \cdot \nabla \dot{\gamma})_{\theta\theta} = (\boldsymbol{v} \cdot \nabla)\dot{\gamma}_{\theta\theta} + \frac{2v_\theta}{r}\dot{\gamma}_{r\theta} - \frac{2v_\phi}{r}\dot{\gamma}_{\theta\phi}\cot\theta$$

$$(\boldsymbol{v} \cdot \nabla \dot{\gamma})_{\phi\phi} = (\boldsymbol{v} \cdot \nabla)\dot{\gamma}_{\phi\phi} + \frac{2v_\phi}{r}\dot{\gamma}_{r\phi} + \frac{2v_\phi}{r}\dot{\gamma}_{\theta\phi}\cot\theta$$

$$(\boldsymbol{v} \cdot \nabla \dot{\gamma})_{r\theta} = (\boldsymbol{v} \cdot \nabla \dot{\gamma})_{\theta r} = (\boldsymbol{v} \cdot \nabla)\dot{\gamma}_{r\theta} + \frac{v_\theta}{r}(\dot{\gamma}_{rr} - \dot{\gamma}_{\theta\theta}) - \frac{v_\phi}{r}(\dot{\gamma}_{\phi\theta} + \dot{\gamma}_{r\phi}\cot\theta)$$

$$(\boldsymbol{v} \cdot \nabla \dot{\gamma})_{r\phi} = (\boldsymbol{v} \cdot \nabla \dot{\gamma})_{\phi r} = (\boldsymbol{v} \cdot \nabla)\dot{\gamma}_{r\phi} - \frac{v_\theta}{r}\dot{\gamma}_{\theta\phi} + \frac{v_\phi}{r}[(\dot{\gamma}_{rr} - \dot{\gamma}_{\phi\phi}) + \dot{\gamma}_{r\theta}\cot\theta]$$

$$(\boldsymbol{v} \cdot \nabla \dot{\gamma})_{\theta\phi} = (\boldsymbol{v} \cdot \nabla \dot{\gamma})_{\phi\theta} = (\boldsymbol{v} \cdot \nabla)\dot{\gamma}_{\theta\phi} + \frac{v_\theta}{r}\dot{\gamma}_{r\phi} + \frac{v_\phi}{r}[\dot{\gamma}_{\theta r} + (\dot{\gamma}_{\theta\theta} - \dot{\gamma}_{\phi\phi})\cot\theta]$$

[a] Reprinted with permission from R. B. Bird, R. C. Armstrong, and O. Hassager, *Dynamics of Polymeric Liquids*, Vol. I, *Fluid Mechanics*, Wiley, New York, 1977.

[b] $(\boldsymbol{v} \cdot \nabla) = v_x \frac{\partial}{\partial x} + v_y \frac{\partial}{\partial y} + v_z \frac{\partial}{\partial z}$.

[c] $(\boldsymbol{v} \cdot \nabla) = v_r \frac{\partial}{\partial r} + \frac{v_\theta}{r}\frac{\partial}{\partial \theta} + v_z \frac{\partial}{\partial z}$.

[d] $(\boldsymbol{v} \cdot \nabla) = v_r \frac{\partial}{\partial r} + \frac{v_\theta}{r}\frac{\partial}{\partial \theta} + \frac{v_\phi}{r\sin\theta}\frac{\partial}{\partial \phi}$

**Table 6.3** *Components of the Vorticity Tensor*[a] $\boldsymbol{\omega}$ *in Three Coordinate Systems*[b]

**Rectangular Coordinates** $(x, y, z)$

$$\omega_{xy} = -\omega_{yx} = \frac{\partial v_y}{\partial x} - \frac{\partial v_x}{\partial y}$$

$$\omega_{yz} = -\omega_{zy} = \frac{\partial v_z}{\partial y} - \frac{\partial v_y}{\partial z}$$

$$\omega_{zx} = -\omega_{xz} = \frac{\partial v_x}{\partial z} - \frac{\partial v_z}{\partial x}$$

**Cylindrical Coordinates** $(r, \theta, z)$

$$\omega_{r\theta} = -\omega_{\theta r} = \frac{1}{r}\frac{\partial}{\partial r}(rv_\theta) - \frac{1}{r}\frac{\partial v_r}{\partial \theta}$$

$$\omega_{\theta z} = -\omega_{z\theta} = \frac{1}{r}\frac{\partial v_z}{\partial \theta} - \frac{\partial v_\theta}{\partial z}$$

$$\omega_{zr} = -\omega_{rz} = \frac{\partial v_r}{\partial z} - \frac{\partial v_z}{\partial r}$$

**Spherical Coordinates** $(r, \theta, \phi)$

$$\omega_{r\theta} = -\omega_{\theta r} = \left(\frac{1}{r}\right)\frac{\partial}{\partial r}(rv_\theta) - \frac{1}{r}\frac{\partial v_r}{\partial \theta}$$

$$\omega_{\theta\phi} = -\omega_{\phi\theta} = \frac{1}{r \sin\theta}\frac{\partial}{\partial \theta}(v_\phi \sin\theta) - \frac{1}{r\sin\theta}\frac{\partial v_\theta}{\partial \phi}$$

$$\omega_{\phi r} = -\omega_{r\phi} = \frac{1}{r\sin\theta}\frac{\partial v_r}{\partial \phi} - \frac{1}{r}\frac{\partial}{\partial r}(rv_\phi)$$

[a] All diagonal components are zero.
[b] Reprinted with permission from R. B. Bird, R. C. Armstrong, and O. Hassager, *Dynamics of Polymeric Fluids*, Vol. I, *Fluid Dynamics*, Wiley, New York, 1977.

## 6.7 Experimental Determination of the Viscosity and Normal Stress Difference Coefficients

This section deals with two common experimental methods for evaluating $\eta$, $\Psi_1$, and $\Psi_2$, as a function of shear rate. The experiments involved are steady capillary flow and steady cone and plate flow. In the first, only the viscosity function can be determined generally for the shear rate region $\dot{\gamma} > 1\ \text{s}^{-1}$. In the second all three viscometric functions can be determined, but only for low shear rates. Both flows are viscometric. Viscosity measurements are much better developed and understood than normal stress difference measurements. It must be emphasized that the results obtained by capillary or cone and plate flow rheometry are *independent of any constitutive equation.* We conduct experiments to *test* the validity of constitutive equations; once one is found valid, the obtained viscometric functions are used in conjunction with it.

### *Capillary Flow Viscometry*

The experimental setup used in capillary viscometry is shown schematically in Fig. 6.1. Care is taken to have a uniform temperature and to eliminate the frictional effects in the reservoir. Either constant pressure or constant flow rate experiments are conducted, depending on the available instrument. At very slow flow rates ($\dot{\gamma} < 1\ s^{-1}$) the surface tension of the emerging extrudate, gravity, and the frictional forces between the piston and the reservoir cannot be neglected; thus the viscosity values obtained are usually high. A capillary viscometer yields viscosity data up to shear rates, where the phenomenon of melt fracture occurs (see Section 13.2). At high shear rates the danger of having a high level of viscous dissipation of energy, thus nonisothermal flow, is very real, as Section 13.1 points out.

The starting point of our analysis is the $z$-component momentum equation arrived at in Example 6.3

$$\frac{dP}{dz} = -\frac{1}{r}\frac{d}{dr}(r\tau_{rz}) \tag{6.7-1}$$

which is valid for all incompressible fluids and is subject to the assumptions of steady and isothermal flow.* Integrating Eq. 6.7-1 we obtain

$$\tau_{rz} = \tau_w\left(\frac{r}{R}\right) \tag{6.7-2}$$

where $\tau_w$ is the shear stress at the "wall" ($r = R$)

$$\tau_w = \left(\frac{P_0 - P_L}{2L}\right)R \tag{6.7-3}$$

$\tau_w$ can be experimentally evaluated by measuring $R$, $L$, and $(P_0 - P_L)$.†

By assuming only that the polymer melt is viscous and time independent, without specifying the viscosity function, we can apply the GNF equation for capillary flow at the wall

$$\tau_w = -\eta\dot{\gamma}_{rz}|_{r=R} \equiv \eta\dot{\gamma}_w \tag{6.7-4}$$

where $\dot{\gamma}_w$ is the shear rate at the wall.

Equation 6.7-4 suggests that if, in some way, $\dot{\gamma}_w$ could be evaluated experimentally, the viscosity function could be determined. This is indeed possible through the measurement of the volumetric flow rate $Q$, which can be expressed independently of any constitutive equation as follows

$$Q = 2\pi\int_0^R rv_z(r)\,dr = 2\pi\left[\left(\frac{r^2 v_z(r)}{2}\right)\Bigg|_0^R - \int_0^R \frac{r^2}{2}\,dv_z\right] \tag{6.7-5}$$

Assuming *no slip* at the wall of the capillary, we note that the first term on the right hand side of Eq. 6.7-5 is zero and it becomes

$$Q = -\pi\int_0^R r^2\left(\frac{dv_z}{dr}\right)dr \tag{6.7-6}$$

* In Example 6.3, as well as here, the entrance and exit effects, which give rise to high axial pressure gradients near the capillary entrance and nonzero gauge pressures at the exit, are neglected. Section 13.2 discusses them, as well as methods used to account for them.

† If "exit" effects are neglected, $P_L = 0$.

From Eq. 6.7-2, $r = \tau_{rz} R/\tau_w$, a relationship that can be utilized to change the integration variable in Eq. 6.7-6,

$$Q = \frac{-\pi R^3}{\tau_w^3} \int_0^{\tau_w} \left(\frac{dv_z}{dr}\right) \tau_{rz}^2 \, d\tau_{rz} \tag{6.7-7}$$

Next, Eq. 6.7-7 is differentiated (37) with respect to $\tau_w$ to give*

$$\frac{1}{\pi R^3}\left[\tau_w^3 \frac{dQ}{d\tau_w} + 3\tau_w^2 Q\right] = -\left(\frac{dv_z}{dr}\right)_{r=R} \tau_w^2 \tag{6.7-8}$$

Equation 6.7-8 can be written as

► $$-\left(\frac{dv_z}{dr}\right)_{r=R} = \dot{\gamma}_w = \frac{3\Gamma}{4} + \frac{\tau_w}{4}\frac{d\Gamma}{d\tau_w} \tag{6.7-9}$$

where $\Gamma$ is the Newtonian shear rate at the wall

$$\Gamma = \frac{4Q}{\pi R^3} \tag{6.7-10}$$

Finally, with Eq. 6.7-3, Eq. 6.7-8 becomes

► $$\dot{\gamma}_w = \frac{1}{\pi R^3}\left[3Q + \Delta P \frac{dQ}{d(\Delta P)}\right] \tag{6.7-11}$$

Either Eq. 6.7-9 or 6.7-11, known as the "Rabinowitsch equations" or "Weissenberg-Rabinowitsch equation" (38), can be used to determine the shear rate at the wall $\dot{\gamma}_w$ by measuring $Q$ and $\Delta P$ or $\tau_w$ and $\Gamma$. Thus in Eq. 6.7-4 both $\tau_w$ and $\dot{\gamma}_w$ can be experimentally measured for *any* fluid having a shear rate dependent viscosity, as long as it does not slip at the capillary wall. Therefore the viscosity function can be obtained.

Experimentally it is found that for polymer melts $\dot{\gamma}_w \geq \Gamma$, with the unequality becoming stronger at higher shear rates. This is consistent with the pluglike velocity profiles predicted by the power law model. Second, the equation $Q = [\pi R^3/(s+3)](-R\Delta P/2mL)^s$ obtained in Example 6.3 with a power law model fluid can be easily shown to be a special case of Eq. 6.7-11 (see Problem 6.8).

Finally, because the results obtained in capillary viscometry, especially for capillaries of small $L/R$, are influenced by both elastic and viscous phenomena associated with the fluid spatial accelerations at the capillary entrance, it is necessary to correct the values of $\tau_w$, given in Eq. 6.7-3. Section 13.2 covers the nature, magnitude, and significance of these corrections.

The "Rabinowitsch equation" has been used in the capillary viscometry data found in Appendix A.

* Use the "Leibnitz formula" for differentiating an integral

$$\frac{d}{dx}\int_{a_1(x)}^{a_2(x)} f(s,x)\,ds = \int_{a_1(x)}^{a_2(x)} \frac{\partial f}{\partial x}\,ds + \left[f(a_2,x)\frac{da_2}{dx} - f(a_1,x)\frac{da_1}{dx}\right]$$

### Cone and Plate Flow Viscometry

The cone and plate flow apparatus is shown schematically in Fig. 6.11. The polymer melt flows in the space formed by the rotating cone and stationary plate. The experimentally measured quantities are:

1. The cone rotational frequency $\Omega$.
2. The resulting torque needed to turn the cone $\mathcal{T}$.
3. The total force normal to the fixed plate (thrust) $F_N$.
4. The pressure distribution on the fixed plate as a function of $r$,

$$\pi_{\theta\theta}(r)|_{\theta=\pi/2} = P + \tau_{\theta\theta}(r)|_{\theta=\pi/2}$$

At shear rates exceeding $10^{-2}$ or $10^{-1}\ \mathrm{s}^{-1}$, fracture of the polymer melt is observed, with the fracture surface initiating at the melt-air interface at the perimeter. This has been attributed to the fact that the elastic energy becomes greater than the energy required to fracture the polymer melt at those shear rates (39). Irrespective of the origin of the fracture, it limits the operation of the cone-and-plate instrument to below the above-mentioned shear rates.

The velocity field between the cone and the plate is considered such that each "liquid cone" described by $\theta$-constant planes, rotates rigidly about the cone axis with the angular velocity that increases from zero at the stationary plate to $\Omega$ at the rotating cone (3). The resulting flow is a unidirectional shear flow. Moreover, because of the very small $\psi_0$ (about 1°–4°), *locally* (at fixed $r$) the flow can be considered to be like a torsional flow between parallel plates (i.e., the "liquid cones" become "disks"). Thus

$$v_\phi = \Omega r \frac{z}{z_0} \tag{6.7-12}$$

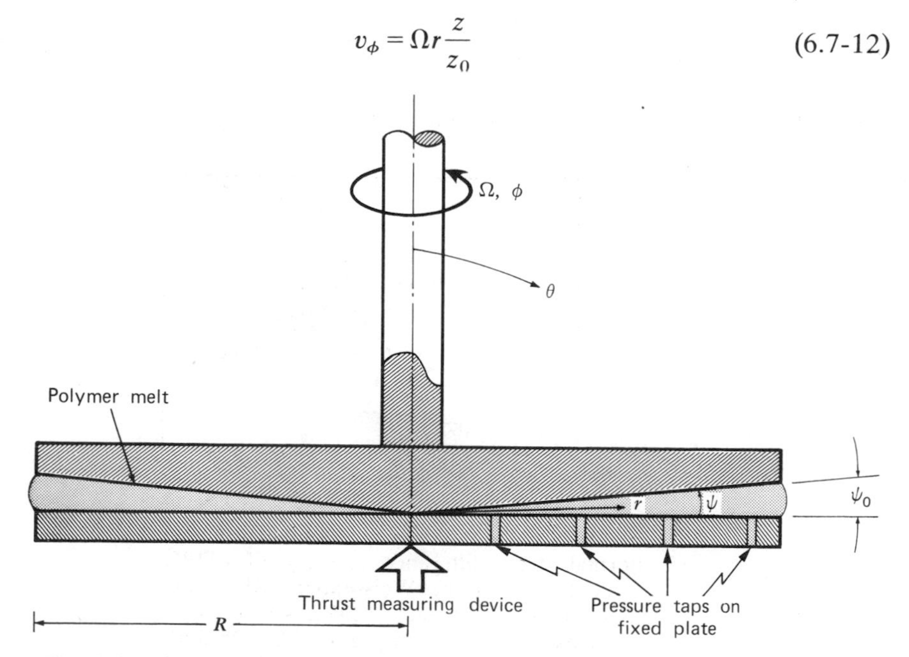


**Fig. 6.11** Schematic representation of the cone and plate viscometer.

where $z$ and $z_0$ can be expressed in terms of the angle $\psi = \pi/2 - \theta$

$$z = r \sin \psi \cong r\psi$$

and $$\tag{6.7-13}$$

$$z_0 = r \sin \psi_0 \cong r\psi_0$$

and Eq. 6.7-12 becomes

$$v_\phi = \Omega r\left(\frac{\psi}{\psi_0}\right) \tag{6.7-14}$$

Accordingly the only nonvanishing component of the rate of deformation tensor is $\dot{\gamma}_{\theta\phi} = \dot{\gamma}_{\phi\theta} = (1/r)(\partial v_\phi/\partial\theta)$, and from Eq. 6.7-14 we obtain

► $$\dot{\gamma}_{\theta\phi} = -\frac{\Omega}{\psi_0} = \text{constant} \tag{6.7-15}$$

The relationship above establishes that the cone and plate flow is viscometric, where $\phi$ is "1," $\theta$ is "2," and $r$ is "3." Furthermore, the flow field is such that shear rate is constant in the entire flow field, as in the flow between parallel plates.

The torque on the shaft of the cone is due to the action of the shear stress $\tau_{\theta\phi}$ on its surface

$$\mathcal{T} = 2\pi \int_0^R (r\tau_{\theta\phi}) r\, dr \tag{6.7-16}$$

where $\tau_{\theta\phi}$ is constant, since $\dot{\gamma}_{\theta\phi}$ is constant throughout the flow field. Upon integration we obtain

► $$\tau_{\theta\phi} = \frac{\mathcal{T}}{(\frac{2}{3})\pi R^3} \tag{6.7-17}$$

This expression suffices to determine experimentally the shear stress. Having evaluated both $\tau_{\theta\phi}$ and $\dot{\gamma}_{\theta\phi}$, we can readily obtain the viscosity function $\eta(\dot{\gamma}_{\theta\phi})$. Figure 6.12 gives such data for low density polyethylene. The data extend beyond

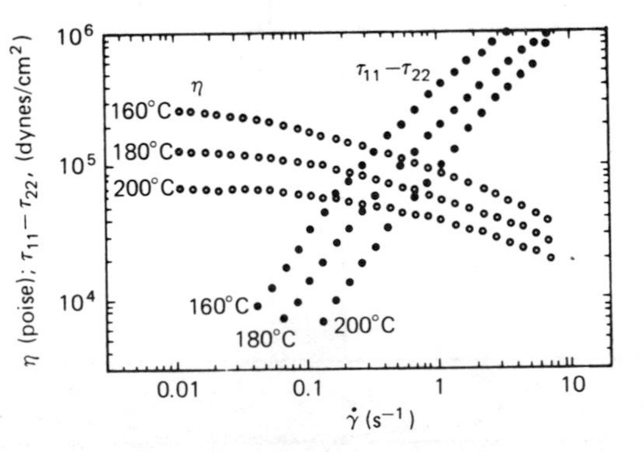


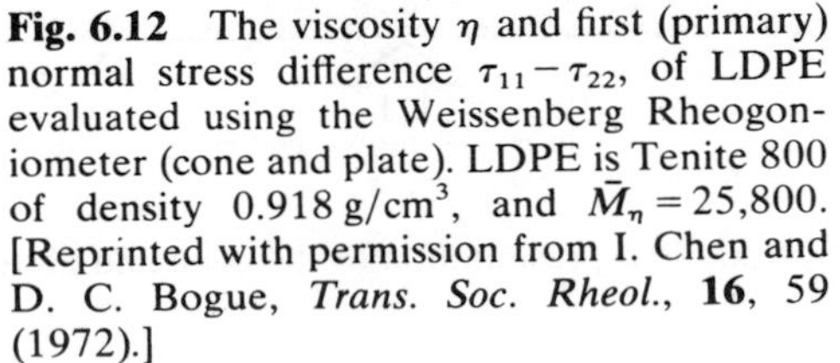

**Fig. 6.12** The viscosity $\eta$ and first (primary) normal stress difference $\tau_{11} - \tau_{22}$, of LDPE evaluated using the Weissenberg Rheogoniometer (cone and plate). LDPE is Tenite 800 of density 0.918 g/cm$^3$, and $\bar{M}_n = 25{,}800$. [Reprinted with permission from I. Chen and D. C. Bogue, *Trans. Soc. Rheol.*, **16**, 59 (1972).]

the commonly accepted upper limit of shear rate for polymer melts, probably because of the low average molecular weight of the polymer.

To obtain experimental information on normal stresses, we employ and mathematically manipulate the $r$-component of the equation of momentum, which (neglecting centrifugal forces) is

$$-\frac{\partial P}{\partial r}-\frac{1}{r^2}\frac{\partial}{\partial r}(r^2\tau_{rr})+\frac{\tau_{\theta\theta}+\tau_{\phi\phi}}{r}=0 \tag{6.7-18}$$

Introducing $\pi_{ii}=\tau_{ii}+P$ (no sum)

$$\frac{\pi_{\theta\theta}+\pi_{\phi\phi}}{r}-\frac{1}{r^2}\frac{\partial}{\partial r}(r^2\pi_{rr})=0 \tag{6.7-19}$$

Upon rearrangement and integration, taking into account that the negative of the secondary normal stress difference, $\pi_{rr}-\pi_{\theta\theta}$, is a constant (since $\dot{\gamma}_{\theta\phi}$ is constant), and that $\pi_{\theta\theta}$ at $\theta=\pi/2$ (the plate) is a function of the radius, we have

$$[\pi_{\theta\theta}(r)-\pi_{\theta\theta}(R)]|_{\theta=\pi/2}=[(\tau_{\phi\phi}-\tau_{\theta\theta})+2(\tau_{\theta\theta}-\tau_{rr})]\ln\left(\frac{r}{R}\right) \tag{6.7-20}$$

The left-hand side of Eq. 6.7-20 can be experimentally evaluated; thus the quantity in brackets on the right-hand side can be determined.

The normal force on the stationary plate can be expressed as

$$F_N=2\pi\int_0^R \pi_{\theta\theta}r\,dr-\pi R^2P_{\text{atm}} \tag{6.7-21}$$

With the help of Eq. 6.7-20 and the relation $P_{\text{atm}}=\pi_{rr}(R)$, we obtain, after integration of Eq. 6.7-21, the simple relation

► $$\tau_{\phi\phi}-\tau_{\theta\theta}=\frac{-2F_N}{\pi R^2} \tag{6.7-22}$$

which yields experimentally the important function $\tau_{\phi\phi}-\tau_{\theta\theta}$.

In summary, and in terms of the viscometric flow notation, we conclude the following about the experimental capabilities of the cone and plate viscometric flow.

1. The viscosity function $\eta$ can be determined with the aid of Eqs. 6.7-15 and 6.7-17.
2. The primary normal stress difference, $\tau_{11}-\tau_{22}=\tau_{\phi\phi}-\tau_{\theta\theta}$, can be calculated through Eq. 6.7-22.
3. The secondary normal stress difference, $\tau_{22}-\tau_{33}=\tau_{\theta\theta}-\tau_{rr}$, can be determined, subsequent to the evaluation of $\tau_{11}-\tau_{22}$, using Eq. 6.7-20.

These conditions are subject to the limitation, for polymer melts, that the applied shear rate $\dot{\gamma}=\Omega/\psi_0$ must be below that which gives rise to fracture in the fluid sample. For solutions of polymers, the upper limit of shear rate (or $\Omega$) is one at which the centrifugal forces become important (40).

Figure 6.12 presents the primary normal stress difference data for low density polyethylene and Fig. 6.13 presents primary and secondary normal stress difference data for a 2.5% polyacrylamide solution, again using a cone and plate viscometer.

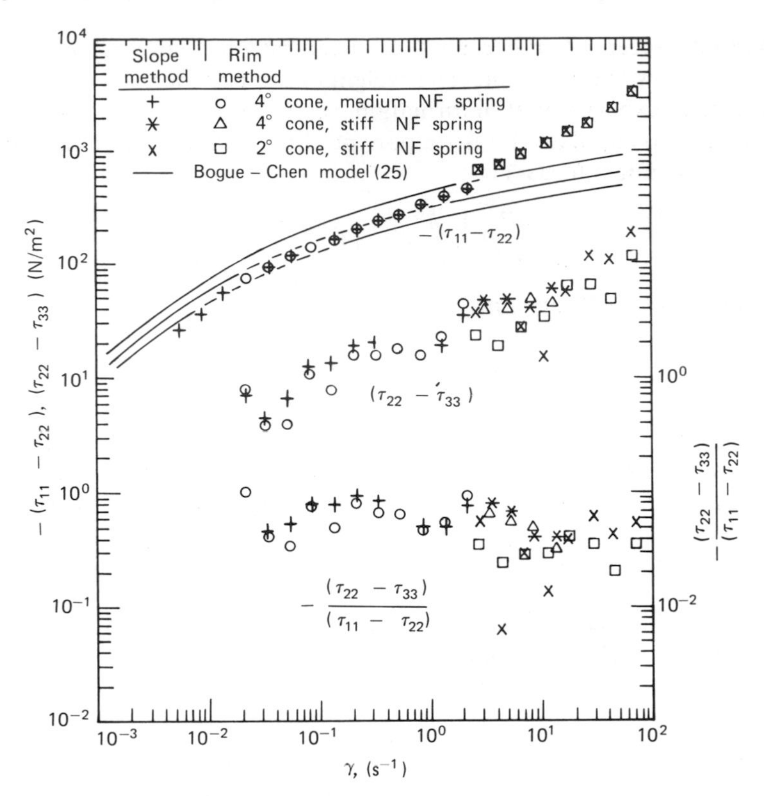


**Fig. 6.13** Values of $-(\tau_{11}-\tau_{22})$, $\tau_{22}-\tau_{33}$, and the ratio $-(\tau_{11}-\tau_{22})/(\tau_{22}-\tau_{33})$ for a 2.5% solution of polyacrylamide measured with a cone and plate rheometer. [Reprinted with permission from E. B. Christiansen and W. R. Leppard, *Trans. Soc. Rheol.*, **18**, 65 (1974).]

We note that the primary normal stress coefficient $\Psi_1$ is positive, whereas the secondary normal stress difference $\Psi_2$ is negative. The scatter of the second normal stress difference data is evident; it is very difficult to measure $(\tau_{22}-\tau_{33})$, and its values are very much in doubt (41). The ratio $-(\tau_{22}-\tau_{33})/(\tau_{11}-\tau_{22})$ is shown to be of the order of $10^{-1}$. The same results, with an equal amount of scatter, were reported for PIB (42).

It is worth noting that if we use Eqs. 6.6-3 and 6.6-4 (i.e., for the CEF fluid) in conjunction with Fig. 6.13, we can reach the conclusions that both $\Psi_1$ and $\Psi_2$ are functions of $\dot{\gamma}$; thus we see that Eqs. 6.6-3 and 6.6-4 state simply that

$$\tau_{11}-\tau_{22}=f_1(\dot{\gamma})$$

$$\tau_{22}-\tau_{33}=f_2(\dot{\gamma})$$

Figure 6.13 reveals that the log-log plots of the normal stress differences versus shear rate have a nearly constant slope, suggesting

$$\tau_{11}-\tau_{22}=-\Psi_1(\dot{\gamma})\dot{\gamma}^2=-m_1'\dot{\gamma}^{n_1'-2}\dot{\gamma}^2$$

$$\tau_{22}-\tau_{33}=-\Psi_2(\dot{\gamma})\dot{\gamma}^2=m_2'\dot{\gamma}^{n_2'-2}\dot{\gamma}^2$$

where $m_1'$ and $m_2'$ are both positive.

The value of the slope for both curves is roughly one-half (i.e., $n'_1 \cong n'_2 \cong 0.5$), leading to the approximate relations

$$\Psi_1(\dot{\gamma}) = m'_1 \dot{\gamma}^{-1.5}$$

$$\Psi_2(\dot{\gamma}) = -m'_2 \dot{\gamma}^{-1.5}$$

That is, the *normal stress difference coefficients* also exhibit *shear thinning.*

Recently Bird and his co-workers (43) have pointed out a simple method of estimating the primary normal stress difference from viscosity data. The method is approximate and has as its origin the Goddard-Miller constitutive equation (Eq. 6.3-7). It predicts that

$$\Psi_1(\dot{\gamma}) = \frac{4K}{\pi} \int_0^{\infty} \frac{\eta(\dot{\gamma}) - \eta(\dot{\gamma}')}{\dot{\gamma}'^2 - \dot{\gamma}^2} \, d\dot{\gamma}' \qquad (6.7\text{-}23)$$

where $K$ is an empirical constant. As predicted by Eq. 6.3-7, $K = 1$ but good fit to data results with $K$ about 2 for solutions and 3 for melts. In Eq. 6.7-23 the viscosity must be known for the entire shear rate range $0 < \dot{\gamma} < \infty$. The viscosity at high shear rates can be either experimentally obtained or predicted using a number of constitutive equations (Bird and his co-workers used the Carreau model), but the viscosity at the Newtonian plateau must be known. Figure 6.14 shows the predictions of Eq. 6.7-23 for the low density polyethylene sample whose viscosity and primary normal stress difference function appears in Fig. 6.12. The agreement is good (excellent for engineering purposes).

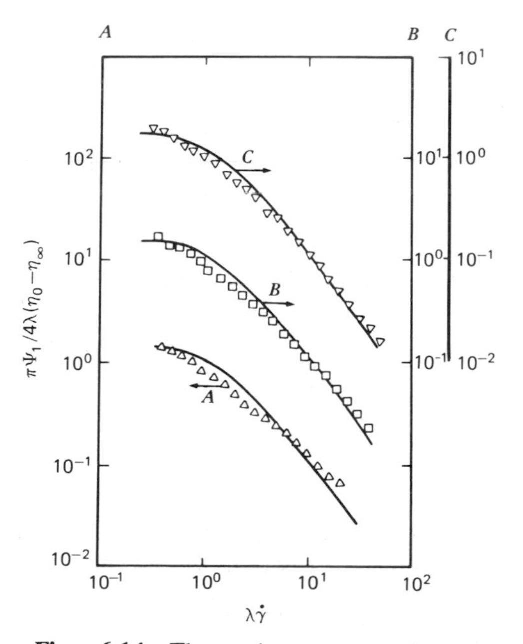


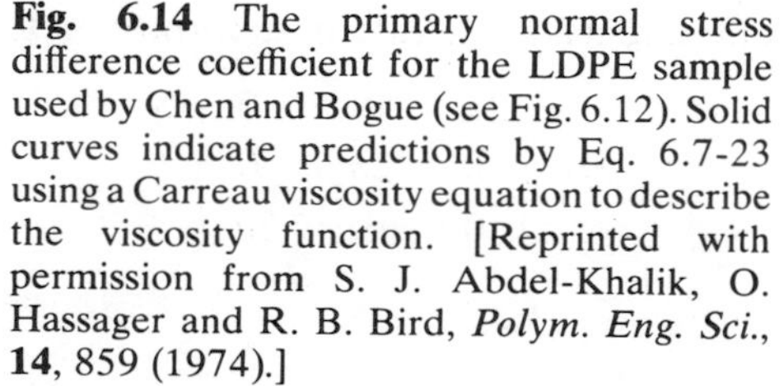

**Fig. 6.14** The primary normal stress difference coefficient for the LDPE sample used by Chen and Bogue (see Fig. 6.12). Solid curves indicate predictions by Eq. 6.7-23 using a Carreau viscosity equation to describe the viscosity function. [Reprinted with permission from S. J. Abdel-Khalik, O. Hassager and R. B. Bird, *Polym. Eng. Sci.*, **14**, 859 (1974).]

The capillary and cone and plate geometries are the most commonly used in the experimental determination of the rheological functions, which can also be measured, in principle, in *any* viscometric flow. The slit (parallel plate pressure) flow has been used by Han and his co-workers (44). The eccentric disk (or Maxwell orthogonal) rheometer (45) is also a flow configuration that is becoming commonly used. Finally, a recent development by Lodge (46), which makes use of the "hole pressure error," permits the measurement of $\eta$, $\Psi_1$, and $\Psi_2$ under steady viscometric flow conditions. Walters in his recent book (40) discusses the subject of viscometry in detail.

## 6.8 Elongational Flows

Although the simple fluid constitutive equation and its general expansion forms are applicable to all types of flows, so far this chapter has emphasized only shear flows, and in particular steady viscometric flows. The reasons are the simplicity of such flows and their predominance in polymer processing. Nevertheless, there is another class of flows known as *elongational*, *extensional*, or *shear free*, which commonly occur in processing. Examples are the stretching of a melt strand in the fiber spinning process, the uniaxial stretching of a molten film exiting a flat film die, the biaxial stretching of a tubular film into a "bubble," the multiaxial stretching of a parison to form an enclosure in the process of blow molding, and finally the flow in converging channels. These examples involve elongational flows that are more complex than those studied for rheological purposes. Whereas rheologists study uniform isothermal elongational flows (experimentally quite difficult in their own right), polymer processing engineers have to deal with nonuniform and quite often nonisothermal elongational flows, since such phenomena are often encountered during forming operations, when solidification occurs. To appreciate the nature of these problems, we briefly discuss the *uniform* elongational flows, which *only approximate* those encountered in polymer processing (47).

### *Kinematics of Steady and Uniform Elongational Flows*

In Cartesian coordinates such flows can be represented by the general flow field

$$v_i = a_i x_i \text{ (no sum)} \tag{6.8-1}$$

where

$$2a_i \equiv \dot{\gamma}_{ii} \tag{6.8-2}$$

For steady flows $a_i \neq f(t)$ and, since they are uniform, all $a_i$ are constant. Furthermore if the fluid is incompressible, the continuity equation dictates that

$$\sum a_i = 0 \tag{6.8-3}$$

Thus in general, the components of the rate of deformation tensor are given by

► $$\dot{\boldsymbol{\gamma}} = \begin{pmatrix} 2a_1 & 0 & 0 \\ 0 & 2a_2 & 0 \\ 0 & 0 & 2a_3 \end{pmatrix} \tag{6.8-4}$$

subject to the incompressibility condition (Eq. 6.8-3), because of which only two of the three normal velocity gradients can be specified. It is evident that these flows are not viscometric, thus constitutive equations such as the CEF do not apply. Three common examples of elongational flows are listed below.

*Simple Extensional or Elongational Flow.* This flow is caused by the uniform stretching of a rectangular bar in the "1" direction. The stretching of a thin filament of circular cross section can be approximated by such flow. The flow is given by

$$a_1 = \dot{\varepsilon}; \qquad a_2 = a_3 = \frac{-\dot{\varepsilon}}{2} \tag{6.8-5a}$$

and

$$\dot{\gamma}_{11} = 2\dot{\varepsilon}; \qquad \dot{\gamma}_{22} = \dot{\gamma}_{33} = -\dot{\varepsilon} \tag{6.8-5b}$$

where $\dot{\varepsilon}$ is the elongation rate.

The scalar invariants of $\dot{\boldsymbol{\gamma}}$ are

$$\mathrm{I}_{\dot{\boldsymbol{\gamma}}} = 0; \qquad \mathrm{II}_{\dot{\boldsymbol{\gamma}}} = 6\dot{\varepsilon}^2; \qquad \mathrm{III}_{\dot{\boldsymbol{\gamma}}} = 2\dot{\varepsilon}^3 \tag{6.8-6}$$

*Planar Extensional or Elongational Flow* (*Pure Shear Flow*). Such flow would be generated by uniformly stretching a film in one direction, allowing its thickness to decrease, but not allowing the other flat dimension to change.

It is described by

$$a_1 = -a_2 = \dot{\varepsilon}_{\mathrm{pl}} \qquad a_3 = 0 \tag{6.8-7a}$$

and

$$\dot{\gamma}_{11} = 2\dot{\varepsilon}_{\mathrm{pl}}; \qquad \dot{\gamma}_{22} = -2\dot{\varepsilon}_{\mathrm{pl}}; \qquad \dot{\gamma}_{33} = 0 \tag{6.8-7b}$$

The scalar invariants of $\dot{\gamma}$ are

$$\mathrm{I}_{\dot{\boldsymbol{\gamma}}} = 0; \qquad \mathrm{II}_{\dot{\boldsymbol{\gamma}}} = 8\dot{\varepsilon}_{\mathrm{pl}}^2; \qquad \mathrm{III}_{\dot{\boldsymbol{\gamma}}} = 0 \tag{6.8-8}$$

*Biaxial Extensional or Elongational Flow.* Such flow would be caused by stretching a film at equal ratios and allowing the thickness to decrease. It is characterized by

$$a_1 = a_2 = \dot{\varepsilon}_{\mathrm{bi}}; \qquad a_3 = -2\dot{\varepsilon}_{\mathrm{bi}} \tag{6.8-9a}$$

and

$$\dot{\gamma}_{11} = \dot{\gamma}_{22} = 2\dot{\varepsilon}_{\mathrm{bi}} \qquad \dot{\gamma}_{33} = -4\dot{\varepsilon}_{\mathrm{bi}} \tag{6.8-9b}$$

The invariants of $\dot{\gamma}$ are

$$\mathrm{I}_{\dot{\boldsymbol{\gamma}}} = 0; \qquad \mathrm{II}_{\dot{\boldsymbol{\gamma}}} = 24\dot{\varepsilon}_{\mathrm{bi}}^2; \qquad \mathrm{III}_{\dot{\boldsymbol{\gamma}}} = -16\dot{\varepsilon}_{\mathrm{bi}}^3 \tag{6.8-10}$$

Turning to the practical question of what kinds of applied dimensional rates of change are needed to create any of these uniform elongational flows, we briefly examine that of simple elongational flow, where

$$v_1 = \frac{dx_1}{dt} = \dot{\varepsilon}x_1 \tag{6.8-11}$$

In terms of the specimen length $l(t)$, this becomes

$$\frac{dl(t)}{dt} = \dot{\varepsilon} l(t) \tag{6.8-12}$$

Upon integration ($l(0) = l_0$), we obtain the following expression for the time dependent length that must be created in order to generate this flow:

$$l(t) = l_0 \exp(\dot{\varepsilon} t) \tag{6.8-13}$$

The resulting reduction in the transverse directions can be similarly obtained

$$x_2(t) = x_{20} \exp\left(-\frac{\dot{\varepsilon}}{2} t\right) \tag{6.8-14}$$

$$x_3(t) = x_{30} \exp\left(-\frac{\dot{\varepsilon}}{2} t\right) \tag{6.8-15}$$

Thus the needed "elongational strain" $\dot{\varepsilon} t$ for simple elongational flow is

$$\ln\left[\frac{l(t)}{l_0}\right] = \dot{\varepsilon} t \tag{6.8-16}$$

Similar expressions for the strains necessary to create the other two uniform elongational flows can be obtained (see Problem 6.13).

### *Elongational Viscosity*

The term "elongational viscosity" has been widely used, almost indiscriminately, to denote the resistance of fluids to any elongational flow, be it uniform (of any kind) or nonuniform, isothermal or not. This, of course, has led to a good deal of confusion, especially when comparing experimental results, as we shall see below.

In steady elongational flows a material function $\bar{\eta}$ called the *elongational viscosity* is defined to describe the normal stress difference. For steady simple elongational flow where $\tau_{22} = \tau_{33}$, it is given by:

► $$\bar{\eta}(\dot{\varepsilon}) = -\frac{\tau_{11} - \tau_{33}}{\dot{\varepsilon}} \tag{6.8-17}$$

The definition expressed in Eq. 6.8-17 implies a Newtonian or a generalized Newtonian fluid. Experimentally,

$$\tau_{11} + P_{atm} = \frac{F_1}{A_1} \tag{6.8-18}$$

$$\tau_{22} = \tau_{33} = -P_{atm} \tag{6.8-19}$$

where $F_1$ is the force needed to cause the simple elongational flow and $A_1$ is the cross-sectional area normal to the flow.

Thus, experimentally, the elongational viscosity becomes

$$\bar{\eta}(\dot{\varepsilon}) = -\frac{F_1/A_1}{\dot{\varepsilon}} \tag{6.8-20}$$

For a Newtonian fluid in a simple elongational flow the constitutive equation becomes

$$\boldsymbol{\tau} = -\mu\dot{\boldsymbol{\gamma}} = -\mu \begin{pmatrix} +2\dot{\varepsilon} & 0 & 0 \\ 0 & -\dot{\varepsilon} & 0 \\ 0 & 0 & -\dot{\varepsilon} \end{pmatrix} \tag{6.8-21}$$

Thus

$$\tau_{11} - \tau_{33} = -\mu(2\dot{\varepsilon} + \dot{\varepsilon}] = -3\mu\dot{\varepsilon} \tag{6.8-22}$$

Combining Eqs. 6.8-17 and 6.8-22 we obtain the so-called Trouton relation, which defines the *Trouton viscosity* (48).

► $$\bar{\eta} = 3\mu \tag{6.8-23}$$

A number of experimental methods have been used to measure $\bar{\eta}$. The most straightforward is the method of Meissner (49), Ballman (50), Stevenson (51), Vinogradov et al. (52) and, more recently, White (53). In its essence, a cylindrical sample is pulled at a rate that increases exponentially with time so that $\dot{\varepsilon}$ = constant (Eq. 6.8-13). When (if) a steady force is attained, $\bar{\eta}$ can be evaluated. Steady state is usually reached (before breaking the sample) only at low $\dot{\varepsilon}$ values. Figure 6.15

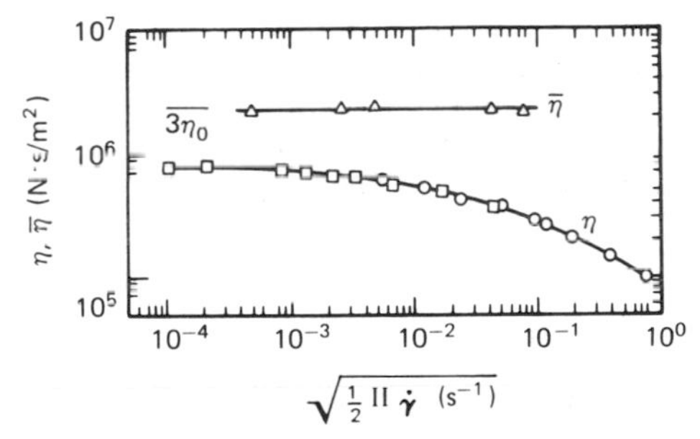


**Fig. 6.15** Comparison between the elongational and shear viscosities, $\bar{\eta}$ and $\eta$, for Butyl 35, a polyisobutylene-isoprene copolymer. Both $\bar{\eta}$ and $\eta$ are plotted versus $\sqrt{\frac{1}{2}\mathrm{II}_{\dot{\gamma}}}$, which is $\sqrt{3}\dot{\varepsilon}$ for simple extension and $\dot{\gamma}$ for shear flows. [Reprinted with permission from R. B. Bird, R. C. Armstrong and O. Hassager, *Dynamics of Polymeric Fluids*, Vol. I., p. 187, Wiley, New York, 1977. Based on data of J. F. Stevenson, *A.I.Ch.E. Journal*, **18**, 540 (1972).]

gives data by Stevenson for a PIB-isoprene copolymer. Two features are worth noting. First, the Trouton relation 6.8-23 holds only between $\bar{\eta}$ and $\eta_0$. Second, $\bar{\eta}$ is fairly independent of $\dot{\varepsilon}$. Thus since $\eta(\dot{\gamma})$ decreases with increasing $\dot{\gamma}$, $\bar{\eta}/\eta > 3$. Very large values of this ratio are reported by Cogswell et al. (54), although the experimental method used does not guarantee that the response measured is steady. At any rate, large values of $\bar{\eta}/\eta$ may be important in the ability of polymers to be spun and partly responsible for the observed entrance pressure losses.

Quite often in experiments where $\dot{\varepsilon}$ = constant, even at low elongation rates, the axial stress $F_1/A_1$ does not stop growing for the duration of the experiment. That is, no steady state is achieved. This fact can be represented by an extensional viscosity $\bar{\eta}^+$, which grows with time, as if the *entire* experiment were a startup experiment. This viscosity can be defined similarly to Eq. 6.8-17 as

$$\bar{\eta}^+ = -\frac{F_1/A_1(t)}{\dot{\varepsilon}_0} = -\frac{\tau_{11}(t) - \tau_{33}(t)}{\dot{\varepsilon}_0} \tag{6.8-24}$$

LPDE melts are the most notable examples of such rheological behavior (Fig. 6.16). Also plotted in Fig. 6.16 are the predictions of the Lodge rubberlike

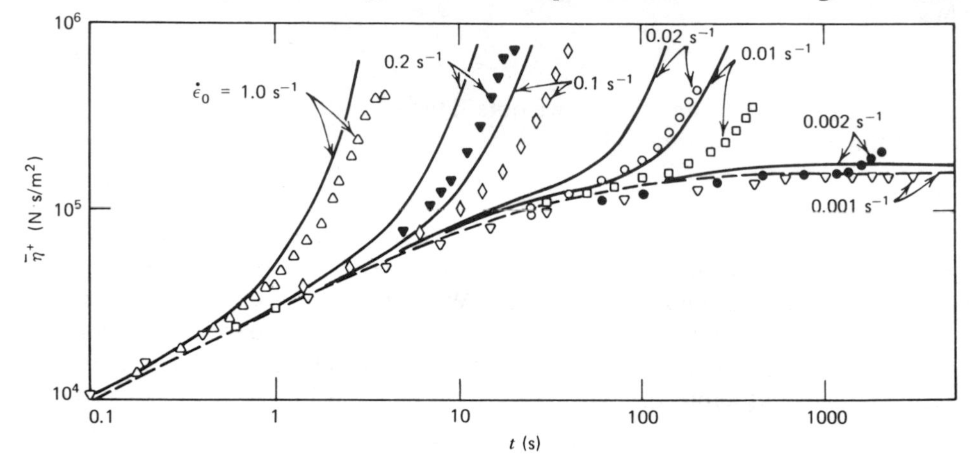


**Fig. 6.16** Comparison of calculated and experimental results of Meissner (49) for $\bar{\eta}^+$ for a LDPE melt. Predictions (solid curves) by Lodge rubberlike liquid constitutive equation with five term memory function (55). (Reprinted with permission from R. B. Bird, R. C. Armstrong and O. Hassager, *Dynamics of Polymeric Liquids*, Vol. I, p. 451, Wiley, New York, 1977.)

liquid equation, Eq. 6.3–13, using the following memory function for small deformations and low deformation rates

$$M(t-t') = \sum_{k=1}^{5} \frac{\eta_k}{\lambda_k^2} e^{-(t-t')/\lambda_k} \tag{6.8-25}$$

where $\lambda_k$ are relaxation times and $\eta_k$ are viscosity constants. The stress growth function is (55) (see also Bird et al. 1d):

$$\bar{\eta}^+(t, \dot{\varepsilon}_0) = \sum_{k=1}^{5} \frac{\eta_k}{\lambda_k^2}\left[\frac{3}{(\lambda_k^{-1} - 2\dot{\varepsilon}_0)(\lambda_k^{-1} + \dot{\varepsilon}_0)} - \frac{2\lambda_k}{(\lambda_k^{-1} - 2\dot{\varepsilon}_0)} \exp -(\lambda_k^{-1} - 2\dot{\varepsilon}_0)t \right.$$
$$\left. - \frac{\lambda_k}{(\lambda_k^{-1} + \dot{\varepsilon}_0)} \exp -(\lambda_k^{-1} + \dot{\varepsilon}_0)t\right] \tag{6.8-26}$$

Outstanding in both the experimental and theoretical results is a critical rate of elongation above which $\bar{\eta}^+$ rises without bound. The theoretical value for the critical rate of elongation is the reciprocal of twice the maximum relation time $\dot{\varepsilon}_0 > (2\lambda_{k,\max})^{-1}$. In physical terms, below this value the stresses relax faster than they grow. Finally we should note that the foregoing results were obtained for relatively *small* levels of total elongational strain $\varepsilon < 5$.

From Eqs. 6.8-24 and 6.8-26, following White (56), two asymptotic expressions can be derived for $\dot{\varepsilon}_0 \to 0$ and $\dot{\varepsilon}_0 \to \infty$. In the former case

$$\bar{\eta}^+(t, \dot{\varepsilon}_0)_{\dot{\varepsilon}_0 \to 0} = 3 \sum_{k=1}^{\infty} \eta_k (1 - e^{-t/\lambda_k}) \tag{6.8-27}$$

Thus at low rates of elongation we find that the linear viscoelastic fluid, subsequent to a transient viscoelastic response, approaches Newtonian behavior. At high rates of elongation we have

$$-(\tau_{11}-\tau_{33})=\sum_{k=1}^{\infty}\left(\frac{\eta_k}{\lambda_k}\right)e^{2\dot{\varepsilon}_0 t} \tag{6.8-28}$$

This is the response of an *elastic solid*. Of course at high elongational rates and large deformations, one should use a nonlinear viscoelastic model to obtain a meaningful asymptote. This was done by White (56), who used a modified BKZ constitutive equation, Eq. 6.3-17, employing deformation rate dependent "effective" relaxation times. Both uniaxial and biaxial deformations were treated, drawing conclusions from the latter about polymer sheet stretching. The results of this nonlinear viscoelastic analysis are as follows. At $\dot{\varepsilon}_0 \to 0$ the nonlinear fluid exhibits a linear viscoelastic response, approaching Newtonian behavior at long times. At nonzero $\dot{\varepsilon}_0$ values, but *below* the critical value, the long duration asymptotes of $\bar{\eta}^+$ can be expressed as a polynomial in powers of $\dot{\varepsilon}_0$, with the Trouton viscosity as the first term. White points out that this is equivalent to the asymptote used by Denson (57, 58), who analyzed biaxial stretching of polymer melt sheets in terms of a non-Newtonian elongational viscosity. Chapter 15 discusses Denson's work. The high rate asymptote $\dot{\varepsilon}_0 \to \infty$ leads to the conclusion that the sheet responds as an elastic solid that is nonlinear. The work of Schmidt and Carley (59) on thermoforming, which Chapter 15 also treats, is in line with this conclusion.

In summary, below the critical elongation rate $\dot{\varepsilon}_0=(2\lambda_{max})^{-1}$ uniaxial and biaxial stretching are due to flow characterized by a non-Newtonian viscosity, whereas above this rate the flow is characterized by nonlinear elastic deformation.

Before closing this chapter, we comment briefly on the important subject of polymer structure–melt rheological properties–processing performance relationships. Stated specifically, how does the macromolecular structure, as measured with available characterization techniques, affect melt flow behavior, as measured with available rheometers? And knowing both variables, can the performance in any polymer processing method (with particular emphasis on shaping and product properties) be predicted? The second question is related to processability.

Although both parts of the question are still open and subject to wide and intense investigation, more is known about the first (the structure–rheological behavior relationships). The reason is that rheological experiments are simpler, free of geometrical complexities and nonuniform temperature effects, and are conducted under controlled conditions, using well-controlled instruments. Thus relationships between $\eta_0$ and MW for monodispersed polymers, and MWD and the shape of $\eta(\dot{\gamma})$ at constant $\bar{M}_w$ have been more or less established. These relationships can be predicted by constitutive equations that have a "molecular" origin. Nevertheless, not much is known about the dependence of $\Psi_1(\dot{\gamma})$ and $\Psi_2(\dot{\gamma})$ on polymer structure. Furthermore, limitations in the analytical characterization methods have not enabled us to have detailed knowledge of polymer structural features, such as the "high MW tail" of the MWD or the number and length distribution of branches. Such features may play an important role in rheological behavior, such as $\Psi_2(\dot{\gamma})$.

Less is known about the influence of structure on polymer processing behavior. Graessley et al. (60) have found that the degree of polydispersity is the primary structural variable affecting extrudate swelling. Although this is satisfying, it is not possible to propose a *specific* quantitative relationship between the two. Furthermore, not even qualitative relationships have been established between polymer structure and processing performance only slightly more complex than extrudate swelling; Miller (61), working with almost identical HDPE samples, obtained quite different parison *diameter* swelling, as Section 15.1 discusses. An IUPAC working party started working in 1967 on the problem of studying the melt rheology, processing behavior (in a blown film process), and end product properties of three practically identical LDPE samples. Their report (62), issued in 1974, concludes the following: (*a*) there is *no* difference in purely viscous and linear viscoelastic behavior; (*b*) there are differences in the values of $\tau_{11}-\tau_{22}$ at low shear rates and the elongational behavior at low strain rates and large strain values; (*c*) there are *notable* differences in the film drawdown behavior, and the optical quality and transverse impact strength of the film. Clearly, this careful and painstaking work suggests that the role of structure with respect to flow and processing behavior is incompletely understood.

We conclude this chapter by emphasizing that the viscometric functions generated during rheological characterization are not necessarily the properties of the melts that control the *processing* flow behavior. This is especially true for fast processing flows that are nonviscometric (have more than one velocity component and numerous velocity gradients) and nonisothermal. Although steady state conditions may prevail, from a Lagrangean point of view a polymer melt element in a processing flow field "sees" rapidly changing conditions; thus its response is different from the steady experience and response in viscometric experiments. Incidentally, it is partly because of the rapidly changing experience of polymer melts in processing flows that "structuring" is possible, as discussed in Chapter 3.

## REFERENCES


1. R. B. Bird, R. C. Armstrong, and O. Hassager, *Dynamics of Polymeric Liquids*, Vol. I, *Fluid Mechanics*, Wiley, New York, 1977: (a) Chapter 8; (b) Chapter 9; (c) Vol. II, Chapter 15; (d) p. 449.
2. S. Middleman, *The Flow of High Polymers*, Wiley-Interscience, New York, 1968.
3. A. S. Lodge, *Elastic Liquids*, Academic Press, London, 1964.
4. B. D. Coleman, H. Markowitz, and W. Noll, *Viscometric Flows of Non-Newtonian Fluids*, Springer-Verlag, New York, 1966.
5. J. A. Brydson, *Flow Properties of Polymer Melts*, Van Nostrand Reinhold, London, 1970.
6. C. D. Han, *Rheology in Polymer Processing*, Academic Press, New York, 1976.
7. S. P. Burrow, A. Peterlin, and D. T. Turner, *Polymer*, **6**, 35 (1965).
8. Z. Tadmor and R. B. Bird, *Polym. Eng. Sci.*, **14**, 124 (1974).
9. N. N. Kapoor, M.S. Thesis in chemical engineering, University of Minnesota, St. Paul; work cited in A. G. Fredrickson, *Principles and Applications of Rheology*, Prentice-Hall, Englewood Cliffs, N.J.; 1964, p. 120.
10. D. F. James, *Nature*, **212**, 754 (1966).
11. J. Eliassaf, A. Silverberg, and A. Katchalsky, *Nature*, **176**, 1119 (1953).
12. W. Noll, *Arch. Rat. Mech. Anal.*, **2**, 197 (1958).
13. J. D. Goddard, *Trans. Soc. Rheol.*, **11**, 381 (1967).
14. A. E. Green and R. S. Rivlin, *Arch. Rat. Mech. Anal.*, **1**, 1 (1957).

15. B. D. Coleman and W. Noll, *Rev. Mod. Phys.*, **33**, 239 (1961).
16. W. O. Criminale, Jr., J. L. Ericksen, and G. L. Filbey, Jr., *Arch. Rat. Mech. Anal.*, **1**, 410 (1958).
17. J. D. Goddard and C. Miller, *Rheol. Acta*, **5**, 177 (1966).
18. J. G. Oldroyd, *Proc. Roy. Soc.*, **A200**, 45 (1950).
19. K. Walters, *Quart. J. Mech. Appl. Math.*, **13**, 444 (1960).
20. A. G. Fredrickson, *Chem. Eng. Sci.*, **17**, 155 (1962).
21. J. D. Ferry, M. L. Williams, and D. M. Stern, *J. Chem. Phys.*, **58**, 987 (1954).
22. B. Bernstein, E. A. Kearsley, and L. J. Zapas, *Trans. Soc. Rheol.*, **7**, 391 (1963).
23. D. C. Bogue, *Ind. Eng. Chem. Fundam.*, **5**, 253 (1966).
24. R. B. Bird and P. J. Carreau, *Chem. Eng. Sci.*, **23**, 427 (1968).
25. I. Chen and D. C. Bogue, *Trans. Soc. Rheol.*, **16**, 59 (1972).
26. A. V. Tobolsky, *Properties and Structure of Polymers*, Wiley, New York, 1960, Chapter 3.
27. E. Catsiff and A. V. Tobolsky, *J. Colloid Sci.*, **10**, 375 (1955).
28. M. L. Williams, R. F. Landel, and J. D. Ferry, *J. Am. Chem. Soc.*, **77**, 3701 (1955).
29. W. Ostwald, *Kolloid-Z.*, **36**, 99 (1925) and A. de Waele, *Oil and Color Chem. Assoc. Journal*, **6**, 23 (1923).
30. P. H. Goldblatt and R. S. Porter, *J. Appl. Polym. Sci.*, **20**, 1199 (1976).
31. Z. Tadmor, *Polym. Eng. Sci.*, **6**, 203 (1966).
32. S. Glasstone, K. J. Laidler, and H. Eyring, *Theory of Rate Processes*, McGraw-Hill, New York, 1941, Chapter 9.
33. M. Reiner, *Deformation, Strain and Flow*, Wiley-Interscience, New York, 1960, p. 246.
34. P. J. Carreau, Ph.D. thesis, Department of Chemical Engineering, University of Wisconsin, 1968.
35. W. P. Cox and E. H. Merz, *J. Polym. Sci.*, **28**, 619 (1958).
36. E. C. Bingham, *Fluidity and Plasticity*, McGraw-Hill, New York, 1922.
37. I. S. Sokolnikoff and R. M. Redheffer, *Mathematics of Physics and Engineering*, McGraw-Hill, New York, 1958, p. 262.
38. K. Weissenberg as cited by B. Rabinowitsch, *Z. Physik-Chemie*, **A145**, 1 (1929) and R. Eisenschitz, *Kolloid-Z.*, **64**, 184 (1933).
39. J. F. Hutton, *Nature*, **200**, 646 (1963).
40. K. Walters, *Rheometry*, Chapman and Hall, London, 1975.
41. E. B. Christiansen and W. R. Leppard, *Trans. Soc. Rheol.*, **18**, 65 (1974).
42. R. I. Tanner, *Trans. Soc. Rheol.*, **17**, 365 (1973).
43. S. J. Abdel-Khalik, O. Hassager, and R. B. Bird, *Polym. Eng. Sci.*, **14**, 859 (1974).
44. C. D. Han, *Trans. Soc. Rheol.*, **18**, 163 (1974).
45. B. Maxwell and R. P. Chartoff, *Trans. Soc. Rheol.*, **9**, 41 (1965).
46. K. Higashitani and A. S. Lodge, *Trans. Soc. Rheol.*, **19**, 307 (1975).
47. J. M. Dealy, *Polym. Eng. Sci.*, **11**, 433 (1971).
48. F. T. Trouton, *Proc. Roy. Soc.*, **A77**, 426 (1906).
49. J. Meissner, *Rheol. Acta*, **10**, 230 (1971); *Trans. Soc. Rheol.*, **16**, 405 (1972).
50. R. L. Ballman, *Rheol. Acta*, **4**, 1938 (1965).
51. J. F. Stevenson, *Am. Inst. Chem. Eng. J.*, **18**, 540 (1972).
52. G. V. Vinogradov, B. V. Radushkevich, and V. D. Fikhman, *J. Poly. Sci.*, A2, **8**, 1 (1970).
53. J. L. White, *Rubber Chem. Technol.*, **50**, 163 (1977).
54. F. N. Cogswell, *Rheol. Acta*, **8**, 187 (1969); Data also in P. C. Penwell, in *Thermoplastics*, R. M. Ogorgiewicz, ed., Wiley, New York, 1974, Chapter 11.
55. H. Chang and A. S. Lodge, *Rheol. Acta*, **11**, 127–129 (1972).
56. J. L. White, *Rheol. Acta*, **14**, 600 (1975).

57. C. D. Denson, *Trans. Soc. Rheol.*, **16**, 377 (1972).
58. D. D. Joye, G. W. Poehlein, and C. D. Denson, *Trans. Soc. Rheol.*, **16**, 421 (1972).
59. L. R. Schmidt and J. F. Carley, *Polym. Eng. Sci.*, **15**, 51 (1975).
60. W. W. Graessley, S. D. Glasscock, and R. L. Crawley, *Trans. Soc. Rheol.*, **14**, 519 (1970).
61. J. C. Miller, *Trans. Soc. Rheol.*, **19**, 341 (1975).
62. Report of IUPAC Working Party on "Structure and Properties of Commercial Polymers," Parts I and II, 1974.

## PROBLEMS

**6.1** ***The Single Maxwell Element LVE Constitutive Equation.*** Consider the single Maxwell mechanical element in Fig. 6.6*a*. A shear strain $\gamma_{12}(t)$ is applied at $t=0$. The element was at rest for $t<0$. By stating that the stress is the same in the dashpot and spring, while the total strain is the sum of those of the spring and the dashpot, obtain Eq. 6.3-9 in shear. Solve the differential equation to obtain Eq. 6.4-2 for a stress relaxation experiment, that is, $\gamma_{12}=\gamma_0$.

**6.2** ***The Boltzmann Superposition Principle.*** Apply the Boltzmann superposition principle to obtain the LVE equation 6.3-8 (in shear) from Eq. 6.4-1. Consider the applied strain $\gamma(t)$ as being applied discretely in a series of small steps $\Delta\gamma$.

**6.3** ***The Single Voigt Element LVE Constitutive Equation.*** In a Voigt mechanical element (Fig. 6.6*b*), the total stress is the sum of the stresses on the dashpot and spring. On the other hand, the strain in each component is equal to the total strain. Use these facts to develop the constitutive equation for a single Voigt element. Solve the differential equation, for a creep experiment ($\tau=0$, $t<0$; $\tau=\tau_0$, $t\geq 0$) to obtain Eq. 6.4-3.

**6.4** ***The Boltzmann Superposition Principle—Alternate form of the LVE Equation.*** Apply the Boltzmann superposition principle for the case of a continuous stress application on a linear viscoelastic material to obtain the resulting strain $\gamma(t)$ in terms of $J(t-t')$ and $d\tau/dt'$, the stress history.

**6.5** ***Creep in Structural Design.*** A pendulum clock manufacturer wants to replace the metal pendulum arm of the clocks with a polymer rod. Is his idea a good one? Use the answer to Problem 6.4.

**6.6** ***Inherent Errors in Using the Power Law Model in Pressure Flows.*** The shear rate during pressure flow between parallel plates varies from zero to $\dot{\gamma}_w$. If for a polymer melt we are to use the power law model, $n \neq 1$, the value of $n$ that is used will persist even near the center, where the shear rate is nearly zero, thus the melt is Newtonian. How would you estimate the error introduced as a function of $\xi^*$, where $\xi^*$ is the position below which the fluid is Newtonian (Fig. P-6.6). [See Z. Tadmor, *Polym. Eng. Sci.*, **6**, 203 (1966).]

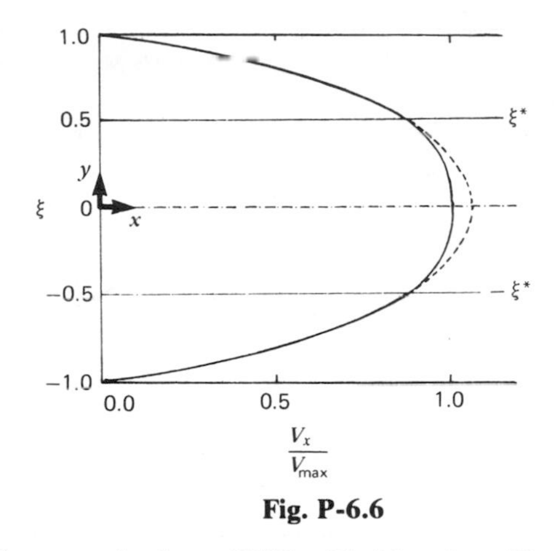


**Fig. P-6.6**

**6.7** ***Stresses Generated by CEF Fluids in Various Viscometric Flows.*** What stresses are necessary to maintain a CEF fluid flowing in the following flows:

(*a*) parallel plate drag.

(*b*) Couette flow with the inner cylinder rotating.

(*c*) parallel plate pressure flow?

Assume the same type of velocity fields that would be expected from a GNF or a Newtonian fluid. The three above-named flows are all viscometric; you should obtain the results in Eqs. 6.6-2 to 6.6-4.

**6.8** ***Special Form of the Rabinowitsch Equation.*** Show that the expression $Q = [\pi R^3/(s+3)](-R\Delta P/2mL)^s$ is a special form of the Rabinowitsch equation 6.7-11 for a power law fluid.

**6.9** ***The Rabinowitsch Equation for Fluids Exhibiting Slip at the Wall.*** Derive the Rabinowitsch equation for the case where the fluid has a "slip" velocity at the wall $V_w$. [See L. L. Blyler, Jr. and A. C. Hart, *Polym. Eng. Sci.*, **10**, 183 (1970).]

**6.10** ***The Flow of Pseudoplastic and CEF Fluids in Flows Between Almost Parallel Plates.*** The lubrication approximation was discussed in Chapter 5 in terms of Newtonian fluids. Considering a nearly parallel plate pressure flow ($H = H_0 - Az$), where $A$ is the "taper," what additional considerations to the lubrication approximation would have to be made for:

(*a*) A pseudoplastic fluid that is forced to flow at *higher* average velocity gradients with increasing $z$.

(*b*) A CEF fluid that exhibits normal stresses in "viscometric" flows such as the nearly parallel plate above. The normal stresses are functions of the shear rate $\dot{\gamma}$.

**6.11** ***Evaluation of GNF Fluid Constants from Viscometric Data.*** Using the viscosity data given on DPDA-6169 in Fig. P-6.11 and Eqs. 6.5-2, 6.5-7, and 6.5-8, obtain the material constants of the three models. Check your results by comparing with the constants given for this material in Appendix A.

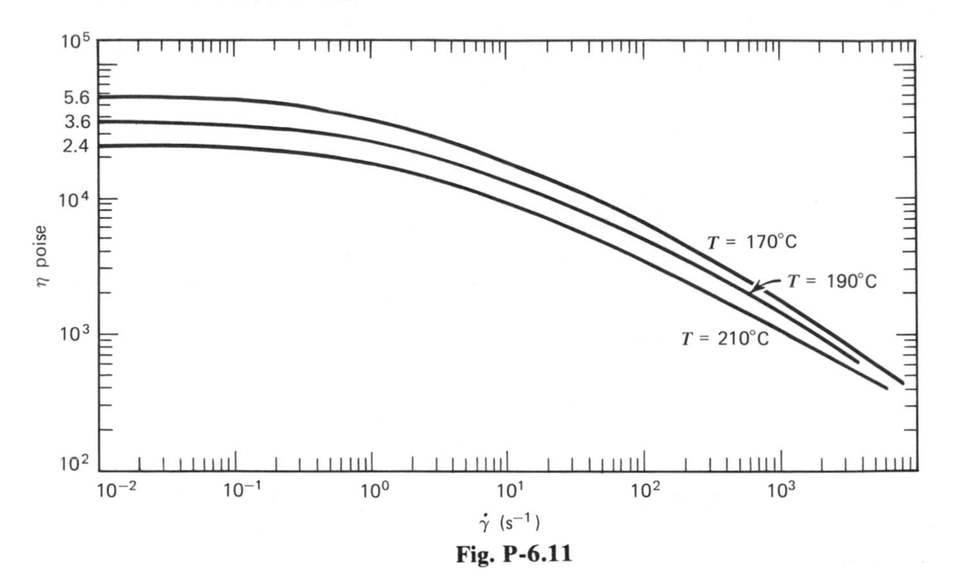


**Fig. P-6.11**

**6.12** ***Helical Annular Flow.*** Consider the helical annular flow brought about by the axial pressure gradient and the rotation of the outer cylinder. Specify the equations of continuity and motion ($z$ and $\theta$ components) and show that if a Newtonian fluid is used, the equations can be solved independently, whereas if $\eta = \eta(\dot{\gamma})$ where $\dot{\gamma}$ is the magnitude of $\dot{\boldsymbol{\gamma}}$, the equations are coupled.

**6.13** ***Dimensional Changes in Planar and Biaxial Extensional Flows.*** Determine the rate of dimensional changes that have to be applied on a flat film in order to generate ($a$) planar extension and ($b$) biaxial extension flows.

**6.14** ***Pressure Flow Calculations Using the Equivalent Newtonian Viscosity.**** Consider fully developed isothermal laminar pressure flow between parallel plates of a non-Newtonian liquid whose rheological behavior is decreased by a polymimial relationship

$$\ln \eta = a_0 + a_1 \ln \dot{\gamma} + a_{11}(\ln \dot{\gamma})^2 + a_2 T + a_{22} T^2 + a_{12} T \ln \dot{\gamma} \qquad \dot{\gamma} \geq \dot{\gamma}_0,$$

$$\eta = \eta(T), \qquad \dot{\gamma} \leq \dot{\gamma}_0$$

The above equations (which are *not* constitutive equations) describe with good agreement the behavior of many polymeric melts. The coefficients $a_{ij}$ can be accurately determined from experimental data by standard multiple regression methods.

($a$) Show that the flow rate per unit width is given by

$$q = \frac{2h^2}{\tau_w^2} \int_0^{\tau_w} \tau \dot{\gamma} \, d\tau$$

where $h$ is half the thickness and $\tau_w$ is the shear stress at the wall.

* E. Broyer, C. Gutfinger, and Z. Tadmor, "Evaluating Flows of Non-Newtonian Fluids by the Method of Equivalent Newtonian Viscosity," *A.I.Ch.E.J.*, **21**, 198 (1975).

(*b*) Show that for a Newtonian fluid the flow rate can be written as

$$q = \frac{2}{3} \frac{h^2 \tau_w}{\mu}$$

(*c*) Show that by defining an equivalent Newtonian viscosity

$$\bar{\mu} = \frac{\tau_w^3}{3 \int_0^{\tau_w} \tau \dot{\gamma} \, d\tau}$$

the flow rate of a non-Newtonian fluid can be calculated with the Newtonian equation in (*b*) with $\mu$ replaced by $\bar{\mu}$.

(*d*) Show that $\bar{\mu}$ can be expressed uniquely in terms of $\tau_w$ and $T$, for example, by an equation such as

$$\ln \bar{\mu} = b_0 + b_1 \ln \tau_w + b_{11}(\ln \tau_w)^2 + b_2 T + b_{22} T^2 + b_{12} T \ln \tau_w$$

and indicate a procedure for evaluating the coefficients of $b_{ij}$ from $a_{ij}$.

(*e*) Using the expression in (*d*) explain how to calculate the flow rate for a given pressure drop, and the pressure drop for a given flow rate.

# 7

# Characterization of Mixtures and Mixing



Mixing is a process that reduces composition nonuniformity. It is a vital step in polymer processing because mechanical, physical, and chemical properties, as well as "appearance," are strongly dependent on composition uniformity. We can cite numerous examples of mixing operations in polymer processing. Indeed, it would be hard to find a processing line without mixing operations. Such operations may involve both solids and liquids: blending polymer particles with color concentrates, fillers, or other additives is a solid-solid mixing operation practiced daily in many plants; the dispersion of carbon black in polyethylene is a typical and important example of solid-liquid mixing operation, and the blending of polymer melts is an example of liquid-liquid mixing.

Among the various mixing operations, those involving polymeric liquids and solid-polymeric liquid systems are perhaps most characteristic of polymer processing. These are dominated by one overriding factor: the *very high viscosity* of the system. Stemming from that is an almost complete lack of eddy and molecular diffusion, which are the main mixing modes commonly encountered by engineers. The significance of this uniqueness can be well appreciated by comparing the mixing of a blob of color in viscous oil paints to the mixing of a spoonful of milk in a cup of coffee.

Viscous mixing raises many interesting practical and theoretical questions. The practical aspects of mixing in general are still referred to, with certain amount of justification, as the "art of mixing," whereas the theoretical aspects are only in a developmental stage. Nevertheless this chapter discusses some fundamental aspects of mixing such as the definition of basic concepts and mechanisms. We shall deal with two main topics: *characterization of quality*, or "*goodness of mixing*," and the *characterization of the mixing process* itself. The two aspects are of course not independent of each other because a certain mixing process leads to a certain mixing quality. However, the quantitative relationship between the two has not yet been properly elucidated.

We shall return to mixing problems in Chapter 11, dealing with the elementary step of "mixing."

## 7.1 Basic Concepts and Mixing Mechanisms

We shall start our discussion with two fundamental concepts: *mixture* and *mixing*. The former defines the nature of the state of the materials we are concerned with, while the latter the mechanism by which we manipulate a property of the former. *Mixture* is defined (1) as "the state formed by a complex of two or more ingredients which do not bear a fixed proportion to one another and which, however comingled, are conceived as retaining a separate existence." This chapter considers systems that have two ingredients. The one having the higher overall concentration is called the *major component*, and the other ingredient is the *minor component*. *Mixing* is an operation that is intended to reduce the nonuniformity of the mixture. This can be accomplished only by inducing physical motion of the ingredients. Three basic types of motion are involved in mixing. Brodkey (2) calls these motions diffusions and classifies them as *molecular diffusion*, *eddy "diffusion*," and *bulk "diffusion*." Molecular diffusion is a process that occurs spontaneously driven by a concentration (chemical potential) gradient. It is the dominant mechanism of mixing in gases and low viscosity liquids. In turbulent mixing, molecular diffusion is superimposed on the gross random eddy motion, which in turn may occur within a larger scale of bulk "diffusion" or *convective flow* process.

In polymer processing, because of the very high viscosities of polymer melts, eddy "diffusion" is rarely reached and molecular diffusion is almost insignificant in that it occurs extremely slowly. Therefore, we are left with convection as the dominant mixing mechanism. The same holds for solid-solid mixing, where convection is the only mixing mechanism. We should note at this point, however, that if one of the components is a low molecular weight material (e.g., certain antioxidants, foaming agents, dyes used for fibers, slip additives) molecular diffusion may be a significant factor in the mixing process. Moreover, the utility of such additives in the finished product must also depend on molecular diffusion. Molecular diffusion plays, of course, an important role in such mass transfer dominated processes as stripping and devolatilization. Nevertheless, our discussion in this chapter centers on viscous systems where molecular diffusion is negligible.

Convection involves movement of fluid particles, blobs of fluid, or clumps of solid from one spatial location in a system to another. Convection results in mixing either if the interfacial area between the minor and the major component increases (3), or if the minor component is distributed throughout the major component without necessarily increasing interfacial area (4). The former criterion is relevant primarily to liquid-liquid mixing, and the latter to solid-liquid and solid-solid mixing. Convective mixing can be achieved by a simple bulk rearrangement of the material that involves a plug-type flow and requires no continuous deformation of the material. Therefore it can be termed *bulk-convective mixing, plug-convective mixing,* or simply, *distributive mixing.* Spencer and Wiley (3) have referred to this kind of mixing as *repetitive mixing,* and McKelvey (5) has used the term *simple mixing.* This kind of mixing, through repeated rearrangement of the minor component, can in principle reduce nonuniformities to the molecular level. The repeated rearrangement in *distributive mixing* can be either *random* or *ordered.* The former is the process that takes place, for example, in V-blenders and many other solid-solid mixers, whereas the latter forms part of the mixing mechanism in certain "motionless" mixers. Figure 7.1 shows schematically these two types of distributive mixing.

Convective mixing also can be achieved by imposing deformation on a system through laminar flow. Hence we term this kind of convective mixing *laminar convective mixing.* Others refer to it as *streamline mixing* (3) or simply *laminar mixing* (5). Liquid-liquid and liquid-solid mixing in processing are accomplished by laminar convective mixing through various types of flow: shear, elongation (stretching), and squeezing (kneading). However, shear flows play the major role in processing. The nature of this kind of mixing is illustrated in Fig. 7.2. Consider a tube filled with viscous liquid at rest into which a pulse of dye is injected (Fig. 7.2*a*). In the absence of molecular diffusion, no mixing occurs. If, however, the fluid is set in motion for a certain period of time $t_1$ (Fig. 7.2*b*), the mean concentration of the dye at the point of injection drops, the interfacial area increases, and the two ingredients move relative to each other. All these changes imply that laminar convective mixing has occurred.

We can generally state that if a liquid-liquid system is to be mixed by a laminar convective mechanism, permanent deformation or strain (3) must be imposed on the system. The term "strain" in the context of laminar convective mixing does not include elastic or delayed elastic strains and certainly not rigid body rotation and translation. Furthermore, it should be obvious that for the interfacial surface to increase, *both* phases must undergo flow. Hence the viscosity ratio (as well as the viscoelasticity of the phases, which would bring about strain recovery) plays an important role in laminar mixing. Chapter 11 deals with these aspects of mixing.

The foregoing discussion implies that in laminar mixing the decisive variable is the strain, whereas the rate of application of strain and stresses play no role. This is indeed the case for all operations that do not involve materials that exhibit a yield point (and are miscible) (1). In these cases, shear stresses are irrelevant as far as the degree of mixing is concerned (it is not irrelevant, of course, to power requirements). However when we do have a component that breaks only upon reaching a certain yield stress, the local stresses do play a very decisive role in the mixing operation. Examples of such components are carbon black agglomerates and viscoelastic polymer blobs. Moreover, with viscoelastic systems in particular, the rate of stress buildup or local stress histories may be very important. This kind of mixing is referred to as *dispersive mixing* (5) for solid-liquid mixtures and *homogenization* for liquid-liquid mixtures. In dealing with dispersive mixing, we shall later use the term *ultimate particle,* which is defined as the smallest particulate piece of the minor component existing in the system.

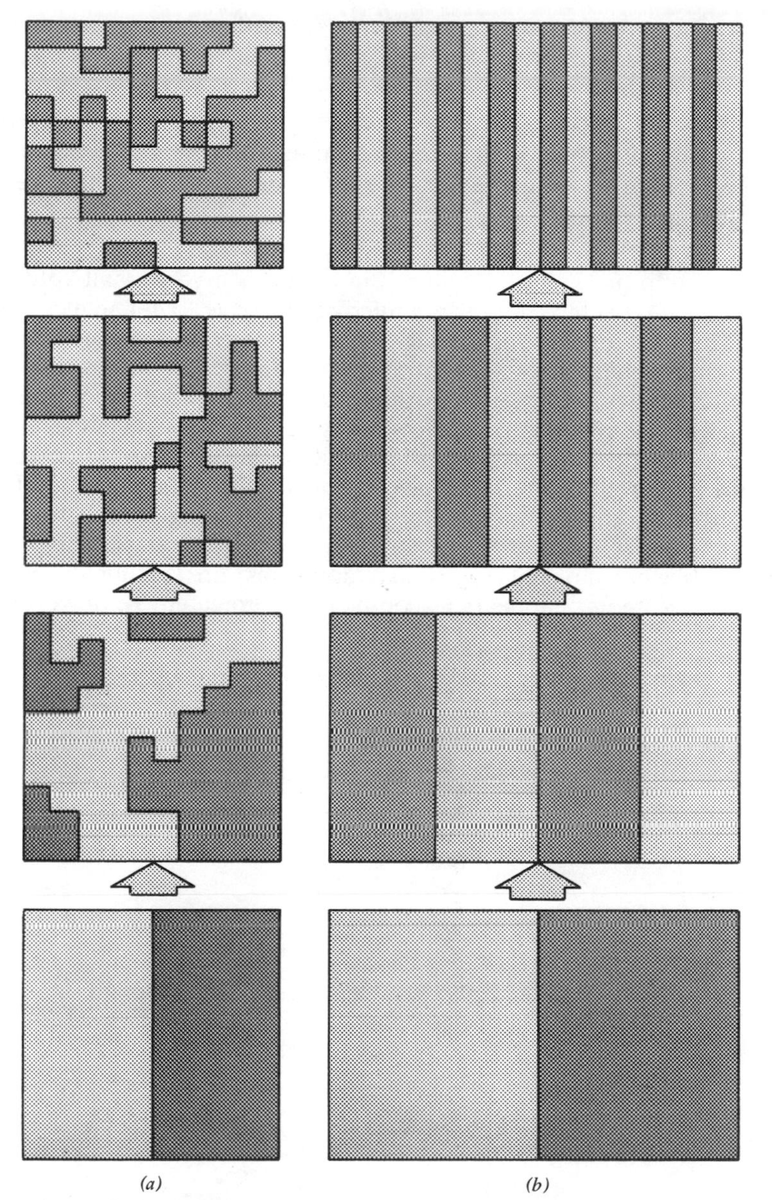


**Fig. 7.1** Schematic representation of distributive mixing. (*a*) Random rearrangement (e.g., a process that takes place in a V-blender). (*b*) Ordered rearrangement (e.g., a process that partially takes place in "motionless mixers").

In summary, we are dealing in polymer processing with both *nondispersive* and *dispersive mixing*, also referred to as *extensive* and *intensive* mixing. The basic type of motion in the former is obtained by *convection*. The type of mixing may be either *distributive* or *laminar*. The mechanism through which the former is imposed on the material may be either an *ordered* or a *random* rearrangement process, whereas the latter (i.e., laminar mixing) is achieved by deforming the material in various laminar flow patterns (e.g., shearing, squeezing, or elongational flows).

We now turn to the two main areas of concern outlined above, namely, how to characterize the state, quality, or "goodness of mixing," and how to characterize the mixing process itself.

It is appropriate at this point, before discussing characterization methods of goodness of mixing, which involve sampling and measurement of concentration of the minor component at various points in the mixture, to discuss the meaning of "concentration" at a "point."

> In continuum mechanics, a "point" in a fluid is a very small volume in macroscopic scale, yet large enough in microscopic scale to define meaningful local averages of temperature, velocity, concentration, and so on. Trying to apply the same definition of concentration in our system, we are bound to encounter difficulties because, as we have concluded earlier, practically all mixing in polymeric systems takes place through convection and in the absence of molecular diffusion. According to this mechanism the process of mixing is nothing but a bulk redistribution of one component within the other. It follows, therefore, that at any "point" in the system, according to the foregoing definition, there will be either the minor component or the major component. In other words, when molecular and eddy "diffusions" are lacking, the mixture at the scale of a "point" appears as completely segregated. If, however, the

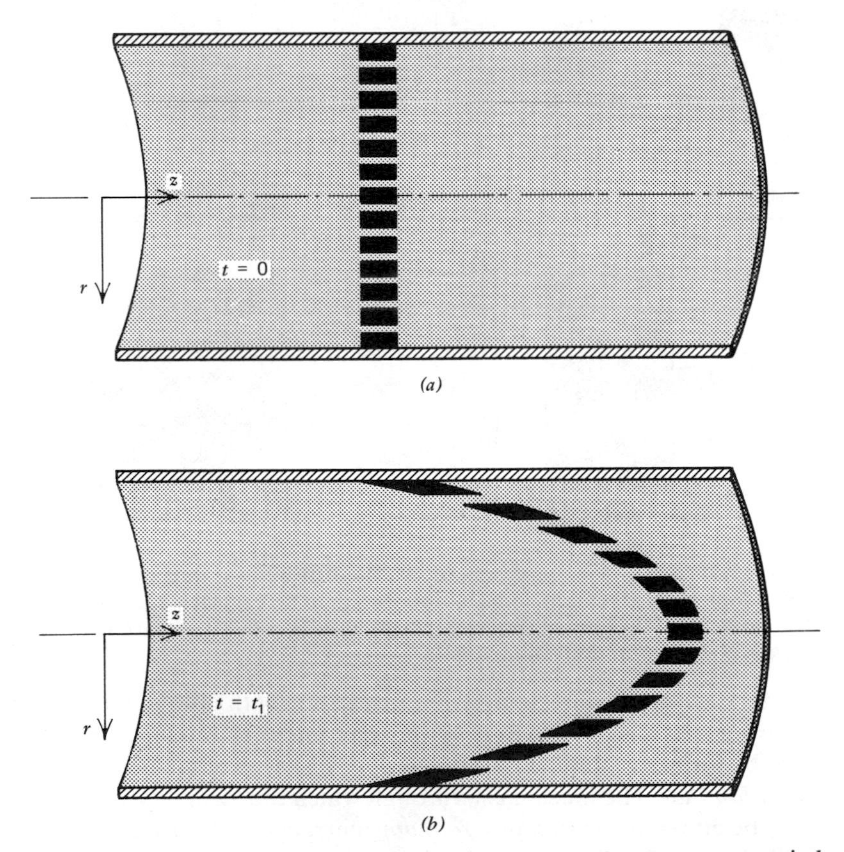


**Fig. 7.2** Laminar mixing in laminar pipe flow. Dark areas are occupied by a tracer consisting of the bulk liquid and a small amount of a dye.

meaning of the "concentration at a point" is expanded to imply a representative concentration of a small local volume, which is much larger than the volume of the ultimate particle or the size of the segregated regions, yet much smaller than the "scale of examination" (see below), a meaningful analysis of goodness of mixing is possible. Of course a concentration so defined cannot be used for evaluating processes such as rates of reaction, determined by molecular events, that require local values on a scale of a "point" much smaller than our "point."

## 7.2 Characterization of Mixtures

A complete characterization of the state of a mixture would require the specification of the size, shape, orientation, and spatial location of every particle, clump, or blob of the minor component. In certain cases, such as the mixture of a uniform sized minor component, the spatial location of every particle would provide complete characterization of the state of the mixture. A somewhat less than complete characterization could be provided by a three-dimensional concentration distribution function, as suggested by Bergen et al. (4). However, for many applications we do not need such a complete characterization of the state of the mixture; in practice, simple methods often suffice. Commonly used methods, for example, are color comparison to a standard for qualitative visual homogeneity, or the measurement of some representative physical property. Every case must be considered on the basis of the nature of the material components, the purpose of the mixing, and the intended application.

Between the two extremes of complete characterization and qualitative or semiqualitative practical evaluation, there is room for a sound quantitative method of characterization. Such a characterization method is described next.

It should be emphasized, however, that unless the reasons for making up a mixture are known, it is impossible to decide whether it is well or poorly mixed.

Generally, in dealing with a mixture we would first examine *overall* or *gross composition uniformity*. By *gross uniformity* we mean some quantitative measure that characterizes the goodness of distribution of the minor component throughout the object or system analyzed. Let us try to clarify this point with an example. Consider an extrusion line for manufacturing blue shopping bags, made out of rolls of blown film. We might take bags from such a roll and examine them for color uniformity. Such an analysis might reveal that all bags appear to be alike, and when examined quantitatively, they contain virtually the same amount of blue pigment; that is, there is a *perfect gross uniformity* throughout the film. Alternatively, the analysis might reveal that although the overall concentration of pigment is virtually the same in each bag, individual bags display nonuniformity in the form of patches, stripes, streaks, and so on; that is, the bags exhibit a certain *texture*. Finally, the analysis might reveal both widely varying pigment concentrations among bags as well as different textures in each one. Indeed, if the mixture fed to the extruder was not uniform in composition, there is a good chance that we would find blue and uncolored regions or a great deal of variation in shades of blue over the roll of film. But not every composition uniformity can be evaluated by visual examination. For example, if the additive is colorless, or if we want quantitative answers on blue pigment distribution in the roll of film, we must take *testing samples*, measure the concentration of the minor component at various points in the film and analyze these for uniformity. As we shall see later, random mixing processes or even pseudorandom processes (such as occur in internal mixers, where the very complex flow pattern is dominated by many uncontrolled events), lead not to a perfect gross uniformity—that is, the same concentration in all samples—but to a binomial distribution of concentration. The binomial distribution may, of course, be very narrow.

In the discussion above two concepts were intuitively introduced and need further clarification. One is the *scale* on which we *examine* for composition nonuniformity, the second is the size of the testing samples.

We are examining in this example the composition uniformity of a *whole* roll of film; this then becomes the *scale* for measuring the composition uniformity. We can, therefore, define a new concept, the *scale of examination*, which is the scale or size of the overall sample we are analyzing for composition uniformity. Clearly the scale of examination is imprecisely defined and can be expressed only as an order of magnitude (length, area, or volume). In testing for gross uniformity, the scale of examination is of the size of the object, system, or overall sample examined for uniformity. To measure the concentration of the minor component in this object, system, or overall sample at various locations, we must withdraw small testing samples. What should their size be? Clearly, they should be small compared to the scale at which we examine for uniformity and large compared to the size of the ultimate particle. If, however, the distribution of the minor component forms a certain *texture*, the sample size should be larger than the "granularity" of the texture.

We have noted before that by *texture* we mean composition nonuniformity reflected in patches, stripes, and streaks. Thus by texture we mean a composition nonuniformity that has some unique *pattern* that can be *recognized* by visual *perception*. Thus a "blind" random sampling of concentration at various points, though it may reveal the existence of compositional nonuniformity and may even suggest the *intensity* of this nonuniformity, will reveal nothing about the character of the texture. *Random* sampling at *individual* points cannot reveal the certain amount of *order* that is retained and exhibited as a texture. Texture is important in polymer processing because (a) laminar and distributive mixing inevitably lead to it, (b) many products are visually examined for lack of texture or for a certain desired texture, and (c) mechanical properties of blends depend on the texture of the mixture.

Since texture has a great deal to do with visual perception, it is interesting to discuss some work done on *texture recognition*. The first question we want to focus on is, What makes two samples with identical overall concentration exhibit different textures?

Perfect gross uniformity implies identical concentration in all samples. Julesz (6), in an interesting paper on visual perception of texture, refers to two samples with the same concentration of a minor component as two samples with the same *first order statistics*. When dealing with visual perception, first order statistics has to do with brightness, or rather luminance. Samples taken from a system having perfect gross uniformity of minor component, or with the same first order statistics, may have nonuniform appearance by way of streaks, stripes and patches, or in general they may have different *granularity*. In other words, each sample may have

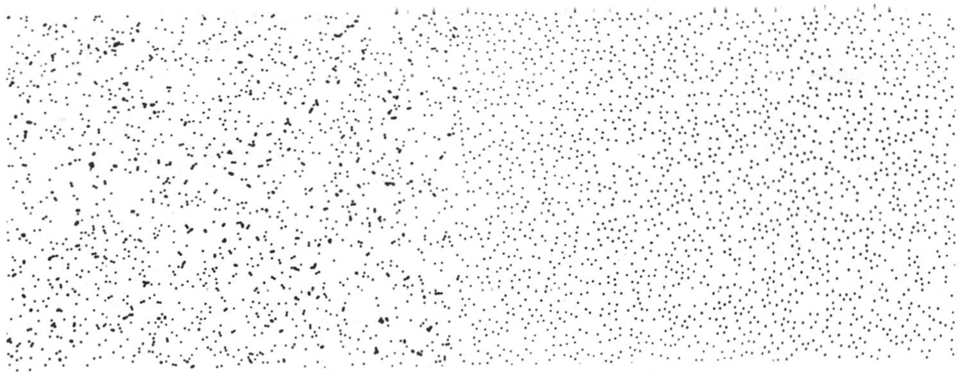

**Fig. 7.3** The two textures (left field and right field) have the same first order statistics (the same number of black dots), but they differ in second order statistics. In the left field the dots fall at random, whereas in the right field there are at least 10 dot diameters between dots. [Reprinted with permission from B. Julesz, *Sci. Am.* **232**, 34 (1975).]

its unique texture, which is determined, of course, by the manner in which the minor component is distributed. Figure 7.3 gives a simple example of two samples sharing the same first order statistics but having different granularity or different second order statistics. There is an apparent difference in granularity (texture) between the left and right fields. In the former, black dots were randomly placed on the field, whereas in the latter there is a distance of at least 10 dot diameters between the black dots. For measuring second order statistics, Julesz (6) suggests dropping a *dipole* (e.g., a needle) on the two textures and observing the frequency with which both ends of the dipole land on black dots. Identical frequencies imply identical second order statistics. His experiments indicate that our visual system can discriminate patterns solely by the perceptual process, only if they differ in second order statistics.* Figure 7.4 presents a somewhat more complex texture; black, dark gray, light gray, and white squares are mixed such that their respective fractions are equal in the left and right fields. The two textures can be discriminated because there is a difference in second order statistics. Third order differences between the texture could no longer be discriminated by perception but require a cognitive process of scrutiny. Neither, of course, can such differences be detected by a "dipole throwing process," but would require a "tripod or triangle throwing process."

Finally, Fig. 7.5 shows a number of film samples extruded under different conditions. They have approximately the same carbon black concentration—that is, they are grossly uniform or have the same first order statistics—but clearly they exhibit different textures or granularities, thus have different second order statistics.

The next section demonstrates that one quantitative measure of texture is the *scale of segregation*, which is calculated by a process of "dipole throwing."

Now we must ask, What is the scale of examination when we analyze texture? In dealing with gross uniformity, the scale of examination was clearly defined, and it equaled the system as a whole. In dealing with texture, the situation is somewhat

* Julesz has made a similar observation with musical "textures," and he found that random melodies could be perceived as being different only if they possessed different second order statistics.

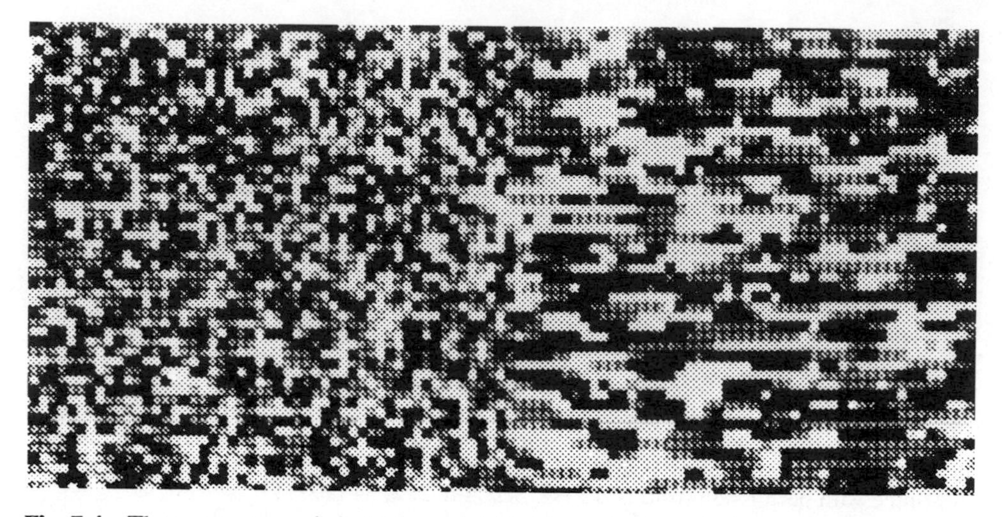

**Fig. 7.4** The two textures (left field and right field) made of black, dark gray, light gray, and white squares have the same first order statistics, but different second order statistics, which appears as a difference in granularity. [Reprinted with permission from B. Julesz, *Sci. Am.* **232**, 34 (1975).]

more complex. We have found that by perception we can discriminate textures differing in second order statistics. These may be due to fine and coarse granularity. The human eye, therefore, through a hierarchy of feature extractors of increasing complexity, can simultaneously analyze texture on a *range of scales of examination.* This range varies according to the distance from which we view the object. Moreover, if in addition to pure perception we start examining and scrutinizing the texture, we commence a cognitive process activating "hypercomplex feature extractors" in our visual system, leading to more detailed and complex texture and pattern recognition (6), which would require a higher than second order statistics evaluation process. Similarly, to the human visual system, in attempting to quantitatively characterize the texture of a sample, the scale of examination *should be*

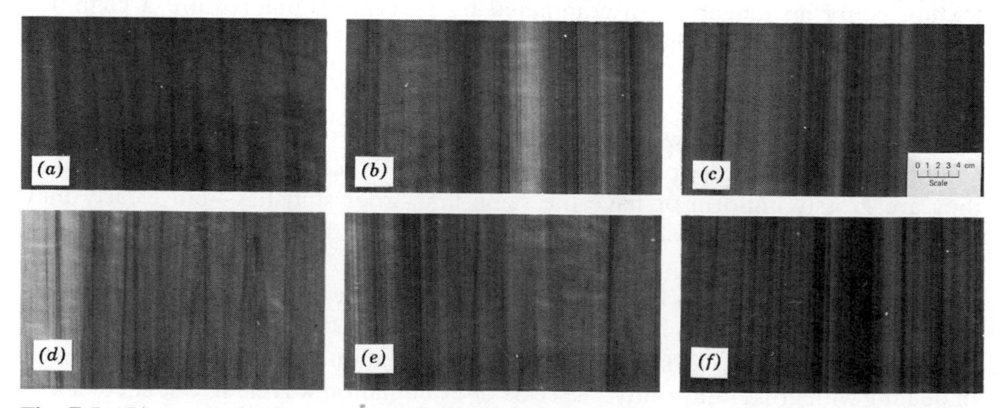


**Fig. 7.5** Photographs of extruded LDPE films with carbon black concentrate extruded at various conditions. The barrel temperature (°C) and screw speed (rpm) are as follows: (*a*) 160°, 40; (*b*) 160°, 60; (*c*) 160°, 80; (*d*) 180°, 40; (*e*) 180°, 60; (*f*) 180°, 80. [Reprinted with permission from N. Nadav and Z. Tadmor, *Chem. Eng. Sci.*, **28**, 2115 (1973).]

*varied over a range*, determined by the requirements. Returning to the example of blue shopping bags, we can decide that the upper limit of the scale on which we shall examine texture is the size of a bag, and the lower limit is the resolution capability of the naked eye or the minimum size of nonuniform regions. This range will suffice if the analysis is carried out for evaluating the appearance of a bag. If, however, we are dealing with a blend of two polymers, the mechanical properties of the mixture are affected by the uniformity and texture down to a microscopic scale throughout the volume of the sample.

At the lower limit of the scale of examination, at the level of ultimate particles, we also must characterize the *local structure*. When dealing, for example, with dispersion of carbon black particles on this scale of examination, we shall determine whether the particles are agglomerated in clumps or individually dispersed. This fact may significantly affect the chemical properties (e.g., resistance to weathering of PE containing carbon black) and the mechanical properties of polymers.

The significance of local structure depends to a large extent on the function of the minor component in the system. It can be investigated by direct microscopic examination or other techniques capable of distinguishing structure on the scale of the ultimate particle (7).

In conclusion, the state of the mixture can be characterized by *gross uniformity*, *texture*, and *local structure*.

Having discussed composition uniformity qualitatively, we now proceed to deal with the quantitative aspects of characterization of mixtures.

## 7.3 Gross Uniformity

A perfect gross uniformity is obtained when there is a uniform concentration of the minor component in all testing samples taken from a specific system. In the example of shopping bags made from a blown roll of film, this implies the same amount of blue pigment in every bag, provided we have chosen a bag as a sample. In most practical cases, however, complete gross uniformity is not achieved. The *maximum* attainable uniformity is controlled by the mixing method, and the *actual* gross uniformity is determined by the conditions and time of mixing. In random mixing processes the maximum attainable uniformity is given by the binomial distribution (8). Consider a system of a mixture of a minor component of uniform size in a major component, which can be solid or liquid. In either case we shall measure the amount of major component in a withdrawn sample as the number of hypothetical particles, each having the same volume as a minor component particle. If the volume fraction of minor component is $p$ and we withdraw samples containing $n$ particles, the fraction of samples containing $k$ minor particles is given by

$$b(k; n, p) = \frac{n!}{k!(n-k)!} p^k (1-p)^{n-k} \qquad (7.3\text{-}1)$$

Equation 7.3-1 shows that the distribution of the minor component in the samples depends both on the average concentration of the minor component $p$ and on the size of the sample $n$. This point becomes more evident by considering the variance of the binomial distribution

$$\sigma^2 = \frac{p(1-p)}{n} \tag{7.3-2}$$

The more particles the samples contain, the narrower the distribution. In samples taken from true solutions, where the ultimate particles are molecules, the number of molecules in the smallest practical withdrawn sample is enormous, and the variance will approach a value of zero and the distribution will be virtually uniform. Figure 7.6 demonstrates the effect of the sample size on the shape of the binomial distribution.

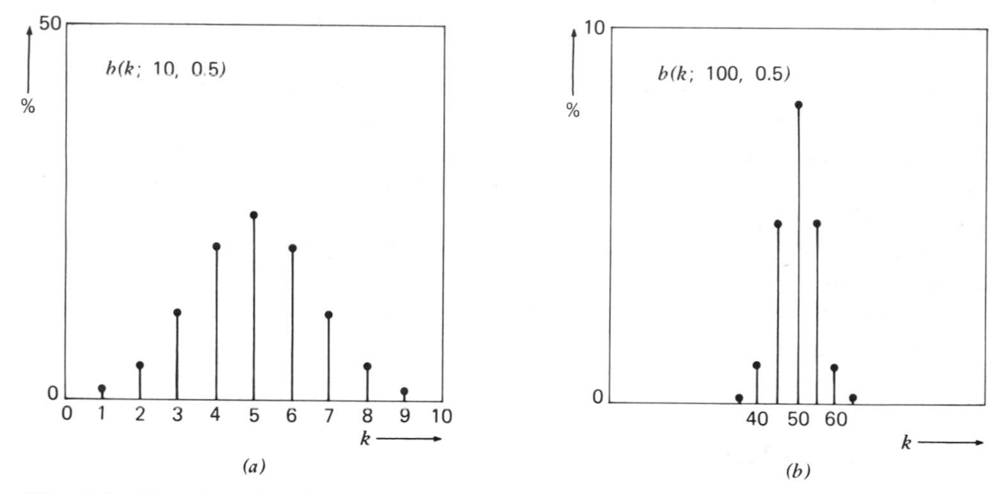


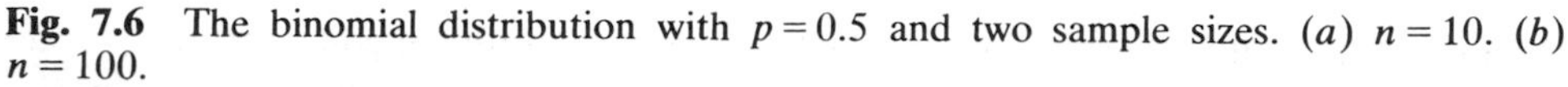


**Fig. 7.6** The binomial distribution with $p = 0.5$ and two sample sizes. (*a*) $n = 10$. (*b*) $n = 100$.

To determine experimentally the closeness of a mixture to random distribution, we must sample the mixture, measure the concentration of the minor component in the withdrawn samples, calculate its volume fractions $x_i$, and compare the resulting distribution to the appropriate binomial distribution. The average volume fraction of the minor in the samples, which is given by

$$\bar{x} = \frac{1}{N} \sum_{i=1}^{N} x_i \tag{7.3-3}$$

where $N$ is the number of withdrawn samples, should not be significantly different from the fraction of the minor component in the mixture $p$. A significant deviation of their values would indicate a faulty testing procedure (9). We can test for the statistical significance of this difference by a $t$-test or $z$-test, for sample numbers below and above 30, respectively. In either case, the statistic $t$ (or $z$) is calculated from the following equation:

$$t \text{ (or } z) = \frac{(\bar{x} - p)}{S/\sqrt{N}} \tag{7.3-4}$$

where $S^2$ is the observed variance (see Eq. 7.3-6) and compared to standard table values. In addition to comparing the means, we also have statistical tests for determining whether the experimentally observed distribution differs significantly from the expected (binomial) distribution, or the difference is only due to chance. The statistic used for this comparison is $\chi^2$.

Unfortunately, this calculation procedure is tedious; in practice, simpler, though less reliable tests are used for evaluating the state of mixing. This is done by calculating certain *mixing indices* that relate representative statistical parameters of the samples, such as the variance and mean to the corresponding parameters of the binomial distribution. One such index is defined as follows:

$$M = \frac{S^2}{\sigma^2} \tag{7.3-5}$$

where $\sigma^2$ is the variance of the binomial distribution, and $S^2$ is the variance of the samples defined as

$$S^2 = \frac{1}{N-1} \sum_{i=1}^{N} (x_i - \bar{x})^2 \tag{7.3-6}$$

where $x_i$ is the volume fraction of the minor component in test sample $i$; clearly, for a random mixture $M = 1$. For an unmixed state that is a completely segregated system (where samples contain either major or minor components only), the variance* is

► $$\sigma_0^2 = p(1-p) \tag{7.3-7}$$

and the mixing index $M$ according to Eq. 7.3-5 attains a value of $n$, where $n$ is the number of particles in the sample.

Numerous other mixing indices were defined using various combinations of $S^2$, $\sigma^2$, $\sigma_0^2$, and $\bar{x}$. These were reviewed by Bourne (10) and analyzed by Fan and Wang (11); Table 7.1 lists a few of them. The last mixing index in Table 7.1 compares breadths of the measured and binomial distribution relative to the mean. The physical reasoning for this is that at low minor concentrations, one requires better (closer to random) mixing for the "same" quality than is needed at higher concentrations. It is perhaps interesting to note in passing that in characterizing molecular weight distribution, we also use a relative measure of the breadth, such as the dispersion index ($D = \bar{x}_w/\bar{x}_n = 1 + \sigma^2/\bar{x}_n^2$), rather than an absolute measure of breadth, such as the variance.

The preceding discussion dealt with mixtures of uniformly sized particulate solids, or uniformly sized particulate solids in a liquid. In the latter case, as we mentioned before, the liquid phase in the sample can be considered to consist of a number of hypothetical particles, each having the same volume as a minor component particle. In the case of two liquid components, the situation may be somewhat different. If we expect a true solution, mixing should lead to virtually uniform composition in the samples. If, however, the two polymers are nonsoluble, we are faced with a more complex problem because one cannot define the theoretical limit for random mixing. The attainable limit will depend in the mixing

* A fraction $p$ of the samples will contain pure minor component and a fraction, $1-p$, pure major component; hence, $\sigma_0^2 = p(1-p)^2 + (1-p)(0-p)^2$. Alternatively, Eq. 7.3-7 can be obtained from Eq. 7.3-2 with $n = 1$, because if the sample size is reduced to the size of the ultimate particle of the minor component, the mixture will be segregated (no mixing is possible on this scale).

**Table 7.1 Mixing Indices Comparing Concentration Distribution of the Minor Component in the Samples to the Perfectly Random Mixed State**

| Mixing Index | Limiting Values | |
|---|---|---|
| | Perfectly Mixed | Completely Segregated |
| $\dfrac{S^2}{\sigma^2}$ | 1 | $n$ |
| $\dfrac{\sigma_0^2 - S^2}{\sigma_0^2 - \sigma^2}$ | 1 | 0 |
| $\dfrac{\ln \sigma_0^2 - \ln S^2}{\ln \sigma_0^2 - \ln \sigma^2}$ | 1 | 0 |
| $\dfrac{S^2 - \sigma^2}{p}$ | 0 | $(1-p)\left(1-\dfrac{1}{n}\right)$ |

method. We can, of course, still sample the mixture and, if the "blobs" of the minor component are much smaller than the sample size, we would expect the variance $S^2$ of the samples to approach zero as gross uniformity is approached.

## 7.4 Texture

We shall now deal with the quantitative characterization of *texture*. Consider a simple, geometrically ordered, checkered texture of dark gray and light gray squares. This texture can be easily and fully characterized by measuring the length of a square and providing in some form the difference in intensity of the light and dark gray colors (e.g., concentration difference). The former will characterize the *scale* of the *granularity* or *segregation* and the latter, the *intensity of segregation*. These two concepts, developed by Danckwerts (12, 13) for characterizing the state of mixing in chemical reactors, were also suggested by him for texture characterization (13). Their statistical definition and physical meaning are discussed later in this section.

Clearly, a state of perfectly textureless compositional uniformity can be obtained by either reducing the scale of segregation to the scale of the ultimate particle (at the lowest limit this will be the molecular scale) or by reducing the intensity of segregation to zero, that is, increasing the concentration of the minor component in the light squares and decreasing it in the dark squares until they match. These processes are depicted in Fig. 7.7. The "goodness of mixing," as far as the texture is concerned, depends on some combination of both scale of segregation and intensity of segregation (perhaps their product). If the scale of segregation is small, a large intensity of segregation might be tolerated, and vice versa. Most textures, however, are not as simple as the previous example. The granularity and the intensity of segregation may spread over broad ranges. Hence some meaningful

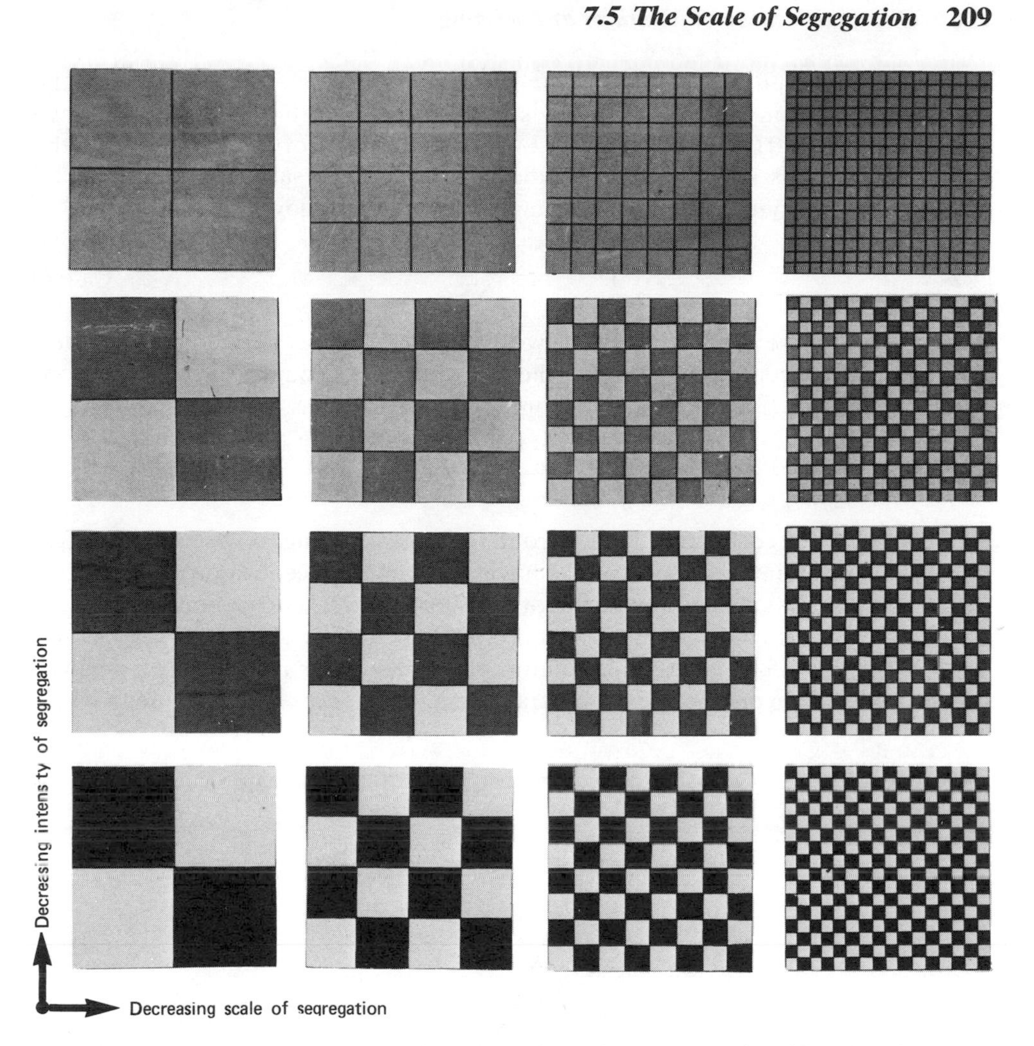


**Fig. 7.7** Schematic representation of scale and intensity of segregation. Note that decreases in both scale and intensity of segregation lead to mixture uniformity.

statistical averaging is needed to render them useful for complex texture characterization.

## 7.5 The Scale of Segregation and the Scale of Segregation Profile

The scale of segregation $\mathscr{s}$ is defined as the integral of the coefficient of correlation $R(r)$ between concentrations (volume fractions) at two points separated by a distance $r$

$$\mathscr{s} = \int_0^{\zeta} R(r)\, dr \qquad (7.5\text{-}1)$$

The integral is taken over values of $r$ ranging from zero, where both points have the same concentration [therefore there is a perfect correlation between the two concentrations $R(0)=1$], to a value $\zeta$ at which there is no correlations between the two concentrations ($R(\zeta)=0$), The dimension of $s$ is the same as that of $r$. The definition of the coefficient of correlation is given in the following equation:

$$R(r)=\frac{\sum_{i=1}^{N}(x'_i-\bar{x})(x''_i-\bar{x})}{NS^2} \tag{7.5-2}$$

where $x'_i$ and $x''_i$ are concentrations at two points at a distance of $r$ from each other, $\bar{x}$ is the mean concentration, and $N$ is the total number of couples of concentrations taken. The variance $S^2$ is calculated from the concentrations at all points

$$S^2=\frac{\sum_{i=1}^{2N}(x_i-\bar{x})^2}{2N-1} \tag{7.5-3}$$

The coefficient of correlation ranges from 1 to −1, denoting, respectively, perfect positive (both points in each couple have the same concentration) and perfect negative (one point is pure minor, the other is pure major) correlations. Figure 7.8 gives some typical correlograms which are plots of $R(r)$. The scale of segregation defined by Eq. 7.5-1 can be calculated along a line, over a surface, or within a volume, depending on how the concentration points are chosen. In all cases a linear

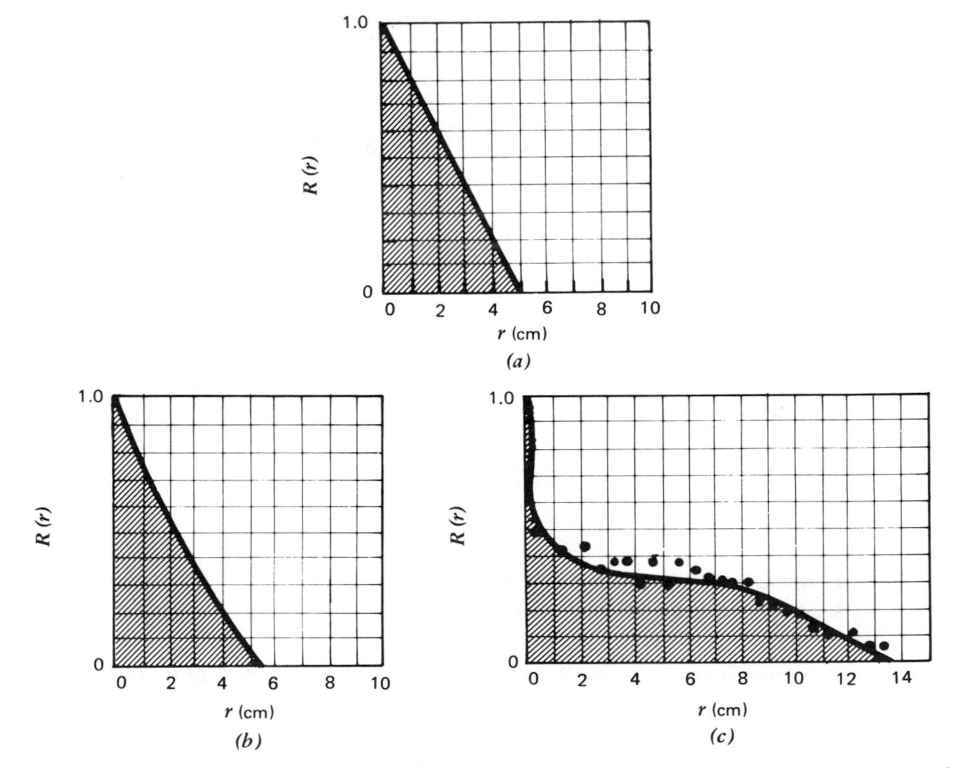


**Fig. 7.8** Typical correlograms. (*a*) Along a line perpendicular to an equally spaced striped texture. (*b*) Over an area of a checkered board texture. (*c*) Along a line of an extruded film, as shown in Fig. 7.5, perpendicular to the extrusion direction.

measure of the scale of segregation is obtained. The evaluation procedure involves random "dipole throwing," recording of the concentrations, and straightforward calculation with the aid of Eqs. 7.5-1 to 7.5-3.

**Example 7.1** ***The Coefficients of Correlation for a Simple Texture in Terms of Probabilities***

A better understanding of the physical meaning of $R(r)$ can be obtained by considering a simple texture consisting of only two types of region, with respective concentrations (volume fractions of the minor component) of $x_1$ and $x_2$, as in Fig. 7.9. By considering a (mental) process of random "dipole throwing," it is easy to show (see Problem 7.5) that the coefficient of correlation can be expressed in terms of probabilities as follows:

$$R(r) = \psi_{11}\frac{\phi_2}{\phi_1} + \psi_{22}\frac{\phi_1}{\phi_2} - \psi_{12} \tag{7.5-4}$$

where $\psi_{11}$ and $\psi_{22}$ are, respectively, the probabilities of both ends of the dipole falling on regions with composition $x_1$ and $x_2$, and $\psi_{12}$ is the probability that each end will fall on different region. The quotient $\phi_1/\phi_2$ is the ratio of

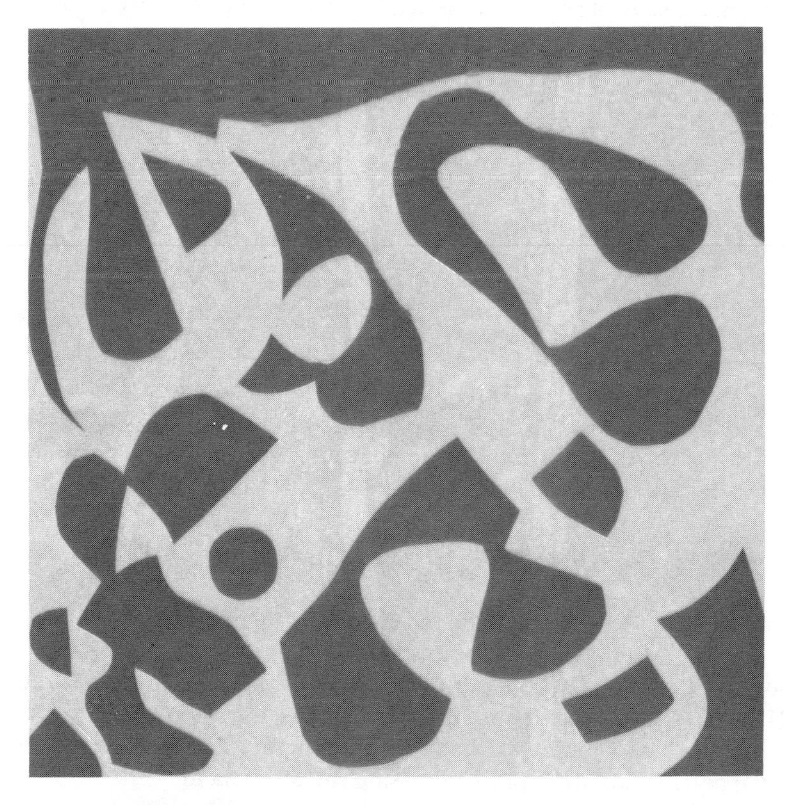

**Fig. 7.9** A texture composed of only two types of regions with compositions $x_1$ and $x_2$ occupying, respectively, area fractions $\phi_1$ and $\phi_2$.

areas [or lengths, when $R(r)$ is calculated along a line] occupied by regions of compositions $x_1$ and $x_2$. For equal areas, Eq. 7.5-4 reduces to

$$R(r) = (\psi_{11} + \psi_{22}) - \psi_{12} \tag{7.5-5}$$

In other words, the coefficient of correlation in this case is the difference between the probability that both ends of a dipole, thrown at random on the texture, will fall on identical regions (of either composition), and the probability that its ends will fall on different regions. Clearly, as $r \to 0$, $\psi_{11} + \psi_{22} \to 1$, $\psi_{12} \to 0$; therefore, $R(0) \to 1$. Conversely, as $r$ increases from zero, a value will be reached at which the probabilities of the dipole ends falling on identical and separate regions become equal and the coefficient of correlation becomes zero. This value of $r$ is, of course, related to the size of the segregated regions, which is the reason for using $R(r)$ in defining scale of segregation.

**Example 7.2** ***Correlograms of Some Simple Patterns and Geometrical Interpretation of the Scale of Segregation***

Figure 7.10 displays a simple linear pattern with variation in one direction. The pattern consists of region I of length $L_1$ with composition $x_1$, followed by region II of length $L_2$ with composition $x_2$. This unit pattern is then repeated.

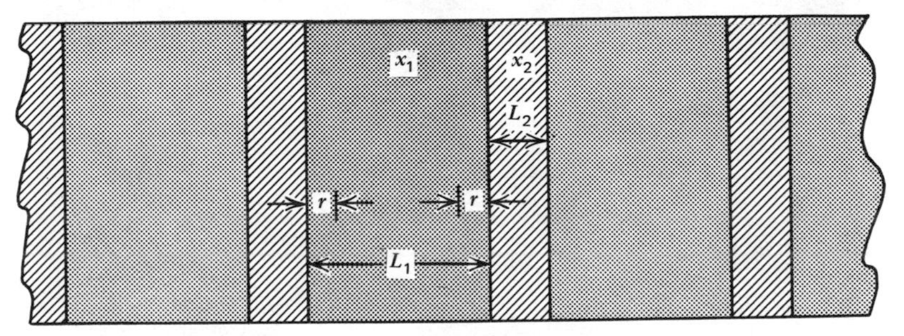


**Fig. 7.10** A simple striped texture.

An expression for the coefficient of correlation along the line perpendicular to the stripes can be derived with the aid of Eq. 7.5-4 (see Problem 7.6).

$$R(r) = 1 - r\left(\frac{L_1 + L_2}{L_1 L_2}\right) \qquad r \leq \frac{2L_1 L_2}{L_1 + L_2} \tag{7.5-6}$$

The correlogram of the pattern can be extended to any $r$. This, for the special case of $L_1 = L_2$, results in a sawtooth structure fluctuating from $+1$ to $-1$, with a period of $2L$. If $L_2 \neq L_1$, the periodic nature of the correlogram is retained, but $R(r) < 1$, with the exception of $R(0)$ which is 1.

Equation 7.5-6 indicates that the correlogram fulfills the requirement that $R(0)$ be 1. It is a straight line crossing the abscissa at

$$r = \frac{L_1 L_2}{L_1 + L_2} \tag{7.5-7}$$

The scale of segregation can now be obtained by integration from Eq. 7.5-1 to give

$$\mathscr{s} = \int_o^{L_1L_2/(L_1+L_2)} R(r)\,dr = \frac{1}{2}\left(\frac{L_1L_2}{L_1+L_2}\right) = \frac{1}{4}\bar{L} \tag{7.5-8}$$

where $\bar{L}$ is the harmonic mean of $L_1$ and $L_2$.

These results reveal a number of interesting characteristics of the scale of segregation. First, it should be noted that $\mathscr{s}$ for the case of only two concentrations becomes independent of concentration. The reason is that $\mathscr{s}$, in effect, is a measure of the sizes of the regions below and above a mean concentration. For such a system, the sizes of the regions will not change with a change in the level of either concentration. If, however, there is a concentration distribution, we would expect a change of the sizes of the regions with a change in the concentration. Results further indicate that the $\mathscr{s}$ accounts for both regions below and above the average concentration. Finally, Eq. 7.5-8 indicates that $\mathscr{s}$ is one-fourth the length of one region when $L_1 = L_2$; it is one-half the length of the minor component when $L_1 \ll L_2$:

$$\mathscr{s} = \frac{L}{4} \quad \text{if} \quad L_1 = L_2 = L \tag{7.5-9}$$

and

$$\mathscr{s} = \frac{L_1}{2} \quad \text{if} \quad L_1 \ll L_2 \tag{7.5-10}$$

For a simple texture, as that analyzed previously, we do not need a statistical measure of the scale of segregation. We could simply specify the value of $L_1$ and $L_2$, or following Mohr et al. (14), we can define the *striation thickness*, which is the linear separation between surfaces of equal concentration $(L_1 + L_2)$*. However the purpose of the discussion is to demonstrate the physical and geometrical meaning of the scale of segregation. The scale of segregation becomes a useful measure of texture for complex textures, as usually encountered in processing, for which the striation thickness cannot be specified.

**Example 7.3** ***The Scale of Segregation Profile***

We concluded earlier that for a complete description of texture, the scale of examination must be varied over a range determined by the uniformity requirements and the texture itself. It is not surprising, therefore, to find that scale of segregation may be a function of the scale of examination, and that by varying the latter over the range of interest and by calculating a *scale of segregation profile*, we obtain a better characterization of texture. Figure 7.11 shows the concentration trace of blown film of low density polyethylene with a carbon black concentrate taken along a line transverse to the "machine"

* The exact definition of the striation thickness is given in Eq. 7.8-1.

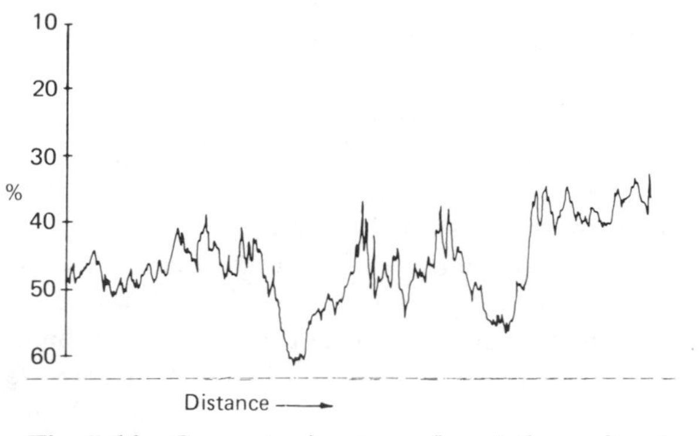


**Fig. 7.11** Concentration trace (in relative units of light transmittance) of a polyethylene film sample (Fig. 7.5), taken perpendicular to the extrusion direction. [Reprinted with permission from N. Nadav and Z. Tadmor, *Chem. Eng. Sci.*, **28**, 2115 (1973).]

direction (15), and Fig. 7.12 gives the scale of segregation profile, which was calculated by varying the scale of examination or the size of the representative sample examined. The actual film samples appear in Fig. 7.5. In this particular example it should be noted that there are two clearly visible characteristic scale of segregation values, about 6 and 0.5 cm. These represent low and medium frequency concentration variations, outstanding in Fig. 7.11, and a less definite lower value that should reflect the high frequency intense texture

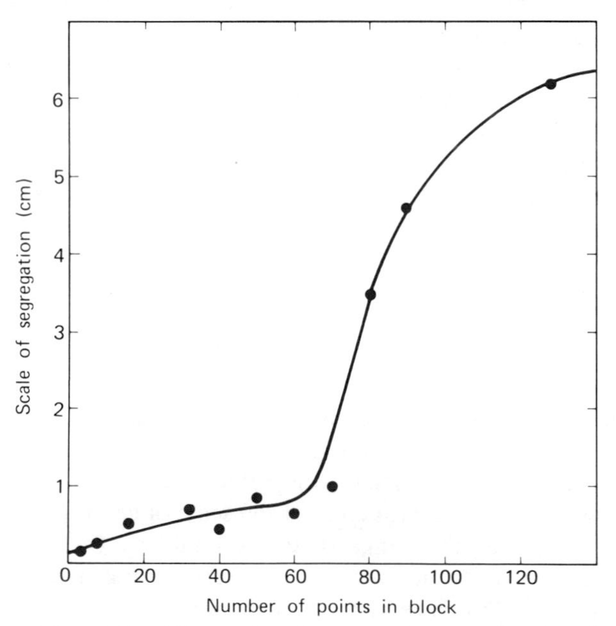


**Fig. 7.12** The scale of segregation profile for the polyethylene film in Fig. 7.5*c*. [Reprinted with permission from N. Nadav and Z. Tadmor, *Chem. Eng. Sci.*, **28**, 2175 (1973).]

visible on samples in Fig. 7.5. The accurate evaluation of the latter was not possible with the experimental method used. Nadav and Tadmor (15) suggested that a relationship may exist between certain mixing modes in extruders and the characteristic scale of segregation values.

## 7.6 The Intensity of Segregation

The intensity of segregation defined as follows

$$I = \frac{S^2}{\sigma_0^2} \tag{7.6-1}$$

expresses the ratio of the measured variance divided by the variance of a completely segregated system. Concentrations are taken as volume fractions of the minor component in the samples. Hence for the completely segregated case, the concentration is either 1 or 0. Thus defined, $I$ has a value of 1 when segregation is complete (i.e., the concentration of any point is either pure minor or pure major) and a value of 0 when the concentration is uniform ($S^2 = 0$). The intensity of segregation, therefore, reflects the departure of the concentration in the various regions from the mean, but not the size of the regions.

The intensity of segregation, as defined in Eq. 7.6-1, reflects to some extent "gross uniformity" on a scale of examination reduced to the scale on which texture is being examined. Referring to the example of blue shopping bags, if we define one bag as the "whole system," the intensity of segregation, as measured by Eq. 7.6-1, will be a measure of the intensity of blue shade variations and also will reflect the gross composition uniformity of blue pigment within a single bag. Thus, $I$ can be viewed as a particular mixing index.

### Example 7.4 *The Intensity of Segregation for a Two Composition System*

A better understanding of the nature of the intensity of segregation can be obtained by analyzing a simple two composition system, as in Fig. 7.9. In this case, the intensity of segregation can be calculated theoretically. Assume that the volume (or area) fraction of regions of composition $x_1$ is $\phi_1$, and that of composition $x_2$ is $\phi_2$ ($\phi_1 + \phi_2 = 1$). The concentrations $x_1$ and $x_2$ themselves are fractions of the minor component in regions 1 and 2, respectively. The average concentration and the variance, assuming that samples taken for examination are smaller than the size of the regions of different compositions (and neglecting the small number of samples falling on border lines), are

$$\bar{x} = \phi_1 x_1 + \phi_2 x_2 \tag{7.6-2}$$

and

$$S^2 = \phi_1 (x_1 - \bar{x})^2 + \phi_2 (x_2 - \bar{x})^2 \tag{7.6-3}$$

The intensity of segregation is obtained by substituting Eqs. 7.6-3 and 7.3-7, with $\phi_1$ replacing $p$, into Eq. 7.6-1 to give

$$I = \frac{\phi_1 \phi_2 (x_1 - x_2)^2}{\phi_1 (1 - \phi_1)} = (x_1 - x_2)^2 \qquad (7.6\text{-}4)$$

Equation 7.6-4 is consistent with the definition of the intensity of segregation to yield $I = 1$ for a completely segregated system ($x_1 = 1$, $x_2 = 0$, or $x_1 = 0$, $x_2 = 1$), and $I = 0$ for uniform composition, $x_1 = x_2$. Clearly, the intensity of segregation is independent of the relative size of the region and reflects the departure of the concentrations from the mean.

We have concentrated our discussion of analyzing texture in terms of scale of segregation and intensity of segregation. This, however, should not imply that this is the only possible approach to the problem. An alternative approach well worth investigating would be a frequency distribution analysis successfully applied to other engineering problems, for example, the wave amplitude distribution in two phase flows (16).

## 7.7 Characterization of the Mixing Process

The previous section discussed quantitative measures for describing certain properties of mixtures related to composition uniformity. The next logical step is to relate these quantitative measures to the mixing process itself: in other words, to find functional relationships between mixer geometry, operating conditions, physical properties of the mixture, and given initial conditions, and these quantitative measures of composition uniformity of mixtures. Unfortunately, this is not an easy step to take. First of all, these quantitative measures are generally based on statistical analysis of withdrawn samples. Therefore, to relate them to the mixing process variables, the exact distribution of the minor component throughout the system should be predictable as a function of the above-mentioned variables. This, however, can easily be done only in relatively simple cases. In ordered distributive mixing, the number of striations, their location, hence the exact distribution of the components, are indeed predictable and can be related to significant variables of the mixing process. Yet in the more common laminar mixing, such a relationship is more difficult to obtain. In this case, to facilitate the quantitative analysis of the mixing process, the interfacial area has been suggested (17, 3) as a variable to be followed during the mixing process. Interfacial area increases can be directly related to the initial orientation and the total permanent strain imposed on the fluid. The latter is, in principle, predictable from a detailed knowledge of the flow pattern. The end result is that in laminar mixing the total strain can be used as one quantitative measure to characterize the mixing process. It can be related, in principle, to mixer geometry, operating conditions, physical properties, and initial conditions. Yet total strain experienced by a fluid cannot be simply measured, and when calculated it does not directly reflect the composition uniformity of the mixture, which depends on the distribution of interfacial elements throughout the system. In relatively simple cases, however, the striation thickness can be calculated from the total strain. Alternatively, empirical correlations are needed for being able to state the total strain needed for a given quality of mixture. Hence strain can be viewed as a very useful link between mixing variables and mixture quality.

We next discuss some of these concepts quantitatively and in some detail.

## 7.8 Distributive Mixing

Figure 7.1*b* represents ordered distributive mixing schematically. Such predictable mixing processes are obtained, in practice, by the various "motionless" mixers, by a combination of distributive and laminar mixing as discussed in more detail in Chapter 11. In such an ordered mixing process, both the quality of the mixture and the mixing process can be well characterized by the *striation thickness.* The striation thickness is defined as the total volume divided by one-half the total interfacial surface (14)

$$r = \frac{V}{(A/2)} \tag{7.8-1}$$

For uniformly spaced and equal sized alternating regions, $r$ simply becomes the linear distance of one repetitive unit. The numerical value of $r$ completely describes the quality of such regular mixtures. The expected value of $r$ in a static mixer is inversely proportional to the predictable number of layers created by the mixer unit. It is worth recalling that for a simple, uniform striped pattern, the scale of segregation was shown to equal one-quarter the thickness of one layer, or $a = r/8$. However, for such simple patterns we do not need a statistically defined measure of texture such as the scale of segregation. The mixing process proceeds by reducing the striation thickness to any desired value; in principle it can be reduced to the molecular level.

Whereas ordered distributive mixing is easy to characterize, random distributive mixing is not. This kind of mixing is common with powders and particulate solids. Valentin (18) reviewed the various models that have been proposed in the literature for characterizing the mixing process. In random distributive mixing we are usually interested in the deviation of the gross uniformity from a completely random state as a function of mixing time and conditions. Furthermore, a common complication involves a process of demixing, which takes place simultaneously with mixing. Various proposed models for mixing are discussed in the literature (19–22).

## 7.9 Laminar Mixing

When two viscous liquids* are mixed, the interfacial area between them increases. Brothman et al. (17) and later Spencer and Wiley (3) have indeed suggested interfacial area as a quantitative measure of the mixing process. This section relates the change of an interfacial area element to the total strain imposed on the viscous fluid.

* Many solid-liquid mixing operations, such as dispersing carbon black, pigments, and fillers in polymers, may be treated as liquid-liquid systems because the mixing operation generally involves the distribution of a small amount of highly filled polymer ("concentrate") within the system.

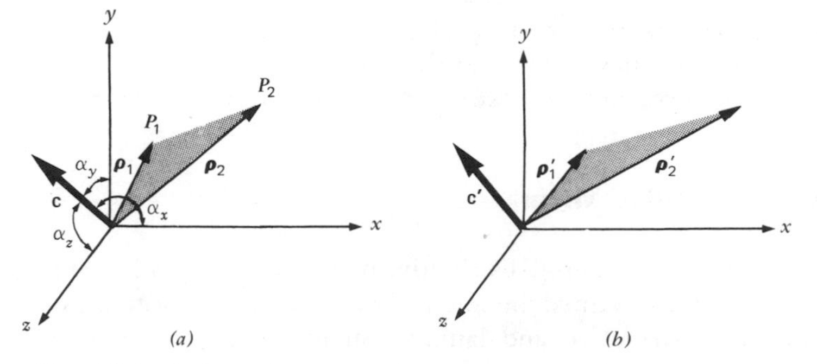


**Fig. 7.13** Surface element confined between position vectors $\boldsymbol{\rho}_1$ and $\boldsymbol{\rho}_2$ in a simple shear flow field $v_x = \dot{\gamma}_{yx}y$. (*a*) At time $t_0$. (*b*) At a later time $t'$.

Consider an arbitrarily oriented surface element in a *simple shear* flow field $v_x = \dot{\gamma}_{yx}y$ (Fig. 7.13*a*). The simple shear flow (which may be steady or time dependent) is homogeneous, and it is the most commonly encountered flow in polymer processing mixing equipment. If the overall shear flow field in the mixing equipment is nonhomogeneous (i.e., the rate of deformation is not constant throughout the field), our discussion holds over a small volume element, where the flow field can be approximated by simple shear. The surface element at time $t_0$ is confined between two position vectors $\boldsymbol{\rho}_1$ and $\boldsymbol{\rho}_2$. The area of the surface element is

$$A_0 = \tfrac{1}{2}|\mathbf{c}| = \tfrac{1}{2}|\boldsymbol{\rho}_1 \times \boldsymbol{\rho}_2| \tag{7.9-1}$$

The vector $\mathbf{c} = \boldsymbol{\rho}_1 \times \boldsymbol{\rho}_2$ is normal to the surface, and its orientation in space is specified by two of the three directional cosines; $\cos\alpha_x$, $\cos\alpha_y$, and $\cos\alpha_z$, which satisfy the following equation:

$$\cos^2\alpha_x + \cos^2\alpha_y + \cos^2\alpha_z = 1 \tag{7.9-2}$$

The angles $\alpha_x$, $\alpha_y$, and $\alpha_z$ (Fig. 7.13) are expressed in terms of the components of the vector $\mathbf{c}$ as follows:

$$\cos\alpha_x = \frac{c_x}{|\mathbf{c}|} \qquad \cos\alpha_y = \frac{c_y}{|\mathbf{c}|} \qquad \cos\alpha_z = \frac{c_z}{|\mathbf{c}|} \tag{7.9-3}$$

The position vectors $\boldsymbol{\rho}_1$ and $\boldsymbol{\rho}_2$ can be expressed in terms of their components as

$$\boldsymbol{\rho}_1 = x_1\boldsymbol{\delta}_x + y_1\boldsymbol{\delta}_y + z_1\boldsymbol{\delta}_z \tag{7.9-4}$$

and

$$\boldsymbol{\rho}_2 = x_2\boldsymbol{\delta}_x + y_2\boldsymbol{\delta}_y + z_2\boldsymbol{\delta}_z \tag{7.9-5}$$

where $x_1$, $y_1$, $z_1$ and $x_2$, $y_2$, $z_2$ are the coordinates of points $P_1$ and $P_2$ in Fig. 7.13. Substituting Eqs. 7.9-4 and 7.9-5 into Eq. 7.9-1, we obtain

$$\mathbf{c} = \boldsymbol{\rho}_1 \times \boldsymbol{\rho}_2 = \begin{vmatrix} \boldsymbol{\delta}_x & \boldsymbol{\delta}_y & \boldsymbol{\delta}_z \\ x_1 & y_1 & z_1 \\ x_2 & y_2 & z_2 \end{vmatrix} = (y_1z_2 - z_1y_2)\boldsymbol{\delta}_x + (z_1x_2 - x_1z_2)\boldsymbol{\delta}_y$$
$$+ (x_1y_2 - x_2y_1)\boldsymbol{\delta}_z = c_x\boldsymbol{\delta}_x + c_y\boldsymbol{\delta}_y + c_z\boldsymbol{\delta}_z \tag{7.9-6}$$

Thus the initial area at $t_0$ is

$$A_0 = \tfrac{1}{2}|\mathbf{c}| = \tfrac{1}{2}(c_x^2 + c_y^2 + c_z^2)^{1/2} \tag{7.9-7}$$

After a certain elapsed time $\Delta t$, the new interfacial area is confined between the position vectors $\boldsymbol{\rho}_1'$ and $\boldsymbol{\rho}_2'$ such that

$$\boldsymbol{\rho}' = \boldsymbol{\rho} + \boldsymbol{v}\Delta t \tag{7.9-8}$$

where the velocity vector $\boldsymbol{v}$ for simple shear flow is

$$\boldsymbol{v} = \dot{\gamma}_{yx} y \boldsymbol{\delta}_x + (0)\boldsymbol{\delta}_y + (0)\boldsymbol{\delta}_z \tag{7.9-9}$$

or

$$\boldsymbol{v}\Delta t = \gamma y \boldsymbol{\delta}_x + (0)\boldsymbol{\delta}_y + (0)\boldsymbol{\delta}_z \tag{7.9-10}$$

where $\gamma$ is the total strain, or the magnitude of the shearing strain $\gamma = \int_0^t \dot{\gamma}_{yx}\ (t')dt'$.

By substituting Eq. 7.9-10 into Eq. 7.9-8, the new position vectors $\boldsymbol{\rho}_1'$ and $\boldsymbol{\rho}_2'$ in terms of their components become

$$\boldsymbol{\rho}_1' = (x_1 + \gamma y_1)\boldsymbol{\delta}_x + y_1\boldsymbol{\delta}_y + z_1\boldsymbol{\delta}_z \tag{7.9-11}$$

and

$$\boldsymbol{\rho}_2' = (x_2 + \gamma y_2)\boldsymbol{\delta}_x + y_2\boldsymbol{\delta}_y + z_2\boldsymbol{\delta}_z \tag{7.9-12}$$

The cross product of $\boldsymbol{\rho}_1'$ and $\boldsymbol{\rho}_2'$ becomes

$$\mathbf{c}' = \boldsymbol{\rho}_1 \times \boldsymbol{\rho}_2 = \begin{vmatrix} \boldsymbol{\delta}_x & \boldsymbol{\delta}_y & \boldsymbol{\delta}_z \\ x_1 + \gamma y_1 & y_1 & z_1 \\ x_2 + \gamma y_2 & y_2 & z_2 \end{vmatrix} = c_x\boldsymbol{\delta}_x + (c_y - \gamma c_x)\boldsymbol{\delta}_y + c_z\boldsymbol{\delta}_z \tag{7.9-13}$$

where $c_x$, $c_y$, and $c_z$ are the components of vector $\mathbf{c}$ defined in Eq. 7.9-6. The new interfacial area $A$ is, therefore,

$$A = \tfrac{1}{2}|\mathbf{c}'| = \tfrac{1}{2}(c_x^2 + c_y^2 + c_z^2 - 2c_x c_y \gamma + c_x^2 \gamma^2)^{1/2} \tag{7.9-14}$$

The ratio of the two areas at $t_0 + \Delta t$ and $t_0$ is obtained by dividing Eq. 7.9-14 by Eq. 7.9-7, and making use of the definition of directional cosines in Eq. 7.9-3.

$$\blacktriangleright \quad \frac{A}{A_0} = (1 - 2\cos\alpha_x \cos\alpha_y \gamma + \cos^2\alpha_x \gamma^2)^{1/2} \tag{7.9-15}$$

This expression, derived by Spencer and Wiley (3), indicates that the increase in interfacial area is a function of initial orientation and total strain. For large deformations, Eq. 7.9-15 reduces to

$$\frac{A}{A_0} = |\cos\alpha_x|\gamma \tag{7.9-16}$$

We, therefore, obtain the important conclusion that the increase in interfacial area is *directly proportional to total strain*. Hence total strain becomes the critical variable for the quantitative characterization of the mixing process.

We further conclude from Eq. 7.9-15 that at low strains, depending on the initial orientation, the interfacial area may increase or decrease with imposed strain. This implies clearly that strain may demix as well as mix two components.

Indeed if the fluid is sheared in one direction a certain number of shear units, an equal and opposite shear will take the fluid back to its original state (no diffusion).

The role of initial orientation is evident from Eq. 7.9-15. The increase in the interfacial area attains a maximum when the initial orientation of the interfacial area is the $yz$ plane ($\cos \alpha_x = 1$, $\cos \alpha_y = 0$). The interfacial area remains unchanged if $\cos \alpha_x = 0$ or $c_x = 0$. From Eq. 7.9-6 we note that this occurs if $y_1 = y_2 = 0$ or $z_1 = z_2 = 0$, implying, respectively, that the surface $A_0$ is either in the $xz$ or the $xy$ plane. In the former case the plane is translated undeformed, whereas in the latter it is deformed, yet retains the same area* $A = A_0$. We must remember, however, that we are dealing with a three-dimensional system in which the interfacial area elements most probably have a random orientation distribution. We are, therefore, interested in calculating the overall increase in interfacial area for such a system. Considering equal initial surface area elements $A_0$, a random orientation distribution implies that we have a collection of vectors **c** pointing randomly in all directions. The fraction of vectors pointing in a certain direction is given by the ratio of an infinitesimal surface area on a sphere of radius $|\mathbf{c}|$ to the total surface area of the sphere

$$f(\theta, \phi)\, d\theta\, d\phi = \frac{1}{4\pi} \sin\theta\, d\theta\, d\phi \tag{7.9-17}$$

where $\theta$ and $\phi$ are spherical coordinates (Fig. 7.14). Therefore $f(\theta, \phi)\, d\theta\, d\phi$ is the probability of finding a vector **c** with orientation between $\theta$ and $\theta + d\theta$ and $\phi$ and

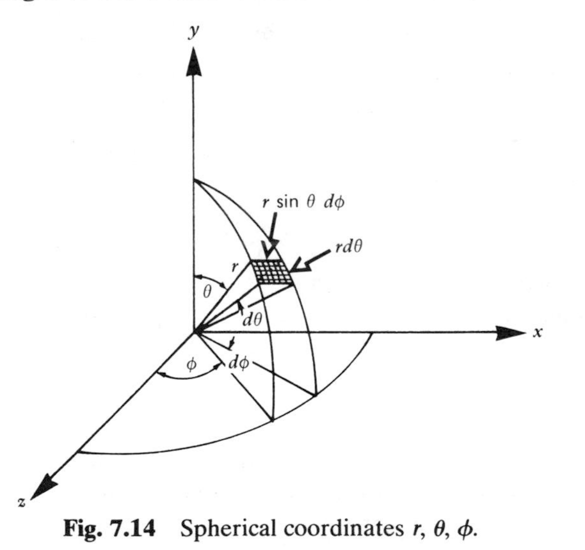


**Fig. 7.14** Spherical coordinates $r$, $\theta$, $\phi$.

$\phi + d\phi$. The angle $\theta$ in the spherical coordinate system is identical to the angle $\alpha_y$, whereas the angle $\phi$ relates to angles $\alpha_x$ and $\alpha_y$ as follows:

$$\cos \alpha_x = \sin \alpha_y \sin \phi \tag{7.9-18}$$

* Considering the latter case, it would appear that mixing does occur despite the constant surface area. Indeed, if we consider two-dimensional mixing over a plane, mixing would occur, but in such a case the criterion for mixing would be the perimeter rather than the area.

Equation 7.9-16 can now be written as

$$\frac{A}{A_0} = |\sin\theta\cos\phi|\gamma \tag{7.9-19}$$

The mean change in interfacial area becomes

$$\blacktriangleright \quad \frac{A}{A_0} = \int_{\phi=0}^{\phi=2\pi}\int_{\theta=0}^{\theta=\pi} |\sin\theta\sin\phi|\gamma\left(\frac{1}{4\pi}\right)\sin\theta\,d\theta\,d\phi = \frac{\gamma}{2} \tag{7.9-20}$$

Hence the ratio of the total final interfacial area to the total initial area in a system of randomly oriented surfaces at large strains in simple shear flow is one-half the total strain imposed on the liquid.

Randomly distributed cubes of a minor component in a major component can be considered to be a system of randomly oriented, equal-sized interfacial areas. The total change in interfacial area for such a system at large deformations (assuming the major and minor components to have the same rheological properties) is given in Eq. 7.9-20. The average striation thickness, defined in Eq. 7.8-1, can be written as

$$r = \frac{2}{(A/A_0)(A_0/V)} \tag{7.9-21}$$

Substituting Eq. 7.9-20 into Eq. 7.9-21 and expressing the ratio of initial surface area to volume in terms of the volume fraction of the minor component $x_v$ and minor cube side $L$, $(A_0/V = 6x_v/L)$ results in

$$\blacktriangleright \quad r = \frac{2}{3}\frac{L}{\gamma x_v} \qquad (\text{large } \gamma) \tag{7.9-22}$$

or

$$\blacktriangleright \quad r = \frac{2r_0}{\gamma} \qquad (\text{large } \gamma) \tag{7.9-23}$$

where $r_0$ is the initial striation thickness given by $L/3x_v$.

The above clearly indicates that striation thickness is inversely proportional to total strain. We also note that the initial striation thickness is proportional to the size of the cube and inversely proportional to the volume fraction of the minor component. Hence for any required final striation thickness, the larger the particles and the smaller the volume fraction of the minor component, the more total strain is required. Hence it is more difficult to mix a small amount of minor components into a major, than to make a 50–50 mixture, and the larger the individual particles of the minor, the more difficult it is to mix. By using Eq. 7.9-22, we can estimate the strain needed to reduce the striation thickness to a level where molecular diffusion or Brownian motion will randomize the mixture for a given strain rate and within the time (residence time) allotted for mixing.

It has been suggested (24) that a good rule of thumb for an adequate mixing in processing equipment is a total $18{,}000 \pm 6000$ shear units. This implies a 4 orders of magnitude reduction in striation thickness. We recall, however, that the term "adequate mixing" is determined by the requirements of the product.

An expression for the increase in interfacial area in tensile elongational flow was derived by Mohr et al. (14). Recently Erwin (25) derived an expression for general homogeneous flow fields

$$\frac{A}{A_0} = \left\{ \frac{\cos^2 \alpha'}{\lambda_x^2} + \frac{\cos^2 \beta'}{\lambda_y^2} + \frac{\cos^2 \gamma'}{\lambda_z^2} \right\}^{1/2} \tag{7.9-24}$$

where $\cos \alpha'$, $\cos \beta'$, and $\cos \gamma'$ are directional cosines with respect to the *principal axes* of the strain tensor in the initial state and $\lambda_x$, $\lambda_y$ and $\lambda_z$ are the principal elongational ratios (see Problem 7.7).

## 7.10 Strain Distribution Functions

If we accept the premise that the total strain is a key variable in the quality of laminar mixing, we are immediately faced with the problem that in most industrial mixers, and in processing equipment in general, different fluid particles experience different strains. This is true for both batch and continuous mixers. In the former, the different strain histories are due to the different paths the fluid particles may follow in the mixer, whereas in a continuous mixer, superimposed on the different paths there is also a different residence time for every fluid particle in the mixer. To quantitatively describe the various strain histories, *strain distribution functions* were defined (26), which are similar in concept to the classical residence time distribution functions. In the following discussion the meaning of strain is restricted to shear strains, and specifically to the magnitude of the shear $\gamma = \int_0^t \dot{\gamma}_{yx}(t')\, dt'$.

### *Batch Mixers*

In a batch mixer the shear rates throughout the volume are not uniform, and neither are the residence times of various fluid particles in the various shear rate regions. Consequently, after a given time of mixing, different fluid particles experience different strain histories and accumulate different shear strains $\gamma$. The strain distribution function (SDF), $g(\gamma)\, d\gamma$, is defined as the fraction of the liquid in the mixer that has experienced a shear strain from $\gamma$ to $\gamma + d\gamma$. Alternatively, it is the probability of a fluid particle fed to the mixer to accumulate a shear strain of $\gamma$ after the given time. The function $g(\gamma)$ changes with time of mixing.

By integrating $g(\gamma)\, d\gamma$, a cumulative SDF can be defined as follows:

$$G(\gamma) = \int_0^{\gamma} g(\gamma)\, d\gamma \tag{7.10-1}$$

where $G(\gamma)$ is the fraction of liquid that experienced a strain of less than $\gamma$. The mean strain of the liquid $\bar{\gamma}$ is given by the following equation:

$$\bar{\gamma} = \int_0^{\gamma_{max}} \gamma g(\gamma)\, d\gamma \tag{7.10-2}$$

The functions $g(\gamma)$ and $G(\gamma)$ depend on mixer geometry operating conditions and liquid rheology.

**Example 7.5** ***The SDF in a "Three Cylinder Mixer"***

The significance of the SDF in batch mixers can be visualized by considering a simple "mixer" consisting of three concentric cylinders (Fig. 7.15). The inner and outer cylinders are stationary, whereas the (zero thickness) middle cylin-

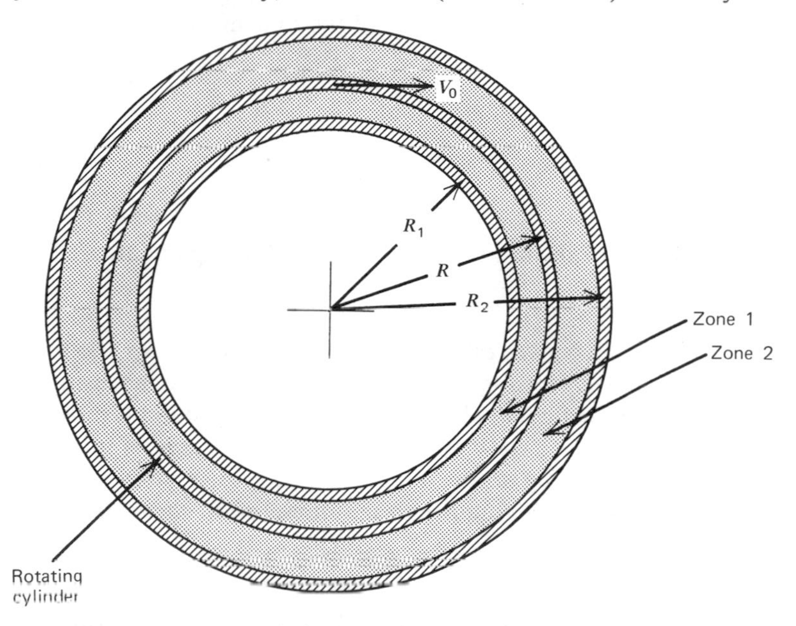


**Fig. 7.15** A "three cylinder mixer" with the middle cylinder rotating at constant tangential velocity $V_0$.

der rotates with tangential velocity $V_0$. The liquid to be mixed is placed between the cylinders. The SDF imposed on the liquid depends on the position of the middle cylinder relative to the others. Neglecting the effect of curvature, the shear rates in zones 1 and 2 are constant, and the strains after time $t$ are given by the following equations:

$$\gamma_1(t) = \frac{\beta}{x(1-\alpha)} \tag{7.10-3}$$

and

$$\gamma_2(t) = \frac{\beta}{(1-x)(1-\alpha)} \tag{7.10-4}$$

where

$$\beta = \frac{V_0 t}{R_2} \tag{7.10-5}$$

and

$$\alpha = \frac{R_1}{R_2} \tag{7.10-6}$$

and $x$ denotes the fractional distance of the middle cylinder from the inner cylinder

$$x = \frac{R - R_1}{R_2 - R_1} \tag{7.10-7}$$

The fraction of the total volume of the "mixer" that zone 1 occupies is

$$\phi_1 = \frac{x^2 + \alpha x(2 - x)}{1 + \alpha} \tag{7.10-8}$$

and that of zone 2 is

$$\phi_2 = 1 - \phi_1 \tag{7.10-9}$$

Finally, the mean strain is given by the following equation:

$$\bar{\gamma} = \phi_1 \gamma_1 + \phi_2 \gamma_2 \tag{7.10-10}$$

Figure 7.16 gives the SDF for a special case of $\alpha = 0.95$ and $\beta = 1$, with $x$ as a parameter. Results show the mean strain to be insensitive to the position of the middle cylinder, whereas the strain distribution is profoundly altered by a change in $x$. For $x = 0.1$, about 10% of the liquid is in zone 1 and 90% in zone 2, undergoing strains of 200 and 22 shear units, respectively. Now if mixing, for example, requires a minimum of 40 shear units to ensure that all

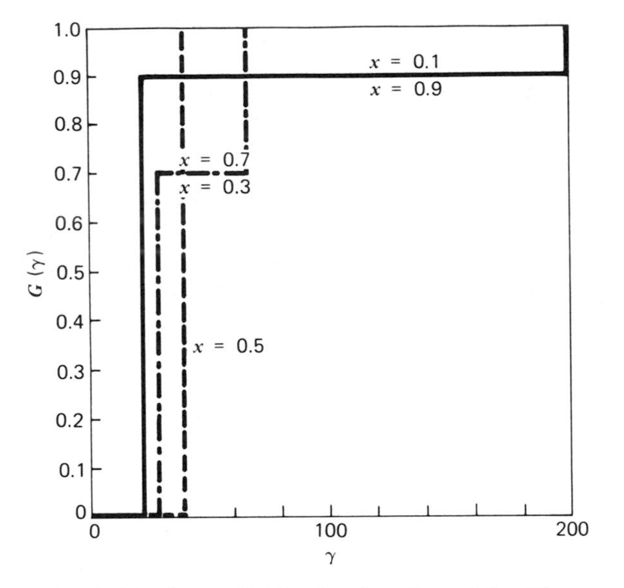


**Fig. 7.16** Strain distribution function of the "three cylinder mixer" with $\beta = 1$ and $\alpha = 0.95$. Full line corresponds to $x = 0.1$ (when $\bar{\gamma} = 39.37$) or $x = 0.9$ (when $\bar{\gamma} = 40.38$). Dashed line corresponds to $x = 0.5$ (when $\bar{\gamma} = 40$). Dashed-dotted line corresponds to $x = 0.3$ (when $\bar{\gamma} = 39.77$) or $x = 0.7$ (when $\bar{\gamma} = 40.18$). Note that mean strain is virtually the same in all cases (small differences are due to curvature), yet the distributions are very different.

fluid particles acquired the minimum strain, mixing time (i.e., $\beta$) would have to be about doubled. However such an increase in mixing time would cause the other 10% of the liquid to be "overmixed," since a strain of 400 shear units would be imposed on them. This waste in time, energy, and overmixing can be eliminated by placing the middle cylinder to $x = 0.5$, where at $\beta = 1$, all liquid acquires a strain of 40 shear units. Clearly, the design objective should be a narrow SDF. This example leads to the further conclusion that in scaling up mixers, the SDF should remain unchanged.

### *Continuous Mixers*

In continuous mixers, exiting fluid particles experience both different shear rate histories and residence times; therefore they have acquired different strains. Following the considerations outlined previously and parallel to the definition of residence time distribution function, the SDF for a continuous mixer $f(\gamma)\,d\gamma$ is defined as the fraction of exiting flow rate that experienced a strain between $\gamma$ and $\gamma + d\gamma$, or it is the probability of an entering fluid particle to exit with that strain. The cumulative SDF, $F(\gamma)$, defined by the following equation:

$$F(\gamma) = \int_{\gamma_0}^{\gamma} f(\gamma)\,d\gamma \tag{7.10-11}$$

is the fraction of exiting flow rate with strain less or equal to $\gamma$, where $\gamma_0$ is the *minimum* strain. The mean strain of the exiting stream is

$$\bar{\gamma} = \int_{\gamma_0}^{\infty} \gamma f(\gamma)\,d\gamma \tag{7.10-12}$$

The SDF, like the residence time distribution (RTD) functions, can be calculated from the velocity distribution in the system; that is, a certain flow pattern determines both functions. The reverse does not necessarily hold. The calculation of the SDF requires a complete description of the flow pattern, whereas RTD functions often may be calculated from a less than complete flow pattern. For example, the RTD of axial annular flow between two rotating concentric cylinders (helical flow) of a Newtonian fluid depends only on the axial velocity, whereas the SDF depends on both the axial and the tangential velocity distributions. Consequently, SDFs cannot be calculated from experimentally measured RTD functions.

#### **Example 7.6** *The Strain Distribution Function in Parallel Plate Flow*

Two parallel plates in relative motion with each other can be viewed as an idealized continuous mixer. A given plane perpendicular to the plates marks the entrance to the "mixer," and another plane at a distance $L$ downstream the exit from the mixer (Fig. 7.17). We assume that a fluid entering the gap between the plate had no previous strain history and that a fully developed pure drag flow exists between the plates. Clearly, although the shear rate is uniform throughout the system, the closer we get to the upper plate, the shorter the residence time will be; hence the fluid particles will experience

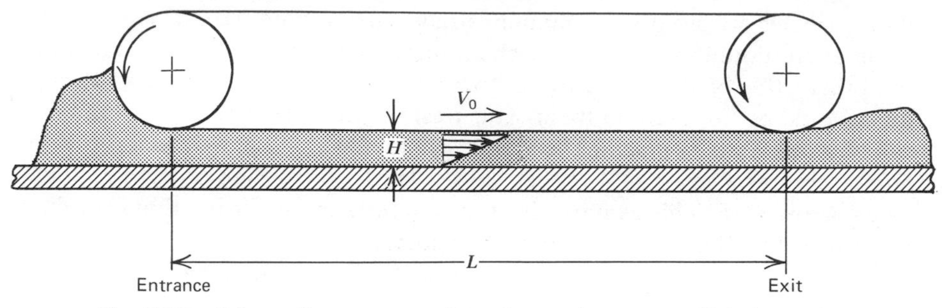


**Fig. 7.17** Schematic representation of a continuous parallel plate mixer.

lower total strains. Moreover, because the velocities are higher in this region, a larger fraction of the existing flow rate will experience the lower strains.

The velocity distribution for a fully developed, isothermal drag flow between parallel plates separated by a distance of $H$ and with the upper plate moving at constant velocity $V_0$ is

$$v_x = \frac{yV_0}{H} \tag{7.10-13}$$

The flow rate per unit width $q$ is

$$q = \frac{V_0 H}{2} \tag{7.10-14}$$

Now the fraction of exiting flow rate between $y$ and $y + dy$ is

$$f(y)\,dy = \frac{dq}{q} = \frac{2y\,dy}{H^2} \tag{7.10-15}$$

The fraction of exiting flow rate in the region greater than $y$ (which is the same as the fraction of flow rate below time $t$, where $t$ corresponds to the time of residence of the fluid at location $y$) is

$$F(y) = \int_y^H f(y)\,dy = 1 - \frac{y^2}{H^2} \tag{7.10-16}$$

The shear rate $\dot{\gamma}$ is

$$\dot{\gamma} = \frac{V_0}{H} \tag{7.10-17}$$

The residence time for a length $L$ is

$$t = \frac{L}{v_x} = \frac{HL}{V_0 y} \tag{7.10-18}$$

The shear strain, as a function of $y$, is obtained from Eqs. 7.10-17 and 7.10-18

$$\gamma = \dot{\gamma} t = \frac{L}{y} \tag{7.10-19}$$

The minimum shear strain $\gamma_0$ is $L/H$.

By substituting Eq. 7.10-19 into Eq. 7.10-16, the SDF is obtained

$$F(\gamma) = 1 - \left(\frac{L}{H\gamma}\right)^2 \tag{7.10-20}$$

and

$$f(\gamma)\,d\gamma = \frac{2L^2}{H^2\gamma^3} \cdot dy \tag{7.10-21}$$

The mean strain can be obtained by either using Eq. 7.10-15 or Eq. 7.10-21

$$\bar{\gamma} = \int_{\gamma_0}^{\infty} \gamma f(\gamma)\,d\gamma = \int_0^H \gamma f(y)\,dy = 2\frac{L}{H} \tag{7.10-22}$$

Thus,

$$F(\gamma) = 1 - \left(\frac{\bar{\gamma}}{2\gamma}\right)^2 \tag{7.10-23}$$

It is worthwhile to note that for a continuous and single valued velocity profile, $v_x(y)$, the mean strain can be shown (see Problem 7.11) to be given by the following expression:

$$\bar{\gamma} = \bar{t}\bar{\dot{\gamma}} = \bar{t}\frac{1}{H}\int_0^H \dot{\gamma}\,dy \tag{7.10-24}$$

where $\bar{t}$ is the mean residence time and $\bar{\dot{\gamma}}$ is the average shear rate. Figure 7.18 shows the SDF and compares it to that of circular tube flow of a Newtonian fluid.

Analysis of the results reveals that the mean strain is proportional to the $L/H$ ratio. Thus for good mixing, plate separation must be small and the length $L$ large. Figure 7.18 reveals a broad strain distribution, with about 75% of the flow rate experiencing a strain below the mean strain.

A better insight into the meaning of the SDF is obtained by following simultaneously the reduction of the striation thickness and the flow rates contributed by the various locations between the plates (Fig. 7.19). The distance between the plates is divided into 10 layers. We assume for the schematic representation of the SDF in Fig. 7.19 that the strain is uniform within each layer. Let us consider in each alternate layer two cubical, minor particles separated by a certain distance such that the initial striation thickness is $r_0$. By following the deformation and separation of the particles with time, Fig. 7.19 shows an arbitrary intermediate and final position of the particles. As pointed out earlier, although the shear rate is uniform everywhere between the plates, since the residence time is different, the total strain experienced by the particle is inversely proportional to the distance from the lower stationary plate (Eq. 7.10-19). Hence the striation thickness at the exit increases with this distance, reaching a maximum value (least mixing) at the moving plate.

The quality of the "product" of such a mixer will not be completely determined by the *range* of strains or striations across the flow field; the flow rate of the various layers also plays a role. As Fig. 7.19 indicates, a sample

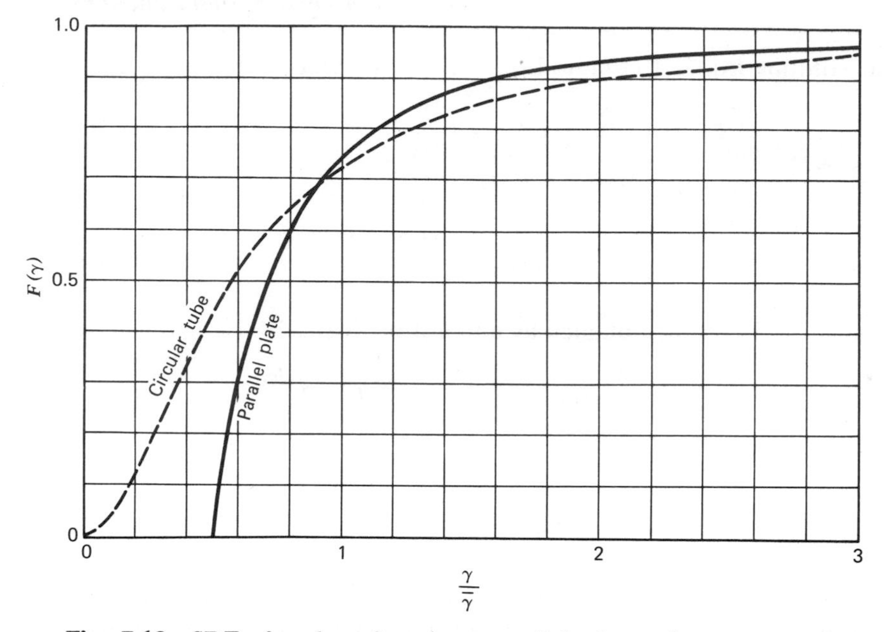


**Fig. 7.18** SDFs for drag flow in a parallel plate mixer compared to Newtonian laminar flow in a circular tube.

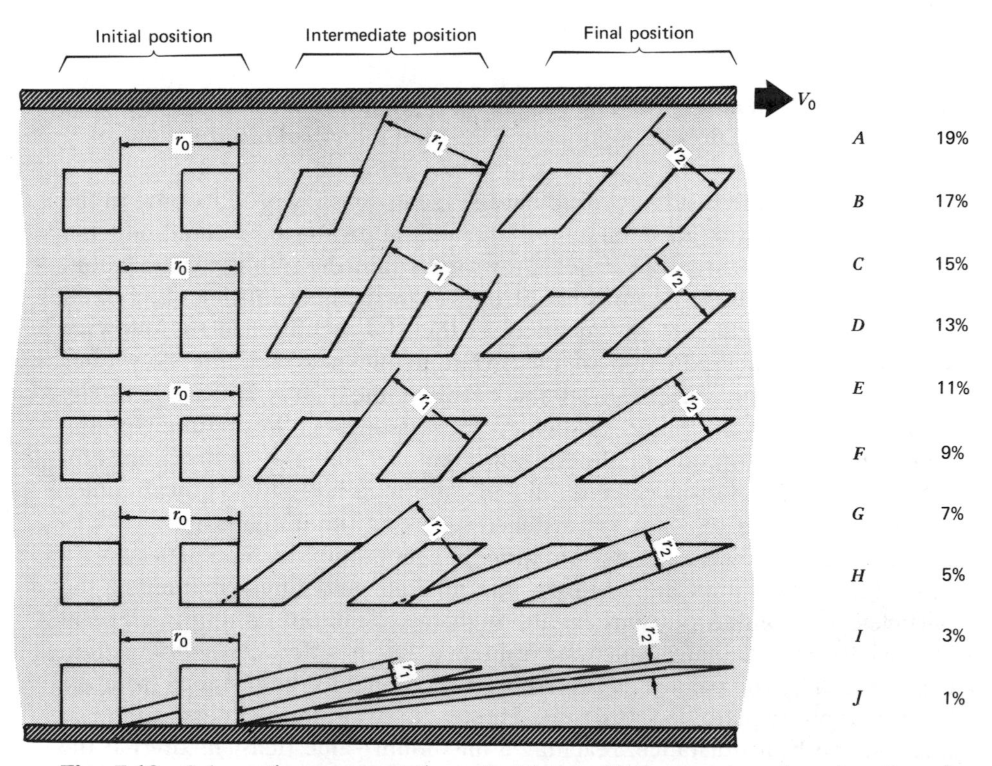


**Fig. 7.19** Schematic representation of striation thickness at various locations in a parallel plate mixer with drag flow. The striation thickness at the entrance to the mixer is $r_0$, somewhere at the middle $r_1$, and at the exit it is $r_2$. Note that $r_2$ is large for particles close to the moving upper plate (because the short residence time) and small for particles close to the stationary lower plate. At the right is given the fraction of flow rate at the exit of layers at different heights.

collected at exit will consist, for example, of 17% of a poorly mixed layer $B$ and only 1% of well-mixed layer $J$. The SDF reflects the combined effect of accumulated strain and local flow rate.

Finally, one would expect that existence of axial pressure gradients, which greatly affect the velocity profile, would also significantly affect the SDF, hence the mixing performance (see Problem 7.10).

## 7.11 Residence Time Distribution

Polymers are temperature sensitive to some degree, and prolonged exposure to high temperatures may result in thermal degradation. The degree of degradation depends on the time-temperature history of the polymer. Often polymer systems are processed with temperature activated reacting additives (foaming agents, cross-linking agents), or the system is reactive per se (reaction injection molding). In these systems, the extent of chemical reaction depends on the time-temperature history. Extrudates of many polymers (e.g., nylon 66) contain varying amounts of "gels", which may be a result of excessive residence time of a small fraction in the vessels. In all the above-mentioned examples, quantitative prediction and design require the detailed knowledge of the residence time distribution functions. Moreover, in operation the time needed to purge a system, or to switch materials, is determined by the nature of this function. Therefore the calculation and measurement of RTD functions in processing equipment have an important role in design and operation in addition to the previously discussed interaction of RTD and SDF.

This section discusses the nature of RTD functions, with special emphasis on laminar flow systems.

The definition of RTD functions is due to Danckwerts (27). We first differentiate between the internal RTD function $g(t)\,dt$ and the external RTD function $f(t)\,dt$. The former is defined as the fraction of fluid volume *in* the system with a residence time between $t$ and $t+dt$, and the latter is the fraction of *exiting* flow rate with a residence time between $t$ and $t+dt$. From these definitions we can define the cumulative functions $G(t)$ and $F(t)$, respectively, as follows:

► $$G(t)=\int_0^t g(t)\,dt \qquad (7.11\text{-}1)$$

and

► $$F(t)=\int_{t_0}^t f(t)\,dt \qquad (7.11\text{-}2)$$

It follows also from the definition that

$$G(\infty)=F(\infty)=1 \qquad (7.11\text{-}3)$$

The mean residence time is given by the following expression

► $$\bar{t}=\int_{t_0}^\infty tf(t)\,dt \qquad (7.11\text{-}4)$$

and it equals the volume of the system divided by the volumetric flow rate.

**Example 7.7** ***The Relationships Among the RTD Functions***

Consider a steady continuous flow system characterized by a certain RTD, $F(t)$ or $f(t)\,dt$, in which a "white" fluid flows at a volumetric rate of $Q$. Suppose that at time $t=0$, we switch to a "red" fluid having otherwise identical properties, without altering the flow rate. Following the switch, after any time $t$ we can make, with the aid of the definitions of the RTD functions, the following material balance of red fluid:

$$Q \quad - \quad QF(t) \quad = \quad \frac{d}{dt}[VG(t)] \tag{7.11-5}$$

| Rate of flow into the system | Rate of flow out of the system | Rate of change within the system |
|---|---|---|

where $V$ is the total volume. Assuming a constant value of $Q$ (i.e. constant density), Eq. 7.11-5 with Eq. 7.11-1 becomes

► $$g(t)=\frac{1-F(t)}{\bar{t}} \tag{7.11-6}$$

where $\bar{t}$, the mean residence time given in Eq. 7.11-4, is also given by

► $$\bar{t}=\frac{V}{Q} \tag{7.11-7}$$

Using Eq. 7.11-6 it is possible to derive all the interrelationships of the RTD functions, which are listed in Table 7.2.

**Table 7.2 Relationships Among the Various RTD Functions**

| | Calculate | | | |
|---|---|---|---|---|
| Given | $f(t)$ | $F(t)$ | $g(t)$ | $G(t)$ |
| $f(t)$ | $f(t)$ | $\int_0^t f(t')\,dt'$ | $\frac{1}{\bar{t}}-\frac{1}{\bar{t}}\int_{t_0}^t f(t')\,dt'$ $\quad t\geq t_0$ | $\frac{t}{\bar{t}}-\frac{1}{\bar{t}}\int_{t_0}^t\int_{t_0}^{t'} f(t'')\,dt''\,dt'$ $\quad t\geq t_0$ |
| $F(t)$ | $\frac{dF(t)}{dt}$ | $F(t)$ | $\frac{1-F(t)}{\bar{t}}$ $\quad t\geq t_0$ | $\frac{t}{\bar{t}}-\frac{1}{\bar{t}}\int_{t_0}^t F(t')\,dt'$ $\quad t\geq t_0$ |
| $g(t)$ | $-\bar{t}\frac{dg(t)}{dt}$ $\quad t\geq t_0$ | $1-\bar{t}g(t)$ $\quad t\geq t_0$ | $g(t)$ | $\int_0^t g(t')\,dt'$ |
| $G(t)$ | $-\bar{t}\frac{d^2G(t)}{dt^2}$ $\quad t\geq t_0$ | $1-\bar{t}\frac{dG(t)}{dt}$ $\quad t\geq t_0$ | $\frac{dG(t)}{dt}$ | $G(t)$ |

The two extreme flow systems, with respect to the RTD are the plug flow system, which exhibits no residence time distribution, and the continuous stirred tank, which exhibits complete back-mixing and has the following RTD function:

► $$F(t)=1-e^{-t/\bar{t}} \tag{7.11-8}$$

In this case, because perfect back-mixing exists, the internal and external RTD functions are identical. This can be easily verified from Table 7.2, recalling that the minimum residence time in this case is zero. This is generally not the case in laminar flow systems where, in principle, the RTD functions can be calculated from the velocity profiles. It should be noted that in complex systems a precise description of the path the fluid follows between subsystems is also needed to calculate the RTD.

In the many cases in which the RTD cannot be calculated theoretically, experimental techniques were developed to measure it. These techniques are carried out by introducing a tracer material into the system and recording its concentration at the exit. These methods are discussed in great detail in the literature. It should be noted, however, that most of these techniques assume a "plug" inlet flow into the system. If this is not the case, special care must be taken in introducing the tracer material (e.g., in an impulse signal, the amount of tracer must be proportional to the local velocity, otherwise complex corrections may be required). In general, a step change in tracer concentration results directly in the $F(t)$ function, and an impulse type of tracer injection results directly in the $f(t)$ function.

**Example 7.8** ***RTD in Fully Developed Laminar Flow of a Newtonian Fluid in a Pipe***

The velocity distribution is

$$v_z = C(R^2 - r^2) \tag{7.11-9}$$

where

$$C = \frac{\Delta P}{4L\mu} \tag{7.11-10}$$

The residence time of a fluid particle depends on its location or its radial position

$$t = \frac{L}{v_z} = \frac{L}{C(R^2 - r^2)} \tag{7.11-11}$$

where $L$ is the pipe length. The minimum residence time at the center of the pipe is

$$t_0 = \frac{L}{CR^2} \tag{7.11-12}$$

The *range* of residence times is therefore from $t = t_0$ to $t \to \infty$. Equation 7.11-11 can now be rewritten as

$$\frac{t}{t_0} = \frac{1}{1 - (r/R)^2} \tag{7.11-13}$$

The fraction of exiting flow rate with a residence time $t$ to $t + dt$ is given by the flow rate between radius $r$ and $r + dr$

$$dQ = 2\pi r v_z \, dr \tag{7.11-14}$$

Therefore, it follows from the definition of $f(t)\,dt$

$$f(t)\,dt = \frac{dQ}{Q} = \frac{2\pi r v_z\,dr}{\int_0^R 2\pi r v_z\,dr} = \frac{2 r v_z\,dr}{R^2 L/2t_0} \tag{7.11-15}$$

Now, to obtain a result with the time $t$, replacing the variable $r$, we use Eq. 7.11-13

$$\begin{aligned} f(t)\,dt &= \frac{2t_0^2}{t^3}\,dt \qquad t \geq t_0 \\ f(t)\,dt &= 0 \qquad t < t_0 \end{aligned} \tag{7.11-16}$$

The mean residence time is obtained with the aid of Eq. 7.11-4 or Eq. 7.11-7

$$\bar{t} = \int_{t_0}^{\infty} t f(t)\,dt = 2t_0 \tag{7.11-17}$$

The other distribution functions can now be obtained, using the relationships in Table 7.2. These are given below.

$$\left.\begin{aligned} F(t) &= \int_{t_0}^{t} \frac{2t_0^2}{t^3}\,dt = 1 - \left(\frac{t_0}{t}\right)^2 \qquad t \geq t_0 \\ F(t) &= 0 \qquad t < t_0 \end{aligned}\right\} \tag{7.11-18}$$

$$\left.\begin{aligned} g(t)\,dt &= \frac{1 - F(t)}{\bar{t}}\,dt = \frac{t_0}{2t^2}\,dt \qquad t \geq t_0 \\ g(t)\,dt &= \frac{1}{\bar{t}}\,dt = \frac{1}{2t_0}\,dt \qquad t \leq t_0 \end{aligned}\right\} \tag{7.11-19}$$

$$\left.\begin{aligned} G(t) &= \frac{t}{\bar{t}} - \frac{1}{\bar{t}} \int_{t_0}^{t} F(t)\,dt = 1 - \frac{t_0}{2t} \qquad t \geq t_0 \\ G(t) &= \frac{t}{2t_0} \qquad t \leq t_0 \end{aligned}\right\} \tag{7.11-20}$$

## 7.12 Generalization of Distribution Functions

Previous sections have demonstrated how strain distribution functions are defined in equivalent terms to the classical residence time distribution functions. Similarly, other useful functions can be defined, using variables or combinations of variables of interest instead of time or strain. Thus, for example, a general function $g(x)\,dx$ can be defined which is the fraction of material within a system having a certain property between values $x$ and $x + dx$, and a function $f(x)\,dx$ which is the fraction of exiting flow rate having a certain property between $x$ and $x + dx$. The variable $x$ may be residence time $t$, total strain $\gamma$, or other variables of interest, such as temperature $T$, where the interest is in critical exposure to certain temperature ranges; the variable may be a product of time and temperature for temperature sensitive materials where the temperature history is critical, or shear stress $\tau$ in dispersive mixing.

## 7.13 Fluctuation of Composition with Time

Up to this point, in dealing with continuous mixing processes, we have assumed a steady inlet composition. Our main concern has been the composition uniformity *across* the emerging stream at any time. Inlet conditions, however, in practice are never uniform in *time*. The performance of a continuous mixer or blender, therefore, is also dependent on how well it can *even out* inlet concentration variations in time. In other words, suppose the entering stream to a mixer has a concentration $C_i$ (averaged over the linlet cross section) that fluctuates with time; the problem is to find the concentration $C_0$ of the emerging stream as a function of time. If not all fluid elements have uniform residence times, elements of fluid that have entered at different times are mixed, so that the outflowing stream shows less variation in composition than the imput. This problem has been dealt with and solved by Danckwerts (27).

The time fluctuations of inlet and outlet concentrations can be expressed as the variation around a time averaged mean concentration $\bar{C}$, which remains constant

$$C_i(t) = \bar{C} + \delta_i(t) \tag{7.13-1}$$

and

$$C_0(t) = \bar{C} + \delta_0(t) \tag{7.13-2}$$

where $\delta_i$ and $\delta_0$ are variations about the mean in the inlet and outlet streams, respectively, and show no trend. It follows from the definition of the variance that

$$\sigma_i^2 = \overline{\delta_i^2} \tag{7.13-3}$$

and

$$\sigma_o^2 = \overline{\delta_o^2} \tag{7.13-4}$$

where $\sigma_i^2$ and $\sigma_o^2$ are the inlet and outlet variances of the concentration fluctuation, which also show no trend. The problem is to find the ratio $\sigma_o^2/\sigma_i^2$.

The concentration $C_o(t)$ is determined by the concentration of all the fluid particles leaving at time $t$, which entered at various times in the past, hence have different concentrations; it is given by the following expression

$$C_o(t) = \int_0^\infty C_i(t-\theta) f(\theta)\, d\theta \tag{7.13-5}$$

where $C_i(t-\theta)$ is the concentration in the inlet stream of fluid particles that entered at time $t-\theta$, thus stayed in the system for a time $\theta$. Equation 7.13-5 can be rewritten as

$$\delta_o(t) = \int_0^\infty \delta_i(t-\theta) f(\theta)\, d\theta \tag{7.13-6}$$

and

$$[\delta_o(t)]^2 = \left[\int_0^\infty \delta_i(t-\theta) f(\theta)\, d\theta\right]^2 \tag{7.13-7}$$

with the aid of the following expression

$$\left[\int_0^\infty f(y)\,dy\right]^2 = 2\int_{y=0}^\infty \int_{r=0}^\infty f(y)f(y+r)\,dy\,dr \tag{7.13-8}$$

where $f(y)$ is any function of $y$; Eq. 7.13-5 can be written as

$$[\delta_o(t)]^2 = 2\int_0^\infty \int_0^\infty \delta_i(t-\theta)\delta_i(t-\theta+\tau)f(\theta)f(\theta+\tau)\,d\theta\,d\tau \tag{7.13-9}$$

Averaging with respect to $t$ results in

$$\sigma_o^2 = \overline{\delta_o^2} = 2\int_0^\infty \int_0^\infty \overline{\delta_i(t-\theta)\delta_i(t-\theta+\tau)}f(\theta)f(\theta+\tau)\,d\theta\,d\tau \tag{7.13-10}$$

The quantity

$$R(\tau) = \frac{\overline{\delta_i(t')\delta_i(t'+\tau)}}{\sigma^2} \tag{7.13-11}$$

(where $t' = t-\theta$) is the coefficient of correlation defined in Eq. 7.5-2 where the pair of points are taken at the same location but at a time difference of $\tau$, and it is known as the *autocorrelation coefficient* or *serial correlation coefficient* of $C_i$ for a time interval $\tau$. Substituting Eq. 7.13-11 into Eq. 7.13-10, we get

► $$\frac{\sigma_o^2}{\sigma_i^2} = 2\int_{\theta=0}^\infty \int_{\tau=0}^\infty R(\tau)f(\theta)f(\theta+\tau)\,d\theta\,d\tau \tag{7.13-12}$$

The autocorrelation coefficient $R(\tau)$ may be calculated from a representative record of inlet concentration as a function of time. It indicates the time scale of fluctuation and its periodicity. The RTD function $f(\theta)\,d\theta$ can be calculated if the flow pattern is known or alternatively, it can be measured experimentally by any of the common tracer method techniques.

Clearly, the magnitude of the exiting composition fluctuation in time is a function of inlet conditions as well as the RTD and their interaction (see Problem 7.15). The extreme condition of plug flow situation only delays and does not reduce the fluctuations. This condition is closely met by a number of types of processing equipment, such as the single and twin extruders. The very narrow RTD in these types of equipment, although conferring great advantages from many points of view, limits the capability of these machines to even out fluctuations in composition of the feed. Hence accurate metering of feed composition to extruders is necessary and is practiced.

## *REFERENCES*

1. H. F. Irving and R. L. Saxton, "Mixing in High Viscosity Materials," in *Mixing*, Vol. 2, V. H. Uhl and J. B. Gray, eds., Academic Press, New York, 1967, Chapter 8.
2. R. S. Brodkey, "Fluid Motion and Mixing," in *Mixing*, Vol. 1, V. H. Uhl and J. B. Gray, eds., Academic Press, New York, 1966, Chapter 2.
3. R. S. Spencer and R. N. Wiley, "The Mixing of Very Viscous Liquids," *J. Colloid Sci.*, **6**, 133–145 (1951).

4. J. T. Bergen, G. W. Carrier, and J. A. Krumbansh, "Criteria for Mixing and Mixing Process," Paper presented at the 14th Society of Plastic Engineers National Technical Conference, Detroit, January 1958.
5. J. M. McKelvey, *Polymer Processing*, Wiley, New York, 1962.
6. B. Julesz, "Experiments in the Visual Perception of Texture," *Sci. Am.*, **232**, 34–43 (1975).
7. W. G. Best and H. F. Tomfohrde, "A Quick Test for Carbon Black Dispersion," *Soc. Plastics Eng. J.*, **15**, 139–141 (1959).
8. W. Feller, *An Introduction to Probability Theory and Its Applications*, Vol. 1, Wiley, New York, 1950.
9. C. W. Clump, "Mixing of Solids," in *Mixing*, Vol. II, V. H. Uhl and J. B. Gray, eds., Academic Press, New York, 1967, Chapter 10.
10. J. R. Bourne, Industrial Research Fellow Report No. 2, "The Mixing of Powders, Pastes, and Non-Newtonian Fluids," The Institute of Chemical Engineers, London, September 1969.
11. L. T. Fan and R. H. Wang, "On Mixing Indices," *Powder Technol.*, **11**, 27–32 (1975).
12. P. V. Danckwerts, "Theory of Mixtures and Mixing," *Research* (London), **6**, 355–361 (1953). See also: P. V. Danckwerts, "General Aspects of Mixtures and Mixing," Paper presented at the Symposium on Mixing of Thick Liquids, Pastes, and Slurries, *British Society of Rheology*, London, September 1953; cf. N. Wooley, *Nature*, **172**, 846–847 (1953).
13. P. V. Danckwerts, "The Definition and Measurement of Some Characteristics of Mixtures," *Appl. Sci. Res.*, Sec. A3, 279–296 (1952).
14. W. D. Mohr, R. L. Saxton, and C. H. Jepson, "Mixing in Laminar Flow Systems," *Ind. Eng. Chem.*, **49**, 1855–1856 (1957).
15. N. Nadav and Z. Tadmor, "Quantitative Characterization of Extruded Film Textures," *Chem. Eng. Sci.*, **28**, 2115–2126 (1973).
16. K. J. Chu and A. E. Dukler, "Statistical Characterization of Thin, Wavy Films," *Am. Inst. Chem. Eng. J.*, **20**, 695 (1974); **21**, 583 (1975).
17. A. Brothman, G. N. Woldan, and S. M. Feldman, *Chem. Metall. Eng.*, **52**, 102 (1945).
18. F. H. H. Valentin, "Mixing of Powders and Particulate Solids," *Chem. Process Eng.*, 181–187, April 1965.
19. S. S. Weidenbaum, Ph.D. thesis, Columbia University, New York, 1953.
20. P. M. C. Lacey, "Developments in the Theory of Particle Mixing," *J. Appl. Chem.*, **4**, 254–268 (1954).
21. W. Weydanez, *Chem. Eng. Technol.*, **32**, 343 (1960).
22. H. E. Rose, *Trans. Inst. Chem. Eng.*, **40**, 272 (1962).
23. W. D. Mohr, "Mixing and Dispersing," in *Processing of Thermoplastic Materials*, E. C. Bernhardt, ed., Reinhold, New York, 1959, Chapter 3.
24. W. C. Brasie, private communication.
25. L. Erwin, "Mixing of Viscous Liquids," *Polym. Eng. Sci.* in press. See also: L. Erwin, "New Fundamental Considerations on Mixing in Laminar Flow," Society of Plastics Engineers 36th Annual Technical Conference, Washington, D.C., 1978, p. 488.
26. Z. Tadmor and G. Lidor, "Theoretical Analysis of Residence Time Distribution Functions and Strain Distribution Functions in Plasticating Screw Extruders," *Polym. Eng. Sci.*, **16**, 450–461 (1976).
27. P. V. Danckwerts, "Continuous Flow Systems—Distribution of Residence Times," *Chem. Eng. Sci.*, **2**, 1 (1953).

## *PROBLEMS*

**7.1** ***Mixing Mechanisms.*** Consider a calendering process for manufacturing polyvinyl chloride floor covering. The line consists of a ribbon-type blender in which the PVC is dry-blended, an internal mixer that feeds a single screw extruder equipped with a static mixer feeding the first nip of the calender.

(*a*) What types of mixing mechanisms does the polymer experience?

(*b*) Specify the locations at which each mechanism occurs.

**7.2** ***Random and Ordered Mixtures.*** Figure P7.2 shows two samples of mixtures of black and white squares. Sample 1 was obtained by tossing a coin for each square; if it showed "heads," the square was set black.

(*a*) If we take a large number of black and white particles of equal numbers and place the mixture in a V-blender, which sample will the mixture resemble?

(*b*) If we take from sample 1 and sample 2 a large number of small testing samples and measure the fraction of black particles in each sample, what type of distribution would be expected in each case?

(*c*) What is the variance of each distribution in part *b*?

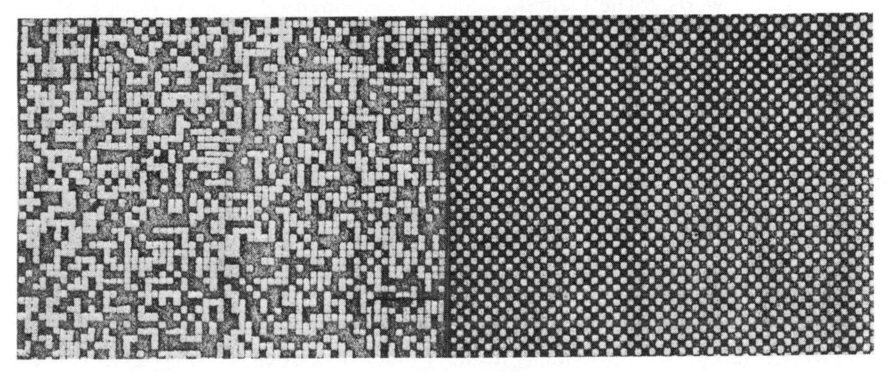

**Fig. P7.2**

**7.3** ***Gross Uniformity and Texture.*** (*a*) Do the two samples in Fig. P7.2 share the same first order statistics (i.e., are they "grossly uniform")?

(*b*) Do they have different textures?

(*c*) Outline a computer program for evaluating the scale of segregation of sample 1.

**7.4** ***The Binomial Distribution.*** Consider a random mixture of minor particles in major particles of equal size. The fraction of minor particles in the mixture is $p$. We withdraw a large number of testing samples from the mixture each containing exactly $n$ particles.

(*a*) Show that distribution of minor particles in the samples is given by Eq. 7.3-1.

(*b*) Derive the mean and the variance of the distribution.

***Answer:*** (*b*) $\sigma^2 = \dfrac{p(1-p)}{n}$

**7.5** ***The Coefficient of Correlation of a Two Composition System in Terms of Probabilities.*** Consider a two-composition texture of concentrations $x_1$ and $x_2$ (Fig. 7.9). Following a (mental) process of "dipole throwing" we find that $k_{11}$ of them fell with both ends on composition $x_1$, $k_{22}$ fell with both ends on composition $x_2$, and $k_{12}$ fell with one end on composition $x_1$ and the other on $x_2$.

(*a*) Show that the mean concentration is given by

$$\bar{x} = \frac{2k_{11}+k_{12}}{2N} x_1 + \frac{2k_{22}+k_{12}}{2N} x_2$$

where $N = k_{11}+k_{22}+k_{12}$.

(*b*) Show that the variance is given by

$$S^2 = \frac{(2k_{11}+k_{12})}{2N} \frac{(2k_{22}+k_{12})}{2N} (x_1-x_2)^2$$

(*c*) Using the results above, show that the coefficient of correlation is given by Eq. 7.5-4, where (for large number of dipoles) $\psi_{11} = k_{11}/N$, $\psi_{22} = k_{22}/N$, $\psi_{12} = k_{12}/N$, and

$$\frac{\phi_1}{\phi_2} = \frac{2k_{11}+k_{12}}{2k_{22}+k_{12}}$$

**7.6** ***The Scale of Segregation of a Striped Texture.*** Using Eq. 7.5-4 show that the linear scale of segregation (perpendicular to the stripes) of the texture in Fig. 7.10 is given by Eq. 7.5-8.

***Hint***: The probabilities in Eq. 7.5-4 are evaluated by a (mental) process of "dipole throwing." Thus the probability that one end of the dipole should fall on region I is $L_1/(L_1+L_2)$. The probability that this point should fall not closer than a distance $r$ from either boundary is

$$\left(\frac{L_1}{L_1+L_2}\right)\left(\frac{L_1-2r}{L_1}\right)$$

The probability that the other end should fall on region I is 1. Therefore the foregoing equation gives the probability that these dipoles fall with both end on region I. (*Note*: there are additional dipoles that fall on region I).

**7.7** ***Increase in Interfacial Area in Homogeneous Flow.*** The increase in interfacial area in a general homogeneous flow is given in Eq. 7.9-24. Prove that for simple shear flow Eq. 7.9-24 reduces to Eq. 7.9-15.

***Hint***: Recall that the angles the principal axes of the strain spheroid form with direction $x$ (direction of shear), $\chi$, $\chi+(\pi/2)$ are related to the total strain $\gamma$ as follows

$$\gamma = 2 \cot 2\chi$$

and the principal elongational ratios are

$$\lambda_x = \cot \chi, \qquad \lambda_y = \frac{1}{\lambda_x}, \qquad \lambda_z = 1$$

The angle the principal axes form with the $x$ direction in the initial state $\chi_0$ is given by

$$\sin \chi_0 = \cos \chi$$

Next, express the angles $\alpha'$ and $\beta'$ in terms of angles $\alpha_x$ and $\alpha_y$ and $\chi_0$ and substitute in 7.9-24.

**7.8** ***The Distribution of Interfacial Area Elements at High Strains in Simple Shear Flow.*** Consider randomly distributed and equal sized interfacial area elements placed in a rheologically uniform medium in simple shear flow. After a given strain, the interfacial area elements *vary* in size; that is, a distribution of interfacial area elements evolves because of the flow.

(*a*) Show that the variance of the distribution is

$$\left(\frac{\sigma}{\bar{A}}\right)^2 = \frac{\overline{A^2}}{(\bar{A})^2} - 1 = \frac{\gamma^2 + 12}{3\gamma^3}$$

where $\overline{A^2}$ is the mean of the square of the area elements and $\bar{A}$ is the mean area

$$\bar{A} = \frac{1}{N} \sum_{i=1}^{N} A_i$$

where $N$ is the number of area elements.

(*b*) It has been shown* that the mean square of interfacial area elements for any homogeneous deformation is given by

$$\overline{\left(\frac{A'}{A_0}\right)^2} = \tfrac{1}{3}\lambda_1^2\lambda_2^2\lambda_3^2(\lambda_1^{-2} + \lambda_2^{-2} + \lambda_3^{-2})$$

where $A_0$ is the initial surface and $\lambda_i$ are the principal elongational ratios. Show that the equation above for simple shear together, with Eq. 7.9-20, reduce to the results in part (*a*).

**7.9** ***Strain Distribution Function in Poiseuille Flow.*** (*a*) Derive the SDF $F(\gamma)$ for fully developed isothermal laminar flow of a Newtonian fluid in a tube.

(*b*) Calculate the mean strain.

(*c*) If the length of the tube is 1 m and its radius 0.01 m, what fraction of the exiting stream experiences a total strain of less than 100?

***Answer*:** (*a*) $F(\gamma) = 1 - 2/(1 + c^2/2 + \sqrt{1 + c^2})$ $\quad c = \gamma R/L$). (*c*) 0.3137.

**7.10** ***Strain Distribution Function in Parallel Plate Flow.*** (*a*) Derive the SDF $F(\gamma)$ for the parallel plate flow in Example 7.6, but with a superimposed pressure gradient. The velocity profile is given by

$$v_x = \left[\xi + 3\xi \frac{q_p}{q_d}(1 - \xi)\right] V_0$$

* A. S. Lodge, private communication, 1976.

where $\xi = y/H$, and $q_p$ and $q_d$ are pressure and drag flow rates per unit width respectively, and their ratio $A$ is

$$A = \frac{q_p}{q_d} = \frac{1}{6}\left(-\frac{dP}{dx}\right)\frac{H^2}{V_0 \mu}$$

(*b*) Calculate the mean strain.
(*c*) Plot the SDF with $q_p/q_d$ as a parameter in the range $-\frac{1}{3} < \frac{q_p}{q_d} < \frac{1}{3}$.

***Answer:***

(*a*) $F(\xi) = 1 - \frac{\xi^2}{1+A}[1 + A(3-2\xi)]$ $\qquad \gamma = \frac{L}{H}\left[\frac{1+3A(1-2\xi)}{\xi + 3\xi A(1-\xi)}\right]$

(*b*) $\bar{\gamma} = 2\frac{L}{H(1+A)}$

**7.11** ***Relationship Between the Mean Strain and the Mean Shear Rate.*** Section 7.10 indicates that for continuous single valued velocity profiles in shear flows of the type $v_x(y)$, the following relationship holds:

$$\bar{\gamma} = \bar{\dot{\gamma}}\bar{t} = \bar{t}\frac{1}{H}\int_0^H \dot{\gamma}\, dy$$

(*a*) Prove this relationship.
(*b*) Verify the validity of this relationship in Problems 7.9 and 7.10.

***Hint:*** Start with the definition of $\bar{\gamma}$ in Eq. 7.10-2, express $\gamma$ in terms of $\dot{\gamma}$ and $t$, and make use of the relationship $f(\gamma)\,d\gamma = dQ/Q$.

**7.12** ***Purging a Tubular Die.*** (*a*) A "red" polymer is pumped over a tubular die. At time $t$, the inlet stream is switched over to a "white" polymer for purging the die. Assuming Newtonian fluids, identical viscosities and densities, and fully developed isothermal laminar flow, calculate the volume fraction of "red" polymer left in the die at the time the first traces of white polymer appear at the exit.

***Answer:*** 0.5

**7.13** ***RTD in Two Systems in Series.*** Figure P7.13 shows two combinations of a tubular vessel and a well-mixed stirred tank.

(*a*) Assuming plug flow in the tube, prove that the RTDs in both combinations are identical.

(*b*) Repeat (*a*), assuming laminar flow in the tube.

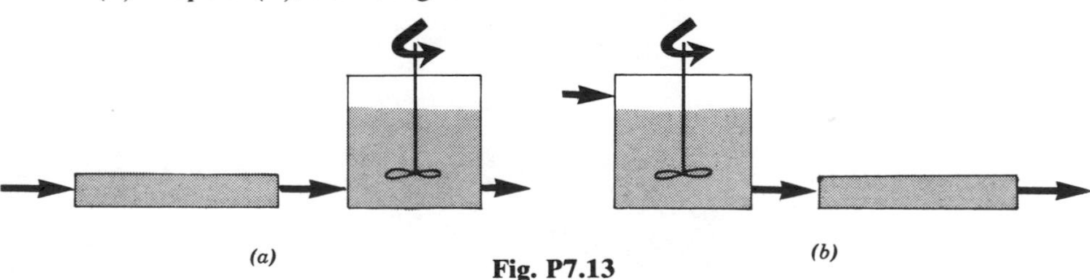


**Fig. P7.13**

**7.14** ***Derivation of RTD in a Continuously Stirred Tank (CST).*** Equation 7.11-8 gives the RTD function $F(t)$ in a CST.

($a$) Calculate $f(t)\,dt$.

($b$) The function $f(t)\,dt$ also expresses the probability that an entering fluid particle will leave at time $t$. Derive this function, using probability considerations.*

($c$) Extend the derivation in ($b$) to $N$ vessels in series.

***Hint*:** Consider the liquid volume in the vessel as consisting of a very large number of fluid particles; then assume that in every infinitesimal time increment, a fluid particle of volume $Q\delta t$, enters the vessel and one is randomly drawn out from the vessel. Each drawing can be considered as a Bernoulli trial.

***Answer*:** ($a$) $f(t)\,dt = (1/t)\,e^{-t/\bar{t}}$;
($c$) $f(t)\,dt = [1/(N-1)!](t/\bar{t})^{N-1}\,e^{-t/\bar{t}}(dt/\bar{t})$.

**7.15** ***Damping of a Sinusoidally Fluctuating Inlet Composition in a Continuous Mixer*** (23). The inlet concentration fluctuation to a continuous mixer is given by

$$\delta_i(t) = \delta_{iM}\cos\frac{2\pi t}{T}$$

where $\delta_{iM}$ is the amplitude about the mean and $T$ is the period.

($a$) Calculate the variance of the inlet composition fluctuation.

($b$) Assuming that the RTD function of the continuous mixer is given by

$$f(t)\,dt = \frac{1}{\bar{t}}e^{-t/\bar{t}}$$

calculate the autocorrelation coefficient. What is the value of $R(\tau)$ at $\tau = 0$, $\tau = T/2$, and $\tau = 1$? Explain.

($c$) Calculate the ratio of exit to inlet variances. How does the mean residence time affect the damping of the fluctuations?

***Answer*:** ($a$) $\dfrac{\sigma_0^2}{\sigma_i^2} = \dfrac{1}{1+\left(\dfrac{2\pi\bar{t}}{T}\right)^2}$

* J. A. Biesenberger, private communication.

*PART THREE*

# ELEMENTARY STEPS IN POLYMER PROCESSING

# 8

# Handling of Particulate Solids



## 8.1 The Role of Particulate Solids in Processing

Most polymers are supplied to processors in the form of particulate solids (Fig. 8.1). This is the most convenient way to handle, transport, blend and compound, store, feed, and more important, to process the polymers on existing machinery. The common forms of the particles are *pellets* in a variety of shapes such as cubes, cylinders, spheres, lenses, and ellipsoids, as well as *powders*, "*beads*," and "*granulates*." Pellets are obtained downstream the polymerization reactor through a special pelletizing step. The choice of the pelletizing system is determined by the nature of the polymer, production rate requirements, and processing characteristics (1). Hot die-face cutting in air or under water, cold cutting of strands, and strip dicing are among the common pelletizing methods. Typical sizes fall in the range of 3–6 mm ($\frac{1}{8}$–$\frac{1}{4}$ in.). The pellets often contain all the additives required for processing. This is generally the form preferred by the processors to feed processing machines such as extruders, injection molding machines, and blow molding machines. Other processing methods may require different forms of feed. Thus, for example, powders are required for rotational molding machines. Powders either are obtained directly from the polymerization reactor, or they are prepared in grinders and pulverizers. Hammer mills, pin mills, fluid energy mills, jet mills, and other highly

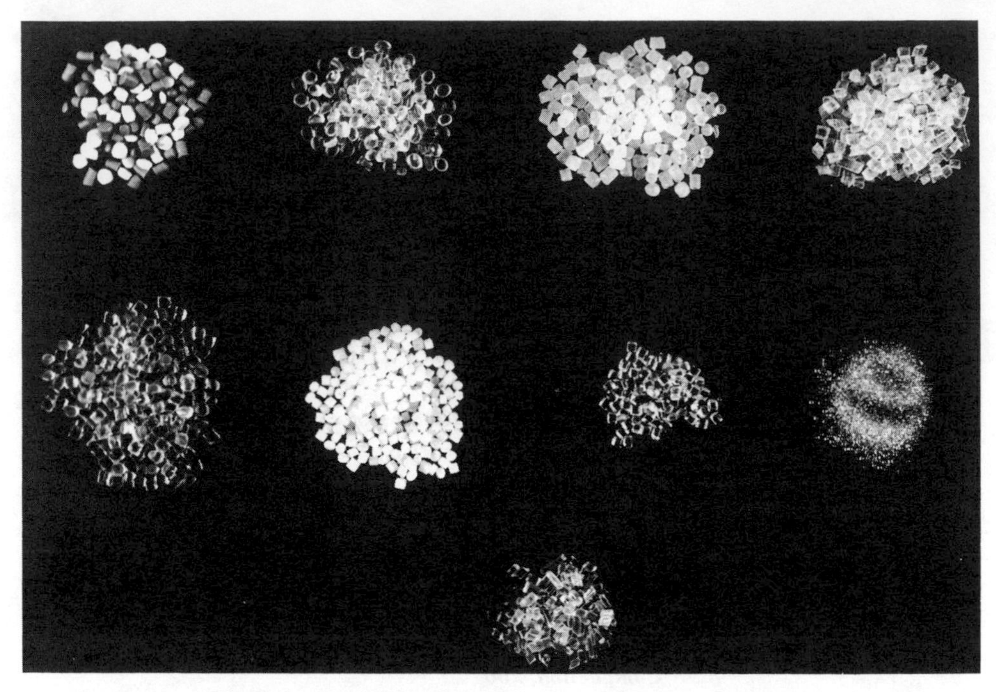

**Fig. 8.1** Particulate forms of some common polymers.

specialized equipment are used for this purpose (2). Scrap material is reduced in size in a variety of granulators, resulting in a "granulate" of nonuniform size and shape, which is often mixed with "virgin" feed material and reprocessed.

The polymer in particulate form is transported, stored, blended, fed through hoppers, conveyed in closed conduits in the processing equipment, pushed, dragged, compressed, and melted. All these intensive manipulations of particulate solids in processing, coupled with the unique properties exhibited by particulate solids (which affect the course and outcome of a processing operation), justify in our judgment the definition of "*handling of particulate solids*" as an elementary unit step of polymer processing.

This chapter discusses the unique properties of particulate solids and their behavior under static and dynamic conditions, relevant to polymer processing. Comprehensive reviews of these subjects are available in the literature (3–5).

## 8.2 Some Unique Properties of Particulate Solids

Scientific and engineering interests into the properties and behavior of particulate solids date back to the early work of Coulomb, who in 1776 developed a theory on "soil pressure and resistance," thus laying sound foundations for important engineering practices. Later, in 1852, Hagen analyzed the flow of sand in an hourglass, and shortly afterward Reynolds, in 1885, made the observation on the dilatancy of a deforming mass of sand.* The latter unique property of particulate

* O. Reynolds, *Phil. Mag.*, Ser. 5, **20**, 469 (1855).

solids can be observed while walking on wet sand at the seashore. The sand "whitens" or appears to dry momentarily around the foot because the pressure of the foot dilates the sand.

The analysis of particulate solids systems, in analogy to fluids, can be divided into *statics* and *dynamics*; it is interesting to note that in spite of the early beginnings of scientific interest into the properties of particulate solids, this field—in particular, the dynamics of particulate solids—has not experienced the same intensive scientific development as fluid dynamics. In most engineering curricula relatively little attention is focused on the analysis of particulate solids. Therefore, as engineers, we are generally ill-equipped to analyze these complex systems and to design equipment for handling them, and we may often be surprised by the behavior of particulate solids.*

Particulate solids are made up of loose, discrete particles of more or less uniform size. Following the definitions of Brown and Richards (4), the term "particulate solids" covers *powders*, which consist of particles up to 0.1 mm (100 $\mu$) in size, *granular solids* consisting of particles from 0.1 to 3 mm, and *broken solids* composed of particles larger than 3 mm. Powders can be further subdivided into *ultrafine* (0.1–1 $\mu$) and *superfine* (1–10 $\mu$), and classified as "*free flowing*" or *cohesive*.

A closer look at the properties of particulate solids and their response to external forces reveals that these are a blend of ($a$) liquidlike behavior, ($b$) solidlike behavior, and ($c$) particle-interface dominated behavior, unique to these systems. Like liquids, particulate systems take the shape of the container they occupy, exert pressure on the container walls, and flow through openings. Yet like solids, they sustain shearing stresses (hence they form piles), may possess cohesive strength, and exhibit a nonisotropic stress distribution upon application of a unidirectional load. But unlike liquids, shearing stress is proportional to *normal load* rather than to *rate of deformation*, and unlike solids, the magnitude of the shearing stress is generally *indeterminate*, and all that can be said is that the following *inequality* holds:

$$\tau \leq f'\sigma \tag{8.2-1}$$

where $f'$ is the interparticle static coefficient of friction and $\sigma$ represents a range of normal forces ("pressures") that can be applied to the particulate system before the value of shear stress $\tau$ is reached that is high enough to start the particles sliding past one another. That is, before particulate solids flow starts, there is a *range of equilibrium states* and a *range of bulk densities* allowable. Only at the inception of flow the frictional forces are fully mobilized (4). At this state the relation 8.2-1 takes the form of Amonton's law, discussed in Section 4.3, which is the defining equation for the coefficient of static friction.

Thus before we consider the response of particulate systems to externally applied stresses, we must know whether the shear and normal stresses at any point and orientation are above the values specified by the *equality* 8.2-1. We do this by considering the Mohr circle of plane stresses. Furthermore, since there are two kinds of particulate solids, the noncohesive (free flowing) and the cohesive, we

* For example, the fact that the drag on the plough is independent of speed has surprised engineers who designed the first tractors (5).

comment on the phenomenon of agglomeration, which transforms the former to the latter. Finally, we must remember that since it is necessary to contain particulate solids, the wall-particle static coefficient of friction and the wall shear and normal forces must be specified. The wall is another location at which flow can be initiated.

## 8.3 Agglomeration

The term *agglomeration* describes the forming of an aggregate from the individual particles. Agglomeration is undesirable when it happens to a free flowing system (caking), but it is desirable for pelleting, tableting, and other similar processes. The pelleting of thermoset molding powders and the extrusion of polytetrafluoroethylene prior to sintering can serve as examples of the latter in polymer processing. In either case it is important to understand the physical mechanisms involved (3, 6). Agglomerates form because of the binding forces between the particles. These forces have been discussed in Chapter 4. Solid-solid forces (see Section 4.3) are significantly amplified by increases in pressure and temperature, which induce simultaneously an increase in contact area, *bridging* due to local heat effects, breakdown of contaminating layers, and generation of "clean" surfaces by fracture. "Liquid-binding" forces due to liquid-solid interaction (Section 4.2) play an important role in agglomeration, and in many "dry" pressing techniques the powder is "moistened" prior to compaction. Finally, electrostatic forces must also be considered in the agglomeration mechanism (Section 4.4), in particular in dealing with fine powders.

## 8.4 The Mohr Circle of Stresses

The theoretical development of statics of particulate solids commences with the work of Coulomb, and it has been oriented primarily toward subjects related to soil mechanics. This section reviews some concepts in elementary statics that are

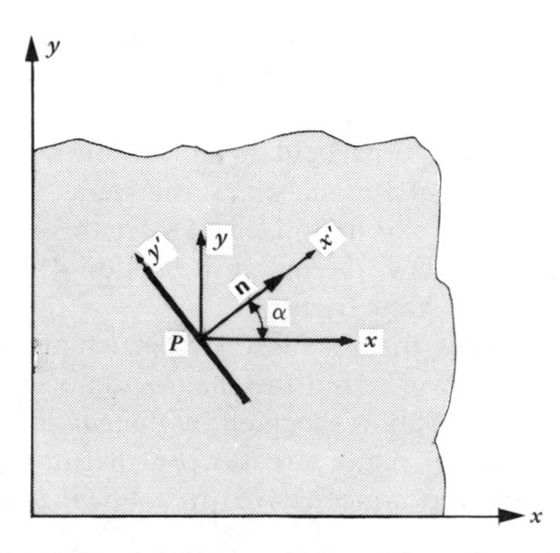


**Fig. 8.2** Arbitrary plane with orientation defined by the unit normal vector **n**.

essential for the understanding of the behavior of particulate solids. We shall be dealing with a condition of static equilibrium that also holds at the limiting condition of *incipient flow*, that is, a condition under which the inequality in Eq. 8.2-1 becomes an equality. Moreover, equilibrium static conditions can also be applied to sufficiently slow steady flows (5). The analysis is limited to a state of *plane stress*.

Section 5.1 pointed out that given the stress field at a point $P$ in a continuum, it is possible to evaluate the resultant stress vector on any arbitrary surface area element containing point $P$. Figure 8.2 shows an arbitrary plane with its orientation defined by the unit normal vector $\mathbf{n}$. Given a stress field $\boldsymbol{\tau}$ the stress vector at point $P$ acting on the above-defined surface is $\mathbf{n} \cdot \boldsymbol{\tau}$. The stress vector can be resolved into two components—a normal and a shear component—given by the following equations:

$$\tau_{x'x'} = \tfrac{1}{2}(\tau_{xx} + \tau_{yy}) + \tfrac{1}{2}(\tau_{xx} - \tau_{yy}) \cos 2\alpha + \tau_{xy} \sin 2\alpha \tag{8.4-1}$$

and

$$\tau_{x'y'} = \tfrac{1}{2}(\tau_{yy} - \tau_{xx}) \sin 2\alpha + \tau_{xy} \cos 2\alpha \tag{8.4-2}$$

where $x'$, $y'$ is a rectangular coordinate system rotated by angle $\alpha$ relative to the system $x$, $y$, as in Fig. 8.2. It is easy to show from Eqs. 8.4-1 and 8.4-2 that $\tau_{x'x'}$ and $\tau_{x'y'}$ acquire extreme values at certain specific angles $\alpha$. Thus $\tau_{x'x'}$ is maximum at an angle $\alpha_m$ given by

$$\tan 2\alpha_m = \frac{2\tau_{xy}}{\tau_{xx} - \tau_{yy}} \tag{8.4-3}$$

and it is minimum at $\alpha_m + \pi/2$. Clearly the value of $\alpha_m$ is a function of the stress field. The corresponding maximum and minimum values of $\tau_{x'x'}$, denoted as $\sigma_{max}$ and $\sigma_{min}$, respectively, are

$$\sigma_{max} = \tfrac{1}{2}(\tau_{xx} + \tau_{yy}) + \tfrac{1}{2}\sqrt{(\tau_{xx} - \tau_{yy})^2 + 4\tau_{xy}^2} \tag{8.4-4}$$

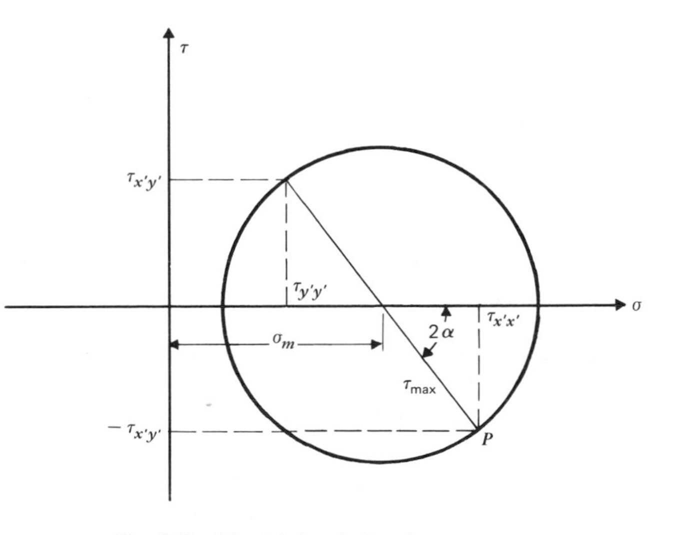


**Fig. 8.3** The Mohr circle of stresses.

and

$$\sigma_{\min} = \tfrac{1}{2}(\tau_{xx} + \tau_{yy}) - \tfrac{1}{2}\sqrt{(\tau_{xx} - \tau_{yy})^2 + 4\tau_{xy}^2} \tag{8.4-5}$$

By substituting Eq. 8.4-3 into Eq. 8.4-2, it can be shown that when the normal stresses attain extreme values, the shear stress vanishes. Hence we can conclude that there is a certain set of perpendicular planes at $\alpha_m$ and $\alpha_m + \pi/2$, where the normal stresses reach maximum and minimum values, respectively, and where the shear stress vanishes. These planes are called the *principal planes*, and the normal stresses are called *principal stresses.* It further follows from this result that the state of stress at point $P$ is fully described by the principal normal stresses and planes. Finally, *any change in the stress loading of the system may change the magnitude of the principal stresses and the orientation of the principal planes, and they both may vary from point to point in the system.*

Similarly, from Eq. 8.4-2 it can be shown that $\tau_{x'y'}$ attains a maximum value at $\alpha'_m$ given by

$$\tan 2\alpha'_m = \frac{\tau_{yy} - \tau_{xx}}{2\tau_{xy}} \tag{8.4-6}$$

and the corresponding maximum shear stress is given by

$$\tau_{\max} = \tfrac{1}{2}\sqrt{(\tau_{xx} - \tau_{yy})^2 + 4\tau_{xy}^2} \tag{8.4-7}$$

A comparison of Eq. 8.4-6 with Eq. 8.4-3 indicates that the shear stress acquires a maximum value in the direction of 45° to the principal planes.

In view of these results, it would be worthwhile to rewrite Eqs. 8.4-1 and 8.4-2 with plane orientation relative to the principal planes. That is, the coordinate system $x$, $y$ will be rotated to coincide with the principal planes, hence $\tau_{xx} = \sigma_{\max}$, $\tau_{yy} = \sigma_{\min}$, $\tau_{xy} = 0$, and the angle $\alpha$ becomes the angle between the normal of an arbitrary plane and the abscissa of the principal planes. This results in the following expressions

► $$\tau_{x'x'} = \tfrac{1}{2}(\sigma_{\max} + \sigma_{\min}) + \tfrac{1}{2}(\sigma_{\max} - \sigma_{\min}) \cos 2\alpha \tag{8.4-8}$$

and

► $$\tau_{x'y'} = \tfrac{1}{2}(\sigma_{\min} - \sigma_{\max}) \sin 2\alpha \tag{8.4-9}$$

This set of equations can be represented graphically by the celebrated Mohr circle (Fig. 8.3). The center of the circle is at point $\sigma_m = (\sigma_{\max} + \sigma_{\min})/2$ on the abscissa representing normal stresses and its radius is $\tau_{\max} = (\sigma_{\max} - \sigma_{\min})/2$. Any point on the circle represents the direction of the plane that is $2\alpha$ to the principal plane. Clearly the shear stress is maximum at 45° to the principal planes.

In conclusion, the Mohr circle represents an equilibrium stress condition at point $P$. Any change in these stress conditions (e.g., an increase in external load), as pointed out earlier, may change the direction of the principal plane and the value of the principal stresses, hence altering the location of the circle on the abscissa as well as its radius.

## 8.5 The Equations of Equilibrium

We have been concerned until now with defining the state of the stresses at a *point* in the system. At equilibrium conditions we can derive certain relationships expressing the manner in which stresses vary from point to point. These are the *equations of equilibrium* that can either be obtained by making a force balance on a small differential element or taking the equation of motion (which is the result of a similar general force balance), and setting all velocity components and the hydrostatic pressure gradients equal to zero. For a *plane stress* situation, we obtain the following two equations of equilibrium:

$$\frac{\partial \tau_{xx}}{\partial x} + \frac{\partial \tau_{yx}}{\partial y} = \rho_b g_x \tag{8.5-1}$$

and

$$\frac{\partial \tau_{xy}}{\partial x} + \frac{\partial \tau_{yy}}{\partial y} = \rho_b g_y \tag{8.5-2}$$

where $\rho_b$ is the bulk density; the sign convention defined in Chapter 5 is followed (i.e., compressive stresses are positive and tensile stresses negative).

In cylindrical coordinates with axial symmetry (i.e., no variations in the $\theta$ direction), we get

$$\frac{\partial \tau_{rr}}{\partial r} + \frac{\partial \tau_{rz}}{\partial z} + \frac{\tau_{rr} - \tau_{\theta\theta}}{r} = \rho_b g_r \tag{8.5-3}$$

and

$$\frac{\partial \tau_{rz}}{\partial r} + \frac{\partial \tau_{zz}}{\partial z} + \frac{\tau_{rz}}{r} = \rho_b g_z \tag{8.5-4}$$

## 8.6 Yield Loci

Shear properties of particulate systems determine their "flowability." This term refers to steady flow conditions as well as "flow–no flow" criteria. When internal sliding or shear failure is just about to occur, the local shear stress is called the *shear strength*. The shear strength is a function of the normal stress. This function is the *yield locus* (YL).

The yield locus of a *noncohesive* particulate system follows from Eq. 8.2-1 at fully mobilized friction conditions

► $$\tau = \tan \beta\, \sigma = f'\sigma \tag{8.6-1}$$

where $\beta$ is the *angle of internal friction.*

*Cohesive* particulate systems gain strength when pressure is applied. Consequently their YL is a function of the *consolidation pressure* and the *consolidation time.* Hence their shear strength behavior can be characterized only by a *family* of yield loci, each curve corresponding to a certain consolidation pressure and time. Often these curves are almost straight lines corresponding to the following equation:

$$\tau = \tan\beta(\sigma + \sigma_a) = c + \sigma \tan\beta \tag{8.6-2}$$

where $\sigma_a$ is an "apparent tensile strength," which is obtained by extrapolating the YL to $\tau = 0$. The actual tensile strength of cohesive particulate systems can be measured, and it is usually less than $\sigma_a$ (4). The value of the shear stress at $\sigma = 0$ is called the *coefficient of cohesion*, $c = \sigma_a \tan\beta$, and it reflects the magnitude of cohesive forces in the particulate system that must be overcome for sliding to occur.

Shear failure, or slip, in a particulate system occurs when in a certain direction the local shear stress (as given by the Mohr circle) exceeds the local shear *strength*. Hence failure at a point will not necessarily occur along the plane of maximum shear stress at that point, because the normal stress perpendicular to that plane may be sufficient to raise the shear strength above the local shear stress. Clearly then, in a system at equilibrium there cannot be any point with a stress state such that the YL crosses the Mohr circle, representing the state of stress at that point, because failure would have preceded such a situation.

In a state of *incipient failure* the YL is *tangent* to the Mohr circle. *The angle at the point of contact indicates the plane orientation along which failure will occur.* Under such conditions it can be shown (see Problem 8.3) that for a noncohesive powder, the following relationship holds:

$$\frac{\sigma_{\max}}{\sigma_{\min}} = \frac{1 + \sin\beta}{1 - \sin\beta} \tag{8.6-3}$$

Thus it is evident that under these conditions the principal stresses bear a fixed relationship to each other. For cohesive systems, with linear YL a Mohr circle can be drawn through the origin and tangent to the YL (Fig. 8.4). The resulting maximum principal stress is called the *unconfined yield strength* $\sigma_c$. This would then be the maximum normal stress, under incipient failure conditions, at a point where the other principal stress vanishes. Such would be the case on an exposed surface of an arch or dome (see Fig. 8.12*b*) at the moment of failure. This unconfined yield strength, therefore, plays an important role in deriving criteria of "flow–no flow" in hoppers and bins. Since $\sigma_c$ depends on the YL and the latter depends on the consolidation pressure, $\sigma_c$ is also a function of consolidation pressure.

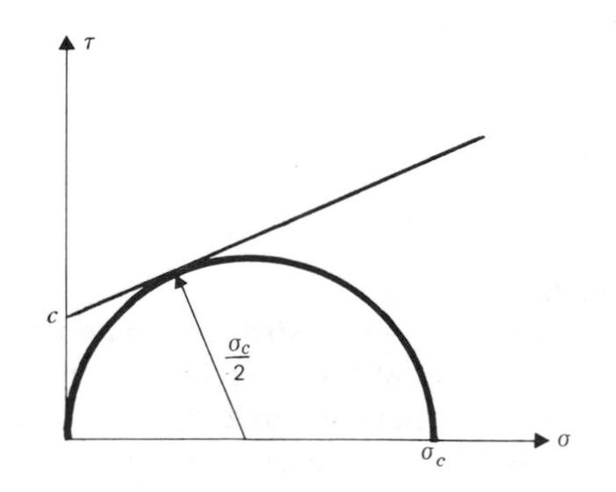


**Fig. 8.4** Unconfined yield strength.

For cohesive particulate systems with YL following Eq. 8.6-2 under incipient failure conditions, the following relationship between principal stresses holds:

$$\sigma_{\max} = \sigma_{\min}\frac{1+\sin\beta}{1-\sin\beta}+\sigma_c \tag{8.6-4}$$

By substituting Eqs. 8.4-4 and 8.4-5 into Eq. 8.6-4, we obtain the *Coulomb yield function*

$$(\tau_{xx}+\tau_{yy})\sin\beta-[(\tau_{xx}-\tau_{yy})^2+4\tau_{xy}]^{1/2}+\sigma_c(1-\sin\beta)=0 \tag{8.6-5}$$

If all points in the particulate solids are under incipient failure conditions, the equations of equilibrium together with Eq. 8.6-5 provide for the stress distribution in the system.

### *Effective Yield Locus*

Each yield locus in a cohesive particulate system terminates at the stress condition where the normal stress equals the consolidation pressure. A higher stress would imply a higher consolidation pressure, hence a different YL. A Mohr circle can now be created such that it is tangent to the terminal point of the particular YL. This process would generate a series of Mohr circles, as in Fig. 8.5. The *envelope* of these Mohr circles is called the *effective yield locus* (EYL), and it is generally a straight line passing through the origin. It forms an angle $\delta$ with the abscissa, which is termed the *effective angle of friction* (7). The EYL represents the shear–compressive stress characteristics of a powder that is consolidated and being sheared under the *same* stress conditions. This is highly relevant to a steady flow situation because during such flow shearing of the powder takes place at all points; hence the Mohr circle representing the stress condition at any given point must also be tangent to the EYL. No shearing flow occurs at any point where the stress field is such that the Mohr circle lies below EYL.

For a noncohesive powder the EYL coincides with the YL. Under steady flow conditions, therefore, as pointed out by Jenike (7), the ratio of the principal stresses

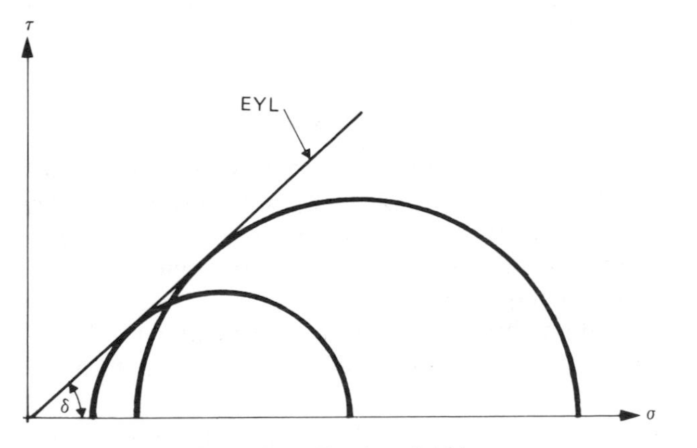


**Fig. 8.5** The effective yield locus.

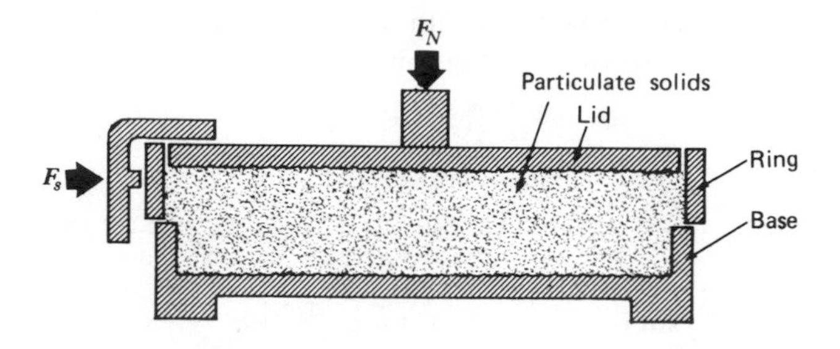


**Fig. 8.6** Shearing cell for measuring the YL of particulate solids, developed by Jenike (7).

at all points is

$$\frac{\sigma_{\max}}{\sigma_{\min}} = \frac{1+\sin\delta}{1-\sin\delta} \tag{8.6-6}$$

and the yield function is given by

$$(\tau_{xx}+\tau_{yy})\sin\delta - [(\tau_{xx}-\tau_{yy})^2 + 4\tau_{xy}]^{1/2} = 0 \tag{8.6-7}$$

An apparatus and methods for measuring the shear properties of particulate solids were developed by Jenike (7). Direct shear measurements are made in disk-shaped shear cells (Fig. 8.6). The shearing force is measured as a function of normal load. Prior to the test the shearing cell is placed in a consolidating bench where the powder is consolidated to the required pressure and time. By measuring the shearing strength of the powder at a number of normal loads below the consolidating pressure, YL is determined. Repeating the test at other consolidating pressures provides the family of yield loci. From these, the internal angles of friction, the effective angle of friction, and the unconfined yield stress can be calculated.

Figure 8.7 shows yield loci of a particulate system of polystyrene pellets resulting in an effective angle of friction $\delta = 34.5°$ and indicating the unconfined yield strength values at various consolidation pressures (8).

The Jenike shear cell can also be used to measure YL curve between the particulate solids and a confining wall. This is called the *wall yield locus* (WYL), and it usually falls below the YL. It is either a curve or a straight line. In the latter case we can write the relationship as

$$\tau_w = c_w + \sigma_w \tan\beta_w = c_w + f'_w\sigma_w \tag{8.6-8}$$

where $\tau_w$ is shear stress at the wall, $\sigma_w$ is the normal stress, $c_w$ is a measure of the adhesion of the solids at the wall, $\beta_w$ is the wall angle of friction, and $f'_w$ is the coefficient of friction at the wall. The WYL was found* to be generally independent of consolidation pressure (7). Chapter 4 gave experimental results for $f'_w$, which according to Eq. 4.3-6 may be a function of $\sigma_w$.

Finally we should mention the *angle of repose*, which reflects the static equilibrium between *unconfined loose solids*. One common way to measure the angle of

* This finding, however, is based on nonpolymeric materials and relatively low pressures compared to those existing in polymer processing machines.

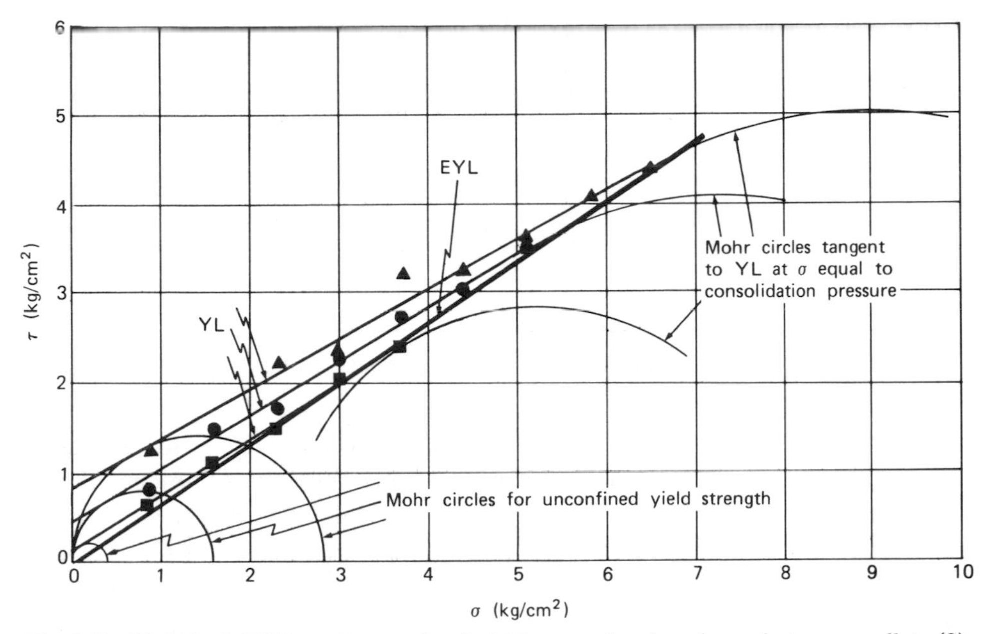


**Fig. 8.7** Yield loci, EYL, and unconfined yield strength values for polystyrene pellets (8).

repose is by allowing the material to flow out from a small opening onto a flat horizontal surface. The angle of inclination to the horizontal of its free surface is the angle of repose. There are a number of other ways to measure angles of repose, all resulting in somewhat different values (4). It should be noted that the angle of repose is generally *not* a measure of flowability of solids, and as Jenike (7) points out, it is strictly useful only to determine the contour of a pile. Its popularity stems from the ease with which it can be measured.

## 8.7 Pressure Distribution in Bins and Hoppers

The static pressure under a liquid column is isotropic and is determined by the height of the column above the point of measurement, $h$, and the density of the liquid $\rho$

$$P = \rho g h \tag{8.7-1}$$

In a column of particulate solids contained in a vertical bin, the pressure at the base will not be proportional to the height of the column because of the friction between the solids and the wall. Moreover, a complex stress distribution develops in the system, which depends on the properties of the particulate solids as well as loading method. The latter affects the mobilization of friction both at the wall and within the powder. Finally, arching or doming may further complicate matters. Hence exact solution to the problem is hard to obtain. Janssen (9) in 1895 derived a simple equation for the pressure at the base of the bin, which is still frequently quoted and used. The assumptions that he made are: the vertical compressive stress is constant over any horizontal plane, the ratio of horizontal and vertical stresses is constant and independent of depth, the bulk density is constant, the wall friction is fully

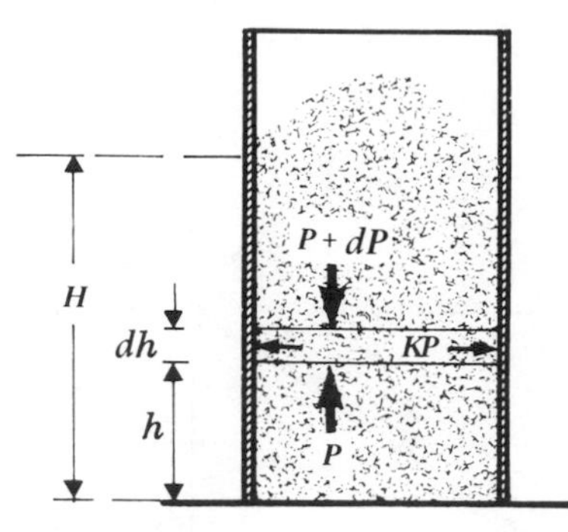


**Fig. 8.8** Vertical bin filled with particulate solids.

mobilized, that is the powder is in incipient slip condition at the wall. A force balance over a differential element (Fig. 8.8) using simply pressure $P$ instead of the compressive stress, with shear stress at the wall from Eq. 8.6-8, gives

$$\underbrace{A\rho_b g\, dh}_{\text{Weight of element}} + \underbrace{(P+dP)A}_{\text{Pressure acting downward}} = \underbrace{(c_w + f'_w KP)\, C\, dh}_{\text{Frictional force supporting the element}} + \underbrace{PA}_{\text{Pressure acting upward}} \qquad (8.7\text{-}2)$$

where $\rho_b$ is bulk density, $A$ is the cross-sectional area, $C$ the "wetted" circumference, and $K$ is the ratio of compressive stress in the horizontal direction to compressive stress in the vertical direction. A rough approximation for this ratio for cohesive particulate solids can be obtained from Eq. 8.6-3, which implies that the maximum principal stress is in the vertical direction and the minimum principal stress is in the horizontal direction, and all points in the system are at incipient flow conditions. Neither of these is strictly correct in this case. If the particulate solids are at steady flow conditions, an approximation for $K$ can be obtained from Eq. 8.6-6 subject to the same restrictions.

Integration of Eq. 8.7-2 results in

► $$P = P_1 \exp\left[\frac{f'_w CK(h-h_1)}{A}\right] + \frac{(A\rho_b g/C) - c_w}{f'_w K}\left\{1 - \exp\left[\frac{f'_w CK(h-h_1)}{A}\right]\right\} \qquad (8.7\text{-}3)$$

where $P_1$ is the pressure at height $h_1$. For the special case of cylindrical bin, with $h_1 = H$, where $P_1 = 0$ and $c_w = 0$ (no adhesion between the solids and the wall), Eq. 8.7-3 reduces to the familiar Janssen equation

► $$P = \frac{\rho_b g D}{4f'_w K}\left\{1 - \exp\left[\frac{4f'_w K(h-H)}{D}\right]\right\} \qquad (8.7\text{-}4)$$

Clearly the pressure at the base approaches a limiting value as $H$ goes to infinity:

► $$P_{\max} = \frac{\rho_b g D}{4f'_w K} \qquad (8.7\text{-}5)$$

Hence most of the weight is *supported frictionally by the walls* of the bin. The maximum pressure is proportional to bin diameter and inversely proportional to

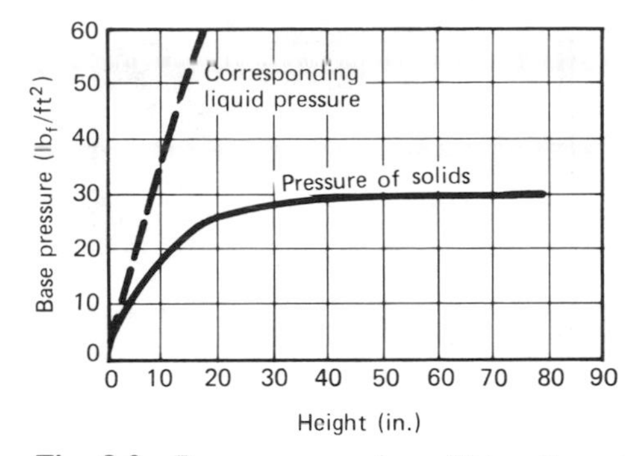


**Fig. 8.9** Base pressure in a 10 in. diameter cylindrical hopper filled with $\frac{1}{8}$ in. PS cubes with $K = 0.521$, $f'_w = 0.523$, and $\rho_b = 39\ \text{lb/ft}^3$. [Reprinted with permission from W. L. McCabe and J. C. Smith, *Unit Operations of Chemical Engineering*, McGraw-Hill, New York, 1956.]

the coefficient of friction at the wall. Figure 8.9 plots the pressure measured under a load of polystyrene pellets in a 10 in. diameter cylindrical bin as a function of solids height (10). Many other attempts to verify the Janssen equation have met with varying success, but the shape of the curve as predicted by the model is usually observed (4).

A more rigorous derivation of the pressure distribution in vertical bins was presented by Walker (11), assuming plastic equilibrium in the solids such that the Mohr circles representing the stress condition at a certain level touch the EYL. The results are only slightly different from those predicted by Eq. 8.7-4, with $f'_wK$ replaced by the product $BD^*$ where $D^*$ is defined as a distribution factor relating the *average* vertical stress with the vertical stress near the wall. The distribution function $D^*$ can be evaluated theoretically by solving the complete stress field (12), but as a first approximation it can be assumed to be unity. The ratio of the shear to the normal stress at the wall $B$ is given by the following equation:

$$B = \frac{\sin\delta \sin\kappa_0}{1 - \sin\delta \cos\kappa_0} \tag{8.7-6}$$

where

$$\kappa_0 = \beta_w + \arcsin\left(\frac{\sin\beta_w}{\sin\delta}\right) \qquad \arcsin > \frac{\pi}{2} \tag{8.7-7}$$

The stress distribution in convergent hoppers is of great interest and has received considerable attention in the literature. Walker (11) derived equations for the vertical and wall stress distribution under mass flow conditions, that is, conditions at which all the solids move toward the exit. For the details of the derivation, the reader is referred to the original paper. The resulting vertical stress or pressure distribution is given as

$$P = \left(\frac{h}{h_0}\right)^{\not{k}} P_0 + \frac{\rho_b g h}{\not{k} - 1}\left[1 - \left(\frac{h}{h_0}\right)^{\not{k}-1}\right] \qquad \not{k} \neq 1 \tag{8.7-8}$$

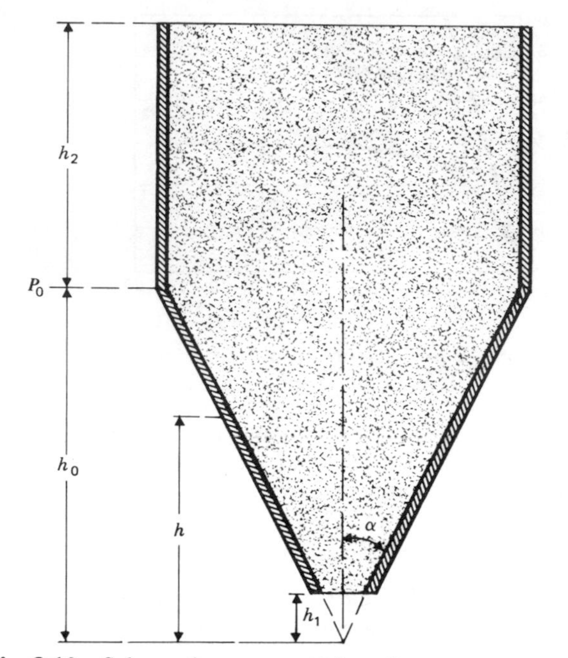


**Fig. 8.10** Schematic representation of a convergent conical hopper.

$$P = \frac{h}{h_0} P_0 + \rho_b g h \ln\left(\frac{h_0}{h}\right) \qquad \not{k} = 1 \tag{8.7-9}$$

where $h_0$ is the height where the vertical pressure is $P_0$ (this can be the pressure at the base of a vertical hopper placed above the convergent hopper, as in Fig. 8.10), the coefficient $\not{k}$ is given for conical and wedge shaped hoppers, respectively, as

$$\not{k} = \frac{2B'D^*}{\tan \alpha} \tag{8.7-10}$$

and

$$\not{k} = \frac{B'D^*}{\tan \alpha} \tag{8.7-11}$$

where $2\alpha$ is the hopper angle, $D^*$ is again the distribution function, which as pointed out earlier can be taken as a first approximation to be unity, and $B'$ is given by

$$B' = \frac{\sin \delta \sin(2\alpha + \kappa_0)}{1 - \sin \delta \cos(2\alpha + \kappa_0)} \tag{8.7-12}$$

where $\kappa_0$ is

$$\kappa_0 = \beta_w + \arcsin\left(\frac{\sin \beta_w}{\sin \delta}\right) \qquad \arcsin < \frac{\pi}{2} \tag{8.7-13}$$

In a conical section both the vertical and wall stresses attain maximum values in the cone (Fig. 8.11). With a cylindrical section added on the top of the cone, instabilities of stress conditions may occur at the transition plane (13).

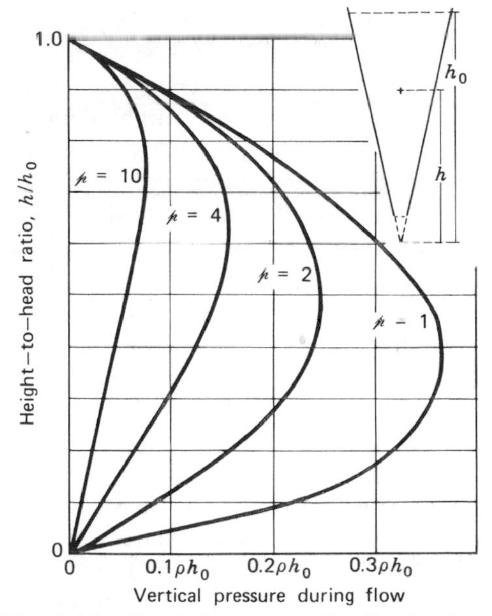


**Fig. 8.11** Vertical pressure distribution in a conical hopper with the coefficient $k$ as a parameter. [Reprinted with permission from D. M. Walker, *Chem. Eng. Sci.*, **21**, 975 (1966).]

## 8.8 Gravitational Flow in Bins and Hoppers

There are three aspects of particular interest in gravitational discharge behavior from bins and hoppers: flow disturbances, flow kinematics, and flow rates.

There are generally two types of gravitational flow in bins and hoppers (Figs. 8.12*a*, 8.12*c*). In "mass" flow the whole mass of particulate solids moves toward the exit, and in "funnel" flow the particles flow out through a central opening. In the former, the main cause for flow disturbance is "doming" or "arching," where all the weight of the solids is supported by the walls (Fig. 8.12*b*), whereas in the latter flow disturbance may occur when the solids can sustain the existence of an empty central tube, called "piping" (Fig. 8.12*d*). These and other flow disturbances were

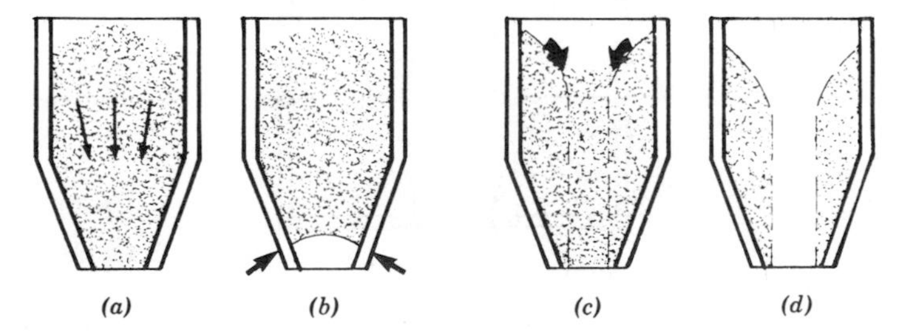


**Fig. 8.12** Schematic representation of (*a*) "mass" flow in hoppers, (*b*) arching, (*c*) "funnel" flow, and (*d*) "piping."

discussed by Johanson (14). In both arching and piping the solids must have consolidated sufficiently to develop the level of strength necessary to sustain the weight of the retained particulate solids. Hence obstruction to flow is acute in cohesive particulate solids (in particular those with high unconfined yield strength); it must also depend, in addition to material properties, on hopper geometry, which determines the stress distribution in the system.

Jenike (7) and his co-workers developed design methods and criteria for building obstruction-free hoppers and bins. We shall not discuss the details of their method but only outline the principles on which it is based. In particular, we consider the problem of arching. The flowability of a hopper can be characterized by a function called "flow factor," defined as

$$ff = \frac{\sigma_1}{\bar{\sigma}_1} \tag{8.8-1}$$

where $\sigma_1$ is the consolidating pressure and $\bar{\sigma}_1$ is the stress that acts in a stable arch. The latter is also the only nonvanishing principal stress, because the arch is assumed to be self-supporting. Both $\sigma_1$ and $\bar{\sigma}_1$ are linear functions of hopper width, and their ratio is constant for a given hopper. The flow factor depends on hopper geometry and material properties; values for a wide range of hoppers and material properties have been computed. The yield strength of the consolidated solids in a stable arch is given by the unconfined yield strength $\sigma_c$, which is a function of the consolidation stress $\sigma_1$. The condition for no arching, therefore, is

$$\sigma_c < \bar{\sigma}_1 \tag{8.8-2}$$

By plotting the unconfined yield strength as a function of consolidating pressure, which was defined by Jenike (7) as the "flow function," and using it together with the flow factor, the critical conditions for arching can be established.

Similar reasoning leads to criteria for no "piping." Further details on arching and piping can be found in the literature (4, 7, 15). Arching criteria were also used by Richmond (16) for designing an optimum hopper shape.

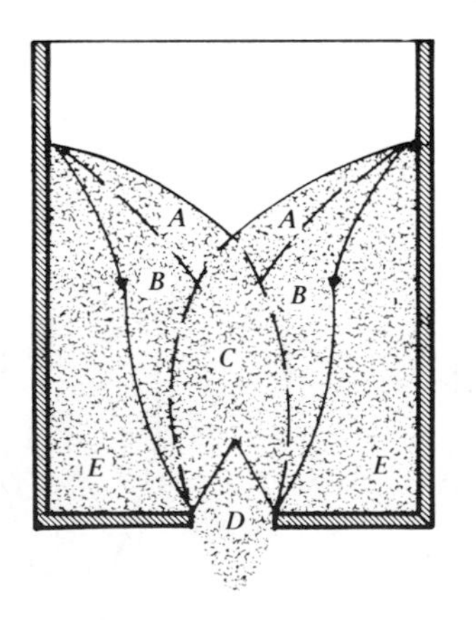


**Fig. 8.13** Schematic representation of the flow regions as observed by Brown and Hawksley (17).

**Fig. 8.14** Regions of different flow behavior in two-dimensional hoppers as observed by radiographic techniques. [Reprinted with permission from J. Lee et al., *Trans. Soc. Rheol.*, **18**, 247 (1974).]

The flow patterns during steady discharge from a rectangular bunker through a slit opening were investigated by Brown and Hawksley (17). They have observed five regions of flow, as illustrated in Fig. 8.13. Region *A* slides more or less as a bulk on region *B*, which slides more slowly and with internal shear on the stagnant region *E*. The angle of inclination of the top surface as well as that of the interfaces remain essentially constant during discharge. Regions *A* and *B* feed into region *C*, where the particles accelerate downward and reach region *D*, where they fall more or less as a free body in gravitational field. The region expands and contracts in a pulsating manner, apparently because of the compression of the particles into tight packing followed by a decompression as the particles fall through the opening.

The flow patterns in hoppers are somewhat different. If the stagnant regions are completely eliminated, the hopper is called a mass flow hopper. Lee et al. (18) determined simultaneously by radiographic techniques the flow field and the porosity field in a two-dimensional hopper. The marks left by tracer particles during discharge permit the evaluation of the local velocity vector and the intensity of the shade, the porosity. On the basis of both velocity and porosity fields, the authors distinguished between four regions (Fig. 8.14). Region *D* is called "plug flow zone" and in region *B* rigid body behavior was observed. Region *A* is called "rupture zone" because of the intensive deformation that occurs there. Finally region $C$ is the free flow zone.

A detailed mapping of flow kinematics in two-dimensional hoppers with stereoscopic technique, developed by Butterfield et al. (19), was done by Resnick et al. (20). According to this technique, photographs of the flow field are taken at short time intervals. Pairs of consecutive photographs under a stereocomparator produce a three-dimensional model of the displacement field, from which the velocity field can be computed. This method requires no tracers and it permits the determination of detailed flow field throughout the system, but it is limited to two-dimensional flows.

Although a great deal of progress has been made in obtaining flow fields of particulate solids, and design criteria for arch-free flow are available, it is not possible yet to calculate discharge rate from first principles. Hence empirical equations are used for this purpose. It should be noted perhaps that in most polymer processing applications, such as in hoppers feeding processing equipment, the maximum open discharge rates are much higher than present processing rates. Therefore we do not list here empirical equations for discharge rates; these are available in the literature (21–23).

## 8.9 Compaction

The response of particulate solid systems, specifically powders, to forced compaction is of great interest in a broad range of processes. Tableting or pelleting of pharmaceutical products, powder pressing in ceramic industries, powder metallurgy, and briquetting of coal can serve as examples. In polymer processing particulate solids are compacted prior to melting *inside* most processing machinery, and the performance of these machines is greatly influenced by the compaction behavior of the solids.

In many of these applications the purpose of compaction is to induce agglomeration, while in others it is needed for efficiency in further processing of the solids (e.g., melting). The compaction is obtained by applying an external force. This force is transmitted within the system through the points of contact between the particles. By a process of elastic and plastic deformation (shear deformation and local failure), the points of contact increase, as do the forces holding the particles together, which were discussed in the section dealing with agglomeration. The externally applied force generates an internal stress field, which, in turn, determines the compaction behavior.

Compaction of powders was discussed by Train and Lewis (6), who point out that Wollaston (24) was the first scientific worker in this field to realize the great pressures needed for compaction of dry powders, leading to his development of the famous toggle press. There are many difficulties in analyzing the compaction process. Troublesome in particular are the facts that the properties of particulate solids vary greatly with consolidation and that stress fields can be obtained in principle only in the limiting cases of steady flow or in a state of incipient flow when the friction is fully mobilized. In compaction these conditions are not necessarily fulfilled.

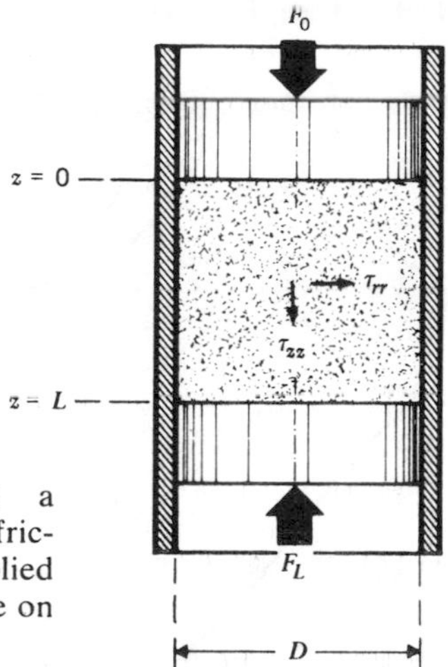


**Fig. 8.15** Compaction in a cylindrical channel, between frictionless pistons. $F_0$ is the applied force, $F_L$ is the resultant force on the lower piston.

Let us consider an apparently simple situation of compaction in a cylinder (Fig. 8.15). A normal force $F_0$ applied to the top ram generates a certain normal stress $\tau_{zz}$, as well as a radial stress $\tau_{rr}$. The frictional shear force due to the latter acts in the opposite direction to the applied force. Hence the transmitted force to the lower ram $F_L$ will be smaller than the applied force. By making a force balance similar to that made in deriving the Janssen equation and assuming that wall friction is fully mobilized, that the ratio of axial to radial stresses is a constant throughout and that the coefficient of friction at the wall is constant, we obtain the following simple exponential relationship between the applied and transmitted force (see details in Section 8.11):

$$\frac{F_0}{F_L} = \exp\left(\frac{4f'_w KL}{D}\right) \tag{8.9-1}$$

Experimental data seem to conform to this relationship (25); yet there are serious doubts about its validity. Both the coefficient of friction and the ratio of normal stresses vary along the compact (although it appears that their product stays approximately constant, explaining the reasonable agreement with experimental data). Experimental measurement of stresses within the compact, however, reveal a rather complex stress distribution (26), which depends very much on conditions at the wall and the geometry of the compact (Fig. 8.16).

Another question of fundamental importance discussed by Long (27) is the nature of the ratio of axial to radial stresses. Provided the axial and radial directions coincide with the principal axes, this ratio was given by Eq. 8.6-3 for noncohesive particulate solids in a state of incipient flow, and we have mentioned that the same equation holds for a steady flow situation of cohesive particulate solids, with the effective angle of friction replacing the internal angle of friction (Eq. 8.6-6). For cohesive particulate solids at the condition of incipient failure, the Coulomb yield function (Eq. 8.6-5) provides a relationship between the principal stresses. Yet as pointed out earlier, the conditions under compaction are not clearly defined because friction may not be fully mobilized throughout the system. There is a

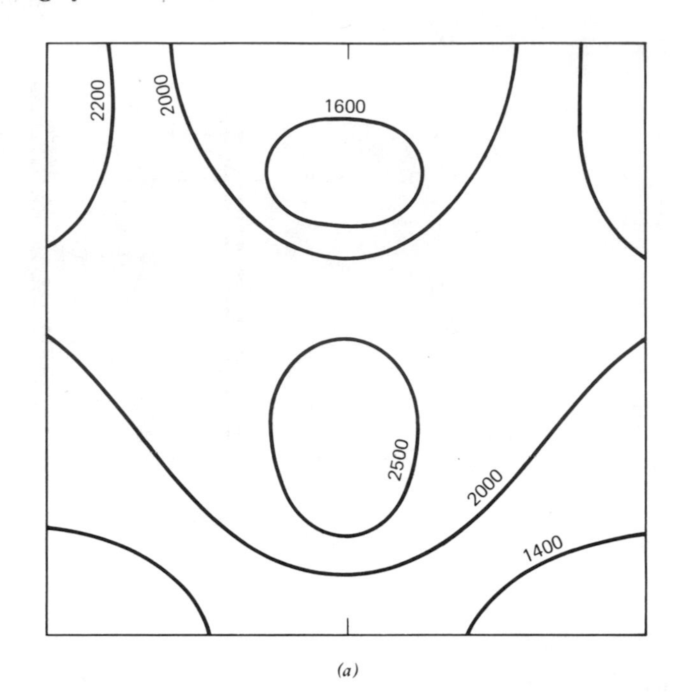


(*a*)

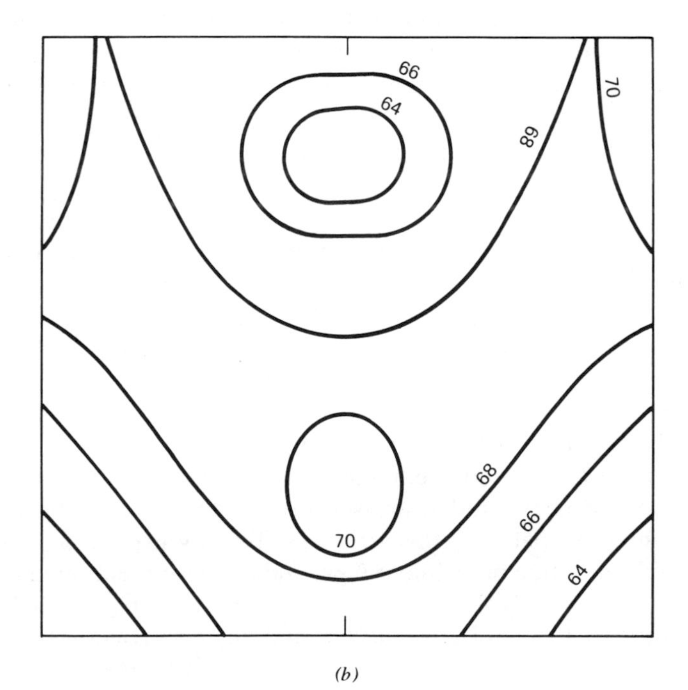


(*b*)

**Fig. 8.16** (*a*) Stress and (*b*) density distribution in a cylindrical compact of magnesium carbonate at an applied pressure of 2040 kg/cm$^2$. [Reprinted with permission from D. Train, *Trans. Inst. Chem. Eng.*, **35**, 262 (1957).]

complex stress distribution, and principal axes may not coincide with the axial and radial directions, respectively. Long (27) investigated this relationship by carrying out "radial stress cycles." The "cycles" are obtained by first increasing the axial stress, then decreasing it. A residual radial stress remains after the axial stress has been reduced to zero. This residual stress is responsible for the necessity of forcing the compacted mass out from the die after removal of the axial stress. According to Long (27), at small axial stresses, before any yield takes place in the powder, the ratio of radial to axial stress will be given by the Poisson ratio $\nu$ ($\nu\sigma_a$ is the stress needed to suppress the radial expansion the compact would undergo if it were free to expand). Once yield takes place, the ratio is determined by some yield criterion such as the Coulomb yield function, and a more or less linear increase of radial stress with axial stress is observed.

The response of polymeric particulate solids to compaction was investigated experimentally by Schneider (28) and Goldacker (29). For polyethylene, for example, a constant radial to axial stress ratio of 0.4 was observed.

The bulk density of particulate solids increases by compaction. (Dilation, mentioned earlier, occurs only in the presence of a free surface, which allows for a loosening of the packing arrangements of the particles.) The increase in density, or decrease in porosity, seems to follow an exponential relationship with the applied pressure

$$\varepsilon = \varepsilon_0 e^{-\beta' P} \tag{8.9-2}$$

where $\varepsilon_0$ is the porosity at $P = 0$, and $\beta'$ is a "compressibility coefficient," which in view of the complex stress distribution in compacts should depend on properties of the particulate system, on compact geometry, and possibly on the loading history. Therefore Eq. 8.9-2 can be viewed as an approximate empirical relationship reflecting some average values.

## 8.10 Flow in Closed Conduits

In polymer processing it is usually necessary to force the particulate solids through some sort of closed conduit or channel. In a ram-type injection molding machine the solids are *pushed* forward by the advancing ram. They move in a channel that becomes an annular gap upon reaching the torpedo. In a screw extruder the solids are *dragged* forward in the helical channel formed between the screw and the barrel. These examples represent the two basic conveying and compaction methods used in polymer processing: external mechanical *positive displacement* conveying and compaction, and *drag induced* conveying and compaction by a solid boundary in the direction of flow. In the former, the friction between the solids and the stationary walls reduces the conveying capacity, whereas in the latter friction between the solids and the moving wall is the *source* of the driving force for conveying. It is perhaps worthwhile to note that the two solids conveying mechanisms are identical in concept to external mechanical pressurization and drag induced viscous pressurization of liquids, discussed in Chapter 10.

Rigorous analysis of particulate solids flow in closed conduits is difficult. The main sources of the difficulty were discussed in Section 8.9 (Compaction). These fundamental difficulties are compounded by the complexities of polymer processing, such as temperature increases as a result of friction and external heating, the viscoelastic response of polymeric particulate systems under deformation, and the usually large ratio of particle-to-channel sizes, casting doubts on the validity of the assumption that the particulate system can be analyzed as a continuum with negligible interaction between internal structure and particle interface. Hence the analysis that follows is limited by the many simplifying assumption that are necessary.

## 8.11 Mechanical Displacement Flow

We shall analyze mechanical displacement flow in straight channels of constant cross-sectional area, as schematized in Fig. 8.17 (with the upper plate at rest). A column of solids of length $L$ is compressed between two rams. The one on the left exerts a force $F_0$ on the solids and it is opposed by a *smaller* force $F_L$ on the right. Thus friction on the channel walls also opposes the applied resultant force.

The variation of axial stress or force with distance can be derived by a force balance in a way very similar to the Janssen equation approach. We shall make the following assumptions: (*a*) The compacted solids are either at a steady motion or in a state of incipient slip on the wall (friction at wall is fully mobilized), (*b*) axial and radial stresses vary only with the axial distance $x$, (*c*) the ratio of the radial to axial stresses is a constant $K$, independent of location, (*d*) the coefficient of friction is constant and independent of compaction, (*e*) temperature effects in the case of steady motion are neglected. A force balance on a differential element as shown in Fig. 8.17 is now written as

$$F_x - (F_x + dF_x) - C\left(\frac{F_x}{A}\right) K f_i \, dx = 0 \qquad (8.11\text{-}1)$$

where $f_i$ is either the static coefficient of friction for the case of incipient motion, or the kinematic coefficient of friction for steady motion; $C$ is the circumference that can be generalized for noncircular cross section to mean the wetted perimeter, and

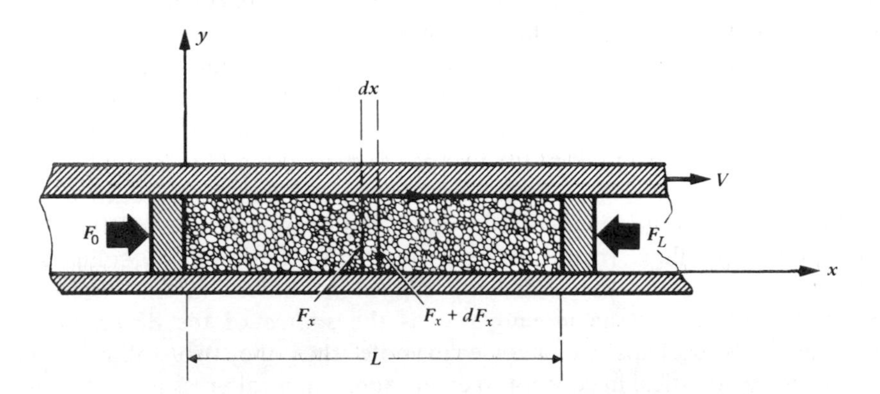


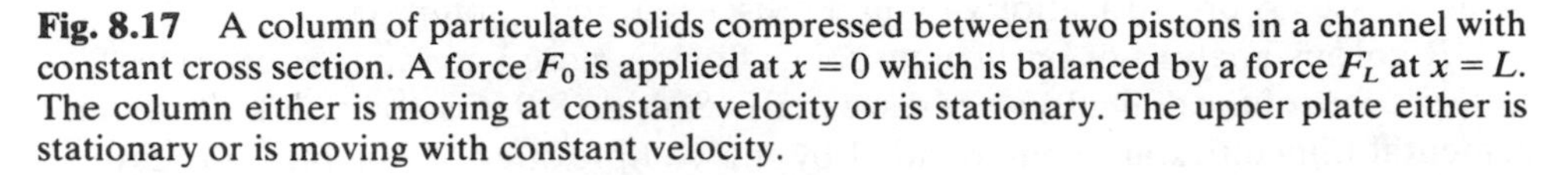
**Fig. 8.17** A column of particulate solids compressed between two pistons in a channel with constant cross section. A force $F_0$ is applied at $x = 0$ which is balanced by a force $F_L$ at $x = L$. The column either is moving at constant velocity or is stationary. The upper plate either is stationary or is moving with constant velocity.

$A$ is the cross-sectional area. Integration of Eq. 8.11-1 gives

$$F_x = F_0\, e^{-f_i KCx/A} \tag{8.11-2}$$

where $F_x$ is the axial force at location $x$. The axial stress may be obtained by multiplying the force with the cross-sectional area. The force at the downstream ram $F_L$ is obtained by setting $x = L$ in Eq. 8.11-2.

Hence in dealing with steady motion of particulate solids it is evident that the axial stress or "pressure" drops exponentially, whereas in the case of liquid flow it drops linearly with distance. This difference stems of course from the fact that frictional forces on the wall are proportional to the absolute local value of the normal stress or pressure. In liquids the pressure *gradient* rather than the *absolute* value of the pressure affects the flow. Furthermore, Eq. 8.11-2 indicates that the pushing force increases exponentially with coefficient of friction and with the geometric dimensionless group $CL/A$, which for a tubular conduit becomes $4L/D$.

Experimental support on the validity of Eq. 8.11-2 was presented by Spencer et al. (25), who also proposed a theoretical derivation based on considering a discrete number of contact points between solids and containing walls. They assumed isotropic stress distribution ($K = 1$) and obtained an expression identical to Eq. 8.11-2

$$\frac{F_L}{F_0} = e^{-4f'_w L_0/D} \tag{8.11-3}$$

where $L_0$ is the *initial* length of the column. The use of initial length of column, even though the column shortens upon compression, is justified by them on the basis of assuming constant number of contact point. Experiments were carried out with a stationary column of saran powders and granular polystyrene, and results confirmed the theoretical derivation within experimental error. Yet in spite of the agreement between experiments and theory, we must recall the drastic simplifying assumptions involved in the theoretical derivation, as discussed in the previous section, and treat the theoretical results with caution. Nevertheless, the investigation just mentioned, as well as other experimental work, seems to confirm the exponential nature of the axial distribution of stresses or pressure.

**Example 8.1** ***Force Requirements of Ram Injection Molding Machines***

A ram injection molding machine consists of a 2 in. diameter barrel in which a well-fitting ram reciprocates. Calculate the maximum length of the solid plug the machine can deliver if the downsteam pressure during injection is 10,000 psi and the barrel can sustain a radial stress of 25,000 psi. The static coefficient of friction is 0.5, and the radial to axial stress ratio is 0.4.

*Solution*: The maximum allowable axial stress is 25,000/0.4 = 62,500 psi. Substituting the appropriate values into Eq. 8.11-3 but not setting $K = 1$, we get

$$\ln(6.25) = (0.5)(0.4)(4)\frac{L}{2}$$

The length $L$ is 4.58 in. Thus with an axial force of about 20,000 $lb_f$ we can only press a 4.58 in. long solids column driving the radial stress to its upper limit! Clearly, if it is necessary for injection molding machines of this type to develop such high downstream pressures, appropriate means must be provided to reduce the coefficient of friction on the wall. This can be achieved, for example, by heating the barrel, generating a liquid film on the wall. This will change the drag mechanism to that of a viscous laminar flow, which is independent of the absolute local normal stresses.

## 8.12 Steady Mechanical Displacement Flow Aided by Drag

Drag aided particulate solids flow occurs when at least one of the confining solid walls moves in the direction of flow parallel to its plane. The friction between the moving wall and the solids exerts a forward dragging force on the solids. Figure 8.17 shows a rectangular channel with the upper plate, forming the top of the channel, moving at a constant velocity in the $x$-direction. Solids are compressed into a column of length $L$ between two rams. We now can differentiate between four possible states of equilibrium:

1. The solids are stationary with friction on the stationary walls fully mobilized and with $F_0 > F_L$.
2. The same as case 1, but with $F_L > F_0$.
3. The solids move at constant velocity (less than the velocity of the upper plate) in the positive $x$-direction.
4. The same as case 3, but the solids move in the negative $x$-direction.

Force balances on a differential element for these four cases appear in Fig. 8.18. The moving plate exerts a force of $C_1 f_{w_1} K(F/A)$ in all cases, where $C_1$ is the portion of the "wetted" perimeter of the moving plate and $f_{w_1}$ is the kinematic coefficient of friction. The stationary channel walls in cases 1 and 2 exert a force $C_2 f'_w K(F/A)$, where $f'_w$ is the static coefficient of friction and $C_2$ is the portion of the "wetted" perimeter of the lower plate and side walls that is stationary. This force points in the direction of increasing force. Thus it points to the left in case 1 and to the right in case 2. Finally in cases 3 and 4 the stationary walls exert a force

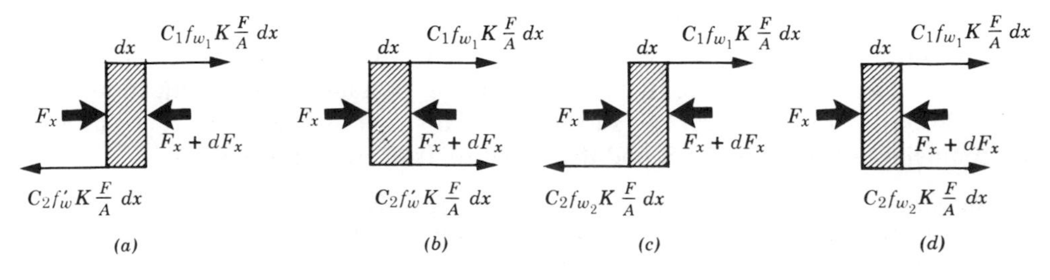


**Fig. 8.18** Force balances on a differential element of solids in Fig. 8.17. (*a*) Stationary solids, $F_0 > F_L$. (*b*) Stationary solids, $F_0 < F_L$. (*c*) Solids move at constant velocity in the positive $x$-direction. (*d*) Solids move at constant velocity in the negative $x$-direction.

$C_2 f_{w_2} K(F/A)$, where $f_{w_2}$ is the kinematic coefficient of friction. This force acts in the direction opposite to the direction of motion of the plug.

Force balances such as Eq. 8.11-1 with the further assumption that the channel is flat and the torque induced by couples of forces can be neglected, lead to the following equations:

### Case 1

*$F_L < F_0$; Stationary Plug; Friction Mobilized*

$$\frac{F_L}{F_0} = \exp\left[(C_1 f_{w_1} - C_2 f'_w)\frac{KL}{A}\right] \tag{8.12-1}$$

### Case 2

*$F_L > F_0$; Stationary Plug; Friction Mobilized*

$$\frac{F_L}{F_0} = \exp\left[(C_1 f_{w_1} + C_2 f'_w)\frac{KL}{A}\right] \tag{8.12-2}$$

### Case 3

Plug moves in the direction of the upper plate:

$$\frac{F_L}{F_0} = \exp\left[(C_1 f_{w_1} - C_2 f_{w_2})\frac{KL}{A}\right] \tag{8.12-3}$$

### Case 4

Plug moves in the direction opposite to the upper plate ($F_L > F_0$):

$$\frac{F_L}{F_0} = \exp\left[(C_1 f_{w_1} + C_2 f_{w_2})\frac{KL}{A}\right] \tag{8.12-4}$$

In the foregoing we have allowed for different kinematic coefficients of frictions on the moving plate $f_{w_1}$ and the stationary walls $f_{w_2}$.

Analysis of these equations reveals the role of drag on the force and stress distribution. We shall consider first the case of a stationary column of solids. Assume that the drag force exerted by the moving plate can be gradually increased by changing $f_{w_1} C_1$ through modifying the surface properties of the plate through coating, roughening and so on, or increasing $C_1$. This is demonstrated graphically in Fig. 8.19. If $f_{w_1} C_1$ is zero, the ratio of the forces is $F_L/F_0 = \exp[(-C_2 f'_w)(KL/A)]$, as given in Eq. 8.11-1. A gradual increase in $f_{w_1} C_1$ increases this ratio, implying that for a given $F_L$, less and less force has to be exerted on the upstream ram, until it reaches a value of 1 (i.e., $F_L = F_0$) when $C_1 f_{w_1} = C_2 f'_w$. At this point the forward dragging force exerted by the upper plate exactly compensates the fully mobilized

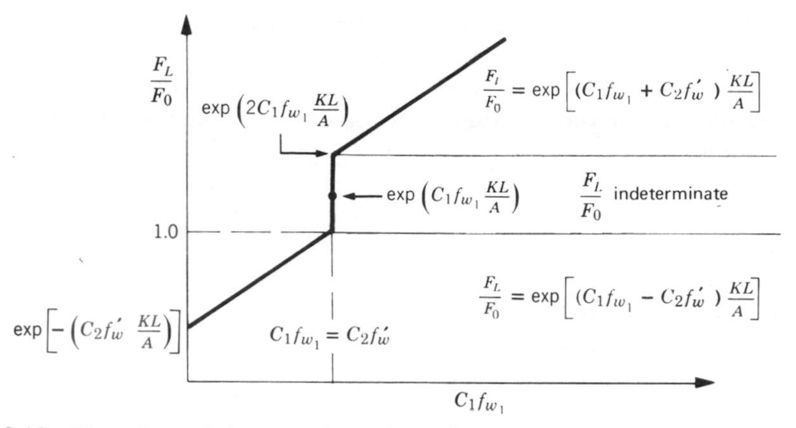


**Fig. 8.19** The effect of drag on the ratio $F_L/F_0$ for a stationary column of solids. (The ordinate is a logarithmic scale.)

frictional forces on the stationary walls. At this point also we can slightly increase $F_L$, thereby demobilizing the friction on the stationary walls. This is indicated by the vertical line in Fig. 8.19. We then reach a point where the frictional forces on the stationary plate are zero and the forward dragging force is fully compensated by the force $F_L$. Under these conditions

$$\frac{F_L}{F_0} = \exp\left(C_1 f_{w_1}\frac{KL}{A}\right) \tag{8.12-5}$$

indicated by the heavy dot in Fig. 8.19. Moreover, the force $F_L$ can be further increased gradually mobilizing the frictional forces on the stationary walls in the opposite direction until they are fully mobilized, where the ratio of forces is

$$\frac{F_L}{F_0} = \exp\left(2C_1 f_{w_1}\frac{KL}{A}\right) \tag{8.12-6}$$

Further increase in $f_{w_1}C_1$ will result in an increase in the ratio $F_L/F_0$ according to Eq. 8.12-2. Hence we have a condition indicated by the vertical line in Fig. 8.19, where the force ratio is indeterminate.

The condition indicated by the heavy dot in Fig. 8.19 can also be interpreted as representing a point where the downstream ram is replaced by a rigid channel block, which responds only to the forces exerted on it by the solids and prevents mobilizing the friction on the stationary walls. This is in agreement with the St. Venant principle, which states that if statically equivalent and opposing surface tractions are applied on a solid, the differences are negligible at far away locations, that is, on the surface of the stationary walls; hence this surface plays no role in the force balance.

The same kind of analysis for the case of steadily moving solids leads to similar conclusions, as Fig. 8.20 demonstrates. We should note, though, that in this case we do not have a continuous transition between the two directions of motion because within the region between the two curves, the solids must come to rest, thus encountering the two previously discussed cases and leading to possible instabilities.

Both cases, however, vividly demonstrate the profound effect the drag forces, induced by a moving boundary, may have on the force distribution. In positive

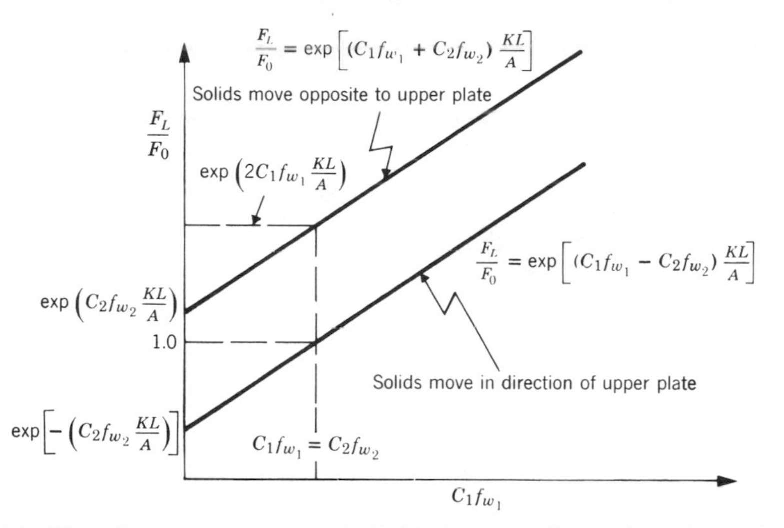


**Fig. 8.20** The effect of drag on the ratio $F_L/F_0$ for a steadily moving column of solids. (The ordinate is a logarithmic scale.)

displacement flow the addition of a drag permits the reduction of the force $F_0$ needed to maintain a certain downstream force $F_L$ to any desired level. Moreover results indicate that drag is capable of generating pressures within the solids *above those applied externally*. The pressure rise is exponential with distance. The same holds for a moving plug. Hence drag, as we shall see in the next section, is a mechanism by which *solids can be compacted as well as conveyed.*

## 8.13 Steady Drag Induced Flow in Straight Channels

We have concluded that frictional drag, when applied to a steadily moving column of solids, can generate stresses or pressures above those applied externally. Consider once more the case of a flat rectangular channel with the upper plate moving and the solids moving in the same direction at constant velocity. The force ratio is plotted in the lower curve of Fig. 8.20. Clearly, for any given $F_0$ (which must be greater than zero except for the frictionless case), we can get any $F_L$ greater than $F_0$, provided $C_1 f_{w_1}$ is large enough. This ratio seems to be independent of either the plate velocity or the velocity of the solids. All that is required is that these velocities be steady. This result was obtained because we have assumed that the frictional force depends *only* on normal stress and is independent of velocity, which as we have seen in Chapter 4 is a reasonable assumption. Yet the velocity of the solids multiplied by the cross-sectional area gives the flow rate. Thus the previous argument implies that in this particular setup *flow rate is indeterminate.* How then can we use the drag induced flow concept to obtain a geometrical configuration in which flow rate not only is not indeterminate but also is predictable? Such a situation would arise if the frictional drag could be made dependent on solids velocity. We can create such a situation by replacing the upper cover with an infinite plate, moving not in the down channel direction but at an *angle* ($\theta$) to this direction, as in Fig. 8.21. The frictional force exerted by the moving plate on the solids remains constant, but the *direction* of this force will be given by the vectorial

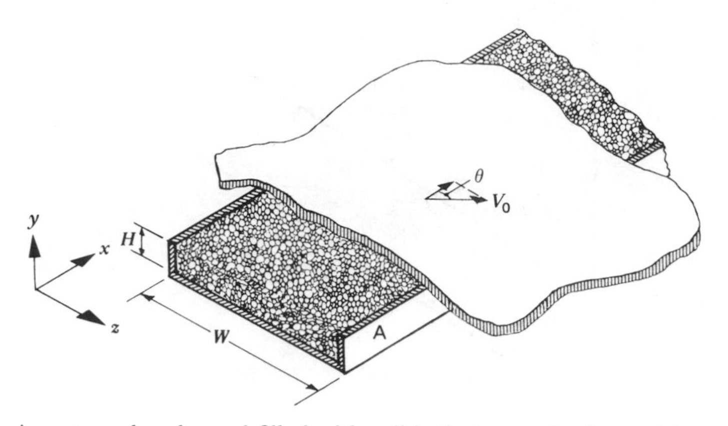


**Fig. 8.21** A rectangular channel filled with solids that move in the positive $x$-direction at constant velocity $u$, covered by an infinite plate moving at constant velocity $V_0$ at an angle $\theta$ to the down channel direction, $z$.

difference between the plate velocity and solids velocity (Fig. 8.22). Hence the velocity component of this force in the down channel direction, which participates in the force balance, becomes a function of both plate velocity and solids velocity (or flow rate). From the velocity diagram in Fig. 8.22 we obtain the following expression for the angle $\phi$, which is the angle between the direction of the force exerted by the moving plate on the solids and the direction of motion of the moving plate (the solids conveying angle)

$$\tan \phi = \frac{u \sin \theta}{V_0 - u \cos \theta} \tag{8.13-1}$$

where $V_0$ is the velocity of the upper plate and $u$ is the velocity of the solids. Note that for stationary solids $\phi$ becomes zero, and it increases with increasing flow rate.

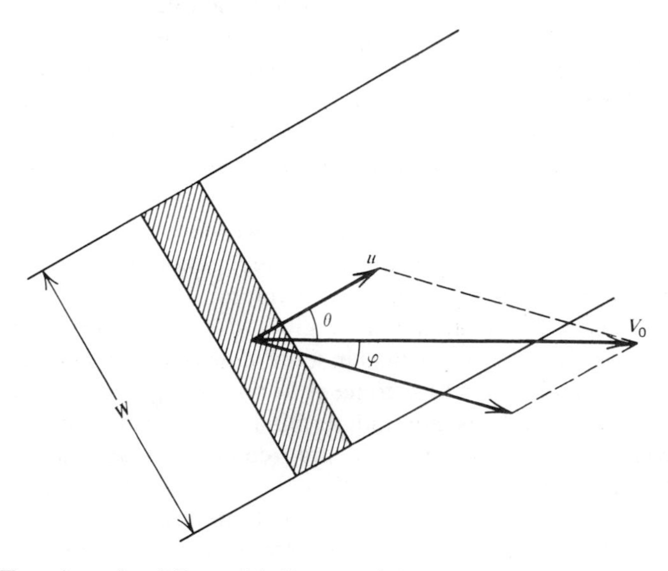


**Fig. 8.22** Top view of a differential element of the column of solids in Fig. 8.21.

Now we can proceed with the force balance. We first concentrate in making a down channel force balance, neglecting the cross channel component of the forces

$$F_x-(F_x+dF_x)+C_1f_{w_1}K\left(\frac{F_x}{A}\right)\cos(\theta+\phi)\,dx-C_2f_{w_2}K\left(\frac{F_x}{A}\right)dx=0 \tag{8.13-2}$$

which upon integration with the initial condition $x=0$, $F=F_0$, gives

$$\frac{P_x}{P_0}=\frac{F_x}{F_0}=\exp\left\{[C_1f_{w_1}\cos(\theta+\phi)-C_2f_{w_2}]\frac{Kx}{A}\right\} \tag{8.13-3}$$

Hence the ratio of forces, which by dividing by the cross-sectional area also equals the ratio of axial stresses, which we shall refer to as "pressures," become a function of the flow rate via the angle $\phi$, determined by Eq. 8.13-1. This implies that for a given inlet pressure $P_0$, a fixed outlet pressure determines the flow rate, or conversely, a given flow rate determines the magnitude of outlet pressure the device can generate. The lower the flow rate, the higher the pressure rise.

The above-described solids conveying mechanism represents in essence the conveying of solids in single screw extruders, although a realistic conveying model for the latter is somewhat more complicated because, as Chapter 12 explains, the channel is curved.

Drag induced flow in a rectangular channel, as in Fig. 8.21, neglecting cross channel forces, resulted in Eq. 8.13-3. We now consider the effect of these forces on the conveying mechanism.

At steady flow conditions the moving plate exerts a force on the solids in the $(\theta+\phi)$ direction. This force is separated into two components—one in the down channel direction, which was used in the force balance, and the other in the cross channel direction, which was neglected. The latter will have the following effects: it will increase the normal stress on the side wall $A$ in Fig. 8.21, and it will alter the stress distribution within the solids. Assume for the sake of simplicity that the St. Venant principle holds; that is, the externally applied force by the plate is completely balanced by the additional force on side wall $A$, and within the solids (which will be considered to be located "far" from the places that these tractions act), there will be no effect. In other words, we neglect the changes in the stress distribution within the solids.

The cross channel force component is

$$F^*=f_{w_1}K\left(\frac{F_x}{HW}\right)(W\,dx)\sin(\theta+\phi)=\frac{f_{w_1}KF_x\sin(\theta+\phi)\,dx}{H} \tag{8.13-4}$$

where $W$ and $H$ are the width and height of the channels, respectively. Now we can write a down channel force balance including the effect of this additional normal force on side wall $A$ on the frictional force along this wall

$$F_x-(F_x+dF_x)+f_{w_1}K\left(\frac{F_x}{WH}\right)(W\,dx)\cos(\theta+\phi)-f_{w_2}K\left(\frac{F_x}{WH}\right)(W+H)\,dx$$

$$-f_{w_2}\left[K\left(\frac{F_x}{WH}\right)H\,dx+F^*\right]=0 \tag{8.13-5}$$

which upon rearrangement and with Eq. 8.13-4, gives

$$\frac{dF_x}{F_x} = \frac{f_{w_1}K}{H}\left[\cos(\theta+\phi) - \frac{f_{w_2}}{f_{w_1}}\left(1+2\frac{H}{W}\right) - f_{w_2}\sin(\theta+\phi)\right]dx \qquad (8.13\text{-}6)$$

Integration of this equation gives

$$\blacktriangleright\ \frac{P_x}{P_0} = \frac{F_x}{F_0} = \exp\left\{\frac{f_{w_1}Kx[\cos(\theta+\phi) - f_{w_2}\sin(\theta+\phi) - (f_{w_2}/f_{w_1})(1+2H/W)]}{H}\right\} \qquad (8.13\text{-}7)$$

Equation 8.13-7 reduces to Eq. 8.13-3 if the second term on the right-hand side vanishes. Clearly the cross channel force induces additional friction on the side wall $A$, which in turn reduces the pressure generation capability for a given flow rate (given angle $\phi$), or it reduces the conveying capacity for a given pressure rise.

## REFERENCES


1. E. W. Schuler, "Shopping for a Pelletizing System?" *Plast. Eng.*, **31** (8), 38–42 (1975).
2. J. K. L. Bajaj, "Cutting Resins to Size," *Plast. Eng.*, **30** (1), 18–23 (1974).
3. C. Orr, Jr., *Particulate Technology*, Macmillan, New York, 1966.
4. R. L. Brown and J. C. Richards, *Principles of Powder Mechanics*, Pergamon Press, Oxford, 1966.
5. K. Weighardt, "Experiments in Granular Flow," *Ann. Rev. Fluid Mech.*, **7**, 89–114 (1975).
6. D. Train and C. J. Lewis, "Agglomerization of Solids by Compaction," Paper presented at the Third Congress of the European Federation of Chemical Engineers, London, June 20–29, 1962. See also A. W. Jenike, P. J. Elsey, and R. H. Woolley, *Proc. Am. Soc. Test. Mater.*, **60**, 1168 (1960).
7. A. W. Jenike, "Gravity Flow of Bulk Solids," Bulletin No. 108 of the Utah Engineering Experimental Station, University of Utah, Salt Lake City, 1961.
8. B. Yavin and J. Ellad, "Properties and Behavior of Polymeric Particulate Systems," Department of Chemical Engineering Report, Technion–Israel Institute of Technology, Haifa, 1975.
9. H. A. Janssen, "Tests on Grain Pressure Silos," *Z. Vereinschr. Dtsch. Ing.*, **39** (35), 1045–1049 (1895).
10. W. L. McCabe and J. C. Smith, *Unit Operations of Chemical Engineering*, McGraw-Hill, New York, 1956, Chapter 5.
11. D. M. Walker, "An Approximate Theory for Pressures and Arching in Hoppers," *Chem. Eng. Sci.*, **21**, 975–997 (1966).
12. J. K. Walters and P. M. Nedderman, "A Note on the Stress Distribution of Great Depth in a Silo," *Chem. Eng. Sci.*, **28**, 1907–1908 (1973).
13. P. L. Bransby and P. M. Blair-Fish, "Wall Stresses in Mass-Flow Bunkers," *Chem. Eng. Sci.*, **29**, 1061–1074 (1974).
14. J. R. Johanson, "Feeding," *Chem. Eng.*, 75–82 (October 13, 1969).
15. O. Richmond and G. C. Gardner, "Limiting Spans for Arching of Bulk Material in Vertical Channels," *Chem. Eng. Sci.*, **17**, 1071–1078 (1962).
16. O. Richmond, "Gravity Hopper Design," *Mech. Eng.*, 46–49 (January 1963).
17. R. L. Brown and P. G. Hawksley, "The Internal Flow of Granular Masses," *Fuel*, **26**, 171 (1947).

18. J. Lee, S. C. Cowin, and J. S. Templeton, "An Experimental Study of the Kinematics of Flow Through Hoppers," *Trans. Soc. Rheol.*, **18**, 247–269 (1974).
19. R. Butterfield, R. M. Harkness, and K. Z. Andrews, "A Stereophotogrammetric Method of Measuring Displacement Fields," *Geotechnique*, **8**, 308 (1970).
20. M. Levinson, B. Shmutter, and W. Resnick, "Displacement Velocity Fields in Hoppers," *Powder Technol.*, **16**, 29–43 (1977).
21. T. Shirai, "Powder Orifice Monograph," *Chem. Eng. (Japan)*, **16**, 86–89 (1952).
22. A. Harmens, "Flow of Granular Material Through Horizontal Apertures," *Chem. Eng. Sci.*, **18**, 297 (1963).
23. H. E. Rose and T. Tanaka, *Engineer*, **208**, 465–469 (1959).
24. W. H. Wollaston, *Phil. Trans.*, **119**, 1 (1829).
25. R. S. Spencer, G. D. Gilmore, and R. M. Wiley, "Behavior of Granulated Polymer Under Pressure," *J. Appl. Phys.*, **21**, 527–531 (1950).
26. D. Train, "Transmission of Forces Through a Powder Mass During the Process of Pelletizing," *Trans. Inst. Chem. Eng.*, **35**, 262–265 (1957).
27. W. M. Long, "Radial Pressures in Powder Compaction," *Powder Metall.*, No. 6, 73–86 (1960).
28. K. Schneider, "Druckausbreitung und Druckverteilung in Schuttgutern," *Chem. Ing. Techn.*, **41**, 142 (1969).
29. E. Goldacker, "Untersuchungen zur inneren Reibung von Pulvern, insbesondere im Hinblick auf die Forderung in Extrudern," Dissertation, Institut für Kunststoffverarbeitung (IKV), Aachen.

## PROBLEMS

**8.1** ***The Stress Vector in Terms of Unit Normal Vector and the Stress Tensor.*** The state of the stress at a point $P$ in a continuum under plane stress conditions is defined by the tensor $\boldsymbol{\tau}$

$$\boldsymbol{\tau} = \begin{pmatrix} \tau_{11} & \tau_{12} & 0 \\ \tau_{21} & \tau_{22} & 0 \\ 0 & 0 & 0 \end{pmatrix}$$

Show that the stress vector, $\boldsymbol{\tau}_n$, acting on a plane defined by the unit normal vector $\mathbf{n}$ inclined by angle $\theta$ to the $x$ axis which contains $P$ is

$$\boldsymbol{\tau}_n = \mathbf{n} \cdot \boldsymbol{\tau} = (\cos\theta\, \tau_{11} + \sin\theta\, \tau_{21})\, \boldsymbol{\delta}_1 + (\sin\theta\, \tau_{22} + \cos\theta\, \tau_{12})\boldsymbol{\delta}_2$$

**8.2** ***The Normal and Shear Stress Components of $\boldsymbol{\tau}_n$.*** Show that the normal and shear components of $\boldsymbol{\tau}_n$ are given respectively by Eqs. 8.4-1 and 8.4-2.

**8.3** ***The Ratio of the Principal Stresses for Noncohesive Powders.*** The stress condition at a point $P$ in a continuum under plane stress is defined by a Mohr circle, as shown in Fig. 8.3. At incipient failure conditions the yield locus is tangent to the Mohr circle. That is, the shear stresses at this particular point $P$, acting on a plane with orientation defined by the tangent point to the Mohr circle, exactly equal the shear strength of the powder at the normal stress acting on the plane.

Prove that for noncohesive powders the ratio of the principal stresses is

$$\frac{\sigma_{\max}}{\sigma_{\min}} = \frac{1+\sin\beta}{1-\sin\beta}$$

**8.4** ***The Yield Locus and the Mohr Circle.*** Explain why the yield locus cannot cross any Mohr circle representing the state of stresses at various points in a particulate solids system in equilibrium.

**8.5** ***Pressure Distribution Cylindrical-Conical Hoppers.*** A cylindrical hopper, as shown in Fig. 8.10, has the following dimensions: Apex angle 40°, $h_0 = 16.5$ in., $h_2 = 18$ in. and the radius of the cylindrical section is 6 in.

The hopper is loaded with particulate solids with the following properties: $\rho_b = 0.028$ lb/in$^3$, $\delta = 50°$, and $\beta_w = 20°$.

(*a*) Calculate the base pressure at the cylindrical section.

(*b*) Plot the vertical stress distribution.

(*c*) Calculate the base pressure for $h_1 = 5$ in.

(*d*) Compare computed results with experimental results obtained by Walker (Ref. 11).

**8.6** ***Stress Distribution During Compaction.*** During compaction of particulate solids in a cylindrical die the applied axial stress induces a certain stress distribution. The ratio of the radial and axial stresses is of particular practical significance.

(*a*) Discuss the theoretical equations suggested for this ratio and list the conditions for their validity.

(*b*) Discuss the difficulties in deriving a general expression.

**8.7** ***The Effect of Drag on the Pressure Distribution in Solids Filling a Rectangular Channel.*** A bed of particulate solids is compressed in a rectangular channel between two freely moving rams, with the upper plate of the rectangular channel moving at a constant velocity. The width of the channel is 5 cm and its height is 0.5 cm. The coefficient of friction on the stationary channel walls is 0.5 and the ratio of axial to perpendicular stresses is 0.4 and can be assumed constant throughout the bed. The force on the downstream ram is 1,000 *N*. The length of the channel is 10 cm.

(*a*) Calculate the force which has to be applied on the upstream ram at equilibrium conditions as a function of the coefficient of friction on the moving wall which can be varied in the range of 0 to 1.0.

(*b*) What effect will the doubling of the velocity of the moving plate have on the results in (*a*).

**8.8** ***Two Dimensional Pressure Distribution in Solids Filling a Rectangular Channel.*** Consider the rectangular channel geometry shown in Fig. 8.21.

Equation 8.13-7 gives the pressure distribution accounting for the cross channel force, but neglecting cross channel pressure distribution.

(*a*) Show that the down channel pressure distribution accounting for cross channel pressure distribution is given by

$$\ln \frac{\bar{P}(x)}{\bar{P}(0)} = (R_2 - R_3)x \qquad \text{P8.8-1}$$

where $\bar{P}(x)$ is the mean pressure over the cross-section at location $x$

$$\bar{P}(x) = \frac{P(x,0)(e^{R_1 W} - 1)}{R_1 W}$$

where $P(x, 0)$ is the axial pressure at $z = 0$ (see Fig. 8.21) and

$$R_1 = f_{w_1} \frac{\sin(\theta + \phi)}{H}$$

and

$$R_2 = \frac{f_{w_1} \cos(\theta + \phi) - f_{w_2}}{H}$$

$$R_3 = \frac{f_{w_3} R_1 (e^{R_1 W} + 1)}{e^{R_1 W} - 1}$$

where $f_{w_1}$, $f_{w_2}$, and $f_{w_3}$ are the coefficients of friction on the moving plate, channel bottom, and channel sidewalls, respectively.

(*b*) Show that for $R_1 W \to 0$, Eq. P8.8-1 reduces to Eq. 8.13-3 with $K = 1$, and $f_{w_3} - f_{w_2}$.

**8.9** ***Flow Rate in a Rectangular Channel.*** The pressure profile for drag induced solids conveying in a rectangular channel is given by Eq. 8.13-7. The channel dimensions are $W = 2.5$ in. and $H = 0.5$ in. The pressure at a certain upstream position is 10 psig and 10 in. downstream it is 55.7 psig. The coefficient of friction on the moving wall is 0.5 and on the stationary walls 0.2. The upper wall moves at an angle of 15° to the downchannel direction and a velocity of 10 in./s. The bulk density is 30 lb/ft$^3$ and $K = 0.5$. Calculate the mass flow rate of solids.

***Answer:*** 101 lb/hr.

# 9

# Melting



Most shaping operations consist of the flow and deformation of *heat-softened* or *melted* polymers; hence the preparation of the polymer for the shaping operation generally includes a "*heating*" or "*melting*" step. In *either* case we classify the process as the "elementary step of melting." Our objectives in this chapter are to elucidate the physical mechanisms of melting, to demonstrate some of the common mathematical tools used in solving them, and to demonstrate how these mechanisms, in conjunction with inherent physical properties of polymers, lead to certain geometrical configurations of melting.

After the polymer has been shaped into the desired form, we are faced with the solidification problem (i.e., the inverse of the melting problem). Solution methods developed in this chapter with regards to melting are also valid for solidification. The chapters dealing with shaping examine specific solidification problems.

The melting step is related primarily to thermoplastics processing, with the exception of cold forming. Yet some of the discussion in this chapter is also relevant to thermosetting processing, where the analogous process of solidification by *cooling* is a process of solidification by *heating*, due to polymerization and crosslinking. The process of heating is usually a heat conduction process coupled with chemical heat generation.

## 9.1 Classification of Melting Methods

A brief analysis of the various terms in the equation of energy (5.1-35) reveals the different possible modes of raising the temperature of a solid: by heat conduction, by compression, by a source term $-(\boldsymbol{\tau}:\nabla \boldsymbol{v})$ that reflects the irreversible conversion of mechanical energy into heat brought about by the irrecoverable deformation of a solid (in liquids this is the viscous dissipation source), and or by the source term $\dot{S}$ that stands for other types of sources, such as chemical and electrical. Among these heat conduction is the most common and the most important mode of raising the temperature of the solid and melting it. The rate controlling factors in conduction melting are the thermal conductivity, the attainable temperature gradients, and the available contact area between the heat source and the melting solid, reflecting, respectively, material, operational, and configurational constraints. Considering the low thermal conductivity of polymers and their temperature sensitivity (limiting attainable temperature gradients), the other melting modes appear, in principle, as very attractive. Thus, ultrasonic heating inducing high frequency cycling deformation in solids and dielectric heating involving an electrical source term in the equation of energy have both been explored as principal melting modes for polymers with limited success. Although a forming method utilizing dielectric heating was developed by Erwin and Suh (1) and dielectric heating of compression molding preforms is being commonly practiced, these melting modes have found broad utilization primarily in localized heating and melting applications such as welding and sealing. The very high pressures needed for "compression melting" also poses severe limitations on utilizing this source of melting as a principal melting mode, although Menges et al. (2) demonstrated the feasibility of an injection molding process based on this mode of melting. Fortunately, as discussed below, the particulate nature of polymeric raw materials broadens the scope of engineering possibilities for melting. Melting methods can be developed in which the particulate system undergoes gross intensive deformation generating heat throughout the volume of the system, both by individual particle deformation as well as interparticle friction. Although the latter is not strictly speaking a homogeneous source, it takes place over particle interfaces that are distributed throughout the system.

The equation of energy provides the basic information on the possible modes of melting. The physical properties of polymers together with the geometrical form of the polymeric raw material suggest ideas on how to utilize these modes of melting, but choosing the actual melting method to be used still represents a substantial engineering design problem.

Let us first examine how other engineering fields have dealt with the "melting" problem. Fusible solids are being extensively dealt with in chemical engineering, and Ross (3) classified the possible melting methods as follows: (*a*) the solid may be added to an already molten material in a heated and agitated vessel, (*b*) the solid may be arranged so that hot gases can be circulated over it and the molten solid drained away, and (*c*) the solid may be placed on a heated surface and the molten material drained away. These melting methods lead to the three common types of industrial melter using mix-melting, convective melting, and contact melting, respectively.

If we attempt to directly apply any of these melting methods to polymers, we are bound to encounter grave difficulties. Let us consider first mix-melting. An attempt to melt a load of polymer pellets in a heated vessel will probably result in a partially degraded nonuniform melt with lots of entrapped gases. Moreover, this unwanted result will take a very long time. These are the consequences of the inherent physical properties of polymers. Outstanding among these are the very

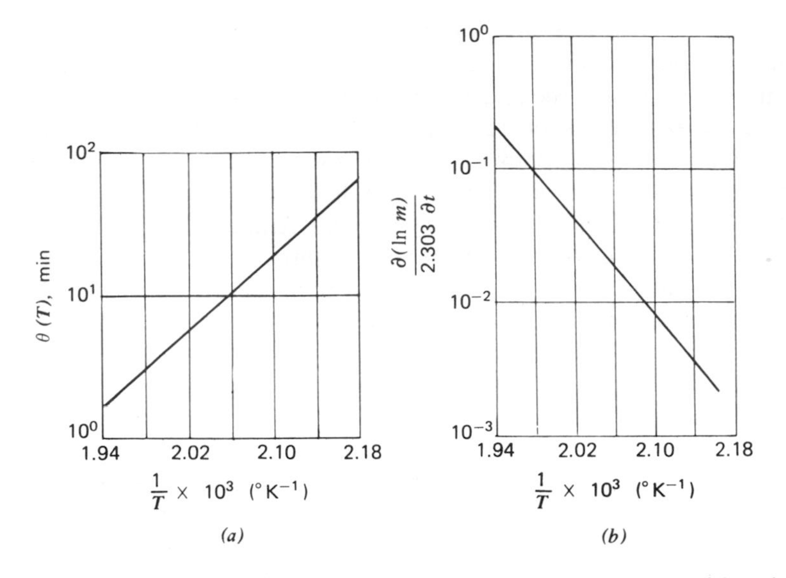


**Fig. 9.1** Thermal degradation of unplasticized PVC, "Geon" 101 EP-F24, as indicated by the time dependence of the consistency index $m$ of the power law fluid model. The two graphs of the above figure indicate that

$$m(t) = m_0 \qquad t \leq \theta(T)$$
$$m(t) = m_0 \exp\left[Ct\, e^{-\Delta E/RT}\right] \qquad t > \theta(T)$$

The induction time $\theta(T)$ is shown on (*a*) and the function $m(t)$ for $t > \theta(T)$ on (*b*). (Reprinted with permission from E. A. Collins, B. F. Goodrich Chemical Co., Avon Lake, Ohio. Paper presented at the 1965 Society of Plastics Engineers Annual National Technical Conference, March 1966.)

low thermal conductivity values of polymers. Furthermore, the thermal instability of polymers, as Fig. 9.1 indicates, places rather low bounds on the maximum temperatures to which polymers can be exposed, as well as the allowable duration of exposure to elevated temperatures. These then limit the attainable temperature gradients and rates of melting. Finally, the high viscosity of the melt eliminates natural and turbulent convection, severely limiting the mixing of the melt, and preventing the elimination of entrapped gases. Clearly then, in order to convert mix-melting into a practical melting method for polymers, intensive agitation, high surface to volume ratio, and, occasionally, repeated exposure to mass transfer surfaces (atmosphere or vacuum) must be provided. Intensive agitation of a highly viscous mixture of molten and partly molten polymer requires very high power inputs. Consequently, the "classical" mix-melting method that relies primarily on conduction melting (with heat being conducted from the molten regions to the solid as well as from the hot vessel walls to the melt) is converted into what could be labeled as a *dissipative mix-melting* method that draws substantially from a mechanical energy source introduced through the shaft and converted into heat by viscous dissipation in the molten regions by mechanical deformation of the solid regions (or particles) and in the initial stages by interparticle friction. Solids fed internal mixers, continuous mixers, roll-mills and certain types of extruders can serve as examples for this melting method.

The other two melting methods are based on providing thermal energy to the solid surface and gravitational draining of the melt. The high viscosity polymeric melts, however, cannot be "drained away" by gravitational forces. Nevertheless, these methods are useful for polymers under two conditions: (*a*) when no melt removal is necessary, and (*b*) when melt removal is achieved by mechanical force. The former relates to such processes as rotational molding, where polymer powder is sintered, and thermoforming, where a sheet is heat softened. In these cases heat energy is provided to the solid by either direct contact with a hot surface or by convection or radiation. Indeed, we can classify this melting method as *conduction melting without melt removal.* Characteristic of this melting method is that we are generally dealing with a finished or semifinished shape. Method *b* is used to generate a mass of melt from a compressed bed of particulate solids for subsequent shaping (e.g., die forming or molding). Continuous *forced* removal of the newly melted polymer leads to the possibility of maintaining continuously a *thin* film of melt between the hot contacting surface and the solid polymer (bed). This in turn implies that *large temperature gradients* (i.e., *large heat conduction rates*) can be obtained while exposing the polymer only to moderate temperatures, thus avoiding thermal degradation. Moreover, quick removal of the polymer from the high temperature regions also results in short residence time of the polymer at elevated temperatures and finally, the forced removal of melt generates heat by viscous dissipation, *aiding* the heat transfer rate. Efficient removal of highly viscous melt is possible by either a *drag* induced flow configuration, whereby the contacting hot surface moves parallel to its plane, or by *pressure* induced flow, whereby the contacting surface moves perpendicular to its plane, toward the solid, generating a squeezing type of flow. Screw extruders and "grid melters" can serve as respective examples of these melting methods. We can classify this type of melting method as *contact or conduction melting with forced melt removal.*

In conclusion then, the available modes of melting, the limitations imposed by the physical properties of polymers, the form of the raw materials, and the intended shaping method lead to a number of well defined melting methods described in Fig. 9.2 and defined as:

1. *Conduction melting without melt removal* where all the heat is provided over contacting or exposed surfaces and the rate of melting is determined only by the conduction term.
2. *Conduction melting with forced melt removal* (drag or pressure) where part of the heat is provided by conduction over a contacting surface and part is provided by converting mechanical energy into heat by viscous dissipation in the molten film. The rate of melting is determined both by the rate of heat conduction as well as by the rate of melt removal and viscous dissipation.
3. *Dissipative mix-melting* where the heat for melting is provided by converting mechanical energy into heat throughout the volume, which is also the rate determining step aided by heat conduction over external walls and over solid melt interfaces in the mixture.
4. Dissipative melting methods utilizing electrical, chemical, or other sources.
5. Compression melting.

The sections that follow discuss some of these melting methods in some detail.

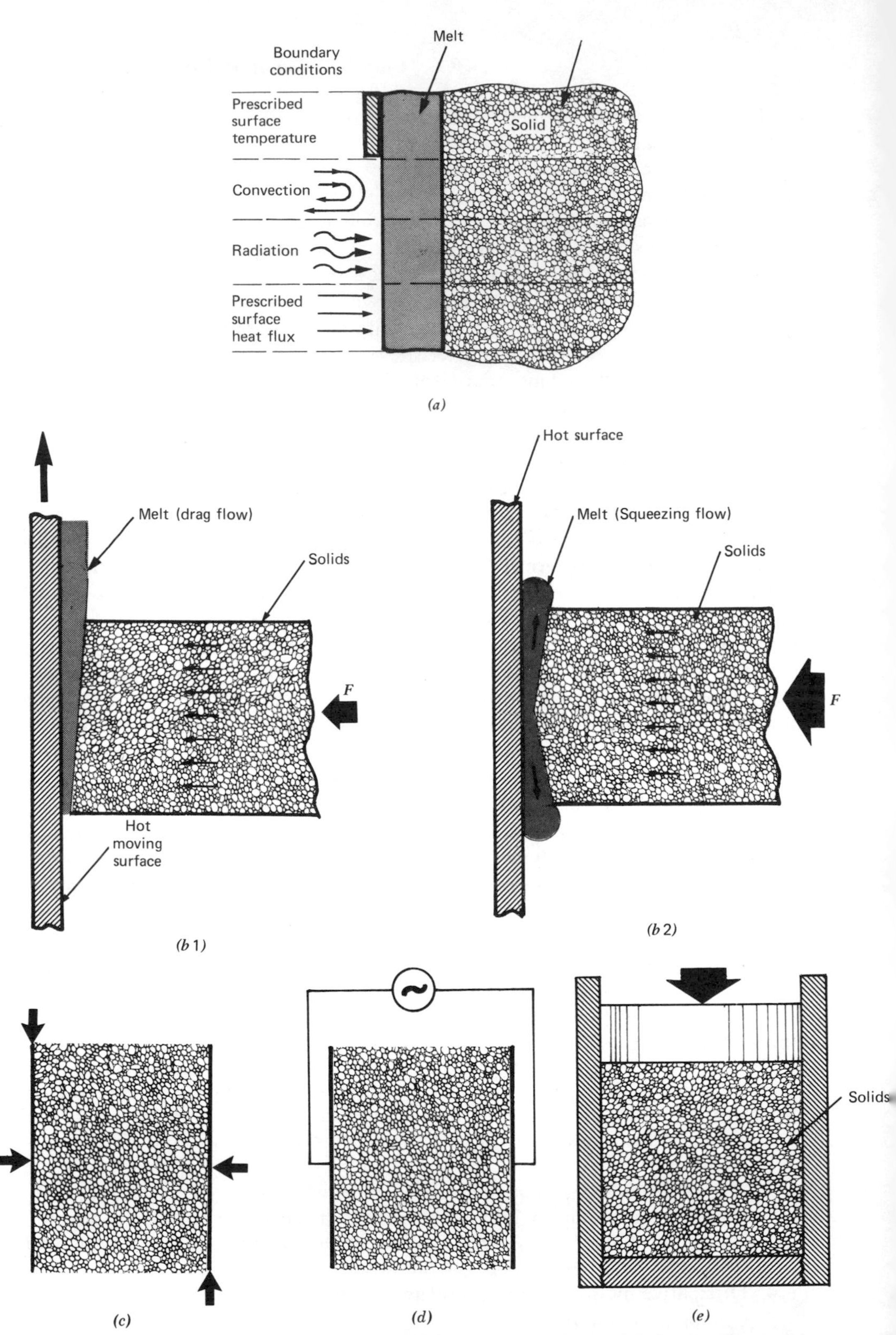


**Fig. 9.2** Schematic representation of the basic melting methods. (*a*) Conduction melting without melt removal indicating the various boundary conditions, (*b*1) conduction melting with melt removal by drag induced flow, (*b*2) conduction melting with melt removal by pressure induced flow. (*c*) Dissipative mix-melting. (*d*) Dissipative dielectric melting. (*e*) Compression melting.

## 9.2 The Roles of Geometry, Boundary Conditions, and Physical Properties in Melting

To isolate and solve a heat transfer problem in polymer processing, the geometrical boundaries of the body must be defined, boundary conditions selected, and the nature of the physical properties of the polymer specified.

Although melting in polymer processing may take place in geometrically complex machinery, the rate determining step often can be modeled in terms of simple geometries, such as semi-infinite bodies, infinite flat slabs, and thin films. Analytical techniques prove to be useful in many of these cases. In solidification, on the other hand, the geometry of the frequently complex finished product coincides with the geometrical boundaries of the heat transfer problem, necessitating the application of *numerical* techniques.

The most important boundary condition in heat transfer problems encountered in polymer processing is the constant surface temperature. This can be generalized to a *prescribed surface temperature* condition; that is, the surface temperature may be an arbitrary function of time $T(0, t)$. Such a boundary condition can be obtained by direct contact with an external temperature controlled surface, or with a fluid having a large heat transfer coefficient. The former is a frequently occurring situation in the heating or melting step in most processing machinery, whereas the latter may be the case in cooling and solidifying, such as in the water trough solidification of extruded products.

A "*prescribed surface convection*" condition mathematically stated as

$$h[T_a(t) - T(0, t)] = -k\frac{\partial}{\partial x}T(0, t) \tag{9.2-1}$$

where $T_a(t)$ is the ambient fluid temperature, and $h$ is the heat transfer coefficient; $T(0, t)$, the exposed surface temperature, is another common boundary condition in heat transfer—in polymer processing, for example, we find it in air cooling of blown films, in oven heating of sheets prior to vacuum forming in cooling of finished injection molding products, and in many other applications.

Yet another boundary condition encountered in polymer processing is a "*prescribed heat flux.*" Surface heat generation via solid-solid friction, as in frictional welding and conveying of solids in screw extruders, can serve as examples. Moreover, certain intensive radiation or convective heating that are weak functions of surface temperature can also be treated as a prescribed surface heat–flux boundary condition. Finally we occasionally must deal with the highly nonlinear, difficult boundary condition of a "*prescribed surface radiation.*" The exposure of the surface of an opaque substance to a radiation source (or sink) at temperature $T_r$ leads to the following heat flux:

$$\sigma\mathscr{F}[T_r^4 - T^4(0, t)] = -k\frac{\partial}{\partial x}T(0, t) \tag{9.2-2}$$

where $\sigma = 5.6697 \times 10^{-8}$ W/m$^2$·°K$^4$ [1.712 × 10$^{-9}$ Btu/hr·ft$^2$·°R$^4$)] is the Stefan–Boltzmann radiation constant, and $\mathscr{F}$ is combined configuration-emissivity factor. As pointed out earlier, if $T_r \gg T$, the boundary condition 9.2-2 reduces to a constant flux condition.

With respect to the thermophysical properties of the polymers to be melted, we recall first from the previous chapter that prior to melting polymers find themselves in the compressed particulate solids form. Because they are compressed, they can be considered for modeling purposes to be *continua*. However in some instances, such as in sintering, their particulate nature must be considered. Nevertheless, for the majority of the melting situations encountered in polymer processing, we can make use of the information on $k$, $\rho$, $C_p$, and $\lambda$ presented in Section 5.5, keeping in mind that these properties are subject to the structuring effects of polymer processing.

But before proceeding with the mathematical treatment of the melting methods already mentioned, let us briefly discuss the aspects of the temperature dependent flow and deformation of amorphous and semicrystalline polymers that are relevant to the elementary step of melting. This discussion is primarily for the clearer understanding of the methods of melting that involve removal of the generated melt. For these methods it is physically important to keep in mind the deformation and flow behavior of polymers just below and above the thermal transition involved.

Amorphous polymers undergo a second order transition and change from brittle to rubbery solids at $T_g$. Although $T_g$ is reported as a single temperature value, it represents practically a range of temperature of the order of 5–10°C. The value of $T_g$ increases with increasing heating rate and applied hydrostatic pressure. Its value (see Table 2.1) depends on the chemical structure of the polymer, additives (plasticizers), and copolymer composition. The shear modulus in the glassy state is of the order of $10^9\ \text{N/m}^2$; it is time independent. Near $T_g$ and especially in the range of $T_g$ to $T_g + 30°\text{C}$, it drops rapidly to values of the order of $10^6\ \text{N/m}^2$, which are characteristic of rubbers. Of course, $10^6\ \text{N/m}^2$ is the *short-term* modulus value; amorphous linear or branched polymers stress relax at $T > T_g$ at rates that increase with increasing temperature (see time-temperature superposition, Section 6.4). The viscosity of amorphous polymers changes from an "infinite" value below $T_g$ to values that are given by the WLF equation in the region $T_g < T < T_g + 100°\text{C}$ (Eq. 6.4-10):

$$\log a_T = \log\left[\frac{\mu(T)}{\mu(T_g)}\right] = \frac{-17.44(T - T_g)}{51.6 + (T - T_g)}$$

Because the viscosity increases *very rapidly* with decreasing temperature near $T_g$, amorphous polymers behave like rubbers. Only much above $T_g$ does the viscosity acquire a constant flow activation energy and behave as a liquid. In the field of linear viscoelasticity, the temperature range $T_g < T < T_g + 100°\text{C}$ is known as the "rubbery plateau" region, and the $T > T_g + 100°\text{C}$ range is called the "flow" (terminal) region (see Refs. 72 and 73 in Chapter 3).

The crystalline portion of semicrystalline polymers undergoes a first order transition from the solid to the liquid state at the melting point $T_m$. Melting of the crystallites occurs over a 10–30°C range, depending on the spectrum of their sizes and "perfection" and on the heating rate. The reported value of $T_m$ is the temperature value at the *end* of this process; it depends on the polymer structure, and in the case of random copolymers, on the copolymer composition. Block copolymers exhibit two melting temperatures characteristic of each of the homopolymers. More important for the melting step in polymer processing, as discussed in Chapter 3, $T_m$ and $\lambda$ are very dependent on the thermal and deformation history in the molten, supercooled, and solid states, as well as the applied hydrostatic pressure in the supercooled state.

The mechanical properties of semicrystalline polymers below $T_m$ depend on the degree of crystallinity. The more crystalline the polymer, the more brittle it is, with a shear modulus approaching $10^9$ N/m$^2$. At such high levels of crystallinity these polymers are also not very time dependent. It is difficult to measure their modulus above $T_m$ because, unlike amorphous polymers just above $T_g$, semicrystalline polymers become liquids with a fairly constant activation energy for flow. Only when their molecular weight is very high, their behavior resembles that of rubbers.

The viscosity of semicrystalline polymers is not "infinite" below $T_m$ because such polymers are somewhat ductile, with ductility increasing with decreasing degree of crystallinity. Above $T_m$, as mentioned earlier, their viscosity has an Arrhenius type of temperature dependence. Some polymers (e.g., nylons) even become liquids of quite low viscosity just a few degree above $T_m$. Therefore a sharp transition is observed in both mechanical and viscous properties of semicrystalline polymers at $T_m$, resulting in a physical situation that is closer to the "classical" melting interface of monomeric crystals where, on one side, there is a viscous liquid and, on the other, an elastic solid. For amorphous polymers, the physical situation is much less "clear-cut." The solid becomes gradually more deformable as its temperature approaches $T_g$, while above $T_g$ it continues to be a rubbery solid that only gradually becomes a polymer melt.

## 9.3 Conduction Melting Without Melt Removal

As pointed out in the previous section, melting can often be modeled in terms of simple geometries. Here we analyze the transient conduction problem in a semi-infinite solid. We compare the solutions of this problem, assuming first constant thermophysical properties, then variable thermophysical properties and finally a phase transition with constant thermophysical properties in each phase. These solutions, though useful by themselves, also help demonstrate the profound effect of the material properties on the mathematical complexities of the solution. Indeed the solution of a melting (or solidification) problem with variable thermophysical properties and phase transition necessitates numerical methods, as Section 9.4 explains.

The equation of energy (5.1-37) reduces for transient conduction in solids without internal heat sources to

$$\rho C_p \frac{\partial T}{\partial t} = \nabla \cdot k \, \nabla T \tag{9.3-1}$$

If the thermal conductivity $k$ and the product $\rho C_p$ are temperature independent, Eq. 9.3-1 reduces for homogeneous and isotropic solids to a linear partial differential equation, greatly simplifying the mathematical difficulties in solving the class of heat transfer problems it describes.*

* For heat conduction in non-isotropic solids, see Ref. 4.

**Example 9.1** ***The Semi-Infinite Solid with Constant Thermophysical Properties and a Step Change in Surface Temperature: Exact Solution***

The semi-infinite solid in Fig. 9.3 is initially at constant temperature $T_0$. At time $t=0$ the surface temperature is raised to $T_1$. This is a one-dimensional transient heat conduction problem. The governing parabolic differential equation

$$\frac{\partial T}{\partial t}=\alpha\frac{\partial^2 T}{\partial x^2} \tag{9.3-2}$$

must be solved to satisfy the following initial and boundary conditions $T(x,0)=T(\infty,t)=T_0$ and $T(0,t)=T_1$.

By introducing a new variable* $\eta$, which *combines* the two independent variables $x$ and $t$, as follows:

$$\eta=Cxt^m \tag{9.3-3}$$

where $C$ and $m$ are constants to be determined, it is possible to reduce Eq. 9.3-2 to

$$m\eta\frac{dT}{d\eta}=\alpha C^2t^{2m+1}\frac{d^2T}{d\eta^2} \tag{9.3-4}$$

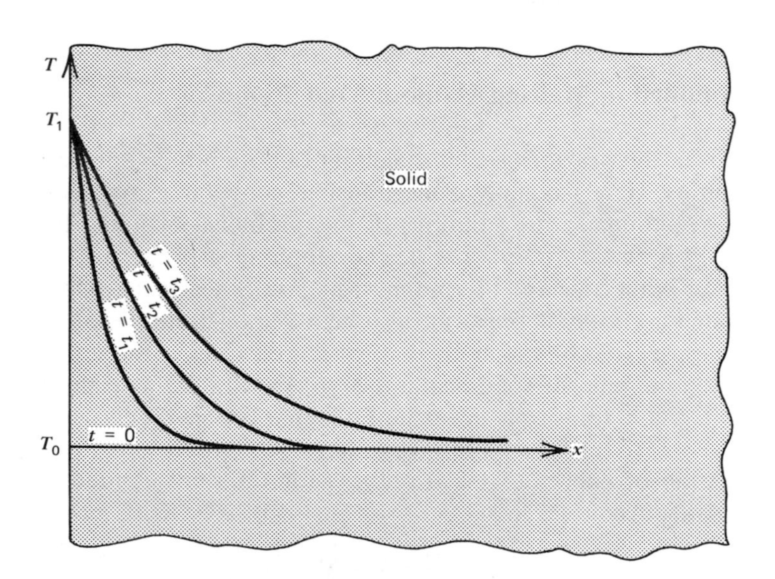


**Fig. 9.3** The semi-infinite solid.

* This transformation follows from general similarity solution methods, and it is a similarity transformation. The term "similar" implies that profiles of the variable $T=T(x,t)$ (at different coordinates $x$) differ only by a scale factor. The profiles can be reduced to the same curve by changing the scale along the axis of ordinates. Problems that lack a "characteristic length" are generally amenable to this solution method.

For Eq. 9.3-4 to be independent of $t$, $2m+1=0$ or $m=-\frac{1}{2}$. Thus the following ordinary differential equation is obtained:

$$\frac{d^2T}{d\eta^2}+\frac{1}{2\alpha C^2}\eta\frac{dT}{d\eta}=0 \qquad (9.3\text{-}5)$$

Next we let $C=1/\sqrt{4\alpha}$, which further simplifies Eq. 9.3-5 to

$$\frac{d^2T}{d\eta^2}+2\eta\frac{dT}{d\eta}=0 \qquad (9.3\text{-}6)$$

Equation 9.3-6 can be easily solved by introducing another variable of transformation, $y=dT/d\eta$. The resulting temperature distribution is

$$T=C_1\frac{\sqrt{\pi}}{2}\operatorname{erf}(\eta)+C_2 \qquad (9.3\text{-}7)$$

where

$$\eta=\frac{x}{\sqrt{4\alpha t}} \qquad (9.3\text{-}8)$$

and erf $(z)$ is the well-known error function defined as

$$\operatorname{erf}(z)=\frac{2}{\sqrt{\pi}}\int_0^z e^{-s^2}\,ds \qquad (9.3\text{-}9)$$

The constants $C_1$ and $C_2$ are obtained from the boundary conditions. Thus boundary condition $T(0,t)=T_1$ is satisfied if $C_2=T_1$, whereas both conditions $T(x,0)=T_0$ and $T(\infty,t)=T_0$ imply $T=T_0$ at $\eta\to\infty$ (which is the direct result of the combination of variables). Thus we get $C_1=2(T_0-T_1)/\sqrt{\pi}$. Substituting these values into Eq. 9.3-7 results in

► $$\frac{T-T_1}{T_0-T_1}=\operatorname{erf}(\eta) \qquad (9.3\text{-}10)$$

which satisfies both the differential equation and the boundary conditions, hence is a solution to the problem.

The heat flux into the solid is obtained by differentiating Eq. 9.3-10 with respect to $x$, and using Fourier's law

► $$q_x=-k\left(\frac{\partial T}{\partial x}\right)_{x=0}=-k\left[\frac{T_0-T_1}{\sqrt{\pi\alpha t}}e^{-\eta^2}\right]_{x=0}=\frac{k}{\sqrt{\pi\alpha t}}(T_1-T_0) \qquad (9.3\text{-}11)$$

The results so far are both interesting and significant. First we have obtained a particular dimensionless combination of the key variables: distance, time, and thermal diffusivity in Eq. 9.3-8. The temperature profile becomes a unique function of this single dimensionless variable $\eta$.

We shall see later that this combination of the key variables is also characteristic of conduction heating with phase transfer. The heat flux is infinite at $t=0$ but quickly drops with inverse of $\sqrt{t}$. Thus after 10 s it is only 30% of the flux at 1 s time, and after 60 s it is only 13% of the heat flux at 1 s time! The obvious conclusion is that *contact melting without melt removal becomes inefficient at anything but short times.*

**Example 9.2** ***The Semi-Infinite Solid with Variable Thermophysical Properties and a Step Change in Surface Temperature: Approximate Analytical Solution***

We have stated before that the thermophysical properties ($k$, $\rho$, $C_p$) of polymers are generally temperature dependent. Hence the governing differential equation 9.3-1 is nonlinear. Unfortunately few analytical solutions for nonlinear heat conduction exist (4a); therefore numerical solutions (finite difference and finite element) are frequently applied. There are, however, a number of useful *approximate analytical* methods available, including the integral method reported by Goodman (5).

The governing differential equation is obtained from Eq. 9.3-1

$$\rho C_p \frac{\partial T}{\partial t} = \frac{\partial}{\partial x} k \frac{\partial T}{\partial x} \tag{9.3-12}$$

which must be solved with the boundary conditions $T(x, 0) = T(\infty, t) = T_0$ and $T(0, t) = T_1$. We now introduce the following variable of transformation for temperature

$$d\Theta(x, t) = \rho C_p \, dT \tag{9.3-13}$$

or in integrated form

$$\Theta(x, t) = \int_{T_0}^{T} \rho C_p \, dT \tag{9.3-14}$$

The variable $\Theta(x, t)$ is the heat added per unit volume at location $x$ and time $t$. Substituting Eq. 9.3-13 into Eq. 9.3-12 gives

$$\frac{\partial \Theta}{\partial t} = \frac{\partial}{\partial x} \alpha(\Theta) \frac{\partial \Theta}{\partial x} \tag{9.3-15}$$

Next we integrate Eq. 9.3-15 over $x$ from the outer surface ($x = 0$) to a certain unknown depth $\delta(t)$, which is the *thermal penetration depth* (to be evaluated later). The penetration depth reflects the time dependent distance from the surface where thermal effects become negligible.

$$\int_0^{\delta} \frac{\partial \Theta}{\partial t} dx = \int_0^{\delta} \left[ \frac{\partial}{\partial x} \alpha(\Theta) \frac{\partial \Theta}{\partial x} \right] dx \tag{9.3-16}$$

Using the Leibnitz formula, we get for the left-hand side of Eq. 9.3-16

$$\frac{d}{dt} \int_0^{\delta} \Theta \, dx - \Theta(\delta, t) \frac{d\delta}{dt}$$

But $\Theta(\delta, t) = 0$, because we defined $\delta$ as the distance at which thermal effects fade away; that is, we assume that $T(\delta) = T_0$. Then the right-hand side of Eq. 9.3-16 simply becomes

$$-\left[ \alpha \frac{\partial \Theta}{\partial x} \right]_{x=0}$$

and Eq. 9.3-16 can be rewritten as

$$\frac{d}{dt}\int_0^{\delta} \Theta\, dx = \left[-\alpha \frac{\partial \Theta}{\partial x}\right]_{x=0} \tag{9.3-17}$$

The advantage of the Goodman transformation is now apparent: the temperature dependent thermophysical properties in the integrated differential equation have to be evaluated only at the constant surface temperature $T_1$. The variation of these properties with temperature appear in the boundary conditions for $\Theta(x, t)$

$$\Theta(0, t) = \Theta_1 = \int_{T_0}^{T_1} \rho C_p\, dT \tag{9.3-18}$$

Boundary conditions $T(x, 0) = T(\infty, t) = T_0$ are both taken care by assuming a time dependent thermal penetration depth of finite thickness.

Next we *assume* a "temperature profile" that a priori satisfies the boundary conditions $[\Theta(0, t) = \Theta_1, \Theta(\delta, t) = 0, (\partial\Theta/\partial x)_{x=\delta} = 0]$ such as

$$\Theta = \Theta_1\left(1 - \frac{x}{\delta}\right)^3 \tag{9.3-19}$$

By substituting Eq. 9.3-19 into Eq. 9.3-17, the time dependence of $\delta$ is obtained

► $$\delta = \sqrt{24\alpha_1 t} \tag{9.3-20}$$

where $\alpha_1$ is $\alpha$ evaluated at $T_1$.

For a polymer with a typical $\alpha$ of $1 \times 10^{-7}$ $m^2/s$, the penetration depth is of the order of 0.1 cm after 1 s and 1 cm after 60 s.

From Eq. 9.3-19 we obtain

► $$\Theta = \Theta_1\left(1 - \frac{x}{\sqrt{24\alpha_1 t}}\right)^3 \tag{9.3-21}$$

The temperature profile at any given time is obtained by calculating $\Theta$ for various $x$ values ($0 < x < \delta$) and obtaining from Eq. 9.3-14 the corresponding temperatures. The latter requires, of course, the knowledge of the temperature dependence of $\rho C_p$.

For constant thermophysical properties Eq. 9.3-21 reduces to

$$\frac{T - T_0}{T_1 - T_0} = \left(1 - \frac{x}{\sqrt{24\alpha t}}\right)^3 \tag{9.3-22}$$

The heat flux at $x = 0$ is

► $$q_x = \frac{k}{\sqrt{(\frac{8}{3})\alpha t}}(T_1 - T_0) \tag{9.3-23}$$

which can be compared to the exact solution in Eq. 9.3-11, showing a small difference between the two solution methods. This difference depends on the selection of the trial function, and in this case it is 8%.

### Example 9.3 *Melting of a Semi-Infinite Solid with Constant Thermophysical Properties and a Step Change in Surface Temperature: The Stefan-Neumann Problem*

The previous two examples investigated the heat conduction problem in a semi-infinite solid with constant and variable thermophysical properties. This example analyzes the same conduction problem with a change in phase.

Interest in such problems started with the early work of G. Lamè and B. P. Clapeyron in 1831 on the freezing of moist solids, and by the work J. Stefan in 1889 on the thickness of polar ice and similar problems. Consequently problems of these types are referred to as "Stefan problems." The exact solution of the phase transition problem in a semi-infinite medium is due to F. Neumann (who apparently dealt with this kind of problem even before Stefan), and it is called the Stefan–Neumann problem. The interest in these problems has grown ever since (6, 7).

The presence of a moving boundary between the phases introduces nonlinearity into the boundary conditions. Hence there are but a few exact solutions, and we turn frequently to approximate analytical or numerical solutions.

In this example we consider the classical Stefan–Neumann solution. The solid is initially at a constant temperature $T_0$. At time $t=0$ the surface temperature is raised to $T_1$, which is above the melting point $T_m$. The physical properties of each phase are different, but they are temperature independent, and the change in phase involves a latent heat of fusion $\lambda$. After a certain time $t$, the thickness of the molten layer is $X_l(t)$, in each phase there is a temperature distribution, and the interface is at the melting temperature $T_m$ (Fig. 9.4). Heat is conducted from the outer surface through the melt to the free interface, where some of the heat is absorbed as heat of fusion, melting some

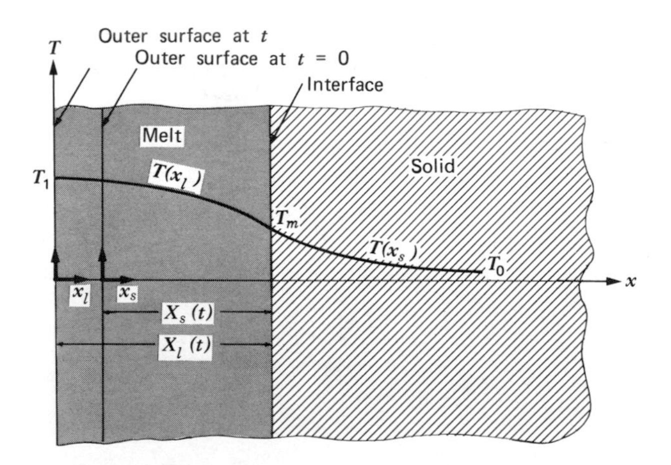


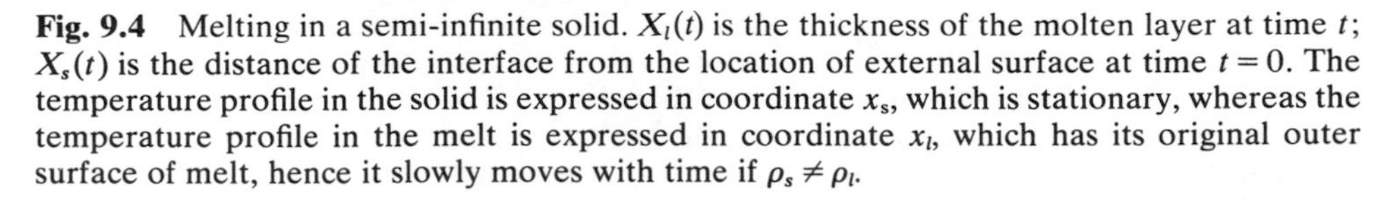
**Fig. 9.4** Melting in a semi-infinite solid. $X_l(t)$ is the thickness of the molten layer at time $t$; $X_s(t)$ is the distance of the interface from the location of external surface at time $t=0$. The temperature profile in the solid is expressed in coordinate $x_s$, which is stationary, whereas the temperature profile in the melt is expressed in coordinate $x_l$, which has its original outer surface of melt, hence it slowly moves with time if $\rho_s \neq \rho_l$.

more solid, and the rest is conducted into the solid phase. The densities of melt and solid are usually different. We shall denote the melt phase with subscript $l$ and the solid with subscript $s$. The thickness of the molten layer increases because of melting, and there is also a slight increase due to a decrease in density as the solid melts. If there were no decrease in density, the thickness of the molten layer would remain $X_s(t)$. Thus the relationship between $X_l(t)$ and $X_s(t)$ is given by

$$\frac{X_l}{X_s}=\frac{\rho_s}{\rho_l}=\beta \tag{9.3-24}$$

The governing differential equation in both phases is Eq. 9.3-2. For the melt phase, it takes the form

$$\frac{\partial^2 T_l}{\partial x_l^2}-\frac{1}{\alpha_l}\frac{\partial T_l}{\partial t}= 0 \tag{9.3-25}$$

and the boundary conditions are

$$T_l(0, t)=T_1 \tag{9.3-26a}$$

$$T_l(X_l, t)=T_m \tag{9.3-26b}$$

It should be noted that the coordinate $X_l$ has its origin at the outer surface of the melt which, if $\rho_l \neq \rho_s$, slowly moves with the melting process. For the solid phase we have

$$\frac{\partial^2 T_s}{\partial x_s}-\frac{1}{\alpha_s}\frac{\partial T_s}{\partial t}=0 \tag{9.3-27}$$

with the boundary conditions

$$T_s(X_s, t)=T_m \tag{9.3-28a}$$

$$T_s(\infty, t)=T_0 \tag{9.3-28b}$$

The coordinate $x_s$ has its origin at the external surface when melting started, and it is stationary. In addition to the boundary conditions above, we can write a heat balance for the interface (this is occasionally referred to as the Stefan condition).

$$k_l\left(\frac{\partial T_l}{\partial x_l}\right)_{x_l=X_l}-k_s\left(\frac{\partial T_s}{\partial x_s}\right)_{x_s=X_s}=\lambda\rho_l\frac{dX_l}{dt}=\lambda\rho_s\frac{dX_s}{dt} \tag{9.3-29}$$

| Heat flux into the interface | Heat flux out from interface | Rate of melting per unit interface |
|---|---|---|

We assume that the temperature profile in each phase has the form of the temperature profile in a semi-infinite solid with a step change in surface temperature as derived in Example 9.1. Thus we get the following temperature profiles for melt and solid phases, respectively:

$$T_l=T_1+A\ \mathrm{erf}\left(\frac{x_l}{2\sqrt{\alpha_l t}}\right) \tag{9.3-30}$$

which automatically satisfies boundary condition 9.3-26a, and

$$T_s = T_0 + B \operatorname{erfc}\left(\frac{x_s}{2\sqrt{\alpha_s t}}\right) \tag{9.3-31}$$

which satisfies boundary condition 9.3-28b. In Eqs. 9.3-30 and 9.3-31 $\operatorname{erfc}(s) = 1 - \operatorname{erf}(s)$. Both equations must satisfy the boundary condition stating that the temperature at the interface is the melting point:

$$T_m = T_1 + A \operatorname{erf}\left(\frac{X_l}{2\sqrt{\alpha_l t}}\right) \tag{9.3-32}$$

$$T_m = T_0 + B \operatorname{erfc}\left(\frac{X_s}{2\sqrt{\alpha_s t}}\right) \tag{9.3-33}$$

Now Eqs. 9.3-32 and 9.3-33 must hold for all times $t$. This is possible *only* if both $X_l$ and $X_s$ *are proportional to the square root of time*. We can, therefore, write

► $$X_s = K\sqrt{t} \tag{9.3-34}$$

and with the aid of Eq. 9.3-24

► $$X_l = \beta K\sqrt{t} \tag{9.3-35}$$

where $K$ is an unknown constant. From Eqs. 9.3-34 and 9.3-35 we conclude, without even having the complete solution, that the thickness of the molten layer grows at a rate proportional to the square root of time. It is interesting to note the similarity between the penetration depth, as obtained in the preceding examples, and the location of the interface. This similarity suggests the application of approximate solution methods to phase transition problems.

The constant $K$ can be evaluated by substituting Eqs. 9.3-30 and 9.3-31 into Eq. 9.3-29. Subsequent to evaluating the constants $A$ and $B$ from the boundary conditions 9.3-26 and 9.3-28, and conditions 9.3-34 and 9.3-35

► $$\frac{(T_m - T_1)k_l e^{-K^2\beta^2/4\alpha_l}}{\sqrt{\pi\alpha_l}\operatorname{erf}(K\beta/2\sqrt{\alpha_l})} - \frac{(T_0 - T_m)k_s e^{-K^2/4\alpha_s}}{\sqrt{\pi\alpha_s}\operatorname{erfc}(K/2\sqrt{\alpha_s})} = \lambda\rho_l\frac{K\beta}{2} \tag{9.3-36}$$

The root of this transcendental equation is $K$, and it is a function of the initial and boundary conditions, as well as the physical properties of the two phases. Tabulated solutions of Eq. 9.3-36 for $\beta = 1$ to four-digit accuracy are given by Churchill and Evans (8). The temperature profiles in the two phases are obtained from Eqs. 9.3-30 and 9.3-31, with the aid of Eqs. 9.3-32 and 9.3-33.

► $$\frac{T_l - T_m}{T_1 - T_m} = 1 - \frac{\operatorname{erf}(x_l/2\sqrt{\alpha_l t})}{\operatorname{erf}(K\beta/2\sqrt{\alpha_l})} \tag{9.3-37}$$

and

► $$\frac{T_s - T_m}{T_0 - T_m} = 1 - \frac{\operatorname{erfc}(x_s/2\sqrt{\alpha_s t})}{\operatorname{erfc}(K/2\sqrt{\alpha_s})} \tag{9.3-38}$$

Equations 9.3-37 and 9.3-38 satisfy the differential equation and the boundary and initial conditions. Therefore they form an exact solution to the problem. In the solution above we neglected heat convection as a result of the "expansion" of the melt phase due to the density decrease.

The rate of melting per unit area as a function of time can be obtained from Eq. 9.3-35.

$$w_A = \rho_l \frac{dX_l}{dt} = \frac{\rho_s K}{2\sqrt{t}} \tag{9.3-39}$$

Again we note the similarity in the solution of the conduction problem with constant thermophysical properties, to those with variable properties, and with phase transition. Clearly, the rate of melting drops with time as the molten layer, which essentially forms a thermal shield, increases in thickness. This result, once again, directs our attention to the advantage accruing from forced removal of the molten layer from the melting site. The average rate of melting is

$$\bar{w}_A = \frac{1}{t}\int_0^t \frac{\rho_s K}{2\sqrt{t}}\,dt = \frac{\rho_s K}{\sqrt{t}} \tag{9.3-40}$$

The foregoing examples discussed the heat conduction problem in a semi-infinite solid with different assumptions regarding the thermophysical properties of the solid. Although the solutions obtained in these examples form useful approximations to the problems at hand, many real cases of polymer melting (and solidification) involve both a change in phase and temperature dependent thermophysical properties. In such cases numerical methods—in particular, finite difference methods—must be used, as the sections that follow explain. Numerical methods have the added advantage that they can also handle complex geometries

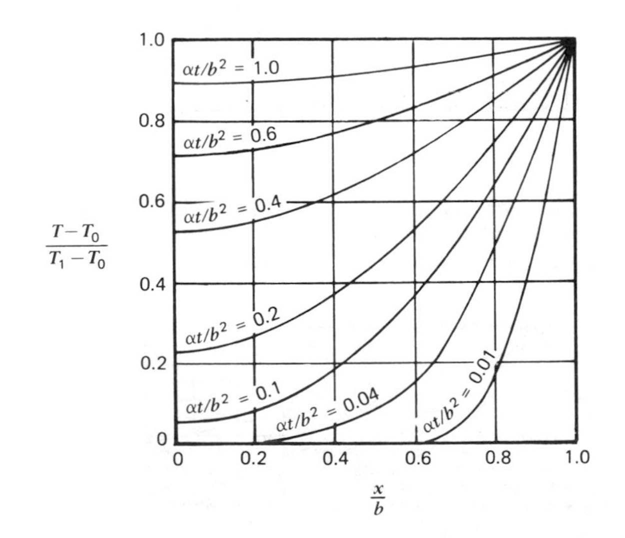


**Fig. 9.5** Temperature profiles for unsteady state heat conduction in infinite flat plates: $T(x, 0) = T_0$, $T(\pm b, t) = T_1$. (Reprinted with permission from H. S. Carslaw and J. C. Jaeger, *Conduction of Heat in Solids*, 2nd ed., Oxford University Press, New York, 1973.)

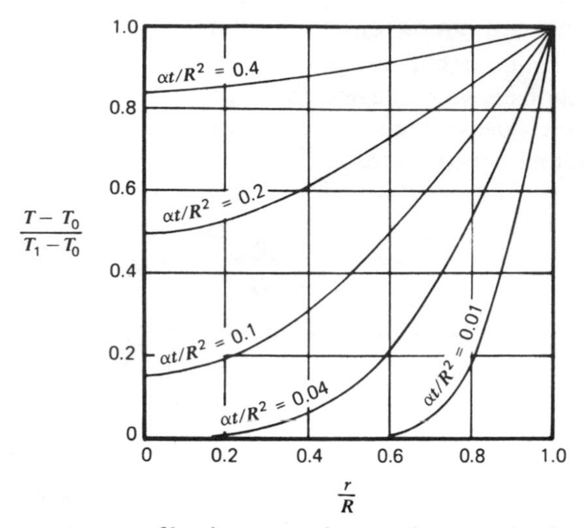


**Fig. 9.6** Temperature profiles for unsteady state heat conduction in an infinite cylinder: $T(r, 0) = T_0$, $T(R, t) = T_1$. (Reprinted with permission from H. S. Carslaw and J. C. Jaeger, *Conduction of Heat in Solids*, 2nd ed., Oxford University Press, New York, 1973.)

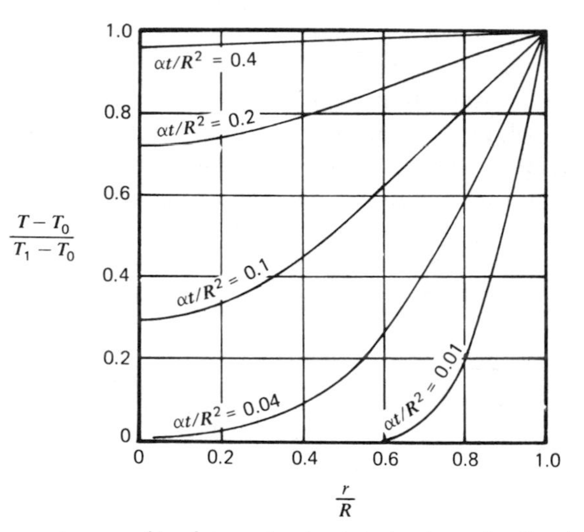


**Fig. 9.7** Temperature profiles for unsteady state heat conduction in a sphere: $T(r, 0) = T_0$, $T(R, t) = T_1$. (Reprinted with permission from H. S. Carslaw and J. C. Jaeger, *Conduction of Heat in Solids*, 2nd ed., Oxford University Press, New York, 1973.)

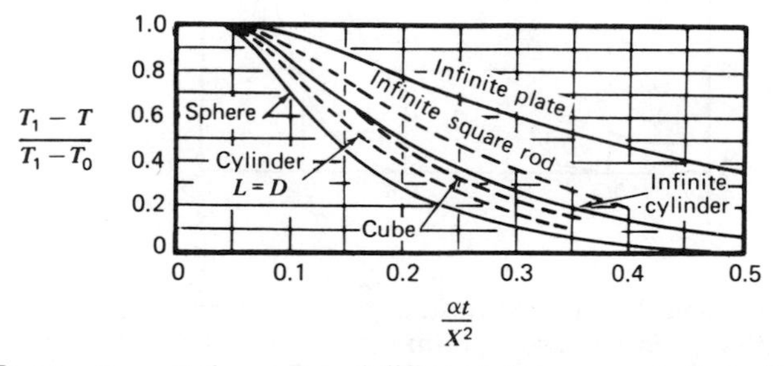


**Fig. 9.8** Temperature at the center of different shapes versus time; $X$ is the thickness, side dimension, or diameter; initial temperature is $T_0$. At $t = t_0$ outside surface temperature is raised to $T_1$. (Reprinted with permission from H. Gröber and S. Erk, *Die Grundgesetze der Wärmeübertragung*, Springer-Verlag, Berlin, 1933, Fig. 28, p. 58.)

and various boundary conditions. As pointed out earlier complex geometries frequently occur in solidification problems. Nevertheless, the vast collection of analytical solutions of heat conduction problems with various geometries and boundary conditions to be found in classical treatises (9, 10), though generally restricted to constant thermophysical properties, are useful in polymer processing. It is, of course, beyond the scope of this text to review these solutions and the mathematical techniques utilized to obtain them. Table 9.1 lists a few well-known and commonly applied solutions, and Figs. 9.5–9.8 show unsteady state temperature profiles for some common geometries.

## 9.4 Solidification or Melting of a Polymer Sheet; Numerical Solution by the Method of Finite Differences

In the finite difference method (FDM) the differential equation is replaced by a *finite difference approximation* and the continuous region in which the solution is desired by *a set of discrete points.* This results in the reduction of the problem to *systems of algebraic equations.* Simultaneous solutions of these equations yields numerical values of the dependent variable (e.g., temperature) at the discrete points. Obtaining only numerical values at discrete locations, rather than a continuous function, is also the main disadvantage of the FDM. This technique involves a large number of rather simple computations, and digital computers are most suitable for this task. Indeed, the great advancements in the computer field have also brought about wide usage of the finite difference method.

The FDM is applicable to both steady and unsteady conduction in one or more dimensions. We shall not attempt to review here the mathematical foundations of this method. Many excellent texts are available for this purpose in the literature (10–13). We derive in detail the solution to the sheet solidification problem.

We consider in this example a polymer melt, initially at $T_0$, being placed between two infinite plates at distance $2b$ from each other, at a temperature $T_1$ below the freezing or solidification temperature (or below $T_g$ for amorphous polymers). The governing differential equation is Eq. 9.3-1, which in one dimension reduces to

$$\rho C \frac{\partial T}{\partial t} = \frac{\partial}{\partial x}\left(k \frac{\partial T}{\partial x}\right) \tag{9.4-1}$$

where for simplicity $C_p$ is written as $C$. The initial and boundary conditions (coordinate system defined in Fig. 9.6) are

$$T(x, 0) = T_0 \tag{9.4-2a}$$

$$\frac{\partial T}{\partial x}(0, t) = 0 \tag{9.4-2b}$$

$$T(b, t) = T_1 \tag{9.4-2c}$$

Following Dusinberre (13), we shall apply Eq. 9.4-1 to the solid as well as the liquid phases. The thermophysical properties $k$, $\rho$, and $C_p$, are functions of

**Table 9.1 Analytical Solutions to Some Common Heat Transfer Problems (Constant Physical Properties)**

| Geometry | Initial and Boundary Conditions | Temperature Distribution |
|---|---|---|
| Semi-infinite solid | $T(x, 0) = T_0$<br>$T(0, t) = T_1$<br>$T(\infty, t) = T_0$ | $\dfrac{T - T_1}{T_0 - T_1} = \operatorname{erf}\left(\dfrac{x}{\sqrt{4\alpha t}}\right)$ |
| Semi-infinite solid | $T(x, 0) = T_0$<br>$T(\infty, t) = T_0$<br>$-k\left(\dfrac{\partial T}{\partial x}\right)_{x=0} = q_0$ | $T - T_0 = \dfrac{q_0}{k}\sqrt{4\alpha t}\,\operatorname{ierfc}\left(\dfrac{x}{\sqrt{4\alpha t}}\right)$ |
| Flat plate | $T(x, 0) = T_0$<br>$T(\pm b, t) = T_1$ | $\dfrac{T_1 - T}{T_1 - T_0} = 2\sum_{n=0}^{\infty} \dfrac{(-1)^n}{(n+\frac{1}{2})\pi} e^{-(n+\frac{1}{2})^2\pi^2(\alpha t/b^2)} \cos\left[(n+\tfrac{1}{2})\pi\dfrac{x}{b}\right]$ |
| Flat plate | $T(x, 0) = T_0$<br>$-\dfrac{\partial T}{\partial x}\Big\vert_{x=-b} = \dfrac{h}{k}[T_1 - T(-b)]$<br>$-\dfrac{\partial T}{\partial x}\Big\vert_{x=b} = \dfrac{h}{k}[T(b) - T_1]$ | $\dfrac{T_1 - T}{T_1 - T_0} = \sum_{n=1}^{\infty} \dfrac{2\left(\dfrac{hb}{k}\right)\cos\beta_n\left(\dfrac{x}{b}\right)}{\left[\beta_n^2 + \dfrac{hb}{k} + \left(\dfrac{hb}{k}\right)^2\right]\cos\beta_n} e^{-\beta_n^2\alpha t/b^2}$<br>$\beta_n \tan\beta_n = \dfrac{hb}{k}$ |
| Cylinder | $T(r, 0) = T_0$<br>$T(R, t) = T_1$ | $\dfrac{T_1 - T}{T_1 - T_0} = \dfrac{2}{R}\sum_{n=1}^{\infty} \dfrac{J_0(rc_n)}{c_n J_1(Rc_n)} e^{-\alpha Cn^2 t}$ $\quad J_0(Rc_n) = 0$ |
| Cylinder | $T(r, 0) = T_0$<br>$-k\left(\dfrac{\partial T}{\partial r}\right)\Big\vert_{r=R} = h[T(R) - T_1]$ | $\dfrac{T_1 - T}{T_0 - T_1} = \dfrac{2}{R}\sum_{n=1}^{\infty} \exp(-\alpha\beta_n^2 t) \dfrac{\dfrac{h}{k}J_0(r\beta_n)}{\left[\left(\dfrac{h}{k}\right)^2 + \beta_n^2\right] J_0(R\beta_n)}$<br>$\beta_n \dot{J}_0(R\beta_n) + \dfrac{h}{k}J_0(R\beta_n) = 0$ |
| Sphere | $T(r, 0) = T_0$<br>$T(R, t) = T_1$ | $\dfrac{T_1 - T}{T_1 - T_0} = \dfrac{2R}{\pi r}\sum_{n=1}^{\infty} \dfrac{(-1)^n}{n} \sin\dfrac{n\pi r}{R} e^{-\alpha n^2\pi^2 t/R^2}$ |
| Sphere | $T(r, 0) = T_0$<br>$-k\left(\dfrac{\partial T}{\partial r}\right)\Big\vert_{r=R} = h[T(R) - T_1]$ | $\dfrac{T - T_1}{T_0 - T_1} = \dfrac{2\dfrac{h}{k}}{r}\sum_{n=1}^{\infty} \exp(-\alpha\beta_n^2 t) \dfrac{R^2\beta_n^2 + \left(R\left(\dfrac{h}{k}\right) - 1\right)^2}{\beta_n^2\left[R^2\beta_n^2 + R\dfrac{h}{k}\left(R\dfrac{h}{k} - 1\right)\right]} \sin(R\beta_n)\sin(r\beta_n)$<br>$R\beta_n \cot(R\beta_n) + R\left(\dfrac{h}{k}\right) - 1 = 0$ |

temperature. Within the solidification range, the latent heat of solidification $\lambda$ is included in the specific heat as follows:

$$C = C_p = \frac{\lambda}{\Delta T_m} \tag{9.4-3}$$

where $\Delta T_m$ denotes the freezing range. This type of representation will result in a fairly sharp change in $C_p$ in the $\Delta T_m$ range for crystalline polymers. In amorphous polymers, of course, there is no fusion, and as discussed in Chapter 5, $C_p$ changes gradually with temperature. The thickness of the solid layer as a function of time extends from the wall to that point inside the melt where $T = T_m + \frac{1}{2}\Delta T_m$.

We first rewrite Eqs. 9.4-1 and 9.4-2 in dimensionless form by defining the following dimensionless variables

$$\Theta = \frac{T - T_1}{T_0 - T_1} \tag{9.4-4}$$

$$\xi = \frac{x}{b} \tag{9.4-5}$$

$$\tau = \frac{\alpha_0 t}{b^2} \tag{9.4-6}$$

$$k' = \frac{k}{k_0},\ \rho' = \frac{\rho}{\rho_0},\ C' = \frac{C}{C_0},\ \text{and } \alpha' = \frac{\alpha}{\alpha_0} \tag{9.4-7}$$

where $k_0$, $\rho_0$, $C_0$, and $\alpha_0$ are the thermophysical properties at $T = T_0$. The resulting equations are:

$$\rho' C' \frac{\partial \Theta}{\partial \tau} = \frac{\partial}{\partial \xi}\left(k' \frac{\partial \Theta}{\partial \xi}\right) \tag{9.4-8}$$

$$\Theta(\xi, 0) = 1 \tag{9.4-9a}$$

$$\frac{\partial \Theta}{\partial \xi}(0, \tau) = 0 \tag{9.4-9b}$$

$$\Theta(1, \tau) = 0 \tag{9.5-9c}$$

In the finite difference formulation, the continuous space is replaced by a set of nodal points $i$ as in Fig. 9.9, with $i = 1$ at the centerline and $i = I$ at the wall ($\xi = 1$). The distance between the nodes is $\Delta\xi$. Before proceeding with the finite difference formulation of the problem, we demonstrate the finite difference formulation of the same problem without phase transition and with constant thermophysical properties. For this case Eq. 9.4-8 reduces to

$$\frac{\partial \Theta}{\partial \tau} = \frac{\partial^2 \Theta}{\partial \xi^2} \tag{9.4-10}$$

The first step in the FDM is to rewrite the differentials in terms of differences. This can be done either by physical considerations—by carrying out heat balances over small spaces about the nodes—or mathematically. In using the latter approach, consider a Taylor series expansion of a continuous function $T(x)$ about

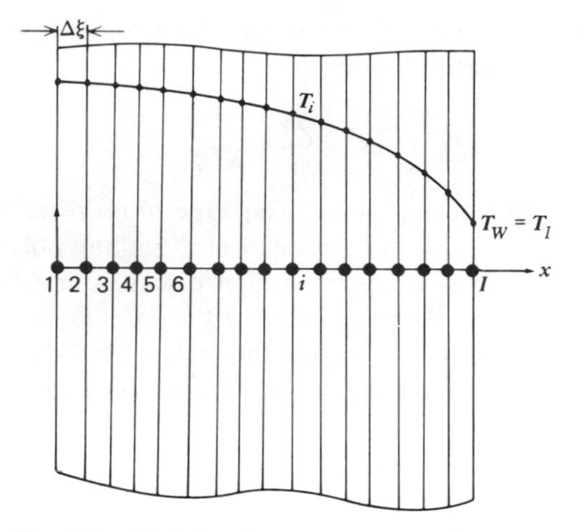


**Fig. 9.9** Nodal point arrangement in one-dimensional heat conduction.

the spatial location $x_i + \Delta x/2$, where $\Delta x$ is the spatial increment from $x_i$, where the temperature is $T_i$, to $x_{i+1}$, where the temperature is $T_{i+1}$:

$$T_{i+1} = T_{i+1/2} + \left.\frac{dT}{dx}\right|_{i+1/2} \frac{\Delta x}{2} + \frac{1}{2}\left.\frac{d^2T}{dx^2}\right|_{i+1/2} \left(\frac{\Delta x}{2}\right)^2 + O[(\Delta x)^3] + \cdots \tag{9.4-11}$$

where $O[\Delta x^3]$ represents the remaining terms of the order of $(\Delta x)^3$. The temperature at $x_i$ can be similarly obtained as

$$T_i = T_{i+1/2} - \left.\frac{dT}{dx}\right|_{i+1/2} \frac{\Delta x}{2} + \frac{1}{2}\left.\frac{d^2T}{dx^2}\right|_{i+1/2} \left(\frac{\Delta x}{2}\right)^2 - O[(\Delta x)^3] + \cdots \tag{9.4-12}$$

Subtracting Eq. 9.4-12 from Eq. 9.4-11, we get

$$T_{i+1} - T_i = \left.\frac{dT}{dx}\right|_{i+1/2} \Delta x + O[(\Delta x)^3] \tag{9.4-13}$$

which may be solved for the temperature derivative to give

$$\blacktriangleright \qquad \left.\frac{dT}{dx}\right|_{i+1/2} = \frac{T_{i+1} - T_i}{\Delta x} + O[(\Delta x)^3] \tag{9.4-14}$$

In a similar way, the second derivative of $T(x)$ can be obtained by expansion about $x_i$ to obtain $T_{i+1}$ and $T_{i-1}$. Thus we have

$$T_{i+1} = T_i + \left.\frac{dT}{dx}\right|_i \Delta x + \frac{1}{2}\left.\frac{d^2T}{dx^2}\right|_i (\Delta x)^2 + \frac{1}{6}\left.\frac{d^3T}{dx^3}\right|_i (\Delta x)^3 + O[(\Delta x)^4] \tag{9.4-15}$$

and

$$T_{i-1} = T_i - \left.\frac{dT}{dx}\right|_i \Delta x + \frac{1}{2}\left.\frac{d^2T}{dx^2}\right|_i (\Delta x)^2 - \frac{1}{6}\left.\frac{d^3T}{dx^3}\right|_i (\Delta x)^3 + O[(\Delta x)^4] \tag{9.4-16}$$

By adding Eqs. 9.4-15 and 9.4-16, followed by some rearrangement, we obtain the required difference form of the second derivative

$$\left.\frac{d^2T}{dx^2}\right|_i = \frac{T_{i-1} - 2T_i + T_{i+1}}{(\Delta x)^2} + O[(\Delta x)^4] \qquad (9.4\text{-}17)$$

Rewriting the right-hand side of Eq. 9.4-10 in terms of finite differences with the aid of Eq. 9.4-17, while omitting the $O[\Delta x^4]$ term, results in

$$\frac{d\Theta_i}{d\tau} = \frac{\Theta_{i-1} - 2\Theta_i + \Theta_{i+1}}{(\Delta\xi)^2} \qquad (9.4\text{-}18)$$

This *ordinary* differential equation is applicable to all interior nodes ($1 < i < I$). Hence the partial differential equation was converted into a *set of ordinary differential equations*. The boundary conditions are accounted for in writing Eq. 9.4-18 for $i = 1$ and $i = I - 1$. In the former we need to know $\Theta_0$, which from boundary condition 9.4-9b is $\Theta_2$; hence we get

$$\frac{d\Theta_1}{d\tau} = \frac{1}{(\Delta\xi)^2}(2\Theta_2 - 2\Theta_1) \qquad (9.4\text{-}19)$$

whereas in the latter, from the boundary condition 9.4-9c $\Theta_I = 0$ and

$$\frac{d\Theta_{I-1}}{d\tau} = \frac{1}{(\Delta\xi)^2}[\Theta_{I-2} - 2\Theta_{I-1}] \qquad (9.4\text{-}20)$$

Thus we have $I - 1$ differential equations for $I - 1$ unknown temperature values at the nodes.

Since the exact analytical solution of a large number of ordinary differential equations is difficult even if the equations are linear, and hardly possible if they are nonlinear, approximate solution methods are being used. Finite difference formulation provides once again the answer to this problem.

The unsteady heat transfer problem is an *initial value problem*. That is, we know the temperature $\Theta_i$ at any node $i$ at time $\tau$, and we seek the temperature $\Theta_i^*$ at any node $i$ and time $\tau + \Delta\tau$, where $\Delta\tau$ is an arbitrarily determined time increment.

The simplest way to estimate the temperature at time $\tau + \Delta\tau$ is to compute the time derivative of the temperature at time $\tau$ and project the change *forward* in time

$$\Theta_i^* = \Theta_i + \left(\frac{d\Theta_i}{d\tau}\right)\Delta\tau \qquad (9.4\text{-}21)$$

By substituting Eq. 9.4-21 into Eqs. 9.4-18 to 9.4-20, we get the following set of *algebraic* equations:

$$\begin{aligned}
\Theta_1^* &= \Theta_1 + \frac{\Delta\tau}{(\Delta\xi)^2}[2\Theta_2 - 2\Theta_1] \\
\Theta_2^* &= \Theta_2 + \frac{\Delta\tau}{(\Delta\xi)^2}[\Theta_1 - 2\Theta_2 + \Theta_3] \\
&\ \vdots \\
\Theta_{I-1}^* &= \Theta_{I-1} + \frac{\Delta\tau}{(\Delta\xi)^2}[\Theta_{I-2} - 2\Theta_{I-1}]
\end{aligned} \qquad (9.4\text{-}22)$$

These equations, for *convenience in notation*, can be written in matrix form as follows:

$$\mathbf{A}\boldsymbol{\Theta} = \boldsymbol{\Theta}^* \tag{9.4-23}$$

where $\mathbf{A}$ is the coefficient matrix, $\boldsymbol{\Theta}$ is the column matrix of the known temperatures, and $\boldsymbol{\Theta}^*$ the column matrix of the temperatures to be calculated, defined respectively as

$$\mathbf{A} = \begin{pmatrix} (1-2p) & 2p & & & \\ p & (1-2p) & p & & \\ & & \cdot & & \\ & & & \cdot & \\ & & & \cdot\; p & (1-2p) \end{pmatrix} \tag{9.4-24}$$

where

$$p = \frac{\Delta\tau}{(\Delta\xi)^2} = \frac{\alpha_0\,\Delta t}{(\Delta x)^2} \tag{9.4-25}$$

$$\boldsymbol{\Theta} = \begin{pmatrix} \Theta_1 \\ \Theta_2 \\ \vdots \\ \Theta_{I-1} \end{pmatrix} \tag{9.4-26a}$$

$$\boldsymbol{\Theta}^* = \begin{pmatrix} \Theta_1^* \\ \Theta_2^* \\ \vdots \\ \Theta_{I-1}^* \end{pmatrix} \tag{9.4-26b}$$

The matrix $\mathbf{A}$ is called a tridiagonal matrix because the only nonvanishing components are three components along the diagonal. The column matrix of the "new" temperatures at time $\tau + \Delta\tau$ is the product of the tridiagonal matrix $\mathbf{A}$ with the column matrix of the "old" known temperatures at time $\tau$. By the matrix multiplication rule we obtain the following set of algebraic equations

$$\begin{aligned} (1-2p)\Theta_1 + 2p\Theta_2 &= \Theta_1^* \\ p\Theta_1 + (1-2p)\Theta_2 + p\theta_3 &= \Theta_2^* \\ p\Theta_2 + (1-2p)\Theta_3 + p\Theta_4 &= \Theta_3^* \\ &\vdots \\ p\Theta_{I-2} + (1-2p)\Theta_{I-1} &= \Theta_{I-1}^* \end{aligned} \tag{9.4-27}$$

The above is equivalent to Eq. 9.4-22.

The tridiagonal matrix $\mathbf{A}$ on the left-hand side of Eq. 9.4-23 is known once we have selected a space increment $\Delta\xi$ and a time increment $\Delta\tau$. The initial value of the column matrix $\boldsymbol{\Theta}$ is known from the initial condition (Eq. 9.4-9), $\Theta_i = 1$. Hence by solving the set of algebraic equations, we obtain the temperatures $\boldsymbol{\Theta}^*$ after time $\Delta\tau$. Next we repeat the procedure replacing $\boldsymbol{\Theta}$ in Eq. 9.4-23 with previously obtained $\boldsymbol{\Theta}^*$ and calculating a new $\boldsymbol{\Theta}^*$ at time $2\Delta\tau$, and so on. This computational procedure is called the *Euler* method and it is *explicit.*

Thus, in conclusion, the partial differential equation has been solved by repeated solution of sets of algebraic equations.

In the Euler method $\Theta_i^*$ at $\tau + \Delta\tau$ were obtained from $\Theta_i$ at $\tau$ by projecting *forward* the time derivative at time $\tau$. Using the mean value of the time derivative between $\tau$ and $\tau+\Delta\tau$, that is, a "central" derivative, would be more accurate. This is being done in the Crank-Nickolson method (14). Thus we write

$$\Theta_i^* = \Theta_i + \frac{1}{2}\left[\frac{d\Theta_i}{d\tau} + \frac{d\Theta_i^*}{d\tau}\right]\Delta\tau \tag{9.4-28}$$

By substituting Eqs. 9.4-18 to 9.4-20 into Eq. 9.4-28, followed by some rearrangement, we obtain the following set of algebraic equations

$$\mathbf{B}\boldsymbol{\Theta} = \mathbf{C}\boldsymbol{\Theta}^* \tag{9.4-29}$$

where **B** and **C** are two tridiagonal coefficient matrices defined, respectively, as

$$\mathbf{B} = \begin{pmatrix} (1-p) & p & & & \\ \frac{p}{2} & (1-p) & \frac{p}{2} & & \\ & & \cdot & & \\ & & & \cdot & \\ & & & \cdot \ \frac{p}{2} & (1-p) \end{pmatrix} \tag{9.4-30}$$

and

$$\mathbf{C} = \begin{pmatrix} (1+p) & -p & & & \\ \frac{-p}{2} & (1+p) & \frac{-p}{2} & & \\ & & \cdot & & \\ & & & \cdot & \\ & & & \cdot \ \frac{-p}{2} & (1+p) \end{pmatrix} \tag{9.4-31}$$

Using this method we do not have an explicit solution for $\Theta_i^*$, but rather an *implicit* tridiagonal system of algebraic equations to be solved simultaneously. Although the Crank–Nickolson implicit method is computationally more involved, larger steps can be taken, and it is less subject to oscillations due to the numerical computation, compared to the explicit Euler method. In addition to these methods, there is the pure implicit O'Brien (15) method, using a *backward difference* for the time derivative that is stable at all conditions.

It is necessary that $\Delta\xi$ and $\Delta\tau$ be assigned such values that a violation of the second law of thermodynamics is avoided (4a). Such a violation would lead to instability. The explicit method may result in unstable oscillations and the Crank–Nickolson method in stable oscillations. The critical value of $p$ beyond which instability occurs depends on the total number of nodal points. The Euler method, for one nodal point, is stable (i.e., has a steady decay) if $p<0.5$, it gets into stable oscillations if $0.5<p<1$, and into unstable oscillation if $p>1$. The Crank–Nickolson method shows a steady decay for $p<1$ and steady oscillations above this value. For increasing nodal numbers, the limiting value of $p$ for stability of the Euler method is reduced. Thus for two nodal points it is 0.586 instead of 1, and for infinite nodal points it is 0.5. The initial value of $p$ for the onset of steady oscillations for the Crank–Nickolson method is also correspondingly reduced with increasing number of nodal points.

The accuracy of the Crank–Nickolson is better than either the explicit or pure implicit methods, except of course for large time intervals (large $p$ values), where only the pure implicit method is stable.

Finally, before returning to our example, let us briefly review the solution methods of Eqs. 9.4-23 and 9.4-29, through a small numerical example. Select $I=5$, $p=0.25$; Eq. 9.4-23 for the first time increment becomes

$$\begin{pmatrix}\Theta_1^*\\ \Theta_2^*\\ \Theta_3^*\\ \Theta_4^*\end{pmatrix}=\begin{pmatrix}0.5 & 0.5 & 0 & 0\\ 0.25 & 0.5 & 0.25 & 0\\ 0 & 0.25 & 0.5 & 0.25\\ 0 & 0 & 0.25 & 0.5\end{pmatrix}\begin{pmatrix}1\\1\\1\\1\end{pmatrix}$$

$$=\begin{pmatrix}(0.5\times 1)+(0.5\times 1)\\ (0.25\times 1)+(0.5\times 1)+(0.25\times 1)\\ (0.25\times 1)+(0.5\times 1)+(0.25\times 1)\\ (0.25\times 1)+(0.5\times 1)\end{pmatrix}=\begin{pmatrix}1\\1\\1\\0.75\end{pmatrix}$$

with $\Theta_5^*=0$ as required by the boundary condition.

The time interval selected by choosing $p=0.25$ is obtained from Eq. 9.4-25

$$\Delta t=b^2\frac{\Delta\tau}{\alpha_0}=p(\Delta\xi)^2\frac{b^2}{\alpha_0}=(0.25)(0.25)^2\frac{b^2}{\alpha_0}=0.0125\frac{b^2}{\alpha_0}$$

During the next time increment the column matrix $\boldsymbol{\Theta}$ becomes

$$\boldsymbol{\Theta}=\begin{pmatrix}1\\1\\1\\0.75\end{pmatrix}$$

and a new column matrix $\boldsymbol{\Theta}^*$ is calculated. Hence the calculation procedure is straightforward, and for large $I$ values it can be quickly carried out by a computer.

The Crank–Nickolson method is more involved. For the same numerical values, we obtain the following set of algebraic equations:

$$\begin{pmatrix}0.75 & 0.25 & 0 & 0\\ 0.125 & 0.75 & 0.125 & 0\\ 0 & 0.125 & 0.75 & 0.125\\ 0 & 0 & 0.125 & 0.75\end{pmatrix}\begin{pmatrix}1\\1\\1\\1\end{pmatrix}=\begin{pmatrix}1.25 & -0.25 & 0 & 0\\ -0.125 & 1.25 & -0.125 & 0\\ 0 & -0.125 & 1.25 & -0.125\\ 0 & 0 & -0.125 & 1.25\end{pmatrix}\begin{pmatrix}\Theta_1^*\\ \Theta_2^*\\ \Theta_3^*\\ \Theta_4^*\end{pmatrix}$$

or after multiplying the matrices, we get the following set of simultaneous equations:

$$\begin{aligned}1.25\,\Theta_1^*-0.25\,\Theta_2^* &= 1\\ -0.125\,\Theta_1^*+1.25\,\Theta_2^*-0.125\,\Theta_3^* &= 1\\ -0.125\,\Theta_2^*+1.25\,\Theta_3^*-0.125\,\Theta_4^* &= 1\\ -0.125\,\Theta_3^*+1.25\,\Theta_4^* &= 0.875\end{aligned}$$

There are a number of methods to solve simultaneous algebraic equations. One standard method is Gaussian elimination, whereby first we eliminate all terms below the main diagonal. In this case the first equation is multiplied by 0.1 and added to the second equation, thereby eliminating from that equation the $\Theta_1^*$ term. The same procedure is continued, resulting in the following set of equations:

$$\begin{aligned}
1.25\,\Theta_1^* - 0.25\,\Theta_2^* \qquad\qquad\qquad\qquad\qquad &= 1 \\
1.225\,\Theta_2^* - 0.125\,\Theta_3^* \qquad\qquad\quad &= 1.1 \\
1.237\,\Theta_3^* - 0.125\,\Theta_4^* &= 1.11224 \\
1.237\,\Theta_4^* &= 0.98739
\end{aligned}$$

This procedure led to a reduction of the unknown variables to two in each equation except the last, where we have one. Hence starting with $\Theta_4^*$ we can calculate all the unknowns in a straightforward manner. The result is:

$$\mathbf{\Theta}^* = \begin{pmatrix} 0.9996 \\ 0.9979 \\ 0.9798 \\ 0.7982 \end{pmatrix}$$

As before, the resulting $\mathbf{\Theta}^*$ becomes the new $\mathbf{\Theta}$ matrix in the following iteration. Clearly this method gives more accurate results, yet involves more labor.

In addition to the Gaussian elimination method, there are other direct methods, such as Cramer's rule and matrix inversion. These schemes result in a solution only after a finite number of steps. If the number of equations is large, indirect or iterative solution methods, such as the Gauss–Seidel iteration method and the method of relaxation, become more efficient (16).

We now return to the original problem with variable thermophysical properties and apply the Crank–Nickolson method of solution.

The right-hand side of Eq. 9.4-8 can be rewritten as

$$\frac{\partial}{\partial \xi}\left(k' \frac{\partial \Theta}{\partial \xi}\right) = \frac{\partial k'}{\partial \xi}\frac{\partial \Theta}{\partial \xi} + k' \frac{\partial^2 \Theta}{\partial \xi^2} \tag{9.4-32}$$

which in terms of finite differences becomes

$$\left(\frac{k'_{i+1} - k'_i}{\Delta \xi}\right)\left(\frac{\Theta_{i+1} - \Theta_i}{\Delta \xi}\right) + k'_i\left[\frac{\Theta_{i-1} - 2\Theta_i + \Theta_{i+1}}{(\Delta \xi)^2}\right] \tag{9.4-33}$$

Then by substituting Eq. 9.4-8, with the right-hand side replaced by 9.4-33 into Eq. 9.4-28 at times $\tau$ and $\tau + \Delta\tau$, we get

$$\mathbf{B}'\mathbf{\Theta} = \mathbf{C}'\mathbf{\Theta}^* \tag{9.4-34}$$

where the matrices $\mathbf{B}'$ and $\mathbf{C}'$ are, respectively,

$$\mathbf{B}'=\begin{pmatrix} \left[\frac{1}{\alpha'_1}-\left(\frac{k'_2-k'_1}{k'_1}\right)\frac{p}{2}-p\right] & \left[\left(\frac{k'_2-k'_1}{k'_1}\right)\frac{p}{2}+p\right] & & \\ \left(\frac{p}{2}\right) & \left[\frac{1}{\alpha'_2}-\left(\frac{k'_3-k'_2}{k'_2}\right)\frac{p}{2}-p\right] & \left[\left(\frac{k'_3-k'_2}{k'_2}\right)\frac{p}{2}+\frac{p}{2}\right] & \\ & & \ddots & \\ & & \left(\frac{p}{2}\right) & \left[\frac{1}{\alpha'_{I-1}}-\left(\frac{k'_I-k'_{I-1}}{k'_{I-1}}\right)\frac{p}{2}-p\right] \end{pmatrix} \tag{9.4-35}$$

and

$$\mathbf{C}'=\begin{pmatrix} \left[\frac{1}{\alpha'_1}+\left(\frac{k'_2-k'_1}{k'_1}\right)\frac{p}{2}+p\right] & -\left[\left(\frac{k'_2-k'_1}{k'_1}\right)\frac{p}{2}+p\right] & & \\ \left(-\frac{p}{2}\right) & \left[\frac{1}{\alpha'_2}+\left(\frac{k'_3-k'_2}{k'_2}\right)\frac{p}{2}+p\right] & -\left[\left(\frac{k'_3-k'_2}{k'_2}\right)\frac{p}{2}+\frac{p}{2}\right] & \\ & & \ddots & \\ & & -\left(\frac{p}{2}\right) & \left[\frac{1}{\alpha'_{I-1}}+\left(\frac{k'_I-k_{I-1}}{k'_{I-1}}\right)\frac{p}{2}+p\right] \end{pmatrix} \tag{9.4-36}$$

where $\alpha'_i$ and $k'_i$ are evaluated at the temperature corresponding to location $i$ at time $\tau$. The solution of Eq. 9.4-34 is not different in principle from that of Eq. 9.4-29, which has been demonstrated. Clearly, for constant thermophysical properties $\alpha'_i=1$ and $k'_i=1$; hence the matrices $\mathbf{B}'$ and $\mathbf{C}'$ of Eqs. 9.4-35 and 9.4-36 reduce to the matrices $\mathbf{B}$ and $\mathbf{C}$ in Eqs. 9.4-30 and 9.4-31.

Figure 9.10 presents some calculated temperature profiles using the Crank–Nickolson method solidification, in a flat injection mold, for $T_0=300°\text{F}$, $T_1=100°\text{F}$, and $T_m=235°\text{F}$ (17).

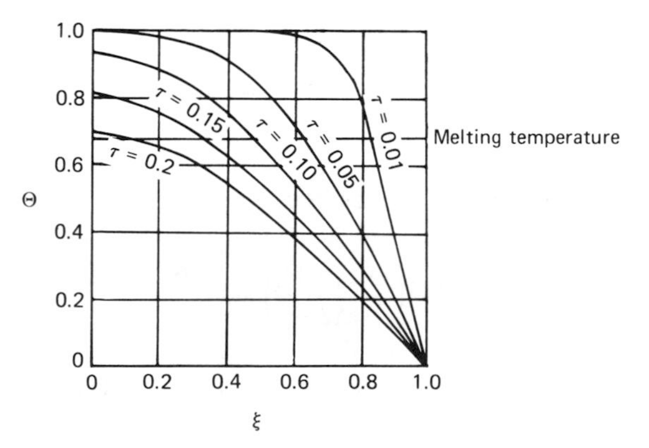


**Fig. 9.10** Temperature profiles during solidification of HDPE in an injection mold. [Reprinted with permission from C. Gutfinger, E. Broyer, and Z. Tadmor, *Polym. Eng. Sci.*, **15**, 515 (1975).]

Finally it should be noted that the same general procedure applies to other types of boundary conditions (e.g., convection and radiation) with the first and last equations in the coefficient matrix assuming different forms.

## 9.5 Moving Heat Sources

Conductive heating with moving heat sources was treated in detail by Rosenthal (18), in particular in relation to metal processing as in welding, machining, grinding, and continuous casting. In polymer processing we also encounter heat conduction problems with moving heat sources as well as heat sinks. The commonly practiced welding of polyvinyl chloride, the continuous dielectric sealing of polyolefins, the heating of films and thin sheets under intense radiation lamps, and in certain cases the heating or chilling of continuous films and sheets between rolls can serve as examples. These processes are usually steady or quasi-steady state, with heat introduced or removed at a "point" or along a "line." We examine one particular case to demonstrate the solution procedure.

### Example 9.4 *Continuous Heating of a Thin Sheet*

Consider a thin polymer sheet infinite in the $x$-direction, moving at a constant velocity $V_0$ in the negative $x$-direction (Fig. 9.11). The sheet exchanges heat with the surroundings, which is at $T = T_a$, by convection. At $x = 0$ there is a plane source of heat of intensity $q$ per unit cross-sectional area. Thus the heat source is moving relative to the sheet. It is more convenient, however, to have the coordinate system located at the source. Our objective is to calculate the axial temperature profile $T(x)$ and the intensity of the heat source to achieve a given maximum temperature. We assume that the sheet is thin, that temperature at any $x$ is uniform, and that the thermophysical properties are constant.

The energy equation for this problem reduces to

$$\rho C_p V_0 \frac{dT}{dx} = k \frac{d^2 T}{dx^2} - Q_v \tag{9.5-1}$$

where $Q_v$ is the heat exchanged with the surrounding per unit volume:

$$Q_v = \frac{hc}{A}[T(x) - T_a] \tag{9.5-2}$$

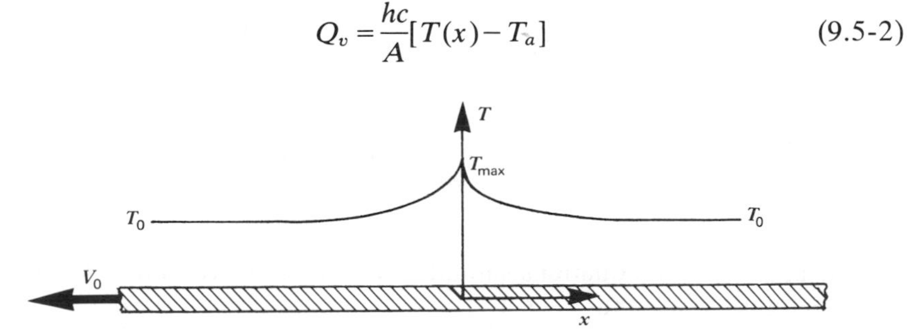


**Fig. 9.11** Heating of a moving thin sheet with a plane heat source.

where $c$ and $A$ are the perimeter and cross-sectional areas, respectively. Substituting Eq. 9.5-2 into Eq. 9.5-1 and using the "excess temperature" $T'(x) = T(x) - T_a$ instead of $T(x)$, we obtain

$$\frac{d^2T'}{dx^2} - \frac{V_0}{\alpha}\frac{dT'}{dx} - m^2T' = 0 \tag{9.5-3}$$

where

$$m = \left(\frac{hc}{kA}\right)^{1/2} \tag{9.5-4}$$

Eq. 9.5-3 is to be solved subject to the.boundary conditions

$$T'(\pm\infty) = 0 \tag{9.5-5}$$

Equation 9.5-3 is a linear second order differential equation that can be conveniently solved by defining a differential operator $D^n \equiv d^n/dx^n$. In the $D^n$ representation it is

$$\left(D^2 - \frac{V_0}{\alpha}D - m^2\right)T' = 0 \tag{9.5-6}$$

in which the differential operator behaves as though it were an algebraic polynominal. Since $T' \neq 0$, the expression in paretheses must equal zero, and solving for $D$ we get as roots:

$$D = \frac{V_0}{2\alpha} \pm \sqrt{m^2 + \left(\frac{V_0}{2\alpha}\right)^2} \tag{9.5-7}$$

The temperature profile is then

$$T'(x) = A_1 \exp\left[\left(\frac{V_0}{2\alpha} + \sqrt{m^2 + \left(\frac{V_0}{2\alpha}\right)^2}\right)x\right] + B_1 \exp\left[\left(\frac{V_0}{2\alpha} - \sqrt{m^2 + \left(\frac{V_0}{2\alpha}\right)^2}\right)x\right] \tag{9.5-8}$$

Since we cannot satisfy both boundary conditions 9.5-5 except for the trivial case $T' = 0$, we split our solution into two regions $x \geq 0$ and $x \leq 0$, resulting in the following solutions:

► $$T'(x) = B_1 \exp\left[-\left(\sqrt{m^2 + \left(\frac{V_0}{2\alpha}\right)^2} - \frac{V_0}{2\alpha}\right)x\right] \qquad x \geq 0 \tag{9.5-9}$$

and

► $$T'(x) = A_1 \exp\left[\left(\sqrt{m^2 + \left(\frac{V_0}{2\alpha}\right)^2} + \frac{V_0}{2\alpha}\right)x\right] \qquad x \leq 0 \tag{9.5-10}$$

Now at $x = 0$ both equations should yield the same, yet unknown maximum temperature $T'_{max}$; thus we get

$$A_1 = B_1 = T'_{max} = T'(0) \tag{9.5-11}$$

The value of $T'_{max}$ depends on the intensity of the heat source. Heat generated at the plane source is conducted in both the $x$ and $-x$ directions. The fluxes $q_1$ and $q_2$ in these respective directions are obtained from Eqs. 9.5-9 and 9.5-10:

$$q_1 = kT'_{max}\left(\sqrt{m^2+\left(\frac{V_0}{2\alpha}\right)^2}-\frac{V_0}{2\alpha}\right) \tag{9.5-12}$$

$$q_2 = -kT'_{max}\left(\sqrt{m^2+\left(\frac{V_0}{2\alpha}\right)^2}+\frac{V_0}{2\alpha}\right) \tag{9.5-13}$$

A heat balance at the interface requires

$$q = |q_1| + |q_2| \tag{9.5-14}$$

Substituting Eqs. 9.5-12 and 9.5-13 into Eq. 9.5-14 and solving for $T'_{max}$ gives:

► $$T'_{max} = \frac{q}{2k\sqrt{m^2+\left(\frac{V_0}{2\alpha}\right)^2}} \tag{9.5-15}$$

Thus the maximum excess temperature is proportional to the intensity of the source, and it drops with increasing speed $V_0$ and increases in thermal conductivity and the heat transfer coefficient. From Eqs. 9.5-9 and 9.5-10 we conclude that the temperature drops quickly in the positive $x$-direction as a result of the convection ($V_0 < 0$) of the solid into the plane source, and slowly in the direction of motion. Later in this chapter we encounter once again exponentially dropping temperatures in solids with convection—a frequent situation in melting configurations.

## 9.6 Sintering

Solid particles, when in contact with each other at elevated temperatures, tend to decrease the total surface area by coalescence. This process is called *sintering* (19). It is usually accompanied by a decrease in the total volume of the particulate bed. As Section 4.1 explained, a decrease in surface area brings about a decrease in (surface) free energy. Thus the surface tension is the driving force for the coalescence process. The sintering process proceeds in two distinct stages, first by developing interfaces and bridges between adjacent particles with little change in density, followed by a stage of *densification* in which the inter-particle cavities are eliminated (Fig. 9.12). It should be noted that sintering is a local phenomenon between adjacent particles involving viscous flow. The rate of the process is, therefore, greatly affected by the local temperature. Hence superimposed on the sintering process we usually have to deal with the overall heat transfer problem within the particulate system, where previously discussed solutions are applicable with the thermophysical properties replaced by "effective" values as discussed in Chapter 5.

The processing of metallic and ceramic powders by sintering is an old and well-developed technological activity. In polymer processing, melting by a sintering

**Fig. 9.12** A monolayer of 700 $\mu$ diameter PMMA beads during a sintering process at 203°C, ×50. (*a*) After 25 min. (*b*) After 55 min. (Reprinted with permission from M. Narkis, D. Cohen, and R. Kleinberger, "Sintering Behavior and Characterization of PMMA Particles," Department of Chemical Engineering, Israel Institute of Technology, Haifa.)

process is practiced in areas such as rotational molding (20, 21) and powder coating. Moreover, it provides the only practical way to process polytetrafluoroethylene, whose very high molecular weight precludes other common processing methods (22). Finally, high pressure compaction followed by sintering has been suggested for melting and shaping high temperature polymers such as polyimides and aromatic polyesters, as well as for physical mixtures of controlled composition distribution of more common polymers (23, 24).

The concept of viscous sintering was developed by Frenkel (25), who derived the following expression for the rate of coalescence of spherical adjacent particles:

$$\frac{x^2}{R} = \frac{2}{3}\frac{\Gamma}{\eta}t \qquad (9.6\text{-}1)$$

subject to $x/R < 0.3$ where $x$ is the neck radius (Fig. 9.13), $R$ is the radius of the particles, $\Gamma$ is the surface tension, and $\eta$ is the viscosity. This expression was applied successfully to glass and ceramic materials, but for polymeric materials Kuczynski

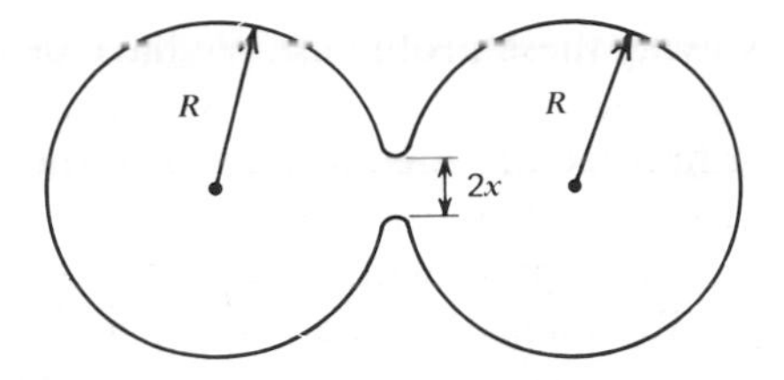


**Fig. 9.13** Schematic view of the first stage in the sintering process.

et al. (19), working with polymethylmethacrylate, found the experimental data to follow the following type of empirical equation:

$$\left(\frac{x^2}{R^{1.02}}\right)^p = F(T)t \tag{9.6-2}$$

where $t$ is sintering time, and $F(T)$ is a function of only the temperature. For $p = 1$, Eq. 9.6-2 reduces to a Frenkel type of equation. Kuczynski et al. derived this equation theoretically by assuming the melt to be non-Newtonian and to follow the power law constitutive equation. The result is

$$\left(\frac{x^2}{R}\right)^{1/n} = \frac{1}{2n}\left(\frac{8n\Gamma}{m}\right)^{1/n} t \tag{9.6-3}$$

where $n$ and $m$ are the power law model constants. Thus the parameter $p$ in Eq. 9.6-2 acquires rheological meaning. For $n = 1$ Eq. 9.6-3 reduces to the Frenkel equation as corrected by Eshelby (26). Yet the flow field during the coalescence process is probably not homogeneous nor isothermal; therefore a complete analysis of the coalescence stage would first require a detailed analysis of the kinematics of the flow field. Then the theoretical analysis should preferably be carried out with a viscoelastic constitutive equation, because viscoelastic effects, as suggested by Lonz (22), may play an important role in sintering of polymeric materials, and accounting for nonisothermal effects.

The coalescence stage is usually considered terminated when $x/R$ reaches a value of 0.5. For the densification stage that follows, Frenkel (25) suggested the following expression

$$\frac{r}{r_0} = 1 - \frac{\Gamma}{2\eta r_0} t \tag{9.6-4}$$

when $r_0$ is the initial radius of the approximately spherical cavity formed by the first stage and $r$ is the radius at time $t$.

As sintering proceeds and coalescence and densification occur, the overall heat conduction problem does not remain unaffected. Clearly the effective thermophysical properties change, influencing the overall temperature distribution, hence the local sintering problem as well.

## 9.7 Conduction Melting with Forced Melt Removal

Preceding sections have discussed the physical mechanisms by which thermal energy can be supplied to a solid polymer and have outlined some of the mathematical tools available for solving these problems. We have dealt with various aspects of "conduction melting without melt removal," which is generally applicable to melting a semifinished or finished product, as well as to the solidification processes following shaping. We have noted in most of the problems analyzed that heat fluxes and rates of melting diminish rapidly with time as the molten layer increases in thickness. It follows logically, then, that the rate of melting can be considerably increased by a *continuous removal of the molten layer formed.* This process, as Section 9.2 pointed out, not only leads to high rates of melting, but is the *essential element in creating a continuous steady source of polymer melt,* which in turn is the heart of the most important shaping methods of die forming, molding, calendering, and coating, as well as for preparing the preshaped forms for the secondary shaping operations. Removal of the melt, also discussed in Section 9.1, is possible in principle by two mechanisms: *drag induced flow* and *pressure induced flow* (Fig. 9.2*b*). In both cases the molten layer must be sheared, leading to viscous dissipation. The latter provides an additional important source of thermal energy for melting, the rate of which can be controlled externally either by the velocity of the moving boundary in drag induced melt removal, or the external force applied to squeeze the solid onto the hot surface, in pressure induced melt removal. In either case we convert external mechanical energy into heat. This source of heat is not negligible; it may even be the dominant or sole source in the melting process—for example, in the case of "autogenous" screw extrusion. Having two alternative sources of heat energy provides the processing design engineer with a great deal of flexibility. Finally, the continuous removal of melt has the added benefit of not exposing polymer melts to high temperature surfaces or regions for long times.

From a mathematical point of view, problems of conduction melting with forced melt removal are far more complex than ordinary conduction melting, because they involve the simultaneous solution of the momentum and energy equations. Moreover boundary conditions are often ill defined.

We analyze next, in some detail, the two important cases of forced drag and pressure melt removal. The former is the dominant melting mechanism in the single screw extruder, which is probably the most important processing machine at the present time. Chapter 10 develops the single screw geometry from first principles, and Chapter 12 analyzes the detailed melting mechanism in extruders, using the model developed in Section 9.8.

## 9.8 Drag Induced Melt Removal

We consider an infinite slab of isotropic homogeneous solid of width $W$, pushed against a moving hot plate (Fig. 9.14). A highly sheared, thin film of melt is formed between the solid and the plate, and this film is continuously removed. After a certain time, steady state conditions evolve; that is, velocity and temperature profiles become time independent. The problem is two-dimensional; that is, the temperature and velocity fields are functions of $x$ and $y$ only. No variations occur in the $z$-direction, which is infinite. The thickness of the melt film is very small at $x = 0$ and it increases in the positive $x$-direction, the shape of the melt film $\delta(x)$ being an a priori unknown function. Heat is conducted from the hot plate, which is at a constant temperature $T_0$, to the solid-melt interface at $T = T_m$, assuming that the

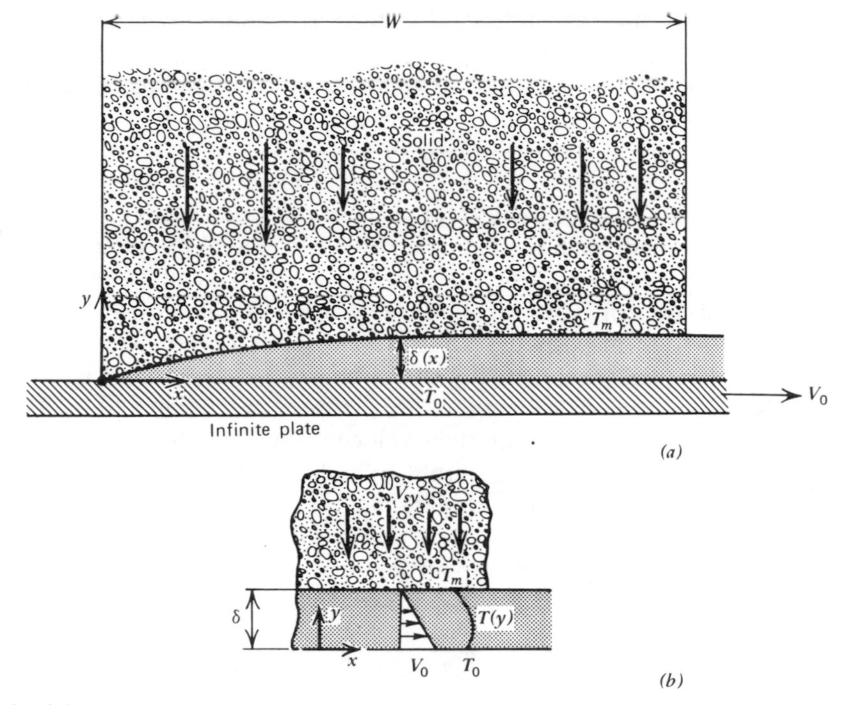


**Fig. 9.14** (*a*) Schematic representation of a slab of polymer melting on a hot moving surface. (*b*) Enlarged view of a portion of the melt film.

polymer is polycrystalline. As discussed in Section 9.2 amorphous polymers at $T_g$ do not change abruptly from brittle solids to viscous liquids. Thus the choice for "$T_m$" is not obvious. One can pretend that the transition is sharp and set an arbitrary level of temperature (larger than $T_g$) at which flow begins to occur. Alternatively, as suggested recently by Sundstrom and Lo (27), the glass transition temperature can be used together with the WLF equation 6.4-10.

We are seeking a solution for the rate of melting and the temperature distribution of the emerging melt. Clearly these variables will be functions of the physical properties of the solid, the plate temperature and velocity, and the width of the solid slab.

Tadmor et al. (28–30) first dealt with this problem in connection to the melting of polymers; they examined the melting mechanism of screw extruders. Later Vermeulen et al. (31), and Sundstrom et al., (32, 27) analyzed the problem both experimentally and theoretically, Mount (33) measured experimental rates of melting, and Pearson (34) analyzed the theoretical problem in detail. We follow Pearson's discussion.

In trying to analyze the detailed mechanism of this melting configuration, we first must consider the nature of the solid. For a perfectly rigid, incompressible body moving toward the interface without rotation, the rate of melting at the interface must be independent of the coordinate $x$, because the bulk velocity of the solid will be uniform across $x$. Hence $\delta(x)$, $P(x)$, and the velocity and temperature fields in the film must assume values that will satisfy this requirement as well as the equations of motion and energy with the appropriate boundary conditions. But in highly sheared thin films of very viscous polymers, a constant pressure assumption in the film is more appropriate. This in turn implies that at steady state conditions *the rate of melting may be generally a function of* $x$, although this variation may be small. *A variable melting rate requires the solid either to deform or rotate or do both.* Solid polymers, in particular in the form of a bed of compressed pellets or powder as encountered in polymer processing, can be considered to be deformable. The melt formed at the interface penetrates some of the voids between the particulate solids forming the bed, enabling sliding and rearrangement in the neighborhood of the interface. Through such a mechanism it is easy to visualize the continuously deforming solid concept. Thus the physical situation in this case would be one of a slowly deformable solid pushed against the moving hot plate. The solid interface has a small velocity in the negative $y$-direction which may slowly vary with $x$. Yet thc solid is rigid enough to sustain the shear stresses in the film and to prohibit the development of an $x$-direction interface velocity. We are now in a position to state the simplifying assumptions to the problem and specify the governing differential equations. The following assumptions are made:

1. Constant thermophysical properties.
2. Incompressible fluid.
3. No slip at the wall.
4. Power law (or Newtonian) fluid with temperature dependent viscosity:

$$m = m_0 \, e^{-a(T-T_m)} \tag{9.8-1}$$

5. Steady state conditions.
6. Gravitational forces are negligible.
7. Laminar flow prevails throughout.
8. The film thickness is much smaller than its width $\delta/W \ll 1$. This, together with the small Reynolds number in the film, justifies the use of the *lubrication approximation*. Moreover, the same considerations lead us to neglect exit effects (at $x = W$), and precise "entrance" conditions (at $x = 0$) need not be specified.

The equations of continuity and motion, respectively, reduce to

$$\frac{\partial v_x}{\partial x} + \frac{\partial v_y}{\partial y} = 0 \tag{9.8-2}$$

and

$$\frac{\partial P}{\partial x} = \frac{\partial \tau_{yx}}{\partial y} \tag{9.8-3}$$

Since we assume a pure drag flow in the film, Eq. 9.8-3 further reduces to

$$\frac{\partial \tau_{yx}}{\partial y} = 0 \tag{9.8-4}$$

Expressing the shear stress in terms of the local velocity gradient, Eq. 9.8-4 becomes

$$\frac{\partial}{\partial y}\left[e^{-a(T-T_m)}\left(-\frac{\partial v_x}{\partial y}\right)^n\right]=0 \tag{9.8-5}$$

Equation 9.8-5 can be integrated with respect to $y$ to give

$$-\left(\frac{\partial v_x}{\partial y}\right)=C_1\, e^{[a(T-T_m)/n]} \tag{9.8-6}$$

Thereby if $a=0$ (i.e., temperature independent viscosity), the velocity profile is linear for both Newtonian and power law fluids. If, however, $a\neq 0$, the local velocity profile becomes a function of the temperature. Since temperature varies sharply over $y$, we expect significant nonlinearity of the profile in the $y$-direction. Moreover, because of convection $T$ is also a (weaker) function of $x$, introducing a corresponding (weak) $x$ dependence of the velocity profile. Hence the equations of motion and energy must be solved simultaneously. The latter reduces to

$$\rho_m C_m\left(v_x\frac{\partial T}{\partial x}+v_y\frac{\partial T}{\partial y}\right)=k_m\frac{\partial^2 T}{\partial y^2}-\tau_{xy}\frac{\partial v_x}{\partial y} \tag{9.8-7}$$

where $\rho_m$, $C_m$, and $k_m$ are the thermophysical properties of the polymer melt, with heat conduction in the $x$-direction assumed to be much smaller than conduction in the $y$-direction, and further assuming that the only significant contribution to viscous dissipation is that originating from the $\tau_{yx}$ component of the stress tensor. Next we specify the boundary conditions in the film

$$T(0)=T_0, \qquad v_x(0)=V_0, \qquad v_y(0)=0 \tag{9.8-8a}$$

$$T(\delta)=T_m, \qquad v_x(\delta)=0 \tag{9.8-8b}$$

The velocity $v_y(\delta)$ at any position $x$ is determined by the rate of melting at the interface, to be obtained from the following heat balance:

$$k_m\left(-\frac{\partial T}{\partial y}\right)_{y=\delta} = \rho_m[-v_y(\delta)]\lambda + k_s\left(-\frac{\partial T_s}{\partial y}\right)_{y=\delta} \tag{9.8-9}$$

| Rate of heat conducted into the interface per unit interface area | Rate of melting at the interface per unit interface area times the heat of fusion | Rate of heat conducted out of the interface per unit interface area |
|---|---|---|

where $\lambda$ is the heat of fusion and $k_s$ and $T_s$ are the thermal conductivity and temperature, respectively, of the solid. The term on the left-hand side is the rate of heat conducted from the hot film into the interface. For melting to take place, $\partial T/\partial y<0$. This term is therefore positive and provides the heat source for melting, which as we see on the right-hand side is used for two purposes: to melt the polymer at the interface $T=T_m$, and to heat the polymer to the melting point.

The last term on the right-hand side can be obtained by solving the temperature profile in the solid bed. Consider a small $x$-direction portion of the film and solid (Fig. 9.14$b$). We assume the solid to occupy the region $y>\delta$ (where $\delta$ is the local film thickness) and to be moving with constant velocity $v_{sy}$ into the interface. The problem thus reduces to a one-dimensional steady heat conduction problem with convection. In the solid, a steady exponentially dropping temperature profile develops. The problem is similar to that in Section 9.5. The equation of energy reduces to

$$\rho_s C_s v_{sy} \frac{\partial T_s}{\partial y} = k_s \frac{\partial^2 T_s}{\partial y^2} \tag{9.8-10}$$

where $\rho_s$, $C_s$, and $k_s$ are the thermophysical properties of the solid polymer. Equation 9.8-10 can be easily solved, with the boundary conditions $T_s(\delta) = T_m$ and $T_s(\infty) = T_{s0}$, to give the following temperature profile:

$$T = T_{s0} + (T_m - T_{s0}) \exp\left[\frac{v_{sy}(y-\delta)}{\alpha_s}\right] \tag{9.8-11}$$

The velocity $v_{sy} < 0$, hence Eq. 9.8-11 satisfies both boundary conditions. The rate of heat conduction out of the interface, noting that $v_{sy}\rho_s = v_y(\delta)\rho_m$, is

$$-k_s\left(\frac{\partial T}{\partial y}\right)_{y=\delta} = -(T_m - T_{s0})v_y(\delta)\rho_m C_s \tag{9.8-12}$$

Thus Eq. 9.8-9 can now be written as

$$k_m\left(\frac{\partial T}{\partial y}\right)_{y=\delta} = \rho_m v_y(\delta)\lambda^* \tag{9.8-13}$$

where

$$\lambda^* = \lambda + C_s(T_m - T_{s0}) \tag{9.8-14}$$

Thus $\lambda^*$ is the total heat energy required to bring *solid* from an initial temperature $T_{s0}$ to $T_m$ and to *melt* it at that temperature. Sundstrom and Young (32) solved this set of equations numerically after converting the partial differential equations into ordinary differential equations by similarity techniques. Pearson (34) used the same technique but obtained a number of useful solutions to simplified cases. He also used dimensionless variables, which aid in the physical interpretation of the results, as shown below:

$$\Theta = \frac{T - T_m}{T_0 - T_m} \tag{9.8-15}$$

$$\xi = \frac{x}{W} \quad \text{and} \quad \eta = \frac{y}{\delta} \tag{9.8-16}$$

$$u_x = \frac{v_x}{V_0} \quad \text{and} \quad u_y = \frac{v_y}{V_0(\delta_0/W)} \tag{9.8-17}$$

where the meaning of $\delta_0$ will be clarified below.

We first rewrite the boundary conditions

$$\Theta(0) = 1, \qquad u_x(0) = 1, \qquad u_y(0) = 0 \tag{9.8-18a}$$

$$\Theta(1) = 0, \qquad u_x(1) = 0 \tag{9.8-18b}$$

The melting condition at the interface (Eq. 9.8-13) reduces to

$$\frac{k_m(T_0 - T_m)W}{\lambda^* \rho_m V_0 \delta_0^2}\left(\frac{\partial \Theta}{\partial \eta}\right)_{\eta=1} = \frac{\delta}{\delta_0} u_y(1) \tag{9.8-19}$$

This relationship provides us with a reasonable choice of $\delta_0$. Since this boundary condition determines the physical process, the dimensionless group $k_m(T_0 - T_m)W/\lambda^* \rho_m V_0 \delta_0^2$ should be of the order of 1. Hence we can choose $\delta_0$ as

$$\delta_0 = \left(\frac{k_m(T_0 - T_m)W}{\lambda^* \rho_m V_0}\right)^{1/2} \tag{9.8-20}$$

As we see later, $\delta_0$ is not merely an arbitrary scaling (normalizing) factor; it is also of the order of the film thickness, provided viscous dissipation or convection is not too significant to the process.

Now we can rewrite the transport equation in dimensionless form as follows. The continuity equation is

$$\frac{\partial u_x}{\partial \xi} - \eta \frac{\dot{\delta}}{\delta} \frac{\partial u_x}{\partial \eta} + \frac{\delta_0}{\delta} \frac{\partial u_y}{\partial \eta} = 0 \tag{9.8-21}$$

where $\dot{\delta} = d\delta/d\xi$.

Details of the derivation of Eq. 9.8-21 are as follows. Substituting $u_x$ and $u_y$ from Eq. 9.8-17 into the equation of continuity results in

$$V_0 \frac{\partial u_x}{\partial x} + \frac{V_0 \delta_0}{W} \frac{\partial u_y}{\partial x} = 0$$

Next we want to rewrite the partial differentials in terms of the new variables $\eta$ and $\xi$. We recall that $u_x(\xi, \eta)$, $u_y(\xi, \eta)$, and $\xi = F_1(x)$ and $\eta = F_2(x, y)$. The $x$ dependence in $\eta$ is due to $\delta(x)$. Hence we can write

$$\frac{\partial u_x}{\partial x} = \frac{\partial u_x}{\partial \xi}\frac{\partial \xi}{\partial x} + \frac{\partial u_x}{\partial \eta}\frac{\partial \eta}{\partial x} = \frac{1}{W}\frac{\partial u_x}{\partial \xi} - \frac{y}{\delta^2}\frac{\partial \delta}{\partial x}\frac{\partial u_x}{\partial \eta}$$

$$= \frac{1}{W}\frac{\partial u_x}{\partial \xi} - \frac{\eta}{W}\frac{\dot{\delta}}{\delta}\frac{\partial u_x}{\partial \eta}$$

Similarly we obtain

$$\frac{\partial u_y}{\partial y} = \frac{\partial u_y}{\partial \xi}\frac{\partial \delta}{\partial y} + \frac{\partial u_y}{\partial \eta}\frac{\partial \eta}{\partial y} = \frac{1}{\delta}\frac{\partial u_y}{\partial \eta}$$

The dimensionless form of the equation of motion is

$$\frac{\partial}{\partial \eta}\left[e^{b\Theta}\left(-\frac{\partial u_x}{\partial \eta}\right)^n\right] = 0 \tag{9.8-22}$$

where

$$b = -a(T_0 - T_m) \tag{9.8-23}$$

Finally the equation of energy using the definition of $\delta_0$ becomes

$$M^{-1}\left[u_x \frac{\partial \Theta}{\partial \xi} - u_x \eta \frac{\dot{\delta}}{\delta}\frac{\partial \Theta}{\partial \eta} + u_y \frac{\delta_0}{\delta}\frac{\partial \Theta}{\partial \eta}\right]$$

$$= \left(\frac{\delta_0}{\delta}\right)^2 \frac{\partial^2 \Theta}{\partial \eta^2} + Br\left(\frac{\delta_0}{\delta}\right)^{n+1} e^{b\Theta}\left(-\frac{\partial u_x}{\partial \eta}\right)^{n+1} \tag{9.8-24}$$

where

$$M = \frac{\lambda^*}{C_m(T_0 - T_m)} \tag{9.8-25}$$

and

$$\mathrm{Br} = \frac{m_0 V_0^{(3n+1)/2} \rho_m^{(n-1)/2} \lambda^{*(n-1)/2}}{(T_0 - T_m)^{(n+1)/2} k_m^{(n+1)/2} W^{(n-1)/2}} \tag{9.8-26}$$

In these equations Br is a modified Brinkman number, which is a measure of the extent to which viscous heating is important, and $M$ measures the ratio of heat energy needed to melt the polymer, as compared to that needed to heat the melt to $T_0$. If the latter is small, $M$ will be large and the convection terms in the energy equation can be neglected. The dimensionless parameter $b$ measures the significance of temperature dependence of the viscosity over the temperature range considered (flow activation energy).

Complete solution of the set of equations above is difficult, as pointed out earlier. In addition to the numerical solution (32), Pearson (34) proposed a heuristic approach. Insight into the nature of melting with drag forced removal can be obtained, however, by considering some special cases that lead to analytical closed form solutions. These simplified cases per se represent very useful solutions to the modeling of processing methods.

### *Newtonian Fluid with Temperature Independent Viscosity and Negligible Convection*

For a Newtonian fluid close to isothermal conditions ($n = 1, |b| \ll 1$) and with convection neglected ($M \gg 1$), the equation of motion becomes

$$\frac{\partial^2 u_x}{\partial \eta^2} = 0 \tag{9.8-27}$$

which for the boundary conditions stated by Eq. 9.8-18 has the solution

$$u_x = 1 - \eta \tag{9.8-28}$$

The equation of energy, which for this case can be solved independently, reduces to

$$\frac{\partial^2 \Theta}{\partial \eta^2} + \mathrm{Br}\left(-\frac{\partial u_x}{\partial \eta}\right)^2 = 0 \tag{9.8-29}$$

Substituting Eq. 9.8-28 into Eq. 9.8-29 followed by integration yield the temperature profile

$$\Theta = 1 - \eta + \frac{\mathrm{Br}}{2}\eta(1 - \eta) \tag{9.8-30}$$

The mean temperature $\bar{\Theta}$ is obtained from Eq. 9.8-30 as follows:

$$\bar{\Theta} = \frac{\int_0^1 u_x \Theta \, d\eta}{\int_0^1 u_x \, d\eta} = \frac{2}{3} + \frac{\mathrm{Br}}{12} \tag{9.8-31}$$

Now we can solve Eq. 9.8-19 for $u_y(1)$ by substituting from Eq. 9.8-30 $(\partial\Theta/\partial\eta)_{\eta=1} = -(1+\mathrm{Br}/2)$ to obtain

$$u_y(1) = -\frac{\delta_0}{\delta}\left(1+\frac{\mathrm{Br}}{2}\right) \tag{9.8-32}$$

We finally turn to the equation of continuity and integrate it over $\eta$, after substituting from Eq. 9.8-28 $\partial u_x/\partial\eta = -1$ and noting that $\partial u_x/\partial\xi = 0$, to obtain

$$u_y(1) = -\frac{1}{2}\frac{\dot{\delta}}{\delta_0} \tag{9.8-33}$$

Combining Eqs. 9.8-32 and 9.8-33 and integrating results in the film profile $\delta(\xi)$

► $$\delta = \delta_0\sqrt{(4+2\mathrm{Br})\xi} \tag{9.8-34}$$

We have obtained the important result that with convection neglected, the film thickness is proportional to square root of the distance. The rate of melting (per unit width) is now given by

$$w_L(x) = \rho_m V_0 \delta \int_0^1 u_x\, d\eta = \frac{V_0\delta}{2}\rho_m \tag{9.8-35}$$

By substituting Eq. 9.8-34 into 9.8-35 with $\xi = 1$; and $\delta_0$ from Eq. 9.8-20 we obtain

► $$w_L = \left[V_0^2\delta_0^2\rho_m^2\left(1+\frac{\mathrm{Br}}{2}\right)\right]^{1/2} = \left[\frac{V_0\rho_m k_m(T_0-T_m)}{\lambda^*}\left(1+\frac{\mathrm{Br}}{2}\right)W\right]^{1/2}$$

$$= \left[\frac{V_0\rho_m[k_m(T_0-T_m)+\mu V_0^2/2]W}{\lambda^*}\right]^{1/2} \tag{9.8-36}$$

The physical meaning of the various terms becomes now evident. The square bracket contains the sum of heat conduction and viscous dissipation terms. The numcrator is the heat energy needed to heat the solid from $T_{s0}$ to melt at $T_m$. The rate of melting also increases proportionally with the square root of the plate velocity and slab width. Yet an increase in plate velocity also increases the viscous dissipation.

In this expression we have neglected convection in the film. Tadmor et al. (29, 30) accounted approximately for convection by including in $\lambda^*$ the heat needed to bring the melt from $T_m$ to the mean melt temperature

$$\lambda^{**} = \lambda + C_s(T_m - T_{s0}) + C_m(T_0 - T_m)\bar{\Theta} \tag{9.8-37}$$

Furthermore, by carrying out a mental exercise of "removing" the newly melted material from the interface, "carrying" it to $\xi = 0$, and allowing it to flow into the film at that point, the film thickness will stay constant and the resulting effect will be a reduction of $w_L$ in Eq. 9.8-36 by a factor of $\sqrt{2}$.

### Power Law Model Fluid with Temperature Dependent Viscosity

Both shear thinning and temperature dependence of viscosity strongly affect the melting rate. Their effect on the rate of melting can be estimated by considering a case in which convection is neglected and viscous dissipation is low enough to permit the assumption that the viscosity variation across the film is determined by a linear temperature profile.

$$\Theta = 1 - \eta \tag{9.8-38}$$

The equation of motion, 9.8-22, reduces to

$$\frac{\partial}{\partial \eta}\left[e^{b[1-\eta]}\left(-\frac{\partial u_x}{\partial \eta}\right)^n\right] = 0 \tag{9.8-39}$$

Equation 9.8-39 can be solved for the local velocity profile $u_x(\eta)$

$$u_x = \frac{e^{b'\eta} - e^{b'}}{1 - e^{b'}} \tag{9.8-40}$$

where

$$b' = \frac{b}{n} = -\frac{a(T_0 - T_m)}{n} \tag{9.8-41}$$

Clearly, $b'$ is a dimensionless number that takes into account both the temperature and shear rate viscosity dependence.

The equation of energy, 9.8-24, reduces in this case to

$$\frac{\partial^2 \Theta}{\partial \eta^2} + \text{Br}\left(\frac{\delta_0}{\delta}\right)^{n-1} e^{b(1-\eta)}\left(-\frac{\partial u_x}{\partial \eta}\right)^{n+1} = 0 \tag{9.8-42}$$

Substituting Eq. 9.8-40 into Eq. 9.8-42, followed by integration, yields

$$\Theta = (1-\eta) + \text{Br}\left(\frac{\delta_0}{\delta}\right)^{n-1}\left(\frac{b'}{1-e^{-b'}}\right)^{n+1}\left(\frac{e^{-b'}}{b'^2}\right)[1 - e^{b'\eta} - \eta(1 - e^{b'})] \tag{9.8-43}$$

As in the Newtonian case, we solve Eq. 9.8-19 for $u_y(1)$, after obtaining $(\partial\Theta/\partial\eta)_{\eta=1}$ from Eq. 9.8-43

$$u_y(1) = -\left(\frac{\delta_0}{\delta}\right)\left[1 + \text{Br}\left(\frac{\delta_0}{\delta}\right)^{n-1}\left(\frac{b'}{1-e^{-b'}}\right)\left(\frac{b' - 1 + e^{-b'}}{b'^2}\right)\right] \tag{9.8-44}$$

Finally, the equâtion of continuity, Eq. 9.8-21, with $\partial u_x/\partial \xi = 0$ and subsequent to substituting $\partial u_x/\partial \eta$ from Eq. 9.8-40, results in

$$-\eta\left(\frac{\dot{\delta}}{\delta_0}\right)\left(\frac{b' e^{b'\eta}}{1 - e^{b'}}\right) + \frac{\partial u_y}{\partial \eta} = 0 \tag{9.8-45}$$

which is integrated to give

$$u_y(1) = \frac{\dot{\delta}}{\delta_0}\left[\frac{1}{b'(1-e^{b'})}\right][e^{b'}(b'-1)+1] \tag{9.8-46}$$

Combining Eqs. 9.8-44 and 9.8-46 results in a differential equation for $\delta$

$$\delta\frac{d\delta}{d\xi}=\frac{-\delta_0^2\left[1+\mathrm{Br}\left(\frac{\delta_0}{\delta}\right)^{n-1}\left(\frac{b'}{1-e^{-b'}}\right)^{n+1}\left(\frac{b'-1+e^{-b'}}{b'^2}\right)\right.}{\frac{e^{b'}(b'-1)+1}{b'(1-e^{b'})}} \tag{9.8-47}$$

An approximate solution of Eq. 9.8-47 can be obtained if a mean $\delta$ value is assumed in the term $(\delta_0/\delta)^{n-1}$. This is a weak dependence of the viscous dissipation term on $\delta$. The resulting melt film profile is

$$\blacktriangleright \quad \delta=\delta_0\left\{\frac{4\left[1+\mathrm{Br}\left(\frac{\delta_0}{\delta}\right)^{n-1}\left(\frac{b'}{1-e^{-b'}}\right)^{n+1}\left(\frac{b'-1+e^{-b'}}{b'^2}\right)\right]\xi}{U_2}\right\}^{1/2} \tag{9.8-48}$$

where

$$\blacktriangleright \quad U_2=2\frac{1-b'-e^{-b'}}{b'(e^{-b'}-1)} \tag{9.8-49}$$

By substituting the expressions of $\delta_0$ and Br from Eqs. 9.8-20 and 9.8-26, respectively, Eq. 9.8-48 can be written as

$$\blacktriangleright \quad \delta=\left\{\frac{2[2k_m(T_0-T_m)+U_1]x}{U_2\rho_m V_0\lambda^*}\right\} \tag{9.8-50}$$

where

$$\blacktriangleright \quad U_1=\frac{2m_0V_0^{n+1}}{(\bar{\delta})^{n-1}}\left(\frac{b'}{1-e^{-b'}}\right)^{n+1}\left(\frac{b'-1+e^{-b'}}{b'^2}\right) \tag{9.8-51}$$

The rate of melting (per unit width) is given by

$$w_L(x)=\rho_m V_0\delta\int_0^1 u_x\,d\eta=\frac{V_0\delta\rho_m}{2}U_2 \tag{9.8-52}$$

And substituting $\delta$ from Eq. 9.8-50 into Eq. 9.8-52 gives

$$\blacktriangleright \quad w_L(x)=\left\{\frac{\rho_m V_0 U_2[k_m(T_0-T_m)+U_1/2]x}{\lambda^*}\right\}^{1/2} \tag{9.8-53}$$

Thus the physical significance of $U_2$ and $U_1$ becomes evident. The former reflects the reduction ($U_2<1$) of the rate of melt removal of the film by drag flow as a result of temperature dependence and shear thinning of the viscosity, whereas $U_1/2$ is the rate of viscous dissipation (per unit width) in the melt film. The relative significance of conduction and dissipation for melting is obtained by comparing the two terms in the square brackets in Eq. 9.8-53.

If convection is to be accounted for by the same approximate method as described in the previous Newtonian case, then $\lambda^*$ in Eq. 9.8-53 is replaced by $\lambda^{**}$, which is given in Eq. 9.8-37, and $w_L(x)$ given in Eq. 9.8-53 is reduced by a factor of $\sqrt{2}$. Finally the mean temperature of the film

$$\bar{\Theta}\equiv\frac{\int_0^1 u_x\Theta\,d\eta}{\int_0^1 u_x\,d\eta} \tag{9.8-54}$$

is obtained by substituting Eqs. 9.8-38 and 9.8-49 into Eq. 9.8-54

$$\bar{\Theta} = \frac{b'/2 + e^{-b'}(1 + 1/b') - 1/b'}{b' + e^{-b'} - 1} \tag{9.8-55}$$

This is an approximate expression because, for the sake of simplicity, a linear temperature profile was used rather than Eq. 9.8-43.

The expressions above were applied to the solution of the melting problem in screw extruders (29, 30) as discussed in Chapter 12.

### Example 9.5 *Drag Induced Melt Removal Melting*

The rate of melting of a 2×2 in. block of solid HDPE on a hot rotating drum was measured by Sundstrom and Young (32). Their results appear in Fig. 9.15.

(*a*) Analyze the effects of drum speed and temperature in light of the previously derived theoretical models.

(*b*) Calculate the rate of melting at a drum speed of 1 in./s on a 168°C drum, using a Newtonian model and compare it to the experimental value.

(*c*) Repeat step *b* with a power law model with a linear temperature profile in the melt film.

The rheological properties of the HDPE used in the experiments follow a power law model (32)

$$\eta = 4.0334 \times 10^3 \, e^{-0.010872(T-127)} \dot{\gamma}^{-0.547}$$

where $\eta$ is the non-Newtonian viscosity ($N \cdot s/m^2$), $T$ is the temperature (°C), and $\dot{\gamma}$ is the shear rate ($s^{-1}$). The power law exponent is $n = 0.453$. The melting point is (32) 127°C. The heat of fusion (Table 5.7) is 218 kJ/kg. The specific heat of the solid polymer (Table 5.8) is 2.3 kJ/kg · °C and that of the

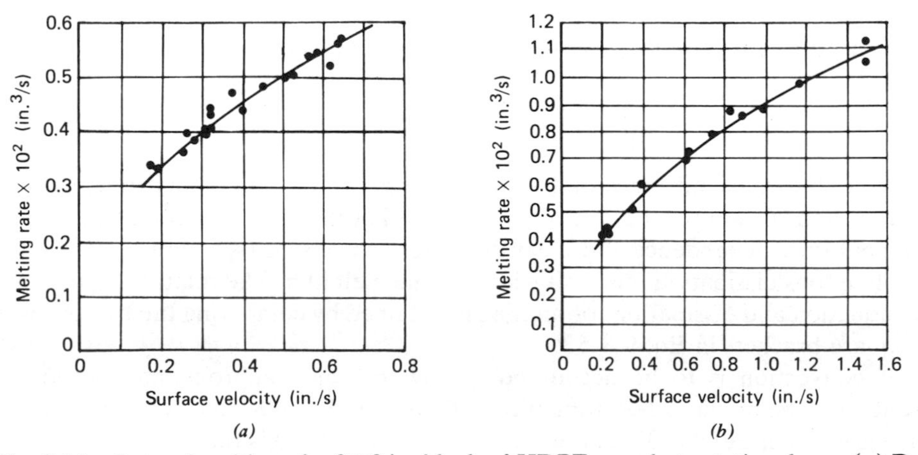


**Fig. 9.15** Rate of melting of a 2×2 in. block of HDPE on a hot rotating drum. (*a*) Drum temperature at 154°C. (*b*) Drum temperature at 168°C. Rate of melting measured in volume of displaced solid. [Reprinted with permission from D. H. Sundstrom and C. Young, *Polym. Eng. Sci.*, **12**, 59 (1972).]

melt (29) is 2.512 kJ/kg · °C. The thermal conductivity of HDPE melt as a function of temperature is (35)

$$k = 0.05736 + 0.0010467T$$

where $k$ is in W/m · °C and $T$ is the temperature (°C). Finally, the density of the solid polymer is 955 kg/m$^3$ and that of the melt (29) is 776 kg/m$^3$.

*Solution*: (*a*) The first step is to evaluate the relative significance of heat conduction and viscous dissipation. This is provided by the Brinkman number in Eq. 9.8-26, which for a Newtonian liquid reduces to

$$\text{Br} = \frac{\mu V_0^2}{k_m(T_0 - T_m)}$$

An estimate of the melt viscosity can be obtained from the power law expression above, assuming a shear rate of 50 s$^{-1}$ and taking a mean temperature of (168 + 127)/2 = 147.5°C (We later check whether these assumptions are acceptable.)

$$\mu = (4.0334 \times 10^3)\, e^{-0.010872(147.5-127)}\, (50)^{-(0.547)} = 379.8 \text{ N·s/m}^2$$

The tangential velocity of the drum selected is 1 in./s, or $V_0 = 0.0254$ m/s, and the thermal conductivity at the mean temperature is 0.212 W/m·°C, thus

$$\text{Br} = \frac{(379.8)(0.0254)^2}{(0.212)(168\text{–}127)} = 0.0282$$

Clearly, viscous dissipation is not significant in the experimental range given for the 168°C drum temperature experiments. Neither is it significant for the lower drum temperature experiments, which were conducted at lower drum speeds. It follows from the theoretical models (Eqs. 9.8-36 and 9.8-53) that the rate of melting in this case is proportional to the square root of drum speed and temperature drop

$$w_L \propto \sqrt{V_0(T_0 - T_m)}$$

It is easy to verify that the curves in Fig. 9.15 follow well the predicted increase in rate of melting with drum speed. For example, the predicted rate of melting at 1.6 in./s from the corresponding value at 0.2 in./s is $0.4\sqrt{1.6/0.2} = 1.13$ in.$^3$/s, which is very close to the measured value. Similarly, selecting a fixed drum speed of 0.5 in./s, the measured rate of melting at 154°C is 0.5 in.$^3$/s. The predicted value at 168°C is $0.5\sqrt{(168-127)/(154-127)} = 0.616$ in.$^3$/s, which once again is very close to the measured value.

(*b*) The rate of melting is evaluated from Eq. 9.8-36. First, however, the viscosity calculation is reexamined. This is done by calculating the film thickness from Eqs. 9.8-20 and 9.8-34. The former gives $\delta_0$ with $W = 0.0508$ m and with $\lambda^*$ calculated from Eq. 9.8-14

$$\lambda^* = 218 \times 10^3 + 2.3 \times 10^3(127 - 25) = 452.6 \times 10^3 \text{ J/kg}$$

Thus

$$\delta_0 = \left[\frac{(0.212)(168-127)(0.0508)}{(452.6\times 10^3)(776)(0.0254)}\right]^{1/2} = 2.225\times 10^{-4}\ \text{m}$$

and the maximum film thickness at $\xi = 1$ from Eq. 9.8-34 is

$$\delta_{max} = 2.225\times 10^{-4}\sqrt{(4)+(2)(0.0282)} = 4.481\times 10^{-4}\ \text{m}$$

The mean film thickness is $3.353\times 10^{-4}$ m and the mean shear rate is $0.0254/3.353\times 10^{-4} = 76\ \text{s}^{-1}$. The mean temperature is obtained from Eq. 9.8-31

$$\bar{\Theta} = \frac{2}{3} + \frac{0.0282}{12} = 0.669$$

Hence, $\bar{T} = 0.669(168-127)+127 = 154.4$. Repeating the calculations with the viscosity evaluated at $76\ \text{s}^{-1}$ and 154°C temperature, and with thermal conductivity of 0.218 W/m · °C, results in a viscosity of 281 N·s/m$^2$, Br = 0.203, $\delta_0 = 2.256\times 10^{-4}$ m, a mean film thickness of $3.495\times 10^{-4}$m, a mean shear rate of 73 s$^{-1}$, and a mean temperature of 154 °C.

Using these values, the rate of melting is calculated from Eq. 9.8-36

$$w_L = [(0.0254)^2(2.256\times 10^{-4})^2(776)^2(1+0.0203/2)]^{1/2}$$
$$= 4.469\times 10^{-3}\ \text{kg/m·s}.$$

The rate of melting for the whole block is $(4.469\times 10^{-3})(0.0508) = 2.27\times 10^{-4}$ kg/s, which is equivalent to 0.0145 in.$^3$/s (note that the volume measured by Sundstrom and Young is the displaced solid). Comparing this result with the measured value of 0.009 in.$^3$/s indicates that the Newtonian model overestimates the rate of melting by about 60%. In the model used, the effect of convection in the film was neglected. By accounting for convection as discussed earlier, the rate of melting is given by

$$w_L = \left[\frac{V_0\rho_m[k_m(T_0-T_m)+\mu V_0^2/2]W}{2[\lambda^* + C_m(T_0-T_m)\bar{\Theta}]}\right]^{1/2}$$

$$= \left\{\frac{(0.0154)(776)[(0.218)(168-127)+(281)(0.0254)^2/2](0.0508)}{2[(452.6\times 10^3)+(2.512\times 10^3)(168-127)(0.669)]}\right\}^{1/2}$$

$$= 2.945\times 10^{-3}\ \text{kg/m·s}$$

which results in a total rate of melting of 0.00956 in.$^3$/s. This is only 6% above the measured value.

(*c*) To calculate the rate of melting from Eq. 9.8-53 we first calculate $b'$, $U_1$, and $U_2$ as follows:

$$b' = -\frac{(0.010872)(168-127)}{(0.453)} = -0.984$$

From Eq. 9.8-49 $U_2$ is obtained

$$U_2 = (2)\frac{(-0.984)-(1)+e^{0.984}}{(-0.984)(1-e^{0.984})} = 0.839$$

which indicates that the reduction in drag removal due to temperature dependence of viscosity is 16%. Finally, $U_1$ is obtained from Eq. 9.8-51 using the previously estimated mean film thickness

$$U_1 = \frac{(2)(4.0334\times 10^3)(0.0254)^{1.453}}{(3.495\times 10^{-4})^{-0.547}}\left(\frac{0.984}{e^{0.984}-1}\right)^{1.453}\left(\frac{(-0.984)-1+e^{0.984}}{(-0.984)^2}\right)$$

$$= 0.1644 \text{ J/s·m}$$

Substituting these values into Eq. 9.8-53, with $\lambda^*$ replaced by $\lambda^{**}$ and a factor of 2 in the demoninator to account for convection, and with $\bar{\Theta}$ from Eq. 9.8-55, gives

$$w_L = \left\{\frac{(0.0754)(776)(0.839)[(0.218)(168-127)+(0.1644)/(2)]0.0508}{(2)[(452.6\times 10^3)+(2.512\times 10^3)(168-127)(0.695)}\right\}^{1/2}$$

$$= 2.6885\times 10^{-3} \text{ kg/m} \cdot \text{s}$$

which is equivalent to a total rate of melting of 0.00872 in.$^3$/s, or about 3% below the measured value.

The good agreements between the predictions and the measured rate of melting is to some degree fortuitous because all the thermophysical properties were selected from the literature rather than measured on the particular grade of HDPE used in the experiments. Thermophysical property data may vary for the same polymer over a relatively broad range. In addition, no doubt, experimental errors were also involved in the measured data, and one cannot expect a perfect agreement. Nevertheless, it is reasonable to conclude that the theoretical models discussed in this section predict correctly the change in melting rate with changing experimental conditions. They provided good estimates of the rate of melting. Incorporating into the model both the effect of convection in the film and the temperature dependence of the viscosity improves the agreement between predictions and experimental measurements. It should be noted, however, that experimental conditions were such that viscous dissipation was insignificant and the temperature drop across the film was relatively small. Consequently, non-Newtonian effects, and effects due to temperature dependence of viscosity, were less significant than were convection effects. This may not be the case in many practical situations, in particular with polymers whose viscosity is more temperature sensitive than is that of HDPE.

## 9.9 Pressure Induced Melt Removal

In the pressure induced process the melt is being removed by the squeezing action of the solid on the melt; hence the force by which the solids are pushed against the hot surface becomes the dominant rate controlling variable. This melting process is less important in polymer processing than the drag removal process. Nevertheless, as Stammers and Beek (36) point out, in manufacturing certain synthetic fibers (e.g., polyester yarns) the polymer is melted on a melting grid; the melting process on such a melting grid is that of pressure removal of the melt. Stammers and Beek

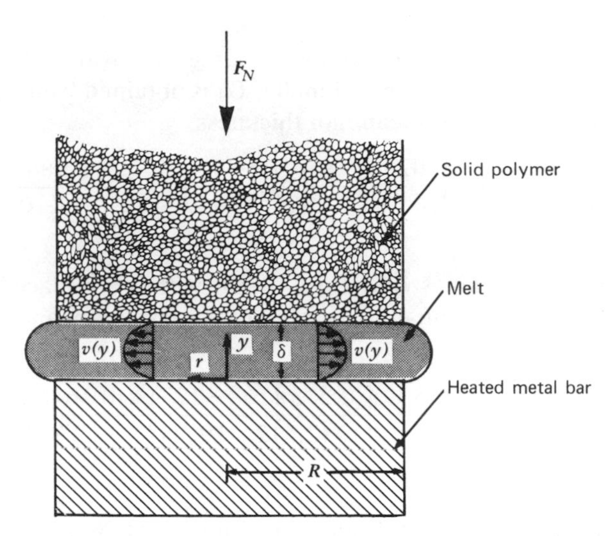


**Fig. 9.16** Schematic representation of a solid polymer melting on a hot metal bar.

(36) developed the following approximate theoretical model for the melting process.

Consider a polymer bar of radius $R$ pressed by force $F_N$ against a hot metal bar at constant temperature $T_b$ of the same radius as in Fig. 9.16. A film of melt is formed which is being squeezed out by radial flow. The following simplifying assumptions are made:

1. The solid is rigid and moves with constant velocity toward the hot bar.
2. The film between the polymer and the hot bar has a constant thickness $\delta$.
3. Flow in the film is laminar.
4. The fluid is Newtonian.
5. Viscosity is temperature independent.
6. Constant thermophysical properties.
7. Steady state.
8. Gravitational forces are negligible.
9. Convection and viscous dissipation on the film are negligible.

Some of these assumptions may be questionable—for example, the assumptions that the solid is rigid and the film thickness constant. In reality, as the preceding section demonstrated, allowing the solid to deform and using an a priori unknown $\delta(r)$ would be more plausible. Nevertheless the foregoing assumptions do allow the "construction" of a simple model for the process, providing insight into its nature. Moreover, the model did show reasonably good agreement with experiments carried out with polyethylene and polyoxymethylene.

With the rigid polymer assumption, the total rate of melting can immediately be written as

$$w_T = \pi(-v_{sy})\rho_s R^2 \tag{9.9-1}$$

where $v_{sy} < 0$ is the velocity of the solid polymer. Our objective is to find a relationship between the velocity $v_{sy}$, the operating conditions (the pushing force $F$, the hot plate and solid temperatures), and the polymer physical properties.

By pressing the bar against the plate, a radial velocity profile will be formed in the melt film, thus removing the newly melted polymer from the location of melting, and draining it. The mean radial velocity at any location $r$ can be expressed in terms of (the yet unknown) velocity $v_{sy}$ by a simple mass balance

$$\rho_s \pi r^2 v_{sy} = 2\pi r \delta \bar{v}_r \rho_m \tag{9.9-2}$$

where $\delta$ is the local separation between the interface and plate. Thus from Eq. 9.9-2 the mean radial velocity is

$$\bar{v}_r = \frac{r v_{sy}}{2\delta} = \frac{1}{\delta}\int_0^\delta v_r \, dy \tag{9.9-3}$$

The radial component of the equation of motion reduces to

$$\frac{dP}{dr} = \mu \frac{d^2 v_r}{dy^2} \tag{9.9-4}$$

We have substituted ordinary differentials for the partial differentials in the equation of motion because the left-hand side is only a function of $r$, whereas we assume the right-hand side to be only a function of $y$ (lubrication approximation). Therefore they simply equal a constant. Equation 9.9-4 can now be integrated over $y$, with boundary conditions $v_r(0)=0$ and $v_r(\delta)=0$, to give

$$v_r = \frac{1}{2\mu}\frac{dP}{dr}(y-\delta)y \tag{9.9-5}$$

An expression for the pressure gradient $dP/dr$ versus $r$ can be obtained by substituting Eq. 9.9-5 into Eq. 9.9-3

$$-\left(\frac{dP}{dr}\right) = \frac{6\mu v_{sy} r}{\delta^3} \tag{9.9-6}$$

Integration of Eq. 9.9-6 with the boundary condition $P(R)=P_0$, where $P_0$ can be the atmospheric pressure, leads to the following pressure profile:

$$P(r) - P_0 = \frac{3\mu v_{sy}}{\delta^3}(R^2 - r^2) \tag{9.9-7}$$

From the pressure profile the total force $F_N$ can be calculated:

$$F_N = \int_0^R 2\pi r P(r)\, dr = \pi R^2 P_0 + \frac{3\mu\pi v_{sy} R^4}{2\delta^3} \tag{9.9-8}$$

Equation 9.9-8 is in effect the relationship we are looking for, and by rearranging it we get a relationship of the velocity $v_{sy}$ in terms of the external total force $F_N$ and a number of other variables.

$$v_{sy} = \frac{2\delta^3}{3\pi\mu R^4}(F_N - \pi P_0 R^2) \tag{9.9-9}$$

We cannot however calculate the melting rate of this geometrical configuration from Eq. 9.9-9 because we do not yet know the value of $\delta$. This is determined by the rate of heat conducted into the solid-melt interface. If we make use of one more of the simplifying assumption above—namely, that viscous dissipation is negligible—the following simple heat balance can be made on the interface (see Eq. 9.8-13):

$$k_m\left(\frac{T_b - T_m}{\delta}\right) = \rho_s v_{sy}[\lambda + C_s(T_m - T_0)] \tag{9.9-10}$$

where $T_0$ is the initial temperature of the solid. Substituting Eq. 9.9-9 into Eq. 9.9-10 results in the final expression—the process design equation

$$v_{sy} = \frac{0.6787}{R}\left[\frac{F_N - \pi P_0 R^2}{\mu}\right]^{1/4}\left[\frac{k_m(T_b - T_m)}{\rho_s[\lambda + C_s(T_m - T_0)]}\right]^{3/4} \tag{9.9-11}$$

The melting capacity of this geometrical configuration can be easily calculated from Eqs. 9.9-11 and 9.9-1.

The results are very revealing and instructive. The rate of melting increases with the total force $F_N$ but only to the $\frac{1}{4}$ power. The physical reason for this is that with increasing force, the film thickness is reduced, thus increasing the rate of melting; but the thinner the film, the larger the pressure drops that are needed to squeeze out the melt. The dependence on the plate temperature is almost linear. The inverse proportionality with $R$ is perhaps the most important result from a design point of view. If viscous dissipation were included, some of these results would have to be modified.

Stammers and Beek (36) have performed some experiments to verify the theoretical model just described. Polyethylene and polyoxymethylene were used. The linear relationship between $v_{sy}/F_N^{1/4}$ and $[(T_b - T_m)^3/\mu]^{1/4}$ as predicted by Eq. 9.9-11 was clearly established, and the slope calculated from this equation agreed well with the experimental data.

## 9.10 Dissipative Mix-Melting

From the foregoing discussion it is evident that a great deal of effort has been invested in elucidating the mechanisms of melting methods based on conduction melting. In particular, the discovery of the relatively ordered conduction melting with drag removal observed in most single screw extruders (see Section 12.1) attracted most of the theoretical effort. This happened because (*a*) single screw extruders play a very important role in polymer processing and (*b*) fortunately this melting method is amenable to a relatively straightforward theoretical formulation. Yet, at the same time, other melting methods were left theoretically unexplored. Thus although dissipative mix-melting is, as Section 9.1 points out, of great practical utility, it has received little or no attention as far as theoretical analysis is concerned. In this section we attempt to discuss in qualitative terms the possible phenomena that may occur in a dissipative mix-melting process, but a great deal more experimental work is needed to verify and fully explore these mechanisms, and ultimately to formulate them into mathematical models.

Consider an internal or continuous mixer (see Section 1.1 and Figs. 11.4, 11.5, and 11.24) fed by solid polymer in particulate form. All the contacting walls are heated, therefore, some polymer melts by conduction melting upon contacting these surfaces. However, the dominant source of energy for melting is mechanical energy introduced through the shafts of the rotors and converted into thermal energy by continuous gross deformation of the particulate charge of material. As Section 9.1 points out, the mechanical energy is converted into heat by a number of mechanisms: individual particle deformation, interparticle friction, and viscous dissipation in the molten regions. As melting progresses, the latter mechanism becomes dominant. The intensive mixing disperses the newly formed melt into the mass. The melt that comes in intimate contact with solid particles cools down and, at the same time, heats and melts the surface layers of the particles. Consequently, the particulate solid charge is gradially converted, first, into a thermally (and rheologically) nonhomogeneous partly fluxed mass, and ultimately into a homogeneous one. In Banbury type mixers, the practice of "seeding" a new charge with a small amount of molten and mixed polymer from the previous charge may serve to accelerate occurrence of the above described phenomena.

One key element of such a melting method is the capability of the system to dissipate mechanical energy at high rates and distribute it uniformly throughout its volume. In internal mixers this is achieved by the geometrical configurations of the rotors (and body) which impose on the polymer charge a variety of deformations—shear, stretching, squeezing, and kneading. In the Banbury type mixer, a vertical ram is used to help force the polymer between the rotor where it undergoes these deformations. The shape of the rotor further ensures such flow patterns are experienced uniformly by all elements of the charge.

An alternative way to induce a similar type of melting process seems to be possible in single screw extruders and similar configurations where the driving force to elementary steps is drag, induced by a moving wall. Specifically, it appears that in screw extruders that are designed and operated in such a way that very high pressure develops in the solids conveying section of the screw (see Section 12.1), higher rates of melting can be obtained than those predicted by melting models based on the conduction melting with drag removal mechanism. It appears that the very high pressure generated serves two functions: (*a*) it brings about particle deformation, interparticle friction and, thus, heating, and perhaps sintering the particulate systems; and (*b*) creates very high shear stresses on the moving wall which are transmitted to the compressed bed and which may bring about internal shear, bed deformation, and ultimately a continuous deformation of the partially fluxed mixture of solids and melt. If a film of melt is created on the moving wall it is possible that under the above conditions the melt is forced into the bed to fill interparticle spaces creating the same type of internal "lubrication" and intimate contact melting mentioned above, which further enhances gross bed deformation. Consequently, in such a mechanism the mechanical energy, rather than being dissipated at limited rates in shearing a thin film, is dissipated at high rates in deforming the whole solid bed. Theoretical analysis of such a melting process is not easy because, in addition to the heat transfer problem, one has to deal with stress distribution and yield in compressed beds of particulate solids at nonuniform temperature, and complex external loading patterns. As Section 8.9 points out even the analysis of a relatively simply loaded isothermal bed is not easy. Nevertheless, the advantages of dissipate mix-melting, which is a melting method of promise and potential (high melting rates and low melt temperatures), dictate that more theoretical analysis efforts be devoted to it in the near future.

## *REFERENCES*


1. L. Erwin and N. P. Suh, "A Method for the Rapid Processing of Thermoplastic Articles," *Polym. Eng. Sci.*, **16**, 841–846 (1976).
2. G. Menges and W. Elbe, "Untersuchungen des Einzugsand Plastifizierverhaltens von Schneckenspritzgiessmachinen," Beitrag zum 5, Kunstofftechnischen Kolloquium des Institut für Kunstoffverarbeitung, Aachen, 1970.
3. T. K. Ross, "Heat Transfer to Fusible Solids," *Chem. Eng. Sci.*, **1**, 212–215 (1952).
4. E. R. G. Eckert and R. M. Drake, Jr., *Analysis of Heat Transfer*, McGraw-Hill, New York, 1972: (a) p. 219; (b) p. 157; (c) p. 12.
5. T. R. Goodman, "Application of Integral Methods for Transient Nonlinear Heat Transfer," in *Advances in Heat Transfer*, Vol. I, T. F. Irvine, Jr., and J. P. Hartnett, eds., Academic Press, New York, 1964, pp. 51–122.
6. S. G. Bankoff, "Heat Conduction or Diffusion with Change in Phase," in *Advances in Chemical Engineering*, Academic Press, New York, 1964, pp. 75–100.
7. L. I. Rubinstein, "The Stefan Problem," in *Translation of Mathematical Monographs*, Vol. 27, American Mathematical Society, Providence, R.I. 5, 1971.
8. S. W. Churchill and L. B. Evans, "Coefficients for Calculation of Freezing in Semi-Infinite Region," *Trans. Am. Soc. Mech. Eng., J. Heat Transfer*, 234–236 (1971).
9. H. S. Carslaw and J. C. Jaeger, *Conduction of Heat in Solids*, 2nd ed., Oxford University Press, New York, 1959.
10. W. M. Rosenhow and J. P. Hartnett, *Handbook of Heat Transfer*, McGraw-Hill, New York, 1973.
11. J. Eisenberg and G. deVahl Davis, "FDM Methods in Heat Transfer," in *Topics in Transport Phenomena*, C. Gutfinger, ed., Wiley, New York, 1975; G. E. Myers, *Analytical Methods in Conduction Heat Transfer*, McGraw-Hill, New York, 1971, Chapter 8.
12. A. M. Clausing, "Numerical Methods in Heat Transfer," in *Advanced Heat Transfer*, B. T. Chao, ed., University of Illinois Press, Urbana, 1969, pp. 157–216.
13. G. M. Dusinberre, *Heat Transfer Calculations by Finite Differences*, 2nd ed., International Textbook, Scranton, Pa., 1961.
14. J. Crank and P. Nickolson, *Proc. Camb. Phil. Soc.*, **93**, 50 (1947).
15. G. G. O'Brien, M. A. Hyman, and S. Kaplan, *J. Math. Phys.*, **29**, 223 (1951).
16. R. D. Kersten, *Engineering Differential Systems*, McGraw-Hill, New York, 1969.
17. C. Gutfinger, E. Broyer, and Z. Tadmor, "Melt Solidification in Polymer Processing," *Polym. Eng. Sci.*, **15**, 515 (1975).
18. D. Rosenthal, "The Theory of Moving Sources of Heat and Its Application Metal Treatment," *Trans. Am. Soc. Mech. Eng.*, **68**, 849–866 (1946).
19. G. C. Kuczynski, B. Neuville, and H. P. Toner, "Study of Sintering of PMMA," *J. Appl. Polym. Sci.*, **14**, 2069–2077 (1970).
20. M. A. Rao and J. L. Throne, "Principles of Rotational Molding," *Polym. Eng. Sci.*, **12**, 237–250 (1972).
21. J. L. Throne, "Rotational Molding Heat Transfer—An Update," *Polym. Eng. Sci.*, **16**, 257–264 (1976).
22. J. F. Lonz, "Sintering of Polymer Materials," in *Fundamental Phenomena in the Material Sciences*, Vol. 1, *Sintering and Plastic Deformation*, L. J. Bonis and H. H. Hausner, eds., Plenum Press, New York, 1964.
23. D. M. Bigg, "High Pressure Molding of Polymeric Powders," Society of Plastics Engineers 33rd Annual Technical Conference, Atlanta, May 5, 1975, pp. 472–476.
24. G. S. Jayaraman, J. F. Wallace, P. H. Geil, and E. Baer, "Cold Compaction Molding and Sintering of Polystyrene," *Polym. Eng. Sci.*, **16**, 529–536 (1976).
25. J. Frenkel, *J. Phys. (U.S.S.R)*, **9**, 385 (1945).
26. J. D. Eshelby, *Trans. Am. Inst. Mech. Eng.*, **185**, 806 (1949).

27. D. H. Sundstrom and J. R. Lo, "Softening Rates of Polystyrene Under Shear Conditions," *Polym. Eng. Sci.* **18**, 422 (1978).
28. Z. Tadmor, "Fundamentals of Plasticating Extrusion—A Theoretical Model for Melting," *Polym. Eng. Sci.*, **6**, 185–190 (1966).
29. Z. Tadmor, I. J. Duvdevani, and I. Klein, "Melting in Plasticating Extruders—Theory and Experiments," *Polym. Eng. Sci.*, **7**, 198–217 (1967).
30. Z. Tadmor and I. Klein, *Engineering Principles of Plasticating Screw Extrusion*, Van Nostrand Reinhold, New York, 1970.
31. J. R. Vermeulen, P. M. Gerson, and W. J. Beek, "The Melting of a Bed of Polymer Granules on a Hot Moving Surface," *Chem. Eng. Sci.*, **26**, 1445–1455 (1971).
32. D. H. Sundstrom and Chi-Chang Young, "Melting Rates of Crystalline Polymers Under Shear Conditions," *Polym. Eng. Sci.*, **12**, 59–63 (1972).
33. E. M. Mount, III, "The Melting of High Density Polyethylene on a Heated, Moving Metal Surface—A Comparison of Experimental and Theoretical Results," M.S. thesis, Rensselaer Polytechnic Institute, Troy, N.Y., May 1976.
34. J. R. A. Pearson, "On the Melting of Solids Near a Hot Moving Interface, with Particular Reference to Beds of Granular Polymers," *Int. J. Heat Mass Transfer*, **19**, 405–411 (1976).
35. T. R. Fuller and A. L. Fricke, "Thermal Conductivity of Polymer Melts," *J. Appl. Polym. Sci.*, **15**, 1729–1736 (1971).
36. E. Stammers and W. J. Beek, "The Melting of a Polymer on a Hot Surface," *Polym. Eng. Sci.*, **9**, 49–55 (1969).

## PROBLEMS

**9.1** ***Solution of Heat Transfer Problems by Combination of Variables.*** Show that the partial differential equation

$$\frac{\partial T}{\partial t} = \alpha \frac{\partial^2 T}{\partial x^2}$$

is reduced to the ordinary differential equation 9.3-4 by defining a new variable $\eta = Cxt^m$, where $C$ and $m$ are constants.

Note that we combine the variables such that $T = f(\eta)$ where $\eta = F(x, t)$. Use the chain rule to obtain expressions for $\partial T/\partial t$, $\partial T/\partial x$ and $\partial^2 T/\partial x^2$, then substitute for $\partial \eta/\partial t$, $\partial \eta/\partial x$ and $\partial^2 \eta/\partial x^2$.

**9.2** ***Cyclic Temperature Boundary Conditions.*** (a) Consider the heat transfer problem involved inside a "semi-infinite solid" of constant properties because of the temperature fluctuation at the solid surface of the cyclic form

$$T(0, t) = T_0 + A \cos(\omega t)$$

(b) Show that, at long times, the relative amplitude of temperature $A_r = A(x)/A(0)$ is given by $A_r = \exp(-x\sqrt{\pi}/x_0)$ where $x_0 = \sqrt{2\pi\alpha/\omega}$. If the heat transfer period equals the fluctuation period, $2\pi/\omega$, then $x_0$ is a good estimate of the penetration thickness.

(c) Find the penetration thickness for a period of 100 $s$ for LDPE, which has thermal diffusivity $\alpha = 7 \times 10^{-8}\ \text{m}^2/\text{s}$.

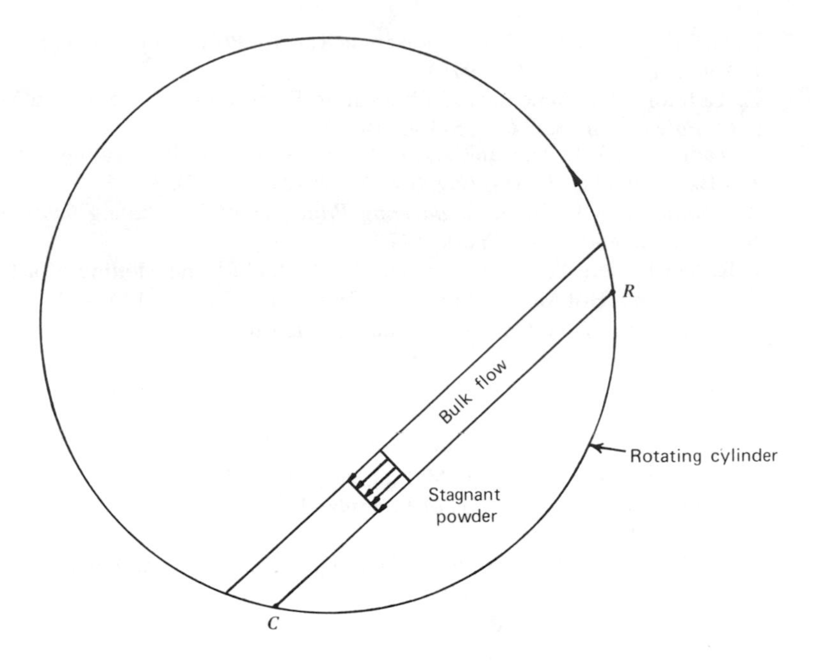


**Fig. P-9.3**

**9.3** ***Rotational Molding.*** Throne† and his co-workers have investigated the heat transfer problems associated with rotational molding of polymer powders, both theoretically and experimentally. One of the simulation models for heat transfer they have considered is depicted on Fig. P-9.3†. The lower area represents a stagnant pool of polymer powder which undergoes rigid body rotation with the rotating mold. When it reaches point $R$ it releases and falls back to $C$ where it is again heated by the hot mold wall. For each "cycle" the time of contact is the time it takes for the mold to rotate from $C$ to $R$. During the flowing stage the powder is considered to mix thermally.

Following their work show that, using the Goodman method (Ref. 5) and a temperature profile

$$T(x, t) = T_s\left(1 - \frac{x}{\delta(t)}\right)^3$$

where $T_s = T(0, t) = T_\infty(1 - e^{-\beta t}) + T^*$,

$T_\infty$ is the oven set point temperature
$\beta$ is the characteristic time of heating of the mold by the external heat transfer medium (steam), experimentally determined, and
$T^*$ is the initial offset temperature,

† See M. Anandha Rao and J. L. Throne, "Principles of Rotational Molding," *Polym. Eng. Sci.*, **12**, 237 (1972).

the penetration thickness $\delta(t)$ is

$$\delta(t)=\frac{2\sqrt{6\alpha_s}}{T_\infty(1-e^{-\beta t_2})+T^*}\Big\{t_c[T_\infty^2+2T_\infty T^*+T^{*2}]$$

$$+\left[\frac{2T_\infty^2}{\beta}+\frac{2T_\infty T^*}{\beta}\right](e^{-\beta t_2}-e^{-\beta t_1})-\frac{T_\infty^2}{2\beta}(e^{-2\beta t_2}-e^{-2\beta t_1})\Big\}^{1/2}.$$

$t_c=t_2-t_1$, is the time of contact, $\alpha_s=\alpha$ at $x=0$.

**9.4** ***Dielectric Heating.*** In dielectric heating the rate of heat generated per unit volume for a field strength $\mathcal{F}$ of frequency $f$ is

$$G=13.3\times 10^{-14} f\mathcal{F}^2 k' \tan\delta$$

where $G$ is in cal/cm$^3$·s, $k'$ is the dielectric constant and $\delta$ the loss tangent (Section 6.4).

Solve the one-dimensional transient heat transfer problem in a slab of constant properties in the presence of dielectric heating of intensity $G$. The slab is initially at a uniform temperature $T_0$ which is kept constant at the two boundaries of the semi-infinite slab, $x=\pm b$.

***Answer:***

$$T-T_0=\frac{G}{2k}\Big\{(b^2-x^2)-\frac{32b^2}{\pi^3}\sum_{n=0}^{\infty}\frac{1}{(2n+1)^3}e^{\frac{-\alpha(2n+1)^2\pi^2 t}{4b^2}}$$

$$\sin\left[\frac{(2n+1)\pi}{2}\left(1+\frac{x}{b}\right)\right]\Big\}$$

**9.5** ***Frictional Welding.*** Two pieces of PMMA are to be welded frictionally. Estimate the normal pressure which has to be applied in order to raise the interface temperature from 25°C to 120°C in 1 s. The relative velocity between the sheets is 10 cm/s. The thermal conductivity of PMMA is $4.8\times 10^{-4}$ cal/cm·°K·s, the thermal diffusivity $9\times10^{-4}$ cm$^2$/s and the coefficient of friction is 0.5.

**9.6** ***A Theoretical Model for Fluidized Bed Coating.*** In fluidized bed coating, a hot metal surface is dipped into a fluidized polymer powder which heats up and melts onto the metal surface. The process is similar in concept to dip coating which utilizes plastisols. Gutfinger and Chen* have considered the one-dimensional transient heat transfer problem involved in the coating of a flat, hot metal body of constant wall temperature $T_w$ by a powder of temperature $T$.

(a) Show that the thermal energy balance, initial and boundary conditions for constant powder properties are

$$\frac{\partial T}{\partial t}=\alpha\frac{\partial^2 T}{\partial x^2}$$

* C. Gutfinger and W. H. Chen, "Heat Transfer with a Moving Boundary. Application to Fluidized Bed Coating," *Int. J. Heat Mass Transfer*, **12**, 1097 (1969).

$T(0, t) = T_w$; $T(X(t), t) = T_m$

$$-k\frac{\partial T}{\partial x}\bigg|_X = h(T_m - T_\infty) + [\rho C_p(T_m - T_\infty) + \lambda]\frac{dX}{dt}$$

where $X(t)$ is the molten coat thickness.

(b) Assuming that the temperature field in $X(t)$ is given by

$$T(x, t) = a(t) + b(t)[X - x] + c(t)[X - x]^2$$

evaluate $a$, $b$, and $c$ using the differential equation, boundary, and initial conditions. This will lead to the *design equation* for the process

$$t = \int_0^{X(t)} \frac{\Delta T_m + \frac{1}{3}\Delta T_w + G/3 + (k\zeta/3)L(\zeta)}{(2\alpha\Delta T_w)/\zeta - \Delta T_w(h/\rho C_p) - (\alpha G/\zeta)} d\zeta$$

where $\Delta T_w = T_w - T_m$; $\Delta T_m = T_m - T$

$$G = \frac{-D + (D^2 + 4\Delta T_w kE)^{1/2}}{2k}$$

$$E = 2\alpha\rho C_p \Delta T_m + \lambda$$

$$D = E - Xh\Delta T_m$$

$$L(\zeta) = h\Delta T_m \zeta\{1 + [(E - \zeta h\Delta T_m)^2 + 4\Delta T_w kE]^{-1/2}(\zeta - E)$$
$$-[-(E - \zeta h\Delta T_m) + [(E - \zeta h\Delta T_m)^2 + 4\Delta T_w kE]^{1/2}]\}$$

Thus the coating thickness as a function of time can be obtained by numerical or graphical integration. Note that when the denominator of the integrant is 0 the time for any $X$ becomes infinite, that is, growth stops.

**9.7** ***Fluidized Bed Coating of an Article.*** A rectangular metal article with dimensions of $0.5 \times 5.0 \times 10.0$ cm is to be coated with PVC powder to a uniform coat thickness of 0.01 cm by the fluidized bed coating process. The fluidized bed temperature is 20°C and the initial metal temperature is 150°C.

(a) Assuming no convective losses to the fluidized bed, what would the metal temperature decrease be to form the desired coat thickness? ($\rho = 7.86$ g/cm$^3$, $C_p = 0.1$ cal/g°K.) Comment on the boundary condition of Problem 9.6,* $T(0, t) = T_w$.

(b) Estimate the effect of convective heat losses on the temperature decrease of the metal.

**9.8** ***Formation of Thick Polymer Sheets.*** Forming thick sheets of unplasticized amorphous polymers (e.g., PVC) is difficult because of the frequency of void formation during cooling. For this reason such products are sometimes made by pressing together a number of thin extruded sheets between hot plates in hydraulic presses.

Using Fig. 9.5 estimate the time required to fuse together twenty sheets of PVC each 0.05 cm thick, initially at 20°C, by pressing them between two

* See N. Abuaf and C. Gutfinger, *Int. J. Heat Mass Transfer* **16**, 213 (1973); and C. Gutfinger and N. Abuaf, "Heat Transfer in Fluidized Beds" in *Advances in Heat Transfer*, Vol. 10, Academic Press, 1974, pp. 167–218.

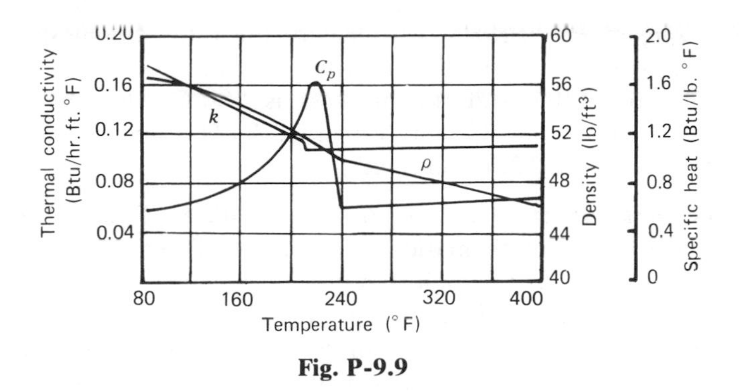


**Fig. P-9.9**

hot plates kept at a constant temperature of 150°C. Use the thermophysical data of Chapter 5 and Appendix A. Discuss the problem of thermal degradation in terms of Fig. 9.1. Specifically, calculate whether any thermal degradation will occur and at any layers of the sheet.

**9.9** ***Cooling of Extruded PE Wire.**** Consider a copper conductor 0.16 in. in diameter coated by extrusion to a 0.62 in. insulated wire (first Transatlantic cable core wire). The conductor is preheated to the extrusion temperature, 412°F, and exits into a water trough maintained at 80°F at 42 ft/min. Assuming a flat temperature distribution in the copper, since its conductivity is about 2,000 times that of PE, solve the heat transfer problem of cooling of the insulated wire in terms of a heat transfer coefficient of 500 (Btu/ft$^2$·hr·°F) and the thermophysical properties for PE shown on Fig. P-9.9.

**9.10** ***Adiabatic Compression Heating.*** Melting of polymers by adiabatic compression has been shown to be feasible for processes such as injection molding (Ref. 2). Discuss this method, in principle, in terms of an order of magnitude analysis of the terms of the thermal energy balance for an amorphous (PS) and a semicrystalline polymer (LDPE). Use the data in Appendix A.

**9.11** ***Necessity of Melt Removal in Conductive Melting.*** There are four reasons for melt removal (from the heat transfer region) in conductive melting. The first is efficiency of melting, Eq. 9.3-39; the second is avoidance of thermal degradation by shortening the residence time of the melt in regions near high temperature surfaces; the third is the further generation of heat in the entire volume of the melt by viscous dissipation of mechanical energy. Finally, the flow that affects melt removal can result in laminar mixing and thermal homogenization.

Compare the melting efficiency and polymer melt stability for the "melting" of PVC in the absence and presence of melt removal. A slab of PVC $8 \times 8 \times 2$ cm at 20°C is to be melted by a hot metal surface at 200°C. Melt removal is accomplished by moving the hot surface at a speed of 1 cm/s. Use data in Fig. 9.1 and Appendix A. Assume an average value for $\rho$, $k$, and $C_p$ below and above $T_g$.

* R. D. Biggs and R. P. Guenther, *Modern Plastics*, 1963, 126 (May 1963).

**9.12** ***Sintering of PS "Pearls."*** Calculate the rate of coalescence of PS "pearls" made from suspension polymerization which are 0.2 cm in diameter. The temperature of the sintering process is 180°C. Use the power law constants of the unmodified PS in Appendix A. The surface tension of the melt can be taken to be 32.4 dyne/cm.*

**9.13** ***Flow and Heat Transfer in the Molten Film During Melt Removal.*** Formulate mathematically the coupled heat transfer and flow problems involved during the melt removal (by a simple shearing flow) in the conductive heating of a polymer sheet. If $x$ is the direction of the melt removal and $y$ the direction of the main temperature gradient, allow both $v_x$ and $v_y$ to be nonzero (because $\delta = \delta(x)$); also allow for a convective heat flux in the $x$-direction. Assume that the polymer is crystalline with constant "average" values for $\rho$, $k$, and $C_p$.

**9.14** ***Heat Transfer in Blow Molding.*** Consider the blow molding process (Fig. 1.17). An HDPE parison 20 cm long, 4 cm in O.D. and 0.3 cm thick exits a reciprocating injection molding machine at 200°C. It is clamped in a 15°C cylindrical bottle mold and blown by cold air of 5°C. The bottle body diameter is 10 cm and its length is 15 cm.

(a) Solve the heat transfer problem involved. Use the $\rho$, $k$, and $C_p$ data given in Appendix A. The air–polymer heat transfer coefficient is 10 Kcal/m$^2$·hr·°C.

(b) Discuss the possible effects of the resulting temperature field on the rate of crystallization (Fig. 3.12).

**9.15** ***Heat Transfer in Underwater Pelletizing.*** In underwater pelletizing the melt strands are extruded directly in a water bath and "chopped" by a rotating high speed knife into short length cylinders called pellets. Such a die is shown in Fig. 12.1. Consider a LDPE extrudate at 200°C chopped into pellets of $L = D = 0.4$ cm in a bath kept at 10°C.

(a) Formulate the complete heat transfer problem.

(b) Estimate the time of bath residence required to bring the centerpoint pellet temperature to 70°C by assuming an infinitely high heat transfer coefficient at the water–polymer interface. Use Fig. 9.8 and data from Appendix A.

* H. Schonhorn, "Theory of Adhesive Joints" in *Adhesion and Bonding*, N. M. Bikales, Ed., Wiley, New York, 1971.

# 10

# Pressurization and Pumping



The polymer melts generated during the melting step of a given processing method must be conveyed or *pumped* out of the processing equipment. They also must be *pressurized*, since they flow *through* a restriction or *into* one in order to assume a useful shape. This chapter deals with the very important *elementary step of pumping and pressurization.* As in Chapters 8 and 9, we first discuss the theoretical principles that can be invoked to achieve various ways of pumping and pressurization. In this case, however, we go one step beyond the principles and commence with the engineering activity of developing and analyzing various pumping geometrical configurations, which are practical for polymer melts. The next chapter explains how these geometrical configurations perform as mixing devices. Finally, Chapter 12 discusses primarily one of them as part of entire polymer processing machines, together with the conveying and melting of polymer solids.

Pumping and pressurization characterizes polymer processing more than any other elementary step. The overall unique features of processing machinery are to a very large extent due to the rheological properties of polymeric melts and, in particular, to their very high viscosities. This together with high production rates lead to the need of developing and working with relatively high pressures. Extrusion pressures of up to $50\,MN/m^2$ and injection pressures of up to 100 $MN/m^2$ are common. Chapter 9 discussed how the very high viscosities inevitably lead to a substantial viscous dissipation during flow. This, coupled with the low polymer thermal conductivities leads in turn to generally *narrow gap configurations* that permit efficient temperature control by conduction to and from external solid walls. Furthermore, the sensitivity of polymers to thermal and shear degradation poses severe restrictions on permissible residence times and indicates the advantage of narrow residence time distributions.

Fortunately many of the flow fields encountered in pumping and pressurization are amenable to theoretical analysis.

## 10.1 Classification of Pressurization Methods

The response of a liquid to external forces is governed by the equation of motion. Therefore, a systematic way to search for all possible pressurization methods is a careful analysis of the equation of motion

$$\rho \frac{D\mathbf{v}}{Dt} = -\nabla P - [\nabla \cdot \boldsymbol{\tau}] + \rho \mathbf{g}$$

We first note that the equation of motion provides information only about the *pressure gradients* in the liquid, not the absolute value of the *hydrostatic pressure.* The latter is determined by *external* conditions. For example, the pressure of a liquid contained in a cylinder equipped with a plunger, is uniform, and is determined by the force applied to the plunger. We classify this pressurization method as *static pressurization,* because the pressure can be *maintained* without flow or without motion of the containing walls. The level of the pressure we can generate by this method is independent of the rheological properties of the liquid. In general, though, the purpose of pressurization is to induce flow that *is* governed by the rheological properties of the liquid. The flow that results from such an external mechanical pressurization is of a *positive mechanical displacement* type, which we encountered with particulate solids in Section 8.11. The outstanding characteristic feature of this flow is that the moving external surface *displaces* part of the liquid in the system. This pressurization method is being extensively used in polymer processing—for example, in injection and compression molding, counterrotating fully intermeshing twin screw extrusion, and gear pump extrusion.

Another way to generate pressure in a liquid is by inducing an internal pressure gradient. The equation of motion generally indicates that a nonzero pressure gradient results if at least one of the three terms $[\nabla \cdot \boldsymbol{\tau}]$, $\rho(D\mathbf{v}/Dt)$, $\rho\mathbf{g}$ has a nonzero value. The first two terms acquire nonzero values only during flow and deformation; hence they give rise to *dynamic pressurization* methods.

The gravitational term leads to potentially a static pressurization method, where an external body (instead of surface) force is utilized, as is the case in casting operations.

Polymer melts are characterized by very high viscosities. It is not surprising, therefore, to find that pressurization methods stemming from the $[\nabla \cdot \boldsymbol{\tau}]$ term (which is proportional to viscosity) have acquired great practical significance in polymer processing. Clearly, the higher the viscosity the larger the pressure gradient that can be generated. Thus the high viscosity of polymeric melts becomes an *asset* in these pressurization methods. The purpose of pressurization is to *generate* pressure as pumps do (as opposed to loose pressure, as in a pipe flow). This can only be achieved by a *moving* external surface that "drags" the melt, leading to drag induced flow fields (Section 8.13). Indeed the outstanding characteristic feature of this *viscous dynamic pressurization* method is an external surface moving *parallel* to itself without displacing the melt. Single screw extrusion, calendering, and roll-milling illustrate the practical significance of this pressurization method.

Viscous dynamic pressurization is not the only pressurization method stemming from the $[\nabla \cdot \boldsymbol{\tau}]$ term. We note from Eq. 6.3-5 that polymeric melts exhibiting a primary normal stress differences may also give rise to nonzero $[\nabla \cdot \boldsymbol{\tau}]$ terms. Indeed, the normal stress extruder utilizes this potential pressurization source.

The acceleration term $\rho(D\boldsymbol{v}/Dt)$, from a practical point of view, is a less important potential source of pressurization. Nevertheless this source did find applications in polymer processing—for example, in centrifugal casting and impact molding.

Finally, we note that in principle, any reduction in density can generate pressure in a closed system. Low pressure structural molding and certain reaction injection molding processes involve foaming during the molding operation, which generates enough pressure to force the polymer melt to fill the mold.

Although the equation of motion provides the information on the *possible* sources of pressurization, the actual multitude of *realistic* geometrical configurations, which can make efficient use of the principles (and also set the boundary conditions for solving the equation of motion) is a matter of creative engineering design that is guided by considerations of the polymer and the entire polymer processing system.

We next analyze the various pressurization methods and the associated characteristic flow fields in some detail. We do this in terms of the practically important geometrical configurations, seeking to achieve two goals: the development of insight into the nature of the flow fields in the various geometries, and the laying of the foundations of the mathematical modeling of associated polymer processing methods.

## 10.2 Dynamic Viscous Pressurization: Parallel Plate Flows

Section 10.1 concluded that dynamic viscous pressurization involves an external surface moving parallel to itself, leading to drag induced flow and pressurization. The simplest possible geometrical configuration that one can envision for analyzing the theoretical possibilities of this kind of pressurization consists of two parallel plates, in relative motion (Fig. 10.1). Since, as a result of the general validity of the lubrication approximation in most cases of interest, this simple geometry contains all the essential features of viscous pressurization (see below), we examine it in detail.

Polymer melt fills the gap between the plates, which are at a distance $H$ apart. As the upper plate is set in motion in the $z$-direction at constant velocity $V_0$, it drags the adjacent fluid layer with it, which in turn "drags" the fluid layer below it. Hence momentum is transported perpendicular to the fluid motion in the negative $y$-direction. After a certain short period of time, a *steady* velocity profile develops. Before we can discuss the velocity profile, the magnitude of the shear stresses, and

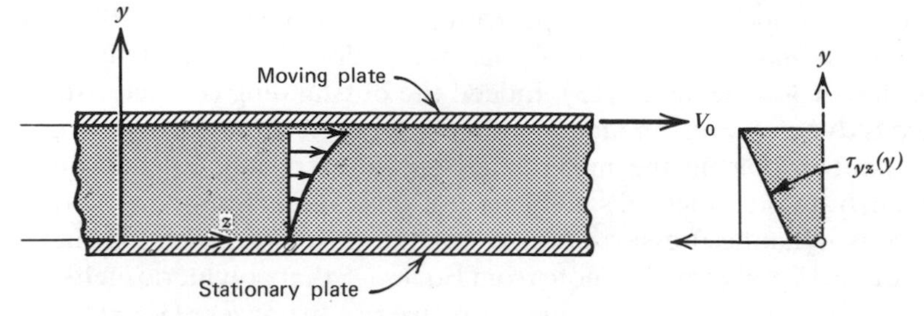


**Fig. 10.1** Schematic representation of the parallel plate geometry. Upper plate moves at constant velocity $V_0$. Lower plate is stationary. The velocity profile for a certain positive pressure gradient ($dP/dz > 0$) is shown schematically between the plates and $\tau_{yz}(y) < 0$ at the right.

the pressure gradients that ensue, we must be more specific about the geometry. We assume that the plates are infinite in the $x$-direction and are long enough in the $z$-direction to render entrance and exit effects negligible. In other words, we assume that over all $z$ there is a fully developed velocity profile. This can be mathematically stated as $\partial v_z/\partial z = 0$. At the exit, located far upstream, we can place a "valve" or a "die." Manipulation of this valve enables us to control the flow rate. If we close the valve, the net flow rate will be reduced to zero, and local pressure gradients as well as pressure at the valve will attain *maximum* values. This condition is referred to as the "*closed discharge*" condition. At the other extreme, the completely open valve condition, the flow rate attains a maximum value and local pressure gradients drop to zero. This is the *open discharge condition;* which in this case is also the *pure drag flow* situation. Pressure at the exit will then equal the pressure at the entrance.

### Newtonian Fluids

Now that the physical flow mechanism has been discussed, we proceed with the mathematical solution of the problem subject to the following assumptions: ($a$) the flow is laminar, ($b$) the flow is isothermal, ($c$) there is no slip at the walls, ($d$) the fluid is Newtonian and incompressible, ($e$) gravitational forces are neglected, and ($f$) the flow is fully developed, $\partial v_z/\partial z = 0$.

We assume no flow in the $x$- and $y$-directions, hence $v_x = v_y = 0$. The velocity component that concerns us is $v_z$, which is a function of $y$ only.

The three components of the equation of motion reduce to

$$\frac{\partial P}{\partial x} = 0 \tag{10.2-1}$$

$$\frac{\partial P}{\partial y} = 0 \tag{10.2-2}$$

$$\frac{\partial P}{\partial z} = -\frac{\partial \tau_{yz}}{\partial y} \tag{10.2-3}$$

From Eqs. 10.2-1 and 10.2-2 we conclude that the pressure is a function of coordinate $z$ only. We notice, therefore, that in Eq. 10.2-3 the left-hand side is a function of $z$ only, whereas the right-hand side is a function of $y$. This is only possible if *both* equal to a constant. Thus we conclude that the pressure gradient is constant; that is, *pressure rises linearly with z*. Equation 10.2-3 can be directly integrated to give $\tau_{yz}(y)$

$$\tau_{yz} = \tau_0 - \left(\frac{dP}{dz}\right)y \tag{10.2-4}$$

where $\tau_0$ is the value of $\tau_{yz}$ at the stationary wall ($\tau_0 < 0$). The shear stress is constant across the gap in the absence of a pressure gradient, and it is a linear function of $y$ in the presence of a pressure gradient.

For a Newtonian fluid, the constitutive equation 6.2-1 reduces in this case to

$$\tau_{yz} = -\mu \frac{dv_z}{dy} \tag{10.2-5}$$

Substitution of Eq. 10.2-5 into Eq. 10.2-4, followed by integration and use of the boundary conditions $v_z(0) = 0$ and $v_z(H) = V_0$, results in the following velocity profile, after evaluation of $\tau_0$:

$$\frac{v_z}{V_0} = \xi - 3\xi(1-\xi)\left(\frac{H^2}{6\mu V_0}\frac{dP}{dz}\right) \tag{10.2-6}$$

where $\xi = y/H$.

The velocity profile consists of two linearly superimposed terms. The first term is due to the motion of the upper plate, and the second term is the result of the pressure gradient in the $z$-direction. The shape of the profile depends on a single dimensionless group, the physical significance of which becomes evident subsequent to deriving the flow rate equation.

The flow rate per unit width $q$ is obtained by integrating the velocity profile over $y$

$$q = \frac{V_0 H}{2} + \frac{H^3}{12\mu}\left(-\frac{dP}{dz}\right)$$

$$= \frac{V_0 H}{2} + \frac{H^3}{12\mu}\frac{(-\Delta P)}{L} \tag{10.2-7}$$

where $L$ is the length of the plates.* The first term on the right-hand side is the flow rate when there is no pressure variation in the direction of flow ($dP/dz = 0$); it is due to the drag exerted by the moving plate on the melt and, as mentioned earlier, it is called *pure drag flow rate*

$$q_d = \frac{V_0 H}{2} \tag{10.2-8}$$

The second term in Eq. 10.2-7 is due solely to the pressure gradient, and in effect it simply expresses pressure flow between two stationary parallel plates; it is referred

* Equation 10.2-7 is the Newtonian fluid parallel plate dynamic viscous pressurization or pump *design* equation. It is a $q$ versus $\Delta P$ relation in terms of $V_0$ (operating variable), $H$ and $L$ (design variables), and $\mu$ (material variable).

to as a *pure pressure flow rate*

$$q_p = \frac{H^3}{12\mu}\left(-\frac{dP}{dz}\right) \tag{10.2-9}$$

The net flow rate is then also the linear superposition of drag and pressure flow rates. It should be noted that this is the consequence of the linearity of the differential equation we started with, and that the differential equation is linear because we have assumed Newtonian fluid behavior and isothermal flow. In terms of flow rates, we can now see that the dimensionless group determining the shape of the velocity profile is the ratio of pressure to drag flow rates

$$\frac{q_p}{q_d} = \frac{q - q_d}{q_d} = -\frac{H^2}{6\mu V_0}\frac{dP}{dz} \tag{10.2-10}$$

A value of $-1$ for this dimensionless group implies a situation of zero net flow rate, or the closed discharge situation. A zero value, as we know, is pure drag flow. A couple of other values of special interest are obtained by calculating from the velocity profile the shear rate distribution $\dot{\gamma}(\xi) = |\dot{\gamma}_{yz}(\xi)|$, where

$$\dot{\gamma}_{yz}(\xi) = \frac{dv_z}{dy} = \frac{V_0}{H}\left[1 + 3(1 - 2\xi)\frac{q_p}{q_d}\right] \tag{10.2-11}$$

From this equation we obtain that if the dimensionless group attains a value of $-\frac{1}{3}$, the shear rate at the stationary plate is zero, and if it attains a value of $\frac{1}{3}$, the shear rate at the moving plate is zero.* When the dimensionless group is in this range, the velocity profile exhibits *no* extremum. In terms of the net flow rate, the condition of no extremum will be

$$\frac{2q_d}{3} < q < \frac{4q_d}{3} \tag{10.2-12}$$

Figure 10.2 plots some typical velocity profiles.

We have embarked on analyzing the parallel plate geometry, having in mind the pressurization problem. Let us return to the original problem. We can obtain further insight into it by rearranging Eq. 10.2-7 with the aid of Eq. 10.2-8

$$\frac{dP}{dz} = \frac{12\mu}{H^3}(q_d - q) \tag{10.2-13}$$

We observe that in parallel plate geometry pressure can be generated, provided $q_d > q$; that is, provided the *moving plate drags more fluid than is actually delivered*. We can further see that the pressure gradient is proportional to the viscosity. Therefore the very high viscosity encountered in polymeric melts increases the system's pressurization capability. We can see that for a constant net flow rate, the pressure gradient rises with the plate velocity, which in an actual machine is an operating variable. Furthermore, for constant $q_d$ it is inversely proportional to $H^3$, which is a design variable in an actual melt pumping device.

* A positive $q_p/q_d$ implies a negative pressure gradient; that is, pressure *drops* in the positive $z$-direction. Such a situation arises if the pressure at the entrance to the plates is higher than at the exit, a common situation in "metering" sections of screw extruders (see Section 10.3).

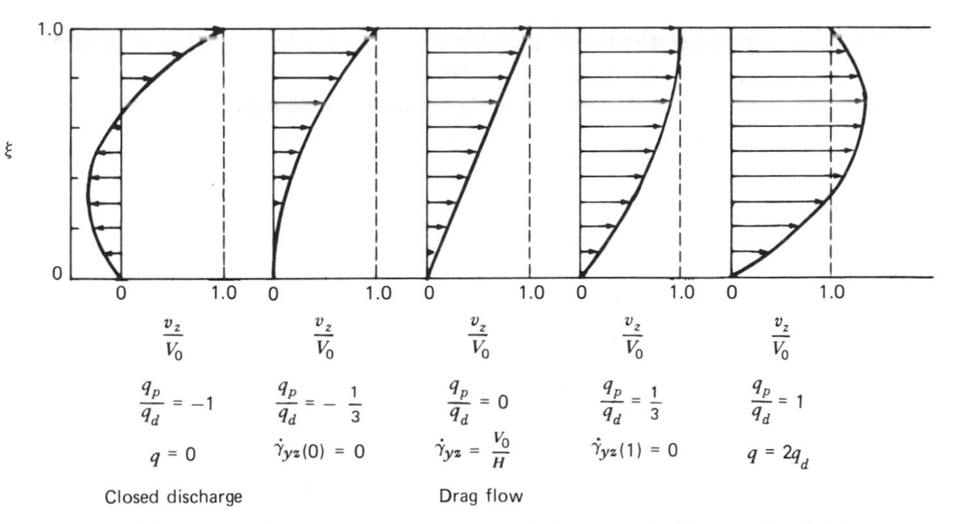


**Fig. 10.2** Velocity profiles between parallel plates of a Newtonian fluid.

The maximum possible pressure rise is obtained by setting $q=0$ (i.e., at the closed discharge situation)

$$\left(\frac{dP}{dz}\right)_{max} = \frac{6\mu V_0}{H^2} \tag{10.2-14}$$

The total pressure rise we can obtain in such a system is the product of the gradient above and the flow length of the plates. It can easily be shown from Eq. 10.2-7 that for a *given* $q$ there is an optimum $H = 3q/V_0$ for a maximum pressure rise, $(dP/dz)_{max} = 6\mu V_0^3/27q^2$.

The shear stress profile $\tau(\xi) = |\tau_{yz}(\xi)|$ can now be obtained from Eqs. 10.2-5 and 10.2-6

$$\tau_{yz}(\xi) = -\mu\frac{V_0}{H} - H\left(\xi - \tfrac{1}{2}\right)\left(\frac{dP}{dz}\right) \tag{10.2-15}$$

The first term in Eq. 10.2-15 is the contribution due to pure drag flow, and the second term is due to the pressure flow. At the moving plate $\tau_{yz}(1)$ is

$$\tau_{yz}(1) = -\mu\frac{V_0}{H} - \frac{H}{2}\left(\frac{dP}{dz}\right) \tag{10.2-16}$$

which with the aid of Eqs. 10.2-13 and 10.2-8 can be written as

$$\tau_{yz}(1) = -\mu\frac{V_0}{H}\left[1 + 3\left(1 - \frac{q}{q_d}\right)\right] \tag{10.2-16a}$$

**Example 10.1** ***Parallel Plate Flow of a Polymer Melt***

Let a parallel plate flow "apparatus" consist of 0.1 m long plates that are 0.005 m apart. The upper plate moves with a velocity of 0.25 m/s. The polymer melt between the plates is nylon (Zytel 31), which at the temperature of the flow and below the shear rate of $100\ s^{-1}$ behaves like a Newtonian fluid with a viscosity of $82.7\ N{\cdot}s/m^2$. Calculate the maximum pressure at the exit, the shear rate and shear stress profiles, and the flow rate at one-half the maximum pressure gradient.

*Solution:* The maximum pressure from Eq. 10.2-14 is

$$P_{max} = \frac{(0.1)(6)(0.25)(82.7)}{(5 \times 10^{-3})^2} = 4.962 \times 10^5\ N/m^2 = 4.897\ atm$$

The pressure gradient at closed discharge conditions is

$$\left(\frac{dP}{dz}\right)_{max} = \frac{(4.962 \times 10^5)}{(0.1)} = 4.962 \times 10^6\ N/m^3$$

The pressure to drag flow ratio from Eq. 10.2-10 at one-half the maximum pressure gradient is

$$\frac{q_p}{q_d} = -\frac{(5 \times 10^{-3})^2 (4.962 \times 10^6)}{(6)(0.25)(82.7)(2)} = -0.5$$

This is an expected result, if we compare Eqs. 10.2-10 and 10.2-14. The shear rate profile $\dot{\gamma}(\xi) = |\dot{\gamma}_{yz}(\xi)|$, where $\dot{\gamma}_{yz}(\xi)$ from Eq. 10.2-11 is

$$\dot{\gamma}_{yz}(\xi) = \frac{(0.25)}{(0.005)}[1 - (3)(0.5)(1 - 2\xi)] = 50(3\xi - 0.5)\ s^{-1}$$

The maximum shear rate at the moving wall is $125\ s^{-1}$; the shear rate reaches a value of zero at $y = 0.1667H$, and at the stationary plate the shear rate is $0.25\ s^{-1}$. Hence the shear rate range in the gap is between zero to $125\ s^{-1}$, which is approximately within the range where the melt behaves as a Newtonian fluid, justifying our assumption.

The shear stress distribution is obtained either by using Eq. 10.2-15 or simply multiplying the shear rate by the viscosity. The maximum shear stress at the moving plate is $1.03375 \times 10^4\ N/m^2$.

Finally the flow rate from Eqs. 10.2-7–10.2-10 can be written

$$q = q_d\left(1 + \frac{q_p}{q_d}\right) = \frac{(0.25)(5 \times 10^{-3})}{(2)}[(1) + (-0.5)] = 3.125 \times 10^{-4}\ m^3/s \cdot m$$

### Non-Newtonian Fluids

We have examined the pressurization principle in the parallel plate geometry with a Newtonian fluid. Yet polymeric melts are generally non-Newtonian. We shall, therefore, consider the effect of the non-Newtonian behavior on this mode of flow and pressurization. Since the most important relevant non-Newtonian property is shear rate dependence, we use the power-law model fluid. This flow problem was treated in detail by Flumerfelt et al. (1) and by Hirshberger (2).

For the flow under consideration, the power law model fluid equation reduces to

$$\tau_{yz} = -m\left|\frac{dv_z}{dy}\right|^{n-1}\frac{dv_z}{dy} \qquad (10.2\text{-}17)$$

Substituting Eq. 10.2-17 into Eq. 10.2-3 and casting it in dimensionless form, we obtain

$$\frac{d}{d\xi}\left(\left|\frac{du_z}{d\xi}\right|^{n-1}\frac{du_z}{d\xi}\right) = 6G \qquad (10.2\text{-}18)$$

where $u_z = v_z/V_0$ and the dimensionless number $G$ is defined as

$$G = \frac{H^{n+1}}{6mV_0^n}\left(\frac{dP}{dz}\right) \qquad (10.2\text{-}19)$$

Note that $G$ for a Newtonian fluid becomes $-q_P/q_d$ (Eq. 10.2-10).

Equation 10.2-18 can be integrated with respect to $\xi$ to give

$$\left|\frac{du_z}{d\xi}\right|^{n-1}\frac{du_z}{d\xi} = 6G(\xi-\lambda) \qquad (10.2\text{-}20)$$

where $-6G\lambda$ is an integration constant. The advantage of writing the integration constant this way is that $\lambda$ acquires a clear physical meaning; it is the location where the shear rate is zero, or the location of the extremum in the velocity profile. We need to know this location in order to rid ourselves of the absolute value in Eq. 10.2-20. Depending on the value of $G$, there are four velocity profiles we must consider (Fig. 10.3). Cases $a$ and $b$ exhibit an extremum in the velocity profile. The location of the extremum is at $\xi = \lambda$. In the former the pressure gradient is positive ($dP/dz > 0$); in the latter it is negative ($dP/dz < 0$). Cases $c$ and $d$ exhibit no extremum in the velocity profile, thus $\lambda$ in these cases lacks physical meaning, although it still is the location of an extremum value of the mathematical function describing the velocity profile. In case $c$, $\lambda < 0$ and in case $d$, $\lambda > 1$. In cases $c$ and $d$ $\dot{\gamma}_{yz} = dv_z/dy$ is positive throughout the flow regime, whereas in cases $a$ and $b$ it changes in sign above and below $\lambda$.

We note in Eq. 10.2-19 that $G$ may be positive or negative depending on the sign of the pressure gradient. It is therefore convenient to introduce at this point a variable accounting for the sign of $G$ defined as (2)

$$\text{sign } G = \frac{G}{|G|} \qquad (10.2\text{-}21)$$

Equation 10.2-20 can now be rewritten as

$$\left|\frac{du_z}{d\xi}\right|^{n-1}\frac{du_z}{d\xi} = 6 \text{ sign } G|G|(\xi-\lambda) \qquad (10.2\text{-}22)$$

It can be easily verified that for regions $\xi \geq \lambda$ for both positive and negative pressure gradients (i.e., both cases $a$ and $b$), Eq. 10.2-22 can be written as follows:

$$\frac{du_z}{d\xi} = |6G|^s(\xi-\lambda)^s \text{ sign } G \qquad (10.2\text{-}23)$$

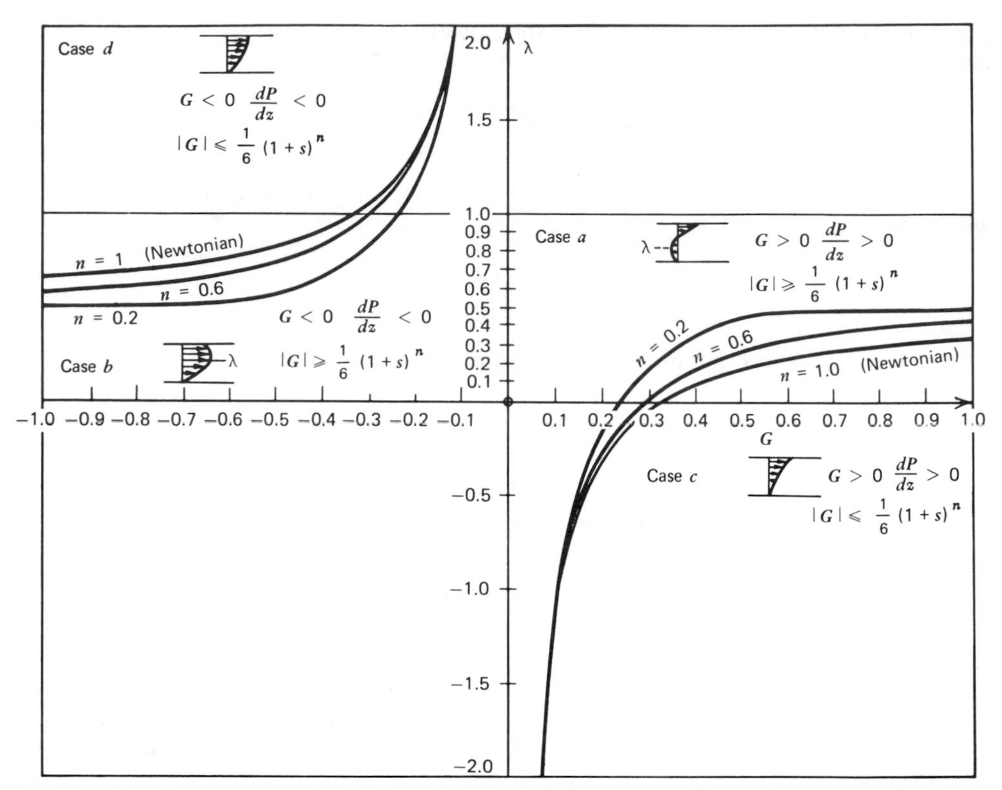


**Fig. 10.3** Four regions of solution of Eq. 10.2-20 corresponding to four types of velocity profiles. In regions (cases) *a* and *b* the velocity profile exhibits an extremum. In the former $dP/dz > 0$, in the latter $dP/dz < 0$. The location of the extremum is $\lambda$. In regions *c* and *d* the velocity profile exhibits no extremum. In the former $dP/dz > 0$, and in the latter $dP/dz < 0$. The curves represent solutions of Eq. 10.2-39 for $n = 1$, $n = 0.6$, and $n = 0.2$.

where $s = 1/n$, and similarly for $\xi \le \lambda$ we get

$$\frac{du_z}{d\xi} = -|6G|^s(\lambda - \xi)^s \operatorname{sign} G \tag{10.2-24}$$

Equations 10.2-23 and 10.2-24 can be integrated subject to the boundary conditions $u_z(1) = 1$ and $u_z(0) = 0$, respectively, to give

$$u_z = 1 - \frac{|6G|^s}{(1+s)}[(1-\lambda)^{1+s} - (\xi - \lambda)^{1+s}] \operatorname{sign} G \qquad \xi \ge \lambda \tag{10.2-25}$$

and

$$u_z = \frac{|6G|^s}{(1+s)}[(\lambda - \xi)^{1+s} - \lambda^{1+s}] \operatorname{sign} G \qquad \xi \le \lambda \tag{10.2-26}$$

Since the velocity is continuous throughout $\xi$, Eqs. 10.2-25 and 10.2-26 are equated at $\xi = \lambda$, resulting in an equation for the unknown $\lambda$

$$\lambda^{1+s} - (1-\lambda)^{1+s} + \frac{1+s}{|6G|^s \operatorname{sign} G} = 0 \tag{10.2-27}$$

Equation 10.2-27 provides the limiting values of $G$, for determining a priori whether the flow corresponds to case $a$ or $b$.

By setting $\lambda=0$ for $G>0$, and $\lambda=1$ for $G<0$, we obtain the following condition for the existence of an extremum within the flow region $0\leq\xi\leq1$:

$$|G|\geq\tfrac{1}{6}(1+s)^n \tag{10.2-28}$$

For a Newtonian fluid Eq. 10.2-28 reduces to $|q_p/q_d|=|G|\geq\frac{1}{3}$, as obtained earlier (cf. Eq. 10.2-12).

By substituting Eq. 10.2-27 into Eq. 10.2-25 we can write the velocity profile in one equation as follows:

$$u_z=\frac{|6G|^s}{(1+s)}(|\xi-\lambda|^{1+s}-\lambda^{1+s})\,\text{sign}\,G \tag{10.2-29}$$

subject to the inequality in Eq. 10.2-28.

Turning now to cases $c$ and $d$, where no extemum occurs and $du_z/d\xi>0$, we note that Eq. 10.2-22 can be written for $G>0$ and $G<0$, respectively, as

$$\frac{du_z}{d\xi}=(6G)^s(\xi-\lambda)^s \qquad G>0 \tag{10.2-30}$$

and

$$\frac{du_z}{d\xi}=(-6G)^s(\lambda-\xi)^s \qquad G<0 \tag{10.2-31}$$

Integration of Eqs. 10.2-30 and 10.2-31, with boundary conditions $u_z(0)=0$ and $u_z(1)=1$ result in the following velocity profiles for each of the cases $c$ and $d$:

$$u_z=\frac{(6G)^s}{(1+s)}[(\xi-\lambda)^{1+s}-(-\lambda)^{1+s}] \qquad G>0 \tag{10.2-32}$$

where $\lambda$ is obtained from

$$(-\lambda)^{1+s}-(1-\lambda)^{1+s}+\frac{1+s}{(6G)^s}=0 \qquad G>0 \tag{10.2-33}$$

and

$$u_z=\frac{(-6G)^s}{(1+s)}[\lambda^{1+s}-(\lambda-\xi)^{1+s}] \qquad G<0 \tag{10.2-34}$$

where $\lambda$ is obtained from the following equation:

$$\lambda^{1+s}-(\lambda-1)^{1+s}-\frac{1+s}{(-6G)^s}=0 \qquad G<0 \tag{10.2-35}$$

By setting $\lambda=0$ in Eq. 10.2-33 and $\lambda=1$ in Eq. 10.2-35, we find the following condition for the flows without an extremum within the flow regime

$$|G|\leq\tfrac{1}{6}(1+s)^n \tag{10.2-36}$$

This result was, of course, predictable from Eq. 10.2-28.

The flow rate per unit width is now obtained by integrating the velocity profiles for each of the cases. Furthermore, the velocity profile and flow rate equations that are obtained subsequent to integrating the velocity profiles, can be fused into single expressions as follows:

$$u_z = \frac{|6G|^s \operatorname{sign} G}{(1+s)} \left( |\xi - \lambda|^{1+s} - |\lambda|^{1+s} \right) = \frac{|\xi - \lambda|^{1+s} - |\lambda|^{1+s}}{|1-\lambda|^{1+s} - |\lambda|^{1+s}} \tag{10.2-37}$$

and

$$q = \frac{V_0 H |6G|^s \operatorname{sign} G}{(1+s)(2+s)} \left[ (1-\lambda)|1-\lambda|^{1+s} + \lambda|\lambda|^{1+s} - (2+s)|\lambda|^{1+s} \right] \tag{10.2-38}$$

where $\lambda$ is obtained from the following generalized expression:

$$|\lambda|^{1+s} - |1-\lambda|^{1+s} + \frac{1+s}{|6G|^s \operatorname{sign} G} = 0 \tag{10.2-39}$$

In solving for $\lambda$ in the last equation, we find multiple solutions but we recall that

if $|G| \geq \frac{1}{6}(1+s)^n$ then $0 \leq \lambda \leq 1$

if $|G| \leq \frac{1}{6}(1+s)^n$ and if $G > 0$ then $\lambda \leq 0$

if $|G| \leq \frac{1}{6}(1+s)^n$ and if $G < 0$ then $\lambda \geq 1$

Figure 10.3, which plots the solution of Eq. 10.2-39 for three $n$ values, also indicates the four solution regions. Figure 10.4 plots dimensionless flow rate versus the dimensionless pressure gradient $G$, as given by Eq. 10.2-38, with $n$ as a parameter. The increasing nonlinearity of this relationship with increasing deviation from Newtonian behavior is clearly noticeable. Of particular interest is the inflection point in the curves.

### *Nonisothermal Flow of Power Law Model Fluids*

So far we have looked at the isothermal parallel plate flow to gain insights into the dynamic viscous pressurization pump. Flow in such systems, though, is rarely isothermal. There are two reasons for this: first, polymer melts are viscous fluids and generate heat during flow; and second, the temperatures of each of the flow channel walls are generally not equal to each other, nor are they uniform. Both sources of nonisothermicity can affect the resulting velocity profile depending on the temperature sensitivity of the viscosity (flow activation energy). For a power law model fluid, this dependence can be expressed as

$$m = m_0 \, e^{-a(T-T_0)} \tag{10.2-40}$$

when $a = 0$, the nonuniform temperature field does not affect the velocity profile. When $a \neq 0$, the equations of motion and energy are coupled by the term above and must be solved simultaneously.

To give a physical insight of the effects of varying temperature on the velocity field, we consider the following simple problem: pure drag and fully developed flow

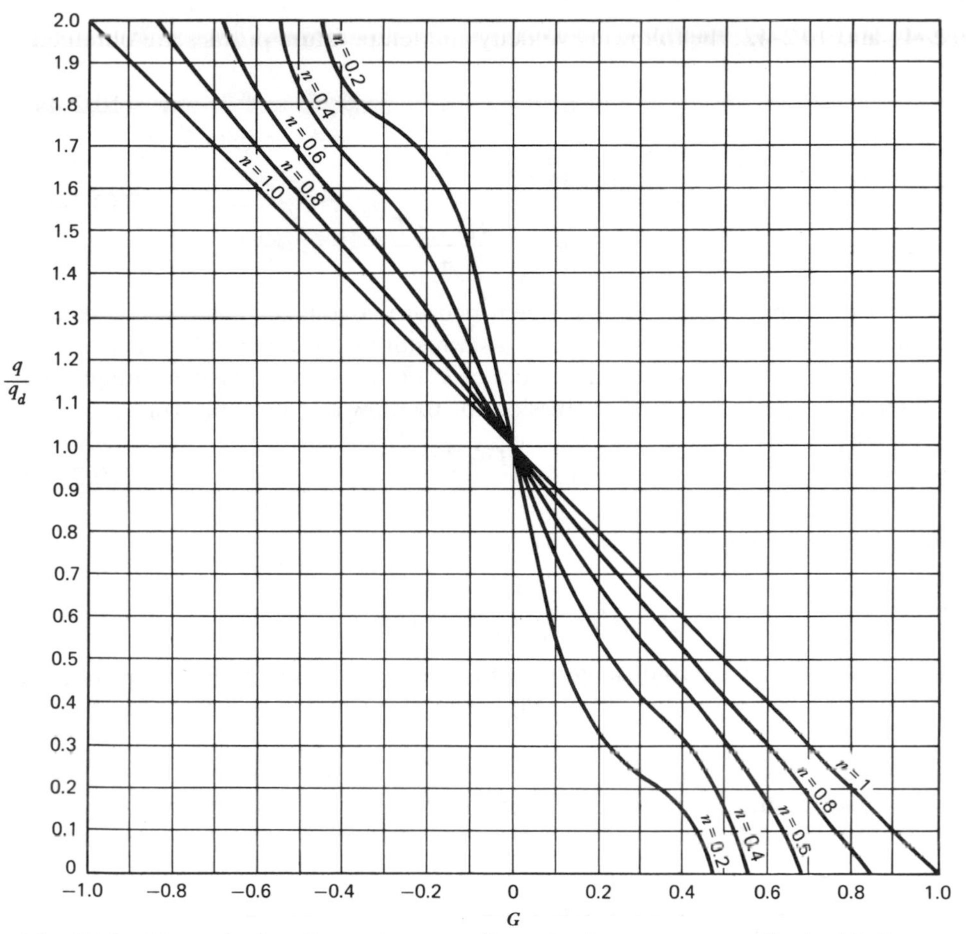


**Fig. 10.4** Dimensionless flow rate versus dimensionless pressure gradient, with the power law exponent $n$ as a parameter, for parallel plate flow as given by Eq. 10.2-38.

of a power law fluid of small $m_0$ (i.e., negligible viscous dissipation, or $\mathrm{Br} \rightarrow 0$) between two parallel plates, one of which is kept at $T_1$ and the other at $T_0$. For this situation, the equations of motion and thermal energy reduce in dimensionless form to

$$\frac{d}{d\xi}\left[\left(\frac{du_z}{d\xi}\right)^n e^{b\Theta}\right]=0 \tag{10.2-41}$$

and

$$\frac{d^2\Theta}{d\xi^2}=0 \tag{10.2-42}$$

where $\Theta=(T-T_0)/(T_1-T_0)$ and

$$b=-a(T_1-T_0) \tag{10.2-43}$$

The boundary conditions are $\Theta(0)=u_z(0)=0$ and $\Theta(1)=u_z(1)=1$. Solving Eqs.

10.2-41 and 10.2-42, the following velocity and temperature profiles are obtained:

$$u_z = \frac{1-e^{-b'\xi}}{1-e^{-b'}} \tag{10.2-44}$$

and

$$\Theta = \xi$$

$$(10.2\text{-}45)$$

where

$$b' = \frac{-a(T_1 - T_0)}{n}$$

Note that the temperature profile is linear because viscous dissipation was neglected, but the velocity profile is nonlinear as a result of the temperature dependent viscosity. Figure 10.5 illustrates the effect of $b'$ on the velocity profile. Integrating the velocity profile provides an expression for the flow rate per unit width

$$q_d = \frac{HV_0}{2} U_2 \tag{10.2-46}$$

where $U_2$ is defined as

$$U_2 = 2\frac{1-b'-e^{-b'}}{b'(e^{-b'}-1)} \tag{10.2-47}$$

and it expresses the quantitative effect of the superimposed temperature profile on the velocity profile in pure drag flow. Figure 10.6 shows this quantity a function of $b'$.

Analyzing Figs. 10.5 and 10.6, we note that a condition of $T_1 > T_0$ ("hot" moving plate) results in a reduction of the conveying capacity by drag (and vice versa). The physical reason for this is that the equation of motion results in a

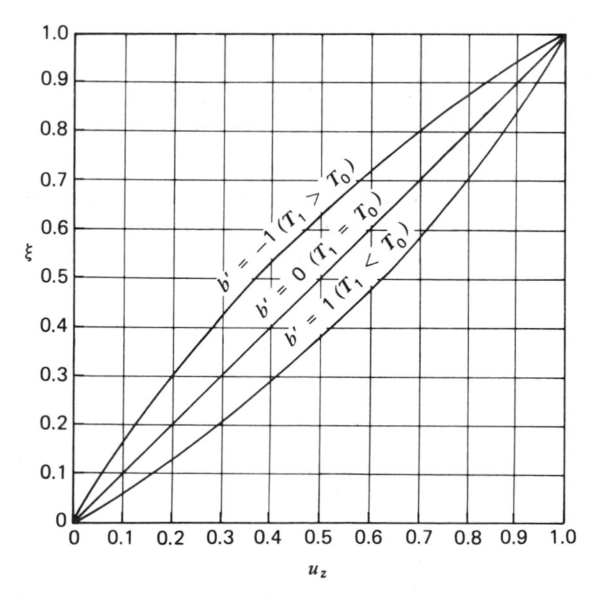


**Fig. 10.5** Velocity profiles in parallel plate flow in the absence of pressure gradient and with a linear temperature profile between the plates: $b' = -a(T_1 - T_0)/n$, where $T_1$ and $T_0$ are the temperatures of the moving and stationary plates, respectively.

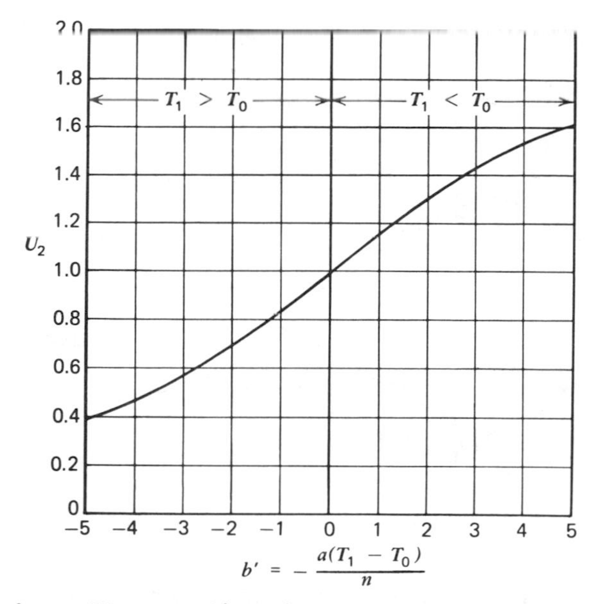


**Fig. 10.6** The factor $U_2$ versus $b'$. $U_2$ is the ratio of drag flow rate with $T_1 \neq T_0$ to isothermal flow rate (Eq. 10.2-47).

constant *shear stress* throughout the gap, and since the *viscosity is lower* at the high temperature plate, the *shear rate must be higher* to maintain the constant shear stress value. The temperature effect may be quite significant in many cases. Indeed if the temperature gradient is very high, the flow rate may decrease so much that the moving plate appears to "slip" past the melt (as a hot knife would slip over butter). Finally, we note that non-Newtonian effects amplify the temperature effect, as is evident from the definition of $b'$.

The coupled flow and heat transfer problem for nonvanishing Brinkman number was solved analytically by Gavis and Laurence (4) for equal plate temperatures and adiabatic stationary plate condition (see Problem 10.6). It is interesting to note in their results that the solution is double valued *in the applied shear stress* at the moving wall (i.e., two different velocity and corresponding temperature profiles satisfy the differential equation and boundary conditions). The solution, however, is unique in the plate velocity or the Brinkman number.

## 10.3 The Screw Pump

We have seen in the previous section how a simple geometrical configuration provides the means for increasing the pressure of a viscous liquid. The next obvious question is, How can this concept be put to practical use? That is, how can we find an engineering solution? After all, we can't build a machine consisting of two infinite plates! Let us therefore start modifying the concept, rendering it more practical, yet without losing the principal advantages provided by it; that is, let us start developing a pump that utilizes the viscous drag pressurization principles.

First, we *confine* the melt by providing side walls on the lower plate, creating a shallow rectangular channel of width $W$ (Fig. 10.7). The moving upper plate slides

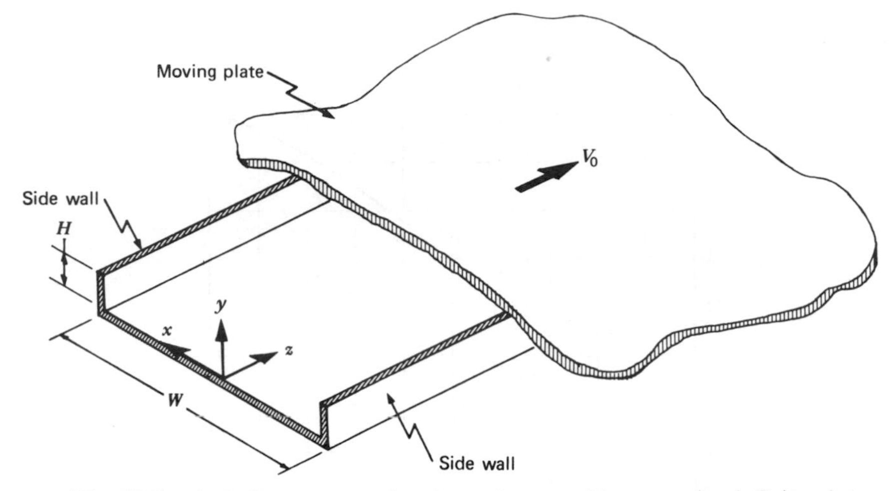


**Fig. 10.7** A shallow rectangular channel covered by a moving infinite plate.

on the channel with a constant velocity in the down channel, $z$-direction. The equations in the previous section hold for this new geometry, provided the channel is shallow, $H/W \ll 1$. Otherwise we must modify the equations (to account for velocity gradients in the $x$-direction), although the basic conclusion will remain the same. Next we make the channel finite in length, close the entrance and exit, equipping the former with a "feeding device" and the latter with a die, as in Fig. 10.8. Clearly, if we can provide for a continuous feed at low pressure $P_1$ at the entrance, the "device" will pump the melt, raise its pressure to $P_2$, and extrude it across the die at the exit. Thus we "almost" have a practical pressure generating device or pump, except that the upper plate is still an "infinite" plate. We could eliminate this last "theoretical" part of the pump by replacing it, for example, with

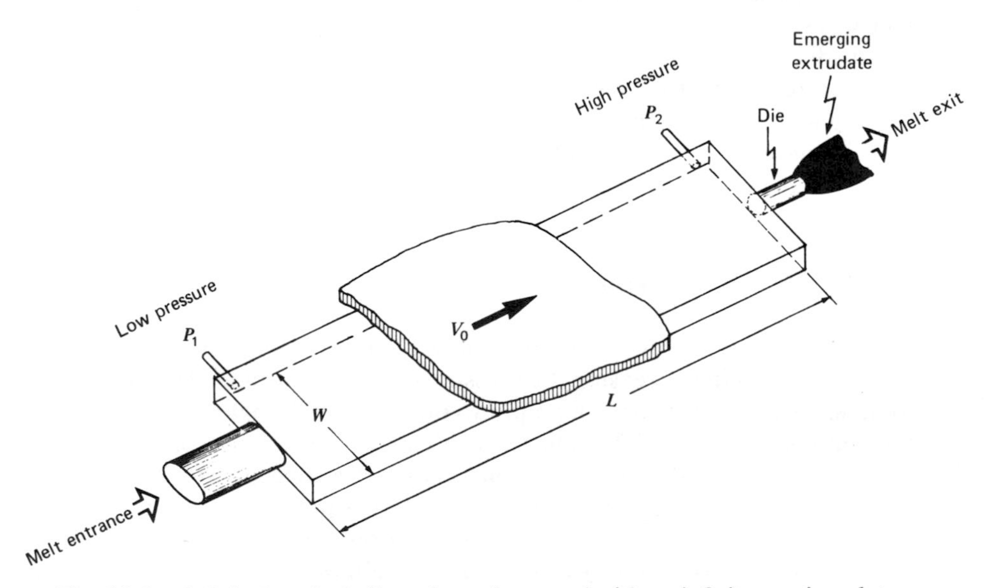


**Fig. 10.8** A finite length shallow channel covered with an infinite moving plate.

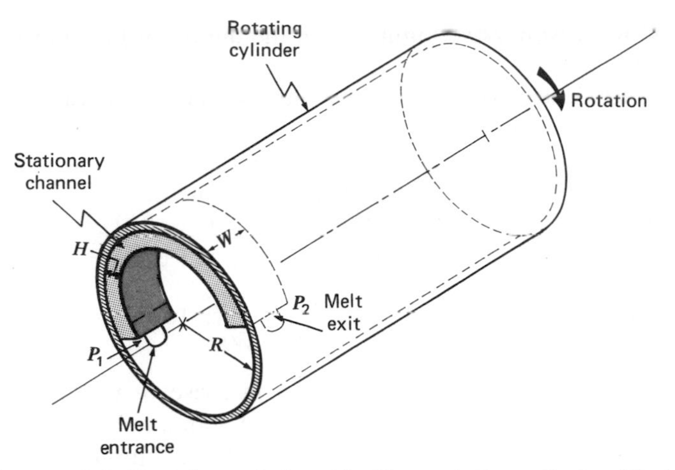


**Fig. 10.9** A curved shallow channel placed inside a rotating cylinder. The inner surface of the cylinder forms the moving infinite surface.

a "practical" infinite belt. Such a solution, however, is hardly practical for a channel filled with hot viscous melt. A *more practical* way to solve this problem would be to bend the channel in the $z$-direction into an arc of circle. Then a rotating cylinder fitted on top of the curved channel would act as the "infinite" plate, as in Fig. 10.9. The curvature will only slightly affect the velocity profile (actually improving the pressure generating capability), without changing the concept.*

The tangential velocity of the moving surface in Fig. 10.9 is given by the following expression:

$$V_0 - \pi ND \tag{10.3-1}$$

where $N$ is the frequency of rotation of the cylinder and $D$ its inner diameter.

With this latter modification in geometry, we have reached a machine configuration that can be built and will "work" (although it still contains serious limitations). Moreover, we also have a theoretical model for the device, which we can use for optimizing the design. Equation 10.2-7 forms the basis for this model. After multiplying it by $W$ to get the total flow rate, and substituting for the pressure gradient the pressure drop from entrance to exit (since the pressure gradient is constant, this is an allowable substitution), we get the "design equation" of this pump:

► $$Q = \frac{V_0 WH}{2} + \frac{WH^3}{12\mu}\left(\frac{P_1 - P_2}{L}\right) \tag{10.3-2}$$

Our objective is to obtain high flow rates and pressures, since high pressures are needed in the shaping step (die flow). By analyzing Eq. 10.3-2, we can explore the possibilities and limitations inherent in each variable appearing in it, to attain this objective. The velocity $V_0$ is proportional to the frequency of rotation. Increasing $N$ will result in an increase in both $Q$ and $P_2$. There is, however, a

* An alternative solution to the problem of providing for a "practical infinite plate" is to use the surface of a flat rotating disk over a spiral channel. The melt is fed at the center of the device and pumped out of an exit port, which can be at any radial position. Such a device has actually been suggested (German Patent DBP 1, 032, 523).

practical limit to increasing $N$ stemming from the nature of polymeric melts. High frequency of rotation implies high shear rates, which may result in mechanical degradation due to shear, overheating and burning due to viscous dissipation, or even slip at the solid boundary. We are, therefore, limited in using this operating variable as a means of increasing the output of the device. We now turn our attention to the design variables. We can increase the pressure $P_2$ for a given output $Q$ by manipulating the channel width $W$, the channel depth $H$, and its length $L$. It will be easier to analyze these variables by rearranging Eq. 10.3-2 as follows:

$$P_2 - P_1 = 12\mu L\left(\frac{V_0}{2H^2} - \frac{Q}{WH^3}\right) \tag{10.3-3}$$

Increasing the channel width will result in an increase in pressure, but we are limited in doing this because soon we will be faced with a major entrance and exit effect. The effect of the channel depth is somewhat more complicated. Section 10.2 indicates that there is an optimum channel depth for a maximum pressure rise for a given flow rate $Q$. The optimum channel depth is

$$H_{\text{opt}} = \frac{3Q}{WV_0} \tag{10.3-4}$$

This is a rather useful result, yet it also serves to indicate the limitations set by manipulating the channel depth in our attempt to reach high pressures.

The last design variable we have at our disposal is the channel length $L$. The geometric configuration we are dealing with sets a physical limit of $\pi D$ on $L$. We could of course increase the diameter, but obviously there is a practical limit to this approach. However we can do something different. By one final slight modification of the geometry, we can increase $L$ to almost any desired value! This we can achieve by *twisting* the channel so that after one full turn it will be displaced in the axial direction by *exactly one channel width* (plus the thickness of the separating wall). Figure 10.10 illustrates this.

The geometrical configuration we have now is equivalent to *a single screw extruder.* All we have to do is to cut out the helical channel from a solid cylinder, as

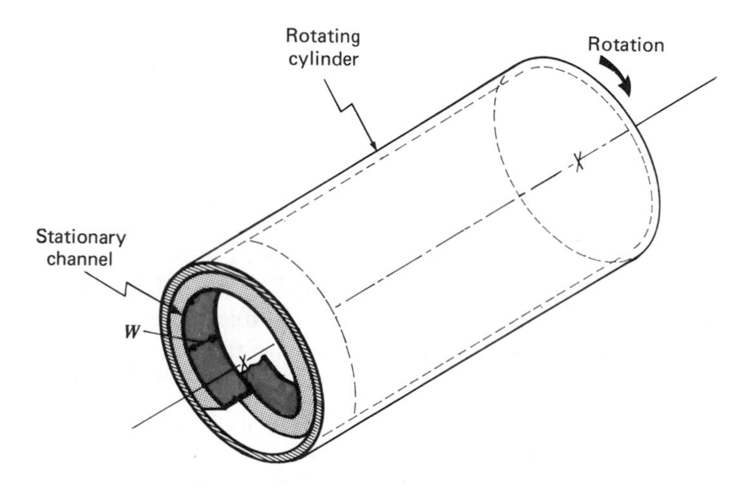


**Fig. 10.10** The helical curved channel placed inside a rotating cylinder.

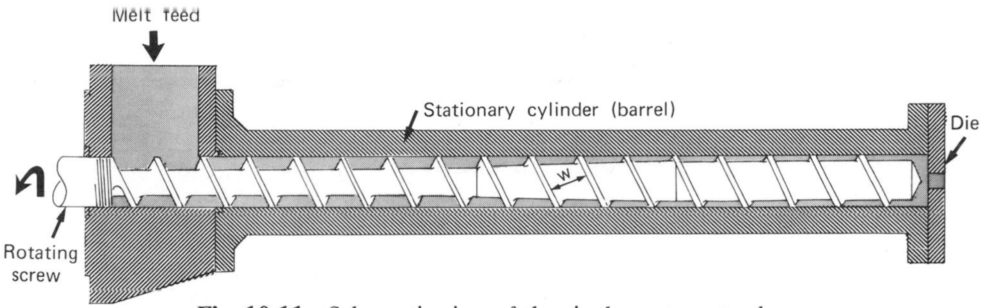


**Fig. 10.11** Schematic view of the single screw extruder.

in Fig. 10.11, and place the screw into a hollow cylinder. Instead of rotating the cylinder, we can rotate the screw in the opposite direction. By doing so we gain two important advantages: both the entrance and exit problems become simple to solve. The former can be achieved through a port in the cylinder (barrel), whereas the latter is solved by itself; the screw is terminated at the required length and the melt is simply pumped into the die.

There are, however, also numerous further advantages to the screw configuration: the stationary barrel can be heated or cooled as needed; the screw can be hollow, permitting heating or cooling; the mechanical energy input for the pumping action is accomplished simply and easily by turning the screw shaft via a motor; the helical channel creates a velocity component perpendicular to the flight, which results in a circulatory flow providing for good mixing action; the final resulting velocity profile gives rise to a very narrow residence time distribution, rendering the screw extruder very effective in handling temperature sensitive polymers; the structure is rugged mechanically and allows for a range of size and operation from minute flow rates to many tons per hour; and finally, the feed may be either melt or particulate solids. As Chapters 8 and 9 demonstrated, the screw extruder, in addition to being an efficient pump, is also a good solids conveying device and an excellent melt generator.

Considering all these advantages, it is indeed no wonder that the single screw extruder has become the most important polymer processing machine.

### *The Screw Pump (single screw extruder "metering" zone) Model*

This section develops a mathematical model of single screw melt pumping. First we briefly discuss some geometrical relations pertaining to screws. The two main geometrical variables characterizing an extruder screw are its diameter $D$ (taken at the tip of the flights) and its axial length $L$. The latter is usually referred to in terms of the length-to-diameter ratio. Typical values of this ratio are 24 to 26, although occasionally we find extruder screws with length-to-diameter ratios as high as 40 and as low as 8. The latter are generally either rubber extruders or early thermoplastic extruders. Diameters range from 2 cm or less to 75 cm or more. The screw cannot be tightly fitted into the barrel because of friction; there is a small radial clearance $\delta_f$ between the tip of the flights and the inner diameter of the barrel

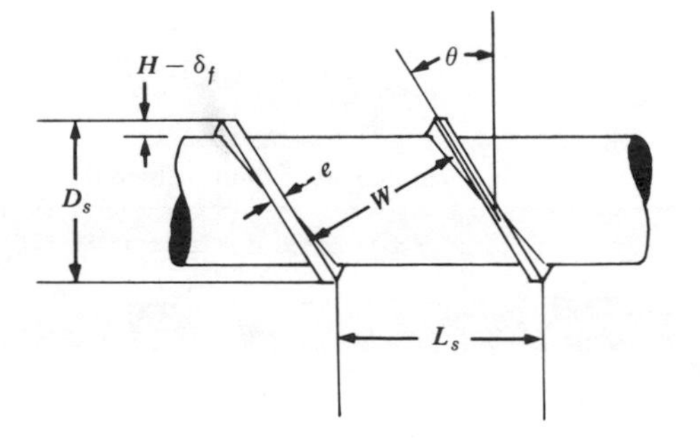


**Fig. 10.12** Geometry of a single flighted extruder screw.

$D_b$ of the order of 0.2–0.5 mm. Polymer melt flows continuously over this clearance, acting as a lubricant. The diameter of the screw at the tip of the flight is $D_s = D_b - 2\delta_f$. The axial distance of one full turn of the flight is called the *lead* $L_s$. Most screws of single extruders are single flighted with $L_s = D_s$. Such a screw is referred to as *squared pitched screw*. A section of such a screw appears in Fig. 10.12. The radial distance between the barrel surface and the root of the screw is called the *channel depth* $H$. The main design variable of screws is the channel depth profile, that is, $H(z)$, where $z$ is the helical, down channel direction. The angle formed between the flight and the plane normal to the screw axis is called the *helix angle* $\theta$. The helix angle can be expressed in terms of the lead and diameter as follows:

$$\tan\theta = \frac{L_s}{\pi D} \tag{10.3-5}$$

The value of the helix angle is therefore a function of the diameter (radial distance). At the tip of the flight it is smaller than at the root of the screw. A square pitched screw with the flight clearance neglected has a helix angle of 17.65° ($\tan\theta = 1/\pi$) at the flight tip.

The *width* of the channel $W$ is the perpendicular distance between the flight, and it is given by

$$W = L_s \cos\theta - e \tag{10.3-6}$$

where $e$ is the flight width. Clearly, since $\theta$ varies with radial location, so does $W$. Finally, the helical distance along the channel $z$ is related to the axial distance $l$ by:

$$z = \frac{l}{\sin\theta} \tag{10.3-7}$$

and it is also a function of the distance from the root of the screw.

We commence the development of the *mathematical model* for screw extrusion by reversing the conceptual process that led to the screw concept. This reverse process (i.e., unwinding of the channel), for a screw with shallow and constant channel depth, leads to a rectangular channel covered by an infinite plate (the barrel) moving at constant velocity of $V_b$

$$V_b = \pi N D_b \tag{10.3-8}$$

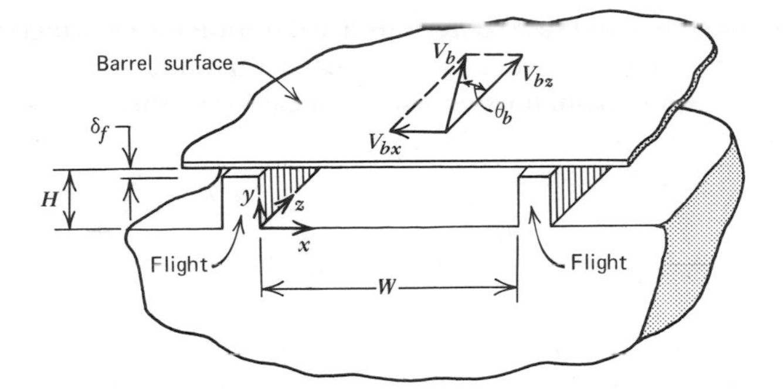


**Fig. 10.13** Geometry of the "unwound" rectangular channel.

at an angle $\theta_b$ to the down channel direction (Fig. 10.13). The surface velocity can be separated into down channel and cross channel components given, respectively, by

$$V_{bz} = V_b \cos\theta_b \tag{10.3-9}$$

and

$$V_{bx} = V_b \sin\theta_b \tag{10.3-10}$$

The former "drags" the polymer melt toward the exit, whereas the latter induces the cross channel mixing. Note that the barrel surface angle $\theta_b$ is used above.

Comparing the present flow situation to the previously discussed simple parallel plate situation, we note two important differences. First, the flow in the down channel, $z$-direction is two-dimensional [i.e., $v_z(x, y)$], and the barrel surface has a velocity component in the $x$-direction that will result in a circulatory flow in the cross channel direction.

The basic assumptions for solving this flow problem are the same as those for the parallel plate flow. Under the above-listed assumptions, the three components of the equation of motion in rectangular coordinates defined in Fig. 10.13 reduce to

$$x\text{-component}\quad \rho\left(v_x\frac{\partial v_x}{\partial x}+v_y\frac{\partial v_x}{\partial y}\right)=-\frac{\partial P}{\partial x}+\mu\left(\frac{\partial^2 v_x}{\partial x^2}+\frac{\partial^2 v_x}{\partial y^2}\right) \tag{10.3-11}$$

$$y\text{-component}\quad \rho\left(v_x\frac{\partial v_y}{\partial x}+v_y\frac{\partial v_y}{\partial y}\right)=-\frac{\partial P}{\partial y}+\mu\left(\frac{\partial^2 v_y}{\partial x^2}+\frac{\partial^2 v_y}{\partial y^2}\right) \tag{10.3-12}$$

$$z\text{-component}\quad \rho\left(v_x\frac{\partial v_z}{\partial x}+v_y\frac{\partial v_z}{\partial y}\right)=-\frac{\partial P}{\partial z}+\mu\left(\frac{\partial^2 v_z}{\partial x^2}+\frac{\partial^2 v_z}{\partial y^2}\right) \tag{10.3-13}$$

where $\rho$ is the density of the melt. In these equations the velocity components are not functions of $z$, since the flow is fully developed. If it is assumed that flow in the cross channel direction is also fully developed (which is a good approximation in shallow channels), then $\partial v_x/\partial x$, $\partial v_y/\partial x$, and $\partial v_z/\partial x$ are zero. Therefore, from the equation of continuity, one obtains that $\partial v_y/\partial y = 0$ and, in combination with the foregoing, that $v_y = 0$. Furthermore, Eq. 10.3-12 reduces to $\partial P/\partial y = 0$, resulting in the pressure $P$ being a function of $x$ and $z$ only. Therefore Eq. 10.3-11 reduces to

$$\frac{\partial P}{\partial x} = \mu\frac{\partial^2 v_x}{\partial y^2} \tag{10.3-14}$$

In Eq. 10.3-13 the left-hand side represents acceleration terms, which in the case of slow motion of a viscous fluid will be much smaller than the terms representing the viscous forces on the right-hand side. In a typical flow situation in extruders the ratio of inertia to viscous forces is of the order of $10^{-5}$ (3b).

Thus Eq. 10.3-13 reduces to

$$\frac{\partial P}{\partial z} = \mu\left(\frac{\partial^2 v_z}{\partial x^2} + \frac{\partial^2 v_z}{\partial y^2}\right) \tag{10.3-15}$$

It is evident from Eq. 10.3-14 that the right-hand side is a function of $y$ only, whereas the left-hand side is a function of $x$ and $z$. Since neither side is dependent on the variables of the other, both are constant. Equation 10.3-14, therefore, can be integrated to give

$$v_x = \frac{y^2}{2\mu}\left(\frac{\partial P}{\partial x}\right) + C_1 y + C_2 \tag{10.3-16}$$

The constants $C_1$ and $C_2$ are evaluated from the boundary conditions

$$\begin{aligned} v_x(0) &= 0 \\ v_x(H) &= -V_{bx} \end{aligned} \tag{10.3-17}$$

Substituting the boundary conditions into Eq. 10.3-16 results in

$$u_x = -\xi + \xi(\xi - 1)\left(\frac{H^2}{2\mu V_{bx}}\frac{\partial P}{\partial x}\right) \tag{10.3-18}$$

where $u_x = v_x/V_{bx}$ and $\xi = y/H$. For the transverse flow in the channel, there is the additional condition of zero flow rate, provided we neglect leakage flow across the flights. Expressing this condition mathematically

$$\int_0^1 u_x \, d\xi = 0 \tag{10.3-19}$$

and substituting Eq. 10.3-18 into Eq. 10.3-19, subsequent to integration, results in

$$\frac{\partial P}{\partial x} = -6\mu\frac{V_{bx}}{H^2} = -6\mu\frac{\pi N D_b \sin\theta_b}{H^2} \tag{10.3-20}$$

This equation gives the cross channel pressure gradient. Note that this gradient is proportional to $N$ and $D_b$ and inversely proportional to the square of the channel depth.

Finally, by substituting Eq. 10.3-20 into Eq. 10.3-18, we obtain the cross channel velocity profile

► $$u_x = \xi(2 - 3\xi) \tag{10.3-21}$$

According to the velocity profile (Fig. 10.14), which of course holds only at some distance away from the flights, the melt circulates around a stagnant layer at two-thirds of the channel height.

In the down channel direction the velocity profile is obtained by solving the partial differential equation 10.3-15. It can be shown (3b) that the pressure gradient $\partial P/\partial z$ is a constant (although $P$ is a function of both $x$ and $z$). The

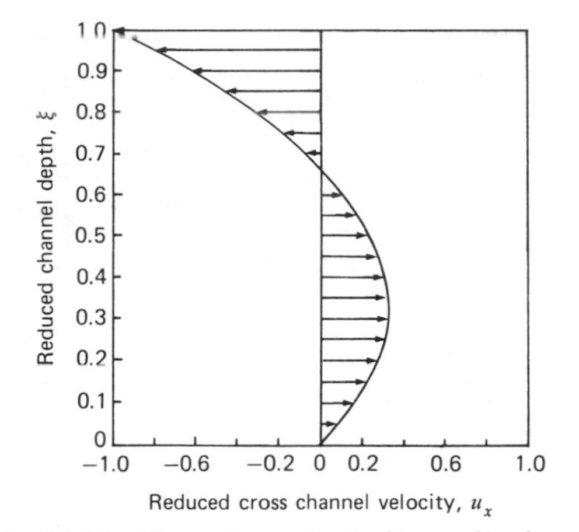


**Fig. 10.14** Cross channel velocity profile, from Eq. 10.3-21.

boundary conditions are

$$\begin{aligned} v_z(x, 0) &= 0 \\ v_z(x, H) &= V_{bz} \\ v_z(0, y) &= 0 \\ v_z(W, y) &= 0 \end{aligned} \tag{10.3-22}$$

The velocity profile in a channel of rectangular cross section has been solved by many authors. The pure pressure flow was first solved by Boussinesq (5). The complete problem was solved by Rowell and Finlayson (6) in 1922, in an effort to develop a mathematical model for screw pumps. The detailed solution by separation of variables is given in Ref. 3c, and the result is

$$u_z = \frac{4}{\pi} \sum_{i=1,3,5}^{\infty} \frac{\sinh(i\pi h\xi)}{i \sinh(i\pi h)} \sin(i\pi\chi) - \left(\frac{H^2}{2\mu V_{bz}} \frac{\partial P}{\partial z}\right) \left[\xi^2 - \xi + \frac{8}{\pi^3} \sum_{i=1,3,5}^{\infty} \frac{\cosh[i\pi(\chi - 0.5)/h]}{i^3 \cosh(i\pi/2h)} \sin(i\pi\xi)\right] \tag{10.3-23}$$

where $u_z = v_z/V_{bz}$, $\chi = x/W$, and $h = H/W$. Figure 10.15 gives the velocity distribution of the pure drag flow ($\partial P/\partial z = 0$), for a number of $h$ values.

The flow rate or throughput in an extruder, the pressure profile along the screw, and the power consumption are the quantities of utmost interest for design, and these are calculated from the velocity profiles in the channel. The net flow rate is obtained by integrating the down channel velocity component $v_z$ given in Eq. 10.3-23 across the channel

$$Q = WHV_{bz} \int_0^1 \int_0^1 u_z \, d\xi \, d\chi \tag{10.3-24}$$

The integration of the infinite series in Eq. 10.3-23 poses no problems. The series is uniformly converging, and termwise integration is permissible. The results

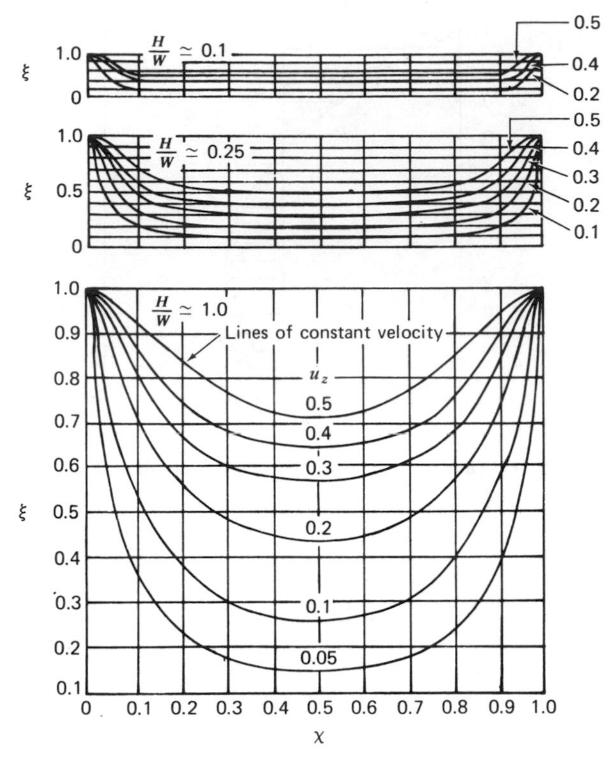


**Fig. 10.15** Down channel velocity distribution for pure drag flow, from Eq. 10.3-23. (Reprinted with permission from E. C. Bernhardt, ed. *Processing of Thermoplastic Materials*, Reinhold, New York, p. 290.)

are given in the following equation:

$$Q = \frac{V_{bz} WH}{2} F_d + \frac{WH^3}{12\mu}\left(-\frac{\partial P}{\partial z}\right) F_p \tag{10.3-25}$$

where $F_d$ and $F_p$ are "shape factors" for the drag and pressure flows, respectively. They assume values that are smaller than 1 and represent the reducing effect of the flights on flow rate between infinite parallel plates. They are given by

$$F_d = \frac{16W}{\pi^3 H} \sum_{i=1,3,5}^{\infty} \frac{1}{i^3} \tanh\left(\frac{i\pi H}{2W}\right) \tag{10.3-26}$$

$$F_p = 1 - \frac{192H}{\pi^5 W} \sum_{i=1,3,5}^{\infty} \frac{1}{i^5} \tanh\left(\frac{i\pi W}{2H}\right) \tag{10.3-27}$$

Note that the two shape factors (Fig. 10.16) depend only on the $H/W$ ratio.

The effect of the flights on pressure flow is stronger than that on drag flow. When the value $H/W$ becomes very small, both $F_d$ and $F_p$ approach unity. In this case, Eq. 10.3-25 reduces to the simplest possible theory of extrusion, that is, flow of isothermal Newtonian fluid between two infinite parallel plates, Eq. 10.3-2. Equation 10.3-25 is the well-known equation for throughput in the isothermal Newtonian extrusion theory. Since this is the solution of a linear differential equation, it is composed of two independent terms, the first representing the contribution of drag flow and the second that of pressure flow. These two terms are independent of each other. This is also true for the velocity equation. An increase in the positive pressure gradient will increase the pressure backflow. This decreases the net flow rate, despite an unchanging drag flow.

The ratio of pressure to drag flow rates is obtained from Eq. 10.3-25

$$-\frac{Q_p}{Q_d}=\frac{H^2}{6\mu V_{bz}}\left(\frac{\partial P}{\partial z}\right)\frac{F_p}{F_d} \quad (10.3\text{-}28)$$

which for shallow channels reduces to Eq. 10.2-10.

The concept of pressure backflow and the shape of the velocity profile in the $z$-direction generated the erroneous concept that actual flow toward the hopper occurs in some part of the channel. *Under no condition,* however, does flow backward along the screw axis occur.* This is clear from the velocity profiles in the axial direction

$$v_l = v_x \cos\theta + v_z \sin\theta \quad (10.3\text{-}29)$$

Substituting Eq. 10.3-21 and Eq. 10.3-23 (which for shallow channels reduces to Eq. 10.2-6) into Eq. 10.3-29, and using Eq. 10.3-28, we get

$$u_l = 3\xi(1-\xi)\left(1+\frac{Q_p}{Q_d}\right)\sin\theta\cos\theta \quad (10.3\text{-}30)$$

where $u_l = v_l/V_b$.

Examining this equation reveals some interesting points. For helix angles between 0 and $\pi/2$, $u_l$ will always be positive. In other words, in no case does a "backflow" exist in the axial direction. For closed discharge ($Q_p+Q_d=0$), $u_l$ becomes zero for all values of $\xi$. The *shape* of the velocity profile is identical for all values of $Q_p/Q_d$. Only the maximum velocity, which always occurs at the mid-channel ($\xi=0.5$), changes from zero in the case of no output to

$$u_{l,\max}=\frac{3\sin\theta\cos\theta}{4} \quad (10.3\text{-}31)$$

for pure drag flow.

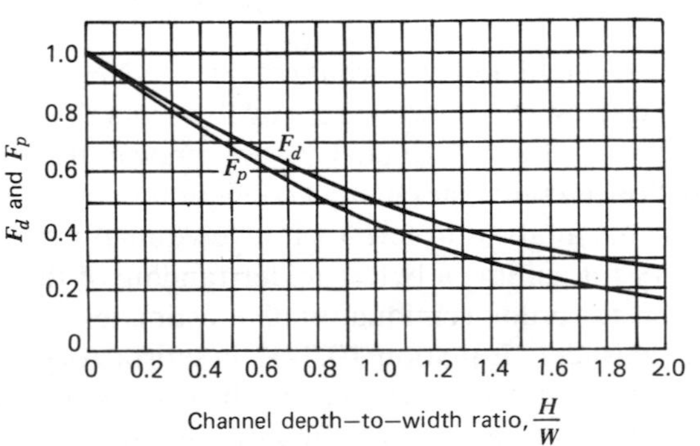


**Fig. 10.16** Shape factors for drag and pressure flow, from Eqs. 10.3-26 and 10.3-27.

* Except if a head pressure higher than the pressure generated by the screw is imposed at the die (e.g., by another extruder in a head on configuration).

Finally, the velocity in the axial direction also depends on $\theta$. For a pure drag flow, the velocity $u_l$ at any $\xi$ will attain a maximum value at $\theta = \pi/4$.

Figure 10.17 shows velocity profiles in various planes as a function of the $Q_p/Q_d$ ratio.

It is worthwhile to note that the velocity profile $u_x$ is independent of the $Q_p/Q_d$ ratio. Another interesting point is that for zero throughput, both $u_x$ and $u_z$ assume a value of zero at the same point, $\xi = \frac{2}{3}$. In other words, under these conditions fluid particles at this position will be stationary. (In the neighborhood of the flight, of course, they will have a $v_y$ component and will circulate.)

From these velocity profiles, one can deduce the possible path of a fluid particle in the channel. For closed discharge, the fluid, making no advance in the axial direction, advances and retreats in the $x$- and $z$-directions. It describes a closed path in a plane perpendicular to the screw axis, as in Fig. 10.18. As the particle comes close to the flight, it starts to move toward the root of the screw. Then the pressure gradients $\partial P/\partial x$ and $\partial P/\partial z$ cause it to start to move backward in the channel toward the other flight.

As the throughput increases from zero, the closed loops open, and the fluid particles travel along a sort of flattened helical path. The closer the flow to pure drag flow, the more open the loops of the helix will be. Thus polymer particles travel in a path that is a flat helix within a helical channel.

It is very difficult to evaluate the effect of leakage flow across the flights on the real flow situation, but we can achieve this quite well for Newtonian isothermal flow in rectangular channels. Such an analysis will lead to a modification of Eq. 10.3-25 (3d)

$$\blacktriangleright \quad Q = \frac{V_{bz} W (H - \delta_f)}{2} F_d + \frac{WH^3}{12\mu} \left( -\frac{\partial P}{\partial z} \right) F_p (1 + f_L) \tag{10.3-32}$$

where

$$f_L = \left( \frac{\delta_f}{H} \right)^3 \frac{e}{W} \frac{\mu}{\mu_f} + \frac{\left(1 + \dfrac{e}{W}\right) \left[ \dfrac{1 + e/W}{\tan^2 \theta} - \dfrac{6\mu V_{bz} (H - \delta_f)}{H^3 (\partial P/\partial z)} \right]}{1 + \dfrac{\mu_f}{\mu} \left( \dfrac{H}{\delta_f} \right)^3 \dfrac{e}{W}} \tag{10.3-33}$$

$\mu_f$ is the viscosity across flight and $\mu$ in channel.* For a Newtonian fluid $\mu_f = \mu$.

We have seen the screw extruder concept evolve from that of the parallel plate. We have also seen how, to be able to analyze the screw extruder, we return to the parallel plate or shallow rectangular channel configuration as a first approximation. We have carried out the analysis in terms of a Newtonian isothermal model. This analysis is satisfactory for gaining a better understanding of the pressurization and flow concept and also for crude semiquantitative estimates of the actual pumping performance of the extruder. Equation 10.3-32 is satisfactory for this purpose. If, however, reliable design equations are needed, it is necessary to eliminate a long series of simplifying assumptions leading, step by step, to more complex solutions. The end result should be a model for nonisothermal flow of a non-Newtonian fluid in the actual helical channel, not disregarding leakage flow across the flights, and

* This is an attempt to approximately account for a non-Newtonian effect in a Newtonian model.

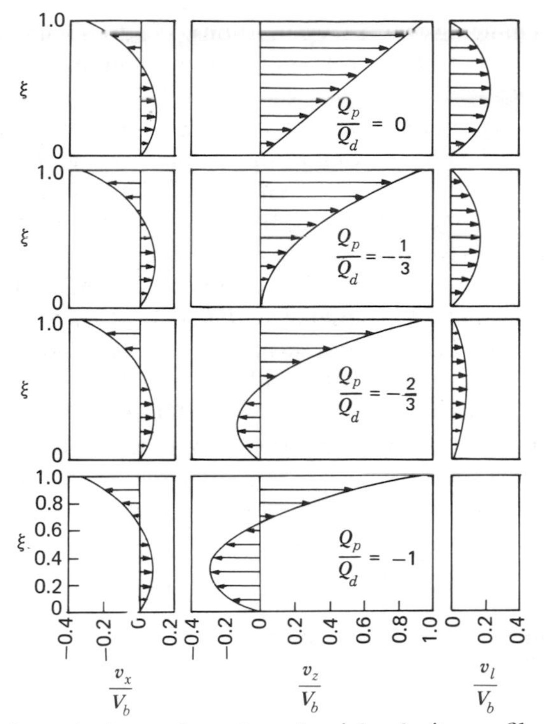


**Fig. 10.17** Cross channel, down channel, and axial velocity profiles for various $Q_p/Q_d$ values, in shallow square pitched screws. (Reprinted with permission from J. M. McKelvey, *Polymer Processing*, Wiley, New York, 1962.)

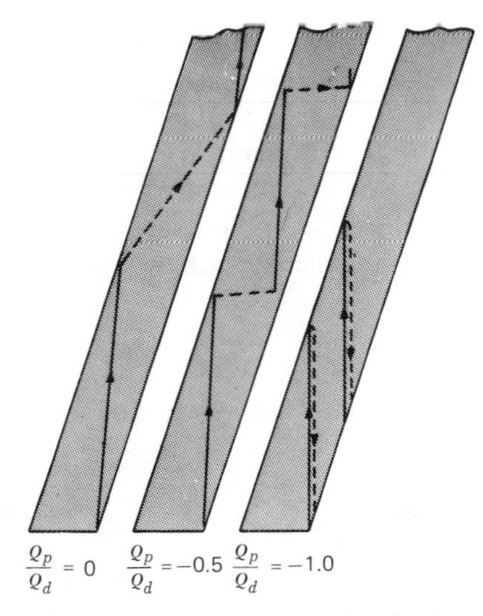


**Fig. 10.18** The path of a fluid particle in the screw channel for $Q_p/Q_d = 0$, $Q_p/Q_d = -0.5$, and $Q_p/Q_d = -1.0$. Solid lines show the path of the fluid in the upper portion of the channel (at $\xi = 0.9$), and the broken curve shows the path of the same fluid particle after it turns over, in the lower portion of the channel (at $\xi = 0.35$).

accounting for the changing boundary conditions. We do not have today a complete and satisfactory solution to this problem, although a great deal of research effort has been invested into it.

In general, two types of approach have been used that are in many ways complementary to each other. Attempting to solve as rigorously as possible the actual flow problems is the approach taken by Griffith (7), Colwell and Nicholls (8), Pearson (9), Zamodits (10), and others. It inevitably leads to numerical solutions. The other approach is to treat idealized systems and try to quantitatively estimate the effects of one particular variable. For example, the effect of channel curvature on flow rate can be evaluated by comparing tangential flow to rectangular channel flow. This can be easily done for isothermal power law fluid (11, 3) for drag and pressure flows separately. The results can be included into the flow rate equation 10.3-32 through curvature correction factors (3e). Other similar correction factors were derived for the various important effects excluded from the simple model.

We return to extrusion modeling, using the results just derived, in Sections 12.1 and 12.2 dealing, respectively, with the engineering systems of the single screw melt pump and single screw plasticating extruder.

## 10.4 Dynamic Viscous Pressurization: Non-Parallel Plate Flow

This section considers another very simple geometry consisting of two non-parallel plates in relative motion. In principle this geometrical configuration is the central element of a number of processing methods, such as knife and roll coating, calendering, and roll-milling (Fig. 10.19). In some of these methods both "plates" are set in motion, in others only one of the "plates" moves. Nevertheless, studying the simplest case (Fig. 10.20), with one plate moving, helps in elucidating the basic mechanism involved. Moreover, this particular case is also applicable to screw extrusion because sections of screw channels are tapered.

We assume narrow gap configurations and moderate tapering and invoke the lubrication approximation, discussed in Section 5.4. Furthermore, assuming steady isothermal flow of an incompressible Newtonian liquid permits direct application of

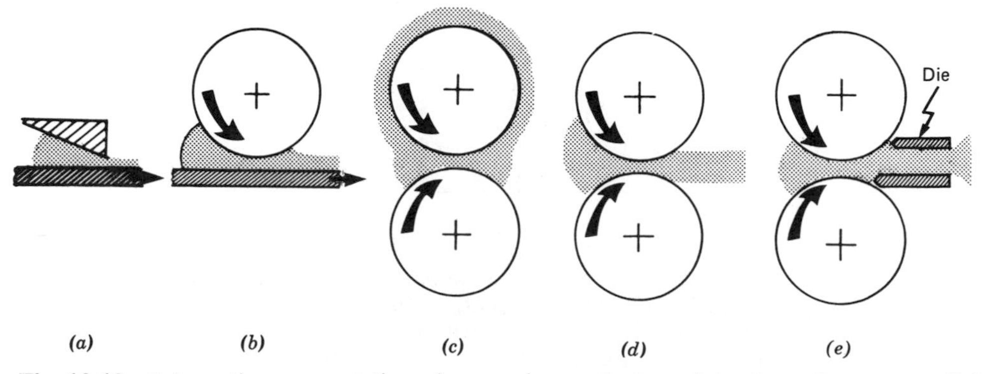


**Fig. 10.19** Schematic representation of processing methods evolving from the non-parallel plate geometry. (*a*) Knife coating. (*b*) Roll coating. (*c*) Roll-mill. (*d*) Calender. (*e*) Twin roll extrusion.

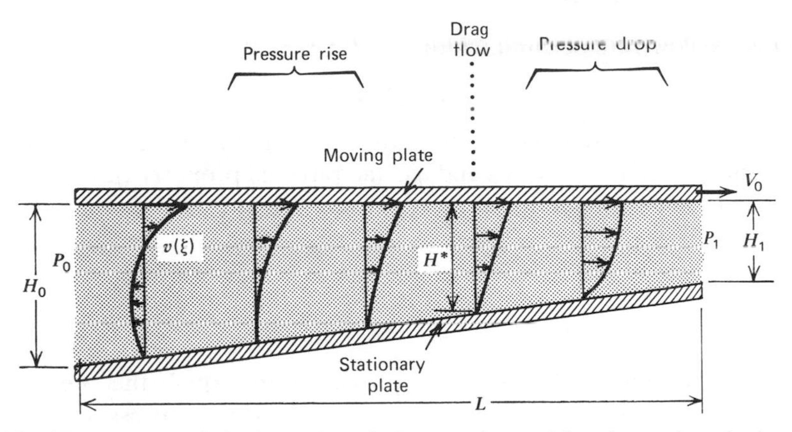


**Fig. 10.20** Two nonparallel plates in relative motion, with schematic velocity profiles corresponding to a pressure rise zone followed by a pressure drop zone ($P_0 = P_1$).

the Reynolds equation 5.4-11, which for one-dimensional flow reduces to

$$\frac{d}{dz}\left(H^3 \frac{dP}{dz}\right) = 6\mu V_0 \frac{dH}{dz} \tag{10.4-1}$$

Equation 10.4-1 can be integrated with respect to $z$ to give

$$H^3 \frac{dP}{dz} = 6\mu V_0 H + C_1 \tag{10.4-2}$$

where $C_1$ is an integration constant, which can be evaluated from the following boundary condition $dP/dz = 0$ at $H = H^*$. If the pressure profile exhibits a maximum within $0 < z \leq L$, then $H^*$ is the separation between the plates at that location; if the pressure profile exhibits no maximum in this range, the mathematical function describing pressure will still have a maximum at $z \geq L$ or at $z \leq 0$, and $H^*$ will be the "separation" between the two plates extended to that point. Equation 10.4-2 can be rewritten as

$$\frac{dP}{dz} = 6\mu V_0 \frac{(H - H^*)}{H^3} \tag{10.4-3}$$

and integrated to give the pressure profile

$$P = P_0 + 6\mu V_0 \int_0^z \frac{H - H^*}{H^3}\, dz \tag{10.4-4}$$

where $P(0) = P_0$. For a constant taper

$$\zeta = \zeta_0 - (\zeta_0 - 1)\frac{z}{L} \tag{10.4-5}$$

where $\zeta = H/H_1$ and $\zeta_0 = H_0/H_1$. Equation 10.4-4 can be integrated to give

$$P = P_0 + \frac{6\mu L V_0}{H_0 H_1}\left[\frac{\zeta_0 - \zeta}{\zeta(\zeta_0 - 1)} - \frac{q}{V_0 H_0}\frac{\zeta_0^2 - \zeta^2}{\zeta^2(\zeta_0 - 1)}\right] \tag{10.4-6}$$

where $q$ is the flow rate per unit width

$$q = \tfrac{1}{2} V_0 H^* \tag{10.4-7}$$

The pressure distribution depends therefore on a number of variables: geometrical ($H_0$, $H_1$, and $L$), operational ($V_0$ and $q$), and physical property ($\mu$). The maximum pressure that can be attained is at the $\zeta = 1$ ($z = L$), at closed discharge conditions ($q = 0$).

$$P_{\max} = P_0 + \frac{6\mu L V_0}{H_0 H_1} \tag{10.4-8}$$

Clearly, if the entrance and discharge pressures are equal, the pressure profile exhibits a maximum value. The location of maximum pressure is at $H^* = 2H_0/(1 + \zeta_0)$. This conclusion, therefore, focuses attention on an important difference between parallel plate and non-parallel plate geometries. In the former, equal entrance and discharge pressures imply *no* pressurization and pure drag flow, in the latter this implies the *existence* of a maximum in pressure profile. Indeed, this pressurization mechanism forms the foundation of hydrodynamic lubrication.

As with the development of the single screw melt extruder from the parallel plate geometry, we can follow the development of a number of processes from the *non*-parallel plate geometry. Roll mills and calenders are examples of such machinery. Furthermore, the two counterrotating roll devices can be converted to a continuous extrusion operation with increased pressurization capabilities, since *both* surfaces are moving parallel to themselves.

A different direction in utilizing the non-parallel plate pressurization concept was taken by Westover (12), who developed the "slider pad extruder." A slider pad (Fig. 10.21) is an "extreme" extension of the non-parallel plate concept. Slider pads are used for hydrodynamic lubrication because of their good load carrying capacity. Westover, by placing stepped slider pads on a disk covered by another rotating disk, converted this pressurization concept into a continuous extruder (Fig. 10.22).

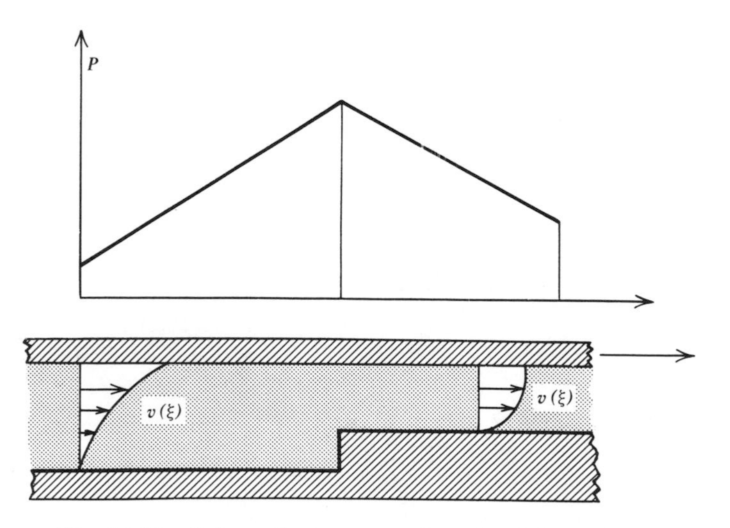


**Fig. 10.21** Schematic representation of a slider pad.

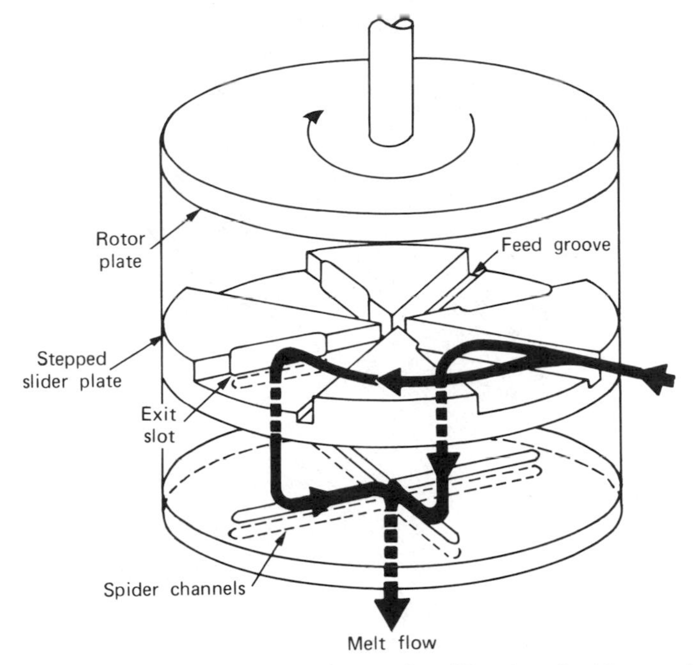


**Fig. 10.22** A rotary sector slider pad extruder. [Reprinted with permission from R. F. Westover, *Soc. Plast. Eng. J.*, 1473 (1962).]

Finally, as pointed out earlier, extruder screws have tapered sections; the various pressure profiles that can be obtained in the non-parallel plate geometry explain the experimentally observed axial pressure profiles in them. We can obtain the flow rate–pressure drop relationship for a tapered extruder section from Eq. 10.4-6 by setting $\zeta = 1$. Following some rearrangements, we have

$$q = \frac{V_0 H_0}{2}\left(\frac{2}{1+\zeta_0}\right) + \frac{H_0^3}{12\mu}\frac{P_0 - P_1}{L}\frac{2}{\zeta_0(1+\zeta_0)} \tag{10.4-9}$$

Equation 10.4-9 reduces to Eq. 10.2-6 for $\zeta_0 = 1$. Hence the terms $2/(1+\zeta_0)$ and $2/\zeta_0(1+\zeta_0)$ can be viewed as "correction factors" when the parallel plate theory is used for tapered geometries.

## 10.5 Two Rotating Rolls: Calenders and Roll-Mills

The two rotating roll geometry of calenders and roll-mills is an important application of the non-parallel plate pressurization concept with both plates moving. There are some differences between the two cases. In a roll-mill the two rolls are generally rotated at different frequencies. The operation is generally batch, with the polymer forming a continuous blanket around one roll. In calenders, on the other hand, there is one pass between any set of two rolls, which occasionally are of different radii, and operate at different speeds. The purpose of the former is to melt and mix the polymer, whereas the aim of the latter is to shape a product. Hence we

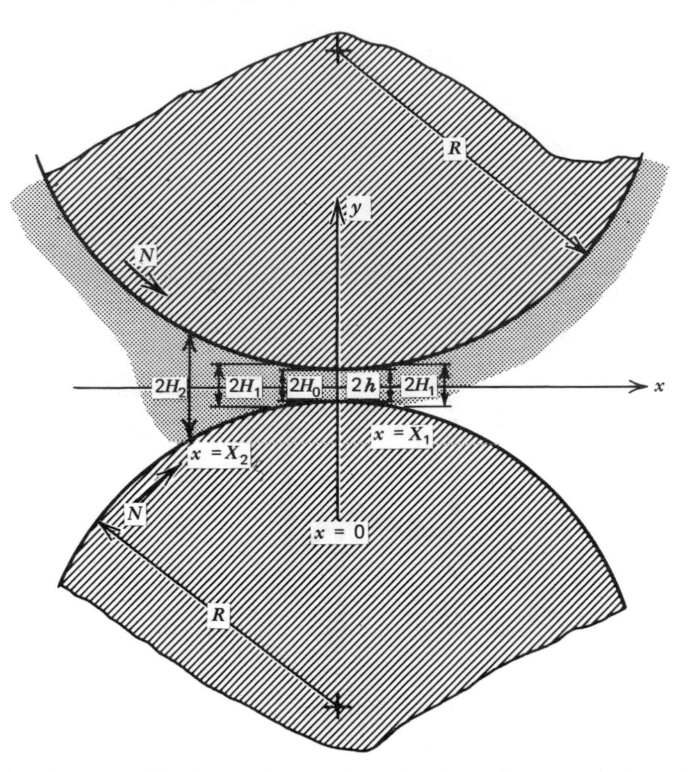


**Fig. 10.23** The nip area of the two roll geometry. A rectangular coordinate system is placed at the midpoint in the gap on the line connecting the two roll centers.

discuss roll-mills in more detail in Chapter 11, dealing with mixing, and calenders in Chapter 16. Nevertheless the nip flow in both cases is based on the same pressurization principle hence we treat it in this chapter.

Figure 10.23 depicts flow configuration schematically. Two identical rolls of radii $R$ rotate in opposite direction with frequency of rotation $N$. The minimum gap between the rolls is $2H_0$. The polymer is uniformly distributed laterally over the roll width $W$. At a certain axial (upstream) location $x = X_2$ ($X_2 < 0$), the rolls start "biting" on the polymer. There the melt contacts both rolls. At a certain axial (downstream) location $x = X_1$, the polymer detaches itself from one of the rolls. Pressure, which is assumed to be atmospheric at $X_2$, rises with $x$, reaches maximum before the minimum gap clearance (recall the pressure profiles in the previous section), then drops back to atmospheric pressure at $X_1$. As a result of this type of pressure profile, there is a force acting on the rolls that tends to increase the clearance between them and even distort them. The location of points $X_1$ and $X_2$ depends on the roll geometry, gap clearance, and the total volume of polymer on the rolls in roll-mills or on the volumetric flow rate in calenders.

We shall first derive a simple Newtonian model following Gaskell's (13) work and McKelvey's (11) treatment. The following assumptions are made: the flow is steady, laminar, and isothermal; the fluid is incompressible and Newtonian; there is no slip at the walls; the clearance-to-radius ratio $h/R \ll 1$ throughout the flow region, which allows us to assume *narrow gap flow with slowly varying gap separation*; that is, we invoke the lubrication approximation where the velocity profile at any location $x$ is considered to be identical to the velocity profile between two infinite parallel plates at a distance $2h$ apart, with pressure gradients and plate velocities equal to the local values between the rolls; finally, gravity forces are neglected, and the polymer melt is uniformly distributed on the rolls. Based on these assumptions, there is only one nonvanishing velocity component $v_x(y)$. Hence the equations of continuity and motion, respectively, reduce to

$$\frac{dv_x}{dx} = 0 \tag{10.5-1}$$

and

$$\frac{\partial P}{\partial x} = -\frac{\partial \tau_{yx}}{\partial y} = \mu \frac{\partial^2 v_x}{\partial y_2} \tag{10.5-2}$$

Equation 10.5-2 can be integrated twice without difficulty because the pressure $P$ is a function of $x$ only. The boundary conditions are $v_x(\pm h) = U$, where $U$ is the tangential velocity of the roll surfaces

$$U = 2\pi NR \tag{10.5-3}$$

The resulting velocity profile is

$$v_x = U + \frac{y^2 - h^2}{2\mu}\left(\frac{dP}{dx}\right) \tag{10.5-4}$$

Note that Eq. 10.5-4 indicates that for a positive pressure gradient (i.e., pressure rise in the positive $x$-direction) $v_x(0) < U$, and for a negative pressure gradient $v_x(0) > U$.

The flow rate per unit width $q$ is obtained by integrating Eq. 10.5-4 to give

► $$q = 2\int_0^h v_x\, dy = 2h\left(U - \frac{h^2}{3\mu}\frac{dP}{dx}\right) \tag{10.5-5}$$

At steady state, $q$ is constant and independent of position $x$. To solve for the pressure profile, we shall require that the velocity be uniform at the exit $v_x(y) = U$. This requirement implies that $\tau_{yx} = 0$, and from Eq. 10.5-2 we conclude that the pressure gradient should also vanish at this point. Hence the flow rate can be expressed via Eq. 10.5-5 in terms of $H_1$ and $U$ as follows:

$$q = 2H_1 U \tag{10.5-6}$$

Combining Eqs. 10.5-6 and 10.5-5, and doing some rearrangement, gives

$$\frac{dP}{dx} = \frac{3\mu U}{H_1^2}\left(1 - \frac{H_1}{h}\right)\left(\frac{H_1}{h}\right)^2 \tag{10.5-7}$$

Equation 10.5-7 implies that the pressure gradient is zero not only at the exit but also at $x = -X_1$, where $h$ also equals $H_1$ and where, as we shall see later, the pressure attains a maximum value. The pressure profile is obtained by integrating Eq. 10.5-7, with the boundary condition $P = 0$ at $x = X_1$. First, however, we have to find a functional relationship between $h$ and $x$. From plane geometry we get

$$h = H_0 + R - \sqrt{R^2 - x^2} \tag{10.5-8}$$

A useful approximation to Eq. 10.5-8 is obtained by expanding $\sqrt{R^2-x^2}$ using the binomial series and retaining only the first two terms. This results in the expressions

$$\frac{h}{H_0}=1+\rho^2 \tag{10.5-9}$$

where

$$\rho^2=\frac{x^2}{2RH_0} \tag{10.5-10}$$

Integration of Eq. 10.5-7, subsequent to substituting Eqs. 10.5-9 and 10.5-10, gives the pressure profile

$$► \quad P=\frac{3\mu U}{4H_0}\sqrt{\frac{R}{2H_0}}\left\{\left[\frac{\rho^2-1-5\lambda^2-3\lambda^2\rho^2}{(1+\rho^2)^2}\right]\rho+(1-3\lambda^2)\tan^{-1}\rho+C(\lambda)\right\} \tag{10.5-11}$$

where

$$\lambda^2=\frac{X_1^2}{2RH_0} \tag{10.5-12}$$

and the constant of integration $C(\lambda)$ is obtained by setting $P=0$ at $\rho=\lambda$

$$C(\lambda)=\frac{(1+3\lambda^2)}{(1+\lambda^2)}\lambda-(1-3\lambda^2)\tan^{-1}\lambda \tag{10.5-13}$$

McKelvey (11) suggested the following approximation for $C(\lambda)$:

$$C(\lambda)\cong 5\lambda^3 \tag{10.5-14}$$

The maximum pressure is obtained by substituting $\rho=-\lambda$ into Eq. 10.5.-11 and it is given by

$$► \quad P_{\max}=\frac{3\mu U}{4H_0}\sqrt{\frac{R}{2H_0}}[2C(\lambda)]\cong\frac{15\mu U\lambda^3}{2H_0}\sqrt{\frac{R}{2H_0}} \tag{10.5-15}$$

Note that the maximum pressure is very sensitive to $\lambda$. An increase in $\lambda$ brings about both a broadening of the pressure profile as well as an increase in the maximum value. This is demonstrated in Fig. 10.24, where $P/P_{\max,\lambda=1}$ is plotted versus $\rho$ with $\lambda$ as a parameter. Results indicate that for any given $\lambda$, there is a particular upstream position at which the pressure drops to zero. Figure 10.23 denoted this point by $X_2$. This unique relationship between $\lambda$ and $X_2$, obtained by setting $P=0$ in Eq. 10.5-11, is plotted in Fig. 10.25 in terms of $\rho_2=X_2/\sqrt{2RH_0}$. Note that both $\rho_2$ and $X_2$ are negative. Finally another property of the pressure profile is that at $x=0$ the pressure equals exactly $P_{\max}/2$.

The velocity profile is obtained by substituting Eq. 10.5-7 into Eq. 10.5-4, with the aid of Eqs. 10.5-9, 10.5-10 and 10.5-12

$$► \quad u_x=1+\frac{3}{2}\frac{(1-\xi^2)(\lambda^2-\rho^2)}{(1+\rho^2)} \tag{10.5-16}$$

where $u_x=v_x/U$ and $\xi=y/H$. Figure 10.26*a* gives velocity profiles for $\lambda^2=0.1$.

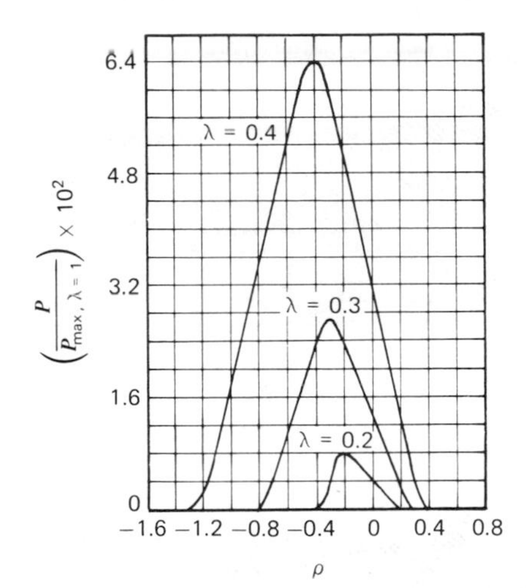


**Fig. 10.24** Pressure profiles between rolls with $\lambda$ as a parameter. (Reprinted with permission from J. M. McKelvey, *Polymer Processing*, Wiley, New York, 1962.)

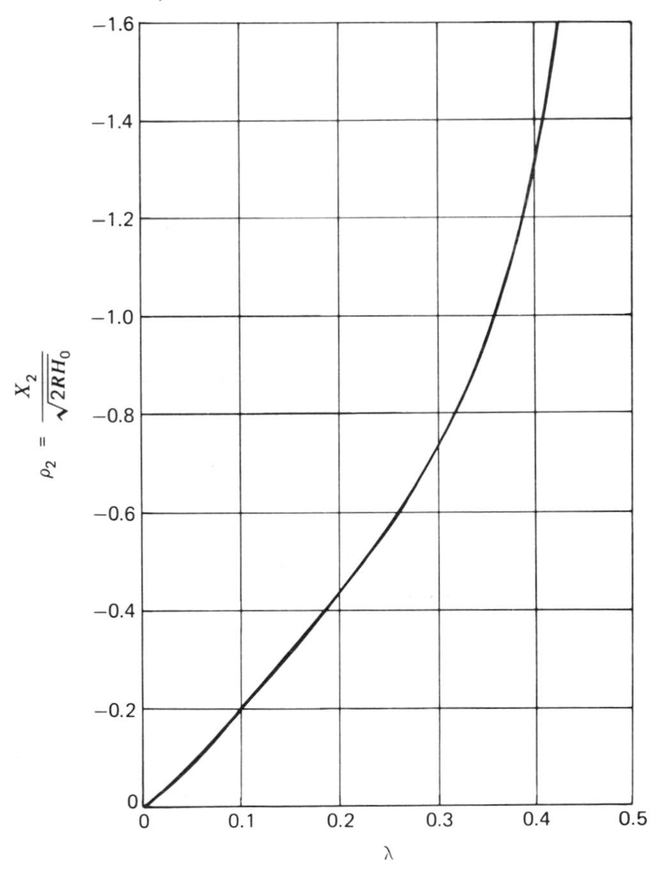


**Fig. 10.25** Relationship between $\rho_2$, where the rolls "bite" on the polymer (equivalent to $X_2$) and $\lambda$ where the polymer detaches (equivalent to $X_1$). Curve based on computations given in G. Ehrman and J. Vlachopoulos, *Rheol. Acta*, **14**, 761 (1975).

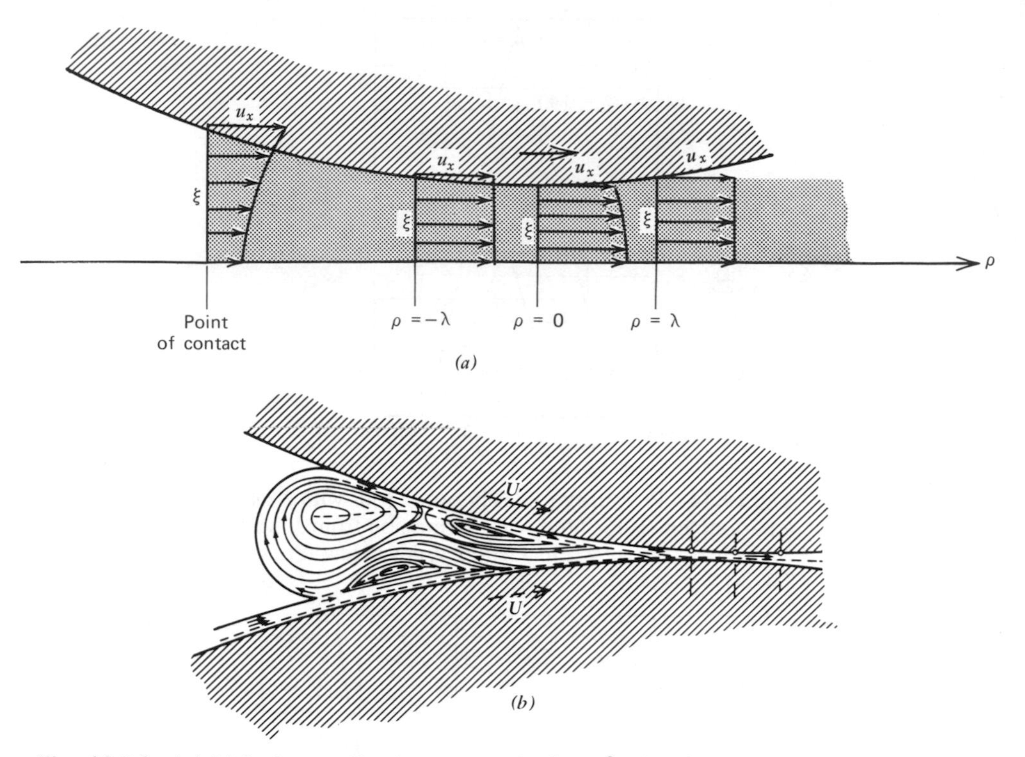


**Fig. 10.26** (*a*) Velocity profiles between rolls for $\lambda^2 = 0.1$ from Eq. 10.5-16. At $\rho = \pm\lambda$ velocity profiles are flat (plug-type flow) because the pressure gradient vanishes at these locations. At $\rho = -2.46\lambda$ the melt comes in contact with the rolls and the velocity profiles indicate a pressure rise in the direction of flow. (*b*) Schematic view of flow patterns obtained by Unkrüer (17) using color tracers. His results indicate circulation patterns not predicted by the Gaskell model.

Equation 10.5-16 indicates that at a certain axial location $\rho^*$ at $\xi = 0$, a stagnation point may occur ($v_x(0) = 0$)

$$\rho^* = -\sqrt{2 + 3\lambda^2} \tag{10.5-17}$$

For $\lambda = 0.425$, the stagnation point is at the contact point; hence for $\lambda > 0.425$ a circulatory flow develops in the entrance region.

The rate of strain and stress distributions can now be obtained from the velocity profile which together with Eq. 10.5-9 give:

$$\blacktriangleright \qquad \dot{\gamma}_{yx}(\xi) = \frac{3U}{H_0} \frac{(\rho^2 - \lambda^2)}{(1 + \rho^2)^2} \xi \tag{10.5-18}$$

and

$$\blacktriangleright \qquad \tau_{yx}(\xi) = \frac{3\mu U}{H_0} \frac{\lambda^2 - \rho^2}{(1 + \rho^2)^2} \xi \tag{10.5-19}$$

An extremum in shear rate $\dot{\gamma} = |\dot{\gamma}_{yx}|$ and shear stress occurs at the roll surface at

$\rho = 0$, where the gap clearance is at minimum

$$\dot{\gamma}_{\text{ext}} = \frac{3U\lambda^2}{H_0} \tag{10.5-20}$$

and

$$\tau_{\text{ext}} = \frac{3\mu U\lambda^2}{H_0} \tag{10.5-21}$$

but the overall maximum value of the shear stress and shear rate occur at $\rho = \rho_2$ if $\rho_2 > -\sqrt{1+2\lambda^2}$, and at $\rho = -\sqrt{1+2\lambda^2}$ if $\rho_2 < -\sqrt{1+2\lambda^2}$ (see Section 11.8). The total power input into both rolls can now be calculated by integrating the product of roll velocity and shear stress at the surface, which is obtained by setting $\xi = 1$ in Eq. 10.5-19

$$P_w = 2UW\sqrt{2RH_0}\int_{\rho_2}^{\lambda} \tau_{yx}(1)\, d\rho \tag{10.5-22}$$

where $W$ is the width of the rolls. Integration of Eq. 10.5-22 yields the following result

► $$P_w = 3\mu WU^2\sqrt{\frac{2R}{H_0}}\, f(\lambda) \tag{10.5-23}$$

where

$$f(\lambda) = (1-\lambda^2)[\tan^{-1}\lambda - \tan^{-1}\rho_2] - \left[\frac{(\lambda-\rho_2)(1-\rho_2\lambda)}{(1+\rho_2^2)}\right] \tag{10.5-24}$$

Figure 10.27 plots the function $f(\lambda)$.

Finally the force separating the two rolls is obtained by integrating the pressure given in Eq. 10.5-11 over the area of the rolls this pressure acts upon

$$F_N = W\sqrt{2RH_0}\int_{\rho_2}^{\lambda} P\, d\rho \tag{10.5-25}$$

which gives

► $$F_N = \frac{3\mu URW}{4H_0}\, g(\lambda) \tag{10.5-26}$$

where

$$g(\lambda) = \left(\frac{\lambda-\rho_2}{1+\rho_2^2}\right)[-\rho_2 - \lambda - 5\lambda^3(1+\rho_2^2)] + (1-3\lambda^2)[\lambda\tan^{-1}\lambda - \rho_2\tan^{-1}\rho_2] \tag{10.5-27}$$

The function $g(\lambda)$ is given in Fig. 10.27*b*. Note that in calculating the force we have neglected roll curvature, which follows from the basic assumption on which the whole model is based, namely, that $h/R \ll 1$.

The treatment of non-Newtonian fluids was outlined by Gaskell (13) in his original paper, and he presented detailed solutions for Bingham fluids. Later McKelvey (11) reported a detailed solution for power law model fluids.

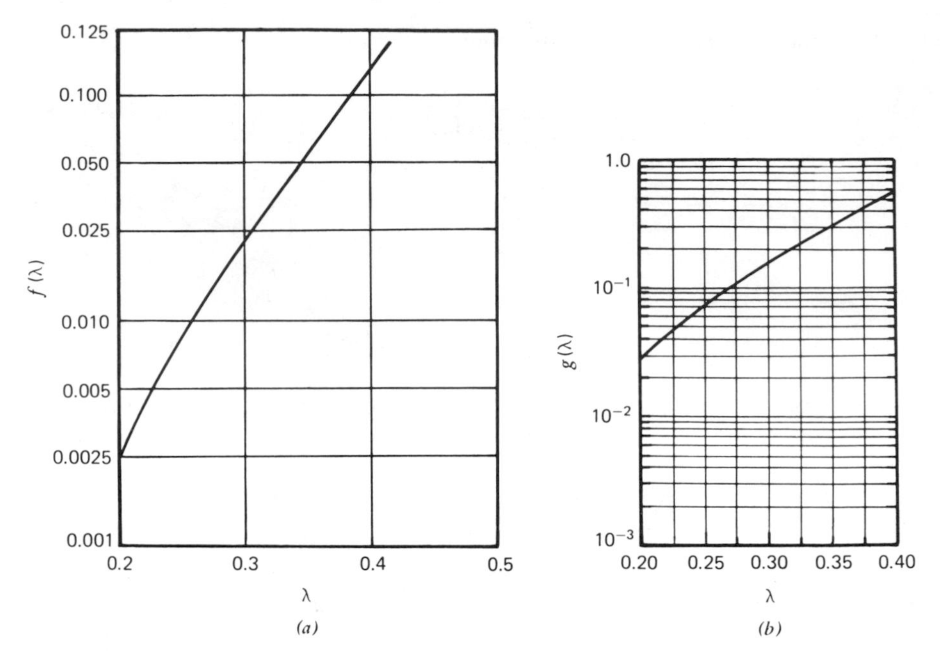


**Fig. 10.27** (*a*) Relation between $f(\lambda)$ and $\lambda$ from Eq. 10.5-24. [Reprinted with permission from G. Ehrmann and J. Vlachopoulos, *Rheol. Acta*, **14**, 761–764 (1975).] (*b*) $g(\lambda)$ and $\lambda$ from Eq. 10.5-27. [Reprinted with permission from J. M. McKelvey, *Polymer Processing*, Wiley, New York, 1962.]

As shown in Fig. 10.26*a* $\dot{\gamma}_{yx}(\xi) \geq 0$ for $\rho < -\lambda$, and $\dot{\gamma}_{yx}(\xi) \leq 0$ for $\rho > -\lambda$, where $-\lambda$ is the yet unknown location, where the pressure profile exhibits maximum value (or $dP/dx = 0$). Moreover, because of symmetry we have the convenient boundary condition that $\tau_{yx} = \dot{\gamma}_{yx} = 0$ at $y = 0$ ($\xi = 0$). Making the same simplifying assumptions as in the Newtonian analysis, the following results are obtained for the velocity profile and flow rate:

$$v_x = U + \frac{\text{sign}(\dot{P})}{(1+s)} \left[\frac{\text{sign}(\dot{P})}{m} \frac{dP}{dx}\right]^s (y^{1+s} - h^{1+s}) \tag{10.5-28}$$

where sign $(\dot{P})$ is defined as

$$\text{sign}(\dot{P}) = \frac{dP/dx}{|dP/dx|} = \begin{cases} +1 & \rho < -\lambda \\ -1 & \rho > -\lambda \end{cases} \tag{10.5-29}$$

and

$$q = 2h\left\{U - \text{sign}(\dot{P}) \left(\frac{h}{2+s}\right)\left[\text{sign}(\dot{P}) \frac{h}{m} \frac{dP}{dx}\right]^s\right\} \tag{10.5-30}$$

By expressing the flow rate in terms of the clearance at the point of detachment, the following expression is obtained for the pressure gradient:

$$\frac{dP}{d\rho} = K \frac{[\text{sign}(\dot{P})(\rho^2 - \lambda^2)]^n}{(1+\rho^2)^{2n+1}} \tag{10.5-31}$$

where

$$K = \text{sign}(\dot{P})m\sqrt{\frac{2R}{H_0}}\left[\frac{(2+s)U}{H_0}\right]^n \qquad (10.5\text{-}32)$$

The pressure profile is obtained by numerical integration of Eq. 10.5-31, where $\lambda$ is given by Eq. 10.5-12 and is determined by the flow rate. The effect of the power law exponent $n$ on the pressure profile is shown in Fig. 10.28.

Experimental measurement of pressure profiles in calenders was conducted by Bergen and Scott (14). A strain gauge pressure transducer was embedded in the surface of one of the 10 in. diameter rolls, and traces were recorded at various conditions corresponding to both calendering and roll-milling. Figure 10.28 compares the experimental pressure profile using a plasticized thermoplastic resin (unfortunately the rheological flow curve was not provided) and theoretical curves using Newtonian and power law fluid models. McKelvey's (11) method of comparison is used, whereby the maximum pressures are forced to coincide by appropriate selection of $\lambda$. With a Newtonian fluid a good agreement between experiments and theory is observed in the region $\rho > -\lambda$. In the region $\rho < -\lambda$ the theoretical prediction falls well below the experimental measurement. The same conclusion is reached when the comparison is made by selecting a viscosity that matches the experimental and theoretical curves at $\rho = 0$, and matching the locations of maximum pressures, as done originally by Bergen and Scott (14). This effective viscosity was lower by 3 orders of magnitude from the measured viscosity. The latter, however, was measured at much lower shear rates than those prevailing

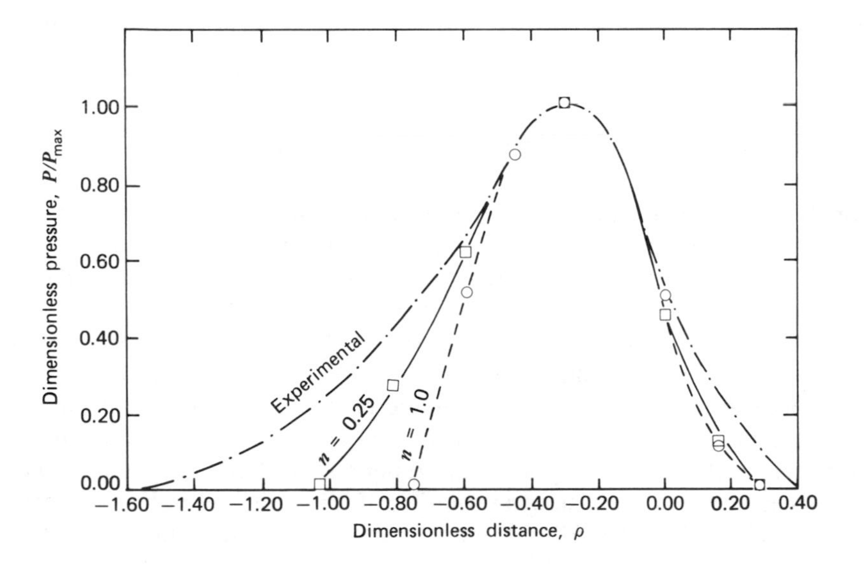


**Fig. 10.28** Comparison between experimental pressure profile for a plasticized thermoplastic resin (14) and theoretical pressure profiles for $n = 1$ and $n = 0.25$ calculated by Kiparissides and Vlachopoulos (15). The theoretical curves were calculated both by finite element method and analytically by way of Gaskell type models, as discussed in this section, giving virtually identical results.

in the nip region. The disagreement between the theory and experiment must be partly due to the Newtonian assumption. This conclusion is indirectly verified by further experimental work of Bergen and Scott (14) with strongly non-Newtonian fluids demonstrating significant deviations from the Gaskell model throughout the nip region. Further support for this conclusion can be derived from the work of Kiparissides and Vlachopoulos (15) who showed that for power law fluids, a decreasing $n$ value considerably reduces the disagreement between theory and experiment. This is illustrated in Fig. 10.28. They used a finite element calculation method that eliminates geometrical inaccuracies involved in the Gaskell model (although it introduces other approximations). Alternatively, the two roll geometry can be better described by using bipolar coordinates as first worked out by Finston (16). This approach, as well as the finite element approach, enables analysis of both equal sized and nonequal sized roll geometry, the latter being referred to as nonsymmetrical calendering. The finite element method appears, however, to be more flexible in dealing with both Newtonian and non-Newtonian fluids. Chapter 16 covers this method in some detail. Recently detailed experimental work on calendering was reported by Unkrüer (17). The calender used had 30 cm diameter 50 cm wide rolls. The material calendered included rigid PVC and PS. Pressure profiles were measured at various axial locations, indicating the existence of cross-machine direction flow superimposed on the main machine-direction flow in the inlet region. This type of flow is neglected in the Gaskell model. Unkrüer, using color tracers, also investigated the flow patterns in the inlet region. Results verified the cross-machine flow and revealed a complex flow pattern with several circulation regions (Fig. 10.26*b*). The results indicate that the incoming stream of melt affects the flow pattern, which may also be affected by the elastic properties of the melt. Both these effects are, of course, neglected in the Gaskell model; therefore it is not surprising that the predictions of the model are at variance with the experimental findings.

## 10.6 Dynamic Normal Stress Pressurization

Section 10.1 noted that the term $[\nabla \cdot \boldsymbol{\tau}]$ in the equation of motion is an important source of pressurization. We have further pointed out that this source may be related either to viscosity or to the normal stress difference coefficient. This section discusses the latter case.

Consider a two disk geometry (Fig. 10.29). Polymer is placed between two disks of radii $R$. The upper disk is attached to a shaft and rotates at frequency $\Omega$. This geometrical arrangement is a schematic representation of the normal stress extruder suggested by Maxwell and Scalora (18) and analyzed by a number of investigators, both theoretically and experimentally (19–21). When a Newtonian fluid is placed between the disks, the rotation, if it is high, will induce a centrifugal force that will tend to "suck" the fluid in through the die and expel it at the circumference, much as a centrifugal pump would do. If a non-Newtonian fluid exhibiting normal stresses is placed between the disks, however, an opposite effect is observed, namely, an inward radial flow *into* the die. Equation 6.7-22 showed that for the cone and plate flow, which is similar to the flow just described, the total thrust on the plate is given by the product of the plate area and one-half the

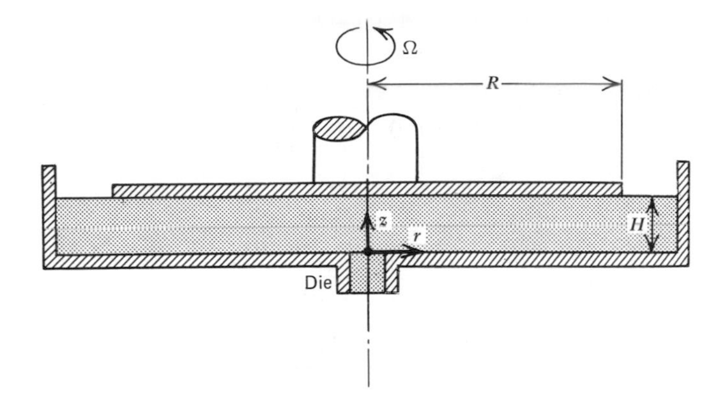


**Fig. 10.29** Schematic view of the normal stress extruder. Polymer is placed between the disks. The upper disk is attached to a shaft and rotates at frequency $\Omega$. Pressure develops at the die and the polymer is extruded.

primary normal stress difference. Furthermore (Eq. 6.7-20), we saw that the experimentally measured pressure distribution on the plate increases logarithmically with *decreasing* radius, the relationship involving both normal stress differences. Here we want to evaluate the pressure at the center as a function of disk geometry, frequency of rotation, and rheological properties. We do this in the absence of radial flow (i.e., for the closed discharge condition), and we neglect any possible secondary flows as well ($v_r = v_z = 0$), although such flows ("cigar rolling") have been observed (19). These impose practical limits on the pressurization capability of the normal stress extruder by forcing upper limits on $\Omega$ and $R$. Yet even under these rather severe simplifying assumptions, the flow field between parallel disks is nonviscometric because the nonvanishing velocity component $v_\theta$ is a function of both $r$ and $z$, $v_\theta(r, z)$. Nevertheless, we use the CEF equation (Eq. 6.3-5), which as Chapter 6 pointed out is applicable to moderately nonviscometric flows with reasonable accuracy. Finally steady isothermal flow with no-slip at the walls will be assumed.

To determine what stresses are generated in the torsional disk flow of a CEF fluid, we first assume that its flow field is that of a purely viscous fluid; then we calculate the tensor quantities $\nabla \boldsymbol{v}$, $\dot{\boldsymbol{\gamma}}$, $\boldsymbol{\omega}$, $\{\dot{\boldsymbol{\gamma}} \cdot \dot{\boldsymbol{\gamma}}\}$, $\{\boldsymbol{\omega} \cdot \dot{\boldsymbol{\gamma}}\}$ and $\{\boldsymbol{v} \cdot \nabla \dot{\boldsymbol{\gamma}}\}$ which appear in the CEF equation. Having these quantities, we substitute them in the constitutive equation to find out which are the nonzero stress components. Recall that this is exactly the procedure we followed in Example 6.4. Finally, these nonzero stress components are inserted into the equation of motion, resulting in the required pressure field.

Assuming that the flow kinematics of the CEF and Newtonian fluids are identical, the velocity profile in a steady torsional disk flow is (see Problem 5.9)

$$v_\theta = \Omega r \frac{z}{H} \tag{10.6-1}$$

where $H$ is the separation between the disks. Thus from Tables 5.3 and 6.3 we obtain

$$\dot{\boldsymbol{\gamma}} = \begin{pmatrix} 0 & 0 & 0 \\ 0 & 0 & \dfrac{\Omega r}{H} \\ 0 & \dfrac{\Omega r}{H} & 0 \end{pmatrix} \tag{10.6-2}$$

and

$$\boldsymbol{\omega} = \begin{pmatrix} 0 & 2\Omega\dfrac{z}{H} & 0 \\ -2\Omega\dfrac{z}{H} & 0 & -\dfrac{\Omega r}{H} \\ 0 & \dfrac{\Omega r}{H} & 0 \end{pmatrix} \tag{10.6-3}$$

From Eqs. 10.6-2 and 10.6-3, the following expressions are obtained:

$$\{\dot{\boldsymbol{\gamma}} \cdot \dot{\boldsymbol{\gamma}}\} = \begin{pmatrix} 0 & 0 & 0 \\ 0 & \left(\dfrac{\Omega r}{H}\right)^2 & 0 \\ 0 & 0 & \left(\dfrac{\Omega r}{H}\right)^2 \end{pmatrix} \tag{10.6-4}$$

$$\{\boldsymbol{\omega} \cdot \dot{\boldsymbol{\gamma}}\} = \begin{pmatrix} 0 & 0 & \dfrac{2\Omega^2 rz}{H^2} \\ 0 & -\left(\dfrac{\Omega r}{H}\right)^2 & 0 \\ 0 & 0 & \left(\dfrac{\Omega r}{H}\right)^2 \end{pmatrix} \tag{10.6-5}$$

and

$$\{\dot{\boldsymbol{\gamma}} \cdot \boldsymbol{\omega}\} = \begin{pmatrix} 0 & 0 & 0 \\ 0 & \left(\dfrac{\Omega r}{H}\right)^2 & 0 \\ -\dfrac{2\Omega^2 rz}{H^2} & 0 & -\left(\dfrac{\Omega r}{H}\right)^2 \end{pmatrix} \tag{10.6-6}$$

Finally, with the aid of Table 6.2 we obtain

$$\{\boldsymbol{v} \cdot \nabla\dot{\boldsymbol{\gamma}}\} = \begin{pmatrix} 0 & 0 & -\dfrac{\Omega^2 rz}{H^2} \\ 0 & 0 & 0 \\ -\dfrac{\Omega^2 rz}{H^2} & 0 & 0 \end{pmatrix} \tag{10.6-7}$$

Substituting the foregoing expressions into the CEF equation 6.3-5 gives

$$\begin{pmatrix} \tau_{rr} & \tau_{r\theta} & \tau_{rz} \\ \tau_{\theta r} & \tau_{\theta\theta} & \tau_{\theta z} \\ \tau_{zr} & \tau_{z\theta} & \tau_{zz} \end{pmatrix} = -\eta \begin{pmatrix} 0 & 0 & 0 \\ 0 & 0 & \dfrac{\Omega r}{H} \\ 0 & \dfrac{\Omega r}{H} & 0 \end{pmatrix}$$

$$-\tfrac{1}{2}(\Psi_1+2\Psi_2)\begin{pmatrix} 0 & 0 & 0 \\ 0 & \left(\dfrac{\Omega r}{H}\right)^2 & 0 \\ 0 & 0 & \left(\dfrac{\Omega r}{H}\right)^2 \end{pmatrix} + \frac{\Psi_1}{2}\begin{pmatrix} 0 & 0 & 0 \\ 0 & -\left(\dfrac{\Omega r}{H}\right)^2 & 0 \\ 0 & 0 & \left(\dfrac{\Omega r}{H}\right) \end{pmatrix} \quad (10.6\text{-}8)$$

Thus the stress components for the assumed flow kinematics and for $\dot{\gamma} = \dot{\gamma}_{\theta z}(r) = (\Omega r/H)$

$$\tau_{rr} = 0 \quad \text{(a)}$$

$$\tau_{\theta\theta} = -(\Psi_1+\Psi_2)\left(\frac{\Omega r}{H}\right)^2 = -(\Psi_1+\Psi_2)\dot{\gamma}^2 \quad \text{(b)}$$

$$\tau_{zz} = -\Psi_2\left(\frac{\Omega r}{H}\right)^2 = -\Psi_2\dot{\gamma}^2 \quad \text{(c)} \qquad (10.6\text{-}9)$$

$$\tau_{\theta z} = \tau_{z\theta} = -\eta\left(\frac{\Omega r}{H}\right) = -\eta\dot{\gamma} \quad \text{(d)}$$

$$\tau_{r\theta} = \tau_{\theta r} = \tau_{rz} = \tau_{zr} = 0 \quad \text{(e)}$$

Hence the normal stress difference functions, keeping in mind the direction convention discussion in Section 6.2, and noting that in this case $\theta$ is direction "1," $z$ is direction "2," and $r$ is direction "3," are

$$\tau_{11} - \tau_{22} = -\Psi_1\left(\frac{\Omega r}{H}\right)^2 = -\Psi_1\dot{\gamma}^2 \qquad (10.6\text{-}10)$$

and

$$\tau_{22} - \tau_{33} = -\Psi_2\left(\frac{\Omega r}{H}\right)^2 = -\Psi_2\dot{\gamma}^2 \qquad (10.6\text{-}11)$$

Having all the stress tensor components, we can proceed with the equation of motion, whose three components reduce to (21)

$$-\rho\frac{v_\theta^2}{r} = -\frac{\partial P}{\partial r} + \frac{\tau_{\theta\theta}}{r} \qquad (10.6\text{-}12)$$

$$\frac{\partial P}{\partial \theta} = 0 \qquad (10.6\text{-}13)$$

and

$$\frac{\partial P}{\partial z} = 0 \qquad (10.6\text{-}14)$$

Hence we find that pressure is a function of only the coordinate $r$. Substituting Eqs. 10.6-9b and 10.6-1 into Eq. 10.6-12, we obtain

$$\frac{dP}{dr}=\rho\Omega^2 r\left(\frac{z}{H}\right)^2-(\Psi_1+\Psi_2)\left(\frac{\Omega}{H}\right)^2 r \qquad (10.6\text{-}15)$$

The first term on the right-hand side is due to centrifugal forces and contributes to an increasing pressure with $r$. Furthermore, we note a certain inconsistency between assumptions and results. For the *assumed* velocity profile, we obtained from the equation of motion that $P \neq f(z)$; yet Eq. 10.6-15 indicates a $z$-dependence. In reality, we should obtain a circulatory flow due to centrifugal forces, resulting in nonvanishing $\partial P/\partial z$, $v_z$, and $v_r$ terms. Our solution should therefore be restricted to conditions under which this flow is negligible; indeed, we are interested in the particular case of centrifugal forces that are small compared to normal stress effects, which are represented by the second term on the right-hand side of Eq. 10.6-15. Therefore we average $P$ over $z$ to obtain

$$\frac{d\bar{P}}{dr}=\rho\frac{\Omega^2 r}{3}-(\Psi_1+\Psi_2)\left(\frac{\Omega}{H}\right)^2 r \qquad (10.6\text{-}16)$$

Chapter 6 gave experimental evidence that $\Psi_1$ is positive, $\Psi_2$ is probably negative and that $-\Psi_2/\Psi_1 \sim 0.1$ for the experimentally investigated shear rate range. Thus $d\bar{P}/dr<0$ and the pressure will increase with decreasing radius, opposing and overcoming centrifugal forces.

Equation 10.6-16 can be integrated to give the pressure at $r=0$

$$\begin{aligned} \blacktriangleright \quad \bar{P}(0) &= \bar{P}(R)+\left(\frac{\Omega}{H}\right)^2\int_0^R(\Psi_1+\Psi_2)r\,dr-\rho\frac{\Omega^2R^2}{6} \\ &= \bar{P}(R)+\int_0^{\Omega R/H}(\Psi_1+\Psi_2)\dot{\gamma}\,d\dot{\gamma}-\rho\frac{\Omega^2R^2}{6} \end{aligned} \qquad (10.6\text{-}17)$$

Making the approximation that $\Psi_1$ and $\Psi_2$ are shear rate independent, we obtain the following expression for $\bar{P}(0)$

$$\blacktriangleright \quad \bar{P}(0)=\bar{P}(R)+\frac{1}{2}\left(\frac{\Omega R}{H}\right)^2(\Psi_1+\Psi_2)-\rho\frac{\Omega^2R^2}{6} \qquad (10.6\text{-}18)$$

which is a design equation for this normal stress pump.

Thus we find the maximum pressure rise at the center of the disk to be proportional to the square of $\Omega R/H$, which is the shear rate at $r=R$. Moreover, comparing Eq. 10.6-18 to Eqs. 10.6-10 and 10.6-11, we find that this pressure rise is the sum of the primary and secondary normal stress difference functions $\{-[(\tau_{11}-\tau_{22})+(\tau_{22}-\tau_{33})]\}$ at $r=R$ less the centrifugal forces. Since $\Psi_2$ is probably negative, it opposes the pressurization; hence the source of pressurization in the normal stress extruder is the *primary* normal stress difference function $\Psi_1$.

### Example 10.2 *The Maximum Pressure in the Normal Stress Extruder*

Calculate the maximum pressure (at closed discharge) in a normal stress extruder composed of two 25 cm radius disks 0.5 cm apart, shearing LDPE at 60 rpm and 200°C.

*Solution*: In view of lack of reliable data on $\Psi_2$ and considering that it is of the order of 10% of $\Psi_1$ we shall assume that $\Psi_2 = 0$. The shear rate range is from zero at the center to the maximum at $R$, $\dot{\gamma}(R) = \Omega R/H = 314.2\ \text{s}^{-1}$.

Figure 6.12 shows experimental data for $\Psi_1$ for LDPE. At 200°C we can calculate the following shear rate dependence of

$$\Psi_1(\dot{\gamma}) \cong 9 \times 10^4\ \dot{\gamma}^{-0.92}$$

Furthermore, we assume that this relationship holds for shear rates greater than $10\ \text{s}^{-1}$. Then, substituting this relationship into Eq. 10.6-17 followed by integration and taking $P(R) = 0$ and melt density of 0.75 g/cm$^3$, we obtain

$$\bar{P}(0) = \frac{(9 \times 10^4)}{(1.08)}(314.2^{1.08}) - \frac{(0.75)(25^2)}{(6)} = 4.17 \times 10^6\ \text{N/m}^2\ (605\ \text{psi})$$

Two facts are worth noting. First, the normal stress pressurization is indeed much greater ($10^5$ times) than that brought about by centrifugal forces. Second, the level of the maximum pressure generated is significant but for

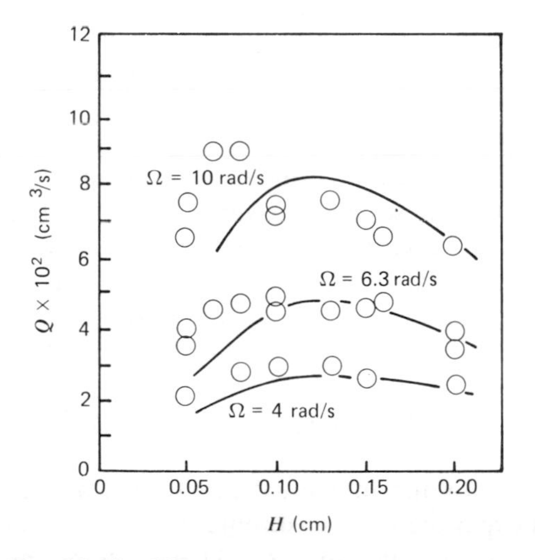


**Fig. 10.30** Flow rate versus gap separation in a 5 cm diameter disk arrangement. Polymer used was a polyacrylamide solution at 28°C. Smooth curves indicate calculated values. Die length was 0.482 cm; 0.244 cm diameter. [Data replotted from P. A. Good, A. J. Schwartz and C. W. Macosko, *Am. Inst. Chem. Eng. J.*, **20**, 67 (1974).]

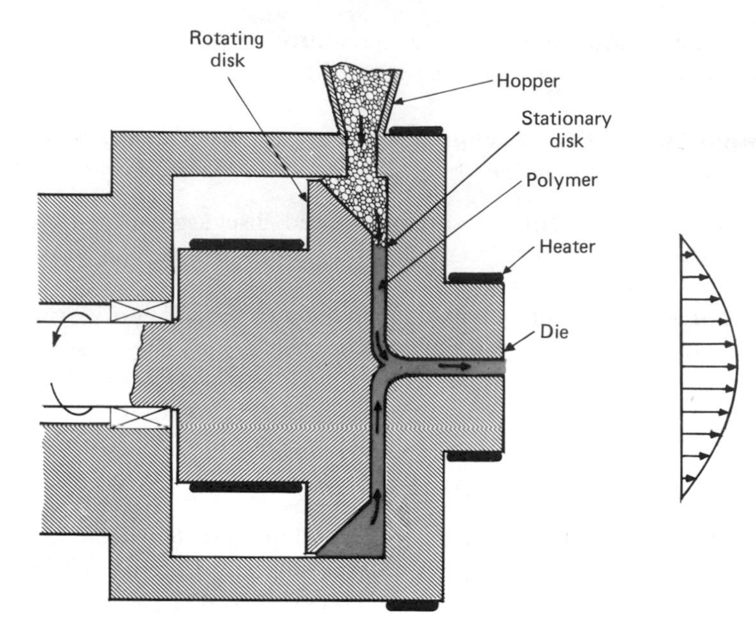


**Fig. 10.31** Schematic representation of a plasticating normal stress extruder.

processing applications not very high. We recall that because of secondary flows ("cigar rolling") there is a limit to $\Omega$ and $R$.

So far, we have neglected the radial flow, but in a normal stress extruder the objective is to extrude the polymer through the die. Such a flow implies a pressure loss in the inward radial direction, consequently reducing the maximum pressure at the die entrance. The ensuing flow rate is determined by die resistance; at steady flow conditions, the pressure rise in the radial direction equals the pressure drop across the die. Rigorous solution of this flow problem is difficult. Macosko et al. (22) have proposed the following approximate analytical solution, which results in good agreement with experiments. They assumed that the pressure rise due to normal stresses ($\Delta P_{NS}$) at no discharge, the pressure loss between the disks due to radial flow ($\Delta P_{rs}$), and the entrance losses into the die ($\Delta P_{DE}$) are related with the available pressure drop for pumping across the die as follows:

$$\Delta P_T = \Delta P_{NS} - \Delta P_{rs} - \Delta P_{DE} \tag{10.6-19}$$

They then solved for the individual pressure drops and compared the calculated results with experimental measurements. Figure 10.30 gives results of this comparison with a polyacrylamide solution. There is an optimum gap separation for maximum output. The physical reason for such an optimum is the $H$ dependence of the pressure rise due to normal stresses and pressure loss due to radial viscous flow.

The normal stress extruder can also be used for melting, as shown in Fig. 10.31. Because of the limited pressurization capability, various modifications have been suggested, one of these being a combination with a screw. In spite of the modifications, it has found limited use as an extrusion system, although it appears that its latest modification (23) has certain advantages in processing hard-to-mix materials.

## 10.7 Static Mechanical Pressurization and Positive Displacement Flows

The next four sections analyze four important practical geometrical configurations (Fig. 10.32) in which the melt is pressurized by an externally imposed (mechanical) hydrostatic pressure; in each case the result is a positive displacement type of flow. These configurations include an axially moving plunger in a cylinder, two axially moving parallel flat disks, two intermeshing rotating gears, and two intermeshing rotating screws. The first two configurations give rise to noncontinuous (batch) operations. In the plunger-cylinder arrangement the external moving surface is normal to main flow direction, whereas in the two-disk arrangement the moving surface is parallel to the main flow direction, giving rise to squeezing type of flow. The intermeshing gear pump and the twin screw extruder provide for continuous operation.

There are a number of potential advantages of positive displacement melt conveying, as compared to drag induced melt conveying. The latter is sensitive to conditions at the moving surface such as wall temperatures, slip at the wall, or apparent slip at the wall due to migration of low viscosity compounds. The former is generally insensitive to these problems. Moreover, flow rate control is generally better and pressure sensitivity lower in positive displacement flows as compared to drag induced flows.

## 10.8 Plunger-Cylinder

As Chapter 1 mentioned, the first recorded application of the plunger-cylinder configuration in the field of polymer processing appears to have been for the extrusion of gutta-percha. The main disadvantage of these extruders, introduced in England in 1845, is their intermittent operation. In an attempt to eliminate this disadvantage, Westover (24) designed a continuous plunger-type extruder consisting of four sets of plunger-cylinders, two for melting and two for pumping. An ingenious shuttle valve connects them all and provides continuous extrusion. But in present day polymer processing practice the plunger-cylinder configuration is applied to injection molding and transfer molding rather than to extrusion. Here the intermittent nature of the operation is of little consequence, since these processing methods are also noncontinuous. Both the old fashioned ram-type injection molding machines and the current reciprocating screw injection molding machines pressurize the melt by the forward motion of the ram or screw, acting as a plunger, and force the melt into the mold cavities by straightforward positive displacement flow. The pressure in front of the plunger is simply the product of the force exerted on the plunger and the cross-sectional area of the cylinder. It is selected on the basis of polymer properties, mold configuration, and product requirements (see Chapter 14). The flow field that develops in the cylinders in front of the advancing plunger, however, is not simple.

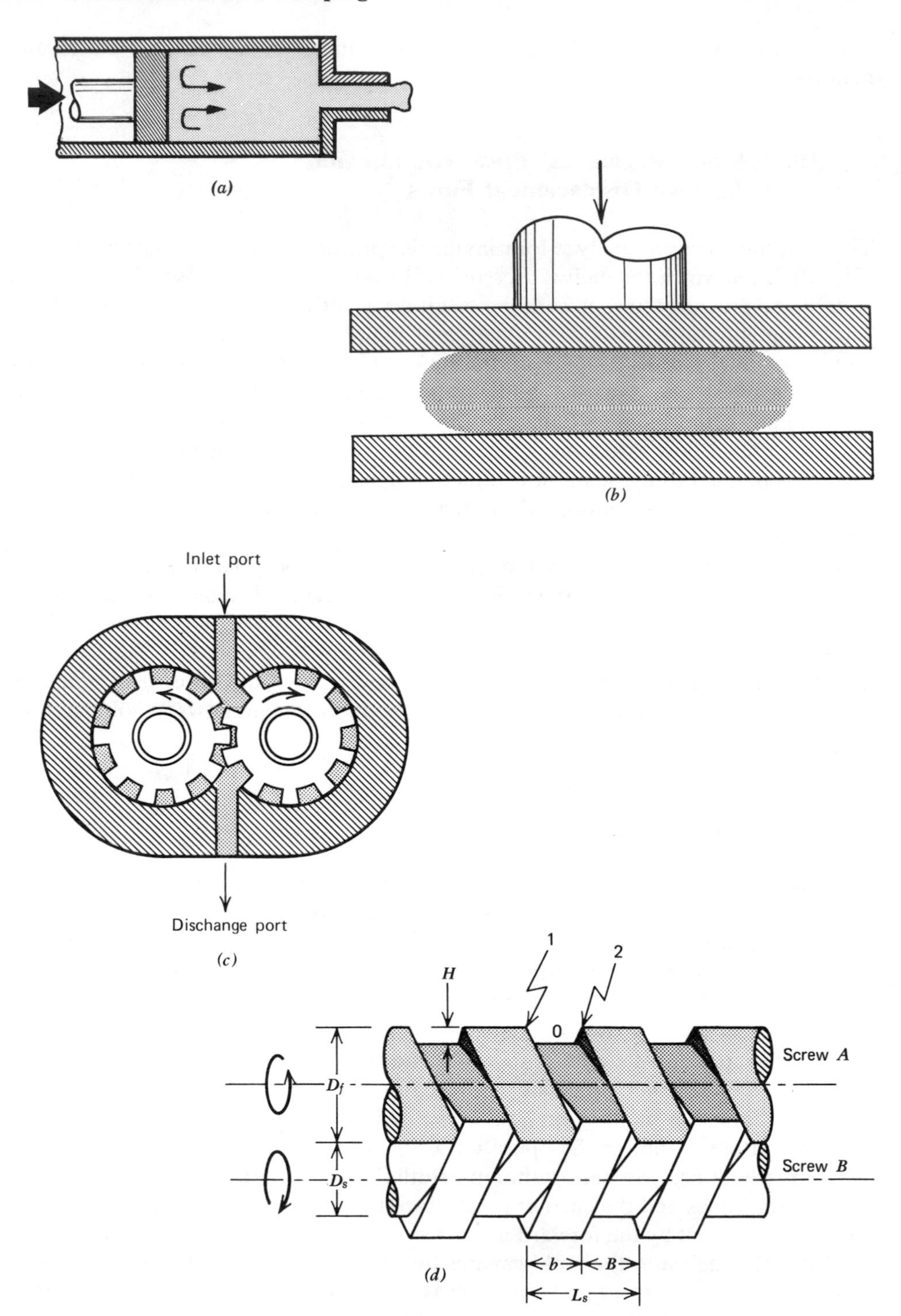


**Fig. 10.32** Schematic representation of four geometrical configurations utilizing external mechanical pressurization and giving rise to positive displacement flow. (*a*) Axially moving plunger in a cylinder. (*b*) Axially moving disks. (*c*) Two rotating intermeshing gears. (*d*) Two counter-rotating intermeshing screws.

It is easier to visualize the problem from a coordinate system placed on the plunger and moving with it (Lagrangean point of view). In this coordinate system the cylinder will be moving at constant velocity $-V_0$. The cylinder by its axial motion drags the adjacent liquid towards the "stationary" plunger. As the liquid approaches the plunger, it must acquire an inward radial velocity, while gradually decelerating to zero axial velocity. As the liquid continues to move radially inward, it acquires a positive axial velocity. Hence the result is an annular "skin" of liquid moving toward the plunger and an inner "core" moving away from it. Such a flow pattern was termed by Rose (25) a "reverse fountain effect." (We encounter a "fountain" type of flow again when dealing with flow into an empty mold cavity: Section 14.1.)

Assuming that the fluid slips at the plunger surface, as well as isothermal flow and incompressible Newtonian fluid; a mathematical formulation of the problem (26) results in the following approximate velocity profiles:

$$v_z = V_0\left(1 - \frac{2r^2}{R^2}\right)\left(1 - e^{-z\sqrt{6}/R}\right) \tag{10.8-1}$$

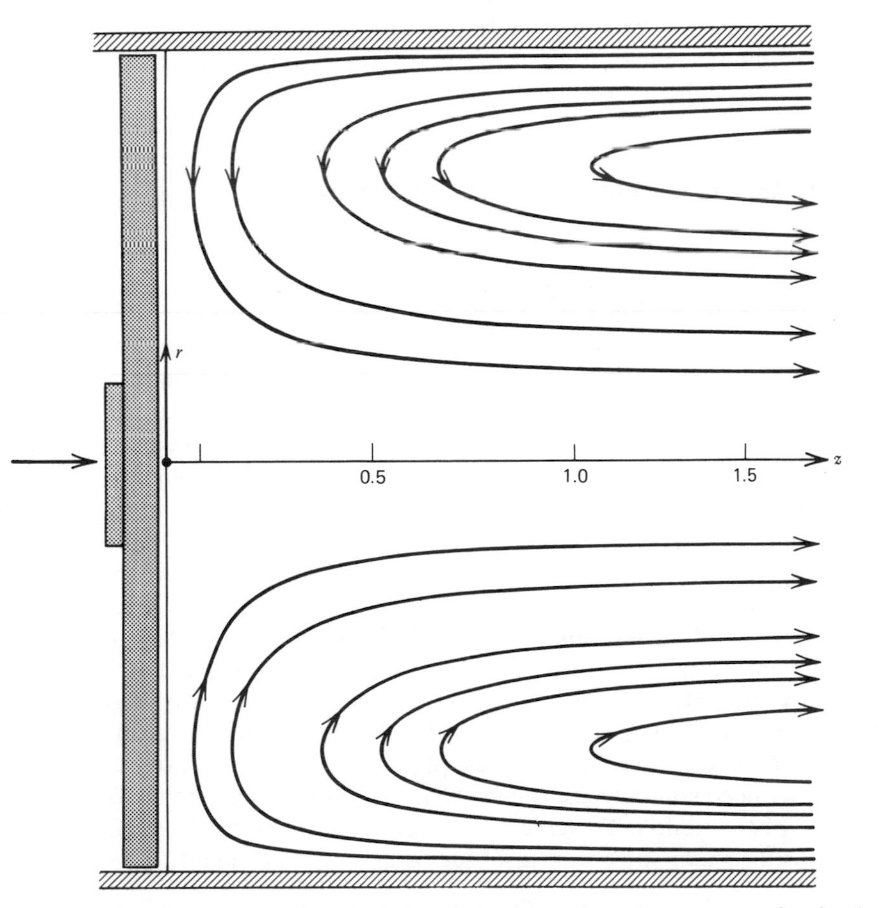


**Fig. 10.33** Computed streamlines in front of the advancing plunger, assuming isothermal flow, Newtonian fluid, and frictionless plunger surface (slip at the plunger wall).

and

$$v_r = -\frac{V_0\sqrt{6}r}{R^3}(R^2 - r^2)\, e^{-z\sqrt{6}/R} \qquad (10.8\text{-}2)$$

We note that if $V_0$ is added to $v_z$ in Eq. 10.8-1, the "laboratory" frame of reference in which the cylinder is stationary and the plunger is in motion is restored. In this frame of reference, at large $z$ values, the axial velocity profiles reduces to the familiar Poiseuille parabolic profile. It is further interesting to note from the foregoing velocity profile expressions that at the axial distances of *one-half* and *one* diameter away from the plunger 91 and 99%, respectively, of the fully developed profile are obtained.

Finally, we note that at $r = R/\sqrt{2}$, the velocity $v_z$ is zero for all $z$ (i.e., in laboratory coordinates, the axial velocity at this radial position equals plunger velocity). For $r > R/\sqrt{2}$ the liquid moves toward the plunger, and for $r < R/\sqrt{2}$, the liquid moves away from the plunger, as shown in Fig. 10.33, where the streamlines are plotted.

The plunger-cylinder flow pattern and circulatory motion (reverse fountain effect) is significant in injection molding in that it determines the residence time distribution as well as the time sequence of exit of the melt produced by the injection molding machine. It also has to be taken into account in explaining time nonuniformities of the injected melt.

## 10.9 Squeezing Flow Between Parallel Disks

Squeezing flows between two disks, as pointed out earlier, has the characteristic features of compression molding. This geometrical configuration and flow type is also used in certain hydrodynamic lubricating systems, as well as in various rheological testing devices for asphalt and other viscous liquids. The Williams Plastometer, based on this geometry, has been used in the rubber industry for many years (27). Recently Leider and Bird (28) pointed out the advantage of this simple geometry for transient nonviscometric rheological testing for polymeric melts.

Specifically, the flow problem considered is that of an incompressible fluid flowing isothermally and radially outward between two disks of radii $R$ and initial gap separation of $2h_0$, after a constant force $F$ has been applied normally to the disks. We are interested in deriving functional relationships between the applied force and the disk separation as well as the applied force and the pressure profile.

This flow problem was solved, with certain assumptions, using a number of constitutive equations, both Newtonian and non-Newtonian. Leider and Bird (28) critically reviewed the various solutions and concluded that a quasi-steady state solution with a power law equation describes the experimental results for "slow" squeezing of melts. However for describing experimental results with "fast" squeezing, we need solutions using constitutive equations accounting for the "stress overshoot" phenomenon. To achieve this Leider and Bird proposed the following behavior for the shear stress during a squeezing flow experiment

$$\tau_{rz} = m\left(-\frac{dv_r}{dz}\right)^n [1 + (b\dot{\gamma}t - 1)e^{-t/an\lambda}] \qquad (10.9\text{-}1)$$

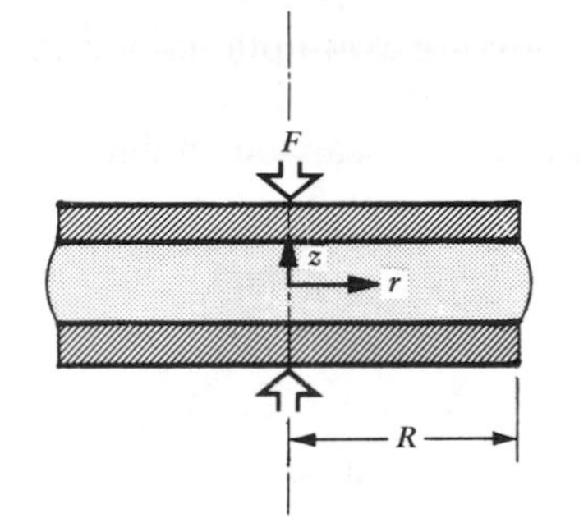


**Fig. 10.34** Schematic diagram of the squeezing flow configuration with a cylindrical coordinate system placed midway between the disks.

where $m$ and $n$ are the familiar power law model parameters, $\lambda$ is a time constant, $\dot{\gamma}$ is the shear rate and $a$ and $b$ are two additional parameters.

We first solve the flow problem for a power law fluid as proposed by Scott (29) and presented by Leider and Bird (28). In addition to the quasi-steady state assumption, the lubrication approximation is invoked. In light of these simplifying assumptions, the power law model, in cylindrical coordinates shown in Fig. 10.34, reduces to

$$\tau_{rz} = m\left(-\frac{dv_r}{dz}\right)^n \tag{10.9-2}$$

and the equation of continuity reduces to

$$\frac{1}{r}\frac{\partial}{\partial r}(rv_r)+\frac{\partial v_z}{\partial z}=0 \tag{10.9-3}$$

which can be integrated to give

$$-\dot{h}\pi r^2 = 2\pi r\int_0^h v_r\,dz \tag{10.9-4}$$

where $\dot{h} = dh/dt$ is the instantaneous disk velocity. The $r$-component of the equation of motion, with inertial terms and normal stress terms omitted, reduces to

$$\frac{\partial \tau_{rz}}{\partial z} = -\frac{\partial P}{\partial r} \tag{10.9-5}$$

The time dependence of $v_r$ is introduced through the boundary condition at $h(t)$ as given by Eq. 10.9-4, where

$$v_r(h)=0 \tag{10.9-6a}$$

furthermore, requirements of symmetry dictate that $v_r$ must satisfy the following condition:

$$\frac{\partial v_r}{\partial z}=0 \quad \text{or} \quad \tau_{rz}=0 \quad \text{at} \quad z=0 \tag{10.9-6b}$$

It follows from the simplifying assumptions that the other two components of the equation of motion reduce to $\partial P/\partial\theta = 0$ and $\partial P/\partial z = 0$, hence the pressure $P$ is a function of $r$ only, and Eq. 10.9-5 can be integrated with respect to $z$, which together with boundary condition 10.9-6b gives

$$\tau_{rz} = \left(-\frac{\partial P}{\partial r}\right)z \tag{10.9-7}$$

Next, Eq. 10.9-7 is substituted into Eq. 10.9-2 and the latter integrated over $z$ to obtain the following radial velocity profile:

$$v_r = \frac{h^{1+s}}{1+s}\left(-\frac{1}{m}\frac{dP}{dr}\right)^s\left[1-\left(\frac{z}{h}\right)^{1+s}\right] \tag{10.9-8}$$

where $s = 1/n$. Note that the velocity profile above is identical to that of a power law fluid in a fully developed flow between parallel plates. By substituting Eq. 10.9-8 into Eq. 10.9-4, followed by integration, a differential equation for the pressure gradient is obtained in terms of the instantaneous disk velocity:

$$\frac{dP}{dr} = -m\left(\frac{2+s}{2h^{s+2}}\right)^n(-\dot{h})^n r^n \tag{10.9-9}$$

which can be integrated to obtain the pressure profile

► $$P = P_a + m\frac{(2+s)^n}{2^n(n+1)}\frac{(-\dot{h})^n R^{1+n}}{h^{1+2n}}\left[1-\left(\frac{r}{R}\right)^{1+n}\right] \tag{10.9-10}$$

where $P_a$ is the atmospheric pressure. The maximum pressure, as expected, is at the center of the disk. The total instantaneous force that must be applied to the disk to maintain the velocity $\dot{h}$ is obtained from Eq. 10.9-10 by integrating $P$ over the disk surface. This results in the following expression:

► $$F_N = m\pi\frac{(2+s)^n}{2^n(3+n)}\frac{(-\dot{h})^n R^{3+n}}{h^{1+2n}} \tag{10.9-11}$$

Equation 10.9-11 is the Scott equation, which for a *constant* applied force predicts the following disk separation as a function of time:

► $$\frac{h(t)}{h_0} = \left[1+\frac{2(1+s)(3+n)^s}{2+s}\left(\frac{F_N}{m\pi R^2}\right)^s\left(\frac{h_0}{R}\right)^{1+s} t\right]^{-n/(1+n)} \tag{10.9-12}$$

where $h_0 = h(0)$.

Finally, the value of the "half-time" that is, the time needed for reducing gap size to one-half the initial value, can be obtained from Eq. 10.9-12:

► $$\frac{t_{1/2}}{n} = K_n\left(\frac{m\pi R^2}{F_N}\right)^s\left(\frac{R}{h_0}\right)^{1+s} \tag{10.9-13}$$

where

$$K_n = \left(\frac{2^{1+s}-1}{2n}\right)\left(\frac{2+s}{1+s}\right)\left(\frac{1}{3+n}\right)^s \tag{10.9-14}$$

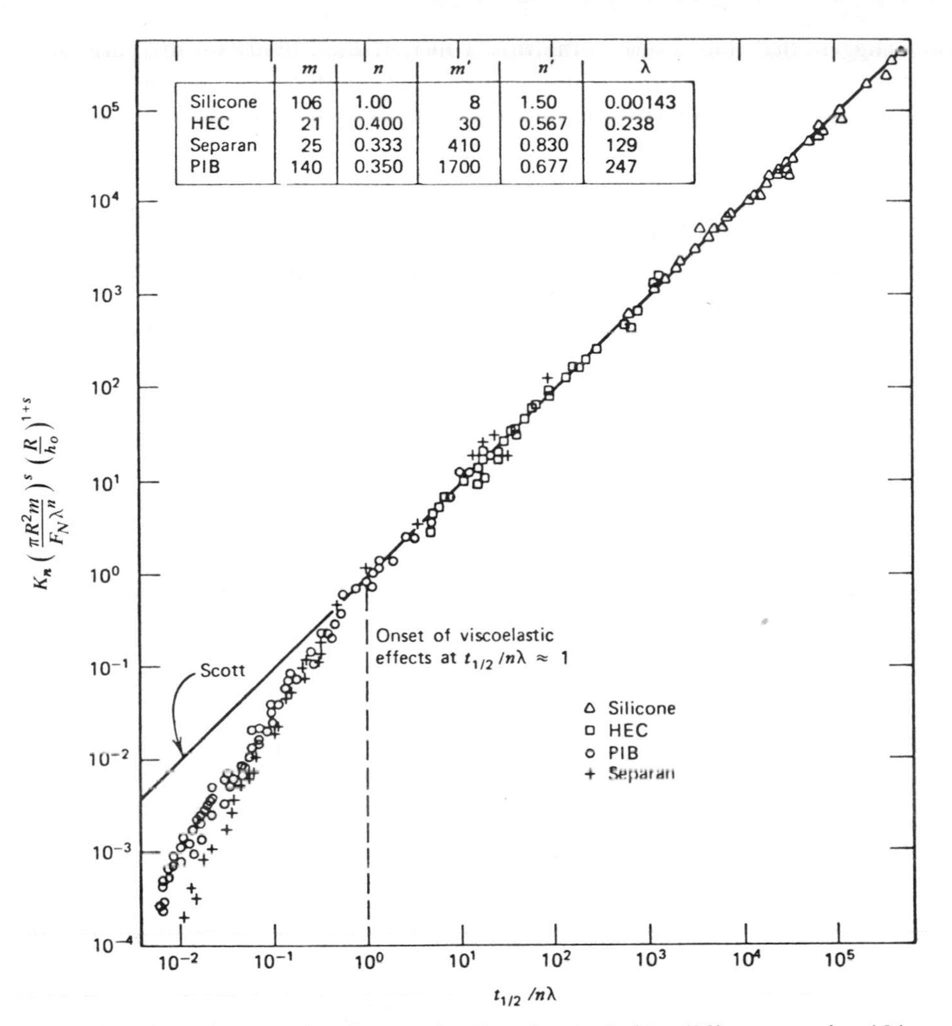


| | $m$ | $n$ | $m'$ | $n'$ | $\lambda$ |
|---|---|---|---|---|---|
| Silicone | 106 | 1.00 | 8 | 1.50 | 0.00143 |
| HEC | 21 | 0.400 | 30 | 0.567 | 0.238 |
| Separan | 25 | 0.333 | 410 | 0.830 | 129 |
| PIB | 140 | 0.350 | 1700 | 0.677 | 247 |

**Fig. 10.35** Dimensionless plot of squeezing flow data by Leider (30) representing 181 runs for four fluids: silicone oil, 1% solution of hydroxyethyl cellulose (HEC), 0.5% solution of Separan (polyacrylamide) in glycerine, and polyisobutylene solution. [Reprinted with permission from P. J. Leider, *Ind. Eng. Chem. Fundam.*, **13**, 342–346 (1974). Copyright by the American Chemical Society.]

This treatment of squeezing flows was tested experimentally by Leider (30) as indicated in Fig. 10.35, which plots Eq. 10.9–13 in a dimensionless form. Leider (30) rendered $t_{1/2}$ dimensionless by dividing by $\lambda$. Clearly the Scott equation describes the experimental results above $t_{1/2}/n\lambda = 1$. Deviations below the value are associated with the viscoelastic nature of polymers as pointed out earlier, in particular with the stress overshoot phenomenon. Leider and Bird (28) proceeded to include this effect empirically by using the relation given in Eq. 10.9–1. The resulting expression for the force $F_N$ using this relation is given by

$$F_N = m\pi \frac{(2+s)^n}{2^n(3+n)} \frac{(-\dot{h})^n R^{3+n}}{h^{1+2n}} \left\{1 + \left[\frac{(2+s)}{2^{1+s}}\left(\frac{-\dot{h} h_0^s R}{h^{2+s}}\right) bt - 1\right] e^{-t/an\lambda}\right\} \quad (10.9\text{-}15)$$

They suggest that the choice of the time constant $\lambda$ be made on the basis of the power law model parameters $m$ and $n$ and a similar power law relationship for the primary normal stress function $\Psi_1(\dot{\gamma}) = m_1 \dot{\gamma}^{n_1 - 2}$ (see Section 6.7), as follows:

$$\lambda = \left(\frac{m_1}{2m}\right)^{1/(n_1 - n)} \tag{10.9-16}$$

and the selection of the parameters $a$ and $b$ so as to give the best fit of the shear stress overshoot data that occur at the beginning of constant shear rate experiment. By following the procedure for evaluating the parameters $\lambda$, $a$, and $b$, Eq. 10.9-15 correctly predicted the direction and order of magnitude of the deviation of the experimental data from the Scott equation in Fig. 10.35.

## 10.10 Gear Pumps

Rotary gear pumps (Fig. 10.32*c*) are widely used with a variety of fluids. The positive displacement flow in these pumps permits accurate flow rate metering, coupled with high discharge pressure—a unique combination in pumping low viscosity oils. The hydraulic systems of many injection molding machines contain gear pumps, although vane-type pumps tend to replace them for this application. Gear pumps have also found application in pumping and pressurizing polymer melts, in particular those of low viscosity. Thus gear pumps are used as "booster" pumps in series with a plasticating screw extruder and with low viscosity polymers such as nylons, for both "boosting" thc pressure and accurately controlling flow rate (e.g., in fiber spinning operations). Gear pumps are also used as high output devices, for pelletizing polyolefins after the polymerization reactor. A solids fed series of three sets of gear pumps was even suggested by Pasquetti (31) for melting and pumping.

The principle of operation of gear pumps is simple. Referring to Fig. 10.32*c*, liquid is fed between the exposed adjacent gear teeth. The liquid is transported by the gear rotation from the inlet to the discharge port. During this time the liquid is enclosed between the adjacent teeth and the housing, with some leakage over the teeth. The relative motion between gear and housing induces circulatory flow, similar to the flow in the cross channel direction in screw extruders discussed in Section 10.3. The inlet and the outlet are sealed from each other by the meshed gears. The meshing action squeezes the melt out from the space. Cycling of both the discharge pressure and discharge rate occurs every time a new set of gear teeth is exposed to the discharge port. The shape of the gear teeth is generally of the involute type (Fig. 10.36). (Such a shape is described by the end of a string wrapped

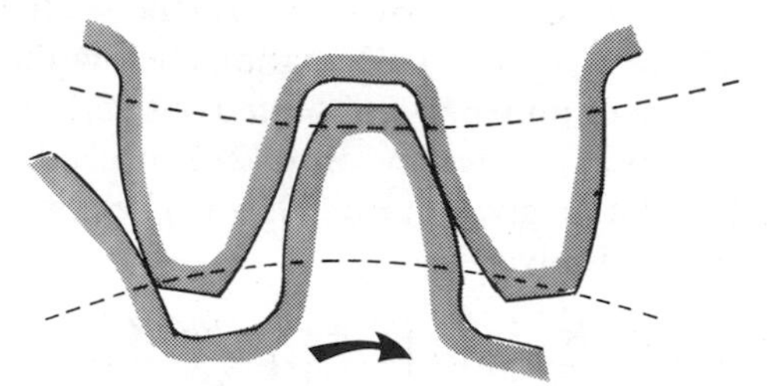

**Fig. 10.36** Tooth configuration for gear pumps.

around the circumference of a cylinder and being unwrapped under tension.) With straight teeth gears liquid may be trapped between the intermeshing teeth resulting in "backlash," excessive noise, and wear. With low viscosity oils this problem is relieved by providing strategically located relief ports. Since this is not possible with high viscosity melts, double helical gears (Fig. 10.37) are used. Upon intermeshing, this geometry of teeth results in a squeezing-out action of the melt, from the center outward.

Ideally the flow rate of gear pumps is determined by the displaced volume; hence it is independent of the rheological properties of the liquid. Yet in reality some leakage does occur between the teeth crest and the housing, between the gear sides and the housing, and between the intermeshing gears, reducing somewhat the pumping efficiency. The leakage flows are sensitive, of course, to liquid viscosity. Discharge pressure is determined by the die resistance at the discharge port.

Gear pumps have a number of limitations in pumping and pressurizing polymer melts. In gravitational or low pressure feeding (as in pelletizing applications), there is an upper limit of melt viscosity beyond which the polymer will not fill the gap exposed to it. This results in "starving." The other limitation is the forces that develop between the intermeshing teeth because of the "squeezing out" process of the melt. These forces tend to separate the gears and bring about wear. The problem is, of course, intensified by high viscosity melts. Finally, for thermally and shear degradable polymers, the gear pump may be inappropriate because of the numerous "dead" spots present.

Theoretical analysis of the gear pump involves the evaluation of the various leakage flows, the hydrodynamic behavior of the melt in the squeezing type of flow upon teeth intermeshing, and the circulatory flow within the gaps. Leakage flows have been analyzed by Ishibashi (32) in reference to low viscosity oils. It appears that no analysis has been attempted on the squeezing type of flow upon intermeshing, where, in addition to simple viscous flow, possible stress overshoot, elastic, and other non-Newtonian effects may play an important role. Finally, the circulatory flow within the gap is amenable to analysis.

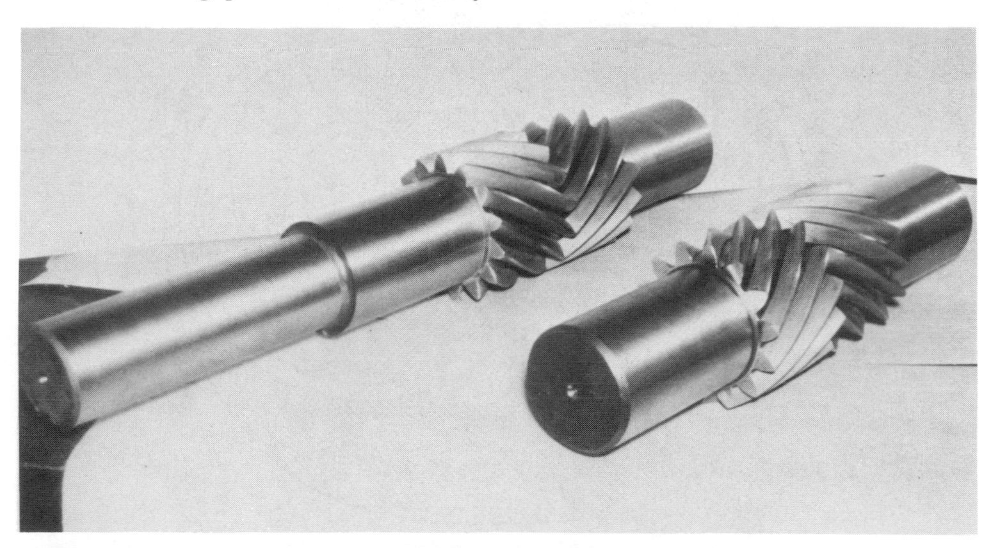

**Fig. 10.37** Double helical gears. (Photograph courtesy of Farrel Co., Ansonia, Conn.)

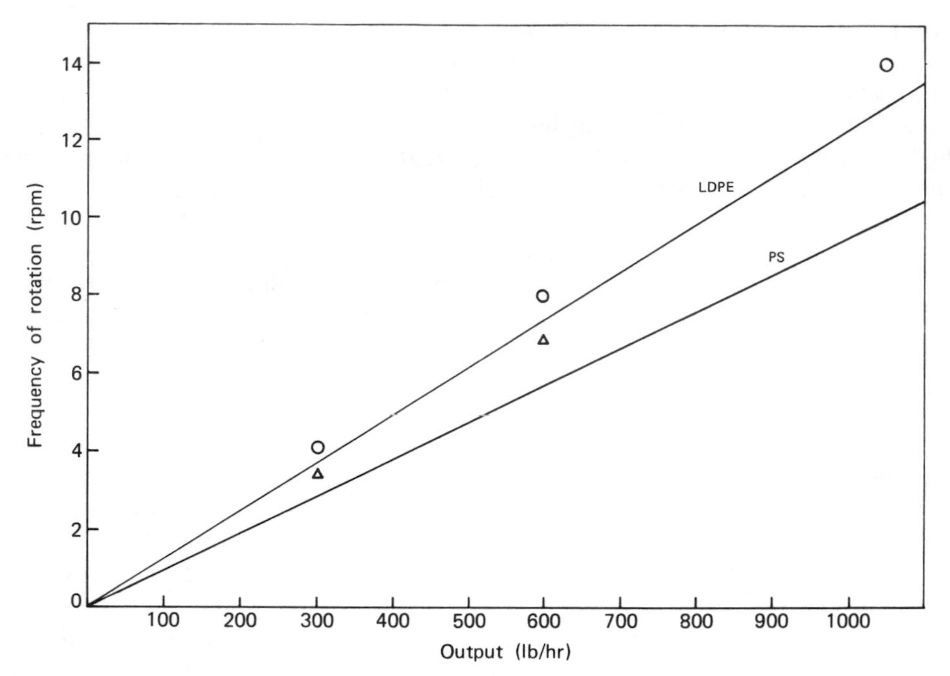


**Fig. 10.38** Output versus gear frquency of rotation for a 5.6 in. diameter double helical gear pump 4.5 in. wide, with LDPE (circles) and PS (triangles). Smooth curves are theoretical curves at the respective densities (33).

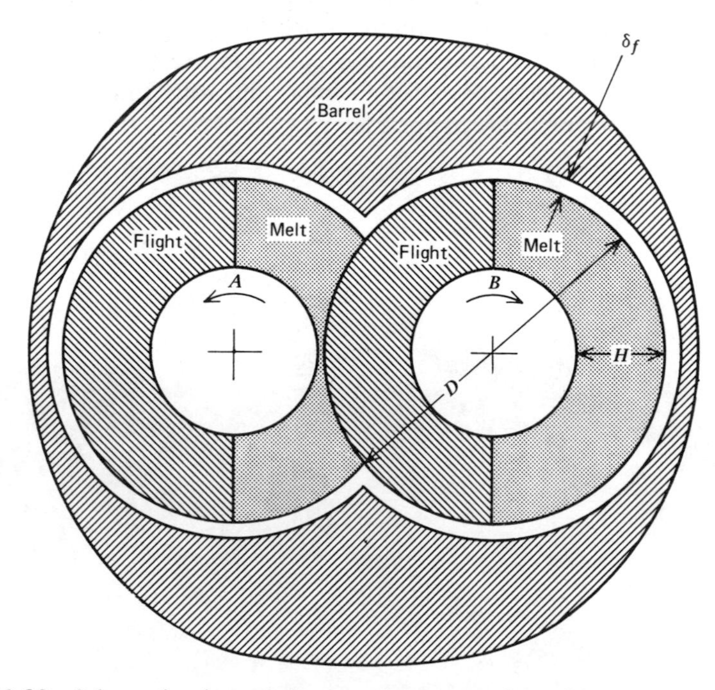


**Fig. 10.39** Schematic view of the barrel cross section with counterrotating screws. Dotted regions indicate polymer melt in screw channels.

Figure 10.38 shows experimental data for PS and LDPE using a 5.6 in. diameter double helical gear pump 4.5 in. wide (33). The gears have 14 teeth, a tooth depth of 0.67 in., and a 30° spiral angle. The smooth curves are theoretical. The experimental points fall above the theoretical line, indicating the existence of leakage flow. This discrepancy increases with increasing flow rate or discharge pressure. Characteristics are the very high flow rates at relatively low frequencies of rotation.

## 10.11 Intermeshing Counterrotating Twin Screws

Twin screw extruders come in a variety of configurations and arrangements. They can be corotating or counterrotating, they can be non-, partially, or fully intermeshing, as well as of further geometrical distinctions. A systematic classification of twin screw configurations and arrangements that is essential for a better understanding of their principle of operation, was recently suggested by Herrmann et al. (34), who also point out that only fully intermeshing counterrotating screws (Figs. 10.32*d*, 10.39) can form truly positive displacement pumps (see Problem 10.13). The theoretical analysis of this latter configuration is the subject of this section.

The easiest way to visualize the conveying mechanism of a twin screw extruder is to place a viewer in the screw channel and let him report his observations (35). We shall first stop the screw rotation and place him at point "0" in Fig. 10.32*d*, and ask him to explore the space around him. He will report that he is standing on the root of screw *A*, that he sees above the barrel surface, and on each of his sides two "walls" (flights 1 and 2) block completely the gap between his "floor" and his "ceiling." He then will start walking in the down channel direction, toward the die; but before he goes too far, a small section of a "cylinder" (screw *B*) blocks his way completely. Unable to move forward, he turns around and moves backward, toward the hopper. After just about one turn he discovers that once again the channel is blocked by a cylinder. He then reports the unpleasant finding that he is completely confined within solid steel walls, and having nothing better to do, he prepares a three-dimensional view of his "living quarters" (Fig. 10.40). From this we conclude that by "meshing" two mirror image (counterrotating type) screws together we break up the otherwise continuous helical channel into *short segments* less than one full turn long. This finding is demonstrated in Fig. 10.41, which shows two intermeshing screws in which the channel segments of one of the screws was filled with silicon rubber. The silicon rubber coating from the second and fourth segments from the left were pulled off the screw. The former appears in its original

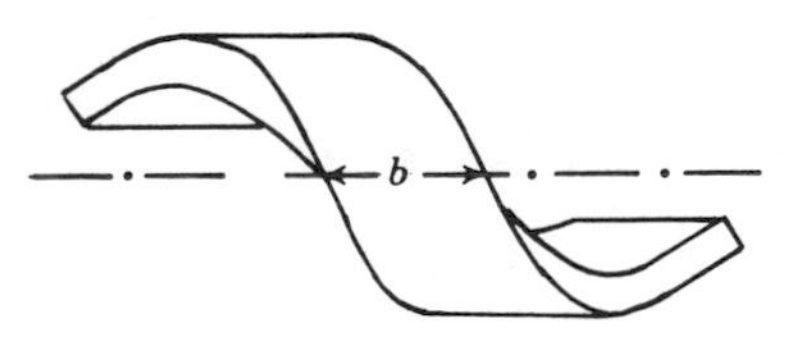


**Fig. 10.40** A three-dimensional schematic view of the "channel segments" formed by intermeshing twin screws.

**Fig. 10.41** Two intermeshing screws: the channel segments in the upper screw were filled with silicone rubber. Two silicone rubber pieces, reproducing the shape of the channel segments—the second and fourth from left—were pulled out: the former shows the shape of the real geometry of the segment, whereas the latter is flattened out into a straight channel.

shape, whereas the latter is shown flattened out, indicating that the segments can be approximated by short rectangular (or trapezoidal) straight channels.

What happens if we start rotating the screws? Turning to our viewer "on the spot," we get his first reaction—that everything moves around him and he constantly bumps into the walls. Then we tell him to move in such a way that the walls will not approach him, although they can move parallel to themselves. This will be rather easy for him, because all he has to do is to acquire an axial velocity of

$$V_l = L_s N \tag{10.11-1}$$

where $N$ is the frequency of screw rotation (usually referred to as screw speed) and $L_s$ is the lead shown in Fig. 10.32*d*. Now, moving at this velocity in the axial direction, what will the viewer observe? First of all, of course, he will say that his whole compartment with the melt in it is being transported at the same velocity toward the die. Second, from his point of view (Lagrangean point of view), it is *the barrel surface that is moving with a velocity* $V_l$ in the opposite direction, and the root of the screw in the *up channel helical direction* with a velocity

$$V_s = NZ_s \tag{10.11-2}$$

where $Z_s$ is the helical length of one full turn on the root of the screw, given by the following equation, analogous to Eq. 10.3-7:

$$Z_s = \frac{L_s}{\sin \theta_s} \tag{10.11-3}$$

where $\theta_s$ is the helix angle on the root of the screw. From the three equations above we get

$$V_s = \frac{V_l}{\sin \theta_s} \tag{10.11-4}$$

The flights, of course, will also be moving in the same direction as the root of the screw. Now that the physical configuration has been clarified, we can consider the details of a simulation model for the process.

Such a model has a number of facets and engineering goals. First we are interested in expressions relating throughput or volumetric flow rate to geometrical and operational variables. This is a relatively simple problem for *totally* isolated channel segments, in which instance the counterrotating twin screw extruder becomes truly a positive displacement pump. Without "communication" between channel segments, the pressure within each segment is independent of the discharge pressure at the die, except during the period when a particular "chamber" reaches the die and exposes itself to it. During this time the "cylinder" or "roll" blocking the channel segment at the upsteam end acts like a plunger, pushing the melt into the die at a constant rate. The discharge pressure depends on die resistance. As in gear pumps, the discharge pressure exhibits pressure and flow rate time fluctuations as successive "chambers" are exposed to the twin screw die. Usually, however, channel segments cannot be totally sealed from each other. Clearances exist between the two screw surfaces, which allow for leakage flows from chamber to chamber. The magnitude of this leakage flow is dependent on the level of pressure differences between chambers, but since there is "communication" between chambers all the way to the die, the leakage flow becomes a function of the die pressure. In practice, the size of the chamber is reduced toward the die (e.g., by a decreasing lead), and only a number of chambers close to the die are full of melt, while the rest are only partially filled. Starve-feeding is a common practice to avoid drive and bearing overloading. The problem of leakage flows was analyzed by Konstantinov and Levin (36, 37) and by Janssen et al. (38). It should be noted at this point that leakage between chambers, although reducing the output, can bring about much intensive mixing, which is of great importance in twin screw extruders. Such a mixing action takes place both between the tip of the flight of one screw and the root of the other, and between the flanks of the intermeshing flight.

Another aspect of a theoretical model we consider is the hydrodynamic behavior of the melt within the chamber. This aspect is related primarily to extensive mixing. Yet the dynamic pressure distribution within the chamber also relates to the leakage flow problem. The hydrodynamic flow problem within the chamber was discussed by Doboczky (39) and Wyman (35).

### *Flow Rate and Leakage Flows*

The theoretical output of the twin screw geometry in the absence of leakage flow can be easily calculated by considering Fig. 10.39 and Eq. 10.11-1. For completely filled and sealed channels, the volumetric flow rate is the product of the axial velocity $V_l$ and the cross-sectional area of melt $A_m$:

$$Q = V_l A_m \tag{10.11-5}$$

For single flighted identical screws, $A_m$ (indicated in Fig. 10.39 by the dotted area) equals the full cross-sectional area of the annular space between the root and barrel of one screw $\pi(D_f - H)H$ less the area corresponding to the overlap of the flanks of the flights. Thus neglecting the flight clearance, we have

$$A_m = \pi(D_f - H)H - \frac{D_f^2}{2}\cos^{-1}\left(1 - \frac{H}{D_f}\right) + \frac{D_f - H}{2}\sqrt{H(2D_f - H)} \tag{10.11-6}$$

The volumetric flow rate (both screws) is given by

$$Q = \pi N\bar{D}L_s H\left[1 - \frac{1}{2\pi}\left(\frac{D_f}{\bar{D}}\right)\left(\frac{D_f}{H}\right)\cos^{-1}\left(1 - \frac{H}{D_f}\right) + \frac{1}{2\pi}\left(\frac{D_f}{\bar{D}}\right)\left(1 - \frac{H}{D_f}\right)\sqrt{\frac{2D_f}{H} - 1}\right] \tag{10.11-7}$$

where $\bar{D}$ is the mean diameter

$$\bar{D} = D_f - H \tag{10.11-8}$$

The actual flow rate is less than the theoretical flow rate because of the leakage flow between adjacent chambers. As stated earlier, there are leakage flows between the screw flights and the barrel, between the tip of the flight of one screw and the root of the second screw, and between the flanks of the flights. Equations for these leakage flows were derived by Doboczky (39) and by Janssen et al. (38), who also carried out experiments with Newtonian liquids, confirming their theoretical results. Power consumption in twin screw geometry is given by Schenkel (40), who also provides detailed information on various twin screws and compares their action to that of single screw pumps.

### *The Plate and Frame Model**

Steady state flow problems within one isolated channel segment or chamber can be analyzed from a Lagrangean point of view, that is, from the point of view of an observer located in the chamber and moving with the velocity given in Eq. 10.11-1. In such a moving coordinate system, channel walls are at fixed positions. Assuming relatively shallow channels, we unwind each channel segment from the "screw" and flatten it out, as in Fig. 10.42. We note the surface of the barrel, which moves at velocity $V_l$ in a direction opposite to the forward axial direction, and the root of the screw, which moves at velocity $V_s$ in the upstream helical direction. The flights, of course, move together with the root of the screw. The blocking cylinder or screw is rotating with a tangential velocity $\pi ND_s$. The end result of the simplified model is as follows: the flights and cylinders (screw $B$) form a parallelepiped frame. The "frame" is placed within two infinite parallel plates—the surface of the barrel and the root of the screw. Figure 10.43 gives top and side views of this plate and frame. Each retaining surface moves parallel to its plane, as explained earlier and shown in the figures. The velocity of the barrel surface can be broken down into two

* A model virtually identical to the "plate and frame" model was recently proposed by U. Burkhardt, H. Herrmann and S. Jakopin, at the Soc. of Plastics Engineers 36th Annual Technical Conference, Washington, D.C., 1978, p. 498. They also proposed a model for certain corotating twin screw extruders.

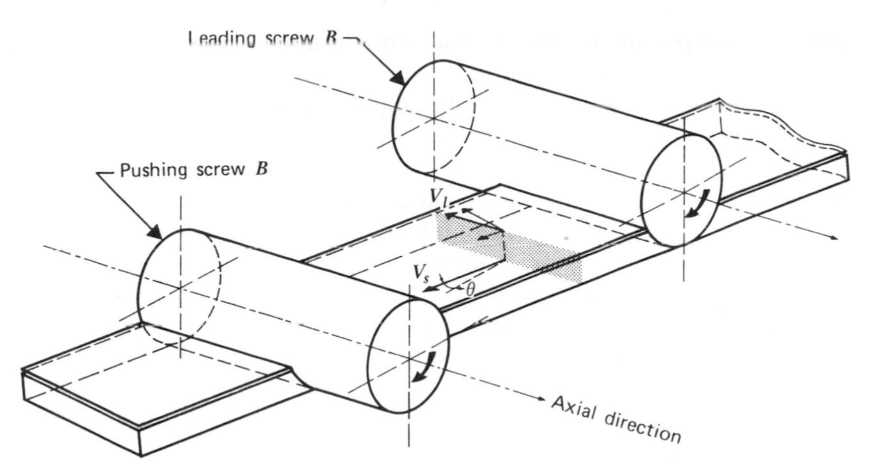


**Fig. 10.42** The unwound helical channel segment of counterrotating screws.

components, down channel $V_l \sin \theta_b$ and cross channel $V_l \cos \theta_b$ toward the "pushing flight." The screw velocity is the vectorial sum of two velocities: the tangential velocity of the root of the screw $\pi N D_s$ and the velocity of the barrel or the viewer $V_l$. Finally, Fig. 10.43*b* gives the first hints on the nature of the flow pattern in the chamber. We note that both the screw and the barrel drag melt toward the pushing screw. Neglecting end effects and considering that the net flow rate is zero (no leakage), the shape of the velocity profile ($v_z$) must be as indicated schematically in Fig. 10.43*b*. This also implies that pressure will build up against the pushing screw.

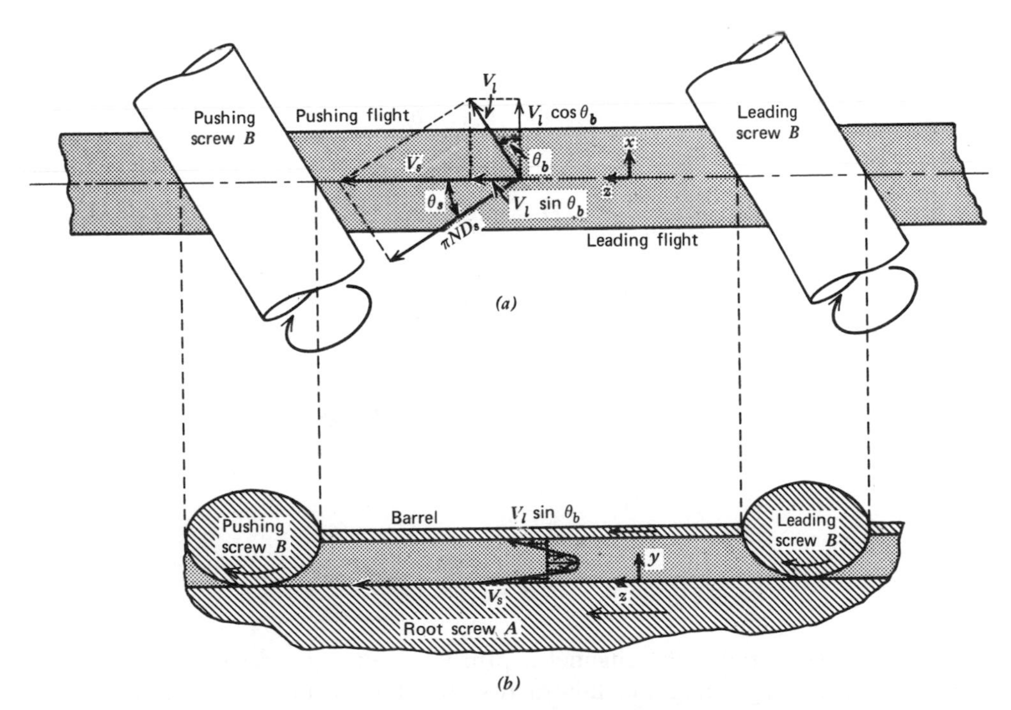


**Fig. 10.43** (*a*) Top and (*b*) side view of the unwound channel in Fig. 10.42.

The velocity profile in the channel compartment, viewed from a moving coordinate system, can be easily derived after making the following assumptions: the fluid is Newtonian, the flow is independent of time (steady state), the flow is laminar, the flow is fully developed, there is no slip at the walls, the fluid is incompressible, gravity forces are neglected, and the flow is isothermal. On the basis of these assumptions, we shall derive velocity profiles that hold for a shallow channel "far" from the frame boundaries (i.e., far from the flights and screw B). From this velocity profile we shall be able to form some ideas about the path of the fluid particles in the channel, the pressure distribution in the channel, and power consumption.

The equation of continuity subject to the foregoing assumptions and in the coordinate system of Fig. 10.43 reduces to

$$\frac{\partial v_y}{\partial y} = 0 \tag{10.11-9}$$

which upon integration yields $v_y = \text{constant}$; but since $v_y$ must be zero at either plate, it must vanish everywhere. Thus $v_y = 0$. The velocity components that remain in the equation of motion are $v_z(y)$ and $v_x(y)$. The equation of motion reduces to

$$\frac{\partial P}{\partial x} = \mu \frac{\partial^2 v_x}{\partial y^2} \tag{10.11-10}$$

and

$$\frac{\partial P}{\partial z} = \mu \frac{\partial^2 v_z}{\partial y^2} \tag{10.11-11}$$

and the boundary conditions for solving these differential equations are

$$v_x(0) = 0 \tag{10.11-12a}$$

$$v_x(H) = V_l \cos \theta_b \tag{10.11-12b}$$

$$v_z(0) = V_s \tag{10.11-13a}$$

$$v_z(H) = V_l \sin \theta_b \tag{10.11-13b}$$

The $y$-component of the equation of motion provides the information that $P \neq f(y)$; thus Eqs. 10.11-10 and 10.11-11 can be integrated, using the boundary conditions 10.11-12 and 10.11-13, to yield the following velocity profiles:

$$v_x = V_l \cos \theta_b \xi + \xi(\xi - 1)\left(\frac{H^2}{2\mu}\frac{\partial P}{\partial x}\right) \tag{10.11-14}$$

and

$$v_z = (V_l \sin \theta_b - V_s)\xi + \xi(\xi - 1)\left(\frac{H^2}{2\mu}\frac{\partial P}{\partial z}\right) + V_s \tag{10.11-15}$$

where $\xi = y/H$, $H$ being the channel depth. By assuming a zero net flow in both directions (i.e., neglecting the flow across the flights and the flow across the blocking screws), we obtain from Eqs. 10.11-14 and 10.11-15, after integration

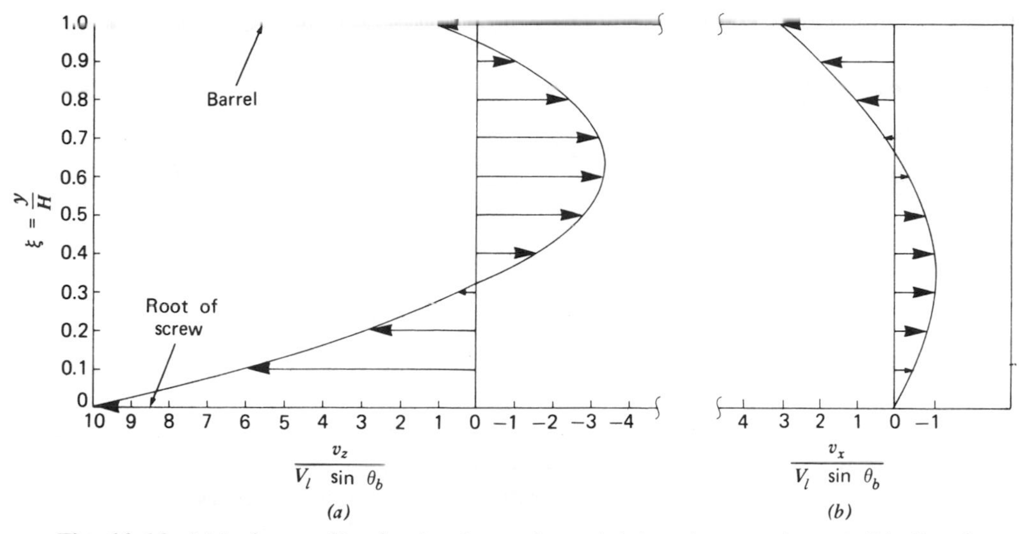


**Fig. 10.44** Velocity profiles in the down channel (*a*) and cross channel (*b*) directions, assuming square pitched screws. Basis: $V_l \sin\theta_b = 1$, $V_s/(V_l \sin\theta_b) = 1/(\sin\theta_s \sin\theta_b) \cong 10$, $v_x(1)/(V_l \sin\theta_b) = v_x(1)/(V_l \cos\theta_b)(\cos\theta_b/\sin\theta_b) \cong v_x(1)/(V_l \cos\theta_b) \cong 3.13$.

between $y = 0$ to $y = H$, the following expressions for the pressure gradients:

$$\frac{\partial P}{\partial x} = \frac{6\mu V_l \cos\theta_b}{H^2} \tag{10.11-16}$$

$$\frac{\partial P}{\partial z} = \frac{6\mu(V_l \sin\theta_b + V_s)}{H^2} \tag{10.11-17}$$

Thus the pressure rises linearly in the directions of the pushing screw and flight, reaching a maximum in the corner between the pushing screw and the pushing flight. However, the absolute pressure cannot be determined from these equations unless the chamber is partly empty, in which case the empty portion can be assumed to be at atmospheric pressure. Otherwise the leakage flow must be considered and the pressure profile along the whole length of the screw determined. By substituting Eqs. 10.41-17 and 10.11-16 into Eqs. 10.11-15 and 10.11-14, respectively, we obtain

$$\frac{v_x}{V_l \cos\theta_b} = \xi(3\xi - 2) \tag{10.11-18}$$

and

$$\frac{v_z}{V_l \sin\theta_b} = \xi(3\xi - 2) + \frac{V_s}{V_l \sin\theta_b}(1 - 4\xi + 3\xi^2) \tag{10.11-19}$$

Figure 10.44 presents the velocity profiles for square pitched screws ($\theta = 17.65°$) with $V_l \sin\theta = 1$. The velocity profiles reveal intense internal circulation, whereby the melt is dragged by the root of the screw toward the pushing screw in the lower portion of the channel, while it flows in the opposite direction (opposing the motion of the barrel surface) in the upper portion of the barrel. At the same time there is also a circulatory flow in the channel width direction where, in the

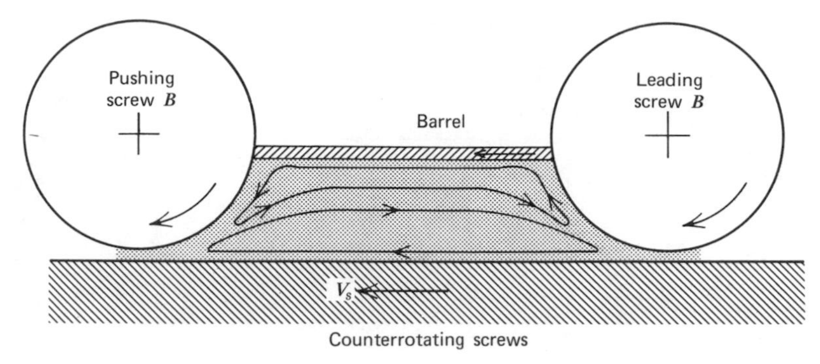


**Fig. 10.45** Schematic representation of streamlines in the neighborhood of the "nips".

upper part of the channel, the melt is dragged by the barrel surface toward pushing flights and flows back in the lower portion of the channel. The interaction of the two velocity profiles eliminates the possibility of any stagnant layers. The paths described by fluid particles will depend on their initial location and will be quite complex. In principle these paths can be calculated from the velocity profiles, and they are expected to have the shape of open loop helices.

The situation is somewhat more complicated in the neighborhood of the nips (Fig. 10.45). Near the pushing flight, both the pushing screw and the root of screw A drag the melt toward the nip. This results in intensive mixing effects with both high rates of shear and high stresses. An opposite effect occurs at the leading screw.

From the foregoing discussion it is evident that the analysis of other types of twin screws will lead to different models and conclusions.

## *REFERENCES*

1. R. W. Flumerfelt, M. W. Pierick, S. L. Cooper, and R. B. Bird, "Generalized Plane Couette Flow of a Non-Newtonian Fluid," *Ind. Eng. Chem. Fundam.*, **8**, 354–357 (1969).
2. M. Hirshberger, "Flow of Non-Newtonian Fluids in Rectangular Channels," M.S. thesis, Department of Chemical Engineering, Technion—Israel Institute of Technology, Haifa, 1970.
3. Z. Tadmor and I. Klein, *Engineering Principles of Plasticating Extrusion*, Van Nostrand Reinhold, New York, 1970: (a) p. 133; (b) p. 190; (c) p. 194; (d) p. 224; (e) p. 397.
4. J. Gavis and R. L. Laurence, "Viscous Heating in Plane and Circular Flow Between Moving Surfaces, *Ind. Eng. Chem. Fundam.*, **7**, 232–239 (1968); "Viscous Heating of a Power Law Liquid in Plane Flow," *ibid.*, **7**, 525–527 (1968).
5. M. Boussinesq, "Sur l'Influence des Frottements dans les Mouvements Reguliers des Fluids," *J. Math. Pures Appl.*, Ser. 2, **13**, 377 (1868).
6. R. S. Rowell and D. Finlayson, "Screw Viscosity Pumps," *Engineering*, **114**, 606 (1922); *ibid.*, **126**, 249 (1928).
7. R. M. Griffith, "Fully Developed Flow in Screw Extruders," *Ind. Eng. Chem. Fundam.*, **1**, 180–187 (1962).
8. R. E. Colwell and K. R. Nicholls, "The Screw Extruder," *Ind. Eng. Chem.*, **51**, 841–843 (1959).
9. J. R. A. Pearson, *Mechanical Principles of Polymer Melt Processing*, Oxford, Pergamon Press, 1966.

10. H. Zamodits, "Extrusion of Thermoplastics," Ph.D. thesis, University of Cambridge, 1964; H. Zamodits and J. R. A. Pearson, "Flow of Polymer Melts in Extruders." Part I. The Effect of Transverse Flow and of a Superimposed Temperature Profile," *Trans. Soc. of Rheol.*, **13** (3), 357 (1969).
11. J. M. McKelvey, *Polymer Processing*, Wiley, New York, 1962.
12. R. F. Westover, "A Hydrodynamics Screwless Extruder," *Soc. Plast. Eng. J.*, 1473–1480 (1962).
13. R. E. Gaskell, "The Calendering of Plastic Materials," *J. Appl. Mech.*, **17**, 334–336 (1950).
14. J. T. Bergen and G. W. Scott, Jr., "Pressure Distribution in the Calendering of Plastic Materials," *J. Appl. Mech.*, **18**, 101–106 (1951).
15. C. Kiparissides and J. Vlachopoulos, "Finite Element Analysis of Calendering," *Polym. Eng. Sci.*, **16**, 712–719 (1976).
16. M. Finston, "Thermal Effects in the Calendering of Plastic Materials," *J. Appl. Mech.*, **18**, 12 (1951).
17. W. Unkrüer, "Beitrag zur Ermittlung des Druckverlaufes und der Fliessvorgange im Walzspalt bei der Kalanderverarbeitung von PVC Hart zu Folien," Doctoral dissertation, Technischen Hochschule, Aachen, 1970.
18. B. Maxwell and A. J. Scalora, "The Elastic Melt Extruder Works Without Screw," *Mod. Plast.*, **37**, 107 (1959).
19. V. L. Kocherov, Yu. L. Lukach, E. A. Sporyagin, and G. V. Vinogradov, "Flow of Melts in a Disk-type Extruder," *Polym. Eng. Sci.*, **13**, 194–201 (1973).
20. J. L. White and A. B. Metzner, "Development of Constitutive Equations for Polymeric Melts and Solutions," *J. Appl. Polym. Sci.*, **7**, 1867 (1963).
21. R. B. Bird, "Macromolecular Hydrodynamics," Rheology Research Center Report 14, University of Wisconsin, 1971.
22. P. A. Good, A. J. Schwartz, and C. W. Macosko, "Analysis of the Normal Stress Extruder," *Am. Inst. Chem. Eng. J.*, **20**, 67–73 (1974).
23. B. Maxwell, "A New Method of Solving the Feeding and Melting Zone Problems in Extruders," 31st Annual Technical Conference, Society of Plastics Engineers, Montreal, Quebec, 1973.
24. R. F. Westover, "Continuous Flow Ram Type Extruder," *Mod. Plast.*, March 1963.
25. W. Rose, "Fluid-Fluid Interfaces in Steady Motion," *Nature*, **191**, 242–243 (1961).
26. S. Bhattacharji and P. Savic, "Real and Apparent Non-Newtonian Behavior in Viscous Pipe Flow of Suspensions Driven by a Fluid Piston," *Proc. Heat Transfer Fluid Mech. Inst.*, **15**, 248 (1965).
27. I. Williams, "Plasticity of Rubber and Its Measurements," *Ind. Eng. Chem.*, **16**, 362–364 (1924).
28. P. J. Leider and R. B. Bird, "Squeezing Flow Between Parallel Disks. I. Theoretical Analysis," *Ind. Eng. Chem. Fundam.*, **13**, 336–341 (1974).
29. J. R. Scott, *IRI Trans.*, **7**, 169 (1931).
30. P. J. Leider, "Squeezing Flow Between Parallel Disks. II. Experimental Results," *Ind. Eng. Chem. Fundam.*, **13**, 342–346 (1974).
31. R. F. Westover, "Melt Extrusion," *Encyclopedia of Polymer Science and Technology*, **8**, 533–587 (1970).
32. A. Ishibashi, "Studies on Volumetric Efficiency and Theoretical Delivery of Gear Pumps," *Bull. Japan Soc. Mech. Eng.*, **13**, 688–696 (1970).
33. C. Y. Cheng, Farrel Company, Ansonia, Conn., private communication, 1972.
34. H. Herrmann, U. Burkhardt and S. Jakopin, "A Comprehensive Analysis of the Multiscrew Extruder Mechanisms," 35th Annual Technical Conference, Society of Plastics Engineers, Montreal, Quebec, 1977.
35. C. E. Wyman, "Theoretical Model for Intermeshing Twin Screw Extruders: Axial Velocity Profile for Shallow Channels," *Polym. Eng. Sci.*, **15**, 606–611 (1975).

36. V. N. Konstantinov and A. N. Levin, "Determination of the Output of Screw Extruders," *Khim. Mashinostr.*, No. 3, 18–22 (1962).
37. V. N. Konstantinov, "Effect of Screw Construction on the Output of Multi-screw Extruders," *Khim. Neft. Mashinostr.*, No. 2, 21–26 (1964).
38. L. P. B. M. Janssen, L. P. H. R. M. Mulders, and J. M. Smith, "A Model for the Output From the Pump Zone of the Double Screw Processor or Extruder," *Plast. Polym.*, 93–98 (June 1975).
39. Z. Doboczky, *Plast. Verarb.*, **16**, 57–67 (1965).
40. G. Schenkel, *Plastics Extrusion Technology and Theory*, Illife Books, London, 1966.

## PROBLEMS

**10.1** ***Pressurization Methods.*** Classify the pressurization methods in the following systems: human heart, centrifugal pump, gear pump, blow molding, volcanos, screw extruders, ram extruders, injection molding and compression molding devices, and normal stress extruders.

**10.2** ***Parallel Plate Flow of Newtonian Fluids.*** A Newtonian fluid is pumped in a parallel plate pump at steady state and isothermal conditions. The plates are 2 in. wide, 20 in. long and 0.2 in. apart. It is required to maintain a flow rate of 50 lb/hr for a polymer melt of viscosity 0.1 $lb_f{\cdot}s/in.^2$ and density 48 $lb/ft^3$.

(a) Calculate the velocity of the upper plate for a total pressure rise of 100 psi.

(b) Calculate the optimum gap size for maximum pressure rise.

(c) Evaluate the power consumption for cases (a) and (b).

(d) Is the isothermal assumption valid in either (a) or (b)?

**10.3** ***Parallel Plate Flow of Power Law Fluids.*** A power law model fluid with $m = 1\ lb_f{\cdot}s^{0.5}/in.^2$ and $n = 0.5$ is pumped in the parallel plate pump of Problem 10.2 with gap separation of 0.1 in. and plate velocity of 10 in./s. The flow rate is maintained at one half the drag flow value. Calculate the pressure rise.

**10.4** ***General Solution to the Parallel Plate Flow of Power Law Fluids.*** Write a computer program to solve Eq. 10.2-38(a) for the flow rate, given the pressure gradient; (b) for the pressure gradient, given the flow rate.

**10.5** ***The Superposition Error.*** It has been suggested in the literature that a combined pressure and drag flow between parallel plates of a non-Newtonian fluid can be solved by superposition of drag flow (which is independent of the nature of the fluid) and of pressure flow of the non-Newtonian fluid between stationary plates.* For a power law model fluid the net flow rate will be given by

$$q = \frac{V_0 H}{2} - \frac{H^{2+s}}{2^{1+s}(2+s)}\left(\frac{\Delta P}{mL}\right)^s$$

(a) Show that the above equation converges to Eq. 10.2-7 for Newtonian fluids.

* H. R. Jacobi, *Screw Extrusion of Plastics*, Illife Books, Ltd. London, 1963.

(b) Explain why this equation is incorrect.

(c) Solve Problem 10.3 with the above erroneous equation and compare the results of the correct and incorrect solutions.

**10.6 *Viscous Heating in Plane Flow Between Infinite Parallel Plates.****† Consider an infinite parallel plate flow of a viscous liquid in the absence of pressure gradient and an exponentially temperature dependent Newtonian viscosity, $\mu = \mu_0 e^{-a(T-T_0)}$.

(a) Derive the velocity and temperature profiles subject to the following boundary conditions: $v_x(0)=0$, $v_x(H)=V_0$, $T(0)=T(H)=T_0$, where $H$ is the plate separation.

(b) Plot the velocity and temperature profiles with $\lambda_1$ as a parameter, where $\lambda_1$ is a modified Brinkman number defined as

$$\lambda_1 = \frac{a\tau_0^2 H^2}{k\mu_0}$$

where $\tau_0$ is the shear stress and $k$ is the thermal conductivity.

(c) Explain the physical reason for the double valued solution in terms of the applied shear stress and the existence of a maximum shear stress that can be applied.

***Answer:*** (a)

$$\theta = a(T-T_0) = \ln\left\{m \operatorname{sech}^2\left[\left(\frac{\lambda_1 m}{8}\right)^{1/2}(2\xi-1)\right]\right\}$$

$$\xi = y/H \qquad m = \cosh^2\left(\frac{\lambda_1 m}{8}\right)^{1/2}$$

$$\frac{v_x}{V_0} = \frac{1}{2}\left\{1 + \frac{\tanh\left[\left(\frac{\lambda_1 m}{8}\right)^{1/2}(2\xi-1)\right]}{\tanh\left(\frac{\lambda_1 m}{8}\right)^{1/2}}\right\}$$

**10.7 *Blade Coating.***‡ A schematic view of a blade coating operation is shown in Fig. P-10.7.

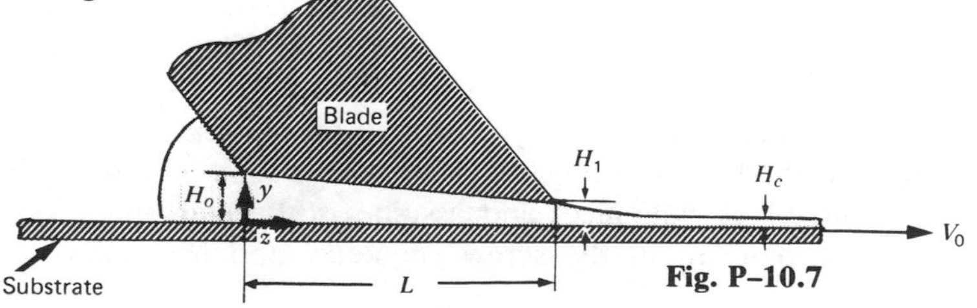


**Fig. P-10.7**

* J. Gavis and R. L. Laurence, "Viscous Heating in Plane and Circular Flow Between Moving Surfaces," *Ind. Eng. Chem. Fundam.*, **7**, 232–239 (1968).

† J. Gavis and R. L. Laurence, "Viscous Heating of a Power Law Liquid in Plane Flow," *Ind. Eng. Chem. Fundam.*, **7**, 525–527 (1968).

‡ Y. Greener and S. Middleman, "Blade Coating of Viscoelastic Fluids," *Polym. Eng. Sci.*, **14**, 791–796 (1974).

(a) Assuming that the pressure at the inlet and outlet is zero, show that for a Newtonian fluid the pressure profile under the blade is

$$P(\zeta)=\frac{6\mu L V_0}{H_0 H_1}\left[\frac{\zeta_0-\zeta}{\zeta(\zeta_0-1)}-\left(\frac{q}{V_0 H_0}\right)\frac{\zeta_0^2-\zeta^2}{\zeta^2(\zeta_0-1)}\right]$$

where $\zeta=\zeta_0-(\zeta_0-1)\frac{z}{L}$, $\zeta_0=H_0/H_1$, $q$ is the volumetric flow rate per unit width, and $V_0$ is the velocity of the substrate.

(b) Show that the coating thickness is

$$H_c=\frac{H_0}{1+\zeta_0}$$

(c) Show that the normal force on the blade $F_N$ is given by

$$F_N=\left(\frac{6\mu L^2 W V_0}{H_0 H_1}\right)\frac{1-\zeta_0(1-\ln\zeta_0)}{(\zeta_0-1)^2}$$

(d) Discuss the possible effects of the viscoelastic nature of the melt and the role of the stress overshoot phenomenon on the blade coating operation. (See Chapter 6.)

**10.8** ***Non-Symmetric Calendering.**** Derive the pressure distribution of a Newtonian fluid in a calender with different size rolls but equal peripheral speed. Make the same simplifying assumptions as used in deriving the Gaskell model in Section 10.5.

***Answer*:** Same as results in Section 10.5 with

$$\frac{1}{R}=\frac{1}{2}\left(\frac{1}{R_1}+\frac{1}{R_2}\right)$$

**10.9** ***Screw Extruders for Pumping Water.*** It is claimed that the screw extruder was invented by Archimedes for pumping water. Is the screw extruder an efficient water pump? Discuss and substantiate your arguments.

**10.10** ***Design of a LDPE Pelletizing Extruder.*** Design a screw extruder for pelletizing LDPE at a rate of 10,000 lb/hr. The head pressure required to pump the melt across the pelletizing plates is 1250 psi. The feed comes from a reactor at 500°F. Assume a Newtonian viscosity of 0.05 $lb_f \cdot s/in.^2$ and density of 48 lb/hr. Inlet opening must be large enough for gravitational feeding, the channel depth in the feed section should be no less than 2 in. deep. Neglect the effect of the flight clearance, and assume isothermal operation. Your answer must be in terms of the screw geometry and frequency of screw rotation.

* R. E. Gaskell, "The Calendering of Plastic Materials," *J. Appl. Mech.*, **17**, 334–336 (1950).

**10.11** ***Screw Extruder With Recycle.*** The constant channel depth screw shown in Fig. P-10.11 has a hollow section connected to the channel such that part of the output can recycle.

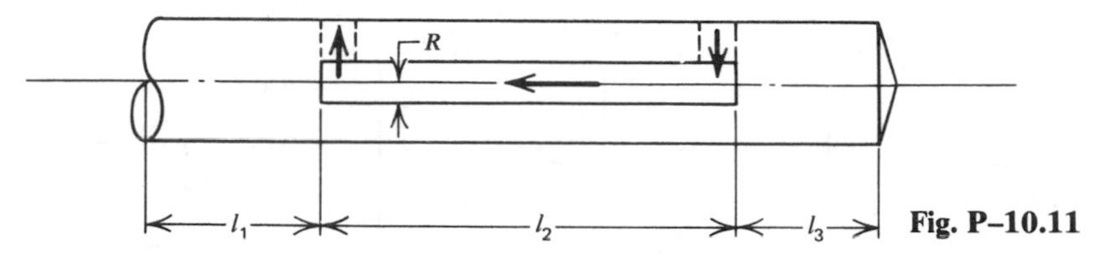


**Fig. P–10.11**

(a) Assuming isothermal Newtonian flow, derive a mathematical model relating flow rate to head pressure.

(b) Derive an expression relating the recycle flow rate to screw geometry and operating conditions.

**10.12** ***Spiral Extruder.*** A spiral flight of height $H$ is placed on a flat stationary disk, as shown in Fig. P-10.12. The flight forms a flat spiral channel of width $W$. The stationary disk is covered by a rotating disk. Inlet and

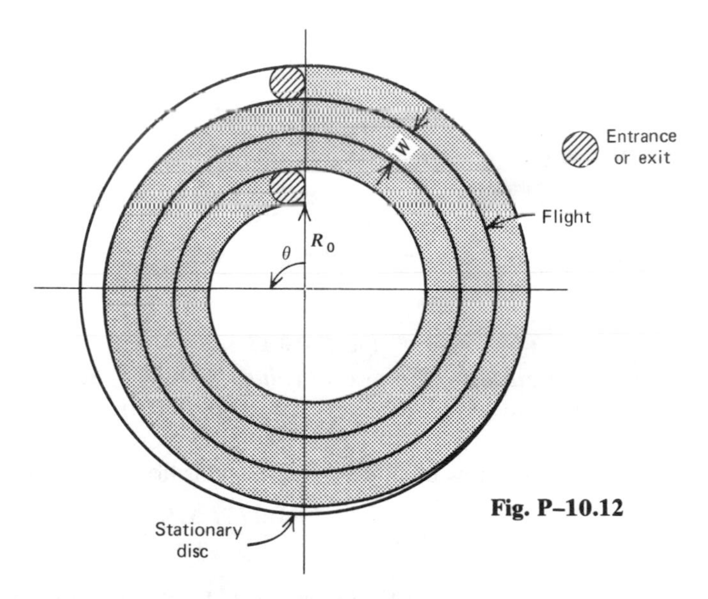


**Fig. P–10.12**

outlet opening are secured through the stationary disk as indicated in the figure. Derive a model for isothermal Newtonian extrusion.

**10.13** ***Intermeshing Twin Screws.*** Intermeshing twin screw extruders are purely positive displacement pumps, provided the flights of one screw fit the channel of the other breaking up the continuous channel into segregated segments. Using Fig. P-10.13 explain why it is possible to have such twin extruders with counterrotating screws but not with corotating screws.

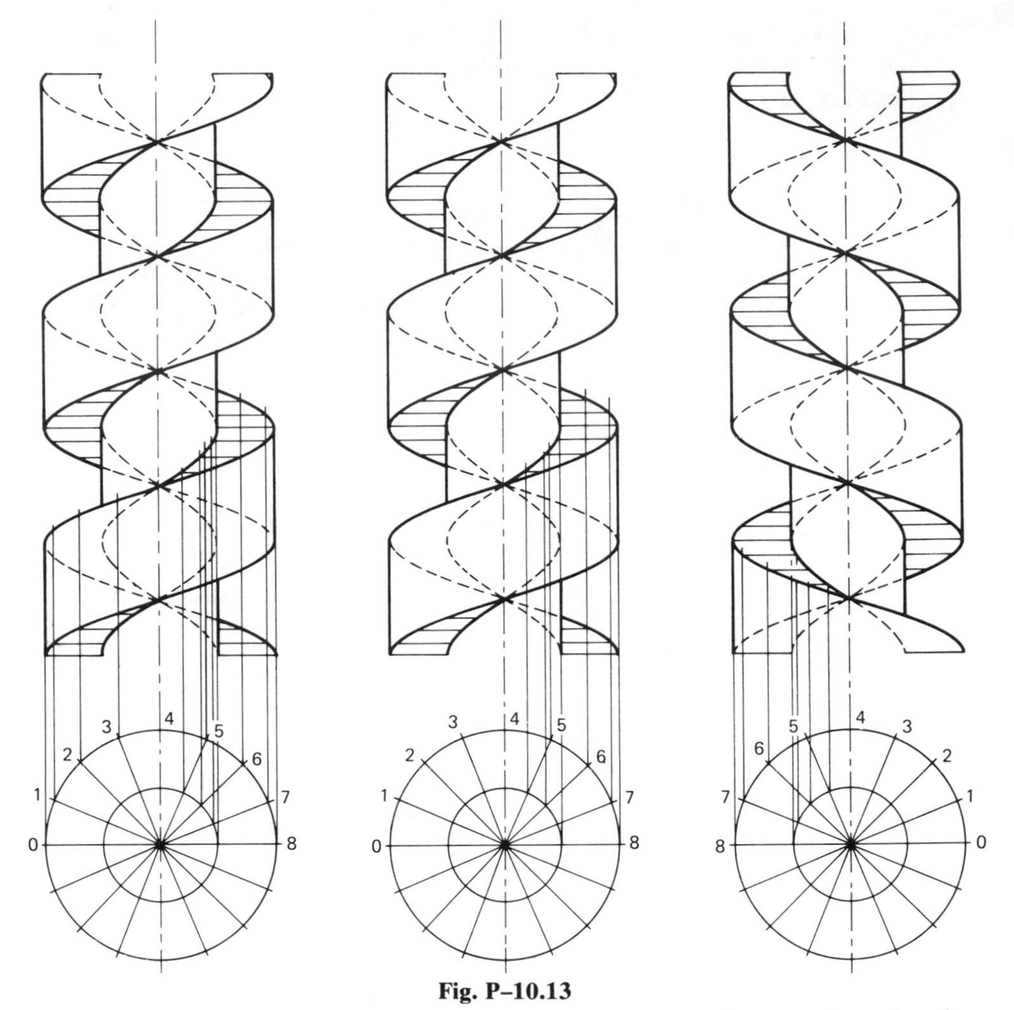


Fig. P–10.13

**10.14 *Nonintermeshing Counterrotating Twin Screw Extruders.**** Nonintermeshing twin screw extruders consist of two identical screws rotating in a barrel, as shown in Fig. P-10.14. Assuming Newtonian isothermal flow and constant depth shallow channels, derive the following relationship for the flow rate

$$Q = \frac{1}{2} WHV_{bz}F_{DTW} - \frac{WH^3}{12\mu}\left(\frac{\Delta P_T}{\Delta Z_T}\right)F_{PTW}$$

where $W$ is the channel width, $H$ channel depth, $V_{bz}$ the down channel velocity of the barrel surface relative to the screw, $\Delta P_T$ the pressure rise over the screw with a helical length of $\Delta Z_T$, and

$$F_{DTW} = \frac{4f}{1+3f}$$

$$F_{PTW} = \frac{4}{1+3f}$$

* A. Kaplan and Z. Tadmor, "Theoretical Model of Non-Intermeshing Twin Screw Extruders," *Polym. Eng. Sci.*, **14**, 58–66 (1974).

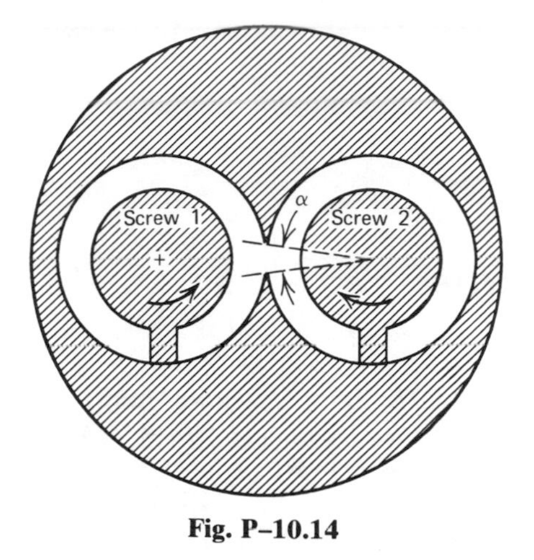


Fig. P–10.14

where $f$ is the fraction of helical channel length which is uninterrupted by an opening to the other screw ($f = \alpha/2\pi$).

**10.15** ***Parallel Plate Flow of a Fluid Obeying the CEF Equation.*** Consider parallel plate flow of a fluid obeying the CEF equation with one plate moving at a constant velocity in the absence of a pressure gradient.

(a) Write expressions for the shear stress, the primary normal stress difference, and secondary normal stress difference.

(b) Compare these expressions with those of a Newtonian fluid and a power law fluid.

# 11

# Mixing



## 11.1 The Scope of Mixing Operations

Most plastics products are not "pure" polymers, but mixtures of the basic polymer with a variety of additives such as pigments, lubricants, stabilizers, antioxidants, flame retardants, "antiblock" agents, slip additives, crosslinking agents, fillers, reinforcing agents, plasticizers, ultraviolet absorbers, and foaming agents. All these additives must be incorporated into the polymer prior to shaping, either during a special postreactor processing step before pelletizing, or just before the shaping operation in conjunction with the other elementary steps. Some of the additives form a significant proportion of the mixture, others only minute amounts. Some are compatible, others are not. The distribution of these additives in the polymer involves the various mixing mechanisms, both extensive and intensive (i.e., dispersive) discussed in Chapter 7. Besides incorporating additives into polymers, we frequently encounter the problem of blending two or more polymeric components. The components may be of the same kind of polymer but differing in molecular weight and molecular weight distribution, in which case they are compatible (miscible), and the mixing mechanism is that of extensive laminar mixing. Alternatively, the components may be different polymers, only partially compatible, in which case the mixing mechanism involves, in addition to laminar mixing, a breaking down of the dispersed liquid phase, leading to what we define as a *homogenization* process. Blending of polymers for obtaining improved properties acquires increased significance with the gradual leveling off of chemically new polymeric systems.

Two additional operations respond to the same variables as mixing does, but involve a single component system. One is MWD alterations by "working" the polymer (e.g., natural rubber), a process discovered by Thomas Hancock, the inventor of the internal mixer (see Chapter 1). The other operation, which is more specific, involves the reduction of elastic properties of LDPE melts coupled with improvements of certain optical and physical properties. The molecular mechanism may involve partial molecular disentanglement. This is achieved by imparting to the melt large total strains in long extruders prior to pelletization.

Finally, there are operations in which components are removed rather than added—for example volatile components such as residual monomer, solvent, or moisture, are sometimes removed. Devolatilization is generally diffusion controlled, and to obtain high removal rates, this operation is often coupled with extensive mixing.

The mixing process in all the above-mentioned cases requires unique and characteristic design considerations that justify its definition as an elementary step in polymer processing.

This chapter reviews mixing, dispersion, and homogenization principles and practices, analyzing mixing geometries by means of the theoretical tools discussed in Chapter 7. For brevity, the term "mixing" is frequently used in an all-embracing sense, covering all the different mixing mechanisms.

## 11.2 Classification of Mixers

The diverse mixing operations mentioned in Section 11.1 are carried out in numerous types of mixing machinery. Some of them are specifically designed for a particular mixing operation, others are "ordinary" pieces of processing equipment designed for other elementary steps as well as for mixing. In attempting to classify mixers, we first distinguish between (*a*) batch and (*b*) continuous mixers.

### *Batch Mixers*

Batch mixers are the oldest type and they are still widely used; only to some extent have they been replaced by continuous mixers. Batch mixers are very versatile units. Operating conditions can be varied during the cycle, additives can be added at an optimal time sequence, temperature control is good; furthermore, they are available over a very broad range of sizes and, if need be, they can be incorporated in continuous lines.

There are no standard engineering classification methods for mixing equipment, and often quite different types of mixers may fulfill a certain mixing job equally well. Nevertheless, we can subdivide batch mixers used in processing into three broad categories: particulate solids mixers, extensive liquid mixers, and intensive liquid mixers. This classification, on the basis of application, is supported by the nature of the primary mixing mechanism taking place in them.

Particulate solids mixers, referred to also as blenders, involve generally a random distributive mixing mechanism (Sections 7.1, 7.8). On the basis of their operation, they can be further subdivided into "tumbling" type, "agitating ribbon" type, and "fluidized bed" mixers (1, 2).

The tumbling-type mixers are the simplest and least expensive, but they cannot handle difficult mixtures. There is a tendency for segregation; stickiness is a problem, and a considerable electrostatic charge may be acquired during rubbing. The latter property may be advantageous, however, as in dry blending of pigments with nonpolar polymers, or during the mixing of two components with opposite electric charges, when charging can greatly improve the mixing (3). Ribbon blenders consist of some moving elements such as a spiral element, which induces convective motion. They are good for sticky mixtures, but they require more power than tumbling blenders and are more difficult to clean. In ribbon-type blenders, as well as some other types, PVC dry blend can be prepared by slowly spraying into the mixture small amounts of liquid additives. Such additives may sometimes generate the formation of small, soft balls, which should be avoided if a free flowing dry blend is desired. Ribbon blenders generate considerable static electricity. Finally, fluidized bed mixers are rapid mixers but cannot, of course, deal with sticky powders; neither are they suitable for powder mixtures with pronounced density and shape variations, because of segregation problems. They generate small static electric charges.

Liquid mixers are dominated by a laminar mixing mechanism and bring about an increase in the interfacial area between the components and the distribution of interfacial elements throughout the mixer volume. These mixers may perhaps be best subdivided on the practical basis of mixture viscosity (4). On one end we have the low viscosity mixers, such as the impeller type and high speed dispersion mixers. In this viscosity range turbulent mixing may still play a significant role. In the medium range we have the various double blade units such as the sigma blade mixer (Fig. 11.1). This design consists of a rectangular trough curved at the bottom to form two half-cylinders. The two blades revolve toward each other at different

**Fig. 11.1** Universal mixer with sigma blades. (Photograph courtesy of Baker Perkins Co., Saginaw, Mich.)

frequencies of rotation. Usually a ratio of 2:1 is used. Mixing is induced by imparting both axial and tangential motion. The clearance between the blades and shell is small, about 1 mm, to eliminate stagnant regions. These mixers handle liquids in the viscosity range of 0.5–500 $N \cdot s/m^2$. Another group of double blade mixers consists of the overlapping blade type, in which the blades rotate at the same frequency of rotation. Double blade mixers are widely used in the preparation of reinforced plastics, as well as for mixing and kneading a great variety of viscous liquids and pastes.

At the high viscosity end of this classification, which is our primary interest, we find among others the high intensity internal mixers such as the Banbury-type mixer and the roll-mill (Chapter 1), both extensively used in the elastomer and plastics industries (5, 6). These mixers, in addition to imparting extensive mixing, are characterized by high shear stress zones where dispersive mixing or homogenization takes place. We return to these mixers for a more detailed analysis later. It is worthwhile to note that dispersive mixing is carried out not only in very high viscosity media but also in lower viscosity ranges—for example, in the paint industry.

### *Continuous Mixers*

In addition to performing the other elementary steps, all continuous processing equipment, such as single and twin extruders, must be able to perform adequate mixing. To enhance this function, as well as to improve temperature uniformity, screw extruders are modified by incorporating into the design "mixing devices" (Fig. 11.2). Similarly, twin screw extruders have also been modified by incorporating special mixing sections (Fig. 11.3).

Substantial modification of the single and twin screw extruders, aimed at improving dispersive mixing capability in particular, led to continuous mixers such as the "Transfermix" and the Buss Ko-Kneader (Fig. 11.4). Another approach in continuous mixer development is to transform batch mixers into continuous ones. Thus the roll-mill can be converted into a continuous mixer by feeding raw material on one side and continuously stripping product on the other side, and the Banbury mixer has been imaginatively transformed into the FCM (Farrel Continuous Mixer), as in Fig. 11.5.

Continuous mixing has the advantages of large output, uninterrupted operation, greater product uniformity, easier quality control, and reduced manpower. It has the disadvantages of generating lower dispersive mixing quality and possessing less flexibility in switching to new mixtures. The feed must be maintained uniform in time, and the order of introducing the components into the mixture is more or less fixed. The design of a continuous mixer is aimed toward a uniform outlet composition across the exiting stream, as well as composition uniformity in time. The former is achieved by imparting to all fluid particles leaving the mixer the same total deformation (i.e., narrow strain distribution function) and by feeding the mixer with a grossly uniform mixture; ample shuffling and rearrangement throughout the mixer also must be ensured. Uniformity in time, as we discussed in Chapter 7, can be obtained either by careful metering of inlet rate of the ingredients or by imparting a certain amount of backmixing. The latter point implies, of

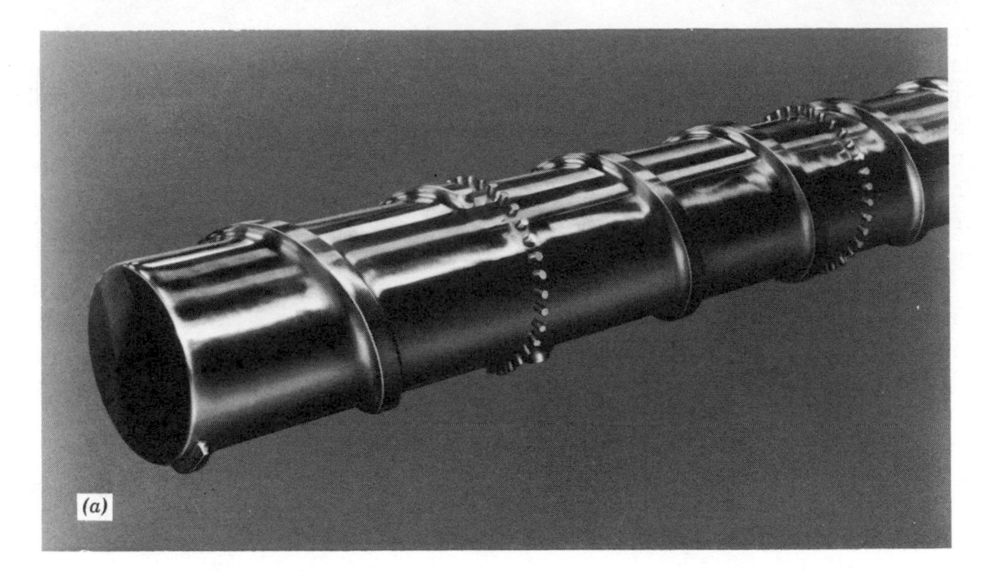


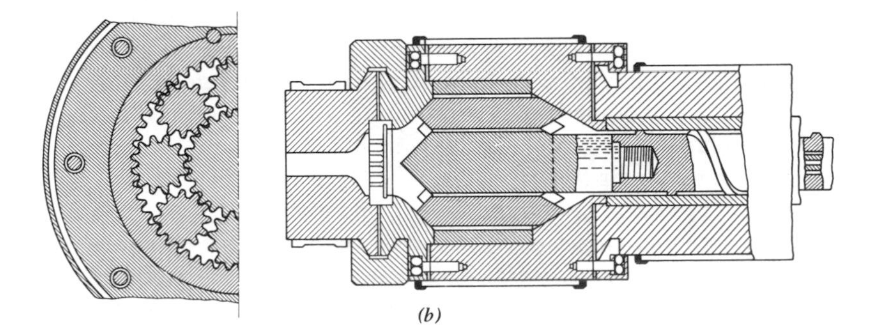


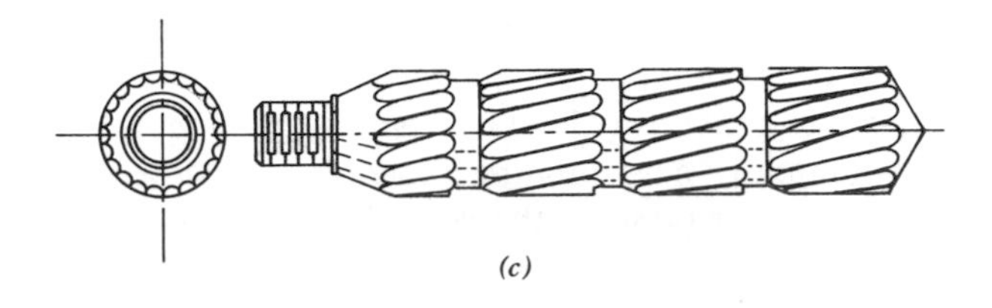


**Fig. 11.2** Mixing devices in single screw extruders. (*a*) Screw with mixing pins (Courtesy of Davis Standard Co., Pawcatuck, Conn.). (*b*) Screw with kneading gears at the end of the screw. (Reprinted with permission from G. Schenkel, *Plastics Extrusion Technology and Theory*, Illife Books Ltd., London, 1966.) (*c*) Torpedo extension to the screw with multiple notched deep thread. (Reprinted with permission from G. Schenkel, *Plastics Extrusion Technology and Theory*, Illife Books Ltd., London, 1966.)

**Fig. 11.3** ZSK corotating intermeshing twin screw extruder with corresponding barrel elements. (Courtesy of Werner and Pfleiderer Co., Waldwick, N.J.)

course, a broadening of the residence time distribution function, with the unwarranted consequence of possible excessive residence times for certain fractions of the material, leading to shear or thermal degradation and longer "switching times" from compound to compound.

Another type of mixers lacking any moving parts are the so called "motionless" mixers (7) discussed in Section 11.7. These are generally amenable to theoretical analysis and their mixing performance is predictable. The mixing mechanism in these mixers is a combination of laminar mixing and ordered distributive mixing.

## 11.3 Laminar Mixing of Rheologically Homogeneous Fluids

Mixing—in particular, viscous liquid mixing—is perhaps the least "scientific" of the elementary steps. Although some of the principles involved are fairly well understood, quantitative characterization methods, discussed in Chapter 7, are not simple. Moreover, as Section 11.2 made evident, the geometrical configurations often required to achieve efficient mixing are quite complex. Hence existing theoretical tools are difficult to use in the analysis, modeling, and design of mixers. Nevertheless, some general guidelines to design and analysis have evolved out of considerations of the underlying principles and relatively simple geometries.

### *Randomization of Interfacial Area Elements*

Chapter 7 pointed out that a quantitative criterion of laminar mixing is the interfacial area between components. The interfacial area, for large strains and simple shear flow, turned out to be directly proportional to the total strain. Thus the first requirement for good mixing is the imposition of large strains on both components. This requirement, however, must be complemented by the equally important

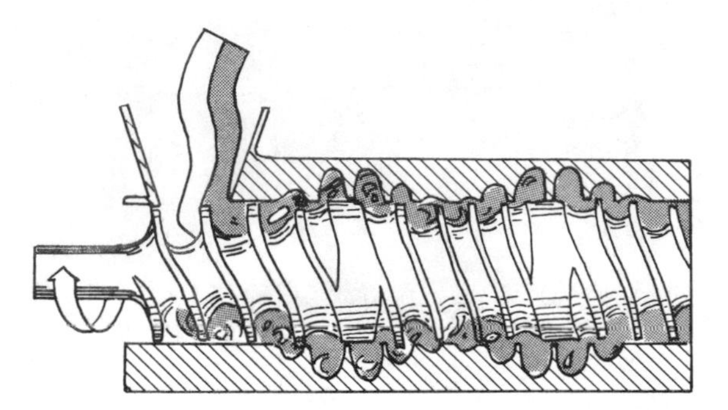

(*a*)

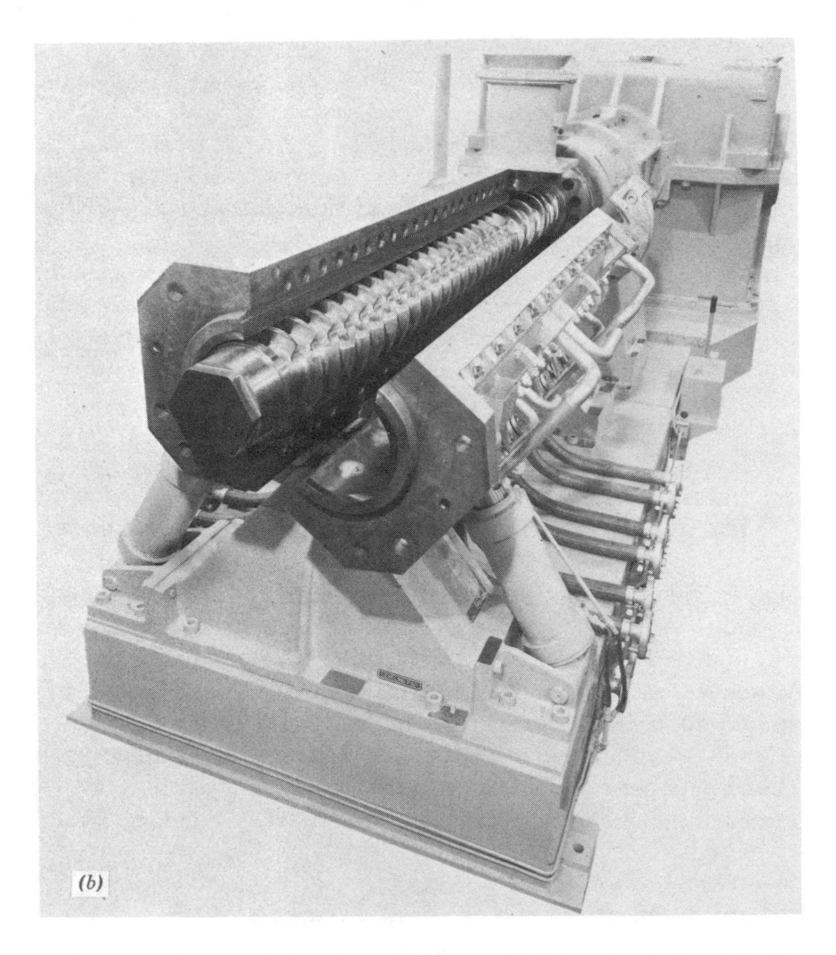

(*b*)

**Fig. 11.4** (*a*) Schematic view of Transfermix (Shearmix) C. M. Parshall and P. Geyer, U.S. Patent 2,744,287 (1956). (*b*) Photograph of the Ko-Kneader. (Courtesy of Baker Perkins Co., Saginaw, Mich.)

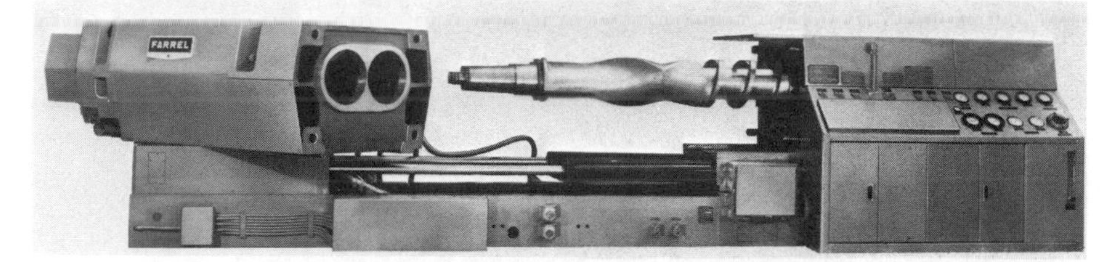

**Fig. 11.5** Size 15 FCM (Farrel Continuous Mixer, P. Hold et al., U.S. Patent 3,154,808, 1969) with chamber opened and rotated hydraulically. (Courtesy of Farrel Co., Ansonia, Conn.)

requirement of *distribution of interfacial elements throughout the system* (9). In other words, good laminar mixing can be achieved only by imposing large strains on the components while approaching (or maintaining) gross uniformity.

In most laminar mixers we can identify the design elements that satisfy the two requirements above. For example, in roll-mills large strains (and stresses) can be imposed on the polymer passing through the nip, satisfying the first requirement for good mixing. The second requirement, however, can be achieved only by "cutting and rolling" of the polymer. Similarly, in a sigma blade mixer, large strains are imposed on the fluid between the blades as well as between the blades and the housing. The blades also have a geometrical configuration that induces axial flow, bringing about the required distribution of interfacial elements throughout the system. Such complex flow patterns as those existing in sigma blade mixers, for example, are dominated by many uncontrollable events and can be viewed as pseudorandom processes. Hence in cases like this, fulfilling the second requirement is tantamount to effecting a *random* distribution of the minor component. Similar results are obtained in "motionless" mixers in an ordered rather than random process. In these mixers the main increase in interfacial area is by laminar mixing, but redistribution of the interfacial area elements also occurs repeatedly and in ordered fashion.

### *The Effect of Interfacial Area Orientation and Initial Arrangement on Mixing Performance*

Both requirements for achieving good mixing are dependent on initial conditions: namely, the *orientation* of the interfacial area and the initial *arrangement* of interfacial area elements. The optimal orientation of the interfacial area elements in unidirectional shear flow is normal to the direction of shear (see Section 7.9). This becomes evident by considering a concentric cylinder mixer (Fig. 11.6), where a single blob of a minor is to be mixed into the major component. In case $a$ we expect no mixing, irrespective of the strain imposed by turning one of the cylinders, whereas case $b$ leads to any desired level of mixing after a sufficient number of revolutions. If, however, we choose the initial orientation and arrangement in Fig. 11.6$c$, we note that the initial orientation provides for an increase in interfacial area to any required level, but the initial arrangement of the minor element, coupled with the nature of the flow field, could *not* bring about a distribution of interfacial

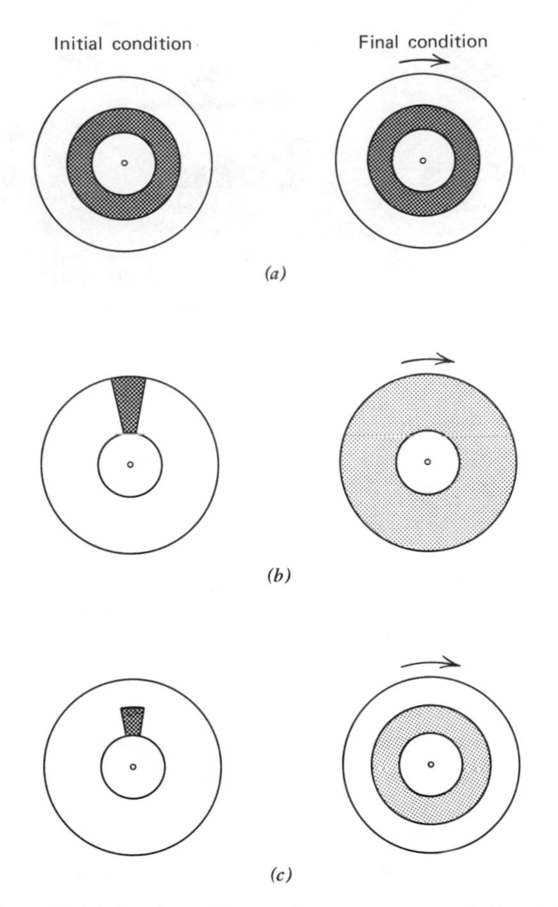


**Fig. 11.6** The role of initial orientation and arrangement of the minor component on mixing in a concentric cylinder mixer. The left-hand column shows initial conditions, whereas the right-hand column shows final conditions after many revolutions of one of the cylinders. (*a*) Minor component does not cut across all streamlines and all interfacial area is parallel to direction of shear. The end result is no mixing at all. (*b*) Minor component cuts across all streamlines at interfacial area perpendicular to the direction of shear. The end result is perfect mixing. (*c*) Mixture of cases *a* and *b*, with partial mixing as the end result.

elements across the system. In this case flow is in closed streamlines; hence the distribution of interfacial elements across the system can be ensured only by arranging the minor component such that it *cuts across all streamlines.* Ensuring such a flow pattern is important because it is hard to control initial orientation and arrangement. In mixers where flow patterns are complex—that is, pseudorandom—initial arrangement and orientation are less important. If the feed is in the form of particulate solids, premixing the feed provides for an initial random arrangement and orientation.

We have considered thus far two extreme flow patterns—closed streamline flow in a concentric cylinder mixer, and the pseudorandom, tortuous flow patterns in certain batch and continuous mixers. Between these two extremes there are many practical cases featuring a complex flow pattern but still amenable to theoretical analysis. Shearer (10) examined a number of such cases. Consider a

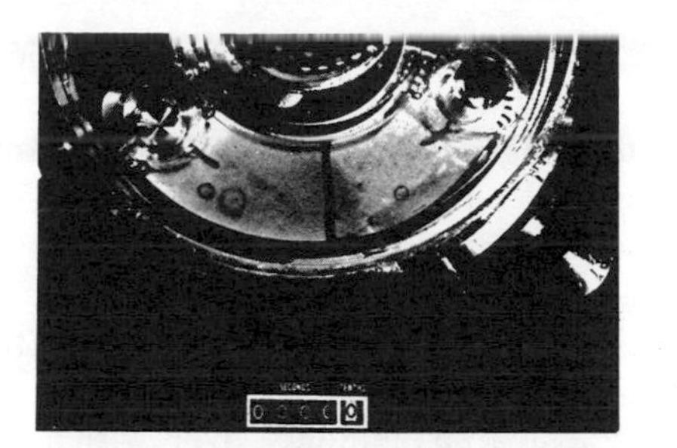

(a)

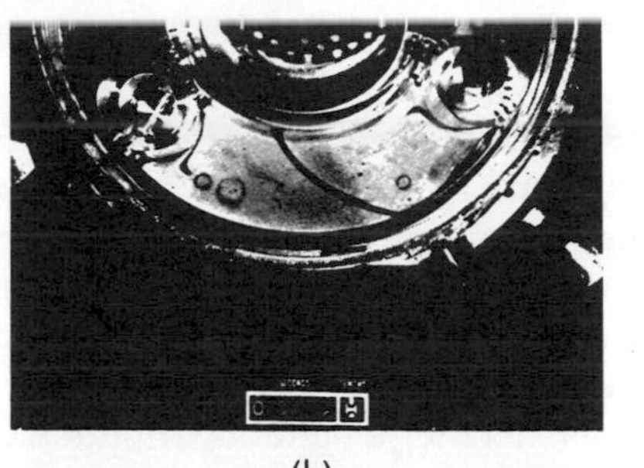

(b)

(c)

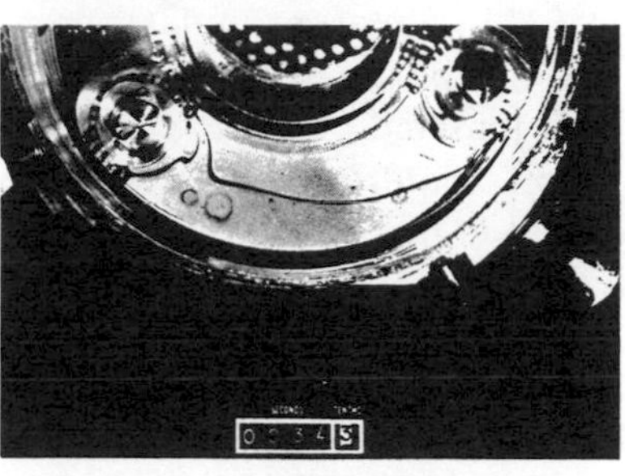

(d)

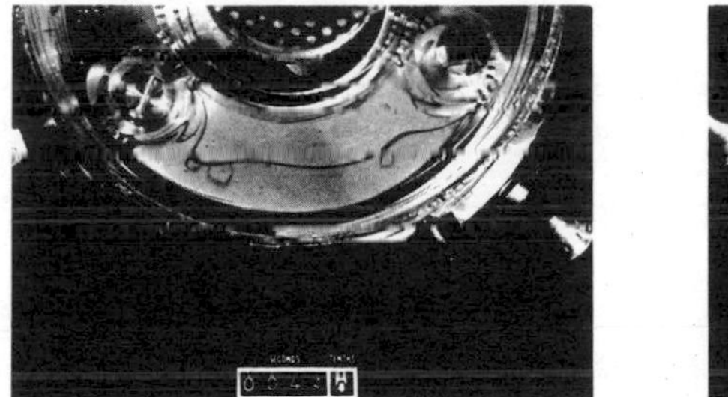

(e)

(f)

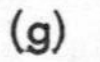

(g)

(h)

**Fig. 11.7** Flow patterns between concentric cylinders equipped with planetary rollers. Wall velocity 10.5 cm/min. Compartment dimensions: outside and inside diameters, 15.3 cm and 8.9 cm respectively, 120° between rollers, 15 cm depth. Liquid: silicone oil, 600 poise. A marker die was injected at time $t = 0$ radially in the middle of the channel, as shown in *a*. (Reprinted with permission from C. J. Shearer, Mixing of Highly Viscous Liquids: Flow Geometries for Streamline Subdivision and Redistribution," *Chem. Eng. Sci.*, **28**, 1091–1098 (1973).)

concentric cylinder arrangement with planetary rollers (Fig. 11.7). During motion, all four walls move. This geometry has some common features with the intermeshing twin screw geometry. It is also used in the single screw extruder as a mixing section, in which case axial flow is superimposed on the tangential flow. A die marker was placed in the center of the cell and followed with time. We observe the increase in interfacial area as well as the distribution of the interfacial elements over the system. Although the initial arrangement of the minor component (the marker die) did cut across all streamlines just as in the concentric cylinder case, we can hope here for a better distribution of interfacial area elements, even with a less favorable initial arrangement of the minor component. Another interesting case investigated by Shearer simulates cross channel flow in single screw extruders (Fig. 11.8). By incorporating into the extruder screw channels mixing sections consisting of rows of pins and other flow obstructions, as mentioned earlier, a more efficient distribution of interfacial area elements is obtained. If interfacial area elements are randomized in the mixing section, Erwin (8) proved that mixing efficiency downstream the mixing section is greatly enhanced.

Until now, in dealing with the orientation of interfacial area elements, we have considered only *initial* orientation. We concluded that for large strains and unidirectional shear, the best initial plane orientation is perpendicular to the direction of shear (i.e., in Fig. 7.13, $\cos\alpha_x = 1$, $\cos\alpha_y = \cos\alpha_z = 0$). If, however, we could continuously readjust the orientation of the interfacial area in a certain preferable way, the increase in interfacial area for a *given strain* could be significantly higher, as the following example demonstrates. Continuous readjustment of interfacial area orientation is tantamount to a continuously changing *direction* of shear. Thus our present discussion is relevant to the optimal flow pattern in mixers discussed previously.

**Example 11.1** ***Optimal Interfacial Element Orientation During Deformation for Maximum Increase in Interfacial Area.***

Equation 7.9-15 provides a relationship between interfacial area $A$ as a function of total strain and initial orientation of the area element. The derivative of $A$ with respect to strain, from Eq. 7.9-15, is

$$\frac{dA}{d\gamma} = A_0 \frac{-\cos\alpha_x \cos\alpha_y + \gamma \cos^2\alpha_x}{(1 - 2\gamma\cos\alpha_x\cos\alpha_y + \gamma^2\cos^2\alpha_x)^{1/2}} \tag{11.3-1}$$

Equation 11.3-1 is correct for any initial orientation and any strain value from 0 to $\gamma$. The initial value of $dA/d\gamma$, at $\gamma = 0$, is therefore

$$\frac{dA}{d\gamma} = -A\cos\alpha_x\cos\alpha_y \tag{11.3-2}$$

where $A = A_0$. First, we note that depending on orientation, the instantaneous interfacial area may *increase* or *decrease*. Next, for the product $\cos\alpha_x\cos\alpha_y$ to be a maximum, with $\cos\alpha_z = 0$, neither $\cos\alpha_x$ or $\cos\alpha_y$ can assume the maximum value of 1 (or $-1$), because, as indicated by Eq. 7.9-2,

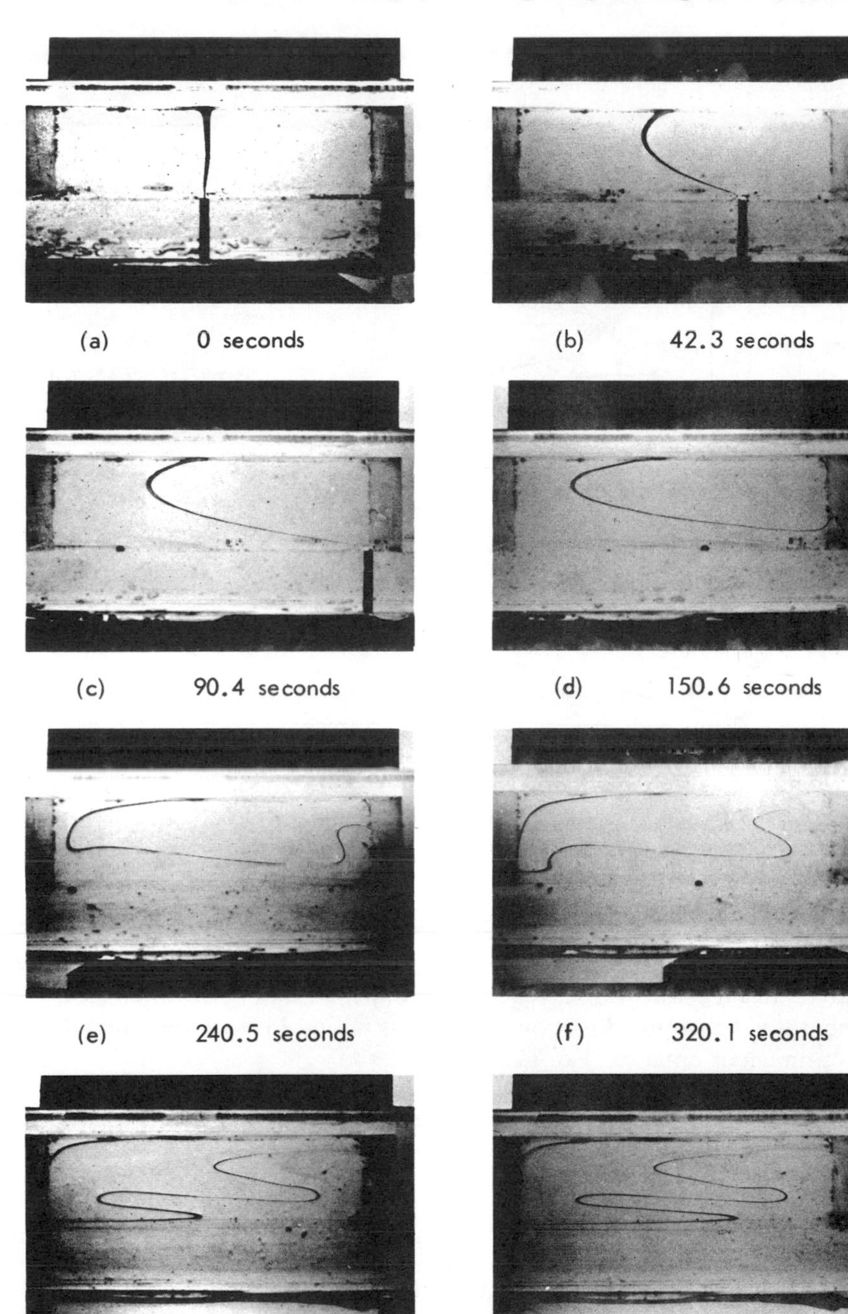

**Fig. 11.8** Flow pattern in a rectangular channel with one wall (lower) moving at constant velocity, 2.4 cm/min. Compartment dimensions: channel width, 9.5 cm; channel height, 1.9 cm; channel depth, 7.6 cm. Liquid: glycerol, 5.7 poise. A marker die was injected at time $t = 0$, as shown in *a*. (Reprinted with permission from C. J. Shearer, Mixing of Highly Viscous Liquids: Flow Geometries for Streamline Subdivision and Redistribution," *Chem Eng. Sci.*, **28**, 1091–1098 (1973).)

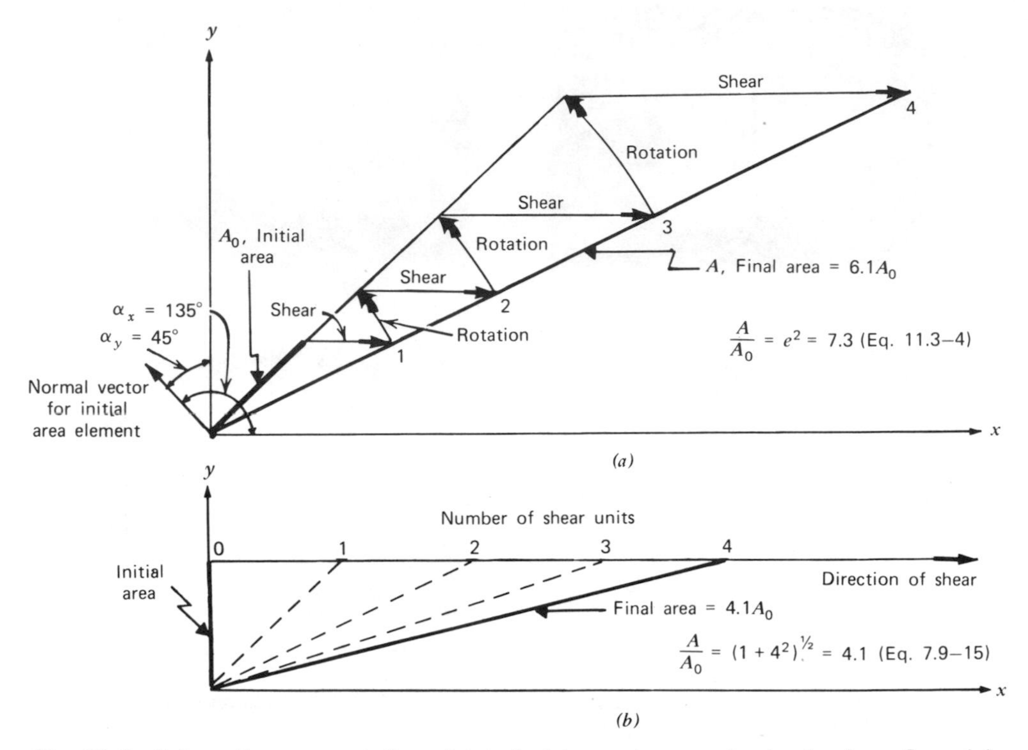


**Fig. 11.9** Schematic representation of interfacial area increase in simple shear flow. (*a*) With initial orientation of 45° to the direction of shear, and after each shear unit the plane is rotated back to 45° orientation. (*b*) With optimal initial orientation and no rotation.

in such a case the other directional cosine assumes a value of zero. It also follows from Eq. 7.9-2 that the product $-\cos \alpha_x \cos \alpha_y$ assumes a maximum positive value if $\cos \alpha_z = 0$, $\alpha_y = 45°$, and $\alpha_x = 135°$, that is, the interfacial area element is at 45° to the direction of shear, as in Fig. 11.9*a*. Consequently, for this optimum orientation, Eq. 11.3-2 reduces to

$$\frac{dA}{d\gamma} = \frac{A}{2} \tag{11.3-3}$$

which upon integration yields

► $$A = A_0 e^{\gamma/2} \tag{11.3-4}$$

or an exponential increase of interfacial area with strain.* Figure 11.9 compares a stepwise increase in interfacial area in simple shear flow with optimal initial orientation, and simple shear flow where at the beginning of each step the interfacial area element is placed 45° to the direction of shear. The figure shows that whereas in the former case the area ratio after four shear units is 4.1, in the latter case the ratio is 6.1, with a theoretical value of 7.3 when the 45° between the plane and direction of shear is maintained at all

* See also L. Erwin, "New Fundamental Considerations of Mixing in Laminar Flows," 36th Annual Technical Conference, Soc. of Plastic Engineers, Washington, D.C., 1978, p. 488.

times. The above-described process of shear and rotation leads to *pure shear flow* or extensional flow (cf. Section 6.8). It should be noted, however, that it is quite difficult to generate steady extensional flows (11) for times sufficiently long to attain the required total elongational strain. As discussed in Section 11.7, the Ross ISG "motionless" mixer provides for an ingenious way to overcome this difficulty.

In a pseudorandom mixing process—needed for distribution of interfacial area elements throughout the system—the direction of shear is also continuously changing, reducing to some extent the disadvantage in steady unidirectional shear.

A different method to overcome the reduction in mixing efficiency in unidirectional shear was suggested by Suh et al. (3). They developed a concentric cylinder mixer in which an electrostatic field can be created between the cylinders. If the viscosity of the components is low enough, a wavy interfacial area is created by the field, enhancing the final stages of mixing.

### *The Strain Distribution Function and Rheological Considerations*

An important factor we have not yet considered is the strain distribution function (SDF; see Section 7.10). Given the most favorable initial conditions, a broad strain distribution function necessarily leads to poor mixing.

We can demonstrate this effect by turning once again to a concentric cylinder mixer. Bergen et al. (9) investigated the mixing of black and white linoleum composition. Figure 11.10 presents the initial condition and mixing resulting after 1 and 20 revolutions. Results indicate that after 20 revolutions a "band" of "uniform" gray mixture is created at the moving inner cylinder, yet a lack of mixing is very evident at the region close to the outer cylinder. The authors relate the nonuniformity to a Bingham fluid response. Although this may partly be the reason, one would expect a mixing nonuniformity even with a Newtonian fluid, and certainly with a power law model fluid, because of a nonuniform SDF, as indicated in the following example.

#### **Example 11.2** *The SDF in Concentric Cylinder Mixer*

Consider a power law model fluid placed between two long concentric cylinders of radii $R_i$ and $R_0$ ($R_0 > R_i$). At a cerain time the inner cylinder is set in motion at constant angular velocity $\Omega$ rads/s. Assuming isothermal laminar flow without slip at the walls, neglecting gravitational and centrifugal forces, and assuming steady state, the velocity profile is given by

$$\frac{v_\theta}{\Omega R_i \rho} = \frac{\beta^{2s} - \rho^{2s}}{\rho^{2s}(\beta^{2s} - 1)} \tag{11.3-5}$$

where $v_\theta$ is tangential velocity, $\rho = r/R_i$, $\beta = R_0/R_i$, and $s = 1/n$, with $n$ being the power law fluid exponent. Taking the derivative of Eq. 11.3-5,

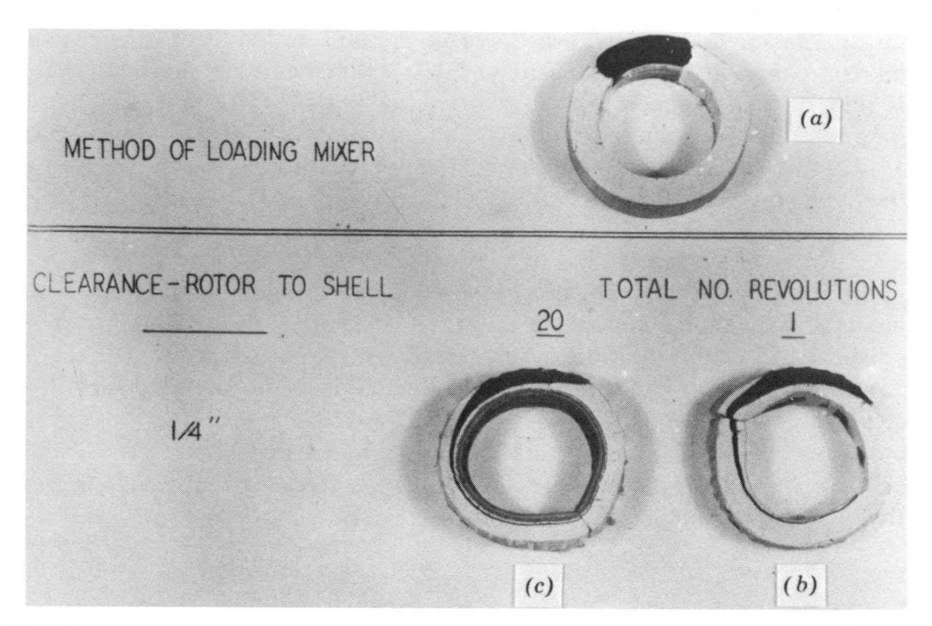


**Fig. 11.10** Mixing of unmatured linoleum compound between concentric cylinder mixers. ($a$) Method of loading—initial conditions. ($b$) After one revolution. ($c$) After 20 revolutions. Diameter of inner cylinder, 1 in; radial clearance, 0.25 in. (Reprinted with permission from J. T. Bergen et al., "Criteria for Mixing and the Mixing Process," paper presented at the 14th National Technical Conference, Society of Plastics Engineers, Detroit, 1958.)

$\dot{\gamma}_{r\theta}(\rho)$ is

$$\dot{\gamma}_{r\theta} = \rho \frac{\partial}{\partial \rho}\left(\frac{v_\theta}{R_i \rho}\right) = -\frac{2s\beta^{2s}\Omega}{\rho^{2s}(\beta^{2s}-1)} \tag{11.3-6}$$

Clearly, the shear rate is maximum at the inner cylinder and minimum at the outer cylinder. The difference between their respective values increases with curvature (increasing $\beta$) and departure from Newtonian behavior (increasing $s$). The strain distribution is obtained by multiplying $|\dot{\gamma}_{r\theta}|$ by the time $t$

$$\gamma(\rho) = |\dot{\gamma}_{r\theta} t| = \frac{2s\Omega t\beta^{2s}}{\rho^{2s}(\beta^{2s}-1)} \tag{11.3-7}$$

Note that $\Omega t$ divided by $2\pi$ is simply the total number of revolutions. The ratio of maximum to minimum strain across the gap is

$$\frac{\gamma_{\max}}{\gamma_{\min}} = \beta^{2s} \tag{11.3-8}$$

The striation thickness is inversely proportional to total strain. This follows from Eqs. 7.8-1 and 7.9-16. Therefore the ratio of the striation thickness at the outer radius (where it is maximum) to the striation thickness at the inner radius (where it is minimum) is given by

$$\frac{r_0}{r_i} = \frac{\gamma_{max}}{\gamma_{min}} = \beta^{2s} \tag{11.3-9}$$

In the particular case in Fig. 11.10, the outer and inner radii are 0.75 and 0.5 in., respectively; hence the curvature $\beta = 1.5$.

It follows from Eq. 11.3-9 that the ratio of the striation thickness for a Newtonian fluid is 2.25; for a power law fluid with $n = 0.5$, it is 5.06, and for a power law fluid with $n = 0.25$, it is 25.6 (!).

A better insight into the nature of this problem can be obtained by calculating the SDF $G(\gamma)$, defined in Section 7.10. The fraction of material that experiences a total strain of $\gamma$ or less is equivalent to the fraction of material found between the radii $\rho$ and $\beta$ (at radius $\rho$ the strain $\gamma$ is given in Eq. 11.3-7):

$$G(\gamma) = \frac{\pi(R_0^2 - r^2)L}{\pi(R_0^2 - R_i^2)L} = \frac{\beta^2 - \rho^2}{\beta^2 - 1} \tag{11.3-10}$$

where $L$ is the length of the "mixer."

Substituting Eq. 11.3-7 into Eq. 11.3-10, we get

$$G(\gamma) = \frac{\beta^2}{\beta^2 - 1}\left[1 - \left(\frac{\gamma_{min}}{\gamma}\right)^n\right] \tag{11.3-11}$$

where $\gamma_{min}$ is the minimum strain (at the outer radius) given by

$$\gamma_{min} = \frac{2s\Omega t}{\beta^{2s} - 1} \tag{11.3-12}$$

The SDF function $g(\gamma)\, d\gamma$, is obtained by differentiating Eq. 11.3-11

$$g(\gamma)\, d\gamma = n\frac{\beta^2}{1 - \beta^2}\frac{\gamma_{min}^n}{\gamma^{n+1}} \tag{11.3-13}$$

Finally, the mean strain $\bar{\gamma}$ is

$$\bar{\gamma} = \int_{\gamma_{min}}^{\gamma_{max}} \gamma g(\gamma)\, d\gamma = \frac{\gamma_{max}}{\beta^2 - 1}\left(\frac{1 - \beta^{2(1-s)}}{s - 1}\right) \tag{11.3-14}$$

Figure 11.11 shows the SDFs for the particular case in Fig. 11.10$a$, with $n$ as a parameter. The value of $\bar{\gamma}$ for each case is marked by a star. We note that even for a Newtonian fluid, 56% of the material experiences a strain less than the mean, with the distribution spreading over a range of 200–450 shear units. It is important to note that the 56% falling below the mean are in a narrower strain range (about 100) than the remaining 44% (which fall in a strain range of about 150 shear units). The nonuniformity substantially increases with decreasing $n$. The mean strain drops, the distribution broadens, and it gets distorted such that higher fractions of material below the mean experience strains in relatively narrower ranges than the smaller fractions above the mean. Thus for $n = 0.25$, 66% of the material falls below the mean in a strain range of about 200 shear units, whereas the remaining 34% experience strain above the mean in a range of about 800 shear units.

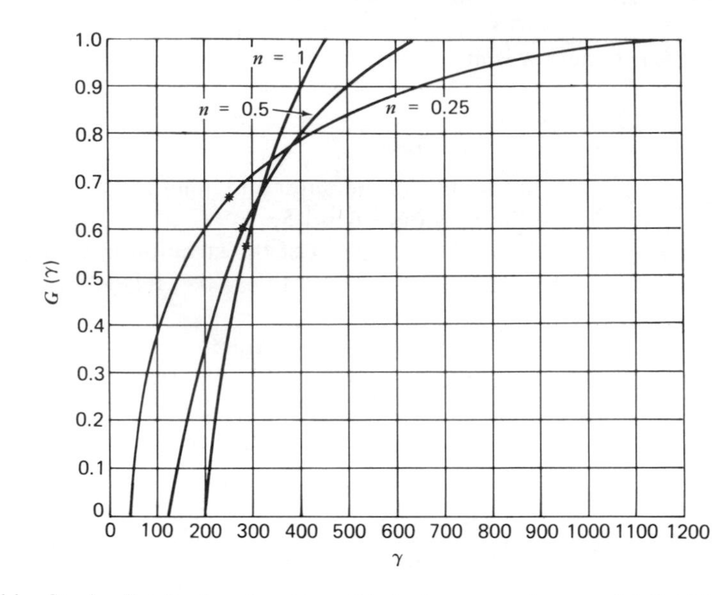


**Fig. 11.11** Strain distribution function $G(\gamma)$ of a power law model fluid in Couette flow between concentric cylinders for case $c$ of Fig. 11.10.

The rheological properties of the polymer used in the experiment of Bergen et al. (9) were not reported. If the polymer behaved like a Bingham plastic, one would certainly expect the nonuniformity as proposed by the authors; but as indicated by the foregoing calculations, one would expect nonuniform mixing even with Newtonian fluids, because of the mixer curvature. Furthermore, the nonuniformity in mixing is considerably amplified by shear thinning, providing an alternative explanation for the experimental observation without the need to assume a Bingham plastic type of flow.

The example above indicates that a mixer characterized by a broad SDF, in spite of favorable initial conditions, leads to a nonuniform mixture. To reduce the nonuniformity to acceptable levels, part of the material must be "overmixed." Hence a narrow SDF is preferable, as we also concluded in Section 7.10 in dealing with the three cylinder mixer.

Finally, Example 11.2 focuses attention on the role of rheological properties in mixing. The physical reasons for the results obtained stem from the nature of the stress distribution between the cylinders. The shear stress is inversely proportional to the square of the radius, $\tau \propto 1/\rho^2$. This is quite different from parallel plate flow, where the stress is constant. (Of course for small curvatures, the variation is negligible.) Thus the shear stress is high at the inner cylinder and low at the outer cylinder, resulting in correspondingly high and low shear rates for Newtonian fluids. If, however, the fluid is non-Newtonian (shear thinning), the viscosity also varies across the gap. It is "low" at the inner cylinder and "high" at the outer cylinder. Therefore, to maintain the required stress distribution, the shear rate must further increase at the inner cylinder and drop at the outer cylinder, with the consequent broadening of the SDF.

**Table 11.1 The Strain Distribution Function for Some Simple Flow Geometries with Newtonian Fluids**

| | Flow Between Parallel Plates[a] | | | |
|---|---|---|---|---|
| Type of Flow | Drag Flow | Pressure Flow | Combined Pressure and Drag Flow | Flow Through Circular Tubes |
| Velocity profiles | $v_z = \xi V_0$ | $v_z = 4\xi(1-\xi)V_{max}$ | $v_z = \left[\xi + 3\xi \frac{q_p}{q_d}(1-\xi)\right] V_0$ | $v_z = \left(1 - \frac{r^2}{R^2}\right) V_{max}$ |
| $F(\gamma)$ | $1 - \left(\frac{\bar{\gamma}}{2\gamma}\right)^2$ | $\frac{C}{1+\sqrt{1+C^2}}\left(1 + \frac{1}{1+\sqrt{1+C^2}}\right)$ <br> $C = \frac{3\gamma}{2\bar{\gamma}}$ | $F(\xi) = 1 - \frac{\xi^2}{1+q_p/q_d}\left[1 + \frac{q_p}{q_d}(3-2\xi)\right]$ <br> $\gamma(\xi) = \frac{1+3(q_p/q_d)(1-2\xi)}{\xi[1+3(q_p/q_d)(1-\xi)]}\left(\frac{1+q_p/q_d}{2}\right)\bar{\gamma}$ <br> $-\frac{1}{3} \le \frac{q_p}{q_d} \le \frac{1}{3}$ | $1 - \frac{2}{1+C^2/2+\sqrt{1+C^2}}$ <br> $C = \frac{8\gamma}{3\bar{\gamma}}$ |
| Mean strain, $\bar{\gamma}$ | $\frac{2L}{H}$ | $\frac{3L}{H}$ | $2\frac{L}{H(1+q_p/q_d)}$ | $\frac{8L}{3R}$ |
| Minimum strain, $\gamma_0$ | $\frac{L}{H}$ | 0 | Varies with $\frac{q_p}{q_d}$ | 0 |

[a] In these equations $\xi = y/H$, where $H$ is the separation between plates, $V_C = v_z(1)$, $V_{max}$ is the maximum velocity, and $q_p/q_d = (H^2/6\mu V_0)(-dP/dz)$ (Eq. 10.2-10).

In continuous mixers the role of the SDF is even more apparent. Section 7.10 examined the SDF in drag flow between parallel plates. In this particular flow geometry, although the shear rate is constant throughout the "mixer," a rather broad SDF results because of the existence of a broad residence time distribution. Consequently, a minor component, even if distributed at the inlet over all the entering streamlines and placed in an optimal orientation, will not be uniformly mixed in the outlet stream. The parallel plate geometry, as we could observe in the preceding chapters, plays a rather dominant role in polymer processing. It is therefore worthwhile to examine an idealized continuous "parallel plate mixer" under more realistic flow conditions, where a pressure gradient is superimposed on the drag flow. The pressure gradient can be considered to be an "operating variable" that can be manipulated during the process to improve mixing performance. The SDFs for a combined pressure and drag flow between parallel plates and pure pressure flow were derived by Lidor and Tadmor (12). Table 11.1 tabulates the results and also lists the corresponding velocity profiles, minimum strain, and mean strain expressions. In addition, the SDFs for pressure flow in a circular pipe are listed for comparison. In the case of combined pressure and drag flow, the SDF cannot be explicitly expressed in terms of $\gamma$, but only in terms of the dimensionless height $\xi = y/H$, which in turn is uniquely related to $\gamma$. The analysis was made for fully developed, isothermal, steady laminar flow of an incompressible Newtonian fluid. The derivation follows the lines of the derivation presented in Section 7.10 for pure drag flow. Results reveal the strong effect of the pressure gradient on the

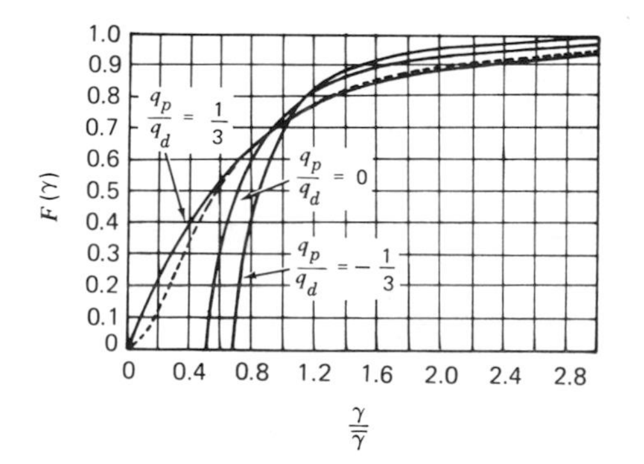


**Fig. 11.12** SDFs for fully developed Newtonian, isothermal, steady flows in parallel plate (solid curves) and tubular (dashed curve) geometries. The dimensionless constant $q_p/q_d$ denotes the pressure gradient. When $q_p/q_d = -1/3$, pressure increases in the direction of flow and shear rate is zero at the stationary plate; $q_p/q_d = 0$ is drag flow, and when $q_p/q_d = 1/3$, pressure drops in the direction of flow and the shear rate is zero at the moving plate. The SDF for the latter case is identical to pressure flow between stationary plates. (Note that in this case the location of the moving plate at $\xi = 1$ is at the midplane of a pure pressure flow with a gap separation of $H' = 2H$.)

SDF, as well as on the mean strain $\gamma$. A positive pressure gradient (pressure rise in the direction of flow, $q_p/q_d<0$) will not only increase the mean strain, it will also reduce the breadth of the distribution, as shown in Fig. 11.12, whereas a negative pressure gradient has the opposite effect. This conclusion is directly relevant to single screw extrusion because the parallel plate flow forms a simple model of melt extrusion (cf. Section 10.3), and it lends theoretical support to the experimental observation that an increase in back pressure (i.e., increase in positive pressure gradient) in the extruder improves mixing. The mean strain, as Table 11.1 indicates, is proportional to the $L/H$ (or $L/R$) ratio, which involves design variables. Thus long and shallow conduits favor good mixing. This ratio, however, does not affect the SDF. We return to analyzing mixing in extruders in more detail in Section 11.10. Finally, pure pressure flows, as shown in Fig. 11.12, are characterized by broad SDF and a minimum strain of zero. Obviously, pressure flow devices are poor laminar mixers.

Pressure gradients introduce shear rate nonuniformity. Therefore we should expect that the SDF, hence the mixing performance, will be affected by the non-Newtonian properties of the polymeric liquids, just as in the case of the concentric cylinder mixer, where curvature introduced a shear rate nonuniformity. This effect is explored in the example that follows.

**Example 11.3** ***The Strain Distribution of a Power Law Model Fluid in Pressure Flow Between Parallel Plates***

Consider two infinitely wide parallel plates of length $L$ and gap $H$. Polymer melt is continuously pumped in the $z$-direction. Assuming isothermal, steady, fully developed flow without slip at the walls, and neglecting gravitational effects, the following SDF is obtained for an incompressible power law model fluid (see Problem 11.4):

$$F(\xi)=\frac{2+s}{1+s}\left(1-\frac{1}{2+s}\xi^{1+s}\right)\xi \qquad 0\le\xi\le 1 \tag{11.3-15}$$

where $\xi=2y/H$, and the strain $\gamma$ is related to $\xi$ as follows:

$$\gamma(\xi)=2(1+s)\frac{L}{H}\frac{\xi^s}{1-\xi^{1+s}} \qquad 0\le\xi\le 1 \tag{11.3-16}$$

The mean strain, using either Eq. 7.10-12 or 7.10-24, is

$$\bar{\gamma}=2\left(\frac{2+s}{1+s}\right)\frac{L}{H} \tag{11.3-17}$$

For a Newtonian fluid these equations reduce* to those given in Table 11.1. Clearly, Eq. 11.3-17 allows us to conclude that the mean strain drops with increasing deviation from Newtonian behavior ($n<1$). Furthermore, by analyzing the SDFs plotted in Fig. 11.13, with $n$ as a parameter, we can

* Note that the coordinate system is placed at the center of the plates, unlike the arrangement in Table 11.1.

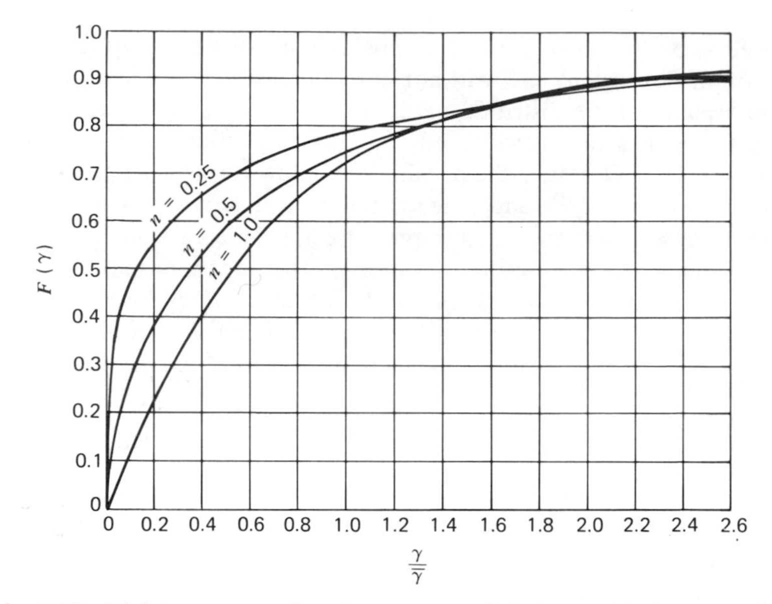


**Fig. 11.13** SDF, $F(\gamma)$ for pressure flow between parallel plates with the power law model exponent $n$ ($s = 1/n$) as a parameter.

conclude that a reduction in the parameter $n$ brings about a broadening of the strain distribution with substantially larger fraction of the exiting flow rate experiencing low total strains.

It appears, therefore, that the SDFs with non-Newtonian (shear thinning) liquids are broader than their Newtonian counterparts.

### *Temperature Effects and Power Considerations*

The temperature of an incompressible fluid element in a deforming medium is governed by the equation of thermal energy (Eq. 5.1-37). This excluding the source and reversible compression terms is

$$\rho C_v \frac{DT}{Dt} = -(\boldsymbol{\nabla} \cdot \mathbf{q}) + (-\boldsymbol{\tau} : \boldsymbol{\nabla} \boldsymbol{v}) \tag{11.3-18}$$

which simply states that the temperature change of a fluid element in a flowing fluid system is determined by the sum of heat gain or loss by conduction to the element and the rate of viscous dissipation within the element. In mixing equipment, practical considerations usually require maintaining a relatively low temperature. This requirement originates from the temperature sensitive nature of polymers, and Section 11.6 shows the need to sustain in certain parts of the mixer large shear stresses required for dispersive mixing. Equation 11.3-18 indicates that this restrictive requirement implies the need for efficient heat removal by conduction. This is not an easy task in polymeric systems characterized by low thermal conductivities. Hence, obviously, geometrical design must ensure not only good surface temperature control but also a large surface-to-volume ratio (i.e., short heat conduction lengths).

The term $(-\boldsymbol{\tau}:\nabla \boldsymbol{v})$ is the rate of irreversible conversion of mechanical energy into heat, that is, the viscous dissipation. It is a scalar function, discussed in Chapter 5. The integral of this term over the volume of the system gives the total rate of conversion of mechanical energy into heat, $E_v$

$$E_v = \int_V (-\boldsymbol{\tau}:\nabla \boldsymbol{v})\, dV \tag{11.3-19}$$

Clearly, Eq. 11.3-19 cannot be evaluated unless the velocity and temperature fields and the rheological behavior of the fluid are known.

In batch mixers the macroscopic energy balance results in

$$\frac{dU}{dt} = \rho C_v M \frac{dT}{dt} = Q_h + P_w \tag{11.3-20}$$

where $U$ is the internal energy, $M$ is the mass of material in the mixer, $Q_h$ is total rate of heat added to the mixer, and $P_w$ is the total rate of work input (power). Note that for incompressible fluids, $C_v = C_p$. Equation 11.3-20 assumes that no change in total kinetic and potential energy occurs. In such a case all the power is dissipated into heat $P_w = E_v$. If a constant temperature is to be maintained, total heat removal must equal the total power input.

In a steady, continuous system, the macroscopic energy balance in terms of the enthalpy $\hat{H}$ (per unit mass) reduces to

$$\Delta \hat{H} = \hat{Q}_h + \hat{P}_w \tag{11.3-21}$$

where $\hat{Q}_h = Q/G$ and $\hat{P}_w = P_w/G$, $G$ being the mass flow rate.

$$\Delta \hat{H} = \int_{T_1}^{T_2} C_p\, dT + \frac{\Delta P}{\rho} \tag{11.3-22}$$

The macroscopic mechanical energy balance (Bernoulli equation) for this case reduces to

$$\hat{P}_w = \hat{E}_v + \frac{\Delta P}{\rho} \tag{11.3-23}$$

or

► $$P_w = E_v + (\Delta P)Q \tag{11.3-24}$$

where $Q$ is the volumetric flow rate.

Hence part of the power is used for increasing the pressure of the liquid and the rest is dissipated into heat (e.g., used for laminar mixing).

The temperature rise over the mixer for a constant $C_p$ follows from Eqs. 11.3-21–11.3-23

$$\Delta T = \frac{1}{C_P}[\hat{Q}_h + \hat{E}_v] \tag{11.3-25}$$

To calculate the temperature rise, in addition to $\hat{Q}_h$, we must know either $\hat{E}_v$, or $\hat{P}_w$ and $\Delta P$.

A quantitative measure of laminar mixing is the total strain $\gamma$, which in simple shear flow is the product of shear rate and time, that is, $\dot{\gamma}t$. Therefore the *same* total deformation can be obtained within different periods of time by appropriate adjustment of the shear rate, which will, consequently, alter the rate of heat generation by viscous dissipation. For a power law fluid, the viscous dissipation (in terms of total strain and the time of shearing for simple shear flow) is given by

$$e_v = \tau\dot{\gamma} = m\left(\frac{\gamma}{t}\right)^{n+1} \tag{11.3-26}$$

Hence imparting a given total shear deformation quickly will raise the viscous dissipation term and, since we are usually limited by the rate of heat removal, this may imply higher ultimate temperatures. The effect is more pronounced in Newtonian fluids than in shear thinning fluids. In either case, the viscous dissipation term is reduced by increasing the temperature level as a result of the lowering of the parameter $m$. If, however, instead of a given total deformation a given shear stress is to be imparted (dispersive mixing), the heat generation term written as

$$e_v = \frac{\tau^{s+1}}{m^s} \tag{11.3-27}$$

indicating that shear thinning will raise the heat generation term, and a temperature drop (resulting in an increase of the parameter $m$) will reduce it. The raised heat generation term occurs because higher shear rates are needed to obtain the same shear stress; the reduction comes about because at lower temperatures smaller shear rates are needed to obtain the same shear stress. Hence from a power consumption point of view, laminar mixing is desirable to carry out at elevated temperatures, whereas dispersive mixing is preferable at lower temperatures.

## 11.4 Laminar Mixing of Rheologically Nonhomogeneous Liquids

Laminar mixing frequently involves systems containing components that have significantly different rheological properties—for example, mixing of different grades of the same polymer, mixing of two components of the same polymer, one of which contains various additives, mixing different polymers, and mixing a lower molecular weight liquid component into a polymer. We could also add to this list thermally nonhomogeneous systems that until temperature differences are reduced during mixing, also form a rheologically nonhomogeneous system. The main question we address in this section is the effect of the difference in rheological properties on the mixing process. To a first approximation, this difference is reflected in the viscosity ratio.

### *Effect of the Viscosity Ratio on Mixing*

It is generally accepted in the field of mixing that it is more difficult to mix a high viscosity minor component into a low viscosity major component than vice versa. This is in agreement with the previously mentioned requirement that to achieve laminar mixing, both components must be deformed; and of course it is harder to

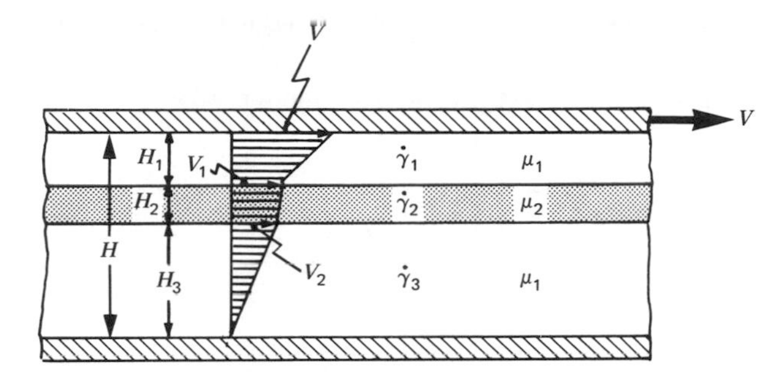


**Fig. 11.14** Velocity profile for drag flow between parallel plates of layers of immiscible Newtonian liquids.

deform a high viscosity minor component placed in an easily deforming low viscosity major component than the other way around. This point can be further amplified by the following example (5).

**Example 11.4** ***Effect of Viscosity Ratio on Shear Strain in Parallel Plate Geometry***

Consider a two parallel plate flow in which a minor component is "sandwiched" between two layers of major component (Fig. 11.14). We assume that the liquids are incompressible, Newtonian, and not miscible in each other. The equation of motion for steady state, using the common simplifying assumption with negligible interfacial tension, indicates a constant shear stress throughout the system. Thus we have

$$\mu_1\dot{\gamma}_1 = \mu_2\dot{\gamma}_2 = \mu_1\dot{\gamma}_3 \tag{11.4-1}$$

and by expressing the shear rates in terms of velocity drop, it is easy to show that

$$\dot{\gamma}_2 = \frac{V}{H}\left[\frac{1}{(1-\phi)\mu_2/\mu_1+\phi}\right] \tag{11.4-2}$$

where $V$ is the velocity of the upper plate, $H$ is the gap between the plates, and $H_2/H = \phi$ is the fraction of the gap occupied by the minor component. Clearly, the shear rate attainable in the minor component is a function of the viscosity ratio $\mu_2/\mu_1$ and $\phi$. If $\phi \ll 1$, $\dot{\gamma}_2 \cong (V/H)(\mu_1/\mu_2)$, indicating little deformation for high viscosity ratios $\mu_2/\mu_1$. Furthermore, the shear rate in the minor component is more sensitive to the viscosity ratio at low $\phi$ values. The relevance of these conclusions to mixing, however, must be treated very cautiously, since the example is highly oversimplified. Moreover, in actual mixing a very low viscosity component may tend to segregate to high shear rate zones, and in other cases one of the phases may break down into a noncontinuous phase leading to a homogenization process, discussed later in this chapter.

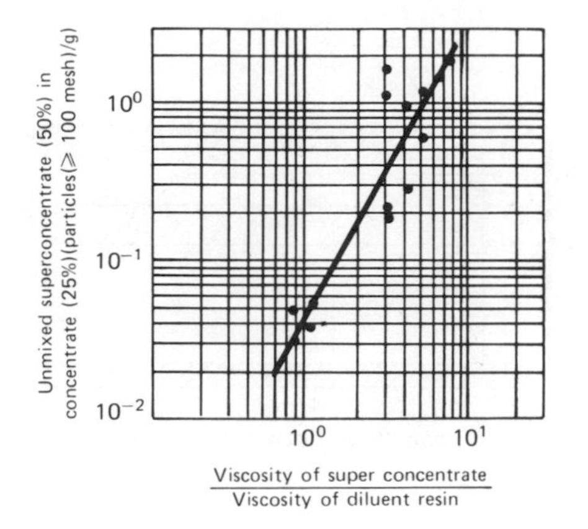


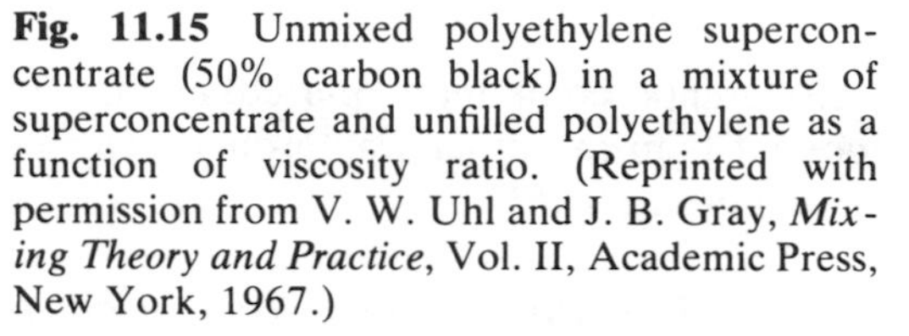
**Fig. 11.15** Unmixed polyethylene superconcentrate (50% carbon black) in a mixture of superconcentrate and unfilled polyethylene as a function of viscosity ratio. (Reprinted with permission from V. W. Uhl and J. B. Gray, *Mixing Theory and Practice*, Vol. II, Academic Press, New York, 1967.)

The general trend of reduction of the mixing quality with increasing the minor to major viscosity ratio was demonstrated experimentally by Irving and Saxton (4). A polyethylene superconcentrate (50% carbon black) was diluted with unfilled polymer to polyethylene concentrate (25% carbon black) in a BR Banbury mixer. The viscosity of the superconcentrate was varied by changing the carrier resin. The population of particles of unmixed superconcentrate was measured as a function of the viscosity ratio. Results plotted in Fig. 11.15 clearly indicate that a high viscosity ratio leads to poor mixing.

However, quantitative predictions from the parallel plate flow cannot be generalized to a more realistic system of large "blobs" of a minor component distributed in a deforming major component. In this case, the hydrodynamic behaviour of the system is far more complex. Bigg and Middleman (13) approached this problem by analyzing the flow of pairs of immiscible liquids of different viscosities in a rectangular channel. This geometrical configuration is relevant to single screw extrusion, and it consists of an infinite rectangular channel (the screw channel) covered by an upper surface (the barrel) moving in the transverse direction. (With a single fluid occupying the channel, a circulatory flow results, which was described in detail in Chapter 10.) The pair of liquids are initially stratified, as in Fig. 11.16. The interfacial tension is assumed to be negligible. The motion of the upper surface induces partial mixing of the fluids, and the interfacial area, which was calculated as a function of time, is used as a quantitative measure of the laminar mixing. The "marker and cell" calculation method, developed by Harlow et al. (14), was used to solve the flow field and calculate the position of the interface. This method is particularly suitable for solving flow problems with free boundaries and interfaces, and it is easy to use computationally in problems involving low viscosity fluids. According to this method, in addition to the numerical solution of the velocity and pressure fields, a number of "marker" particles, initially placed one or

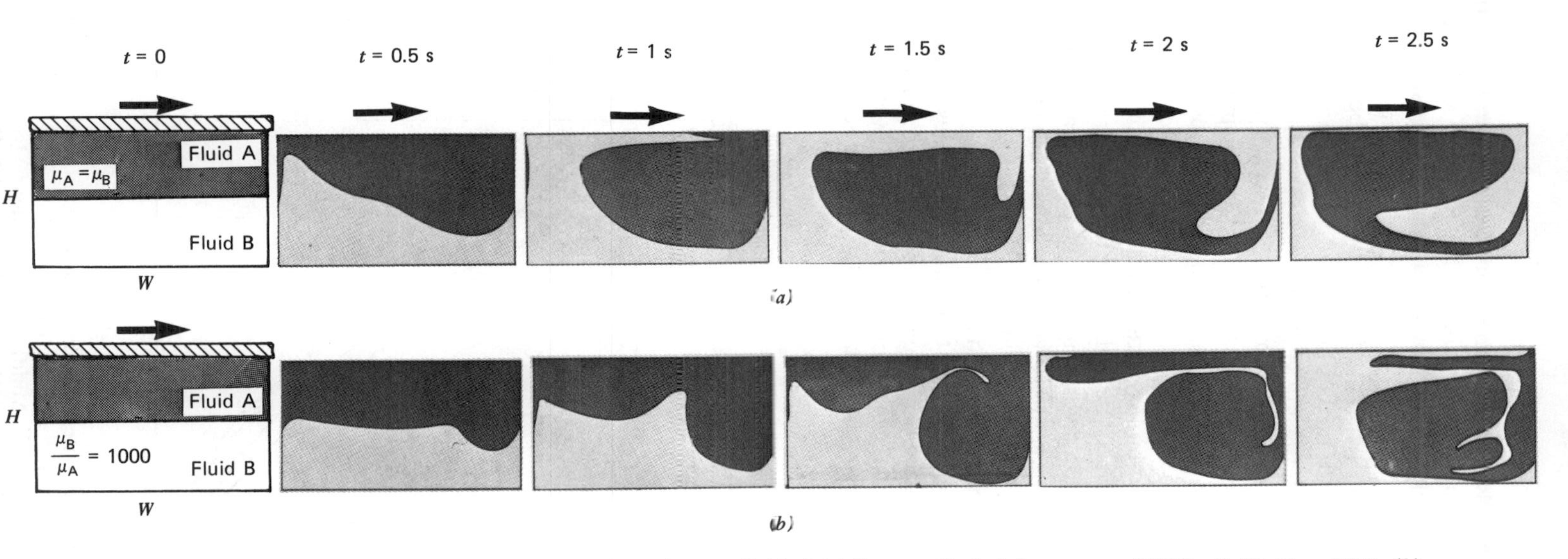


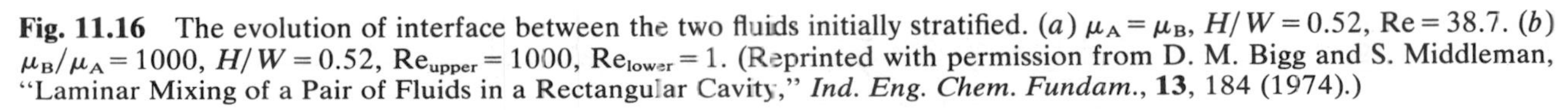

**Fig. 11.16** The evolution of interface between the two fluids initially stratified. (*a*) $\mu_A = \mu_B$, $H/W = 0.52$, $Re = 38.7$. (*b*) $\mu_B/\mu_A = 1000$, $H/W = 0.52$, $Re_{upper} = 1000$, $Re_{lower} = 1$. (Reprinted with permission from D. M. Bigg and S. Middleman, "Laminar Mixing of a Pair of Fluids in a Rectangular Cavity," *Ind. Eng. Chem. Fundam.*, **13**, 184 (1974).)

more in every cell, are followed with time; thus the positions of free boundaries and interfaces can be determined at any time. The evolution of the interface of two fluids of equal viscosities and densities in a channel with an aspect ratio of 0.52, and a Reynolds number defined as $V_0 W\rho/\mu$ of 38.7 appears in Fig. 11.16*a*. Note that after 2.5 s distinct striations between the two fluids have been formed. Additional striations will be formed with further mixing until the striation thickness is reduced to the desired level. Next the effect of the viscosity ratio was investigated on interface evolution. Figure 11.17 shows that for viscosity ratios up to 30, the evolution of the interfacial area was reduced. In these simulations the Reynolds number in the upper layer was maintained at the previous level. If, however, the viscosity ratio was increased to 1000, together with an increase in Reynolds number of the upper layer, a more complex picture evolved, as indicated in Fig. 11.16*b* and the broken curve in Fig. 11.17. The high Reynolds number in the upper region with vortex formation may be somewhat misleading, yet it draws attention to the fact that with variation in viscosity, the Reynolds number also changes and flow conditions may be altered. Bigg and Middleman (13) have also presented experimental verification to their theoretical calculations.

One could perhaps then conclude that equal viscosities of the two components should be favored, at least when equal proportions are mixed. Yet we cannot generalize this result to mixing conditions with significantly different proportions of components, and further work is needed for better understanding of rheologically nonhomogeneous systems.

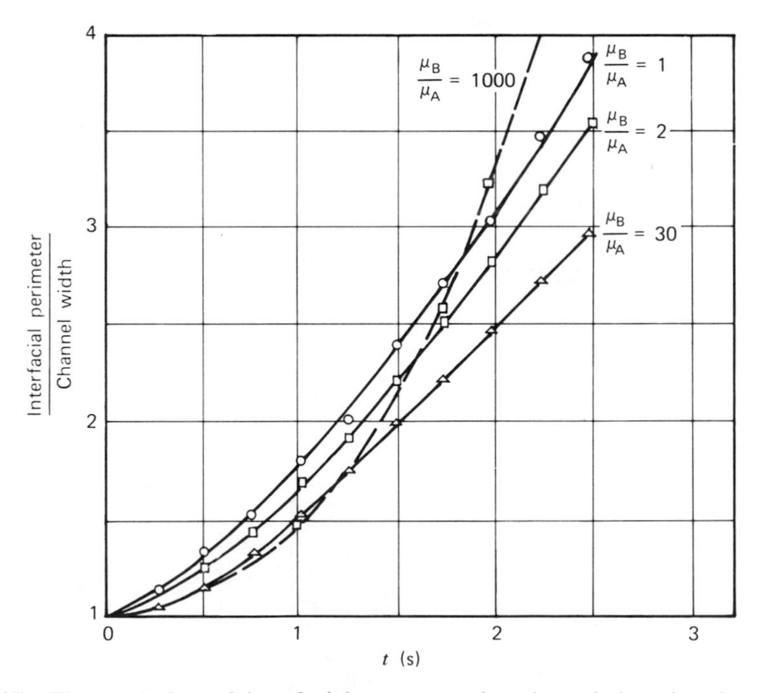


**Fig. 11.17** The evolution of interfacial area as a function of time for the rectangular channel circulatory flow shown in Fig. 11.16, at various viscosity ratios. (Reprinted with permission from D. M. Bigg and S. Middleman, "Laminar Mixing of a Pair of Fluids in a Rectangular Cavity," *Ind. Eng. Chem. Fundam.*, **13**, 184 (1974).)

### *Miscibility of Polymeric Liquids*

The foregoing discussion and examples assume immiscible or incompatible liquids. This is not a highly restrictive assumption, at least for the initial stages of mixing. Before discussing this point, however, let us briefly review the thermodynamic elements of miscibility.

First, we note that miscibility and compatibility mean the same thing. The former refers generally to liquid systems, whereas the latter usually designates solid systems. There are two aspects of the question of miscibility: will the two liquids mix (thermodynamics), and how long would this process take (kinetics). The second aspect is important in polymer–polymer and polymer–monomer systems because of the low diffusivities involved. Thermodynamically, mixing will take place at a temperature $T$ when

$$\Delta G = \Delta H - T\Delta S \leq 0 \tag{11.4-3}$$

where $\Delta G$, $\Delta H$, and $\Delta S$ are the Gibbs free energy, heat, and entropy of mixing. Increasing temperature tends to favor the thermodynamic conditions of mixing, since $\Delta S$ is almost always positive. Naturally, the rate of mixing (if $\Delta G < 0$) also increases with temperature because of increased diffusivities. To predict when mixing is possible, one must be able to calculate $\Delta H$ and $\Delta S$. The heat of mixing can be estimated using either the solubility parameter $\delta$ (cohesive energy density) of the liquids (15, 16) or the parameter $\chi_1$, which represents the interaction energy per solvent molecule divided by $kT$ as follows:

$$\Delta\tilde{H} = v_1 v_2(\delta_1 - \delta_2)^2 \tag{11.4-4}$$

where $\Delta\tilde{H}$ is the heat of mixing per unit volume, and

$$\Delta H = \chi_1 kTN_1 v_2 \tag{11.4-5}$$

where $v_1$ and $v_2$ are the volume fractions of the solvent and solute, $N_1$ is the number of solvent moles and $k$ is the Boltzmann constant. In a polymer–polymer system, it is not clear which is the solvent and which the solute.

The entropy of mixing can be evaluated approximately from statistical mechanics, as in the Flory–Huggins theory (17, 18), where

$$\Delta S = -k(N_1 \ln v_1 + N_2 \ln v_2) \tag{11.4-6}$$

Using these expressions, which hold for flexible monodispersed polymer-solvent systems of normal heat of mixing, critical conditions for phase separation can be obtained. Such predictions indicate that miscibility over the entire composition range occurs between solvents and polymers when $|\delta_1 - \delta_2| < 1.7$, and between polymer melts when $|\delta_1 - \delta_2| < 0.1$, for $\bar{M} \approx 10^5$. The more restrictive requirement for polymer–polymer systems stems from the relatively small entropy increase of such systems upon mixing.

Let us turn now to the following practical laminar mixing situation. A minor liquid component is introduced into the mixer containing the major, which is a melt, in the form of a "blob" of macroscopic dimensions. We maintain that what happens to the "blob" in the mixer flow field does not depend *initially* on whether the minor and the major components are miscible. This is because, at best, a thin interfacial volume will form if the solution kinetics are fast. The blob will deform with time, undergoing internal flow subject to the local stress field, which is *inhomogeneous* because the minor and major components have different rheological properties, both viscous and elastic. The role of the surface tension (therefore the existence of a sharp interface) should not matter. Viscous forces will dominate over interfacial tension. As the "blob" deforms and interfacial area increases, the degree of miscibility of the two components may play an increasing role. In miscible systems, interdiffusion aids in achieving mixing down to a molecular level, whereas in immiscible systems the minor component breaks into small domains. These domains, subject to viscous drag and surface tension forces, will probably attain a state of constant deformation. Thus in immiscible systems, mixing commences with an extensive mixing mechanism and gradually proceeds toward a mechanism of homogenization. The morphology of such domains in blends as well as copolymers is the subject of intensive investigations (19).

## 11.5 Homogenization

The deformation of a spherical liquid droplet in a homogeneous flow field of another liquid was studied in the classical work of G. I. Taylor (20) on emulsions. Taylor showed that for simple shear flow, a case in which interfacial tension dominates, the drop would deform into a spheroid with its major axis at an angle of 45° to the flow, whereas for the viscosity dominated case it would deform into a spheroid with its major axis approaching the direction of flow. He expressed the deformation $D$ as follows

$$D = \frac{L-B}{L+B} \qquad (11.5\text{-}1)$$

where $L$ and $B$ are the major and minor axes of the spheroid (Fig. 11.18). At equilibrium, a steady flow field ensues in the droplet, and for small deformations the deformation $D$ is given by (21)

► $$D = \frac{\mu_0 \dot{\gamma} r}{\Gamma} \frac{19\lambda + 16}{16(\lambda + 1)} \qquad (11.5\text{-}2)$$

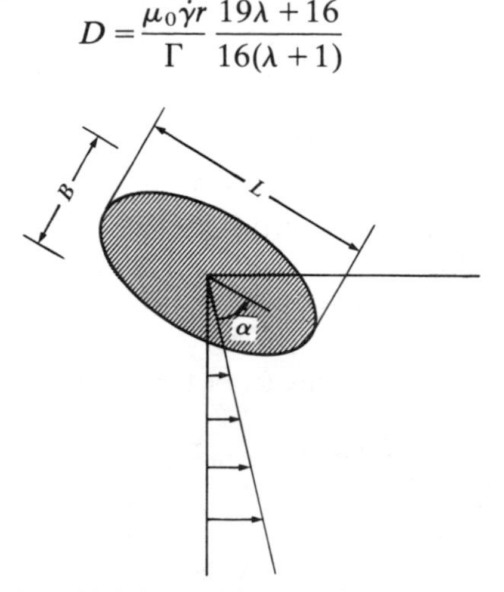


**Fig. 11.18** A single droplet of liquid deformed into a spheroid in a homogeneous shear flow field.

where $\mu_0$ is the viscosity of the continuous phase, $\dot{\gamma}$ the shear rate, $r$ the initial droplet radius, $\Gamma$ the interfacial tension, and $\lambda = \mu^*/\mu_0$ the ratio of viscosities of the dispersed and continuous phases, respectively. With $\dot{\gamma}$ replaced by $2\dot{\varepsilon}_{pl}$, where $\dot{\varepsilon}_{pl}$ is the rate of elongation, Eq. 11.5-2 holds also for two-dimensional steady planar elongational flow, $v_x = \dot{\varepsilon}_{pl}x$, $v_y = -\dot{\varepsilon}_{pl}y$, $v_z = 0$. Taylor also showed that for high $\lambda$ values the final deformation becomes a function of $\lambda$ only

$$D = \frac{5}{4}\frac{\mu_0}{\mu^*} \tag{11.5-3}$$

Equation 11.5-3 is consistent with the previously discussed problem of mixing a high viscosity minor into a low viscosity major.

The problem of droplet breakup was also investigated by Taylor (21) and recently by Rumscheidt and Mason (22) in an extensive study on particle motion in suspensions. Following the suggestion of Taylor, Rumscheidt and Mason assumed that the droplet will burst when the pressure drop generated across the interface exceeds the surface tension force, which tends to hold it together. This condition can be shown to occur at a critical deformation of $D_b = \frac{1}{2}$ for shear and planar extension flows. Experimental evidence has revealed a complex bursting behavior, but the critical deformation at burst agreed reasonably well with the theoretical predictions up to a viscosity ratio of about 4. Above this value, no droplet bursting was observed. Karam and Bellinger (23) have also verified this condition, but they have also found that in addition to the upper limit, there is a lower viscosity ratio limit of 0.005, below which no breakup occurs. Moreover, they discovered that breakup takes place most readily when the viscosity ratio is in the range of 0.2–1 and $D \geq 0.5$. Rumscheidt and Mason (22) have also experimented with extremely high viscosity ratio of $2 \times 10^4$ with a system of sucrose acetate isobutyrate (about $10^5$ poise) dispersed in silicone oil. The nearly spherical drops rotated like rigid bodies. When pulled out into threads and then immersed in the medium, they tended to develop helical and coiled rotational orbits. In this case, the viscosity of the dispersed phase was so high that the drops never had sufficient time to enable the interfacial tension to bring the drop into equilibrium configuration. This behavior demonstrates the complications involved in real polyblend systems.

The foregoing discussion leads us to the conclusion that droplet break-up, when attainable for a given system, occurs at a certain critical shear stress level ($\mu_0\dot{\gamma}$), and that for a given stress level a droplet size $r$ is reached below which droplet breakup is unlikely to occur.

The effect of the elastic properties of a viscoelastic continuous phase on the deformation and breakup of a Newtonian dispersed phase was investigated by Flumerfelt (24). He used a polyacrylamide solution in water as the continuous phase and a low molecular weight polystyrene solution in dibutylphthalate as the dispersed phase. His results indicated that there is a minimum drop size, which varies with each fluid system, below with breakup cannot be achieved. The elasticity of the continuous phase tends to increase the minimum drop size as well as the critical shear rate when breakup does occur, since the resulting shear stress is $\dot{\gamma}$ dependent. Consistent with previous results, an increase in the viscosity of the continuous phase had an opposite effect. Flumerfelt also reported the interesting finding that under unsteady shear conditions (a step change application of shear), both the minimum drop size and the critical shear rate significantly *decrease* as compared to those under steady shear. He suggested that in light of this finding, dispersion in viscoelastic media might be best accomplished under transient conditions. Indeed, such transient conditions exist in the narrow gaps between rotor and housing in mixers as well as in the "barrier" type of mixing devices in extruders.

Heterogeneous blends of polymers containing comparable fractions of each polymer show, of course, a complex behavior. Depending on the rheological properties of the phases and mixing conditions, one or the other component may form a continuous phase, generally with the low viscosity phase tending to encapsulate the high viscosity phase (25). Alternatively, conditions may arise when both components form a "continuous" phase. The domain, size, and morphology of the dispersed phase were shown to be dependent on mixing conditions and related to the rheology of both phases (26). Finally, as observed by White and Tokita (27), one or both polymers may break up into crumbs, and this determines the continuous phase and globule or domain size. This is an important rubber processing problem when treating polybutadienes. It is a solidlike break up, as in dispersive mixing.

## 11.6 Dispersion

Dispersive mixing in polymer processing involves the rupture of clumps and agglomerates of solid particles such as pigments and carbon black in a deforming viscous liquid. It is accomplished by forcing the mixture to pass in high shear zones generated in narrow clearances such as the gap between the rolls of a roll-mill or in the clearance between the blades and the shell in internal mixers.

Following Bolen and Colwell (28), we assume that the agglomerates break when internal stresses, induced by viscous drag on the particles, exceed a certain threshold value. We consider the forces acting on a single agglomerate in the form of a *rigid dumbbell* (Fig. 11.19), consisting of two unequal beads of radii $r_1$ and $r_2$, whose centers are a distance $L$ apart, in a homogeneous velocity field of an incompressible Newtonian fluid. As a result of the viscous drag on each of the beads, a certain force develops in the connector which depends on the magnitude of the viscous drag and on dumbbell orientation. When these forces exceed a certain critical value, which equals the attractive cohesive forces, the beads break apart. The mathematical formulation of this problem was proposed and solved in detail by Bird et al. (29) in connection with molecular interpretation of macroscopic flow phenomena of polymer solutions. Their solution was adopted (30) with two minor modifications: terms due to Brownian motion, which are irrelevant on our scale, are neglected, and bead radii are assumed to be unequal.

It is assumed that the presence of the dumbbell does not alter the flow field of the liquid in the neighborhood of the dumbbell (i.e., that the dumbbell does not "drift") and that the flow field is homogeneous, meaning that the rate of deformation is the same at all points. Hence the velocity field is given by

$$v = (\boldsymbol{\kappa} \cdot \boldsymbol{\rho}) \tag{11.6-1}$$

where $\boldsymbol{\kappa}$ is a tensor that specifies the flow field and may be time dependent, and $\boldsymbol{\rho}$ is a position vector as indicated in Fig. 11.19. The rate of deformation tensor $\dot{\boldsymbol{\gamma}}$ in

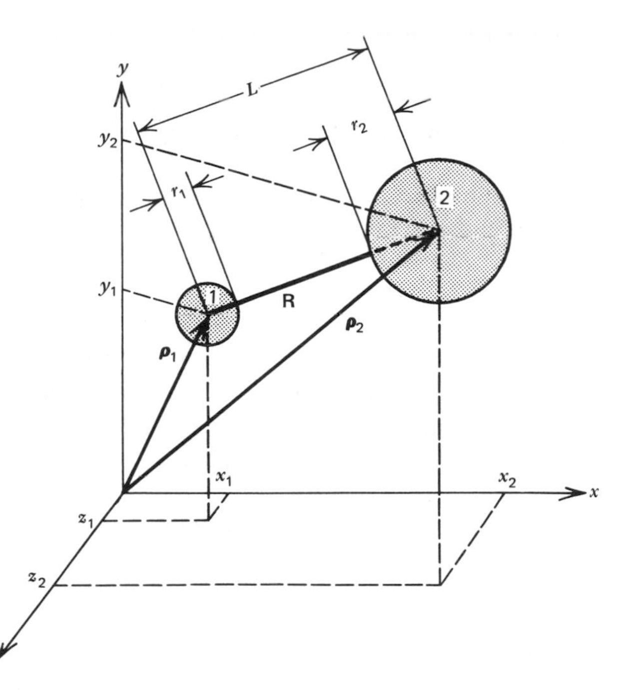


**Fig. 11.19** A rigid dumbbell with beads of radii $r_1$ and $r_2$, respectively, and connector of length $L$; $\boldsymbol{\rho}_1$ and $\boldsymbol{\rho}_2$ are position vectors of the bead centers.

terms of $\boldsymbol{\kappa}$ is:

$$\dot{\boldsymbol{\gamma}} - \boldsymbol{\kappa} + \boldsymbol{\kappa}^\dagger - \nabla \boldsymbol{v} + (\nabla \boldsymbol{v})^\dagger \tag{11.6-2}$$

where $\boldsymbol{\kappa}^\dagger$ is the transpose of $\boldsymbol{\kappa}$. The nonvanishing velocity component for steady simple shearing flow is

$$v_x = \kappa_{xy} y \tag{11.6-3}$$

and those for steady elongational flow are

$$\begin{aligned} v_x &= \kappa_{xx} x \\ v_y &= -\tfrac{1}{2}\kappa_{xx} y \\ v_z &= -\tfrac{1}{2}\kappa_{xx} z \end{aligned} \tag{11.6-4}$$

For each bead of the dumbbell, an equation of motion can be written indicating that mass-time-acceleration equals the sum of forces acting on it. Among the forces, viscous drag forces, and the force in the connector are included

$$m_1 \ddot{\boldsymbol{\rho}}_1 = -\zeta_1(\dot{\boldsymbol{\rho}}_1 - \boldsymbol{v}_1) - \mathbf{F} \tag{11.6-5}$$

$$m_2 \ddot{\boldsymbol{\rho}}_2 = -\zeta_2(\dot{\boldsymbol{\rho}}_2 - \boldsymbol{v}_2) + \mathbf{F} \tag{11.6-6}$$

where $\mathbf{F}$ is the unknown force in the connector, $\boldsymbol{v}_i$ is the local fluid velocity at bead $i$, $\dot{\boldsymbol{\rho}}_i$ is the velocity of bead $i$, and $\zeta_i$ is the viscous drag on bead $i$, which according to Stokes' law is given by the following expression

$$\zeta_i = 6\pi\mu r_i \tag{11.6-7}$$

The acceleration term is assumed to be small compared to the other terms, and it is neglected. Subtracting Eq. 11.6-6 from Eq. 11.6-5 gives

$$\dot{\boldsymbol{\rho}}_1-\dot{\boldsymbol{\rho}}_2=\boldsymbol{v}_1-\boldsymbol{v}_2-\mathbf{F}\left(\frac{\zeta_1+\zeta_2}{\zeta_1\zeta_2}\right) \tag{11.6-8}$$

Substituting Eq. 11.6-1 into Eq. 11.6-8 and defining a vector $\mathbf{R}=\boldsymbol{\rho}_1-\boldsymbol{\rho}_2$, which is a vector pointing from bead 1 to bead 2, as in Fig. 11.19, gives

$$\dot{\mathbf{R}}=\boldsymbol{\kappa}\cdot\mathbf{R}-\mathbf{F}\left(\frac{\zeta_1+\zeta_2}{\zeta_1\zeta_2}\right) \tag{11.6-9}$$

Now by forming a scalar product of $\mathbf{R}$ with Eq. 11.6-9 noting that $\mathbf{R}\cdot\dot{\mathbf{R}}=d(\mathbf{R}\cdot\mathbf{R})/2dt=d(L^2)/2dt=0$, and that $\mathbf{F}$ has a nonvanishing component only in the direction of the rigid connector, thus $\mathbf{R}\cdot\mathbf{F}=LF$, the following equation for the magnitude of the force in the connector is obtained

$$F=\frac{\zeta_1\zeta_2}{(\zeta_1+\zeta_2)L}(\boldsymbol{\kappa}:\mathbf{RR}) \tag{11.6-10}$$

Equation 11.6-10 indicates that the force in the connector is proportional to the harmonic mean of the viscous drags on the beads and proportional to the term $(\boldsymbol{\kappa}:\mathbf{RR})/L$, which depends on the flow field, the orientation of the dumbbell, and its size. By evaluating the latter term for some specific flow fields, we can obtain a better insight into the nature of the dispersive mixing process.

### Example 11.5 *Breakup of Rigid Agglomerates in Steady Simple Shear Flow and in Steady Elongational Flow*

The term $(\boldsymbol{\kappa}:\mathbf{RR})$ is a scalar that in term of its components, is given as follows:

$$\begin{aligned}(\boldsymbol{\kappa}:\mathbf{RR}) &=\kappa_{11}R_1R_1+\kappa_{12}R_2R_1+\kappa_{13}R_3R_1\\ &+\kappa_{21}R_1R_2+\kappa_{22}R_2R_2+\kappa_{23}R_3R_2\\ &+\kappa_{31}R_1R_3+\kappa_{32}R_2R_3+\kappa_{33}R_3R_3\end{aligned} \tag{11.6-11}$$

For simple steady shear flow given in Eq. 11.6-3, Eq. 11.6-11 reduces to

$$(\boldsymbol{\kappa}:\mathbf{RR})=\kappa_{xy}R_yR_x \tag{11.6-2}$$

The vector $\mathbf{R}$ was defined earlier as the difference of the two position vectors $\boldsymbol{\rho}_1$ and $\boldsymbol{\rho}_2$. Expressing these position vectors in terms of unit vectors in rectangular coordinate systems results in

$$\mathbf{R}=\boldsymbol{\rho}_1-\boldsymbol{\rho}_2=\boldsymbol{\delta}_x(x_1-x_2)+\boldsymbol{\delta}_y(y_1-y_2)+\boldsymbol{\delta}_z(z_1-z_2) \tag{11.6-13}$$

Substituting the appropriate components of $\mathbf{R}$ from Eq. 11.6-13 into Eq. 11.6-12 gives

$$(\boldsymbol{\kappa}:\mathbf{RR})=\kappa_{xy}(y_1-y_2)(x_1-x_2) \tag{11.6-14}$$

where $x_1$ and $y_1$ are the coordinates of bead 1 and $x_2$ and $y_2$ those of bead 2. If the coordinate system is placed at the center of bead 1 (and recalling that for this flow situation, according to Eq. 11.6-2, $\kappa_{xy}$ is simply the shear rate $\dot{\gamma}$), Eq. 11.6-10, with the aid of Eq. 11.6-7, reduces to

$$F = \frac{6\pi\mu\dot{\gamma}x_2y_2}{L}\left(\frac{r_1r_2}{r_1+r_2}\right) \tag{11.6-15}$$

Equation 11.6-15 indicates that the force in the connector vanishes if the dumbbell is either parallel to the flow field ($y_2=0$) or perpendicular to the flow field ($x_2=0$). The length of the connector $L$ is related to the position of bead 2 by

$$L^2 = x_2^2 + y_2^2 + z_2^2 \tag{11.6-16}$$

Inserting Eq. 11.6-16 into Eq. 11.6-15 results in

$$F = \frac{6\pi\mu\dot{\gamma}}{L}\left(\frac{r_1r_2}{r_1+r_2}\right)x_2\sqrt{L^2 - x_2^2 - z_2^2} \tag{11.6-17}$$

Equation 11.6-17 indicates that the maximum force in the connector will be obtained when the dumbbell is placed in the $x$–$y$ plane (i.e., $z_2=0$) and its orientation is at 45° angle to the direction of shear (i.e., $x_2=y_2=L/\sqrt{2}$), thus

$$F_{\max} = 3\pi\mu\dot{\gamma}L\left(\frac{r_1r_2}{r_1+r_2}\right) \tag{11.6-18}$$

Finally, Eq. 11.6-18, for the special case of two beads in contact with each other, where $L=r_1+r_2$, reduces to

$$F_{\max} = 3\pi\mu\dot{\gamma}r_1r_2 \tag{11.6-19}$$

Equation 11.6-19 indicates that the maximum force acting on a clump or dumbbell, which tends to separate the beads, is proportional to the *shear stress* ($\mu\dot{\gamma}$) and the product $r_1r_2$. Thus dispersing mixing is improved by increasing the shear stress; it is also easier to break apart two large beads than to break a small bead from a large bead or two small ones from each other.

For steady elongational flow, the maximum force in the connector is obtained when the dumbbell is aligned in the direction of flow, and for the case of the beads in contact with each other is given by

$$F_{\max} = 6\pi\mu\dot{\varepsilon}r_1r_2 \tag{11.6-20}$$

where $\dot{\varepsilon}$ is the rate of elongation.

Comparing Eqs. 11.6-20 and 11.6-19 indicates that at the *same* rate of deformation the force in the connector in elongational flow is twice the force in shear flow. But in practice, very large shear rates *are* obtainable, whereas large rates of elongation rates are hard to obtain. Hence virtually all dispersive mixers are based on shear dispersion in narrow clearances.

The validity of the foregoing derivation is restricted by the simplifying assumptions made—in particular, that the beads do not affect the local flow field and that dumbbell interaction can be neglected.

It is instructive to compare Eqs. 11.6-19 and 11.5-2 and to note that in both agglomerate and liquid breakup, the stress and the particle size play similar roles.

In both cases, mixer design should incorporate high shear zones and ensure that all fluid particles pass the high shear zone repeatedly. Among all mixing operations dispersive mixing is perhaps the most difficult and costly. Therefore it is common practice to prepare *master batches*, that is, mixtures containing a high proportion of given additives. For example, in preparing mixtures of polyethylene with carbon black, a "*superconcentrate*" containing about 50% carbon black is prepared, then diluted in an internal mixer to 25%, and once more diluted to the final low concentration in a processing extruder. The high intensity deagglomeration operation takes place in the "superconcentrate." The dilution of the master batch is a "simple" extensive mixing operation. Hence by preparing the master batch, the difficult and costly dispersive mixing procedure has to be applied to only a small fraction of the final product. Moreover, it is easier to break up clumps and agglomerates at high concentration levels because the high viscosity of the system leads to high local shear stresses and the high concentration facilitates agglomerate breakup by particle interactions. Finally, dilution of master batches makes it easier to maintain uniform product quality than when using a direct mixing process. The color of the product, for example, depends on whether the pigment particles are deagglomerated, and, by direct mixing, this would be more difficult to achieve on a uniform level than by diluting master batches. Yet the preparation of good master batches is not a simple task, and sometimes special precautions must be taken to ensure good dispersion. Thus in dispersing finely divided powders with large surface areas, it is sometimes necessary to wet the surface. For example, water may be added to carbon black before dispersing it in polyethylene.

Finally, it is worthwhile to mention that rigid agglomerates can be also dispersed in particulate solid form in centrifugal impact mixers (31).

## 11.7 Motionless Mixers

The name, "motionless" mixer, is derived from the fact that these devices achieve mixing without any moving parts. Instead, through ingenious construction, they rearrange the flow field and reshuffle the fluid streams in such a way that the interfacial area increases appreciably and predictably every time the fluid mixture flows through each one of the repetitive mixing elements making up motionless mixers. Although the exact mixing pattern is specific to the particular type of the motionless mixer used, it is generally true that the interfacial area between the major and minor components is increased in two ways: by shear or extensional flow and by splitting and recombining fluid streams. Both involve pressure losses. Thus there is a practical limit to the number of motionless mixer elements that can be used, hence the quality of mixing that can be achieved by these devices. We briefly describe two such mixers here, the Ross ISG (Interfacial Surface Generator) and the Kenics Static Mixer. Schott et al. (7) have reviewed these motionless mixers commonly used in the polymer processing industry.

The Ross ISG is shown in Fig. 11.20*a*. In each element, the four circular entrance holes form a line perpendicular to that formed by the exit holes. This is achieved by drilling the holes obliquely with respect to the axis of the element and in such a way that an outside entrance has an inside exit, thus achieving radial mixing. This is shown on Fig. 11.20*b*. It is evident that during the flow *inside* each

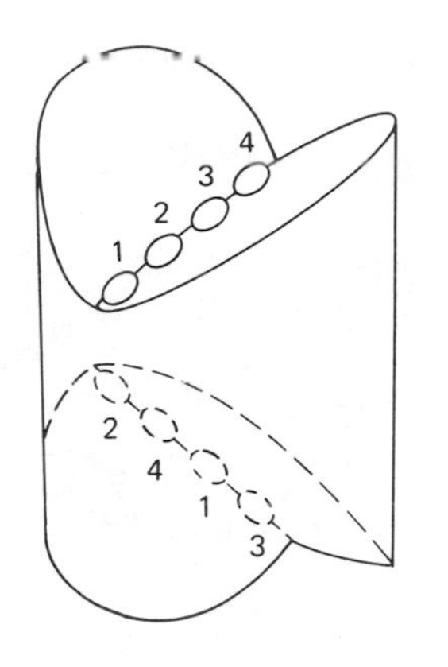

4
3
2
1
2
4
1
3


(*a*)

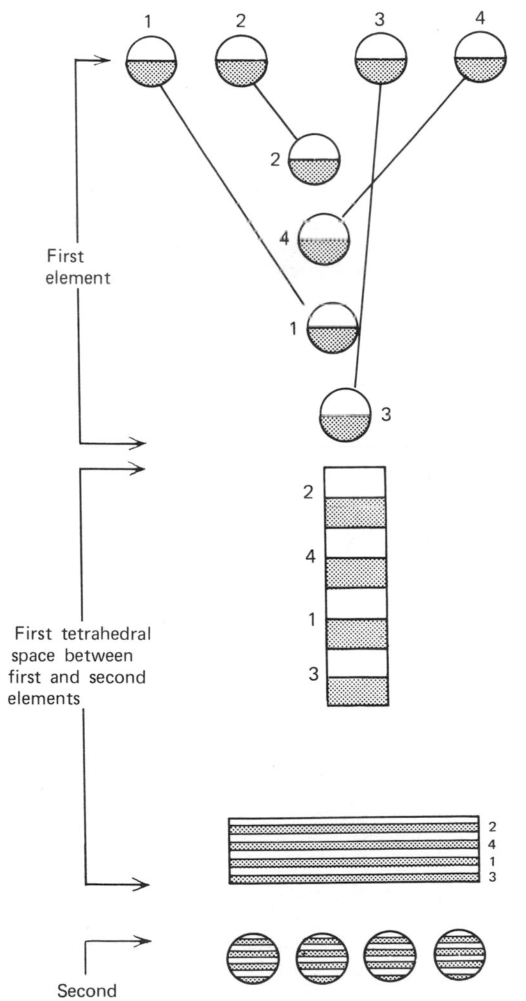

1
2
3
4
2
4
1
3
First element
2
4
1
3
First tetrahedral space between first and second elements
2
4
1
3
Second element


(*b*)

(c)

(d)

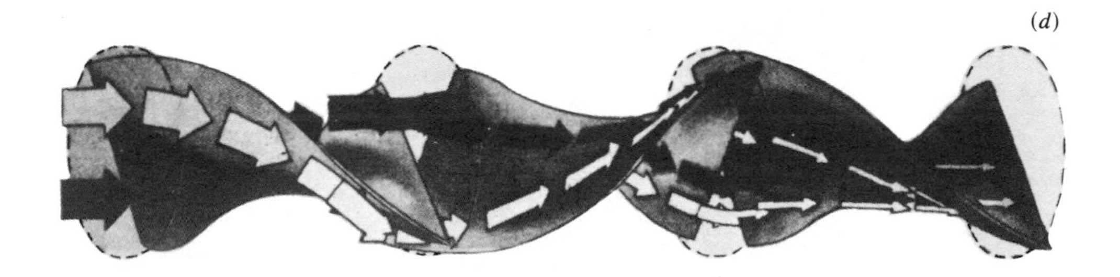

First element

In

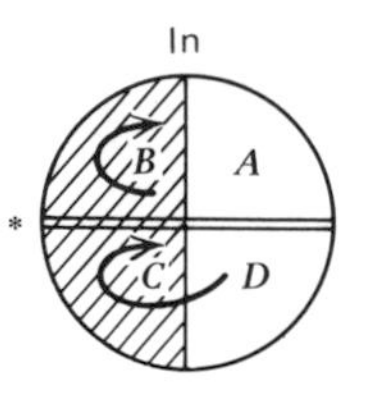


Out

D C
A B
*

(e)

Second element

In

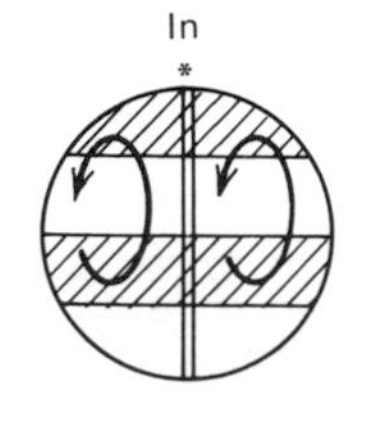


Out

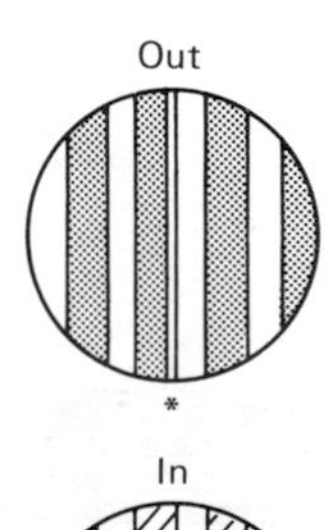


Third element

In

*

Out

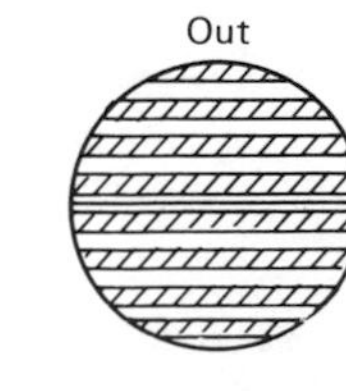

**Fig. 11.20** Motionless mixers. The Ross ISG: (*a*) Schematic representation of one element, (*b*) Schematic representation of the mixing involved in each element and the tetrahedral space between elements, (*c*) The resulting increase of the number of striations for a pair of polyester resins. The Kenics Static Mixer: (*d*) Schematic representation, (*e*) Representation of the "ideal" results of laminar mixing inside each of the helical elements, (*f*) Experimental results showing the real progression of mixing. [(*c*) Courtesy of the Charles Ross and Son Co., Hauppauge, N.Y.; (*f*) courtesy of the Kenics Corp., Danvers, Mass.]

element there is practically no mixing, except for the radial redistribution of the four streams. On the other hand, by construction, a tetrahedral space is formed between two consecutive elements. The four streams from the first element join together in this space, creating new interfaces (striations), as shown on the last part of Fig. 11.20*b* and on Fig. 11.20*c*. The flow that occurs in the tetrahedral spaces is divergent-covergent in nature and results in an effective stretching of the interfacial area elements. In this stretched state the fluid is divided into four streams by the holes of the second element where radial redistribution will occur again. The net result of the fluid stream combination and flow is the fourfold increase of the number of striations $N_s$. Following the fluid through consecutive elements one can easily show that the relation between $N_s$ and the number of elements $E$ is

$$N_s = 4^E \tag{11.7-1}$$

The Kenics Static Mixer, Fig. 11.20*d*, consists of a series of helical elements fixed within a tubular housing. Consecutive helices are of opposite sense. They are welded together so that adjacent edges are perpendicular to each other. The fluid is thus split every time it leaves one element and enters the next. Within each element the fluid flows in two semicircular helical channels where the flow field consists of a combination of down-channel and a significant cross-channel velocity component. To understand the effects of the flow within each element and the stream splitting from element to element, we turn to Fig. 11.20*e*. Segregated black and white semicircular streams enter the first element with the interface perpendicular to the helical separator. At the end of the element, because of the cross-channel component of flow, the interfaces become parallel to the separator in both compartments. Thus they have not only increased in area, but are favorably oriented with respect to the stream splitting by the helix of the second element. Upon entering the second element, the fluid is split perpendicularly to the striations. This initial orientation is of course most favorable relative to the circulating flow in the new semicircular channels (see Section 11.3). In summary, then, during each splitting the number of striations is doubled leading to the relationship

$$N_s = 2^E \tag{11.7-2}$$

and during the flow within each element the interfacial area is increased while the number of striations remains constant. This process is evident in Fig. 11.20*f*, where the first six frames represent the first element.

From the foregoing we can conclude that in motionless mixers stream splitting and recombination results in a predicted and repeated increase in the number of striations (ordered distributive mixing). On the other hand, the flow configuration involved brings about a most favorable orientation of the interfacial area elements with respect to the particular flow (laminar mixing): in the Kenics Mixer they are perpendicular to the predominately shear flow and in the Ross ISG parallel to the predominately elongational flow.

## 11.8 Roll-Mills

Chapter 1 noted the historical significance of the roll-mill with two rolls and the calender in the development of polymer processing machinery, and Section 10.5 analyzed, in some detail, the hydrodynamics of this important geometrical configuration, pointing out that it involves a nonparallel plate viscous drag pressurization mechanism. This section further explores the same configuration with respect to its mixing performance.

As Section 10.5 pointed out, in a roll-mill the two rolls may turn at different frequencies of rotation, inducing further shear between the rolls and facilitating, together with temperature control, the formation of a band on one of the rolls. The operation of the open roll-mill depends on the adhesion of the masticated mass to one of the rolls as a clinging "blanket." Certain materials have an "affinity" for one of the rolls, thus butyl rubber, for example, has an affinity for the faster roll. White and Tokita (27) describe and explain this and other phenomena associated with roll-milling elastomers in terms of rheological properties of the material. During the operation, cutting of the blanket and rolling are carried out, facilitating the distribution of interfacial elements throughout the system. On small roll-mills this is done manually, rendering the mixing process dependent on the skill of the operator, whereas on large mills the operator knife is replaced by "curling wheels" or "ploughs," which continuously roll up strips of stock and redistribute them. This redistribution is needed because the roll-mill lacks lateral motion almost completely. The roll-mill can also be operated continuously. For example, polymeric feed in the form of pellets or powder can be fed on one end of the rolls, while "ploughs" turn and redistribute the molten layer as it moves along the mill with mixed material continuously stripped off at the far end. During the milling operation, additives may be added to the mill. Thus fillers and oils are added to rubber. With polyvinyl chloride it is preferable to prepare a dry-blend and feed the mill with a uniform mix of PVC powder, plasticizer, and additives. Rolls range in diameter up to 1 m and may be as wide as 2.5 m. Mills with three and five rolls are also manufactured. In the case of mills with more than two rolls, the material passes from one nip to the next.

The pressure distribution in the nip region of a roll-mill with equal size rolls and the same roll speed (32) was given in Eq. 10.5-11 for Newtonian fluids and in Eqs. 10.5-31 and 10.5-32 for power law model fluids. Solution of the pressure profile requires the knowledge of $\lambda$, defined in Eq. 10.5-12, which is a transformed variable expressing the distance $X_1$, where the polymer detaches from one of the rolls. The axial location where the rolls "bite" the polymer is uniquely related to $X_1$ as shown in Fig. 10.25. The axial locations where the rolls "bite" the polymer and the polymer detaches from one of the rolls depend on the total volume of polymer on the rolls, and the roll and gap sizes. Clearly, if the "blanket" thickness equals the minimum gap separation, $X_1 = X_2 = 0$, and pressure will be zero throughout. Hence the total volume of polymer on the rolls must be more than this minimum amount (per unit width) of $2\pi(R+H_0)(2H_0)$. If at constant roll speed more polymer is added to the rolls, $|X_2|$ will increase as well. This in turn results in higher pressures developed between the rolls, leading to an increased flow rate out of the roll, later detachment, and a thicker "blanket." The relationship between the total volume of polymer per unit width on the roll-mill $V$ and the variables $X_1$ and $X_2$, assuming constant "blanket" width, is given approximately by

$$V - 4\pi H_1(R+H_1)\phi + 2\int_{X_2}^{X_1} h\,dx \tag{11.8-1}$$

where $\phi$ is the fraction of circumference covered by the free "blanket" and given approximately by

$$\phi = 1 - \frac{\alpha+\beta}{2\pi} = 1 - \frac{\tan^{-1}\left(\dfrac{\lambda\sqrt{2RH_0}}{R+H_0}\right) + \tan^{-1}\left(\dfrac{|\rho_2|\sqrt{2RH_0}}{R+H_0}\right)}{2\pi} \tag{11.8-2}$$

where $\alpha$ and $\beta$ are the angles in Fig. 11.21.

The integral in Eq. 11.8-1 can also be expressed in terms of $\rho_2$ and $\lambda$ by substituting Eqs. 10.5-9 and 10.5-10 into Eq. 11.8-1 and expressing $H_1$ in terms of $\lambda$ via Eqs. 10.5-9 and 10.5-10.

$$\blacktriangleright \quad V = 4\pi H_0(1+\lambda^2)[R + H_0(1+\lambda^2)]\phi + \frac{2H_0\sqrt{2RH_0}}{3}(3\lambda + \lambda^3 - 3\rho_2 - \rho_2^3) \tag{11.8-3}$$

Clearly, if $H_1 = H_0$, then $\lambda = \rho_2 = 0$, $\phi = 1$, and $V_0 = 4\pi H_0(R+H_0)$; that is, the total volume is simply a free blanket of thickness $2H_0$, covering the circumference of one roll. Equation 11.8-3, which incorporates minor geometrical approximations that are not more severe than the approximation made for the whole derivation and valid for $h/R \ll 1$, permits the theoretical evaluation of $\lambda$ from the additional information of the total volume of polymer placed on the rolls together with only roll diameter and minimum gap separation. By dividing Eq. 11.8-3 by the minimum amount of polymer $V_0$, we obtain a unique relationship between $V/V_0$ and $\lambda$, with $H_0/R$ as a parameter (Fig. 11.22). The validity of the relationship above has not

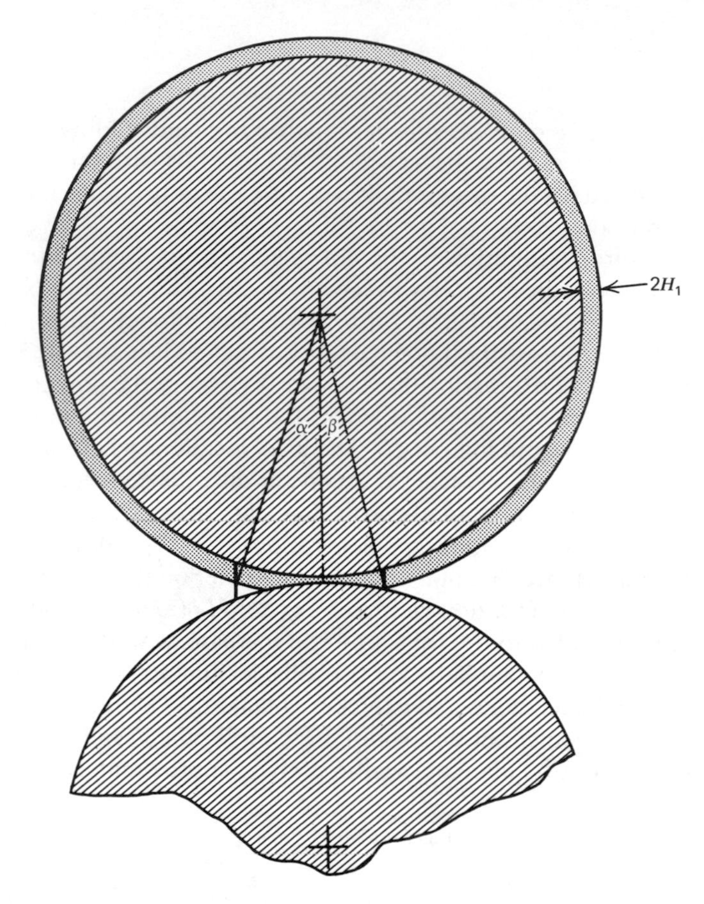


**Fig. 11.21** The evaluation of the total polymer volume on roll-mills. Angles $\alpha$ and $\beta$ represent the locations $\rho_2$ and $\lambda$, respectively.

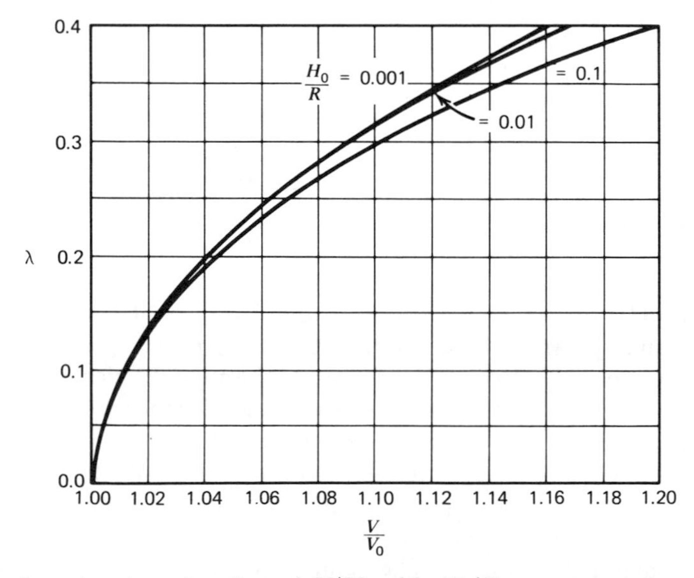


**Fig. 11.22** Location $\lambda$ as function of $V/V_0$ with $H_0/R$ as a parameter; $V$ is the total volume of polymer on the roll-mill, $V_0$ is the minimum volume, $2H_0$ is the minimum gap size, and $R$ is roll radius.

been verified experimentally, but Bergen and Scott (33), who carried out a detailed experimental investigation of the pressure distribution on the rolls in both calendering sheets and milling polymer, have observed that in a series of milling runs, when only the speed was varied, both $\lambda$ and $\rho_2$ remained "remarkably constant." This finding is consistent with the prediction of Eq. 11.8-3, that is, roll speed should not affect the value of $\rho_2$ and $\lambda$. Nevertheless, there are not enough experimental data to verify that total volume of polymer was maintained constant. In another set of data the same investigators have varied gap separation and observed a variation in $\rho_2$ and $\lambda$, which once again is in qualitative agreement with Eq. 11.8-3. Quantitative verification based on these data is unavailable, not only because of lack of adequate experimental information, but also because the theoretical prediction of the relationship between $\lambda$ and $\rho_2$ (Fig 10.25) was not observed experimentally. The reasons for disagreement were discussed in Section 10.5.

Roll-mills are efficient dispersive mixers. As Section 11.6 discussed, in dispersive mixing fluid particles must attain a certain critical shear stress to bring about agglomerate breakup. It is evident from the hydrodynamic analysis of roll-mills in Section 10.5 that depending on their radial location, fluid particles experience different maximum shear stresses as they pass across the nip region. A quantitative evaluation of dispersive mixing requires, therefore, the evaluation of the maximum shear stress distribution function. This is the object in the example that follows.

**Example 11.6** ***The Maximum Shear Stress Distribution Function for Two Roll Mills of Equal Radii and Rotational Frequency***

The maximum shear stress distribution function $F(\tau_m)$ is defined as the fraction of polymer leaving the nip that experiences a maximum shear stress equal to or less than $\tau_m$ (see Section 7.12). Equation 10.5-19 indicates that the maximum shear stress along any flow line is experienced at $\rho = \rho_2$, if $\rho_2 > -\sqrt{1+2\lambda^2}$ and at $\rho = -\sqrt{1+2\lambda^2}$ if $\rho_2 < -\sqrt{1+2\lambda^2}$. Thus the maximum stress profile across the clearance at this location is given by

$$|\tau_m| = \frac{3\mu U}{H_0}\left[\frac{1}{4(1+\lambda^2)}\right]\xi = \tau_{\max}\,\xi \qquad \rho_2 < -\sqrt{1+2\lambda^2}$$

$$\text{(i.e., } \lambda > 0.38 \quad \text{or} \quad \rho_2 < -1.135) \qquad (11.8\text{-}4)$$

and

$$|\tau_m| = \frac{3\mu U}{H_0}\left[\frac{\lambda^2 - \rho_2^2}{(1+\rho_2^2)^2}\right]\xi = \tau_{\max}\,\xi \qquad \rho_2 > -\sqrt{1+2\lambda^2}$$

$$\text{(i.e., } \lambda < 0.38 \quad \text{or} \quad \rho_2 > -1.135)$$

where $\tau_{\max}$ is the overall maximum shear stress in the system at $\xi = 1$. The shear stress profile, as given in Eq. 11.8-4, is a simple linear profile increasing with increasing $\xi$. Hence the fraction of flow rate leaving the nip region that experiences a maximum shear stress of equal or less to $\tau_m$, is obtained by integrating the velocity profile from 0 to $\xi$, provided* $\rho_2 > -\sqrt{2+3\lambda^2}$ (i.e., $\rho_2 > -1.6$ or $\lambda < 0.425$)

$$F(\tau_m) = \frac{\int_0^\xi u_x \, d\xi}{\int_0^1 u_x \, d\xi} \tag{11.8-5}$$

Substituting Eq. 10.5-16, into Eq. 11.8-5 and integrating the resulting expression provides the following equations

$$F(\tau_m) = \tfrac{1}{2}\xi(1+\xi^2) \qquad -\sqrt{2+3\lambda^2} < \rho_2 < -\sqrt{1+2\lambda^2}$$

$$(-1.6 < \rho_2 < -1.135 \quad \text{or} \quad 0.38 < \lambda < 0.425)$$

$$F(\tau_m) = \xi\left(\frac{1+\rho_2^2}{1+\lambda^2}\right)\left[1 + \frac{\lambda^2 - \rho_2^2}{2(1+\rho_2^2)}(3-\xi^2)\right] \qquad \rho_2 > -\sqrt{1+2\lambda^2} \tag{11.8-6}$$

$$(\rho_2 > -1.135 \quad \text{or} \quad \lambda < 0.38)$$

which subsequent to substituting Eq. 11.8-4 provides the required distributions

$$F(\tau_m) = \frac{1}{2}\frac{\tau_m}{\tau_{max}}\left[1 + \left(\frac{\tau_m}{\tau_{max}}\right)^2\right] \qquad -\sqrt{2+3\lambda^2} < \rho_2 < -\sqrt{1+2\lambda^2}$$

► 

$$F(\tau_m) = \frac{\tau_m}{\tau_{max}}\left(\frac{1+\rho_2^2}{1+\lambda^2}\right)\left\{1 + \frac{\lambda^2 - \rho_2^2}{2(1+\rho_2^2)}\left[3 - \left(\frac{\tau_m}{\tau_{max}}\right)^2\right]\right\} \qquad \rho_2 > -\sqrt{1+2\lambda^2} \tag{11.8-7}$$

Figure 11.23 plots $F(\tau_m)$ with $\lambda$ as a parameter. Results indicate broad stress distribution. As $\lambda$ approaches zero, the distribution becomes linear where 50% of the leaving polymer experiences a maximum stress of less than $\tau_{max}$. As $\lambda$ increases, the distribution narrows. But even for $\lambda > 0.38$, about 30% of the leaving polymer experiences a maximum stress of less than $\tau_{max}$. The value of $\tau_{max}$ also increases with $\lambda$ in the range $\lambda < 0.38$.

A suggested procedure for utilizing the maximum shear stress distribution function for dispersive mixing is to calculate the critical shear stress needed to break up the agglomerates with the aid of Eq. 11.6-19. This will be the value of $\tau_m$ in Eq. 11.8-7. After deciding on the fraction of polymer we are willing to permit to experience a maximum stress level below the critical value, we set $F(\tau_m)$ in Eq. 11.8-7 equal to this value, and substitute $\tau_{max}$ from Eq. 11.8-4 into Eq. 11.8-7. We then have one equation with the following variables: $\lambda$, $\mu$, $U$, and $H_0$. The variable $\lambda$, as we have seen earlier, is determined by the total amount of polymer on the rolls (which we assume to be a given quantity) $R$ and $H_0$. If we set a given temperature (fixing $\mu$), we are left with one unknown $U$. If $U$ turns out to be unreasonable, we either reduce $F(\tau_m)$, recalling that the polymer passes many times through the nips and, each time it is shuffled by the cutting and rolling process, or we readjust $H_0$ to a new value (which also changes $\lambda$) and change the operating temperature, which changes the viscosity $\mu$.

* According to Eq. 10.5-16 at $\rho = -\sqrt{2+3\lambda^2}$ and $\xi = 0$, a stagnation point occurs. For $\lambda = 0.425$ the stagnation point is at the contact point. For this $\lambda$, $\rho_2 = -1.6$. If $\rho_2 < -1.6$, circulatory flow occurs in the entrance region and Eq. 11.8-5 does not hold.

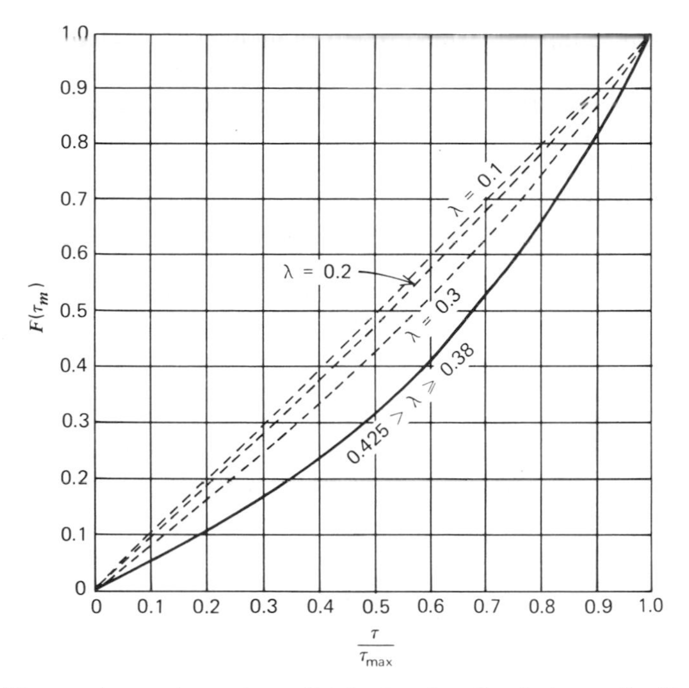


**Fig. 11.23** The maximum shear stress distribution function in two roll mills, $F(\tau_m)$, for a Newtonian fluid. Solid curve for $-\sqrt{2+3\lambda^2}<\rho_2<-\sqrt{1+2\lambda^2}$, broken curves for $\rho_2>-\sqrt{1+2\lambda^2}$ with $\lambda$ as parameter. The $\tau_{max}$ is given in Eq. 11.8-4.

The analysis presented in this section is limited to the validity of the Gaskell model (32) discussed in Section 10.5. Experimental evidence indicates a more complex flow behavior in the inlet region. Moreover, viscoelastic melts may show further deviations from this model (27, 34, 35). For example, they may exhibit difficulty entering the shearing region.

## 11.9 High Intensity Internal Batch Mixers

One of the earliest and most common high intensity internal batch mixers, still widely used in the plastics and rubber industries is the Banbury* type mixer.

The Banbury mixer (Fig. 11.24) consists of a mixing chamber shaped like a figure eight with a spiral lobed rotor in each chamber. The shape of the rotor is such that it induces axial mixing along the rotors toward the center. The mixture is fed to the mixing chamber through a vertical chute in which an air or hydraulic driven ram is located. The lower face of the ram is part of the mixing chamber. The homogenized mixture is discharged through a slide or "drop-door" at the bottom. There is a small clearance between the rotors, which usually operate at different speeds (e.g., 100 and 80 rpm in polyethylene master batching), and between the rotors and the chamber wall. In these clearances dispersive mixing takes place. The shape of the rotors and the motion of the ram during operation ensure that all fluid particles undergo high intensive shearing flow in the gaps (clearances). Both rotors and

* "Banbury" is a registered trade mark of the Farrel Co., Ansonia, Conn.

chamber walls are temperature controlled. Banbury mixers vary in size from a capacity of a couple of pounds up to thousands of pounds. A 60 lb capacity mixer for plastics is equipped with a 600 hp motor. A Banbury polymer compounding line usually consists of storage vessels, automatic scales, the Banbury, a pelletizing extruder, and a product conveying system. The solid or partly melted polymer is dropped into the Banbury and the ram squeezes it against the rotors for a period of time $t_1$. Then the ram is raised and the mixing of the polymer is continued for a period of "stress relaxation" $t_2$. Finally the ram drops once more and mixing continues for another period $t_3$. Hence operating conditions include, in addition to rotor speeds and temperature, the time intervals $t_1$, $t_2$, and $t_3$, as well as ram pressure. The physical properties of the mixture change during the mixing process and intensive mixing is often used to improve certain properties. Meyuhas and Tadmor (36) investigated the effect of Banbury operating conditions on optical and rheological properties of low density polyethylene and on the polymer temperature profile. By standard statistical methods of regression analysis and response surface analysis developed by Box et al. (37), the optimal operating conditions can be

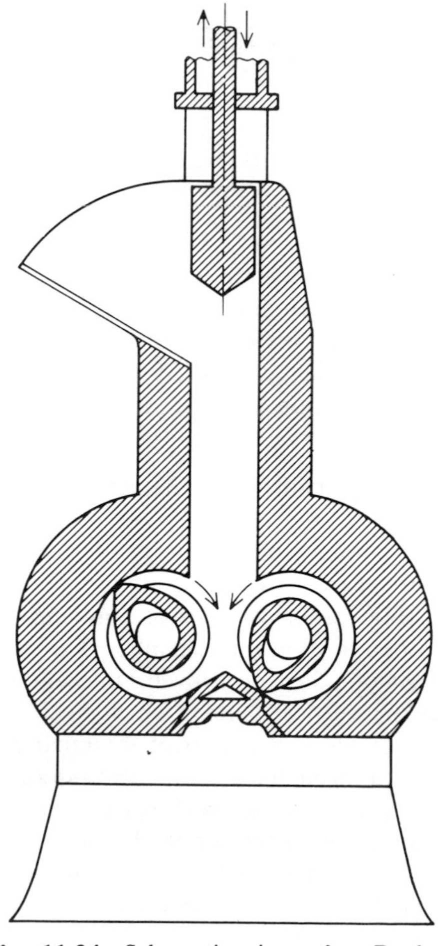

**Fig. 11.24** Schematic view of a Banbury mixer.

determined for a desirable combination of properties. In practice, however, often a simple criterion of constant total mixing time or constant final temperature is used.

In addition to the above-mentioned operating conditions, the feeding schedule of the additives may also be varied. There are a number of commonly practiced procedures: in "dump mixing" all ingredients are added at once; in "upside down mixing" the solid additives are added first, followed by the polymer; finally in "seeding" a small amount of previously well-mixed batch is added to the new batch. Usually hard pigments like carbon black are added as early in the cycle as possible after the polymer has softened. Diluents on the other hand are added as late as possible. Difficulties are experienced sometimes with the mixer when slip on the blades takes place. Another frequent phenomenon with rubbers is "ram bouncing" related to non-Newtonian properties when internal normal stresses exceed ram pressure. These and other interesting phenomena were discussed in detail by White (34).

Complete hydrodynamic analysis of a Banbury mixer is difficult, although useful modeling attempts necessarily leading to large computer programs have been made (38). In this section, following Bolen and Colwell (28) and McKelvey (5), only a highly idealized concentric cylinder geometry (Fig. 11.25*a*) is analyzed. This geometry provides some insight into the nature of dispersive mixing in all common intensive mixers.

**Example 11.7** ***Hydrodynamic Analysis of a Simple Internal Mixer***

In high intensity dispersive mixers the maximum shear stress distribution function is important, whereas in low intensity mixers the strain distribution function is of interest. In both cases power consumption in the clearance is an important factor in design. This example, however, discusses only maximum and mean stress.

Consider a highly simplified idealized internal mixer consisting of two infinitely long concentric cylinders with a short low clearance section (Fig. 11.25*a*). If curvature is negligible ($H/R \ll 1$), we can analyze the flow problem in rectangular coordinates, as in Fig. 11.25*b*. The flow takes place between an infinite upper plate moving at constant velocity over a lower plate, with a step change in the clearance between them. This geometrical configuration is very similar to the "slider pad" extruder (Section 10.4). Assuming the familiar simplifying assumptions of laminar isothermal steady flow, incompressible Newtonian fluid, no slip at the walls, negligible entrance and exit effects at the step, and neglecting gravitational forces, we can write expressions for the flow rate, which is the same in the deep and shallow sections, in terms of local conditions (see Section 10.2)

$$q = \frac{V_0 H}{2} - \frac{H^3(P_1 - P_2)}{12\mu_L L} = \frac{V_0 h}{2} - \frac{h^3(P_2 - P_1)}{12\mu_l l} \tag{11.9-1}$$

where we have allowed for different viscosities in the two regions, to account approximately for the non-Newtonian melt behavior. By

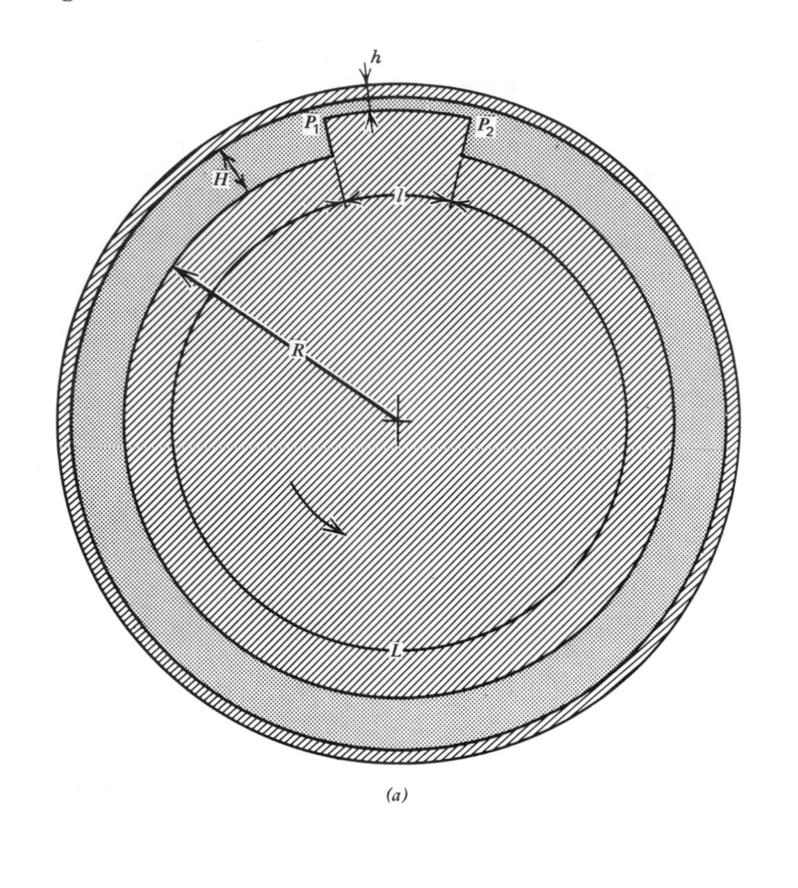


(a)

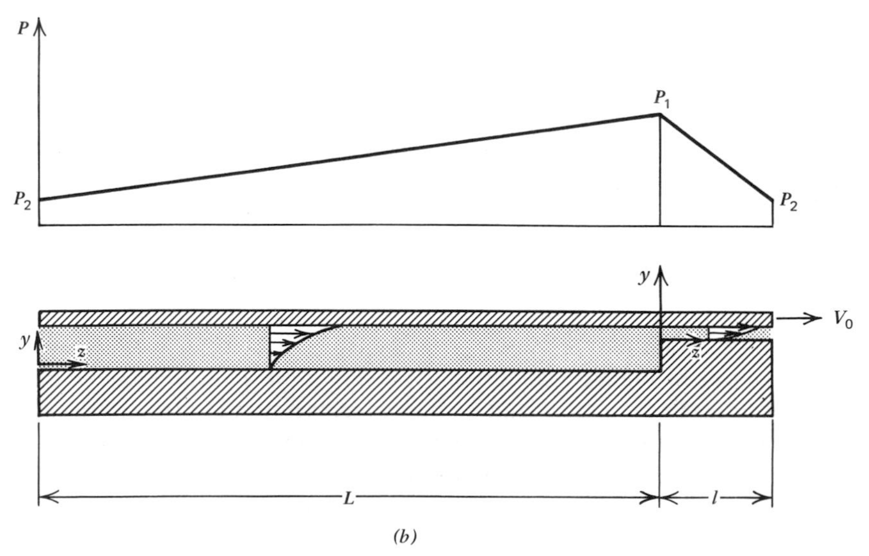


(b)

**Fig. 11.25** (*a*) A simple two concentric cylinder intensive mixer. The inner cylinder rotates at constant frequency of rotation. (*b*) Unwound channel with the coordinate system placed on the inner cylinder and the outer cylinder, which became in the unwinding process a flat plate, moving with velocity $V_0$ in the opposite direction.

rearranging Eq. 11.9-1, we get for the pressure drop over the "step"

$$\frac{P_1 - P_2}{l} = \frac{6\mu_l V_0}{h^2} K \tag{11.9-2}$$

where

$$K = \frac{(H/h) - 1}{(H/h)^3 (l/L)(\mu_l/\mu_L) + 1} \tag{11.9-3}$$

The velocity profile between two parallel plates was given in Eq. 10.2-6. By substituting the pressure drop from Eq. 11.9-2 into Eq. 10.2-6 in terms of the dimensionless variable $\xi = y/h$, we get

$$\frac{v_z}{V_0} = \xi + 3\xi(1-\xi)K \tag{11.9-4}$$

By taking the derivative of Eq. 11.9-4 with respect to $y$, we obtain $\dot{\gamma}_{yz}(\xi)$ in the narrow clearance

$$\dot{\gamma}_{yz}(\xi) = \frac{V_0}{h}[1 + 3K(1 - 2\xi)] \tag{11.9-5}$$

Comparing Eqs. 11.9-5 and 10.2-11, we notice that $K$ is simply the ratio of pressure to drag flow in the narrow gap. Finally, by multiplying Eq. 11.9-5 with the viscosity $\mu_l$, we can get the shear stress distribution. The maximum stress from Eq. 11.9-5 is at $\xi = 0$

$$\tau_{max} = \frac{\mu_l V_0}{h}(1 + 3K) \tag{11.9-6}$$

The mean stress in the gap defined as

$$\bar{\tau} = \mu_l \bar{\dot{\gamma}} = \mu_l \frac{\int_0^h |\dot{\gamma}_{yz}| v_z \, dy}{\int_0^h v_z \, dy} \tag{11.9-7}$$

is obtained with the aid of Eqs. 11.9-5 and 11.9-4 to give

► $$\bar{\tau} = \frac{\mu_l V_0}{h(1+K)} \qquad -\tfrac{1}{3} \leq K \leq \tfrac{1}{3} \tag{11.9-8}$$

and

► $$\bar{\tau} = \frac{V_0 \mu_l}{h}\left(\frac{1+3K}{6K}\right)^2 \frac{1}{1+K}\left[\frac{(1+3K)^2}{2} - 1\right] \qquad K \geq \tfrac{1}{3} \tag{11.9-9}$$

In the deep section of the channel the same equations hold with $\mu_l/h$ replaced by $\mu_L/H$ and $K$ replaced by $K'$, which is defined as follows:

$$K' = \frac{(h/H) - 1}{(h/H)^2 (L/l)(\mu_L/\mu_l) + 1} \tag{11.9-10}$$

The stresses in the gap, both the maximum and the mean, are functions of only two variables $\mu_l V_0/h$, which is the drag flow stress, and $K$, which is a function of $h/H$ and $l/L$. Figure 11.26 plots the ratios of the maximum to drag stresses and mean to drag stresses versus $K$ calculated from Eqs. 11.9-6, 11.9-8, and 11.9-9, respectively. We note that the former increases linearly with $K$, whereas the latter drops moderately until $K = \frac{1}{3}$ (zero shear stress at the moving plate), then starts to increase. Hence large maximum stresses can be obtained by selecting short steps (low $l/L$) and small gaps (low $h/H$). The mean stress, on the other hand, responds in a more complex way to $l/L$, yet it increases with decreasing clearance because the drag flow stress is inversely proportional to $h$. In making these considerations, however, we should account for two other important effects in the clearance region, namely, the effect of shear rate and temperature on viscosity. A higher shear rate in that region will result in a lower apparent viscosity, which is only slightly compensated by an increase in $K$. The great sensitivity of the viscosity on temperature may completely alter the picture. Indeed Bolen and Colwell (28) showed that if a certain reasonable mean temperature rise with velocity $V_0$ is assumed, the mean stress rises quickly to a maximum, then gradually drops off, instead of increasing linearly with $V_0$ as predicted previously.

Finally, in addition to the absolute level of the shear stress, its duration and rate of buildup may be critical factors in mixer performance as discussed earlier (24), because of the viscoelasticity of melts. Also as shown by Strauss and Datta (39), elastic properties of melts may lead to vortices in the region before the moving blade.

## 11.10 Single Screw Extruders

Section 10.3 arrived at single screw geometry in a step-by-step procedure from parallel plate geometry. We recall that one of the last steps in this deductive process was to "twist" the channel placed on the inside of a rotating barrel, so that in one turn it is displaced one channel width, as in Fig. 10.10. We pointed out that this modification provides for a circulatory motion of the polymer in the channel, leading to good laminar mixing and narrow residence time distribution. Having narrow residence time distribution implies that inlet time fluctuations of composition should be *avoided* because the extruder is ineffective in "washing" out these fluctuations (see Section 7.13). Good laminar mixing implies that a large total strain can be imposed on the polymer melt in this device. Thus compositional nonuniformities across the channel can be effectively reduced. But the nature of the velocity fields in the extruder is such that the total strain imparted to fluid particles is a function of position, hence the degree of mixing is nonuniform across the channel. Consequently we should expect a certain composition nonuniformity across the extrudate stream. Quantitative evaluation of this nonuniformity is provided by the strain distribution functions $F(\gamma)$ and $f(\gamma)\,d\gamma$.

We shall analyze these distribution functions for melt extruders with constant channel depth screws utilizing the simple Newtonian isothermal model developed in Sections 10.2 and 10.3. Chapter 12 discusses mixing in plasticating extruders, where the melting process profoundly affects these functions.

The velocity field in the cross channel direction is given in Eq. 10.3-21

$$u_x = \xi(2 - 3\xi) \qquad (11.10\text{-}1)$$

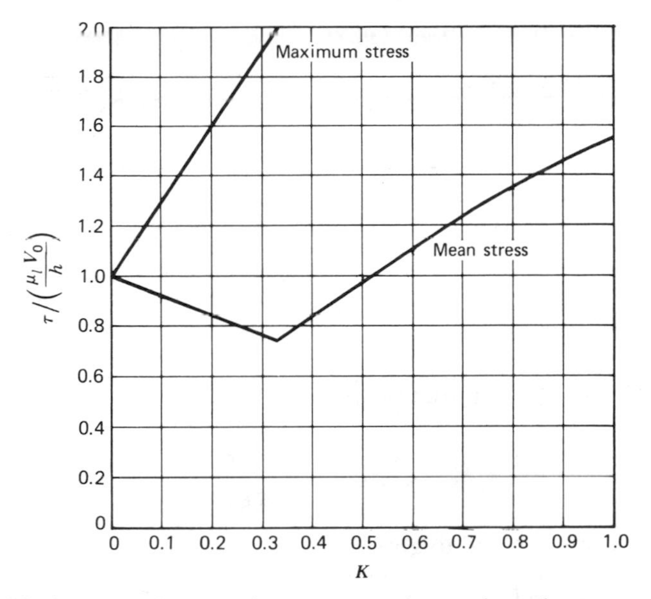


**Fig. 11.26** Maximum and mean shear stress ratios to drag shear stresses in the simple concentric mixer (Fig. 11.25) versus the dimensionless group $K$.

and that in the down channel direction by Eq. 10.3-23, which for shallow channels reduces to Eq. 10.2-6. The latter with $V_0$ replaced by $V_{bz}$ is

$$u_z = \xi - 3\xi(1-\xi)\left(\frac{H^2}{6\mu V_{bz}}\frac{\partial P}{\partial z}\right) \tag{11.10-2}$$

The latter equation, with the aid of Eq. 10.3-28, can be written as

$$u_z = \xi + 3\xi(1-\xi)\frac{Q_p}{Q_d} \tag{11.10-3}$$

Under closed discharge conditions, $Q_p/Q_d = -1$ the cross and down channel velocity profiles become similar differing only in sign. Operating conditions, melt viscosity, and channel depth affect the down channel but not the cross channel velocity profile.

The path of a fluid particle in the extruder channel can be evaluated from Eqs. 11.10-1 and 11.10-3 (see Section 10.3). Consider a fluid particle at location $\xi$ in the upper position of the channel ($\xi > \frac{2}{3}$) shown in Fig. 11.27. Equation 11.10-1 indicates that it will have a certain constant velocity in the negative $x$-direction. As the particle approaches the leading edge of the advancing flight (the pushing flight), it will "turn over" and start moving in the positive $x$-direction at a certain location $\xi_c$. The particle will cross the channel and upon reaching the "trailing" flight, will "return" to its previous location $\xi$. There is a unique relationship between $\xi$ and $\xi_c$ given by the following balance equation and reflecting the circulatory flow pattern just described:

$$\int_0^{\xi_c} u_x \, d\xi = -\int_{\xi}^{1} u_x \, d\xi \tag{11.10-4}$$

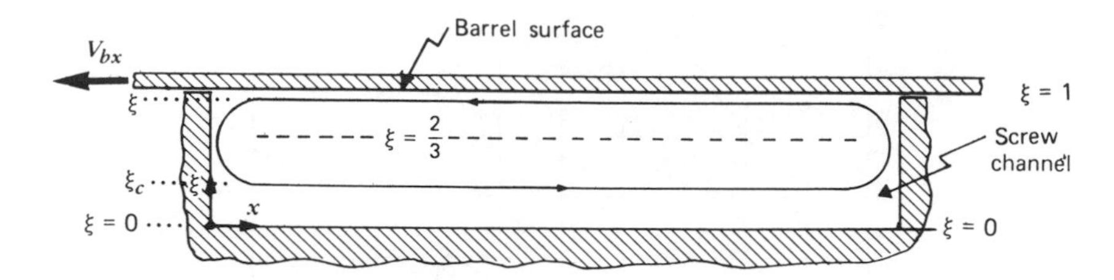


**Fig. 11.27** The path of a fluid particle in the extruder channel. Particle circulates between $\xi$ and $\xi_c$. Fluid particles at $\xi = 2/3$ have no velocity component in the $x$-direction. It is assumed that there is no leakage over the flights.

which upon substituting Eq. 11.10-1 yields

$$\xi_c^2 - \xi_c^3 = \xi^2 - \xi^3 \qquad \begin{array}{c} 0 \le \xi_c \le \frac{2}{3} \\ \frac{2}{3} \le \xi \le 1 \end{array} \tag{11.10-5}$$

Equation 11.10-5 can be solved to give

$$\xi = \tfrac{1}{2}\left(1 - \xi_c + \sqrt{1 + 2\xi_c - 3\xi_c^2}\right) \qquad 0 \le \xi_c \le \tfrac{2}{3} \tag{11.10-6}$$

and

$$\xi_c = \tfrac{1}{2}\left(1 - \xi + \sqrt{1 + 2\xi - 3\xi^2}\right) \qquad \tfrac{2}{3} \le \xi \le 1 \tag{11.10-7}$$

Figure 11.28 plots the relationship between $\xi$ and $\xi_c$. In this analysis we disregarded the complex flow fields in the neighborhood of the flights, which are restricted to regions of the order of one height distance from the flight. A complete two-dimensional solution of this flow problem can be obtained by standard numerical methods (e.g., Refs. 40, 41). These solutions also lead to circulatory-type flow

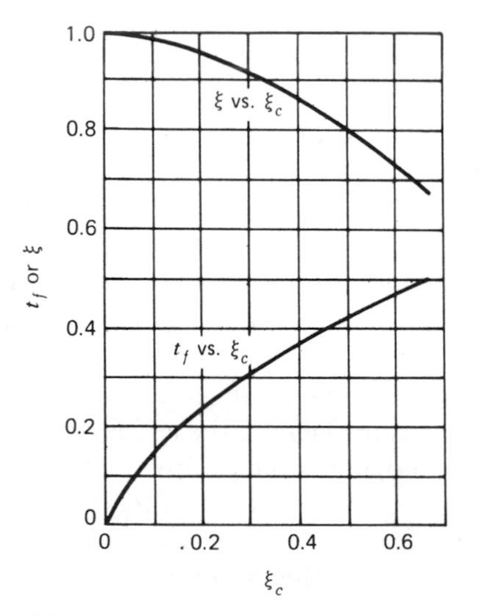


**Fig. 11.28** Fluid particle position in the upper part of the channel $\xi$ versus its position in the lower part of the channel $\xi_c$, and the fraction of time $t_f$ versus position $\xi_c$.

(although in the lower corners small stagnant areas of reverse circulation appear). In shallow channels Eqs. 11.10-1 and 11.10-3 provide good representation of the flow field throughout most of the channel. Thus cross channel pressure rise calculations as discussed in Section 10.3 are valid approximations for shallow channels. However residence time distribution and strain distribution, calculated below, are more sensitive to the exact nature of the flow in the flight neighborhood. Other complications of real systems, such as leakage flow across the flight, non-Newtonian, and nonisothermal effects, further reduce to validity of this simple model in these calculations. These limitations of the model must be kept in mind, and care should be exercised in deriving conclusions and making quantitative calculations.

### *The Residence Time Distribution Function*

From Eq. 11.10-1 the residence time of a fluid particle in the upper part of the channel is $W/(V_{bx}|u_x(\xi)|)$, and in the corresponding lower part of the channel it is $W/(V_{bx}|u_x(\xi_c)|)$, where $W$ is the channel width. The fraction of time fluid particle spends in the upper portion of the channel at $\xi$, $t_f(\xi)$ is therefore given by

$$t_f(\xi) = \frac{1}{1 + \left|\dfrac{u_x(\xi)}{u_x(\xi_c)}\right|} = \frac{1}{1 - \dfrac{\xi(2-3\xi)}{\xi_c(2-3\xi_c)}} \tag{11.10-8}$$

Figure 11.28 shows $t_f(\xi)$ as a function of position. Results indicate that the further away a fluid particle is from $\xi = \frac{2}{3}$, the shorter is the fraction of time it spends in the upper portion of the channel. The fluid particle, while alternating between layers $\xi$ and $\xi_c$, moves also in the down channel direction as dictated by Eq. 11.10-3. The residence time of this fluid particle in an extruder of axial length $l$ is

$$t = \frac{l}{V_b \bar{u}_l(\xi)} \tag{11.10-9}$$

where $\bar{u}_l(\xi)$ is the average velocity of the particle in the axial direction given by

$$\bar{u}_l(\xi) = u_l(\xi)t_f(\xi) + u_l(\xi_c)[1 - t_f(\xi)] \tag{11.10-10}$$

The velocity profile in the axial direction for shallow channels is obtained directly from Eqs. 11.10-1 and 11.10-3, as shown in Section 10.3 (Eq. 10.3-30).

► $$u_l(\xi) = 3\xi(1-\xi)\left(1 + \frac{Q_p}{Q_d}\right)\sin\theta\cos\theta \tag{11.10-11}$$

Substituting Eqs. 11.10-10 and 11.10-11 into Eq. 11.10-9, together with Eqs. 11.10-6 and 11.10-7, we obtain the residence time as a function of $\xi$

► $$t(\xi) = \left[\frac{l}{3V_b(1 + Q_p/Q_d)\sin\theta\cos\theta}\right]\frac{3\xi - 1 + 3\sqrt{1 + 2\xi - 3\xi^2}}{\xi[1 - \xi + \sqrt{1 + 2\xi - 3\xi^2}]} \qquad \frac{2}{3} \le \xi \le 1 \tag{11.10-12}$$

An identical expression is obtained in terms of $\xi_c$ with $\xi$ replaced by $\xi_c$ in Eq. 11.10-12. The minimum residence time $t_0$ at $\xi = \frac{2}{3}$ is

$$t_0 = \frac{3l}{2V_b(1+Q_p/Q_d)\sin\theta\cos\theta} = \frac{3Z}{2V_{bz}(1+Q_p/Q_d)} \tag{11.10-13}$$

where $Z$ is the helical distance and $V_{bz}$ the down channel component of the barrel surface velocity. Figure 11.29 plots the relative $t/t_0$ as a function of $\xi$. It indicates that over a broad region of the channel core, the residence time is close to the minimum; but as we approach the top or bottom of the channel, the residence times increases significantly. The importance of this, however, cannot be appreciated unless we derive the residence time distribution function, which tells what *fraction of the flow rate* stays longer than a given time $t$ in the metering section of the extruder. This we can obtain without difficulty from the velocity fields above (42).

The fraction of flow rate in the upper portion of the channel between $\xi$ and $\xi + d\xi$, corresponding to residence times $t$ and $t + dt$, is

$$\frac{dQ}{Q} = \frac{V_{bz}WH}{Q} u_z(\xi)\, d\xi = \frac{V_{bz}WH}{Q}\left[\xi + 3\xi(1-\xi)\frac{Q_p}{Q_d}\right] d\xi \tag{11.10-14}$$

But in the lower portion of the channel there is a corresponding fraction of flow between $\xi_c$ and $\xi_c - d\xi_c$ which is also characterized by residence times $t$ and $t + dt$:

$$\frac{dQ_c}{Q} = \frac{V_{bz}WH}{Q} u_z(\xi_c)|d\xi_c| = \frac{V_{bz}WH}{Q}\left[\xi_c + 3\xi_c(1-\xi_c)\frac{Q_p}{Q_d}\right] |d\xi_c| \tag{11.10-15}$$

The total fraction of flow rate corresponding to residence times $t$ and $t + dt$ is the sum of Eqs. 11.10-14 and 11.10-15, which is also the definition of the residence time distribution function $f(t)\, dt$.

$$f(t)\, dt = \frac{dQ + dQ_c}{Q} \tag{11.10-16}$$

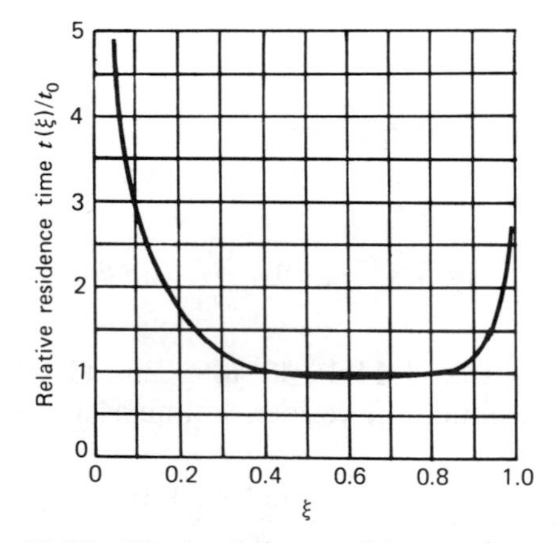


**Fig. 11.29** The relative residence time as a function of location in the channel $\xi$.

By substituting Eqs. 11.10-14 and 11.10-15 into Eq. 11.10-16 and further deriving a relationship between $d\xi$ and $d\xi_c$ from Eq. 11.10-7, we have

$$d\xi_c = \frac{1-3\xi-\sqrt{1+2\xi-3\xi^2}}{2\sqrt{1+2\xi-3\xi^2}}\,d\xi \tag{11.10-17}$$

and noting that $|d\xi_c| = -d\xi_c$, we get

$$f(t)\,dt = \frac{3\xi[1-\xi+\sqrt{1+2\xi-3\xi^2}]}{\sqrt{1+2\xi-3\xi^2}}\,d\xi \tag{11.10-18}$$

Equation 11.10-12 provides a unique relationship between $\xi$ and $t$, thus the RTD function $f(t)\,dt$ can be easily evaluated. We note that the RTD function is dependent on only one dimensionless group $l/[3V_b(1+Q_P/Q_d)\sin\theta\cos\theta]$, which is a simple multiplier and does not alter its shape. The mean residence time $\bar{t}$ can be calculated from Eqs. 11.10-16 and 11.10-12

$$\bar{t} = \int_{t_0}^{\infty} tf(t)\,dt = \frac{4t_0}{3} \tag{11.10-19}$$

where $t_0$ is given in Eq. 11.10-13.

The cumulative RTD function $F(t)$ is obtained by integrating Eq. 11.10-18, recalling that the fraction of flow rate with a residence time less than $t$ is located in the region between $\xi = \frac{2}{3}$ and $\xi$.

$$F(t) = F(\xi) = \tfrac{1}{2}[3\xi^2 - 1 + (\xi-1)\sqrt{1+2\xi-3\xi^2}] \tag{11.10-20}$$

Once again $F(t)$ can be calculated from Eq. 11.10-20 in conjunction with Eq. 11.10-12. Figure 11.30 plots the RTD function $F(t)$ versus reduced time $t/\bar{t}$ and compares it to the RTD function of Newtonian laminar flow in a pipe and that in a well-stirred vessel. The RTD function in melt extruder is quite narrow, approaching plug-type flow. Only about 5% of the flow rate stays more than twice the mean residence time in the extruder.

Wolf and White (43) verified the theoretical RTD function experimentally with radioactive tracer methods. Figure 11.31 gives some of their results, indicating excellent agreement with theory. RTD functions in extruders using non-Newtonian power law model fluids have also been derived (44, 45).

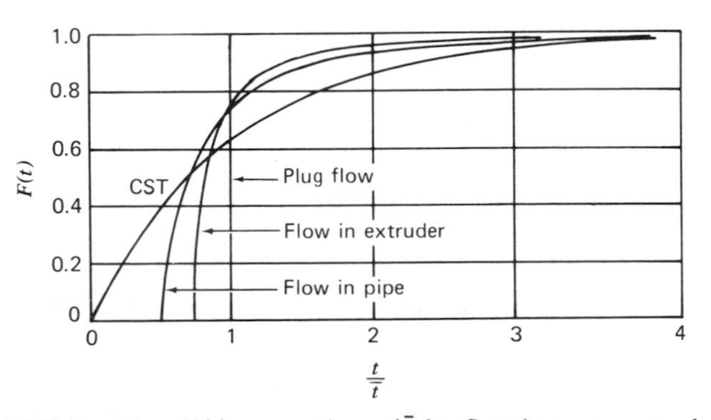


**Fig. 11.30** The RTD function $F(t)$ versus time $t/\bar{t}$ for flow in screw extruder compared to plug flow, isothermal flow of Newtonian fluids in pipes, and a continuously stirred tank vessel (CST).

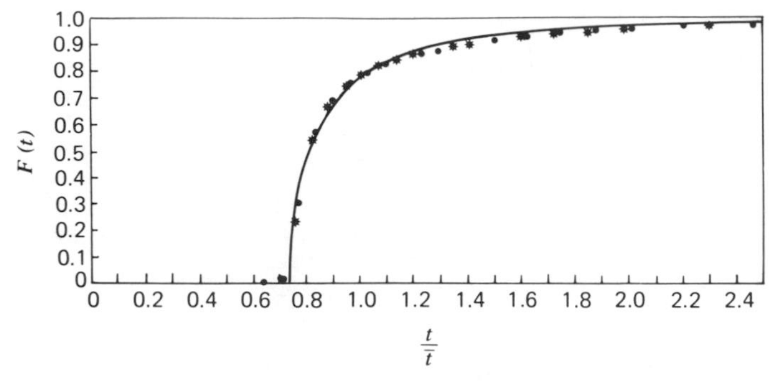


**Fig. 11.31** Experimental verification of the RTD function in extruder by radioactive tracer techniques with a 44.2 mm diameter, 24:1 $L/D$ extruder, liquid polyester resin, and a radioactive manganese dioxide tracer: $*$, experiment 1; ●, experiment 2; smooth curve indicates theoretical prediction. [Reprinted with permission from D. Wolf and D. H. White, "Experimental Study of the Residence Time Distribution in Plasticating Screw Extruders," *Am. Inst. Chem. Eng. J.*, **22**, 122–131 (1976).]

### *Strain Distribution Functions*

The rate of strain components $\dot{\gamma}_{yx}(\xi)$ and $\dot{\gamma}_{yz}(\xi)$ can be calculated from the velocity profiles in Eqs. 11.10-1 and 11.10-3

$$\dot{\gamma}_{yx}(\xi) = \frac{V_{bx}}{H}\frac{du_x}{d\xi} = \frac{2V_b \sin\theta}{H}(1-3\xi) \qquad (11.10\text{-}21)$$

and

$$\dot{\gamma}_{yz}(\xi) = \frac{V_{bz}}{H}\frac{du_z}{d\xi} = \frac{V_b \cos\theta}{H}\left[1+3(1-2\xi)\frac{Q_p}{Q_d}\right] \qquad (11.10\text{-}22)$$

According to our model, these are the only components of the rate of strain tensor. Thus we can write an expression for the magnitude of the rate of strain tensor (cf. Eqs. 5.1-29 and 6.5-1)

$$\dot{\gamma}(\xi) = (\dot{\gamma}_{yx}^2 + \dot{\gamma}_{yz}^2)^{1/2} = \frac{V_b}{H} R\left(\xi, \theta, \frac{Q_p}{Q_d}\right) \qquad (11.10\text{-}23)$$

where

$$R\left(\xi, \theta, \frac{Q_p}{Q_d}\right) = \left\{4(1-3\xi)^2 \sin^2\theta + \left[1+3(1-2\xi)\frac{Q_p}{Q_d}\right]\cos^2\theta\right\}^{1/2} \qquad (11.10\text{-}24)$$

A corresponding expression for $\dot{\gamma}(\xi_c)$ is obtained by replacing $\xi$ with $\xi_c$ in Eqs. 11.10-23 and 11.10-24. The direction of shear of a fluid particle differs at $\xi$ and at $\xi_c$. This poses difficulties in evaluating the total strain experienced by a fluid particle circulating at positions $\xi$ and $\xi_c$, because depending on the specific position of $\xi$, and the exact nature of the turnover at the flight, partial demixing may occur. Exact solution of the problem would require following the fluid particle in the true three-dimensional flow patterns, with the prerequisite of relating the increase of interfacial area to the invariants of the strain tensor. As a first approximation, however, we assume that total strains acquired in the upper and lower parts of the channel are additive. Thus the total strain acquired by the fluid particle after time $t$ is

$$\gamma(\xi)=\dot{\gamma}(\xi)t_f(\xi)t(\xi)+\dot{\gamma}(\xi_c)[1-t_f(\xi)]t(\xi) \qquad (11.10\text{-}25)$$

Substituting Eqs. 11.10-12 and 11.10-23 into Eq. 11.10-25 gives

$$\blacktriangleright \quad \gamma(\xi)=\frac{l}{3H}\left(\frac{1}{1+Q_p/Q_d}\right)\left[E_1\left(\xi,\theta,\frac{Q_p}{Q_d}\right)+E_2\left(\xi,\theta,\frac{Q_p}{Q_d}\right)\right] \qquad (11.10\text{-}26)$$

where

$$E_1\left(\xi,\theta,\frac{Q_p}{Q_d}\right)=\frac{2}{\cos\theta}\,\frac{t_f(\xi)\left[(1-3\xi)^2+\dfrac{\cot an^2\theta}{4}\left(1+3\dfrac{Q_p}{Q_d}-6\xi\dfrac{Q_p}{Q_d}\right)^2\right]^{1/2}}{\xi_c(1-\xi_c)+[t_f(\xi)](\xi-\xi_c)(1-\xi-\xi_c)} \qquad (11.10\text{-}27)$$

and

$$E_2\left(\xi,\theta,\frac{Q_p}{Q_d}\right)=\frac{1}{\sin\theta}\,\frac{[1-t_f(\xi)]\left[4(1-3\xi_c)^2\tan^2\theta+\left(1+3\dfrac{Q_p}{Q_d}-6\xi_c\dfrac{Q_p}{Q_d}\right)^2\right]^{1/2}}{\xi_c(1-\xi_c)+[t_f(\xi)](\xi-\xi_c)(1-\xi-\xi_c)} \qquad (11.10\text{-}28)$$

The time fraction $t_f(\xi)$ as a function of $\xi$ and $\xi_c$ is given in Eq. 11.10-8, and the relationship between $\xi$ and $\xi_c$ appears in Eq. 11.10-6. Thus the total strain as a function of position $\xi$ can be evaluated subject to the assumption above, without difficulties. Figure 11.32 presents the distribution of total strain as a function of $\xi$ for various $Q_p/Q_d$ values. It is interesting to note that for pure drag flow the minimum strain is obtained (as expected) at $\xi=\frac{2}{3}$, but when back pressure is applied ($Q_p/Q_d<0$), the minimum strain is obtained elsewhere. Yet, like residence time, the total strain is rather uniform over a significant portion of the channel.

The strain distribution function can be calculated by the same procedure used for the residence time distribution function. For pure drag flow, where the minimum strain is obtained at $\xi=\frac{2}{3}$, it is given by Eq. 11.10-20, with $\xi$ uniquely related to $\gamma$ in Eq. 11.10-26. For the more general case, however, the fraction of flow rate between $\xi=\frac{2}{3}$ and $\xi$ is *not* the fraction of flow rate experiencing a total strain of $\gamma$ or less, as is evident from Fig. 11.32. First the location of minimum $\gamma$ must be established, then Eqs. 11.10-14 and 11.10-15 must be integrated over the appropriate limits. An alternative approach, which is also applicable to a plasticating extruder, in which the channel is broken up into small height increments and fluid particles are followed with time, was described by Lidor and Tadmor (12). Figure 11.33 gives results of such computations for a 6 in. diameter plasticating extruder at 500 lb/hr. At high screw speeds melting is rapid, and the strain distribution function is essentially that of a melt extruder. Increasing the screw speed at constant flow rate implies an increase in back pressure, and we note a consequent shift of the SDF to higher strain ranges. Once again we observe that SDF in screw

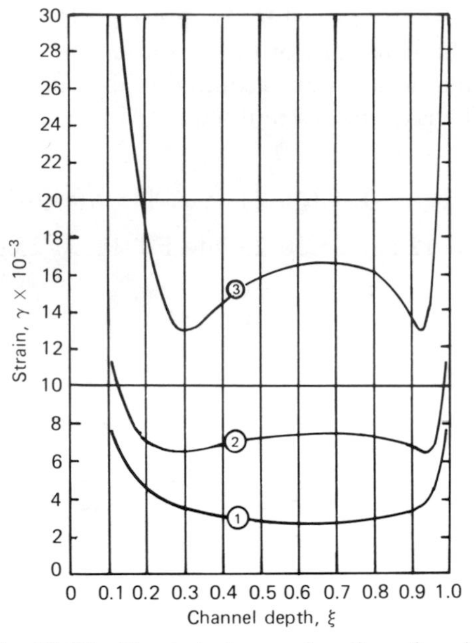


**Fig. 11.32** Total strain as a function of position $\xi$ in a screw channel. Calculations are based on a total axial length of 100 in., channel depth of 0.2 in., and 20° helix angle. Curve 1, $Q_p/Q_d = 0$; curve 2, $Q_p/Q_d = -0.5$; Curve 3, $Q_p/Q_d = -0.75$.

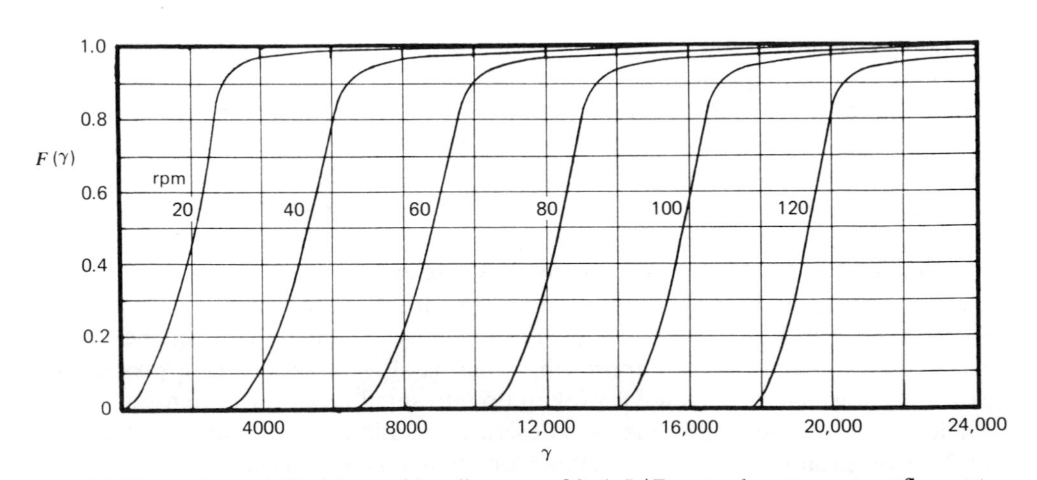


**Fig. 11.33** Values of $F(\gamma)$ for a 6 in. diameter, 20:1 $L/D$ extruder at constant flow rate (500 lb/hr) with screw speed as a parameter. Simulation was made for a square pitched screw with a constant channel depth of 0.6 in. [Reprinted with permission from G. Lidor and Z. Tadmor, "Theoretical Analysis of Residence Time Distribution Functions and Strain Distribution Functions in Plasticating Extruders," *Polym. Eng. Sci.*, **16**, 450–462 (1976).]

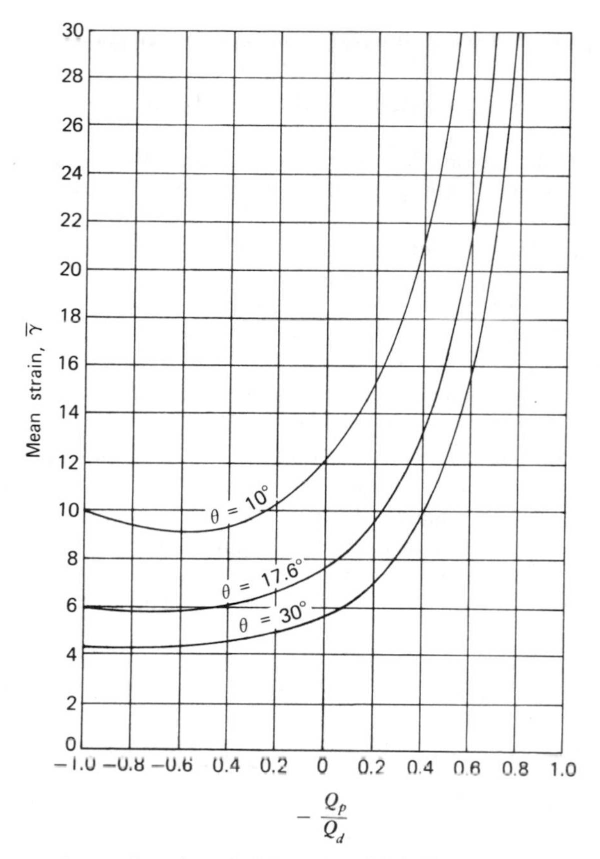


**Fig. 11.34** Mean strain as a function $Q_p/Q_d$ ratio with helix angle as a parameter for $l/H = 1$. When $Q_p/Q_d = -1$, $\gamma$ goes to infinity; this condition corresponds to a closed discharge operation.

extruders is quite narrow. Hence a good indication of the mixing performance can be obtained by calculating the mean strain $\bar{\gamma}$(46)*. The mean strain is proportional to $l/H$ and a function of $Q_p/Q_d$ and the helix angle $\theta$. Figure 11.34 is a generalized plot of $\bar{\gamma}$ as a function of $Q_p/Q_d$ with the helix angle as a parameter. That the mean strain is proportional to $l/H$ is borne out by engineering practice, as we pointed out earlier in this chapter when dealing with SDF in parallel plate flow. Similarly, increasing the mean strain with increasing back pressure is also well supported by experience, and the same holds for the limited effect of helix angle within the practical range of $\theta$ values.

## *REFERENCES*

1. F. H. H. Valentin, "Mixing of Powders and Particulate Solids," *Chem. Process Eng.* 181–187 (April 1965).
2. C. W. Clump, "Mixing of Powders," in *Mixing—Theory and Practice*, Vol. II, V. H. Uhl and J. B. Gray, eds., Academic Press, New York, 1967, Chapter 10.

* The mean strain in Ref. 46 is called "weighted" average total strain.

3. C. L. Tucker and N. P. Suh, "Electrostatic Powder Mixing," *Polym. Eng. Sci.*, **16**, 657–663 (1967). C. A. Rotz and N. P. Suh, "New Techniques for Mixing Viscous Reacting Liquids. Part II. Electrical-Mechanical Hybrid Mixer," *ibid.*, **16**, 672–678 (1976).
4. H. F. Irving and R. L. Saxton, "Mixing of High Viscosity Materials," in *Mixing—Theory and Practice*, Vol. II, V. W. Uhl and J. B. Gray, eds., Academic Press, New York, 1967, Chapter 8.
5. J. M. McKelvey, *Polymer Processing*, Wiley, New York, 1962, Chapter 9.
6. J. T. Bergen, "Mixing and Dispersing Processes," in *Processing of Thermoplastic Materials*, E. C. Bernhardt, ed., Reinhold, New York, 1960, Chapter 7.
7. N. R. Schott, B. Weinstein, and D. LaBombard, "Motionless Mixers in Plastics Processing," *Chem. Eng. Prog.*, **71**, 54–58 (1975).
8. L. Erwin, "New Fundamental Considerations on Mixing in Laminar Flow," 36th Annual Technical Conference, Soc. of Plastics Eng., Washington, D.C., 1978, p. 488.
9. J. T. Bergen, G. W. Carrier, and J. A. Krumhansl, "Criteria for Mixing and the Mixing Process," Paper 97, presented at the 14th National Technical Conference, Society of Plastics Engineers, Detroit, January 1958, p. 987.
10. C. J. Shearer, "Mixing of Highly Viscous Liquids: Flow Geometries for Streamline Subdivision and Redistribution," *Chem. Eng. Sci.*, **28**, 1091–1098 (1973).
11. J. M. Dealy, "Extensional Flow of Non-Newtonian Fluids—A Review," *Polym. Eng. Sci.*, **11**, 433–445 (1971).
12. G. Lidor and Z. Tadmor, "Theoretical Analysis of Residence Time Distribution Functions and Strain Distribution Functions in Plasticating Extruders," *Polym. Eng. Sci.*, **16**, 450–462 (1976).
13. D. M. Bigg and S. Middleman, "Laminar Mixing of a Pair of Fluids in a Rectangular Cavity," *Ind. Eng. Chem. Fundam.*, **13**, 184 (1974).
14. F. H. Harlow and J. E. Welch, "Numerical Calculation of Time Dependent Viscous Incompressible Flow of Fluid with Free Surface," *Phys. Fluids*, **8**, 2182–2189 (1965).
15. J. H. Hildebrand and R. L. Scott, *The Solubility of Non-Electrolytes*, 3rd ed., Reinhold, New York, 1950, reprinted by Dover, New York, 1964.
16. H. Burrell and B. Immergut, "Solubility Parameter Values," IV-341 in *Polymer Handbook*, J. Brandrup and E. H. Immergut, eds., Wiley-Interscience, New York, 1966.
17. P. J. Flory, "Thermodynamics of High Polymer Solutions," *J. Chem. Phys.*, **10**, 51 (1942).
18. M. L. Huggins, "Theory of Solutions of High Polymers," *J. Am. Chem. Soc.*, **64**, 1712 (1942).
19. D. J. Meier, in *Block Copolymers*, J. Moacanin, G. Holden, and N. W. Holden, eds., Interscience, N.Y., 1969.
20. G. I. Taylor, "The Viscosity of a Fluid Containing Small Drops of Another Fluid," *Proc. Roy. Soc.*, **A138**, 41 (1932).
21. G. I. Taylor, "The Formation of Emulsions in Definable Fields of Flow," *Proc. Roy. Soc.*, **A146**, 501–523 (1934).
22. F. D. Rumscheidt and S. G. Mason, "Particle Motion in Sheared Suspensions. XII. Deformation and Burst of Fluid Drops in Shear and Hyperbolic Flow," *J. Colloid Sci.*, **16**, 238–261 (1961).
23. H. J. Karam and J. C. Bellinger, "Deformation and Breakup of Liquid Droplets in a Simple Shear Field," *Ind. Eng. Chem. Fundam.*, **7**, 576–581 (1968).
24. R. W. Flumerfeld, "Drop Breakup in Simple Shear Fields of Viscoelastic Fluids," *Ind. Eng. Chem. Fundam.*, **11**, 312–318 (1972).
25. B. L. Lee and J. L. White, "Interface Deformation and its Mechanisms in Two-Phase Stratified Flow," *Trans. Soc. Rheol.*, **18**, 467 (1974).

26. G. N. Avgeropoulos, F. C. Weissert, P. H. Biddison, and G. G. A. Bahm, "Heterogeneous Blends of Polymers Rheology and Morphology," *Rubber Chem. Technol.*, **49**, 93–109 (1976).
27. J. L. White and N. Tokita, "Instability and Failure Phenomena in Polymer Processing," *J. Appl. Polym. Sci.*, **12**, 1589–1600 (1968).
28. W. R. Bolen and R. E. Colwell, *Soc. Plast. Eng. J.*, **14** (8), 24–28 (1958).
29. R. B. Bird, H. R. Warner, Jr., and D. C. Evans, "Kinetic Theory and Rheology of Dumbbell Suspension with Brownian Motion," *Fortsch. Hochpolymerenforch.*, Springer-Verlag, Berlin, **8**, 1–90 (1971).
30. Z. Tadmor, "Forces in Dispersive Mixing," *Ind. Eng. Chem. Fundam.*, **15**, 346–348 (1976).
31. W. K. Maddern, "Centrifugal Impact Mixing," in *The Encyclopedia of Plastics Equipment*, H. R. Simonds, ed., Reinhold, New York, 1969.
32. R. E. Gaskell, "The Calendering of Plastic Materials," *J. Appl. Mech.*, **17**, 334–336 (1950).
33. J. T. Bergen and G. W. Scott, "Pressure Distribution in the Calendering of Elastic Materials," *J. Appl. Mech.*, **18**, 101–106 (1951).
34. J. L. White, "Elastomer Rheology and Processing," *Rubber Chem. Technol.*, 257–338 (1969).
35. N. Tokita and J. L. White, "Milling Behavior of Gum Elastomers," *J. Appl. Polym. Sci.*, **10**, 1011–1026 (1966).
36. G. Meyuhas and Z. Tadmor, "The Effect of Operating Conditions on Product Properties in Banbury Mixers," Handasah and Adrichalut (Hebrew), 1970; also, G. Meyuhas, M.S. thesis, Department of Chemical Engineering, Technion—Israel Institute of Technology, Haifa, 1970.
37. G. E. P. Box and J. S. Hunter, "Multifactor Experimental Design for Exploring Response Surfaces," *Ann. Math. Statist.*, **28**, 195 (1957).
38. P. Hold and C. Y. Cheng, "Analytical Investigation on the Performance of the Banbury Mixer," Farrell Co. Report 71RP-445 (1972).
39. K. Strauss and K. B. Datta, "Slow Flow of Viscoelastic Fluids Through a Contraction," *Rheolog. Acta*, **15**, 403–410 (1976).
40. O. R. Burggaf, "Analytical and Numerical Studies of the Structure of Steady Separated Flows," *J. Fluid Mech.*, **24**, 113–151 (1966).
41. D. Greenspan, P. C. Jain, R. Manolar, B. Noble, and A. Sakurai, "Numerical Studies of the Navier–Stokes Equation," M.R.C. TR., No. 482, University of Wisconsin, May 1969.
42. G. Pinto and Z. Tadmor, "Mixing and Residence Time Distribution in Melt Extruders," *Polym. Eng. Sci.*, **10**, 279–288 (1970).
43. D. Wolf and D. H. White, "Experimental Study of the Residence Time Distribution in Plasticating Screw Extruders," *Am. Inst. Chem. Eng. J.*, **22**, 122–131 (1976).
44. M. Hirshberger, "Two-Dimensional Non-Newtonian Flow in Rectangular Channels," M.S. Thesis, Department of Chemical Engineering, Technion—Israel Institute of Technology, Haifa, 1972.
45. D. Bigg and S. Middleman, "Mixing in Screw Extruders: A Model for Residence Time Distribution and Strain," *Ind. Eng. Chem. Fundam.*, **13**, 66 (1974).
46. Z. Tadmor and I. Klein, *Engineering Principles of Plasticating Extrusion*, Van Nostrand Reinhold, New York, 1970.

## *PROBLEMS*

**11.1** ***Processing Equipment as Mixers.*** Good extensive mixing is obtained by increasing the interfacial area and the distribution of interfacial elements throughout the system whereas dispersive or intensive mixing requires high shear stresses. In light of these mixing requirements, discuss the following processing equipment: the roll mill,. the Banbury mixer, the single screw extruder, and the intermeshing counterrotating twin screw extruder.

**11.2** ***The Two Cylinder Mixer.*** A mixture of two viscous liquid components with identical rheological properties ($n = 0.5$) is placed in a Two Cylinder Mixer with $\beta = 1.2$, as shown in Fig. 11.6b. Two experiments are carried out: (a) The outer cylinder rotates clockwise at 10 rpm for 1 min.; (b) The outer cylinder rotates alternately clockwise and counter-clockwise at 20 rpm, 15 s in each direction for a total of 2 min. Which mode of motion leads to better mixing?

**11.3** ***Strain Distribution Function in Parallel Plate Combined Pressure and Drag Flow.*** (a) Derive the SDF $F(\gamma)$ and the mean strain $\bar{\gamma}$ in a combined drag and pressure flow between parallel plates. Assume isothermal laminar flow and a Newtonian liquid.

(b) Prove that for $q_p/q_d = \frac{1}{3} F(\gamma)$ reduces to the SDF of pressure flow between parallel plates.

***Answer*:** See Table 11.1.

**11.4** ***The Strain Distribution Function of a Power Law Fluid in Pressure Flow Between Parallel Plates.*** Consider two infinitely wide parallel plates of length $L$ gap $H$. Polymer melt is continuously pumped in the $x$ direction. Assuming isothermal steady, fully developed flow:

(a) Show that $F(\xi)$ is given by

$$F(\xi) = \frac{2+s}{1+s}\left(1 - \frac{1}{2+s}\xi^{1+s}\right)\xi \qquad 0 \le \xi \le 1$$

where $\xi = 2y/H$ (the coordinate system is placed at the center of the gap).

(b) Show that the total strain is related to $\xi$ via

$$\gamma(\xi) = 2(1+s)\frac{L}{H}\frac{\xi^s}{1-\xi^{1+s}} \qquad 0 \le \xi \le 1$$

(c) Show that the mean strain $\bar{\gamma}$ is

$$\bar{\gamma} = \int_0^1 \gamma(\xi) f(\xi)\, d\xi = 2\left(\frac{2+s}{1+s}\right)\frac{L}{H}$$

(d) Show that the mean strain can also be calculated by using the following relationship

$$\bar{\gamma} = \bar{t}\int_0^1 \dot{\gamma}(\xi)\, d\xi$$

(c) Prove that for the Newtonian case the SDF converges to the expression in Table 11.1. (Note the different definitions of $\xi$ here and in Table 11.1.)

**11.5** ***Cohesive Forces in Dispersive Mixing.*** Estimate the cohesive forces holding together carbon black particles on the basis of the following information. Breakup of the agglomerates was obtained in an idealized internal mixer, as shown in Fig. 11.25*a* at $V_0 = 25$ cm/s. The carrier is LDPE at 150°C. The gap clearance is $h = 0.1$ cm, $H = 1.0$ cm; the clearance length is $l = 0.5$ cm and $L = 10.0$ cm. The particle size is 5 $\mu$ in diameter. Assume a constant viscosity 0.05 $\text{lb}_f\cdot\text{s/in}^2$.

**11.6** ***Roll-Mill for Dispersive Mixing.*** A laboratory roll-mill with 5 in. diameter rolls and 0.05 in. minimum clearance between the rolls is used for dispersive mixing of carbon black agglomerates in LDPE. Calculate the roll speed needed to break up 5% of the particles per pass assuming that the critical shear stress needed for breakup is that obtained in Problem 11.5 in the narrow clearance and that the amount of polymer on the rolls is 50% above the minimum. Assume the same constant viscosity as in Problem 11.5.

**11.7** ***Deformation of a Sphere in Various Types of Flows.*** A spherical liquid particle of radius 0.5 in. is placed in a liquid medium of identical physical properties. Plot the shape of the particle (a) after 1 s and 2 s in simple shear flow with $\dot{\gamma} = 2\ \text{s}^{-1}$; (b) after 1 s and 2 s in steady elongational flow with $\dot{\varepsilon}_{pl} = 1\ \text{s}^{-1}$; and (c) after 1 s in simple shear flow with $\dot{\gamma} = 2\ \text{s}^{-1}$ in the perpendicular direction; (d) in each case the ratio of the surface area of the deformed particle to the initial one can be calculated. What does this ratio represent?

**11.8** ***The Rate of Mixing in Intensive Mixers.*** An intensive internal mixer can be viewed as consisting of two regions—a high shear region at the tip of the blade, and the rest of the mixer volume which can be assumed to be well mixed. This is schematically shown in Fig. P-11.8.

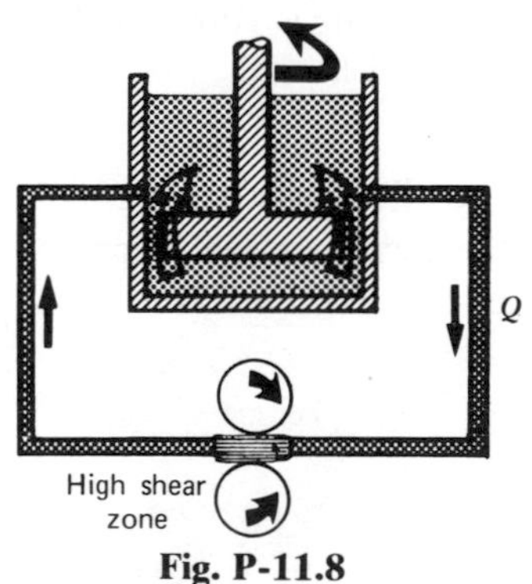


**Fig. P-11.8**

(a) Using the idealized internal mixer developed in Section 11.9 show that the rate of flow over the blade is

$$Q = \frac{hV_0W}{2}(1+K)$$

where $W$ is a characteristic width.

(b) Show that the volume fraction which at time $t$ has passed through the high shear region is

$$G = 1 - e^{-t/\bar{t}}$$

where $\bar{t}$ is the mean residence time of the polymer between successive passes

$$\bar{t} = V/Q$$

(c) What is the mixing time needed to assure that 90% and 99% of the polymer in the mixer have passed through the high shear zone?

(d) What is the effect of reducing the clearance to one half its value on the shear stress level and the mixing time?

**11.9** ***Mixing in Single Screw Extruders.*** A 24:1 $L/D$, 2.5 in. diameter, square-pitched screw with constant channel depth of 0.4 in., flight width of 0.25 in., and negligible clearance is used to extrude a "Newtonian" polymer melt of viscosity, 0.1 $lb_f \cdot s/in^2$. and density of 45 $lb/ft^3$ at a rate of 180 lb/hr. The screw speed is 100 rpm. Calculate the mean strain imposed on the polymer.

**11.10** ***Design of a "Modifying" LDPE Extruder.*** Hanson* investigated the effect of prolonged shear on the rheological and physical properties of LDPE. He observed certain (reversible) changes in these properties possibly related to a process of chain disentanglement. The rheological changes included, for example, an increase in draw-down which permits manufacturing of thinner films at higher rates. The rheological changes are often accompanied by improvements in optical properties as well. It was further observed that these improvements are a function of total strain. The strain levels needed are of the order of 10,000 shear units. Finally, it was found that if the LDPE undergoes such a treatment under the reactor and the melt is pelletized the improvements are retained in the downstream processing by film blowing extruders. Design a constant channel depth melt extruder to pelletize 25,000 lb/hr LDPE. The head pressure needed for pelletizing is 1000 psi. Assume a constant viscosity of 0.05 $lb_f \cdot s/m^2$ and a density of 48 $lb/ft^3$.

***Note*:** If the shear rate imposed on the polymer is too high, viscous dissipation may exceed the heat removal capacity with a consequent detrimental temperature rise of the polymer.

* D. E. Hanson, "Shear Modification of Polythene," *Polym. Eng. Sci.*, **9**, 405–414 (1969).

# 12

# Modeling of Processing Machines with Elementary Steps



By dissecting polymer processing into elementary steps, we were able to focus attention on the underlying principles and basic mechanisms of each individual step. Having understood these principles and having derived the governing equations, we are now in a position to analyze and model logically and systematically real machines. We recall from Section 5.2 that modeling a complex process is done by breaking it down into subsystems. The elementary step concept is most useful in defining these subsystems. The machine selected for this analysis, the single screw extruder, was chosen not only because it enjoys a dominant role in the processing industry, and therefore ought to be studied by all processing engineers, but also because of its inherent capability to perform all the elementary steps. Indeed, this inherent versatility of the screw extruder is the main reason for its great practical importance and dominant role.

The elementary step approach, however, is useful not only in studying and gaining a better insight into existing machines, but also in inventing new ones. An example of this was given in Chapter 10 in the logical evolution of the screw extruder from the parallel plate model. This chapter continues this process and points to the development of a new machine, the "Diskpack Polymer Processor."

## 12.1 The Single Screw Melt Extrusion Process

Melt fed screw extruders are widely used in the polymer processing industry—for example, in postreactor compounding and pelletizing of LDPE and PS. In these extruders the elementary steps of pressurization and mixing take place simultaneously and interact with die forming shaping methods. The discussion of the melt extrusion process in this section also introduces the plasticating extrusion process covered in Section 12.2.

## *Modeling Considerations of the Melt Extrusion Process*

In the general approach to the modeling of the melt extruder (Fig. 12.1), after the identification of the subsystems and the development of the subsystem modeling in terms of equations based on transport phenomena, certain conditions are specified *linking* the subsystems into the integral process. For the melt extrusion process, the

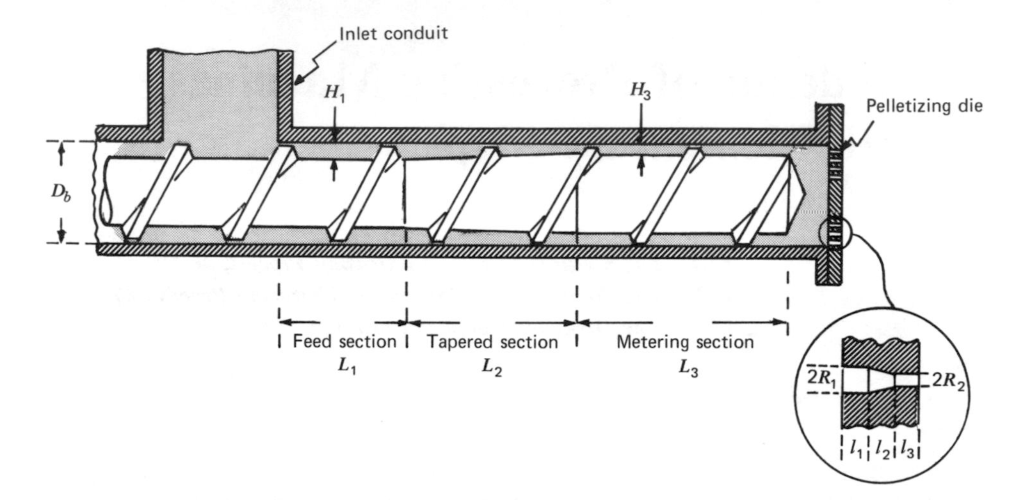


**Fig. 12.1** Schematic view of a single screw melt extruder. The extruder is equipped with a pelletizing die plate and a metering-type screw (i.e., a screw with three geometrical sections feed, tapered and metering).

subsystems are (a) gravitational melt flow in the inlet conduit, (b) drag induced pressurization and laminar mixing in the screw channel, and (c) pressure flow in the extruder die. Some of these must be further broken down into additional subsystems. Thus in the screw channel there are parallel and tapered sections, and the "die" consists of a combination of conduits: a breaker plate and screen pack, the die entrance section, and the die itself. The subsystems in this particular case are in *series*; therefore for a steady state operation with atmospheric inlet and outlet conditions, we can state that the mass flow rate in each subsystem $G_i$ is constant

$$G_i = G_0 \tag{12.1-1a}$$

and the sum of the pressure changes over the entire process is zero

$$\sum_i \Delta P_i = 0 \tag{12.1-1b}$$

Note that the foregoing condition also implies that pressure *rise* in the extruder equals the pressure *drop* over the die. But mass flow rate and pressure changes are not the only operational variables of interest. Linked to the pressurization step are also temperature changes and power input through the screw shaft. Finally, we are interested in the mixing step characterized by the SDF and RTD functions or, in simpler terms, the mean strain and residence time.

The assembled models of the subsystems provide relationships between the main operational variables of interest (i.e., flow rate, pressure profile, temperature profile, power requirements, mean strain, and residence time) and all the relevant geometrical variables (i.e., design variables), rheological and thermophysical properties of the melt, and process variables (i.e., frequency of screw rotation, barrel screw, and die temperature settings). These relationships can, therefore, be used for both new machine design or the analysis (e.g., optimization) of an existing one. In addition to the main operational variables, other variables may be desirable to study, such as cross die variations in temperature and flow rate, extrudate uniformity, swelling and shape stability, and parameter sensitivity of the process. Chapter 13, on die forming, deals with some of these.

### *The Isothermal Newtonian Model for a Constant Channel Depth Screw*

We now proceed with the main task of subsystem modeling. The inlet flow to the extruder is simple gravitational flow through (generally) tubular conduits. In such slow flows the shear rate range is very low and the isothermal Newtonian assumption is valid. For a circular entrance channel the flow rate is given by the Hagen–Poiseuille law

$$Q = \frac{\pi(\mathcal{P}_0 - \mathcal{P}_L)R^4}{8\mu L_0} \tag{12.1-2}$$

where $\mathcal{P} = P + \rho g z$, $z$ being the upward distance in the inlet conduit of height $L_0$.

Drag induced pressurization and laminar mixing in shallow screw channels were discussed in Sections 10.3 and 11.10, respectively. Equation 10.3-32, derived for isothermal flow of an incompressible Newtonian liquid, can be written as

► $$Q_s = \tfrac{1}{2}\pi N D_b \cos\theta_b W(H - \delta_f)F_d - \frac{WH^3}{12\mu}\frac{\Delta P_s}{L}\sin\bar{\theta}(1 + f_L)F_p \tag{12.1-3}$$

where $f_L$ is given in Eq. 10.3-33, the geometry in Fig. 10.12; $Q_s$ is the screw volumetric flow rate, $L$ is the axial length of the screw, and $\Delta P_s$ is the pressure *rise* over that length. It should be noted that this equation, when applied to a screw of finite length, contains in addition to all the simplifying assumptions listed in Section 10.3, the further geometrical assumption of disregarding the oblique ends of the unwound channel (Fig. 12.2), as well as the inlet and outlet effects (1).

Equation 12.1-3 can be well represented by plotting the flow rate $Q_s$ versus the pressure rise $\Delta P_s$. Such plots, called *screw characteristics*, appear in Fig. 12.3. The intersection with the ordinate gives the drag flow value and that with the

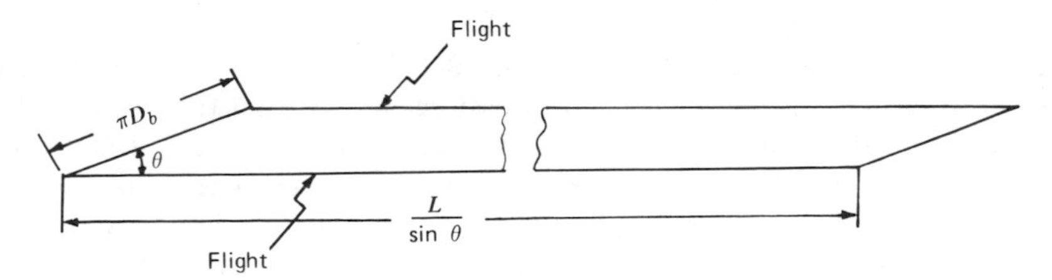


**Fig. 12.2** Top view of an "unwound" screw channel indicating the oblique ends of the channel. The geometrical contribution of this end effect is of the order of: $[\pi \sin\theta \cos\theta/(L/D)]$.

abscissa the maximum pressure at closed discharge. For isothermal flow of a Newtonian fluid in the absence of leakage flow, the screw characteristics are straight lines with a negative slope of $-(WH^3/12\mu)(\sin\bar{\theta}/L)F_p$. Figure 12.3 illustrates the effect of two important variables on screw characteristics: frequency of screw rotation, or "screw speed," as it is commonly referred to, and channel depth. Changing the former shifts the screw characteristic, whereas the latter affects both its level and slope.

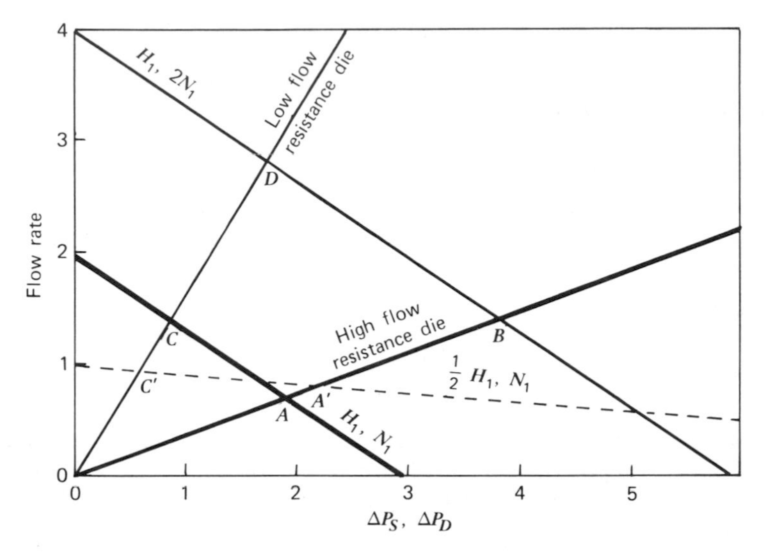


**Fig. 12.3** Schematic view of screw and die characteristics lines for Newtonian fluids and isothermal flow. The points where screw and die characteristics lines cross are the operating points. The effect of screw speed and channel depth on the location of the operating point is demonstrated.

A general equation for the isothermal pressure flow of an incompressible Newtonian fluid in a die, can be written as

$$Q_D = \frac{K}{\mu}\Delta P_D \tag{12.1-4}$$

where $K$ is the *die constant* or "flow conductance" term, determined by the die geometry. For relatively simple dies it can be calculated, but for the more complex ones it must be determined experimentally. The subscript $D$ refers to the die, and $\Delta P_D$ is the die pressure *drop*.

We now use the subsystem linking Eqs. 12.1-1a and 12.1-1b. Recalling that the flow is isothermal and the fluid incompressible, we see that the former implies that $Q_s = Q_D$, and the latter implies that $\Delta P_s = \Delta P_D$. Thus Eq. 12.1-4 can *also* be plotted in Fig. 12.3, resulting in the *die characteristics* lines. The figure gives two such plots, one for a high resistance (low $K$) and another for a low flow resistance (high $K$). The points where the screw characteristics intersect the die characteristics are *operating points* (i.e., they specify the flow rate and die pressure drop of a given extruder and die, operating at a particular $N$, pumping a "Newtonian melt" of a particular viscosity). Analytically, the operating point is obtained by solving simultaneously Eqs. 12.1-3 and 12.1-4, where $\Delta P = \Delta P_s = \Delta P_D$

$$Q = \frac{\frac{1}{2}\pi N D_b \cos\theta_b W(H-\delta_f)F_d}{1+[WH^3 \sin\bar{\theta}(1+f_L)F_p]/12\mu LK} \tag{12.1-5}$$

$$\Delta P = \frac{\frac{1}{2}\mu\pi N D_b \cos\theta_b W(H-\delta_f)F_d}{K+[WH^3 \sin\bar{\theta}(1+f_L)F_p]/12L} \tag{12.1-6}$$

For the simple case under consideration, Eqs. 12.1-5 and 12.1-6 form the main part of the melt extrusion model. Additional insight into the screw-die interaction can be obtained by turning to Fig. 12.3. Point $A$ denotes the operating point of a screw characteristic curve of channel depth $H_1$ and screw speed $N_1$, and a die characteristic of constant $K$. Doubling the screw speed shifts the operating point, along the same die characteristic line, to point $B$. As a result, flow rate and head pressure (which for atmospheric inlet and outlet conditions equals $\Delta P_s$ or $\Delta P_D$) double in value. This result is a consequence of the Newtonian and isothermal assumptions. In the case of a non-Newtonian fluid and nonisothermal flow, flow rate and head pressure increase with increasing screw speed, of course, but not in the same proportion. Considering the low flow resistance die characteristic line, we find exactly the same behavior as with the high flow resistance die. Equipping the extruder with a valve is tantamount to having a "die" with continuously variable $K$. Hence the die characteristics become an infinite family of lines, and the operating point can be shifted at will to any point on the screw characteristic curve.

It is interesting to note in Fig. 12.3 the effect of channel depth on the operating point. With the high flow resistance die, a reduction in channel depth brings about an increase in output (the operating point shifts from $A$ to $A'$), whereas with the low flow resistance die it results in a reduction in output (the operating point shifts from $C$ to $C'$). We recall from Chapter 10 that there is an optimum channel depth for maximum output or pressure, and in this example the two cases happen to fall on opposite "sides" of the optimum.

This general behavior of the extruder-die combination was based on a simple model, but qualitatively it represents the behavior of any melt extruder-die combination.

Mechanical work is introduced into the extruder through the screw shaft. This work is partly used for pressurizing the melt and partly dissipated into heat according to Eq. 11.3-24. The work needed for inducing viscous mixing originates from the latter part of the total work input. In our simplified model, the total rate of work input (i.e., power) into the fluid is given by the following expression:

$$P_w = \int_S (-[\mathbf{n}\cdot\boldsymbol{\tau}]\cdot\boldsymbol{v})\, dS \tag{12.1-7}$$

where $S$ is the surface area of the barrel, $\mathbf{n}$ is the outward directed normal unit vector, $\boldsymbol{\tau}$ is the stress tensor at the barrel surface, and $\boldsymbol{v}$ is the barrel velocity. Equation 12.1-7 for the simple model discussed above reduces to

$$P_w = -[\tau_{yz}(H)V_{bz} + \tau_{yx}(H)V_{bx}]ZW \tag{12.1-8}$$

where $Z$ is the total "unwound" channel length. The shear stress components at the moving surface are obtained from Eq. 10.2-16$a$. The $\tau_{yz}(H)$ term is obtained from Eq. 10.2-16$a$ by replacing $V_0$ with $V_{bz}$; whereas the $\tau_{yx}(H)$ term is found by replacing $V_0$ with $V_{bx}$ and setting $q = 0$, since there is no net flow in the cross channel direction. Substituting these equations into Eq. 12.1-8, noting that $q/q_d = Q/Q_d$, results in

$$P_w = \mu \frac{\pi^2 N^2 D_b^2 WL}{\sin\bar{\theta} H}\left(4 - 3\cos^2\theta_b \frac{Q}{Q_d}\right) \tag{12.1-9}$$

The total power is maximum at closed discharge condition $Q = 0$ and minimum (nonzero) at open discharge condition $Q = Q_d$. In addition to the power consumption by the channel flow, we must account for power consumption in the flight clearance, which is by no means negligible (2a).

The total rate of viscous dissipation can be obtained from Eq. 11.3-24 by subtracting from $P_w$ the power used for pressurizing the melt $Q\Delta P$, which results in

$$E_v = \mu \frac{\pi^2 N^2 D_b^2 WL}{\sin\bar{\theta} H}\left[4 - 6\cos^2\theta_b\left(\frac{Q}{Q_d}\right) + 3\cos^2\theta_b\left(\frac{Q}{Q_d}\right)^2\right] \tag{12.1-10}$$

This expression can also be obtained by integrating the term $\frac{1}{2}\mu(\dot{\boldsymbol{\gamma}} : \dot{\boldsymbol{\gamma}})$ (see Eq. 5.1-38) over the volume of the channel.

The validity of the isothermal assumption can be tested* by calculating the adiabatic temperature rise from Eq. 11.3-25

$$\Delta T = \frac{1}{C_p} \frac{E_v}{\rho Q} \tag{12.1-11}$$

Finally, evaluation of the mixing performance is obtained directly from the results derived in Section 11.10.

So far, we have assembled from the various elementary steps a "complete" model for the process under consideration. We can extend the model without too much difficulty to any screw design consisting of combinations of constant channel depth and moderate tapers by using Eq. 12.1-3 separately for each section and calculating the pressure rise over the screw using Eq. 12.1-1b. For tapered sections we can use the correction factors given in Eq. 10.4-9. Thus the drag flow and pressure flow terms, respectively, are multiplied by the expressions $2/(1+\zeta_0)$ and $2/\zeta_0(1+\zeta_0)$ where $\zeta_0 = H_0/H_1$ and $H_0$ is the channel depth at the inlet of the section and $H_1$ at the outlet; $H$ in Eq. 12.1-3 is then replaced by $H_0$.

### Non-Newtonian and Nonisothermal Models

Both non-Newtonian properties and nonisothermal flow conditions significantly affect the melt extrusion process. We analyzed certain aspects of these effects (independently of each other) in dealing with parallel plate flow in Section 10.2. Figure 10.4, which plots flow rate versus the pressure gradient with the power law exponent as a parameter, illustrates the effect of the shear dependent viscosity on flow rate using a power law model fluid. These curves are equivalent to screw characteristics curves with the cross channel flow neglected. The Newtonian straight

* Note that if $\Delta T$ is significant the isothermal assumption is rejected, but if $\Delta T$ is small, *local* temperature effects may still be important.

lines are replaced with S-shaped curves. The cross channel flow, induced by the cross channel component of the barrel surface velocity, however, affects further the screw characteristic curve as a result of the *coupling* of the down channel and cross channel velocity profiles. The coupling is, of course, absent in Newtonian fluid flow, and the down channel velocity profile (which determines the flow rate) is unaffected by the cross channel circulatory flow. The effect of this coupling in non-Newtonian fluid flow becomes apparent when considering "pure" drag flow (i.e., $\partial P/\partial z = 0$). Under such a condition the down channel velocity profile with Newtonian fluids (in shallow channels) is linear. This is no longer the case with non-Newtonian fluids, as the following example demonstrates.

### Example 16.1 *Coupling of Down and Cross Channel Flows for Non-Newtonian Fluids*

Derive the differential equation governing isothermal "open discharge" flow of an incompressible power law model fluid in shallow screw channels.

*Solution*: Making the common simplifying assumptions, the equation of motion reduces to

$$\frac{\partial \tau_{yz}}{\partial y} = 0 \qquad (12.1\text{-}12)$$

and

$$\frac{\partial \tau_{yx}}{\partial y} = -\frac{\partial P}{\partial x} \qquad (12.1\text{-}13)$$

Expressing the stress components in terms of velocity gradients, we obtain the following two *coupled* differential equations:

$$\frac{dv_z}{dy} = \frac{C_1}{m}\left[\left(\frac{dv_x}{dy}\right)^2 + \left(\frac{dv_z}{dy}\right)^2\right]^{(1-n)/2} \qquad (12.1\text{-}14)$$

and

$$\frac{dv_x}{dy} = \frac{1}{m}\left[\left(\frac{dv_x}{dy}\right)^2 + \left(\frac{dv_z}{dy}\right)^2\right]^{(1-n)/2}\left(\frac{dP}{dx}\right)(y - y_3) \qquad (12.1\text{-}15)$$

where $C_1$ and $y_3$ are integration constants to be evaluated from the boundary conditions $v_x(0) = v_z(0) = 0$, $v_x(H) = -V_b \sin\theta$, $v_z(H) = V_b \cos\theta$. The constant $C_1$ is the shear stress component in the down channel direction, and $y_3$ is the height where the cross channel velocity profile exhibits an extremum point. Clearly, the down channel velocity profile $v_z(y)$, in spite of the constant shear stress component value or absence of down channel pressure gradient, is no longer linear. In physical terms, the cross channel shear rate distribution affects the non-Newtonian viscosity, which varies with $y$. Hence in the down channel direction the liquid responds as a rheologically nonhomogeneous liquid. Consequently, the deviation from linearity becomes a function of the helix angle; as the helix angle approaches a value of zero, the parallel plate model is regained.

For solving such a combined flow problem, one must turn to numerical methods. This has been done by Griffith (3) and later by Pearson et al. (4, 5).

Figure 12.4 shows some of Griffith's results of isothermal power law model fluids in shallow channels, exhibiting the typical S-shaped curves, with increasing

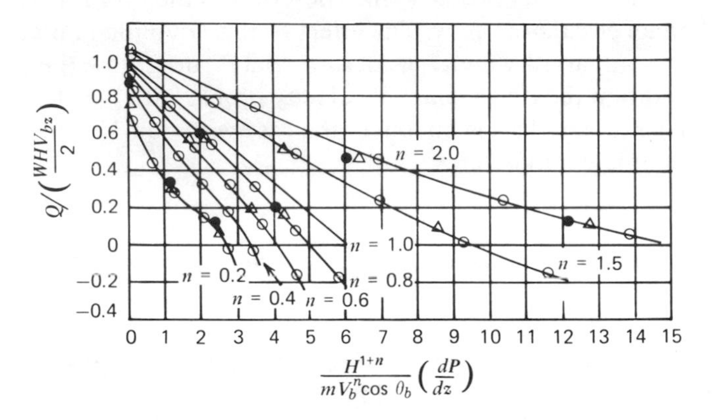


**Fig. 12.4** Computed curves of dimensionless flow rate versus dimensionless pressure gradient for isothermal flow of a power law model fluid in shallow screw channels with the power law exponent $n$ as a parameter: helix angles $\theta_f$ as follows: ○, 30°; △, 20°; ●, 10°; solid curves are for a helix angle 30°. Note that for $n<1$ the reduced flow rate is less than 1, with the deviation diminishing with decreasing the helix angle. [Reprinted with permission from R. M. Griffith, "Fully Developed Flow in Screw Extruders," *Ind. Eng. Chem. Fundam.*, **1,** 180–187 (1962).]

deviation from Newtonian behavior ($n<1$), as well as a downward shift of the curves as a result of the cross channel–down channel coupling effect. This downward shift diminishes with diminishing helix angle. Griffith, who tested these results experimentally with a 1% solution of a carboxyvinyl polymer in water extruded in a 2 in. extruder, reported good agreement with theoretical predictions.

The discussion above was restricted to shallow rectangular channels. Attempting to solve such a flow in curved deep channels greatly increases the difficulty of the numerical solution. An appreciation of these effects, however, can be obtained by considering separately the pressure and drag flow configurations. We found that with Newtonian fluid flow, the flights restrict both the drag and pressure flows. The same is true for a non-Newtonian (e.g., power law model) fluid, but the amount of the reduction is a function of both the shape $H/W$ and the power law exponent $n$. Moreover, generalized curves (i.e., shape factors) can be derived only for the separate pure drag flow and pure pressure flow in the absence of cross channel flow (6). An appreciation of curvature effects on drag flow can be obtained in a similar way by comparing tangential drag flow between concentric annuli to drag flow between parallel plates (2b). The ratio of these flow rates provides for a curvature correction factor that can be plotted as a function of $R_o/R_i$ with $n$ as a parameter (Fig. 12.5). For pure pressure flow, it can be shown (2c) that if the length of the channel is taken along the mean diameter $D_b-H$, curvature effects can be neglected.

These individual correction factors, in addition to providing some quantitative insight into the magnitude of the effects considered, can be made useful for

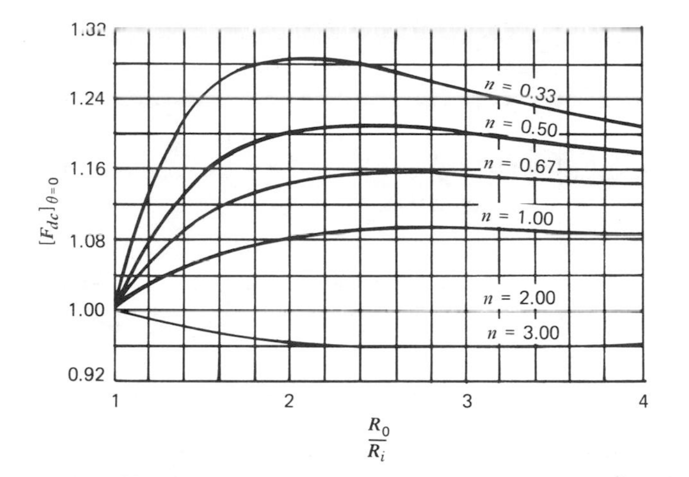


**Fig. 12.5** Curvature correction factor for drag flow for a non-Newtonian fluid $[F_{\mathrm{dc}}]_{\theta=0}$ denotes the ratio of drag flow between parallel plates to drag (Couette) flow between concentric cylinders at equal gaps and moving surface velocities. The subscript $\theta = 0$ indicates that in screw extrusion this correction factor is rigorously valid only in the limit of zero helix angle. (Reprinted with permission from Z. Tadmor and I. Klein, *Engineering Principles of Plasticating Extrusion*, Van Nostrand Reinhold Book Co., New York, 1970.)

practical calculations by an appropriate modeling scheme. Such a procedure was developed by Tadmor et al. (2d, 7).

Temperature effects in extrusion are not negligible if either the barrel temperature or the screw temperature is significantly different from the melt temperature, or if viscous dissipation brings about a substantial temperature rise in the melt. Rigorous solution of nonisothermal flow requires the simultaneous solution of the energy and momentum equations. Moreover, to account for the temperature effects due to the circulatory flow, the complete flow field, including $v_y(x, z)$, has to be solved. This becomes a difficult three-dimensional non-Newtonian and nonisothermal flow problem.

Therefore, simplifying assumptions were introduced into numerical solution methods. The first to treat the nonisothermal flow problem were Colwell and Nicholls (8), who completely neglected cross channel flow and assumed fully developed temperature and velocity profiles in parallel plate geometry. For a power law model fluid then, analysis parallels the nonisothermal effects considered in Section 10.2, with pure drag flow replaced by combined pressure and drag flow. Figure 12.6 summarizes some of their results, plotting the "screw" characteristic (curve $A$). We note the typical S-shaped curve, but at nonisothermal conditions, the mean extrudate temperature varies slightly along the characteristic curve. There is interaction among heat transfer through the constant temperature walls, viscous dissipation, and the velocity profile, coupled through the temperature dependent viscosity. Colwell and Nichols (8) also calculated similar curves under adiabatic conditions, shown in Fig. 12.6 for high and low inlet temperatures. It is worth noting that the adiabatic screw characteristic curves exhibit maximum attainable pressures and a region of doubled values of flow rate for a given pressure rise (cf. Section 10.2).

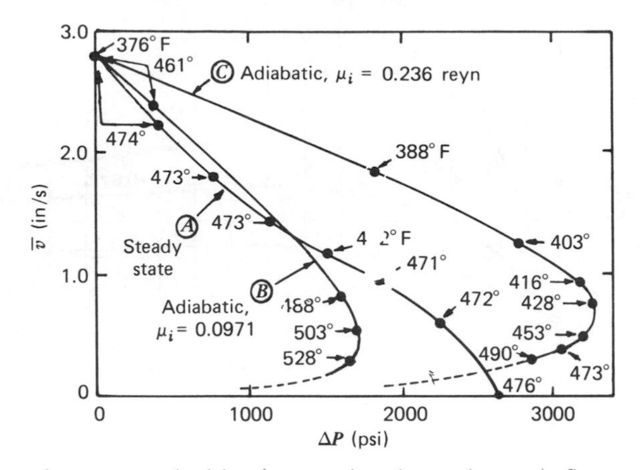


**Fig. 12.6** Computed average velocities (proportional to volumetric flow rate) as a function of pressure drop. Length, 50 in.; $V_{bz}$, 5.66 in./s. Curve *A*, steady state; curve *B*, adiabatic high inlet temperature; curve *C*, adiabatic low inlet temperature. Note the double valued flow rates at a given pressure rise in the adiabatic operation and the maximum pressure rise at finite flow rate values. [Reprinted with permission from R. E. Colwell and K. R. Nicholls, "The Screw Extruder," *Ind. Eng. Chem.*, **51**, 841–843 (1959).]

Numerical solutions for nonisothermal flow including cross channel flow were carried out by Griffith (3), who assumed constant temperature along a streamline, Zamodits and Pearson (4), who assumed fully developed flow in both cross and down channel directions, and by Pearson et al. (5), who analyzed the various boundary conditions. The main difficulty in treating rigorously the nonisothermal problem arises from the cross channel circulatory nature of the flow. In real cases this difficulty is amplified by the existence of leakage flow and curvature.

The previously mentioned modeling approach of superimposing and "correcting" drag and pressure flow terms (2d), however, permits us once again to account in a nonrigorous way for certain nonisothermal effects. From a practical point of view we are primarily interested in the effect of the nonisothermal conditioning on flow rate and on the average extrudate temperature. In many practical situations the screw is "neutral"; that is, it is neither heated nor cooled. In such cases it has been found (2e) that the screw temperature is very close to the melt temperature. Hence the main nonisothermal effect on flow rate stems from a significant difference between the barrel and melt temperatures. This may strongly affect the rate of drag flow as implied by the expression in Eq. 10.2-46. The mean extrudate temperature rise, on the other hand, is the result of a gradual temperature change in the direction of flow. By applying the "lubrication approximation," dividing the screw into small finite elements, carrying out the detailed calculation within each element, assuming a uniform mean temperature, and then performing a heat balance that accounts for both the heat transfer across the walls (barrel) and viscous dissipation, the mean temperature change across the element can be obtained. This calculation procedure also permits the evaluation of the power law parameters at local conditions within each element from a general flow curve ($\eta(\dot{\gamma}, T)$) as well as the handling of screw geometries with variable channel depth. Thus the model that evolves can be characterized as a *lumped parameter model*, where within each element the various subsystems are either lumped or distributed parameter models.

Furthermore, the non-Newtonian and nonisothermal nature of the flow in extruders also must be considered when dealing with the die. Section 13.1 deals with this problem.

Melt extrusion modeling incorporating the above-named effects leads inevitably to large computer programs, which have been described in the literature (7).

The foregoing approach becomes particularly useful when dealing with the plasticating extrusion process, where in addition to pressurization and mixing, we are faced with the elementary steps of solids handling and melting. Both these steps can be handled well by the approach to screw extrusion modeling outlined here.

## 12.2 The Single Screw Plasticating Extrusion Process

Most single screw extruders used in the plastics industry are "plasticating extruders"; that is, they are fed by polymer in the particulate solids form. The solids flow gravitationally in the hopper and into the screw channel, where they are conveyed and compressed by a drag induced mechanism, then melted or "plasticated" by a drag induced melt removal mechanism. Side by side with the melting step, pressurization and mixing take place. Hence we have the plasticating extrusion process (Fig. 12.7) consisting of all four elementary steps: handling of

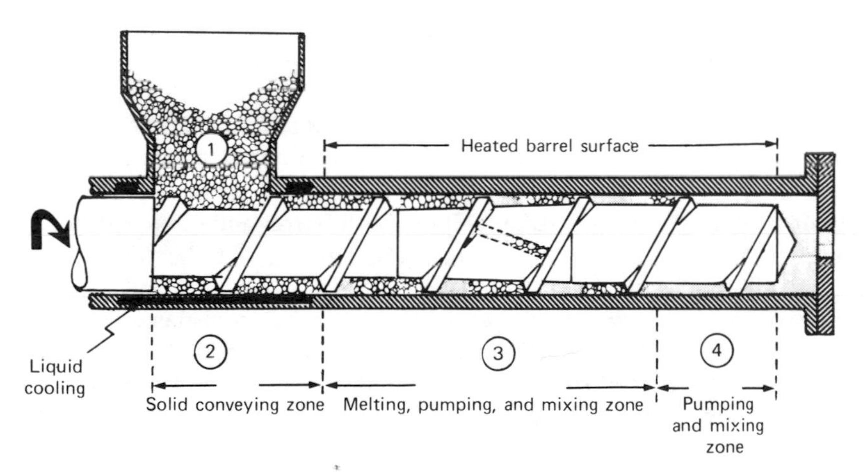


**Fig. 12.7** Schematic representation of a plasticating screw extruder. The barrel is cooled in the hopper region and heated downstream.

particulate solids in Regions 1, 2, and 3; melting, pumping, and mixing in Region 3, and pumping and mixing in Region 4. Devolatilization may also occur in Regions 3 and 4 by appropriate screw design and operating conditions.

### *The Polymer "Experience" in the Extruder*

The plasticating extrusion process is a relatively complex one, and unlike the melt extrusion process, the detailed physical mechanisms—in particular, the melting

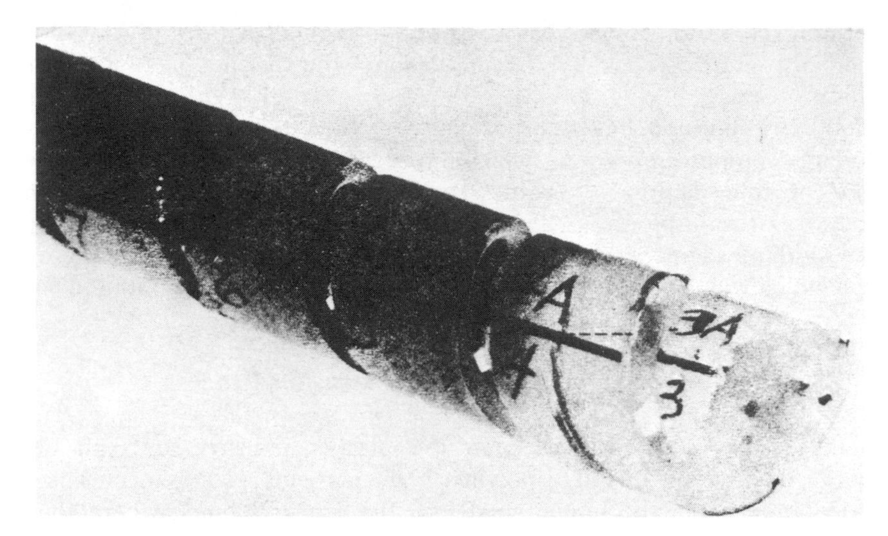


**Fig. 12.8** Helical ribbon of LDPE, after it was taken off the screw following a cooling experiment. The numbers indicate turns from the hopper and cross sections for examination obtained by slicing it perpendicular to the flights as shown by the broken line. (Reprinted with permission from Z. Tadmor and I. Klein, *Engineering Principles of Plasticating Extrusion*, Van Nostrand Reinhold Book Co., New York, 1970.)

mechanism—cannot be easily visualized, predicted, and modeled from basic principles without some experimental investigation. Indeed, qualitative understanding of the melting process was obtained only after Maddock (9) and Street (10) developed a simple and ingenious experimental technique that permitted a visual analysis of the process. This experimental technique consists of abruptly stopping an extruder operating at steady state, chilling both barrel and screw (consequently solidifying the polymer in the screw channel), pushing out the screw from the barrel, unwinding the polymer from the screw (Fig. 12.8), and slicing thin representative sections perpendicular to the flights. To better visualize the details of the process, a small fraction (3–5%) of colored polymer is added as a tracer. This helps to distinguish between solid filled and molten regions. It also provides some information on flow patterns. Figures 12.9–12.14 give some results of such "cooling" experiments obtained by Tadmor et al. (11). The slices sectioned every half-turn from the hopper to the die are shown for each experiment. Next to the slices the turn number (starting downstream from the hopper) is given. Figures 12.9–12.13 were obtained from a 2.5 in. diameter extruder with a metering-type screw having a feed section of 12.5 turns with a channel 0.370 in. deep, a compression section of 10 turns, and a 4 turn metering section, 0.127 in. deep. Results in Fig. 12.14 were obtained from an 8 in. extruder with a metering-type screw. The polymers used and the operating conditions accompany each figure. Figure 12.8 illustrates the relative position of a typical "slice" or "cross section." Thus the "pushing" flight is on the left and the "trailing" flight on the right; the barrel surface on the top and root of the screw on the bottom.

Analyzing the experimental results, we note that throughout most of the extruder, the solid and liquid phases coexist, but rather clearly segregated from each other, with the melt phase accumulating and the pushing flight in a *melt pool*

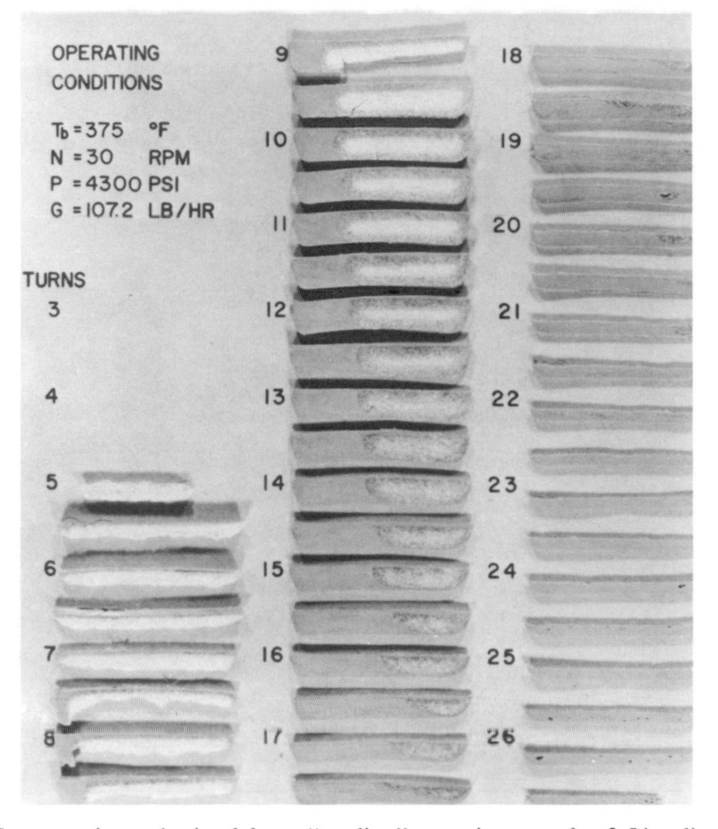


**Fig. 12.9** Cross sections obtained from "cooling" experiments of a 2.5 in. diameter 26.5 length-to-diameter ratio screw extruder. Material: rigid PVC. Operating conditions are listed in the figure: $T_b$ is the barrel temperature, $N$ the screw speed, $P$ the pressure at the die, and $G$ the mass flow rate. Numbers denote turns from the beginning (hopper side) of the screw. The screw was of a metering type with 12.5 turn feed section, 0.37 in. deep, a 9.5 turn transition section, and a 4.5 turn metering section 0.127 in. deep. (Reprinted with permission from Z. Tadmor and I. Klein, *Engineering Principles of Plasticating Extrusion*, Van Nostrand Reinhold Book Co., New York, 1970.)

and the solids segregated at the trailing flight as a *solid bed*. The width of the melt pool gradually increases in the down channel direction, whereas that of the solid bed generally decreases. The solid bed, shaped as a continuous long, helical ribbon of varying width and height, slowly turns in the channel (much like a nut on a screw) sliding toward the exit, while gradually melting. Upstream from the point, where melting starts, the whole channel cross section is occupied by the solid-bed, which is composed, as the hopper is approached, of less compacted solids. The continuity of the solid bed provides an explanation for the capability of the screw extruder to generate melt that is free of air bubbles: the porous continuous solid bed provides uninterrupted air-filled passages from deep in the extruder all the way back to the hopper. Thus particulate solids forming the solid bed move down channel while the air is stationary.

Although the melting behavior in extruders just described appears to be quite general for both amorphous and crystalline polymers, small and large extruders,

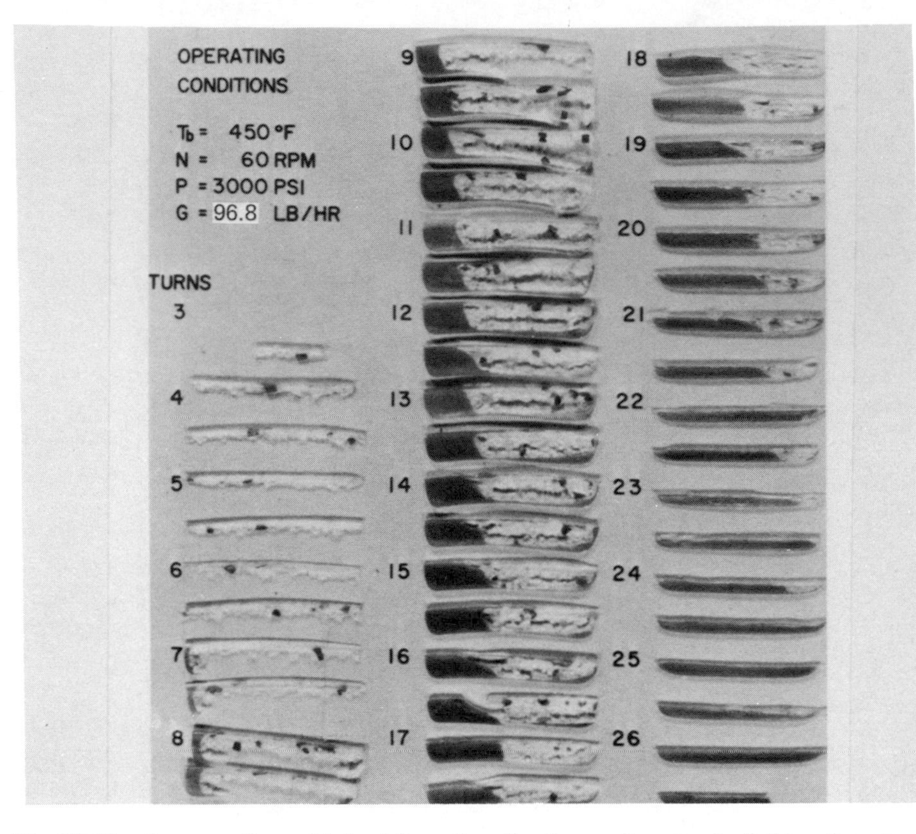


**Fig. 12.10** Cross sections obtained from "cooling" experiments of a 2.5 in. diameter, 26.5 length-to-diameter ratio screw extruder. Material: PP. Operating conditions are listed in the figure. Symbols and screw descriptions as in Fig. 12.9.

and diverse operating conditions, it appears that with certain PVC compounds the melt pool accumulates at the trailing flight (12). Moreover, with large extruders there was no segregated melt pool at the channel side but rather a thickening layer of melt at the barrel surface was observed (13). Finally, as Section 9.10 point out, dissipative mix-melting may take place in screw extruders under conditions which lead to high pressures in the feed zone. In this section however we concentrate on the commonly observed melting mechanism. It is noteworthy that melting takes place along most of the extruder. Indeed, the production capacity of plasticating extruders is frequently determined by their plasticating capacity.

Further visual analysis of the experimental results reveals a tendency of the melt pool to penetrate "under" the solid bed and occasionally to completely surround it; frequently the continuity of the solid bed is broken, and a melt filled gap appears (e.g., turn 15.5, Fig. 12.12). This tendency of solid bed breakup seems to occur in the tapered sections of the extruder, and it appears to be a source of "surging" (i.e., fluctuation in time of temperature, pressure, and flow rate) of the extrudate at the die as well as a source of entrapping some air bubbles into the melt stream.

A close analysis of an individual cross section (Fig. 12.15) suggests further details on the physical mechanisms taking place in the screw channel. We observe a thin film of melt between the surface of the barrel and the solid bed. The relative motion of the barrel surface in the cross channel direction drags the melt in the film into the melt pool, generating a cross channel pressure gradient and a circulatory flow. This hydrodynamically generated pressure in the melt pool no doubt brings about the segregation of the solids at the trailing flight, and since melt is continuously removed by drag from the film, the solid bed must acquire a velocity component toward the barrel surface. But at the same time it also slides down channel; consequently the size of the solid bed at a fixed position in the bed is continuously reduced until, at the end of melting, it completely disappears. At a fixed position in the screw channel, on the other hand, the size of the solid bed remains constant in time. *Thus all the elements have emerged for a drag induced melt removal, steady state, conduction melting mechanism*, discussed in Section 9.8. Moreover, the film region at the barrel surface is the only place where such a mechanism can develop. Recalling the significant difference between the rates of melting in conduction melting with and without melt removal, we can conclude that the melting at the root of the screw (even when there is melt penetration under the

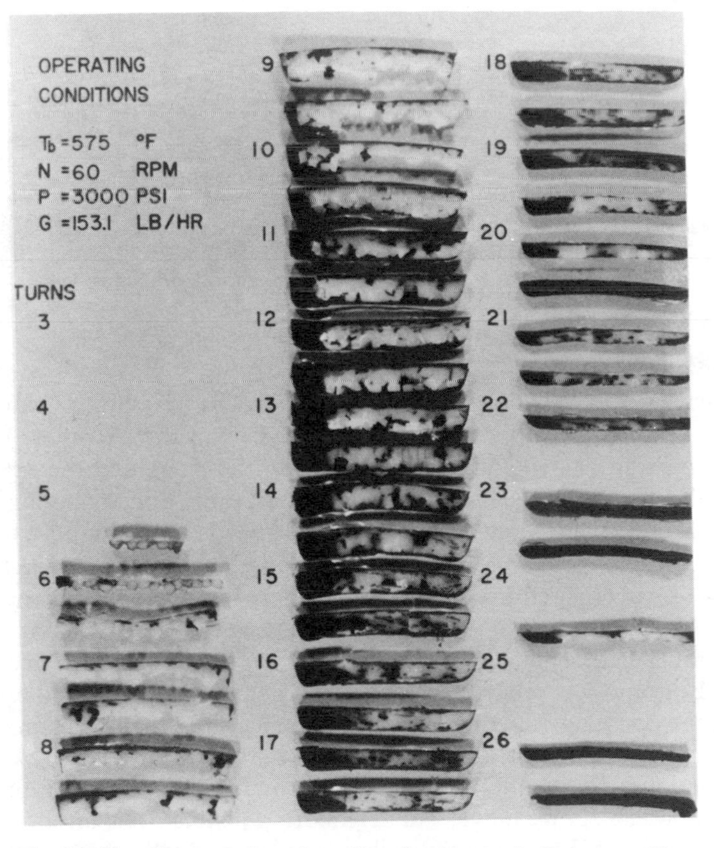


**Fig. 12.11** Cross sections from "cooling" experiments—see Fig. 12.10. Material: nylon.

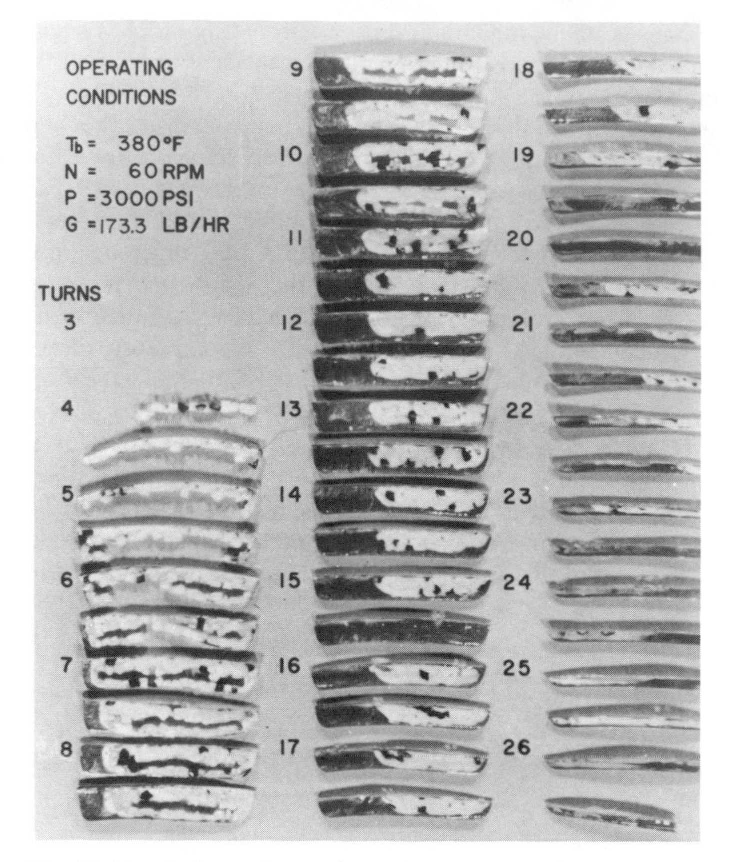


**Fig. 12.12** Cross sections from "cooling" experiments—see Fig. 12.10. Material: ABS.

solid bed) as well as at the melt pool–solid bed interface, are second order effects in most of the melting region.

Having this overview on the physical mechanisms in mind, we can proceed to examine the "complete" experience of the polymer in the extruder. We shall do this by "embarking" a pellet and "traveling" with it throughout the extruder.

In the hopper, where only the elementary step of solids handling occurs, we commence a slow, somewhat erratic motion downward repeatedly bumping into neighboring pellets and occasionally being hung up in a stable arch for short periods, until we reach the "throat" area. Here we observe the helical flight, "sweeping" the pellets from underneath and pushing them forward. At the moment we are caught up by the flight and start rotating, our coordinate system changes. We now record our motion relative to the screw; hence the barrel will appear to be rotating in the opposite direction. We find ourselves in a shallow channel confined between the flights, the roots of the screw, and the barrel surface. We commence a slow motion down the channel, while generally maintaining our position relative to the confining walls. As we move, the neighboring pellets exert an increasing force on our pellets, the void between the pellets is being gradually reduced. Most pellets have the same experience except those in contact with the barrel and those in contact with the screw. The former experience an intense frictional drag by the moving barrel surface, whereas the latter experience a frictional drag force in the up channel direction by the screw surfaces. We know from Section 8.13 that this frictional drag at the barrel is the driving force of the solids conveying mechanism in the screw channel. Both these frictional processes result in heat generation, raising the polymer temperature, in particular the surface layer at the barrel surface. At some point, this temperature may exceed the melting point or softening range of the polymer, converting the frictional drag mechanism into a viscous drag mechanism. That is, the solids are conveyed forward in the channel by the shear stresses generated in the melt film. A more common situation, however, is that before any significant frictional heating, an axial position is reached where the barrel is heated to well above the melting point, forcing the creation of a film of melt. In either case, this marks the end of that portion of the process in the extruder called the *solids conveying zone*, where only solids are present and the only elementary step that occurs is handling of solids. By this time we find our pellet somewhat deformed by the neighboring pellets in close contact with it, forming together a rather sturdy, but deformable, solid bed moving in pluglike fashion down channel. The thin film separating the bed from the barrel is sheared intensely. Heat is generated by this shearing action and heat is conducted from the barrel to the

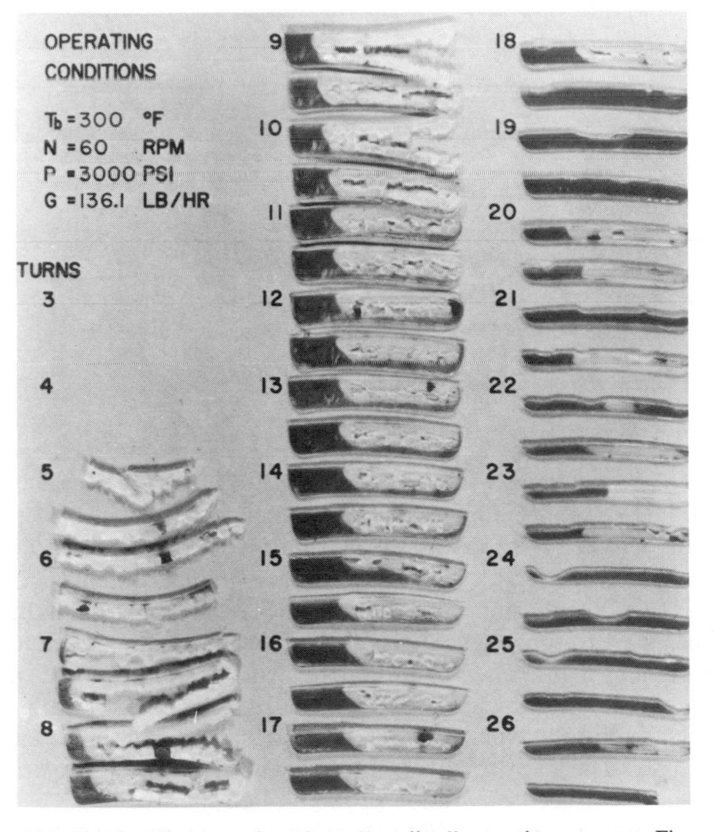


**Fig. 12.13** Cross sections from "cooling" experiments—see Fig. 12.10. Material: LDPE (DFDW 0173, Union Carbide Co.).

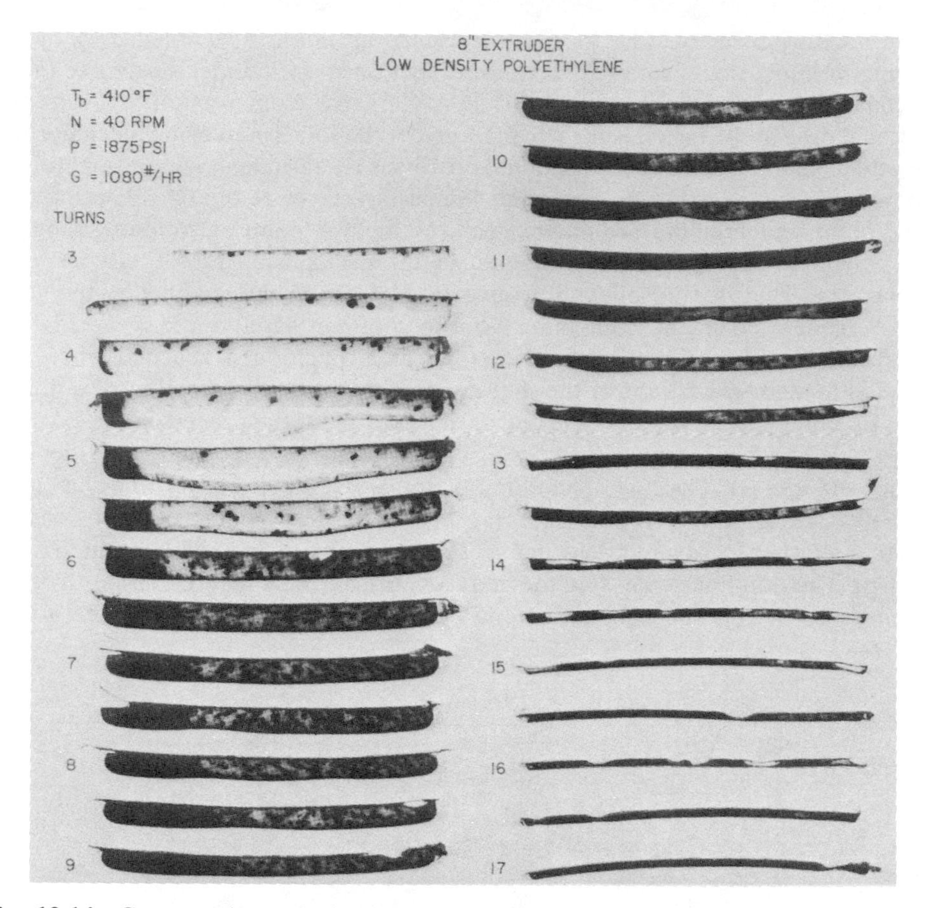


**Fig. 12.14** Cross sections obtained from a "cooling experiment" of an 8 in. diameter extruder. Material and operating conditions indicated in the figure. (Reprinted with permission from Z. Tadmor and I. Klein, *Engineering Principles of Plasticating Extrusion*, Van Nostrand Reinhold Book Co., New York, 1970.)

solid bed. The temperature gradient is large because the barrel temperature drops to the melting point over a very thin film. As a result of this heat transfer, our pellet experiences from this point on a gradual rise in temperature. Since there is a small radial clearance between the tip of the flights and the barrel surface, until the melt film is thinner than the clearance, nothing "drastic" happens. This condition may continue for a few turns, during which the film thickens beyond the flight clearance. The flights then start to scrape the melt off the barrel and the melt starts accumulating at the pushing flight. The portion of the process from the end of the solids conveying zone to the point where the melt pool first appears is called the *delay zone* (14). In this zone the elementary step of melting occurs simultaneously with handling of solids. The melting mechanism however, is contact melting without melt removal but with heat generation in the molten film. In Fig. 12.9, the solids conveying zone ends at turn 3, where barrel heating starts, and the delay zone, which starts at this location, ends at turn 7, where the melt pool begins to form.

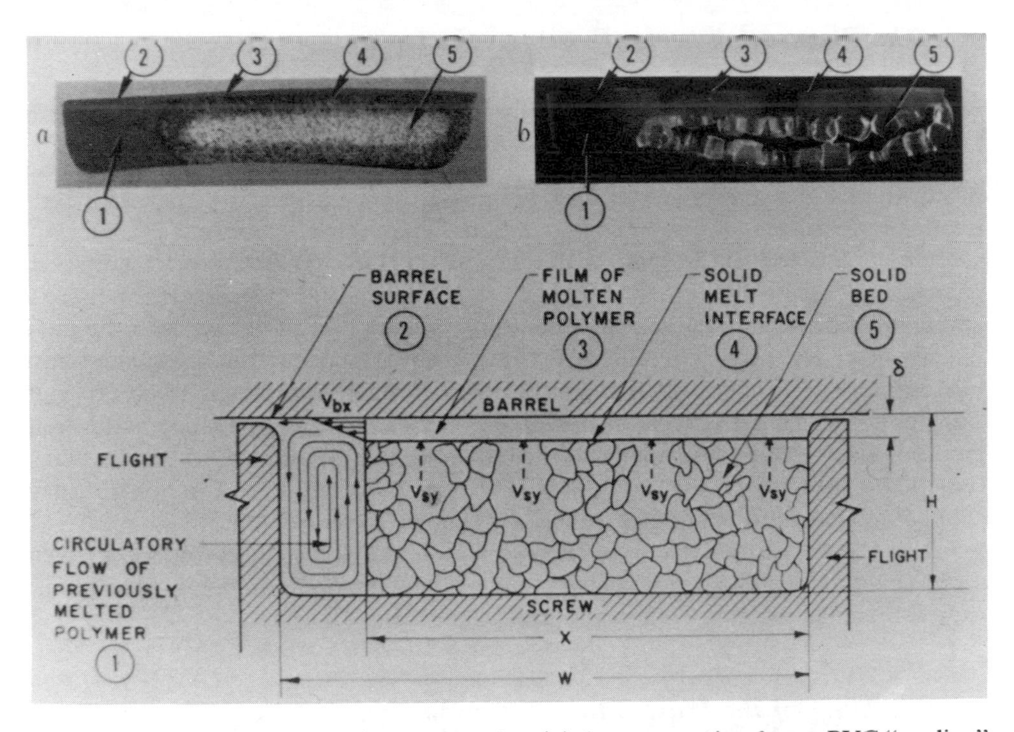


**Fig. 12.15** Idealized cross section compared to (*a*) the cross section from a PVC "cooling" experiment and (*b*) the cross section from an LDPE "cooling" experiment. (Reprinted with permission from Z. Tadmor and I. Klein, *Engineering Principles of Plasticating Extrusion*, Van Nostrand Reinhold Book Co., New York, 1970.)

Returning to "our" pellet, we note the end of the delay zone when the solid bed has acquired a small upward velocity toward the barrel surface. At some point in the extruder our pellet will reach the melt film–solid bed interface, experiencing toward the end of this approach a quick (exponential) rise in temperature up to the melting point. After being converted into melt, our fluid particle is quickly swept into the melt pool. (For amorphous polymers, as the polymer softens it moves both toward the barrel as well as toward the pushing flight.) Once in the melt pool, the fluid particle settles at some position in the channel and commences the circulatory flow alternating between two positions. In the upper portion of the channel it moves toward the pushing flight and down channel relatively fast, whereas in the lower portion of the channel it moves toward the solid bed (which also slides down channel) or trailing flight (if melting is completed) and down channel relatively slowly. This continues until it leaves the screw channel. In the melt pool both the temperature and the pressure change; they generally increase. The portion of the process where melting takes place is called the *melting zone*, which lies side by side with the *melt conveying zone*. The latter extends to the end of the screw. Clearly, then, in the melting zone *all* elementary steps occur simultaneously; whereas in the melt conveying zone, as discussed in the preceding section, only pumping and mixing take place.

### *The Plasticating Screw Extrusion Operation*

The screw characteristic of a plasticating extruder retains the general nonlinear shape of the melt extruder screw characteristic discussed in the preceding section. But other factors appear in addition to pressurization and mixing because the plasticating screw has a number of functional zones and all elementary steps involved are affected by changing conditions. For example, above a certain flow rate the solids conveying may be insufficient resulting in "starved" feeding. A change in flow rate changes the length of the melting zone; hence along the screw characteristic not only the melt temperature varies, as was the case with the melt extruder (see Fig. 12.6), but also the extrudate may start to contain unmelted solids. Furthermore, the mean melt temperature is no longer determined solely by heat transfer to the melt pool from the boundaries and the viscous dissipation in the melt pool but also by the melting performance (i.e., the conditions of the melt film being fed to the melt pool). Finally, the delay zone, its location, and its length, may change; affecting the melting as well as melt conveying zones that follow.

For a plasticating extruder–die combination, as with melt extruders, a given screw speed results in a certain output rate given by the operating point. The latter, as before, is the crossing of die and screw characteristic curves.

### *Modeling of the Plasticating Extrusion Process*

In addition to the modeling objectives listed in melt extruders, which remain valid for plasticating extrusion, we add the following: gravitational flow behavior of particulate solids in hoppers—in particular pressure distribution, arching, and bridging, and stress and temperature distribution in the solids conveying zone; length of the delay zone, stress and temperature distribution in the solid bed, rate of melting, mean width profile of the solid bed (solid bed profile, SBP), mean temperature of the melt film flowing into the melt pool, power consumption in the solids conveying delay and melting zones, as well as surging conditions. We could generalize our modeling objectives to velocity, temperature, and stress fields in both solid and liquid phases, from which we could calculate all the other variables of interest. But in plasticating extrusion, more than in melt extrusion, it is very difficult to obtain a complete solution to this problem. We shall follow the modeling approach outlined for the melt extrusion process. Namely, we shall assume steady state conditions and a given mass flow rate; then starting from the hopper, calculations are made in finite steps, ending up at the die with an extrudate pressure, mean temperature, and solids content. If the flow rate at these conditions does not match the die, calculations are repeated at a new mass flow rate.

### *The Solids Conveying Zone*

The conveying mechanism in screw extruders is one of drag induced flow discussed in Section 8.13. Indeed for shallow channels we could turn directly to Eq. 8.13-7, which would form the solids conveying model. The feed section for screw extruders, however, is generally deep and curvature effects are not negligible. Following the

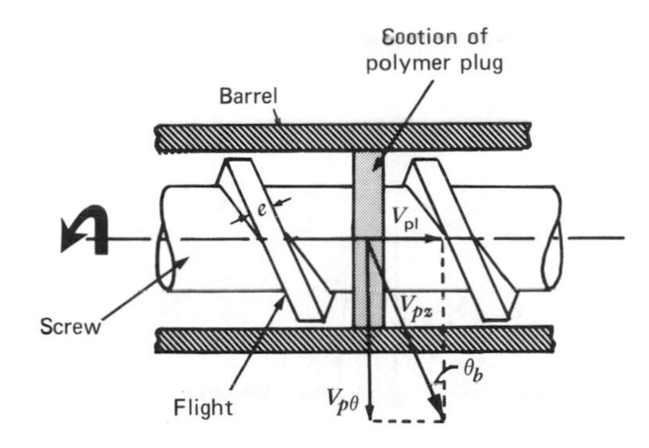


**Fig. 12.16** Axial increment of the solid plug. Velocities given relative to a stationary screw: $V_{pl}$ is the axial velocity of the plug, which is independent of the radial position, $V_{p\theta}$ and $V_{pz}$ are the tangentical and down channel components of the plug surface velocity.

original work of Darnell and Mol (15), we therefore derive a solids conveying model in deep screw channels subject to the same simplifying assumptions made in Section 8.13 and following the same procedure.

At steady state the solid plug has a constant axial velocity $V_{pl}$ and constant angular velocity $V_{p\theta}$ (Fig. 12.16). The former is related to the mass flow rate by the following equation:

$$G = V_{pl}\rho_b\left[\frac{\pi}{4}(D_b^2 - D_s^2) - \frac{eH}{\sin\bar{\theta}}\right] \tag{12.2-1}$$

where $D_b$ is the inside diameter of the barrel, $D_s = D_b - 2H$, $H$ is the channel depth, $e$ is the flight width, and $\bar{\theta}$ the mean helix angle. The down channel velocity of the solids is $V_{pl}/\sin\theta$, which varies with channel depth. (Note that this velocity is equivalent to velocity $u$ in Section 8.13.) It is more convenient to express the flow rate $G$ in terms of the angle $\phi$ formed between the velocities of the solids and the barrel surface because force and torque balances provide an expression for this angle (cf. Eq. 8.13-7). The relationship between $V_{pl}$, $V_b$ and the angle $\phi$ can be easily obtained, as shown in Fig. 12.17

$$V_{pl} = V_b\frac{\tan\phi\tan\theta_b}{\tan\phi + \tan\theta_b} \tag{12.2-2}$$

where $V_b = \pi N D_b$ is the tangential velocity of the barrel surface. Clearly at closed discharge conditions $\phi = 0$ and $V_{pl} = 0$. Substituting Eq. 12.2-2 into Eq. 12.2-1, followed by rearrangements, results in

$$G = \pi^2 NHD_b(D_b - H)\rho_b\frac{\tan\phi\tan\theta_b}{\tan\phi + \tan\theta_b}\left[1 - \frac{e}{\pi(D_b - H)\sin\bar{\theta}}\right] \tag{12.2-3}$$

Equation 12.2-3 can be used either to calculate $\phi$ from $G$, or vice versa. If bulk density changes cannot be neglected, the calculations should be performed in small axial increments. Next we proceed with the force and torque balances. Since

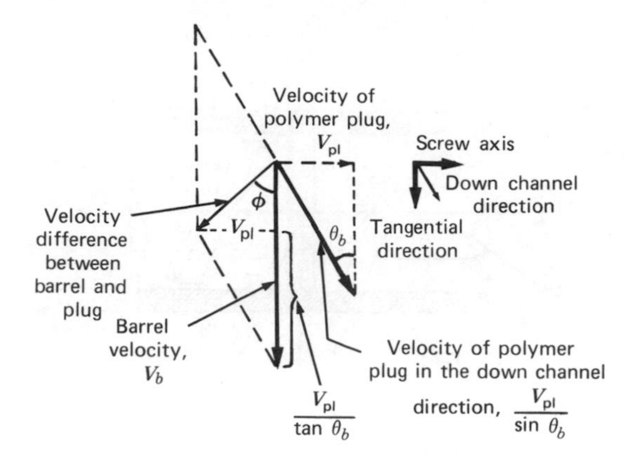


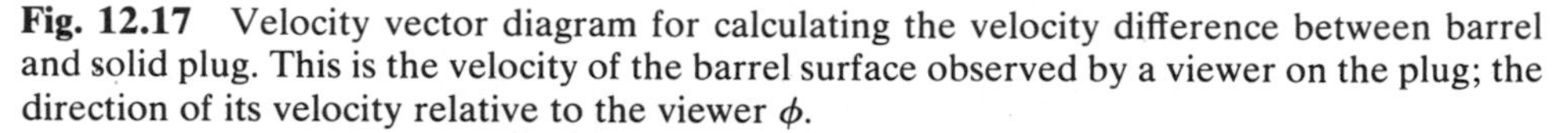

**Fig. 12.17** Velocity vector diagram for calculating the velocity difference between barrel and solid plug. This is the velocity of the barrel surface observed by a viewer on the plug; the direction of its velocity relative to the viewer $\phi$.

pressure builds up in the down channel direction, the force and torque balances are made on a differential increment in the down channel direction; this is illustrated in Fig. 12.18, where the various forces acting on the element are also depicted. These can be expressed in terms of the coefficients of friction, local geometry, and the differential pressure increment, which compensates for the other forces and torques. For an isotropic stress distribution, these are

$$
\begin{aligned}
F_1 &= f_b P W_b \, dz_b \\
F_6 - F_2 &= H\bar{W} \, dP \\
F_8 &= PH \, d\bar{z} \\
F_7 &= PH \, d\bar{z} + F^* \\
F_3 &= f_s F_7 \\
F_4 &= f_s F_8 \\
F_5 &= f_s P W_s \, dz_s
\end{aligned}
\tag{12.2-4}
$$

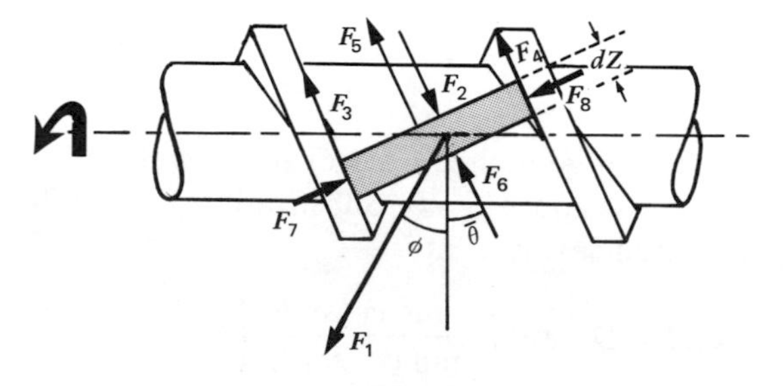


**Fig. 12.18** Forces acting on a down channel increment of the solid plug: $F_1$ is the forward dragging frictional force exerted on the plug by the barrel surface, $F_6 - F_2$ is the net force resulting from the down channel pressure gradient, $F_3$, $F_4$ and $F_5$ are the frictional retarding forces of the screw, and $F_7$ and $F_8$ are the normal forces by the flights on the plug.

where the subscripts $b$ and $s$, respectively, denote the surfaces of the barrel and the root of the screw, and the bar denotes the mean value over the channel depth. The motion of the plug consists of a pure translation in the axial direction and pure rotation in the angular direction. Hence by calculating the components of all the forces in the axial and tangential directions, a force balance can be written in the former direction and a torque balance in the latter direction. By solving the two equations simultaneously the force $F^*$ is eliminated and, subsequent to considerable algebraic rearrangements, the following simple expression is obtained:

$$\cos\phi = K_s \sin\phi + M \tag{12.2-5}$$

or

$$\sin\phi = \frac{\sqrt{1+K_s^2-M^2}-K_sM}{1+K_s^2} \tag{12.2-6}$$

where

$$K_s = \frac{\bar{D}}{D_b}\,\frac{\sin\bar{\theta}+f_s\cos\bar{\theta}}{\cos\bar{\theta}-f_s\sin\bar{\theta}} \tag{12.2-7}$$

and

► $$M = 2\frac{H}{W_b}\frac{f_s}{f_b}\sin\theta_b\left(K_s+\frac{\bar{D}}{D_b}\operatorname{cotan}\bar{\theta}\right)+\frac{W_s}{W_b}\frac{f_s}{f_b}\sin\theta_b\left(K_s+\frac{D_s}{D_b}\operatorname{cotan}\theta_s\right) + \frac{\bar{W}}{W_b}\frac{H}{Z_b}\frac{1}{f_b}\sin\bar{\theta}\left(K_s+\frac{\bar{D}}{D_b}\operatorname{cotan}\bar{\theta}\right)\ln\frac{P_2}{P_1} \tag{12.2-8}$$

where $P_1$ is the initial pressure at $z=0$ and $P_2$ is the pressure at any down channel distance $Z_b$, where solids conveying is the only elementary step taking place. For a given flow rate, $\phi$ is obtained from Eq. 12.2-3, $M$ is then calculated from Eq. 12.2-5, and the pressure rise from Eq. 12.2-8. If a given pressure rise is needed, the process is reversed, with the angle $\phi$ calculated from Eq. 12.2-6. Finally, the total power consumption in the solids conveying zone is obtained by taking the product of the force between the barrel surface and solid plug $F_1$ and the barrel velocity in the direction of the force $\pi ND_b\cos\phi$

$$dP_w = \pi ND_b\cos\phi\, f_bW_bP\,dz_b \tag{12.2-9}$$

Integrating Eq. 12.3-9 after substituting the exponential relationship between $P$ and $Z_b$ as expressed in Eq. 12.2-8 results in (16)

► $$P_w = \pi ND_bW_bZ_bf_b\cos\phi\frac{P_2-P_1}{\ln(P_2/P_1)} \tag{12.2-10}$$

Perhaps the most severe assumption in the Darnell and Mol model is the isotropic stress distribution. Recalling the discussion on compaction in Section 8.9, the stress distribution in the screw channel is expected to be complex. The first attempt to account for the nonisotropic nature of the stress distribution is due to Schneider (17). By assuming a certain ratio between compressive stresses in perpendicular directions and accounting for the solid plug geometry, he obtained a more realistic stress distribution, where the pressure exerted by the solids on the flights, the root of the screw, and the barrel surface are all different and less than the down channel pressure. The ratio between the former and the latter is of the order of 0.3–0.4.

Another questionable assumption is that of constant temperature. Frictional forces lead to surface heat generation. The total power introduced through the shaft is partly dissipated into heat at the barrel, flights, and root of the screw surfaces and is partly used to generate pressure. However most of the power is dissipated into heat of the barrel surface (Fig. 12.19). This quantity is given by the

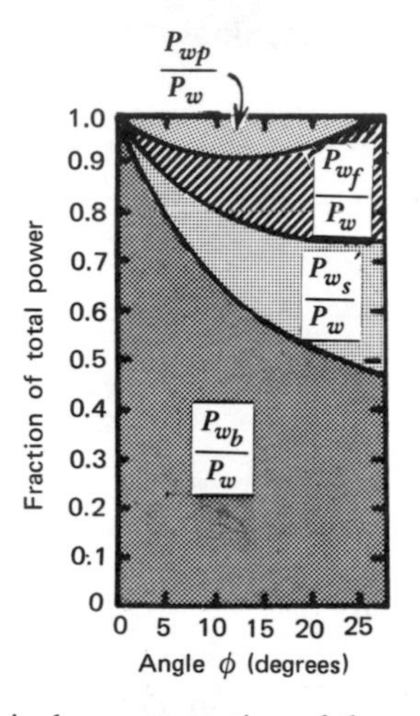


**Fig. 12.19** Graphical representation of the various components of the total power consumption in the solids conveying zone of screw extruders for $H/D_b = 0.15$ and a constant $f_s/f_b$ ratio as a function of the angle $\phi$: $P_{wb}$ is the power dissipated into heat of the barrel plug interface. This is the major component of the power. $P_{ws}$ and $P_{wf}$ are the power dissipated on the root of the screw and flights respectively. $P_{wp}$ is the power consumed for the pressurization of the plug. [Reprinted with permission from E. Broyer and Z. Tadmor, "Solids Conveying in Screw Extruders, Part I. A Modified Isothermal Model," *Polym. Eng. Sci.*, **12**, 12–24 (1972).]

product of the force $F_1$ and the relative velocity between barrel surface and solid plug (16)

$$P_{wb} = \pi N D_b W_b Z_b f_b \frac{\sin \theta_b}{\sin (\theta_b + \phi)} \frac{P_2 - P_1}{\ln (P_2/P_1)} \tag{12.2-11}$$

This heat generated at the interface is partly conducted through the cooled barrel and partly conducted into the solid plug. Consequently a temperature profile develops in the plug with a maximum temperature at the interface. If we neglect heat generation at the other surfaces, the problem reduces to a one-dimensional heat conduction problem soluble by methods discussed in Section 9.3. Since the rate of heat generation varies with axial location, numerical solution methods are needed. This was done by Tadmor and Broyer (18). The results indicate that the solids surface temperature at the barrel increases exponentially. Clearly, once the melting point is reached, the frictional drag conveying mechanism changes into a viscous drag mechanism (14). This nonisothermal solids conveying mechanism explains the need for efficient barrel cooling in the solids conveying zone if high pressure generation is desired.

Finally, all solids conveying models require an estimate of the inlet pressure $P_1$. One approach of evaluating $P_1$ is to assume that it equals the pressure under the granular material at the hopper base (16), which can be evaluated from the equations given in Section 8.7. This approach neglects the complex transition between the gravitational flow in the hopper and the drag induced flow of the pluglike solid in the closed screw channel. It connects, however, the extruder performance to hopper design and loading level. The need for such a connection follows from experimental observations that, under certain conditions, small variations in solids height in the hopper bring about significant variation in the extruder performance—for example, pressure variations at the die. In such cases keeping the solids height above a certain level eliminates pressure surging at the die. One possible reason for this behavior is the effect of solids height on the inlet pressure.

Another approach to the problem, proposed by Lovegrove and Williams (19, 20), is to disregard the hopper and assume that the initial pressure in the solids conveying zone is the result of local gravitational and centrifugal forces. This is a reasonable assumption, considering the generally low level of pressure values in the inlet region. However it fails to account for the effect mentioned previously relating the hopper design and loading level and extrusion performance. Clearly, there is a need for additional experimental observation and a detailed mathematical model encompassing the hopper, the portion of the screw under the hopper, and the inlet region in the extruder, where the Darnell and Mol model does not apply.

For good solids conveying performance, pressure should rise over this zone. However the maximum theoretical conveying capacity is obtained by setting $P_2 = P_1$. Analysis of the solids conveying equations indicates that there is an optimum helix angle as well as an optimum channel depth for maximum conveying capacity or maximum pressure rise. We pointed out before that $P_1$ is low; consequently $P_2/P_1$ must be very high to obtain a substantial pressure level $P_2$. Increasing $P_1$ by forced feeding (e.g., with a feeding screw in the hopper) will proportionately increase $P_2$. Equation 12.2-8 indicates that the pressure profile in the solids conveying zone of screw extruders in exponential as is in shallow rectangular channels discussed in Section 8.13. Solids conveying is improved by increased $f_b/f_s$ and by increasing the screw speed ($\phi$ is decreased for a given $G$), provided the isothermal conditions are maintained and the coefficients of friction remain constant. An accurate measurement of the latter involves many experimental difficulties, as discussed in Section 4.3.

**Example 12.2** *Solids Conveying in Screw Extruders*

LDPE is extruded in a $6.35 \times 10^{-2}$ m (2.5 in.) diameter, 26.5 turn long single screw extruder, with a square pitched ($L = D_b$) metering type of screw. The feed section is 12.5 turns long and $9.398 \times 10^{-3}$ m (0.37 in.) deep, the transition section is 9.5 turns long, and the metering section is $3.22 \times 10^{-3}$ m (0.127 in.) deep. Flight width is $6.35 \times 10^{-3}$ m (0.25 in.), and flight clearance is negligible. Hopper diameter is 0.381 m (15 in.), with a converging conical section of 90° and discharge opening of 0.127 m (5 in.), as in Fig. 12.20. The barrel temperature is maintained at 149°C (300°F) and heating starts 3 turns from the beginning of flights, with the hopper opening occupying the first two turns, leaving one turn for solids conveying.

At a screw speed of 60 rpm, a mass flow rate of 67.1 kg/hr is obtained with polymer feed at 24°C (screw design and operating conditions correspond to the experiment reported in Fig. 12.13).

Calculate (*a*) the base pressure of the hopper, (*b*) pressure at the end of the solids conveying zone, and (*c*) power consumption in the solids conveying zone.

Assume isothermal operation and that the inlet pressure to the solids conveying zone equals the base pressure of a fully loaded hopper. Bulk density of the feed is 595 kg/m$^3$, the static coefficient of friction in the hopper is 0.3, the effective angle of friction is 33.7°, and the dynamic coefficients of friction on barrel and screw are 0.45 and 0.25, respectively.

*Solution*: First we compute some geometrical data (needed both in this and the following example) summarized below.

| Variable | Feed Section | Transition Section | Metering Section |
|---|---|---|---|
| Helix angle at the barrel surface, $\theta_b$ | 17.65° | 17.65° | 17.65° |
| Mean helix angle, $\bar{\theta}$ | 20.48° | Varies linearly | 18.54° |
| Helix angle at the root of the screw, $\theta_s$ | 24.33° | Varies linearly | 19.51° |
| Mean channel width, $\bar{W}$ | $5.314\times10^{-2}$ m | Varies linearly | $5.358\times10^{-2}$ m |
| Channel width at the barrel surface, $W_b$ | $5.416\times10^{-2}$ m | $5.416\times10^{-2}$ m | $5.416\times10^{-2}$ m |
| Channel width at the root of screw, $W_s$ | $5.151\times10^{-2}$ m | Varies | $5.350\times10^{-2}$m |
| Axial length, $l$ | 10.5 turns, 0.666 m | 9.5 turns, 0.603 m | 4 turns, 0.286 m |
| Mean helical length, $\bar{z}$ | 2.270 m | — | 0.800 m |

a. We commence calculations in the solids conveying zone. The initial pressure in the solids conveying zone $P_1$ is assumed to equal the pressure under the solids in the hopper. We can calculate it using Eq. 8.7-8. The value $P_0$ is evaluated assuming that the height of solids in the vertical part of the hopper is sufficient to result in at least 99% of the maximum pressure. Thus from Eq. 8.7-5 we have

$$P_0=\frac{(0.99)(595)(9.806)(0.381)}{(4)(0.3)(0.286)}=6.412\times10^3\ \mathrm{N/m^2}$$

where the value of $K=0.286$ was obtained from Eq. 8.6-6

$$K=\frac{\sigma_{\min}}{\sigma_{\max}}=\frac{1-\sin(33.7)}{1+\sin(33.7)}=0.286$$

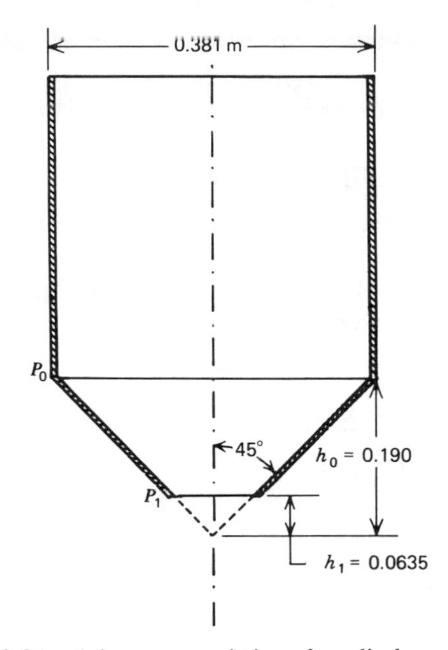


**Fig. 12.20** A hopper consisting of a cylinder over a truncated cone.

As Section 8.7 pointed out, a somewhat more accurate result can be obtained by replacing $f'_w K$ in Eq. 8.7-5 by $BD^*$, where $D^* = 1$ and $B$ is given in Eq. 8.7-6. The angle $\beta_w = \arctan(0.3) = 16.7°$; the angle $\kappa_0$ is given by Eq. 8.7-7,

$$\kappa_0 = 16.7 + \arcsin\left(\frac{\sin 16.7}{\sin 33.7}\right) = 16.7 + 148.8 = 165.5 \qquad \arcsin > \pi/2$$

and $B$ is given by

$$B = \frac{\sin(33.7)\sin(165.5)}{1 - \cos(165.5)\sin(33.7)} = 0.0904$$

We note that $B$ is very close to $f'_w K = 0.086$. The more accurate value of $P_0$ is $5.78 \times 10^3$ N/m$^2$.

The pressure $P_1$ is calculated by Eq. 8.7-8 with $h_1 = 0.0635$ and $h_0 = 0.19$ m (see Fig. 8.10). The angle $\kappa_0$ is given in Eq. 8.7-13.

$$\kappa_0 = 16.7 + \arcsin\left(\frac{\sin 16.7}{\sin 33.7}\right) = 16.7 + 31.2 = 47.9 \qquad \arcsin < \pi/2$$

From Eq. 8.7-12 we obtain $B'$:

$$B' = \frac{\sin(33.7)\sin[(2)(45) + (47.9)]}{1 - \sin(33.7)\cos[(2)(45) + (47.9)]} = 0.2635$$

The value of $\not{h}$ is obtained from Eq. 8.7-10 with $D^* = 1$:

$$\not{h} = \frac{(2)(0.2635)}{\tan(45)} = 0.527$$

Finally $P_1$ from Eq. 8.7-8 is

$$P_1 = \left(\frac{0.0635}{0.190}\right)^{0.527} (5.78\times 10^3) + \frac{(595)(9.806)(0.0635)}{(0.527-1)}\left[1-\left(\frac{0.0635}{0.190}\right)\right]^{0.527-1}$$

$$= 3.244\times 10^3 + 532 = 3.776\times 10^3 \text{ N/m}^2 \ (=0.55 \text{ psi})$$

We therefore note that the conical section results in a 35% reduction in pressure.

(b) The axial velocity of the solid plug $V_{pl}$ from Eq. 12.2-1 neglecting the effect of pressure on bulk density is

$$V_{pl} = \frac{\dfrac{61.7}{3600}}{(595)\left\{\dfrac{\pi}{4}[(0.0635)^2-(0.0447)^2] - \dfrac{(0.00635)(0.003226)}{\sin(20.48)}\right\}}$$

$$= 0.0187 \text{ m/s}$$

The velocity of the barrel surface $V_b = \pi N D_b = 0.1995$ m/s. Hence from Eq. 12.2-2 we obtain

$$\tan\phi = \frac{\tan\theta_b}{(V_b/V_{pl})\tan\theta_b - 1} = \frac{\tan(17.65)}{(0.1995/0.0187)\tan(17.65)-1} = 0.13288$$

and $\phi$ is 7.57°. Next $K_s$ is evaluated from Eq. 12.2-7

$$K_s = \frac{(0.0541)}{(0.0635)}\,\frac{\sin(20.48)+(0.25)\cos(20.48)}{\cos(20.48)-(0.25)\sin(20.48)} = 0.5859$$

and $M$ is evaluated from Eq. 12.2-5

$$M = \cos(7.57) - 0.5859\sin(7.57) = 0.9141$$

The pressure rise ratio $P_2/P_1$ over one turn of solids conveying section ($Z_b = 0.0635/\sin 17.6° = 0.209$ m) from downstream the hopper to the location where barrel heating starts, is obtained from Eq. 12.2-8.

$$0.9141 = (2)\frac{(0.009398)}{(0.05416)}\frac{(0.25)}{(0.45)}\sin(17.65)\left[(0.5859)+\frac{(0.0541)}{(0.0635)}\text{cotan}(20.48)\right]$$

$$+\frac{(0.05151)(0.25)}{(0.05416)(0.45)}\sin(17.65)\left[(0.5859)+\frac{(0.0447)}{(0.0635)}\text{cotan}(24.33)\right]$$

$$+\frac{(0.05314)(0.009398)\sin(20.48)}{(0.05416)(0.209)(0.45)}\left[(0.5859) + \frac{(0.0541)}{(0.0635)}\text{cotan}(20.48)\right]\ln\frac{P_2}{P_1}$$

$$= 0.1676 + 0.34328 + 0.09813\ln\frac{P_2}{P_1}$$

and

$$\frac{P_2}{P_1} = 60.9$$

Thus the exist pressure from the solids conveying zone is $3.776 \times 10^3 \times 60.9 = 2.3 \times 10^5$ N/m$^2$ (33 psi). This result indicates that the solids conveying section functions properly and that higher outputs could be obtained at this screw speed before solids conveying limitations (e.g., "starving") were encountered. We should note, however, that the analysis of the solids conveying zone is very sensitive on the values of the coefficients of friction.

(c) The power consumption is calculated from Eq. 12.2-10

$$P_w = (\pi)(1)(0.0635)(0.05416)(0.209)(0.45)\cos(7.57)$$

$$\frac{(2.3\times10^5)-(3.776\times10^3)}{\ln 60.9} = 56.7 \text{ W } (=0.076 \text{ hp})$$

## *The Melting Zone*

As we pointed out earlier, from the axial location where a melt film is formed at the barrel surface (either as a result of barrel heating or as a result of heat generation due to friction) to the axial location where a melt pool appears at the "pushing" flight, lies the *delay zone*. The conveying mechanism in this zone is one of viscous drag at the barrel surface determined by the shear stresses in the melt film and generally frictional (retarding) drag on the root of the screw and the flights (14, 21). Thickness of the melt film increases with down channel distance and attains a value of several times that of the flight clearance at the end of the zone. There is no reliable mathematical model available to predict the length of the delay zone. Figure 12.21 gives an approximate empirical correlation based on limited experimental data, relating the length of the zone expressed in turns and a dimensionless group $\psi$ related to melting rate, discussed below. The correlation is crude and takes no account of the mechanical properties of the solid bed, which probably play a role in determining the length of the delay zone.

The melting mechanism in screw extruders was first formulated by Tadmor (22) on the basis of the previously described visual observations. The model assumes Newtonian fluids and shallow channels. The channel cross section and that of the solid bed are assumed to be rectangular as in Fig. 12.15. The width of the solid bed is denoted by $X$. The prediction of the solid bed profile $X(z)$ is one of the

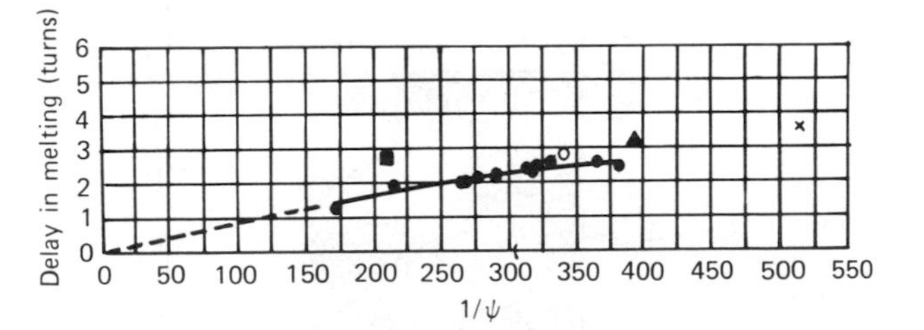


**Fig. 12.21** Delay in melting expressed in "turns" from the point the barrel temperature exceeds the melting point (melt film formed) to the point where a melt pool is segregated at the pushing flight as obtained from cooling experiments versus calculated dimensionless group $1/\psi$. Solid curve for LDPE, ●; HDPE, ○; PP, ▲; rigid PVC, ■; nylon, ×. (Reprinted with permission from Z. Tadmor and I. Klein, *Engineering Principles of Plasticating Extrusion*, Van Nostrand Reinhold Book Co., New York, 1970.)

main objectives of the model. This prediction is amenable to easy and direct experimental verification (as in Figs. 12.9–12.14). The product of $X(z)$ with the solid bed height, solids velocity and density gives the local solids mass flow rate. Furthermore, by subtracting $X(z)$ from the channel width $W$, we obtain the width of the melt pool, which is needed to model the pressurization step in this zone.

A basic assumption of the model is that a steady state condition develops in the extruder. It is further assumed that melting takes place *only* at the barrel surface, where a drag induced melt removal mechanism exists. The solid bed is assumed homogeneous, deformable and continuous. Next we assume that the *local* down channel velocity of the solid bed is constant. Slow variations in this velocity as well as those of physical properties (e.g., density of the solid bed), of operating conditions (e.g., barrel temperature), and of geometry (e.g., channel depth) can be accounted for by a calculation procedure involving small finite down channel increments. This can be viewed as an "extended lubrication approximation," whereby changes in the direction of the main flow are assumed to be small as compared to changes in the perpendicular direction to this flow, and local changes are functions of local conditions only. Finally local physical and thermophysical properties are assumed to be constant and the solid bed–melt film interface is assumed to be a sharp interface existing at a specified temperature $T_m$, the melting point.

The change in size of the solid bed over a small down channel increment will depend on the rate of melting at the solid bed–melt film interface. Consider a small differential volume element, perpendicular to the solid-melt interface (Fig. 12.22). The solid bed has a local down channel velocity $V_{sz}$ and a local velocity component into the melt film of $V_{sy}$. The barrel surface velocity $V_b$ is resolved into down channel, and a cross channel components $V_{bz}$ and $V_{bx}$. The relative velocity between barrel surface and solid bed is

$$\mathbf{V}_j = \mathbf{V}_b - \mathbf{V}_{sz} \tag{12.2-12}$$

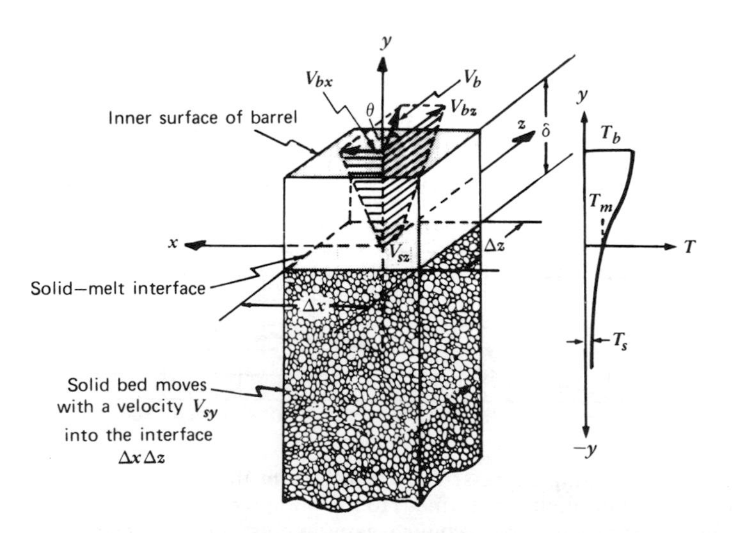


**Fig. 12.22** A differential volume perpendicular to the melt film–solid bed interface. Schematic view of temperature profile in the film and solid bed shown at right. Schematic view of velocity profiles (isothermal model) in the $x$ and $z$ directions are also shown.

or

$$|V_j| = (\mathbf{V}_j \cdot \mathbf{V}_j)^{1/2} = (V_b^2 + V_{sz}^2 - 2V_b V_{sz} \cos\theta)^{1/2} \tag{12.2-12a}$$

which determines the rate of viscous dissipation. For a linear velocity profile, the shear rate is $V_j/\delta$, where $\delta$ is the local film thickness. The melting mechanism, however, is a drag removal mechanism. The melt in the film is primarily removed in the *cross channel* direction, resulting in a reduction of solid bed width. No such effective removal mechanism is possible in the down channel direction; indeed the film thickness variation in the down channel direction is small. Therefore it is the cross channel velocity component $V_{bx}$ which determines the rate of melt removal.

In view of these assumptions, neglecting small changes in the down channel direction over the differential element $dz$, we deal with a thin slice of solid bed that has a molten film, whose thickness varies in the $x$-direction, *and we regain the basic melting mechanism discussed in detail in Section* 9.8. In particular, we can turn directly to Eq. 9.8-36, with $X$ replacing $W$, $T_b$ replacing $T_0$, $V_{bx}$ replacing $V_0$ in the first term, and $V_j$ replacing $V_0$ in the second term, to obtain the rate of melting per unit down channel distance

$$w_L(z) = \left\{ \frac{V_{bx}\rho_m[k_m(T_b - T_m) + (\mu/2)V_j^2]X}{\lambda + C_s(T_m - T_{s0})} \right\}^{1/2} \tag{12.2-13}$$

In this equation the effect of convection on the temperature distribution in the melt film was neglected. This effect, however, is hardly negligible, and since the equation with convection terms are hard to solve, we must turn to approximate methods as pointed out in Section 9.8. Consider an imaginary model in which the newly melted polymer at the melt-solid interface is removed (by a Maxwell-type "demon") brought to the location $x = 0$, heated up to the local melt temperature and fed to the melt film. Thus with convection into the film taken care of in this fashion, the film thickness remains constant with fully developed velocity and temperature profile. The heat needed to bring the "removed" melt from the melting point to the local melt film temperature can be added to the heat of fusion and is given by $C_m\bar{\Theta}(T_b - T_m)$; here $\bar{\Theta}$ is given in Eq. 9.8-31

$$\bar{\Theta} = \frac{2}{3} + \frac{\text{Br}}{12} \tag{12.2-14}$$

where Br given in Eq. 9.8–26 reduces for a Newtonian fluid to

$$\text{Br} = \frac{\mu V_j^2}{k_m(T_b - T_m)} \tag{12.2-15}$$

If we derive an expression under the foregoing conditions (constant $\delta$) we finish with an expression that is different from Eq. 9.8-36 by a factor of $\sqrt{2}$ and the term $C_m\bar{\Theta}(T_b - T_m)$ added to $\lambda$

$$w_L(z) = \left\{ \frac{V_{bx}\rho_m[k_m(T_b - T_m) + (\mu/2)V_j^2]X}{2[\lambda + C_s(T_m - T_{s0}) + C_m\bar{\Theta}(T_b - T_m)]} \right\}^{1/2} \tag{12.2-16}$$

Thus, as expected, neglecting convection would result in overestimation of the rate of melting. The procedure above is approximate, but as frequently seen before, including such approximations enables us to avoid larger errors that would arise if the effects of convection were neglected completely.

The change in solid bed width is obtained by a differential mass balance

$$\rho_s V_{sz}(H-\delta)X|_z - \rho_s V_{sz}(H-\delta)X|_{z+\Delta z} = w_L(z)\,\Delta z \tag{12.2-17}$$

which at the limit $\Delta z \to 0$, neglecting the change in film thickness in the down channel direction, reduces to

$$-\frac{d(HX)}{dz} = \frac{w_L(z)}{\rho_s V_{sz}} \tag{12.2-18}$$

By substituting Eq. 12.2-16 into 12.2-18 we get

$$-\frac{d(HX)}{dz} = \frac{\Phi\sqrt{X}}{\rho_s V_{sz}} \tag{12.2-19}$$

where

► $$\Phi = \left\{\frac{V_{bx}\rho_m[k_m(T_b - T_m) + (\mu/2)V_j^2]}{2[C_s(T_m - T_{s0}) + C_m\bar{\Theta}(T_b - T_m) + \lambda]}\right\}^{1/2} \tag{12.2-20}$$

For a constant channel depth, Eq. 12.2-19 can be integrated to give

► $$\frac{X_2}{W} = \frac{X_1}{W}\left[1 - \frac{\psi(Z_2 - Z_1)}{2H}\right]^2 \tag{12.2-21}$$

where $X_1$ and $X_2$ are the widths of the solid bed of locations $Z_1$ and $Z_2$, respectively, and the *dimensionless* group $\psi$ is defined as

► $$\psi = \frac{\Phi}{V_{sz}\rho_s\sqrt{X_1}} \tag{12.2-22}$$

Thus when performing calculations in small increments $\Delta z = Z_2 - Z_1$, the width of the solid bed at the exit $X_2$ can be calculated via Eq. 12.2-21 from its value at the entrance $X_1$.

For tapered section with a constant taper

$$-\frac{dH}{dz} = A \tag{12.2-23}$$

Eq. 12.2-19 can be rewritten as

► $$\frac{d(HX)}{dH} = \frac{\Phi\sqrt{X}}{A\rho_s V_{sz}} \tag{12.2-24}$$

which can be integrated to give

► $$\frac{X_2}{W} = \frac{X_1}{W}\left[\frac{\psi}{A} - \left(\frac{\psi}{A} - 1\right)\sqrt{\frac{H_1}{H_2}}\right]^2 \tag{12.2-25}$$

where $X_2$ and $X_1$ are the widths of the solid bed at down channel locations corresponding, respectively, to heights $H_2$ and $H_1$. These locations can be obtained from solving Eq. 12.2-23.

Equations 12.2-21 and 12.2-25 are the basic equations for the melting model. We note that the solid bed profile in both cases is a function of one *dimensionless* group $\psi$, which in physical terms expresses the ratio of the local rate of melting per unit solid-melt interface $(\Phi\sqrt{X_1}/X_1)$ to the local solid mass flux $(V_{sz}\rho_s)$, where $\rho_s$ is the local mean solid bed density. The solid bed velocity at the beginning of melting is obtained from the mass flow rate

$$V_{sz} = \frac{G}{\rho_s HW} \tag{12.2-26}$$

In Eq. 12.2-26 we neglected the effect of the melt film. The solid bed velocity was found experimentally (2f) to remain virtually constant in the feed and moderately tapered sections. In extreme cases, however—in particular, large tapers and for low rates of melting—solid bed acceleration is possible, as discussed below.

A better insight into the nature of the melting model can be obtained by considering first melting in a constant depth channel, with constant $\psi$ throughout the melting zone. The latter implies both constant physical properties and constant solid bed velocity. Equation 12.2-21 with $Z_1 = 0$ and $X_1 = W$ reduces to

$$\frac{X}{W} = \left(1 - \frac{\psi}{2}\frac{z}{H}\right)^2 \tag{12.2-27}$$

where $\psi$ reduces in this case to

$$\psi = \frac{\Phi}{V_{sz}\rho_s\sqrt{W}} = \frac{\Phi\sqrt{WH}}{G} \tag{12.2-28}$$

Equation 12.2-27 indicates that the solid bed profile in constant depth channel is parabolic. The total (down channel) length of melting is obtained from Eq. 12.2-27 by setting $X = 0$:

$$Z_T = \frac{2H}{\psi} \tag{12.2-29}$$

We therefore note that the length of melting is inversely proportional to $\psi$; that is, it is proportional to mass flow rate and inversely proportional to the rate of melting. Clearly, through $\Phi$ in Eq. 12.2-20, the effect of the various operating conditions on the length of melting can be evaluated. Thus an increase in screw speed at constant flow rate brings about an increase in the rate of melting, both because melt removal is improved ($V_{bx}$ increases) and because viscous dissipation increases. An increase in barrel temperature initially brings about an increase in the rate of melting because the conduction term $k_m(T_b - T_m)$ increases. But because further increases in the barrel temperature decrease the melt film viscosity and the amount of viscous dissipation, there is an optimum barrel temperature for maximum melting rate. (There is, however, an additional reason for the existence of the optimum, as we shall see below.) Finally an increase in solids feed temperature $T_{s0}$, increases the rate of melting and reduces $Z_T$.

Similar conclusions are drawn by considering melting in a tapered section with initial channel depth $H$ and taper $A = dH/dz$. In this case Eq. 12.2-25 reduces to

$$\frac{X}{W} = \left[\frac{\psi}{A} - \left(\frac{\psi}{A} - 1\right)\sqrt{\frac{H}{H - Az}}\right]^2 \tag{12.2-30}$$

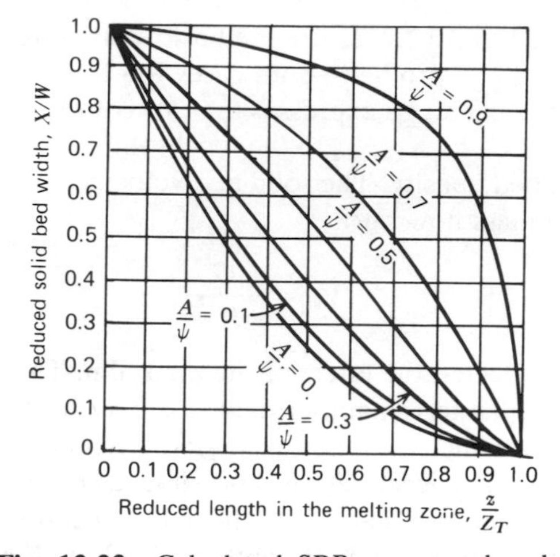


**Fig. 12.23** Calculated SBP versus reduced length of melting in a single section screw: $A/\psi = 0$ denotes a constant channel depth section. The SBP becomes increasingly concave as $A/\psi$ increases, either as a result of an increasing taper (i.e., increasing $A$) or a decreasing rate of melting (i.e., decreasing $\psi$). (Reprinted with permission from Z. Tadmor and I. Klein, *Engineering Principles of Plasticating Extrusion*, Van Nostrand Reinhold Book Co., New York, 1970.)

and the length of melting

$$Z_T = \frac{H}{\psi}\left(2 - \frac{A}{\psi}\right) \tag{12.2-31}$$

Comparing Eq. 12.2-31 to Eq. 12.2-29, we note that the length of melting in a tapered channel is always shorter than in a channel of constant depth.

Furthermore, the higher the taper, the shorter the melting, $Z_T$. But there is a limit to the taper that can be allowed because a high taper may lead to conditions under which the solid bed width will tend to increase instead of decrease (the cross-sectional area, of course, always decreases), which may lead to plugging of the screw channel, solid bed acceleration, and surging conditions in general. It is common practice to characterize the tapered sections of screws by "compression ratio," that is, the ratio of the channel depth in the feed section to that in the metering sections, although from the foregoing discussion, screw taper instead of compression ratio should be the value by which transition sections are characterized. Figure 12.23 illustrates the effect of taper on the shape of the calculated SBP. The width of the solid bed drops if $A/\psi < 1$, it stays constant if $A/\psi = 1$, and it increases if $A/\psi > 1$. All these cases have been experimentally observed. An increase in solid bed width means that the reduction in channel depth is faster than the rate of melting. This condition frequently occurs in a tapered section following a constant depth feed section. Thus at the beginning of the taper $X < W$, and an increase in $X$ is possible without generating either surging conditions or a breakdown of the drag removal melting mechanism. If, however, melting starts in a tapered section and conditions are such that $A/\psi > 1$, a stable drag-removal melting mechanism as described in this section may not be attainable. Conceivably under these conditions other melting mechanisms may be triggered into action, such as the previously mentioned dissipative mix-melting. Unfortunately, there is no sufficient information on these alternative melting mechanisms, neither were theoretical tools developed to a priori predict the nature of the melting mechanism in a particular operation.

Conditions that result in approximately constant solid bed width in the tapered section are desirable and frequently used. Even moderate solid bed width increases may often be tolerated. The experimental SBPs of Figs. 12.9, 12.11, 12.13 appear in Figs. 12.24–12.26. As predicted by the model, in all cases the solid bed drops continuously in the feed section (up to turn 12), it changes slope upon entering the tapered section, with plugging type conditions observed with nylon and stable constant width conditions observed with LDPE. In the metering section experimental measurements are inaccurate because solid bed breakup occurs, or because the bed is too narrow. These particular melting conditions are the combined result of operating conditions, screw design, and polymer properties. The calculated SBPs in Figs. 12.24–12.26 are based on a model that is no different in concept from the one discussed previously except that some of the simplifying assumptions were eliminated. In particular, the Newtonian constant viscosity fluid assumption is replaced by a power law model fluid, with temperature dependent parameter. Moreover, flight clearance and curvature effects were also accounted for. Figure 12.26 indicates that in this particular case, the simple Newtonian model provides a reasonable estimate, although it overestimates the rate of melting (cf. Example 9.5). Note that the predicted curve should approach the closed circles and triangles, which are the measured solid bed width at the melt film, rather than the open circles and

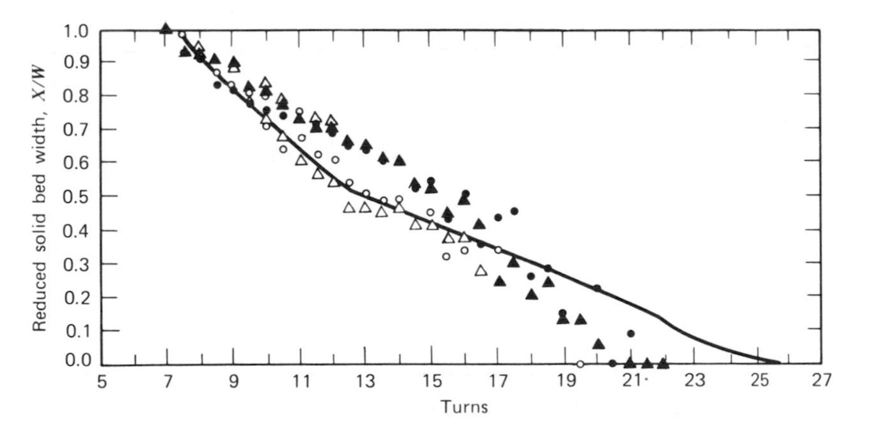


**Fig. 12.24** Experimentally measured SBPs by "cooling" experiments as in Fig. 12.9 (PVC) and theoretically calculated SBP (solid curve). Circles and triangles denote two identical experiments. Solid circles and triangles denote solid bed width at the barrel surface (maximum); open circles and triangles represent the solid bed width at the root of the screw (minimum). Operating conditions as follows: $T_b$, 375°F; $N$, 30 rpm; $P$, 4300 psi; $G$, 107.2 lb/hr. (Reprinted with permission from Z. Tadmor and I. Klein, *Engineering Principles of Plasticating Extrusion*, Van Nostrand Reinhold Book Co., New York, 1970.)

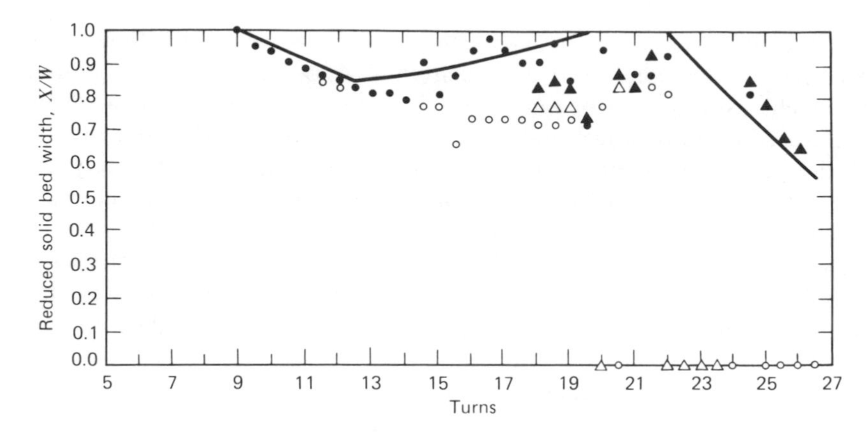


**Fig. 12.25** Experimentally measured SBPs, by "cooling" experiments as in Fig. 12.11 (nylon) and theoretically calculated SBP (solid curve). Circles and triangles as in Fig. 12.24. Note the increasing solid bed width in the tapered section. Operating conditions as follows: $T_b$, 575°F; $N$, 60 rpm; $P$, 3000 psi; $G$, 153.1 lb/hr. (Reprinted with permission from Z. Tadmor and I. Klein, *Engineering Principles of Plasticating Extrusion*, Van Nostrand Reinhold Book Co., New York, 1970.)

triangles, which are the corresponding values at the root of the screw. As observed experimentally, the width near the root of the screw is reduced as a result of melt pool circulation.

In the melting model already described, the assumption of constant viscosity is particularly troublesome, since very large variations of the viscosity are expected because of the large temperature variations. The rate of drag removal melting for a power law model fluid with temperature dependent viscosity is given in Eq. 9.8-53.

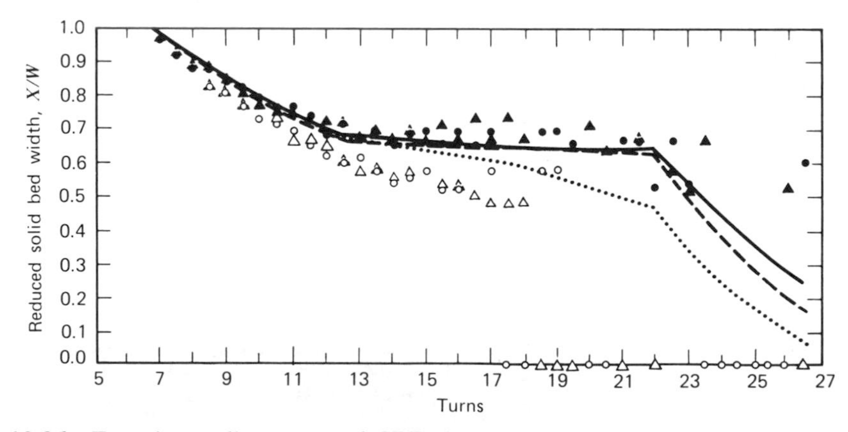


**Fig. 12.26** Experimentally measured SBPs by "cooling" experiments as in Fig. 12.13 (LDPE) and theoretically calculated SBP (solid curve). Circles and triangles as in Fig. 12.24. Upper broken curve is the result of the calculation of the SBP in Example 12.3. Lower broken curve is the result of a simplified Newtonian model. Operating conditions as follows: $T_b$, 300°F; $N$, 60 rpm; $P$, 3000 psi; $G$, 136.1 lb/hr. (Reprinted with permission from Z. Tadmor and I. Klein, *Engineering Principles of Plasticating Extrusion*, Van Nostrand Reinhold Book Co., New York, 1970.)

The corresponding value of $\Phi$, accounting once again for convection, is

$$\Phi=\left\{\frac{V_{bx}\rho_m U_2[k_m(T_b-T_m)+U_1/2]}{2[C_s(T_m-T_{s0})+C_m\bar{\Theta}(T_b-T_m)+\lambda]}\right\}^{1/2} \tag{12.2-32}$$

and the film thickness $\delta$ is

$$\delta=\left\{\frac{[2k_m(T_b-T_m)+U_1]X}{V_{bx}U_2\rho_m[C_s(T_m-T_{s0})+C_m\bar{\Theta}(T_b-T_m)+\lambda]}\right\}^{1/2} \tag{12.2-33}$$

where $U_2$ is given by Eq. 9.8-49

$$U_2=2\frac{1-b'-e^{-b'}}{b'(e^{-b'}-1)} \tag{12.2-34}$$

and $b'$ is defined* in Eq. 9.8-41:

$$b'=\frac{-a(T_b-T_m)}{n} \tag{12.2-35}$$

and $U_1$ is given by (cf. Eq. 9.8-51):

$$U_1=2m_0V_j^{n+1}\bar{\delta}^{1-n}\frac{(e^{-b'}+b'-1)}{(b')^2}\left(\frac{b'}{1-e^{-b'}}\right)^{n+1} \tag{12.2-36}$$

Equation 12.2-36 for constant viscosity Newtonian fluid ($n=1$, $\mu=m_0$, and $a=0$) reduces to $\mu V_j^2$. Finally the mean temperature of the film $\bar{\Theta}$ is given in Eq. 9.8-55:

$$\bar{\Theta}=\frac{b_1'/2+e^{-b'}(1+1/b')-1/b'}{b'+e^{-b'}-1} \tag{12.2 37}$$

Modifications of this model, including a nonlinear temperature profile in the melt film, channel curvature effects, and an approximate method to account for the flight clearance effect, are presented in Ref. 2, together with expressions for power calculations. Numerous other improvements of the melting model have been suggested in the literature (13, 23–28). A detailed discussion of these is, however, beyond the scope of this text.

### Example 12.3 *Melting in Screw Extruders*

The screw geometry and operating conditions for the LDPE extrusion experiment (Figs. 12.13 and 12.26) were given in Example 12.2. Calculate the SBP, using the power law model with temperature dependent viscosity and linear temperature profile.

The polymer melt follows the power law model in the shear rate and temperature ranges of interest

$$\eta=5.6\times10^4\,e^{-0.01(T-110)}\dot{\gamma}^{-0.655}$$

* Note that for $T_b>T_m$, $b'<0$ and is identical to $A_4$ in Ref. 2.

where $\eta(\mathrm{N \cdot s/m^2})$ is the non-Newtonian viscosity, $T$ (°C) is the temperature, and $\dot{\gamma}$ ($\mathrm{s^{-1}}$) the shear rate. Polymer melt density as a function of pressure and temperature follows the empirical relationship

$$\rho_m = 852.7 + 5.018 \times 10^{-7} P - 0.4756T$$

where $\rho_m$ ($\mathrm{kg/m^3}$) is the density, $P$ ($\mathrm{N/m^2}$) the pressure, and $T$ (°C) the temperature. The melting point of the polymer $T_m$ is 110°C, its thermal conductivity $k_m$ is 0.1817 W/m · °C, and heat capacity $C_m$ is 2.596 kJ/kg · °C. The density of the solid polymer is 915.1 $\mathrm{kg/m^3}$, its heat capacity $C_s$ is 2.763 kJ/kg · °C, and the bulk density of the feed at atmospheric pressure is 595 $\mathrm{kg/m^3}$. The heat of fusion is 129.8 kJ/kg.

*Solution*: In this example we know the location of the beginning of the melting zone from experimental data. As Fig. 12.13 indicates, melting starts at turn number 7. Hence we can proceed with the SBP calculation without evaluating the length of the delay zone. The first step is calculating $\Phi$ from Eq. 12.2-32. In the expression for $\Phi$, we have the variables $U_1$, $U_2$, and $\bar{\Theta}$, which we calculate from Eqs. 12.2-36, 12.2-34, and 12.2-37, respectively, with $b'$ evaluated from Eq. 12.2-35:

$$b' = \frac{-(0.01)(149 - 110)}{(0.345)} = -1.1304$$

hence $U_2$ is

$$U_2 = (2)\frac{1 + (1.1304) - e^{1.1304}}{(-1.1304)(e^{1.1304} - 1)} = 0.814$$

The physical meaning of this result is that the drag removal action of the barrel is reduced by a factor of 0.814 as a result of the temperature profile in the film on which the shear thinning effect is superimposed.

The down channel velocity of the solid bed is obtained from Eq. 12.2-26:

$$V_{sz} = \frac{(67.1/3600)}{(915.1)(0.009398)(0.05314)} = 0.0408 \text{ m/s}$$

The velocity of the barrel surface is

$$V_b = \pi(1)(0.0635) = 0.1995 \text{ m/s}$$

The absolute value of the velocity difference $\mathbf{V}_b - \mathbf{V}_{sz}$ from Eq. 12.2-12a is

$$V_j = [(0.1995)^2 + (0.0408)^2 - (2)(0.1995)(0.0408)\cos(17.65)]^{1/2}$$
$$= 0.161 \text{ m/s}$$

To calculate $U_1$ from Eq. 12.2-36, we must solve simultaneously Eqs. 12.2-36 and 12.2-33:

$$U_1 = (2)(5.6 \times 10^4)(0.161)^{1.345}\, \delta^{0.655} \frac{[e^{1.1304} - (1.1304) - (1)]}{(1.1304)^2}\left(\frac{-1.1304}{1 - e^{1.1304}}\right)^{1.345}$$

$$= 3163.8\, \delta^{0.655}$$

The mean dimensionless temperature $\bar{\Theta}$ from Eq. 18.2-37 is

$$\bar{\Theta} = \frac{e^{1.1304}[(1)-(1.1304)^{-1}]+(1.1304)^{-1}-(1.1304)/(2)}{e^{1.1304}-1.1304-1} = 0.700$$

In Eq. 12.2-33 for $\delta$ we face a difficulty with the density, whose value is a function of pressure and temperature. The pressure varies with the down channel location, which couples the melting with melt conveying. This is a *weak* coupling, however, and we shall use a constant density at a mean temperature of $\bar{T} = (0.7)(149-110)+110 \cong 137°\text{C}$ and estimated mean pressure of $6.89 \times 10^6\ \text{N/m}^2$ ($\cong$1000 psi). Thus with $\rho_m = 791\ \text{kg/m}^3$, Eq. 12.2-33 results in

$$\delta = \left\{ \frac{[(2)(0.1817)(149-110)+U_1]X}{(0.1995)\sin(17.65)(0.814)(791) \times [(2763)(110-24)+(129.8\times 10^3)+(2596)(0.638)(149-110)]} \right.$$

$$= 2.438 \times 10^{-4}[(14.17+U_1)X]^{1/2}\ \text{m}$$

Next, we solve $U_1$ and $\delta$ simultaneously for a few $X$ values. Results are given in the following table.

| $X$ (m) | $U_1$ (N/s) | $\delta$ (m) |
|---|---|---|
| 0.055 | 16.07 | $3.144 \times 10^{-4}$ |
| 0.035 | 13.46 | $2.398 \times 10^{-4}$ |
| 0.025 | 11.81 | $1.965 \times 10^{-4}$ |

We note that the value of $\delta$ is very small compared to the channel depth.

From Eq. 12.2-32 we obtain

$$\Phi = \left\{ \frac{(0.1995)\sin(17.65)(791)(0.814)[(0.1817)(149-110)+(0.5)U_1]}{(2)(2763)(110-24)+(129.8\times 10^3)+(2596)(0.638)(149-110)]} \right\}^{1/2}$$

$$= 4.7474 \times 10^{-3}(14.17+U_1)^{1/2}\ \text{kg/s}\cdot\text{m}^{1.5}$$

By comparing $U_1$ values from the table to the value of 14.17, we note that viscous dissipation and heat conduction are about equal in this case.

Since melting starts in the feed section, we use Eq. 12.2-21 to compute the SBP. Melting starts at turn 7, and the constant channel feed section ends at turn 12.5. There are, therefore, 5.5 turns of constant channel depth where melting takes place. Since $\Phi$ depends on $U_1$, which is a function of $X$, we make the calculation in increments. For this example, we take increments one turn long and evaluate $\delta$, $U_2$, and $\Phi$ at conditions prevailing at the entrance of the increment. A more accurate calculation procedure would involve the evaluation of these variables at conditions prevailing in the middle of each increment. In this example this involves the solid bed width $X$ appearing in the expression for $\delta$. Hence iterative calculation procedures are needed for this purpose. Such a procedure can easily be carried out on a computer. We calculate $\psi$ from Eq. 12.2-22 subsequent to solving for $U_1$ and $\Phi$ by interpolation in the table (for the interpolation, we use $\delta_1/\delta_2 \cong \sqrt{X_1/X_2}$). Results of the computations are tabulated below. When the SBP is calculated with Eq. 12.2-21, in the first increment we use $X_1/W = 1$.

| Increment | | Inlet Conditions | | | | $X/W$ at |
|---|---|---|---|---|---|---|
| Start (turn) | End (turn) | $X_1$ (m) | $U_1$ (N/s) | $\Phi$ (kg/s · m$^{1.5}$) | $\psi$ | End of Increment |
| 7 | 8 | $5.416\times10^{-2}$ | 15.99 | 0.0261 | $3.0006\times10^{-3}$ | 0.934 |
| 8 | 9 | $5.06\times10^{-2}$ | 15.64 | 0.0259 | $3.0862\times10^{-3}$ | 0.871 |
| 9 | 10 | $4.717\times10^{-2}$ | 15.28 | 0.0257 | $3.1173\times10^{-3}$ | 0.810 |
| 10 | 11 | $4.389\times10^{-2}$ | 14.71 | 0.0255 | $3.2616\times10^{-3}$ | 0.752 |
| 11 | 12 | $4.074\times10^{-2}$ | 14.14 | 0.0253 | $3.3521\times10^{-3}$ | 0.697 |
| 12 | 12.5 | $3.774\times10^{-2}$ | 13.79 | 0.0251 | $3.4617\times10^{-3}$ | 0.670 |

In the tapered section we follow in principle the same procedure but use Eq. 12.2-25 to calculate the SBP. The mean taper is

$$A = \frac{(9.398\times10^{-3})-(3.226\times10^{-3})}{(9.5)(6.35\times10^{-2})/\sin(19.51)} = 3.4169\times10^{-3}$$

Results are tabulated on p. 507.

From turn 22 to turn 26.5 we have once again a constant channel section, which is computed by the same procedure as in the feed section. The channel depth in Eq. 12.2-21 is that of the metering section.

| Increment | | Inlet Conditions | | | | |
|---|---|---|---|---|---|---|
| Start (turn) | End (turn) | $X_1$ (m) | $U_1$ (N/s) | $\Phi$ (kg/s · m$^{1.5}$) | $\psi$ | $X/W$ at End of Increment |
| 22 | 23 | $3.482\times10^{-2}$ | 13.44 | 0.0249 | $3.580\times10^{-3}$ | 0.502 |
| 23 | 24 | $2.720\times10^{-2}$ | 21.14 | 0.0243 | $3.955\times10^{-3}$ | 0.381 |
| 24 | 25 | $2.065\times10^{-2}$ | 11.10 | 0.0239 | $4.448\times10^{-3}$ | 0.279 |
| 25 | 26.5 | $1.511\times10^{-2}$ | 10.02 | 0.0233 | $5.087\times10^{-3}$ | 0.158 |

The fraction of unmelted polymer leaving the screw is obtained from the last $X/W$ value. The fraction of unmelted polymer in the extrudate is $G_s/G$ is

$$\frac{G_s}{G} = \frac{(X/W)\bar{W}HV_{sz}\rho_s}{G} = \frac{(0.158)(5.385\times10^{-2})(3.226\times10^{-3})(0.0408)(915.1)}{67.1/3600}$$

$$= 0.055$$

Such a level of unmelted polymer can be tolerated frequently because the screen pack and the die itself provide additional opportunities for melting and mixing.

| Increment | | Channel Depth | | | Inlet Conditions | | | | | |
|---|---|---|---|---|---|---|---|---|---|---|
| Start (turn) | End (turn) | Start, $H_1$ (m) | End, $H_2$ (m) | $H_1/H_2$ | $X_1$ (m) | $U_1$(N/s) | $\Phi$ (kg/s · m$^{1.5}$) | $\psi$ | $\psi/A$ | $X/W$ at End of Increment |
| 12.5 | 14.5 | $9.3980 \times 10^{-3}$ | 8.0986 | 1.1604 | $3.628 \times 10^{-2}$ | 13.61 | 0.025 | $3.519 \times 10^{-3}$ | 1.030 | 0.667 |
| 14.5 | 16.5 | $8.0986 \times 10^{-3}$ | 6.7993 | 1.1911 | $3.612 \times 10^{-2}$ | 13.60 | 0.025 | $3.525 \times 10^{-3}$ | 1.032 | 0.663 |
| 16.5 | 18.5 | $6.7993 \times 10^{-3}$ | 5.4999 | 1.2362 | $3.592 \times 10^{-2}$ | 13.57 | 0.025 | $3.534 \times 10^{-3}$ | 1.034 | 0.658 |
| 18.5 | 20.5 | $5.4999 \times 10^{-3}$ | 4.2005 | 1.3093 | $3.563 \times 10^{-2}$ | 13.54 | 0.025 | $3.546 \times 10^{-3}$ | 1.038 | 0.651 |
| 20.5 | 22.0 | $4.2005 \times 10^{-3}$ | 3.2260 | 1.3020 | $3.525 \times 10^{-2}$ | 13.49 | 0.025 | $3.562 \times 10^{-3}$ | 1.042 | 0.643 |

Figure 12.26 plots the calculated SBP as a broken curve compared to experimental data obtained (Fig. 12.13). We note the agreement is generally good except in the metering section where, as a result of the solid bed breakup, experimental data are scattered. The parabolic drop in constant depth sections, the sharp break in the slope of the SBP upon entering the tapered section, as well as the approximately constant solid bed width (which is the combined result of channel taper and operating conditions) are all clearly evident in the experimental results and were predicted by the model. The solid curves are the result of computations with a more accurate model accounting for flight clearance and channel curvature effects (approximately offsetting each other in this example) as well as nonlinear temperature distribution in the film of melt and reevaluation of the rheological parameters at the local conditions. Nevertheless, the example demonstrates that calculations with a relatively simple model provide useful results.

## *Melt Conveying*

In a plasticating extruder two distinct melt conveying regions may be found. One is downstream the melting zone after the completion of melting, where the models derived in the preceding section can be applied without modifications. In addition, melt conveying occurs in the melt pool, which extends side by side with the solid bed profile. Here the width of the melt pool changes in the flow direction.

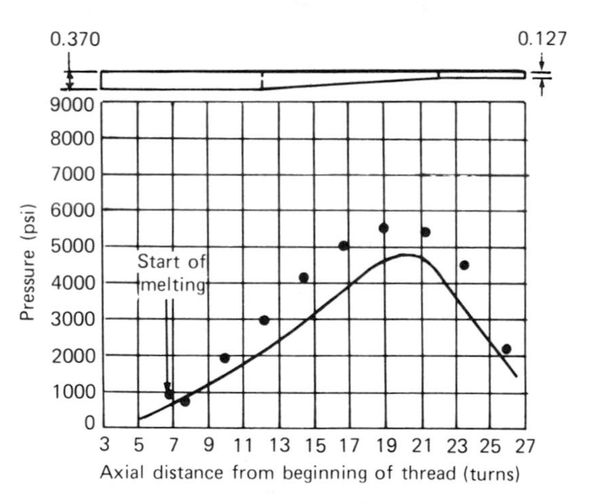


**Fig. 12.27** Simulated and measured pressure profiles for an LDPE extruded in a 2.5 in. diameter, 26.5 length-to-diameter ratio extruder. For details of the experiment, see Fig. 12.13. The model used to simulate the pressure profile is discussed in detail in Ref. 2. (Reprinted with permission from Z. Tadmor and I. Klein, *Engineering Principles of Plasticating Extrusion*, Van Nostrand Reinhold Book Co., New York, 1970.)

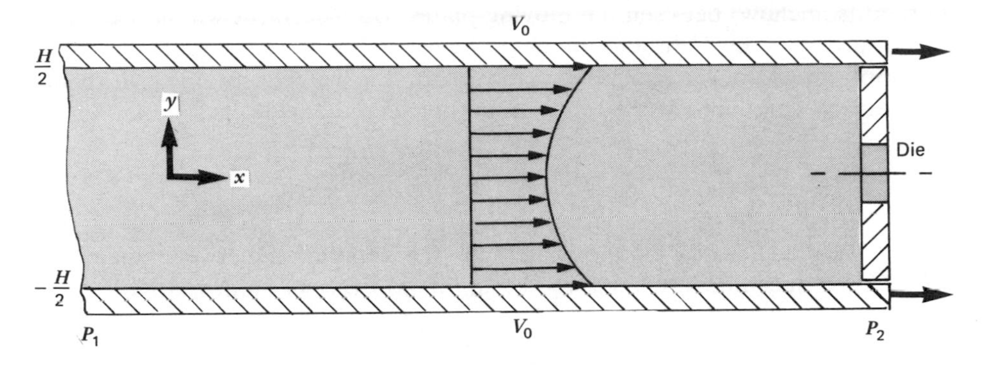


**Fig. 12.28** Two parallel plate geometry with both plates moving of constant velocity $V_0$.

Moreover, the mass flow rate of melt also changes as a result of the influx from the melt film. Both these quantities, as well as the mean melt film temperature, are obtained from the melting model. Hence the melt conveying model can be applied approximately to calculate local pressure gradients and temperature changes over small finite axial increments using the mean local flow rate and melt pool size (2, 27). Figure 12.27 presents results for such a calculation. In this method one assumes that the melting step affects strongly the melt pumping step, but the possible effect of the latter on the former is neglected. In effect the melt in the pool exerts pressure and viscous drag on the solid bed which, together with the viscous drag over the melt film, the root of the screw, and the trailing flight, determine the stress distribution in the solid bed. Such an analysis coupling the two phases has been attempted (13, 28), and in principle, it may provide the way to predict solid deformation, acceleration, and breakup.

## 12.3 The Diskpack* Polymer Processor

Sections 12.1 and 12.2 assembled the relevant elementary steps into the important and widely used screw extrusion process. The common and outstanding feature of all elementary steps taking place in the screw channel is that they are all induced by drag brought about by a single moving surface—that of the barrel. Solids are conveyed and compressed by the frictional drag of the barrel surface; they are melted by a drag removal melting step, and the melt is pressurized and mixed by drag induced flow. The other main surface of the channel, the stationary root of the screw has no "useful" operational function but to serve as a geometrical boundary. The efficiency of the elementary steps is even retarded by it. In search of a better processing configuration, while keeping in mind the nature of the elementary steps, one is, therefore, forced to abandon the traditional screw, and investigate the improvements possible with a geometrical configuration that has *two* moving drag inducing surfaces.

We consider first the elementary step of pumping in the parallel plate geometry in Fig. 12.28, where both the upper *and* lower plates move in the positive $x$-direction. As with the parallel plate geometry with one moving surface, we assume that somewhere downstream there is a restriction to flow, such as a die, through which the pressurized polymer is shaped. Consider a viscous Newtonian

*Trade name applied for by the Farrel Company.

liquid fed (somehow) between the moving plates. Both surfaces will now drag the melt toward the die. Without much difficulty, making the usual simplifying assumptions, we can derive a simple model for the velocity profile, which in a coordinate system placed at the center point between the plates is

$$v(\xi) = V_0 - (1-\xi^2)\left(\frac{H^2}{8\mu}\frac{dP}{dx}\right) \tag{12.3-1}$$

where $V_0$ is the velocity of the plates, $H$ the separation between them, and $\xi = 2y/H$, which can be integrated to obtain the flow rate per unit width $q'$

$$q' = q'_d + q'_p = V_0 H - \frac{H^3}{12\mu}\left(\frac{dP}{dx}\right) \tag{12.3-2}$$

where $q'_d$ and $q'_p$ are the drag and pressure flow rates per unit width, respectively, and their ratio is

$$\frac{q'_p}{q'_d} = -\frac{H^2}{12\mu V_0}\frac{dP}{dx} \tag{12.3-3}$$

Comparing Eqs. 12.3-2 and 10.2-7, we note that as expected, physically, the *drag flow* term is doubled. Consequently there is a difference of a factor of 2 between Eqs. 12.3-3 and 10.2-10. The increased drag flow implies a substantial increase in the pressurization capability. A quantitative estimate of this increased capability is obtained by comparing the pressure rises predicted by Eqs. 10.2-7 and 12.3-2, at equal net flow rate $q = q'$, equal viscosity, equal plate velocity $V_0$, and *optimum* corresponding $H$. The maximum pressure gradient under these conditions with one plate moving was shown in Section 10.2 to be $(dP/dx)_{\max} = 6\mu V_0^3/27q^2$. Similarly, from Eq. 12.3-2 we obtain that for a given $q'$ and $V_0$ a maximum pressure gradient is obtained with $H = 3q'/2V_0$ (i.e., $q'/q'_d = 2/3$), resulting in a maximum pressure gradient of $(dP/dx)'_{\max} = 48\mu V_0^3/27q^2$. Thus

$$\frac{\left(\frac{dP}{dx}\right)'_{\max}}{\left(\frac{dP}{dx}\right)_{\max}} = \frac{48\mu V_0^3/27q^2}{6\mu V_0^3/27q^2} = 8 \tag{12.3-4}$$

indicating an eightfold increase in pressurization capability.

There is, however, a further fundamental difference between the two cases (i.e., one moving plate vs. two moving plates), and it is reflected in the shear rate distributions. These are obtained from Eq. 12.3-1 and 10.2-1, respectively, for both plates moving

$$\dot{\gamma}'(y) = \left|\frac{y}{\mu}\frac{\Delta P}{L}\right| \tag{12.3-5}$$

and for one plate moving

$$\dot{\gamma}(y) = \left|\frac{V_0}{H} + \frac{y}{\mu}\frac{\Delta P}{L}\right| \tag{12.3-6}$$

(Note, $y = 0$ is at the center point of the gap.) Clearly, shear rate is *independent* of plate velocity when both plates move at the same velocity, whereas shear rate increases with plate velocity when one plate moves relative to the other. In the former case, when $\Delta P = 0$—that is, at pure drag flow (plug flow) conditions—shear rate, consequently viscous heat dissipation, both vanish, whereas in the latter case, at pure drag flow conditions, mechanical energy continues to be dissipated into heat at a rate of $\mu(V_0/H)^2$. The practical consequences of these differences between the two cases are that machines based on relative motion between surfaces, such as screw extruders, have upper bounds on plate velocity set by thermal and shear degradation sensitivity of the polymer. Indeed, in developing the screw extruder geometry we recall that this limitation led to the need to increase channel length (by wrapping it around a cylinder) to obtain sufficient pressurization capability. Therefore, a machine that is based on the motion of both surfaces at equal velocities, not only generates pressure much more efficiently than its counterpart with one plate moving, at equal $V_0$ and $q$, but its pressurization capability is greatly expanded by the fact that no inherent upper theoretical limit on $V_0$ exists.

Turning now to the elementary step of mixing, we rewrite Eq. 12.3-1 with the aid of Eq. 12.3-3 as follows:

$$\frac{v(\xi)}{V_0} = 1 + \frac{3}{2}\frac{q'_p}{q'_d}(1-\xi^2) \tag{12.3-7}$$

Figure 12.29 shows some calculated velocity profiles with $q'_p/q'_d$ as a parameter. The ratio $q'_p/q'_d < 0$ decreases ($|q'_p/q'_d|$ increases) with increasing pressure rise, as indicated by Eq. 12.3-3. At closed discharge ($q'_p/q'_d = -1$), Fig. 12.29 indicates that a vigorous circulatory flow arises between the parallel plates reminiscent of that in intermeshing twin screw extruders. Fluid is dragged toward the exit of both surfaces and flows backward in the center zone ($-\sqrt{\frac{1}{3}} < \xi < \sqrt{\frac{1}{3}}$). At $q'_p/q'_d = -\frac{2}{3}$, the velocity in at the center is zero and no back flow takes place. The velocity profile at maximum pressure rise for a *given* net flow rate $q$, and velocity $V_0$ where $q'_p/q'_d = -\frac{1}{3}$ (or

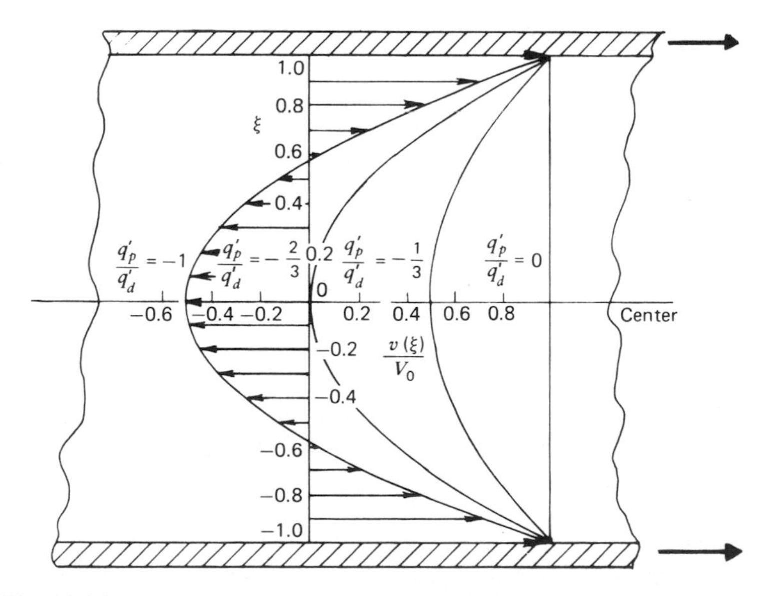


**Fig. 12.29** Calculated velocity profiles from Eq. 12.3-7 at various $q'_p/q'_d$.

$q'/q'_d = \frac{2}{3}$) also appears in Fig. 12.29. Finally, at $q'_p/q'_d = 0$ the expected plug-shaped velocity profile is obtained.

The discussion above indicates that although good laminar mixing potential does exist in this configuration, it is obtained at operating conditions at which the pumping step is less efficient.

Turning to the other elementary steps, similar advantages are evident. In solids conveying, much as in pressurization, the two forward dragging surfaces enhance conveying capacity. With regard to the elementary step of melting, we know that one practical and efficient melting mechanism is conduction melting with forced drag melt removal. Considering the geometry of two moving parallel plates, it is reasonable to assume that such a mechanism can be obtained with both surfaces removing melt (Fig. 12.30). Such a mechanism has the added advantage of a short

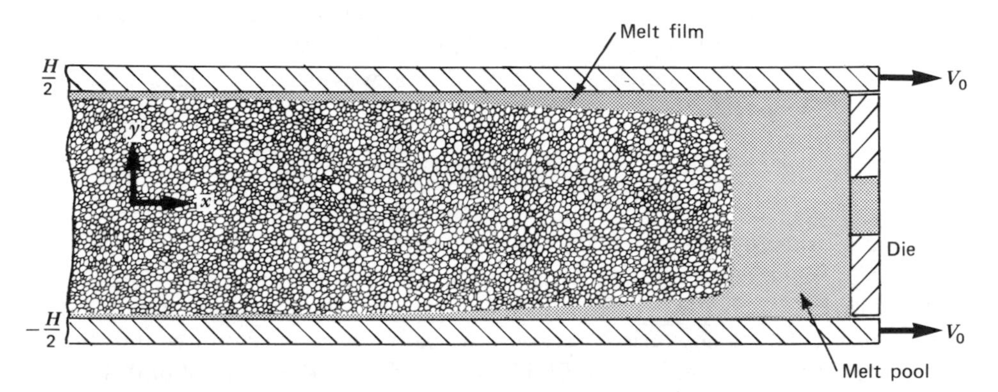


**Fig. 12.30** Schematic representation of the melting mechanism that may take place in a parallel plate geometry with both plates moving at equal velocities.

residence time of the melt because the newly melted polymer is quickly brought to the die. Recall that in screw extruders, melt generated at the beginning of the melting zone spends a relatively long time in the extruder.

Thus by considering the basic mechanisms of the individual elementary steps as discussed in preceding chapters, it was possible to conclude that a parallel plate geometry with both surfaces moving has a good potential for performing all these steps efficiently. The question that arises, then, involves the means of converting this simplified geometry into a *useful machine* configuration. That is, to find a practical design solution utilizing the geometrical "building block" of two moving parallel plates. This is not an obvious step and as all design problems calls for inventive engineering. However, as Section 10.3, discussing the evolution of the screw extruder from the parallel plate geometry, stresses, the understanding of the basic mechanisms may be a most useful aid to this process.

Repeating the metal exercise in Section 10.3 it can be visualized that one possible outcome of such a process would be a "screw extruder" with the flights detached from the screw: that is, a freely turning spiral, closely fitted inside by a cylinder and outside by the barrel. The barrel and inner cylinder would then rotate in unison, or preferably, the flight would rotate in the opposite direction amid the stationary barrel and cylinder. Theoretically this would form an excellent machine with greatly improved performance over the screw extruder. Mechanically, however, great difficulties would arise in transmitting the power through such a flight, which might deform and twist. Nevertheless, solids feeders of such design are being manufactured (29).

Another possible solution to the problem is to use the outside surfaces of cylinders of rolls, creating essentially a two roll mill as described in Section 10.5. Indeed, as Fig. 10.19 shows, die forming by this process is feasible. Inherent in this geometry, however, is the limitation on the length along which pressure can be built up, and in addition, the hot polymer is open to the atmosphere.

Yet a further possible way to create continuously moving surfaces in a practical way is through rotating disks. We have already encountered three processing machines using this geometry, namely, the normal stress extruder, the sliding pad extruder, and the flat spiral extruder. In all these cases, however, we have one surface *moving relative* to the other. But two surfaces moving together can also be obtained by *two* disks. They can be attached to a shaft and enclosed in a barrel, as in Fig. 12.31. We now have our basic configuration with the processing channel completely enclosed. As with screw extruders, an opening in the barrel can provide for inlet and feeding, and another opening for discharge. By placing a "*channel block*" attached to the stationary barrel separating the inlet from the exit, pressure will build up against the back of the block and the polymer will be diverted toward the exit to a die (Fig. 12.31*b*). Such a geometry was proposed by Tadmor (30) and it is the structural element of the Diskpack Polymer Processor. Its development into a processing machine and its theoretical analysis was discussed by Tadmor, Hold and Valsamis (31, 32).

The present geometry limits the (circumferential) length of the channel to somewhat less than the circumference of the disks. We cannot "twist" the channel and extend it indefinitely as we did with the screw extruder. It is possible, however, to have a *pack of disks* connected in series, with a stagewise pressure buildup. Alternatively, a pack of chambers in parallel arrangement provides great flexibility, increasing melting capacity.

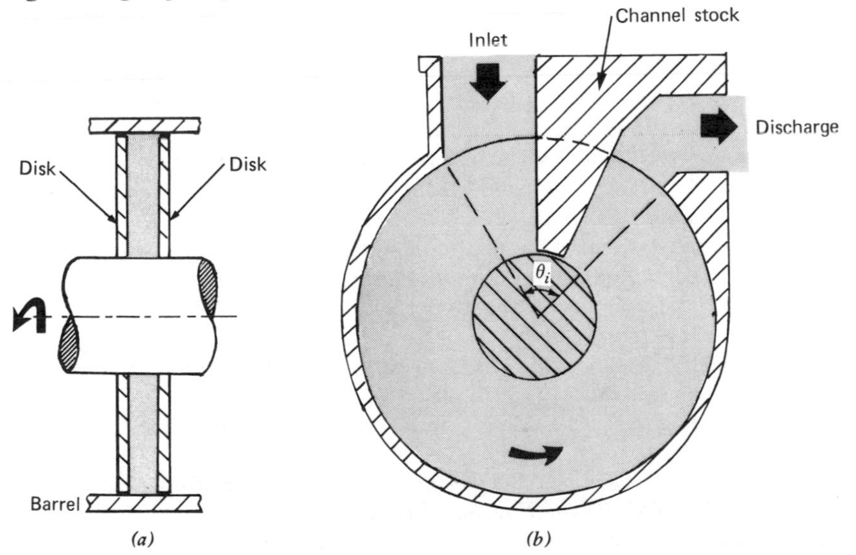


**Fig. 12.31** Schematic view of a structural element of the Diskpack Polymer Processor (30–32) showing (*a*) two disks attached to the shaft forming one annular channel, with (*b*) an inlet and an outlet opening through the barrel separated by a channel block.

As pointed out earlier, extensive mixing can be attained by restricting the output, inducing double circulatory flow patterns, whereas intensive mixing can be induced by allowing a small clearance between the channel block on disk surfaces (conductive to either batch or continuous mixing operations), or by inserting various auxiliary "mixing blocks" into the channel upstream the channel block. Such auxiliary channel blocks can also be used for creating the possibility for venting, much as in screw extrusion (30–32).

## REFERENCES


1. M. L. Booy, "Influence of Oblique Channel Ends on Screw Pump Performance," *Polym. Eng. Sci.*, **7**, 5–16 (1967).
2. Z. Tadmor and I. Klein, *Engineering Principles of Plasticating Extrusion*, Van Nostrand Reinhold, New York, 1970: (a) p. 234; (b) p. 310; (c) p. 328; (d) Chapters 6, 8; (e) p. 186; (f) p. 90; (g) p. 133.
3. R. M. Griffith, "Fully Developed Flow in Screw Extruders," *Ind. Eng. Chem. Fundam.*, **1**, 180–187 (1962).
4. H. Zamodits and J. R. A. Pearson, "Flow of Polymer Melts in Extruders. Part I. The Effect of Transverse Flow and of Superposed Steady Temperature Profile," *Trans. Soc. Rheol.*, **13**, 357 (1969).
5. B. Martin, J. R. A. Pearson, and B. Yates, "On Screw Extrusion. Part I. Steady Flow Calculations," University of Cambridge, Department of Chemical Engineering, Polymer Processing Research Center, Report No. 5, 1969.
6. S. Middleman, "Flow of Power Law Fluids in Rectangular Ducts," *Trans. Soc. Rheol.*, **9**, 83 (1965).
7. I. Klein and Z. Tadmor, *Computer Programs for Plastics Engineers*," I. Klein and D. I. Marshall, *eds.*, Reinhold, New York, 1968, Chapter 6.2.
8. R. E. Colwell and K. R. Nicholls, "The Screw Extruder," *Ind. Eng. Chem.*, **51**, 841–843 (1959).
9. B. H. Maddock, "A Visual Analysis of Flow and Mixing in Extruder Screws," *Technical Papers*, Vol. V, 15th Annual Technical Conference, Society of Plastics Engineers, New York, January 1959; also *Soc. Plast. Engs. J.*, **15**, 383 (1959).
10. L. F. Street, "Plastifying Extrusion," *Int. Plast. Eng.*, **1**, 289 (1961).
11. Z. Tadmor, I. J. Duvdevani, and I. Klein, "Melting in Plasticating Extruders—Theory and Experiments," *Polym. Eng. Sci.*, **7**, 198–217 (1967).
12. G. Menges and P. Klenk, "Melting and Plasticating of Unplasticized PVC Powder in the Screw Extruder," *Kunststoffe* (*German Plastics*), **57**, 598–603 (1967).
13. J. T. Lindt, "A Dynamic Melting Model for Single Screw Extruders," *Polym. Eng. Sci.*, **16**, 284–291 (1976).
14. L. Kacir and Z. Tadmor, "Solids Conveying in Screw Extruders. Part III. The Delay Zone," *Polym. Eng. Sci.*, **12**, 307–395 (1972).
15. W. H. Darnell and E. A. J. Mol, "Solids Conveying in Extruders," *Soc. Plast. Engs. J.*, **12**, 20–28 (1956).
16. E. Broyer and Z. Tadmor, "Solids Conveying in Screw Extruders. Part I. A Modified Isothermal Model," *Polym. Eng. Sci.*, **12**, 12–24 (1972).
17. K. Schneider, "Technical Report on Plastics Processes in the Feeding Zone of an Extruder" (in German), Institute of Plastics Processing (IKV), Aachen, 1969.
18. Z. Tadmor and E. Broyer, "Solids Conveying in Screw Extruders. Part II. Non-Isothermal Model," *Polym. Eng. Sci.*, **12**, 378–386 (1972).

19. J. G. A. Lovegrove and J. G. Williams, "Solids Conveying in Single Screw Extruders—The Role of Gravity Forces," *J. Mech. Eng. Sci.*, **15**, 114–122 (1973); also, *ibid.*, 195–199.
20. J. G. A. Lovegrove and J. G. Williams, "Pressure Generation Mechanism in the Feed Section of Screw Extruders," *Polym. Eng. Sci.*, 589–594 (1974).
21. C. I. Chung, "New Ideas About Solids Conveying in Screw Extruders," *Soc. Plast. Engs. J.*, **26** (5), 32–44 (1970).
22. Z. Tadmor, "Fundamentals of Plasticating Extrusion." I. A. Theoretical Model for Melting," *Polym. Eng. Sci.*, **6**, 185–190 (1966).
23. C. I. Chung, "A New Theory for Single Screw Extrusion. Part I and Part II. *Mod. Plast.*, **45**, 178 (1968).
24. D. R. Hinrich and L. U. Lilleleht, "A Modified Melting Model for Plasticating Extruders," *Polym. Eng. Sci.*, **10**, 268–278 (1970).
25. R. C. Donovan, "A Theoretical Melting Model for Plasticating Extruders," *Polym. Eng. Sci.*, **11**, 247–257 (1971).
26. J. R. Vermeulen, P. G. Scargo, and W. J. Beek, "The Melting of a Crystalline Polymer in a Screw Extruder," *Chem. Eng. Sci.*, **26**, 1457–1465 (1971).
27. I. R. Edmundson and R. T. Fenner, "Melting of Thermoplastics in Single Screw Extruders," *Polymer*, **16**, 48–56 (1975).
28. J. Shapiro, A. L. Halmos, and J. R. A. Pearson, "Melting in Single Screw Extruders," *Polymer*, vol. 17, No. 10, 905–918 (Oct. 1976).
29. G. Schenkel, *Plastics Extrusion Technology and Theory*, Illife Books, London, 1966.
30. Z. Tadmor, "Method and Apparatus for Processing Polymeric Materials," Patents to be issued for Farrel Co. Division of USM.
31. Z. Tadmor, P. Hold and L. Valsamis, "Method and Apparatus for Processing Polymeric Materials," pending patent applications for Farrel Co. Division of USM.
32. Z. Tadmor, P. Hold, and L. Valsamis, "A Novel Polymer Processing Machine Theory & Experiments" — 37th Soc. of Plastics Eng., Annual Technical Conference, New Orleans, 1979, pp. 193-204. P. Hold, Z. Tadmor, and L. Valsamis, "Application and Design of a Novel Polymer Processing Machine," *ibid.*, pp. 205-211.

## PROBLEMS

**12.1** ***Analysis of a Melt Extrusion Process.*** Consider the extrusion process shown in Fig. 12.1 for pelletizing 8,000 Kg/hr of polymer melt. The 40 cm diameter, and 12 $L/D$ extruder, has a square pitched metering type screw. The feed section is 3 turns long and 7.5 cm deep and the metering section is 6 turns long and 2.5 cm deep. The flight width is 3 cm and the flight clearance is negligible. Neglecting the breaker plate and screen pack, the die consists of a 3 cm thick pelletizing plate with 1000 holes over its surface of geometry, shown in Fig. 12.1, with $l_1 = l_2 = l_3 = 1$ cm, $R_1 = 0.5$ cm and $R_2 = 0.25$ cm. The extruded polymer is an incompressible Newtonian fluid with a viscosity of $10^3\ \mathrm{N \cdot s/m^2}$ and a density of $0.75\ \mathrm{g/cm^3}$.

Calculate (a) The screw speed needed to obtain the required output and the resulting head pressure, (b) the power and (c) the mean strain and residence time in the extruder. Furthermore, specify if, (d) the isothermal assumption is valid and (e) estimate the minimum size of a tubular inlet conduit for gravitational feeding.

***Answer*:** (a) 34.7 rpm; (b) 116,711 $W$; (c) $\bar{t} = 70.9$ s, $\bar{\gamma} = 3750$; (e) $R_0 = 0.179$ m.

**12.2 *The Superposition Correction Factor.**** Combined drag and pressure flow between parallel plates (or concentric cylinders†) of a Newtonian fluid at isothermal conditions leads to a flow rate expression which is the linear sum of two independent terms, one for drag flow and another for pressure flow

$$Q = Q_d + Q_p$$

The former vanishes when the velocity of the moving plate is zero, and the latter term vanishes in the absence of a pressure gradient.

(a) Explain on physical and mathematical grounds why the solution of the same flow problem with a non-Newtonian fluid, for example a power law model fluid, no longer leads to the same type of expressions.

(b) It is possible to define a superposition correction factor $\varepsilon$ as follows

$$Q = \varepsilon(Q_d^* + Q_p^*)$$

where $Q_d^*$ and $Q_p^*$ are hypothetical drag and pressure flow terms, each calculated assuming the other vanishes with a power law model. Thus, $Q_d^* = Q_d$, the pure drag flow term, and $Q_p^*$ is the pure pressure flow term of a power law model fluid. Show that $\varepsilon$ in parallel plate flow is a function of only $Q/Q_d$ and $n$. Restrict your analysis to a positive pressure gradient and no extremum in the velocity profile.

(c) Explain how by using a generalized plot of $\varepsilon$ one can calculate in a straightforward manner the pressure gradient for a given geometry, plate velocity, and net flow rate requirement.

**12.3 *The Isothermal Newtonian Screw Extrusion Model.*** Use Eq. 12.1-3 assuming constant depth shallow channels and negligible flight clearance to derive expressions for (a) maximum pressure rise at closed discharge, (b) optimum channel depth for maximum pressure rise for a given flow rate, (c) optimum channel depth and helix angle for maximum flow rate at constant screw speed (assume the flow rate through the die is given by $Q = K_d(\Delta P/\mu)$, (d) channel depth for lowest screw speed for a given flow rate, and (e) what is the $Q/Q_d$ ratio in (b)?

**12.4 *"Viscous" Seals.*** Vertical extruders with the feed end of the screw protruding into the hopper at the top, and the drive attached to the discharge end of the screw at the bottom, have many advantages (e.g., good feeding and high torque input capability), but have leakage problems at the high pressure end—the screw at the discharge end becomes a shaft attached to a drive. The shaft rotates in a slider bearing where substantial leakage may occur depend-

* Z. Tadmor and I. Klein, "Engineering Principles of Plasticating Extrusion," Van Nostrand Reinhold Book Co., New York, p. 323, 1970.

† Z. Tadmor, "Non-Newtonian Tangential Flow in Cylindrical Annuli," *Polym. Eng. Sci.*, **6**, 203 (1966).

ing upon the clearance. One way to reduce the leakage or stop it completely is to cut on the shaft a reverse flight. Thus, the bearing with the shaft turns into a screw extruder which pumps the leaking polymer melt back into the high pressure discharge region. This is called a viscous dynamic seal. The design can be viewed as two extruders pointing head-on on each other. The main extruder has a certain throughput and generates a pressure $P$; the dynamic seal, if no leakage is allowed, generates the same $P$ at zero flow rate condition. Design a dynamic seal on a 2 in. shaft to prevent leakage for $P = 1000$ psi. Assume a Newtonian fluid of viscosity 0.05 $\text{lb}_f \cdot \text{s/in}^2$ and isothermal conditions.

**12.5** ***Solids Conveying of Nylon in Screw Extruders.*** Consider a 1.991 in diameter screw with 2 in lead, 1.375 in root diameter, an 0.2 in flight width conveying nylon pellets with bulk density of 0.475 $\text{g/cm}^3$ and a coefficient of friction of 0.25. Assuming no pressure rise, calculate the solids conveying rate (g/rev) at the following conditions: (a) no friction between the screw and the solids, (b) no friction on the screw flights, (c) no friction on the trailing flight, (d) friction on all contacting surfaces, and (e) compare the experimentally measured value by Darnell and Mol* of 14.9 g/rev with your results. Discuss.

**12.6** ***Solid Bed Profile in Screw Extruders.*** Determine the solid bed profile and length of melting of LDPE extruded in a 2.5 in diameter, single flighted screw extruder with the following screw geometry and operating conditions. Feed section 3.2 turns and 0.5 in channel depth; compression section 12 turns with linear taper; metering section 12 turns and 0.125 in channel depth. The flight width is 0.25 in and flight clearance is negligible. The operating conditions are 82 rpm screw speed, 150°C constant barrel temperature and 120 lb/hr flow rate. Use polymer physical property data in Example 12.3 and assume that melting starts one turn before the end of the feed section.

***Answer:*** At the end of the feed section $\frac{X}{W} = 0.905$, at the end of the compression $\frac{X}{W} = 0.023$.

**12.7** ***The Rotating Flight Extruder.*** The flight on the screw forms a helical spiral. Consider a processing machine consisting of a detached flight freely rotating on a smooth shaft in a lightly fitted barrel. The difference between this "extruder" and the conventional one is that the root of the screw is stationary relative to the flights just like the barrel.

(a) Derive a theoretical model for pumping (equivalent to the simple Newtonian melt single screw extrusion model).

* W. H. Darnell and E. A. J. Mol, "Solids Conveying in Extruders," *Soc. Plast. Engs. J.*, **12**, 20–28 (1956).

(b) Derive a theoretical model for melting (equivalent to the simple Newtonian melting model in screw extruders).

(c) Derive a theoretical model for solids conveying.

(d) How do the theoretical predictions of the rotating flight extruder compare to an equivalent size screw extruder?

(e) Discuss the engineering feasibility of the rotating flight extruder.

**12.8** ***Modeling the Reciprocating Screw Injection Molding Process.**** Analyze the reciprocating screw injection molding process in terms of elementary steps and suggest a procedure for building a theoretical model for this process.

* R. C. Donovan, D. E. Thomas and L. D. Leversen, "An Experimental Study of Plasticating in a Reciprocating Screw Injection Molding Machine," *Polym. Eng. Sci.*, **11**, 353 (1971); S. D. Lipshitz, R. Lavie, and Z. Tadmor, "A Melting Model for Reciprocating Screw Injection Molding Machines," *Polym. Eng. Sci.*, **14**, 553 (1974).

# *PART FOUR*

# SHAPING

# 13

# Die Forming



Dies, as used in polymer processing, are metal flow channels or restrictions that serve the purpose of imparting a specific *cross-sectional shape* to a stream of polymer melt that flows through them. They are primarily used in the extrusion process to *continuously* form polymer products such as tubes, films, sheets, fibers, and "profiles" of complex cross-sectional shapes. Dies are positioned at the exit end of the melt generating or conveying equipment and, generally speaking, consist of three functional and geometrical regions:

1. The die *manifold*, which serves to distribute the incoming polymer melt stream over a cross-sectional area similar to that of the final product but different from that of the exit of the melt conveying equipment.
2. The *approach channel*, which streamlines the melt into the final die opening.
3. The *die* "*lip*," or final die opening area, which is designed to give the proper cross-sectional shape to the product and to allow the melt to "forget" the generally nonuniform flow experience in regions 1 and 2.

Figure 13.1 shows these regions schematically for a sheet forming die. The shape of both the manifold and approach channels may vary in the cross-die $x$-direction, to permit achievement of design goals just outlined, that is, melt distribution and delivery to the die lips area at uniform pressure. Minor adjustments of the die lip opening are often required to correct for temperature gradients along the die, as well as bending of the die under the applied pressure.

The engineering objectives of die design are to achieve the desired shape within set limits of dimensional *uniformity* and at the *highest* possible *production rate*. This chapter discusses both objectives, but the question of die formed product uniformity deserves immediate amplification. To understand the problem, we must distinguish between two types of die formed product nonuniformity:

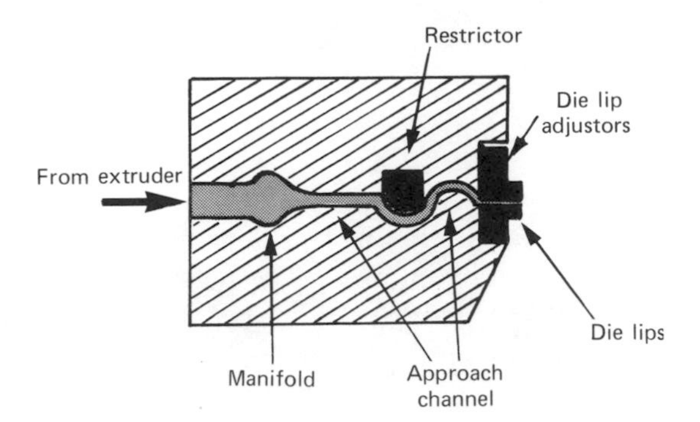


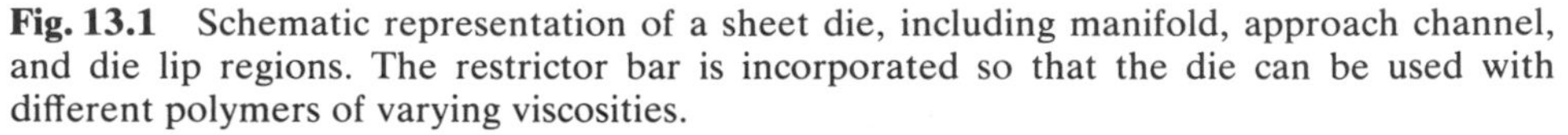

**Fig. 13.1** Schematic representation of a sheet die, including manifold, approach channel, and die lip regions. The restrictor bar is incorporated so that the die can be used with different polymers of varying viscosities.

(a) nonuniformity of product in the machine direction, direction $z$ (Fig. 13.2*a*), and (b) nonuniformity of product in the cross-machine direction, direction $x$ (Fig. 13.2*b*).

These dimensional nonuniformities originate generally from entirely different sources. The main source of the former is the time variation of the inlet stream temperature, pressure, and composition (when mixtures are extruded through the die). The latter is generally due to improper die design. Section 7.13 noted that a system's capability of eliminating inlet concentration variations depends on the

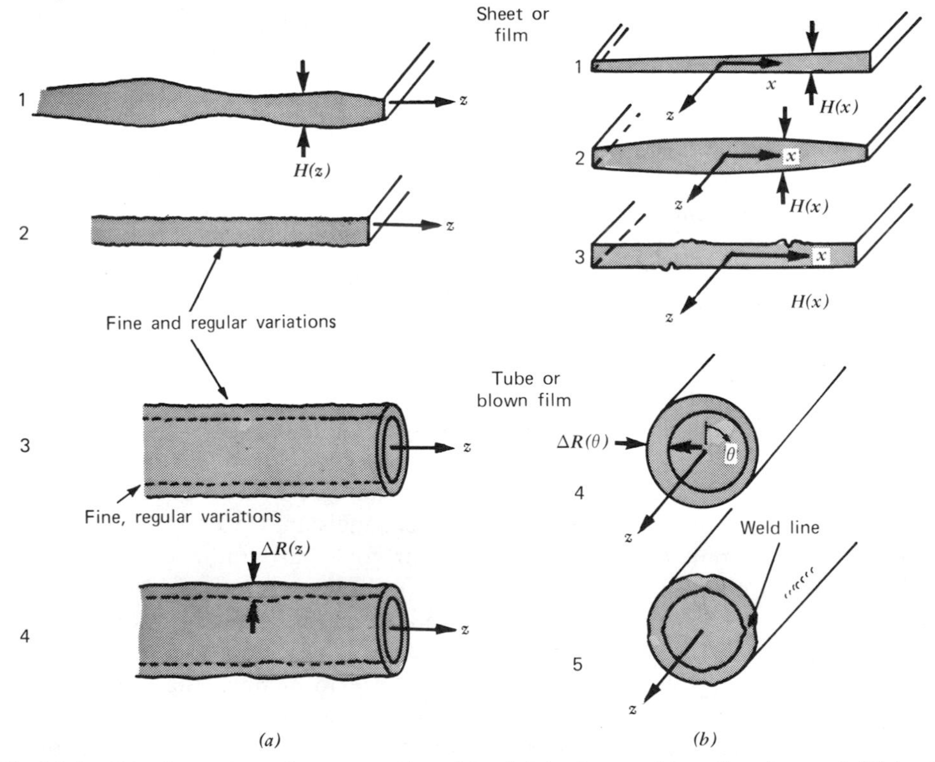


**Fig. 13.2** Die-formed product nonuniformities (*a*) in the machine direction and (*b*) in the cross-machine direction.

RTD function. Narrow RTD functions, as in pressure-type flow dies, cannot be expected to eliminate concentration or temperature variation by mixing. Hence we must ensure adequate inlet uniformity in time, of both temperature and pressure. These depend on the melt generating and conveying equipment preceding the die. In terms of the extrusion operation, improper solids conveying, solid bed breakup, incomplete melting, inefficient mixing due to deep metering channel (all for set extrusion conditions), or the absence of a mixing or barrier device, may result in a *time variation* of the pressure and temperature of the melt being delivered to the die; this will certainly cause machine direction nonuniformities. Figure 13.3 gives examples of acceptable and unacceptable temperature and die inlet pressure variations for a melt stream of LDPE.

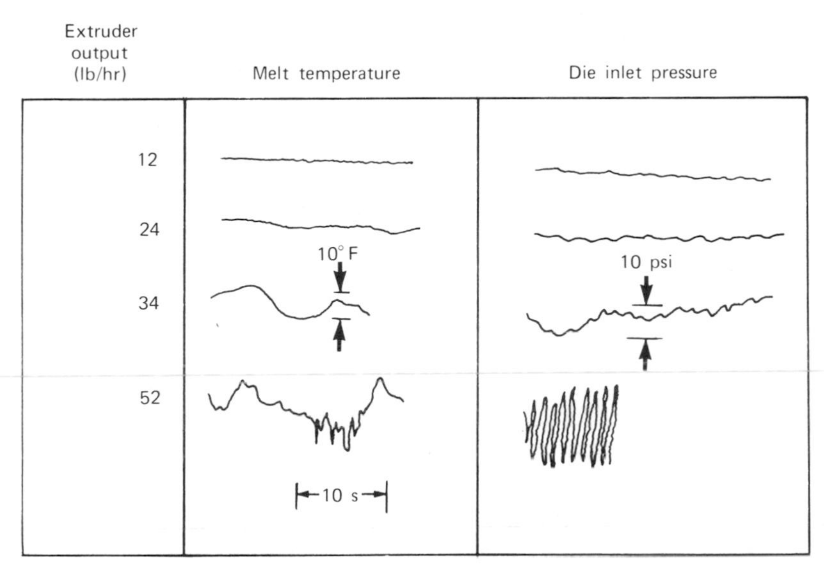


**Fig. 13.3** Types of actual melt temperature and die inlet pressure variations obtained with LDPE. The last two would result in product nonuniformities in the machine direction. [Reprinted with permission from H. B. Kessler, R. M. Bonner, P. H. Squires and C. F. Wolf, *Soc. Plast. Eng. J.*, **16**, 267 (1960).]

The level of variations that can be tolerated depends on both the product specifications and the temperature sensitivity of the viscous and elastic responses of the melt. It is also worth noting in Fig. 13.3 that the two engineering objectives of die forming—namely, uniformity of product and maximum throughput rate—are competing ones. That is, a high throughput rate can be achieved at the expense, in general, of machine direction product uniformity, at set processing conditions. Additionally, there exists an intrinsic upper limit in the throughput rate set by the phenomenon of melt fracture, observed with all polymer melts and discussed later in this chapter (Section 13.2). Machine direction product nonuniformities always accompany melt fracture, and this is why the phenomenon marks the throughput upper limit to die forming. These nonuniformities can be intense or mild, depending on the die *streamlining*, but they are always high frequency disturbances in the product thickness. Other causes for machine direction nonuniformities can be found in the die design that creates stagnation areas where the melt gets trapped and periodically surges forward, and in the post-die forming, cooling, and stretching operations, which may vary with time or may be subject to periodic instabilities.

Product nonuniformities in the cross-machine direction are as pointed out above due to *poor die design* or are intrinsic to particular types of die. Any number of die-related causes can be responsible for the types of nonuniformities in Fig. 13.2*b*: inappropriate design of any one of the three die regions (Fig. 13.2*b*, 1–4), inadequate temperature control of the die walls (Fig. 13.2*b*, 1, 2, 4); bending of the die walls by the flow pressure (Fig. 13.2*b*, 2), and finally the presence of obstacles in the flow channels for die support purposes (Fig. 13.2*b*, 5). In principle, all types of cross-machine direction nonuniformity can be remedied by proper die design, which can be achieved in part through the development of *die design equations* which form the mathematical *model* of the die. This chapter discusses the origin, form, and limitations of such models as they apply to dies of various types.

Before doing so, however, we examine in detail the flow of polymer melts in capillaries. There are two reasons for revisiting capillary flow, after dealing with it in Chapter 6. (*a*) Capillary flow is characterized by all the essential problems found in any die forming flow: flow in the entrance, fully developed, and exit regions; therefore the conclusions reached from the study of capillary flow can, and will be, generalized for all pressure flows. (*b*) Capillary flow has been more widely studied by both rheologists and engineers than has any other flow configuration. We pay particular attention to the problems of entrance pressure drops, viscous heat generation, extrudate swelling, and melt fracture.

Finally, in discussing die flow models we take into account the results presented in connection with capillary flow, but we also look at the details of the flow in each particular die. Die flow models should provide quantitative answers to questions of the following type: (a) If a tube of given dimensions, uniform in the cross-machine direction, is to be extruded at a given rate and with a specific polymer, what should the die design (or designs) be, and what would be the resulting pressure drop? (b) How do the design and pressure drop depend on the processing variables and melt rheology.

## 13.1 Capillary Flow

Figure 13.4 depicts schematically the experimental setup used in capillary flow studies. The primary application of the discussion that follows is in capillary viscometry, useful to die design. The ratio $R_r/R$ should be greater than 10, so that the pressure drop due to the flow in the reservoir can be neglected.* The reservoir radius cannot be too large, though, because the time required for uniform heating of the solid polymer load would be too long (see Table 9.1 for an estimate). Long heating cannot be used for sensitive polymers such as PVC, which readily degrade thermally.

It is useful at this point to consider the "polymer melt experience" as it flows from the reservoir into, through, and out of the capillary tube. In the entrance region (see Fig. 13.4) the melt is forced into a converging flow pattern and

* In careful viscometric studies this pressure drop is never neglected; the pressure reading at zero reservoir height, as found by extrapolation, is used for the value of $P_0$.

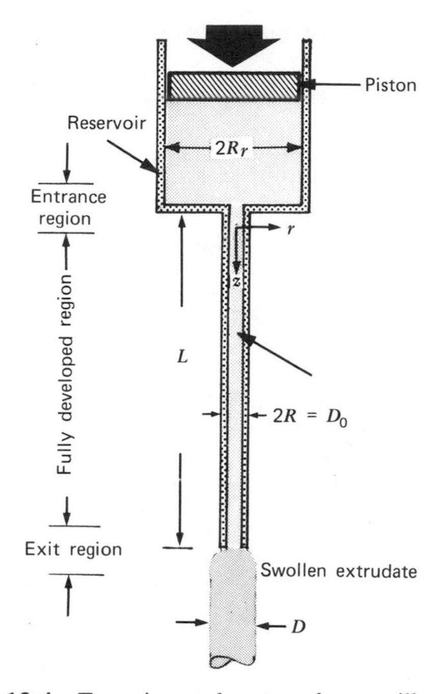


**Fig. 13.4** Experimental setup for capillary flow showing reservoir, entrance, fully developed, and exit regions.

undergoes a large axial acceleration, that is, it stretches. As the flow rate is increased, the axial acceleration also increases, and as a result the polymer melt becomes more elastic, with the possibility of rupturing much like "silly putty" would, when stretched fast. Barring any such instability phenomena, a fully developed velocity profile is reached a few diameters after the geometrical entrance to the capillary. The flow in the capillary, which for pseudoplastic fluids is characterized by a "flat" velocity profile, imparts a shear strain on the melt near the capillary walls. The core of the melt, if the capillary $L/R$ is large and the flow rate is small, can undergo a partial strain recovery process during its residence in the capillary. At the exit region the melt finds itself under the influence of no externally applied stresses. It can thus undergo delayed strain recovery, which, together with the velocity profile rearrangement to one that is pluglike, results in the phenomenon of extrudate swelling.

The "polymer melt experience" briefly described above is complex and varied: it involves steady flow, accelerating flow, and strain recovery. It is not surprising then that this apparently simple experiment is used to study not only the viscous but also the elastic nature of polymer melts.

**Example 13.1** ***The Handling of Entrance Losses in Capillary Viscometry (the "Bagley Correction")***

In the fully developed flow region, the assumptions of steady and isothermal flow, constant fluid density, and independence of the pressure from the radius

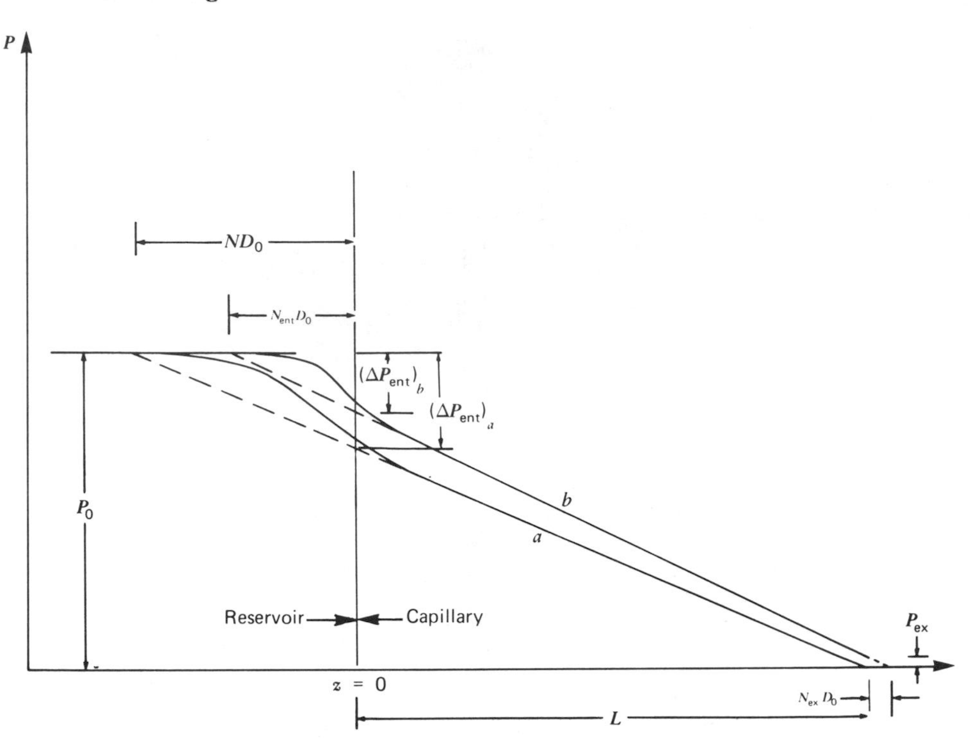


**Fig. 13.5** Schematic representation of the capillary pressure along its axis: curve *a*, without $P_{ex}$; curve *b*, in the presence of $P_{ex}$.

resulted in the conclusion that $-dP/dz = \Delta P/L$. This conclusion is obviously not correct near the capillary entrance where, because of the converging flow, extra velocities ($v_r$) and velocity gradients ($dv_z/dz$, $dv_r/dr$, etc.) are present. Although the flow pattern is not known precisely, we know that a higher pressure drop is needed to support the additional velocity gradients for any viscous or viscoelastic fluid. Schematically then, the pressure profile can be represented as in Fig. 13.5 (curve *a*). Thus $-dP/dz = \Delta P/L^* = \Delta P/(L + ND_0)$, where $N > 0$ is the entrance loss correction factor, which must be evaluated experimentally. The entrance, or Bagley correction (1), $N$ must be considered in the calculation of the shear stress at the wall

$$\blacktriangleright \qquad \tau_w^* = \frac{(P_0 - P_L)D_0}{4L^*} = \frac{D_0}{4}\left(\frac{\Delta P}{L + ND_0}\right) \qquad (13.1\text{-}1)$$

where $\tau_w^*$ is the corrected shear stress at the wall for fully developed flow. The stress at the capillary wall is given by

$$\tau_w^* = -\eta \dot{\gamma}_{12}(R) = \eta \dot{\gamma}_w^* \qquad (13.1\text{-}2)$$

From the Rabinowitsch equation 6.7-9, $\dot{\gamma}_w^*$ is given by

$$\blacktriangleright \qquad \dot{\gamma}_w^* = \tfrac{3}{4}\Gamma + \frac{\tau_w^*}{4}\frac{d\Gamma}{d\tau_w^*},$$

where $\Gamma = 32Q/\pi D_0^3$.

Combining the equations for $\dot{\gamma}_w^*$ and $\tau_w^*$, we see that $\tau_w^* = g(\Gamma)$. Equation 13.1-1 can be rearranged as follows:

$$\frac{L}{D_0} = -N + \frac{\Delta P}{4g(\Gamma)} \tag{13.1-3}$$

where $g(\Gamma)$ is a function of the Newtonian shear rate at the wall. Equation 13.1-3 forms the basis of the so-called Bagley plots (1) (Fig. 13.6) through

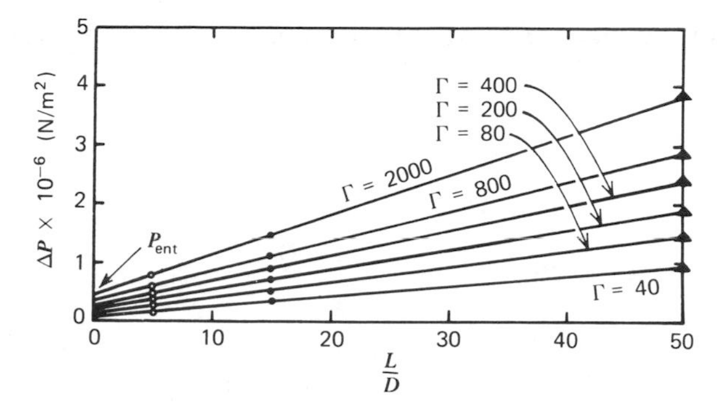


**Fig. 13.6** Bagley plots for a polystyrene melt at 200°C, from which $N(\Gamma)$ can be evaluated; $\Delta P$ at $L/D_0 \to 0$ is the entrance pressure drop $\Delta P$. [Reprinted with permission from J. L. White, *Appl. Polym. Symp.*, No. 20, 155 (1973).]

which $N(\Gamma)$ is evaluated graphically and the corrected wall shear stresses are determined using Eq. 13.1-1. Experimental evidence indicates that the $\Delta P$ versus $L/D_0$ curves at constant $\Gamma$ are not straight, but curve upward at high $L/D_0$ values. This phenomenon has been attributed in the literature both to the relaxation of normal stresses (2) and, probably to a large extent, to the effect of the hydrostatic pressure on melt viscosity (3–5), which can be expressed as

$$\mu(P) = \mu(P_0) \exp [\beta(P - P_0)] \tag{13.1-4}$$

where $\beta$ is related to polymer melt compressibility and is of the order of $5 \times 10^{-9}$ (N/m$^2$)$^{-1}$. It should be emphasized that although the treatment above may give better values for the fully developed region, the same is not true for its entrance region, where the actual $\tau_w^*$ values are large. Despite the above-mentioned experimental facts, the Bagley correction $N(\Gamma)$ is *functionally useful* in that it eliminates the effect of $L/D_0$ from the capillary flow curves. Problem 13.1b indicates the error involved in neglecting $\Delta P_{ent}$ with various $L/D_0$ capillaries.

Also worth noting in connection with the foregoing discussion is the experimental observation of a nonzero gauge pressure at the capillary exit $P_{ex}$ (curve *b* of Fig. 13.5). Thus one must worry about *end corrections*, both entrance and exit, instead of entrance corrections alone, as was done by Bagley. Sakiades (6) was the first to report the existence of $P_{ex}$ for polymer solutions. Han and his co-workers have studied it extensively using polymer melts (7–9). It has been found that $P_{ex}/\Delta P_{ent}$ is between 0.15 and 0.20 and that although $\Delta P_{ent}$ does not depend on $L/D_0$, $P_{ex}$ decreases up to $L/D_0 = 10$, then remains constant (9).

With these observations in mind (see curve *b*, Fig. 13.5), we can rewrite Eq. 13.1-1.

$$\tau_w^* = \frac{\Delta P D_0}{4L^*} = \frac{D_0}{4}\left(\frac{\Delta P}{L + N_{ent} D_0 + N_{ex} D_0}\right) \tag{13.1-1a}$$

where $N_{ent}$ is the entrance capillary length correction and $N_{ex}$ the exit capillary length correction. If $(P_{ex}/\Delta P_{ent}) \ll 1$, then $N = N_{ent}$; if $P_{ex}$ is included it is found that $N_{ent} < N$. Problem 13.2 indicates that for HDPE, the data improvements obtained with the inclusion of $P_{ex}$ are slight. $\Delta P_e$ is the *sum* of $P_{ex}$ and $\Delta P_{ent}$.

### Viscous Heat Generation

One of the assumptions made in solving the flow in the fully developed region of the capillary was that of constant fluid temperature throughout the flow region. This is not a valid assumption for the flow of very viscous fluids at high rates of shear in which a nonuniform temperature field is created. As we have already mentioned in connection with the thermal energy balance (Section 5.1), the rate of viscous heating per unit volume $e_v$ is

$$\blacktriangleright \qquad e_v = -(\boldsymbol{\tau} : \nabla \boldsymbol{v}) = -\tfrac{1}{2}(\boldsymbol{\tau} : \dot{\boldsymbol{\gamma}} + \cancelto{0}{\boldsymbol{\tau} : \boldsymbol{\omega}}) \tag{13.1-5a}$$

Thus for a Newtonian fluid

$$\blacktriangleright \qquad e_{v,N} = \frac{\mu}{2} \dot{\boldsymbol{\gamma}} : \dot{\boldsymbol{\gamma}} \equiv \mu \Phi_v \tag{13.1-5b}$$

while for a power law fluid

$$\blacktriangleright \qquad e_{v,PL} = \frac{m}{2}\left[\sqrt{\tfrac{1}{2}(\dot{\boldsymbol{\gamma}} : \dot{\boldsymbol{\gamma}})}\right]^{n-1}(\dot{\boldsymbol{\gamma}} : \dot{\boldsymbol{\gamma}}) = m \Phi_v^{(n+1)/2} \tag{13.1-5c}$$

In the fully developed region of the capillary the flow is described by

$$\dot{\boldsymbol{\gamma}} = \begin{pmatrix} 0 & 0 & \dfrac{dv_z}{dr} \\ 0 & 0 & 0 \\ \dfrac{dv_z}{dr} & 0 & 0 \end{pmatrix}$$

Thus Eqs. 13.1-5b and 13.1-5c become

$$e_{v,N} = \mu\left(\frac{dv_z}{dr}\right)^2 \tag{13.1-6a}$$

$$e_{v,PL} = m\left[\left(\frac{dv_z}{dr}\right)^2\right]^{(n+1)/2} \tag{13.1-6b}$$

where

$$\mu = \mu_0 \, e^{\Delta E'/RT} \tag{13.1-7a}$$

$$m = m_0 \, e^{\Delta E/RT} \tag{13.1-7b}$$

Figure 13.7 gives the velocity gradients of a Newtonian and a power law fluid with $\mu = m$; isothermal flow is assumed. It is clear that for power law fluids, viscous

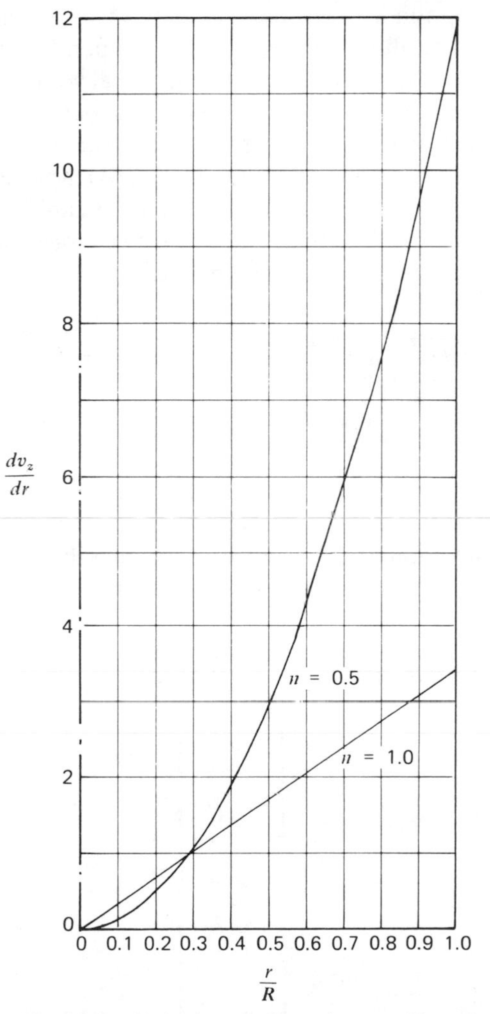


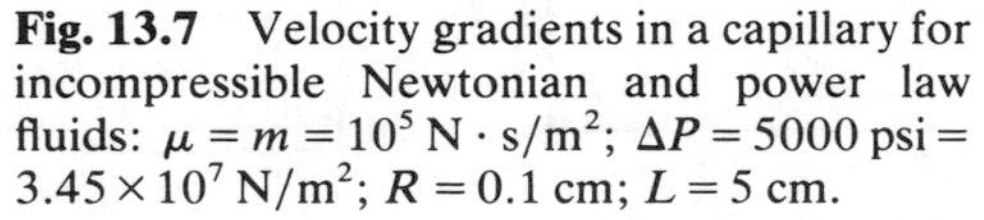

**Fig. 13.7** Velocity gradients in a capillary for incompressible Newtonian and power law fluids: $\mu = m = 10^5\ \text{N} \cdot \text{s/m}^2$; $\Delta P = 5000$ psi $= 3.45 \times 10^7\ \text{N/m}^2$; $R = 0.1$ cm; $L = 5$ cm.

heating may be intense near the capillary wall, whereas the central portion of the fluid is relatively free of this effect.

The mathematical solution of the nonisothermal flow problem in the fully developed region of the capillary, even with the simplifying assumption of constant fluid density, involves the simultaneous solution of the momentum and energy balances, which have the general forms discussed in Section 5.1, subject to the appropriate boundary conditions. The two equations must be solved simultaneously because they are coupled through the temperature dependent viscosity (Eq. 13.1-7). Nahme (10) was the first investigator to look into this coupled transport problem. Brinkman (11) made the first significant contributions to the understanding of the problem (the Brinkman number), although his solution suffers from the severe assumption of constant viscosity. The problem of viscous heat generation of a Newtonian fluid with temperature dependent transport properties was solved by Turian and Bird (12), the first numerical solution is due to Gerrard, Steidler, and Appeldoorn (13). Turian solved by a perturbation procedure the problem of a power law fluid with temperature dependent viscosity and conductivity, flowing in Couette and cone and plate flows (14). Morrette and Gogos (15), using a numerical solution, solved the same problem for capillary flow of PVC melts that are subject to thermal degradation, and viscosity changes because of it. For this system the coupled momentum and energy balances are

$$-\frac{dP}{dz}+\frac{1}{r}\frac{d}{dr}\left[rm_0\,e^{\Delta E/RT}\left|\frac{dv_z}{dr}\right|^{n-1}\frac{dv_z}{dr}\right]=0 \qquad (13.1\text{-}8)$$

$$\rho C_p v_z\frac{dT}{dz}=\frac{1}{r}\frac{\partial}{\partial r}\left(rk\frac{\partial T}{\partial r}\right)+m_0\,e^{\Delta E/RT}\left|\frac{dv_z}{dr}\right|^{n-1}\left(\frac{dv_z}{dr}\right)^2 \qquad (13.1\text{-}9)$$

Results obtained by the latter investigators indeed show what is expected from Eqs. 13.1-6a and 13.1-6b; temperature rises are intense and significant for the layer of the PVC melt near the capillary wall, where $e_{v,\mathrm{PL}}$ has high values. As Fig. 13.8 indicates, about 50% of the temperature rise occurs in the first tenth of the capillary length when $m$ has a high value, since the temperature of the entering melt is low. Two cases were considered, those of isothermal and adiabatic capillary walls, because actual flows lie between these two extremes. There has not been a good

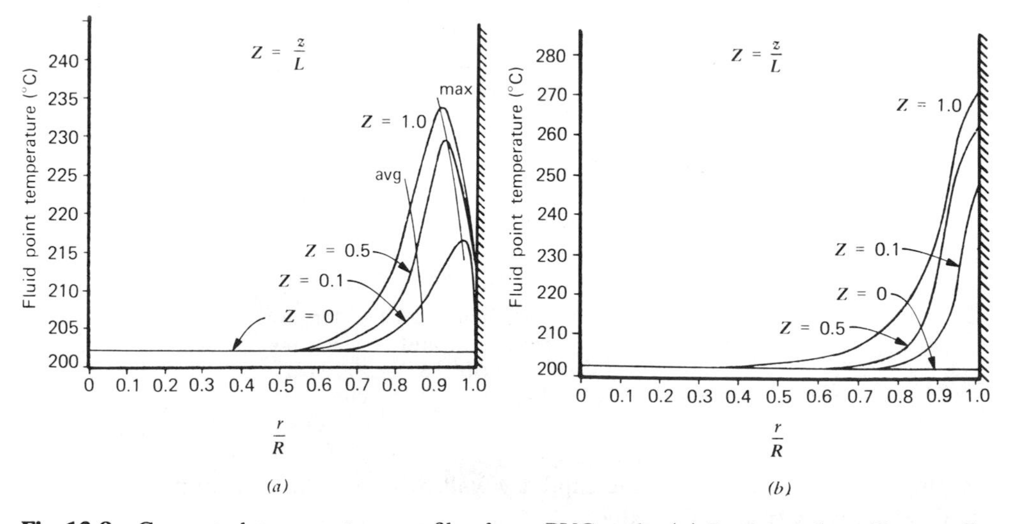


**Fig. 13.8** Computed temperature profiles for a PVC melt. (*a*) Isothermal capillary wall. (*b*) Adiabatic wall. [Reprinted with permission from R. A. Morrette and C. G. Gogos, *Polym. Eng. Sci.*, **8**, 272 (1968)].

experimental method yet devised to measure the temperature of highly viscous fluids flowing at high flow rates. Thermocouple measurements (16–18) have not been successful because they disrupt the flow field and become heated by the viscous fluid flowing past their surface. Cox and Macosko (19) have reported experimental results on measurements of the melt surface temperature upon exit from the capillary using infrared pyrometry. The infrared pyrometer used senses the radiation emitted by the hot polymer melt surface. Their work also included the numerical simulation of viscous heating in a capillary, a slit, and an annular die. Their method resembles that of Gerrard, Steidler, and Appeldoorn (13). They used a boundary condition at the die wall in between the isothermal and adiabatic case, $-k(\partial T/\partial r) = h(T - T_0)$ at the wall where $T_0$ is the die temperature "far" from the melt-die interface as well as the inlet melt temperature. Some of their results for ABS Cycolac T appear in Figs. 13.9–13.11.

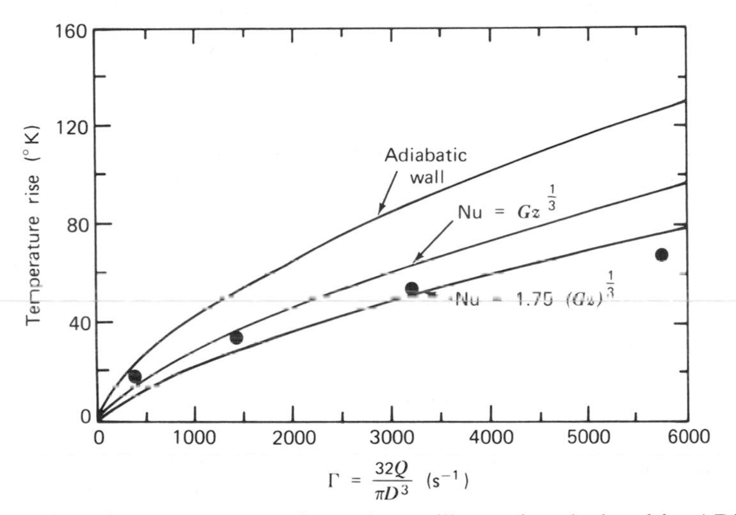


**Fig. 13.9** Melt surface temperature rise at the capillary exit, calculated for ABS Cycolac T and measured (●) with an infrared pyrometer: $T_0 = 505°K$, $D_0 = 0.319$ cm, $L/D_0 = 30$. The relationships $Nu = C(G_z)^{1/3}$ are used to estimate $h$. [Reprinted with permission from H. W. Cox and C. W. Macosko, *Am. Inst. Chem. Eng. J.*, **20**, 785 (1974)].

The data and numerical results depicted in these figures suggest the following:

1. Large temperature rises due to viscous heating do indeed occur in melt capillary flow at moderate and high shear rates. These must be estimated and taken into account whenever temperature sensitive polymers are extruded and whenever the extrudate surface quality and properties are of critical importance.
2. At least as far as estimating the surface temperature rise, the simple dimensionless relationship, concerning heat transfer at the wall suffices:

$$Nu = C(G_z)^{1/3} \tag{13.1-10}$$

where $Nu = hD_0/k$, $G_z = \dot{m}C_p/kL$, $\dot{m}$ is the mass flow rate, and $C = 1.75$ for satisfactory fit with the data. The temperature at the capillary wall is never the maximum; this occurs at about $\frac{9}{10}$ the capillary radius. The

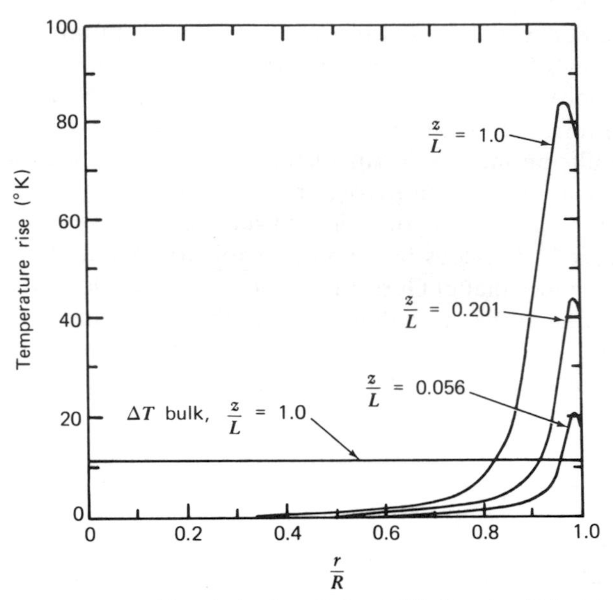


**Fig. 13.10** Temperature profiles (calculated) for ABS Cycolac T in tube flow using $\mathrm{Nu} = 1.75(Gz)^{1/3}$ to estimate $h$; $D_0 = 0.319$ cm; $L/D_0 = 30$; $T_0 = 505°\mathrm{K}$; $\Gamma = 5730\ \mathrm{s}^{-1}$. [Reprinted with permission from H. W. Cox and C. W. Macosko, *Am. Inst. Chem. Eng. J.*, **20**, 785 (1974)].

extrudate temperature field at the exit $T(r, L)$ will influence the swelling and drawing behavior of the extrudate.

3. About 50% of the temperature rise occurs near the capillary entrance $z \leq 0.2L$. Thus shortening the capillary length does not decrease the temperature rise due to viscous heating, proportionally.
4. The bulk temperature rise $\Delta T_b$ (Fig. 13.10) does not seem to serve any useful purpose in that it is much smaller than the maximum and is heavily influenced by the central core of the fluid, which does not significantly heat

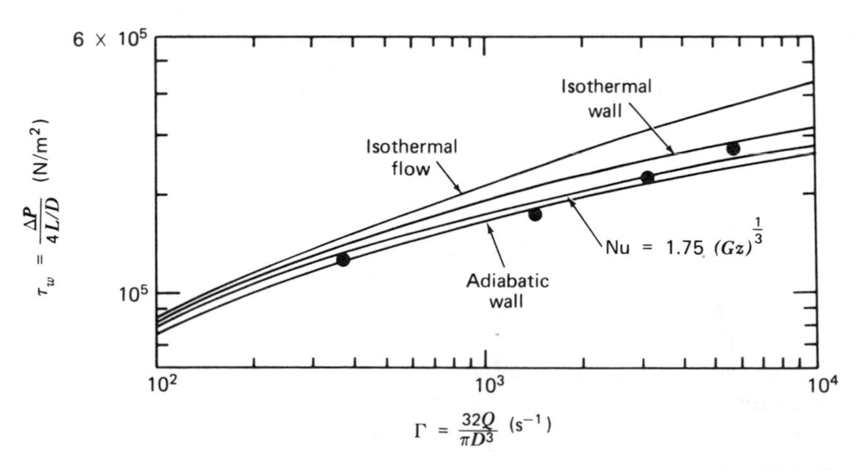


**Fig. 13.11** Uncorrected shear stress versus Newtonian wall shear rate for ABS Cycolac T; measured and calculated using various thermal boundary conditions. $D_0 = 0.319$ cm, $L/D_0 = 30$, $T_0 = 505°\mathrm{K}$. [Reprinted with permission from H. W. Cox and C. W. Macosko, *Am. Inst. Chem. Eng. J.*, **20**, 785 (1974)].

up; $\Delta T_b$ is a quantity often calculated and presented as the reason for *not* having to worry about viscous heating. A simple estimate of $\Delta T_b$ is obtained by assuming that the entire mechanical energy degenerates to heating the melt, as discussed in Section 11.3. It, therefore, seems that one must consider and deal with the nonisothermal nature of any type of pressure flow when the calculated value of $\Delta T_b$ exceeds 4–5°C. Galili and Takserman-Krozer (20) have proposed a simple criterion that signifies when nonisothermal effects must be taken into account. The criterion is based on a perturbation solution of the coupled heat transfer and pressure flow isothermal wall problem of an incompressible Newtonian fluid.

The pressure drop calculated assuming the relationship $\text{Nu} = 1.75(Gz)^{1/3}$ for estimating $h$, is smaller than the calculated $\Delta P$ assuming isothermal flow. For the conditions depicted in Fig. 13.11, at $\Gamma = 10^3\ \text{s}^{-1}$ the isothermal pressure drop is about 30% higher than the measured value. This fact must be taken into account in the design of extrusion dies so that gross die overdesign can be avoided, as well as in capillary viscometry.

## 13.2 Elastic Effects in Capillary Flows

So far in this chapter we have looked into the viscous phenomena associated with the flow of polymer melts in capillaries. We now turn to the phenomena that are related to melt elasticity, namely:

1. Swelling of polymer melt extrudates.
2. Large pressure drops at the capillary entrance, compared to those encountered in the flow of Newtonian fluids.
3. Capillary flow instabilities accompanied by extrudate defects, commonly referred to as "melt fracture."

These phenomena have been the subject of intensive studies during the last 25 years and still represent major problems in polymer rheology. From a processing point of view they are very important, since melt fracture represents an upper limit to the rate of extrusion, and swelling and the large pressure drops must be accounted for in product considerations and in the design of the die and processing equipment.

### *Extrudate Swelling*

Extrudate swelling refers to the phenomenon observed with polymer melts and solutions that when extruded, emerge with cross-sectional dimensions appreciably larger than those of the flow conduit. The ratio of the final jet diameter to that of the capillary, $D/D_0$, for Newtonian fluids varies only from 1.12 at low shear rates to 0.87 at high rates. Polymer melts exhibit the same low shear rate $D/D_0$ value in the Newtonian plateau region, but swell 2–4 times the extrudate diameter at high shear rates (21, 22). Figure 13.12 gives the shear rate dependent $D/D_0$ for melts, together with $\eta(\dot{\gamma})$. Extrudate swelling increases are accompanied by $\eta(\dot{\gamma})$ decreases.

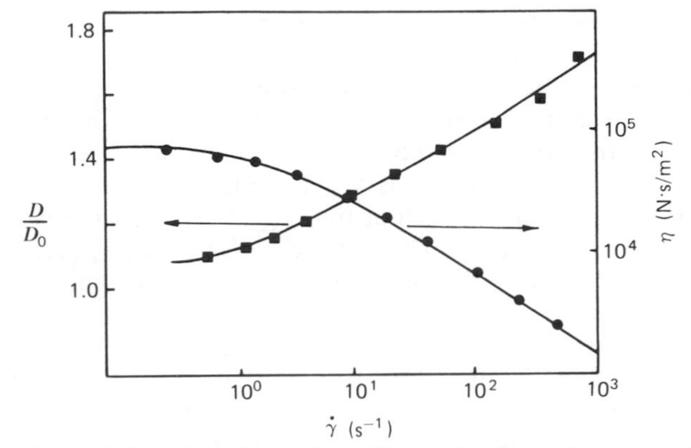


**Fig. 13.12** Comparison of the viscosity and swelling ratio dependence on shear rate for a polystyrene melt of $\bar{M}_w = 2.2 \times 10^5$ and $\bar{M}_w/\bar{M}_n = 3.1$. [Reprinted with permission from W. W. Graessley, S. D. Glasscock and R. L. Crawley, *Trans. Soc. Rheol.*, **14**, 519 (1970)].

Experimentally $D/D_0$ depends on the following variables (Fig. 13.13): the shear stress at the wall $\tau_w$ (a flow variable) and the molecular weight distribution (a structural variable) (22). The length-to-diameter ratio of the capillary (a geometric variable) also influences $D/D_0$. The swelling ratio at constant $\tau_w$ decreases exponentially with increasing $L/D_0$ and becomes constant for $L/D_0 > 30$. The reason for this decrease is qualitatively the following. Extrudate swelling is related to the ability of polymer melts and solutions to undergo delayed elastic strain recovery, as discussed in Section 6.1. The more strained and the more entangled the melt is at the capillary exit, the more it will swell.* From this point of view, the decrease of swelling with increasing $L/D_0$ is due to two causes. First, in a long capillary the melt recovers from the tensile deformations suffered at the capillary inlet, which is due to the axial acceleration in that region. Second, the shear strain imposed on the melt while in the capillary may bring about disentanglements.

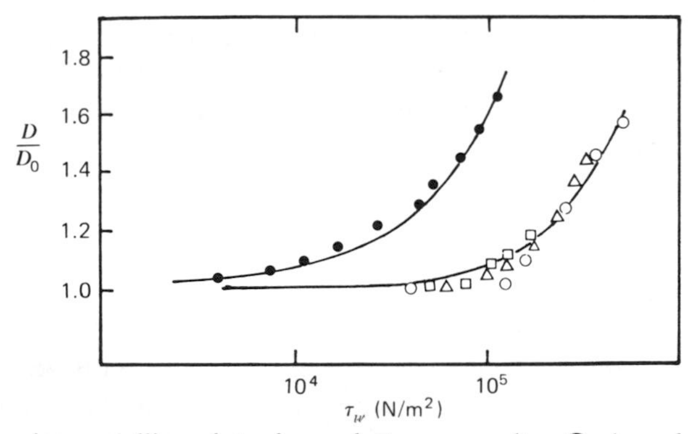


**Fig. 13.13** Extrudate swelling data for polystyrene melts: ●, broad molecular weight sample; ○, □, △, narrow distribution sample data at various temperatures. [Reprinted with permission from W. W. Graessley, S. D. Glasscock and R. L. Crawley, *Trans. Soc. Rheol.*, **14**, 519 (1970)].

* Solutions of *rigid* polymer molecules (e.g., poly-*p*-phenylene terephthalate) may also exhibit extrudate swelling because they too are entropy elastic: molecules exit the capillary in a fairly oriented state and become randomly oriented downstream.

Polymer melts and solutions are entangled and under quiescent conditions are characterized by a high entanglement density value. In this sense they possess a "structure." During shear flow the entanglement density is reduced, and so is the ability of the fluid to undergo strain recovery. Thus the value of $D/D_0$ at very long $L/D_0$ values reflects the ability of the viscoelastic liquid to recover from *shear* strains only. The melt issuing from a very short capillary is much more entangled and recovers from both shear and tensile strains.

In view of the foregoing discussion, it is not surprising that the magnitude of the first normal stress difference, which measures the extra tension in the flow direction during the flow in a long capillary, reflects the magnitude of extrudate swelling. Such a relationship has indeed been suggested by Tanner (23), who applied Lodge's theory of free recovery following steady shearing flows (24). For long capillaries Tanner obtains

$$\frac{D}{D_0} = 0.1 + (1 + S_R^2)^{1/6} \qquad (13.2\text{-}1)$$

The constant 0.1 is empirical; $S_R$, the recoverable shear strain, is

$$S_R = \frac{\tau_{11} - \tau_{22}}{2\tau_{12}} = \frac{\Psi_1 \dot{\gamma}_w}{2\eta} \qquad (13.2\text{-}2)$$

where the stresses are evaluated at the wall shear rate $\dot{\gamma}_w$. The shear stress and normal stress difference can be measured experimentally or they can be calculated with the aid of indirect experimental measurements and either continuum or molecular theories. For example, Tanner used the BKZ theory, and Bird and his co-workers (25) the Goddard–Miller theory. When the Rouse molecular theory is used, the steady state shear compliance $J_R$ is (26).

$$J_R = 0.4 \frac{\bar{M}_w}{\rho R T} \cdot \frac{\bar{M}_z \bar{M}_{z+1}}{\bar{M}_w^2} \qquad (13.2\text{-}3a)$$

for PS. Graessly et al., (22) found

$$J_0 = \frac{2.2 J_R}{1 + 2.1 \times 10^{-5} \rho \bar{M}_w} = \frac{S_R}{\tau_w^*} \qquad (13.2\text{-}3b)$$

where $\rho$ is in g/cm$^3$. Thus, for high $\bar{M}_w$

$$J_0 \cong \frac{0.4 \times 10^5}{\rho^2 R T} \frac{\bar{M}_z \bar{M}_{z+1}}{\bar{M}_w^2} \qquad (13.2\text{-}3c)$$

These relationships indicate that $S_R$ and, according to Eq. 13.2-1, $D/D_0$, depend on the molecular weight distribution. We recall that this was observed by Graessley.

Equation 13.2-1 has been semiquantitatively successful in predicting extrudate swelling (25). However recently White and Roman (27) have shown experimentally with a number of polymers that $D/D_0$ is *not a function of* $S_R$ only. Furthermore, they demonstrated that the success of the Tanner equation depends on the method of measurement of $D/D_0$. As Fig. 13.14 reveals, extrudate swelling values can differ by as much as 30%, depending on the measurement method. The Tanner equation compares best with the results obtained on extrudates frozen in air. The work of White and Roman, to which we return in Section 15.1, is important to

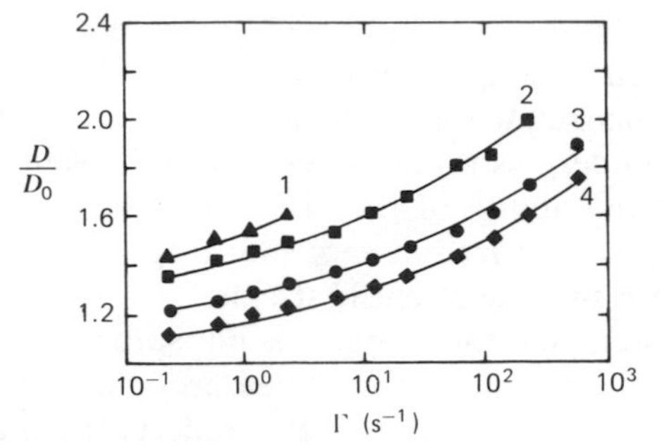


**Fig. 13.14** Effect of the method of measurement on the value of $D/D_0$ for HDPE. Curve 1, frozen extrudates; curve 2, extrudates annealed at 160°C in hot silicon oil; curve 3, photographs of extrudates emerging from capillary; curve 4, photographs of extrudates in hot silicon oil. [Reprinted with permission from J. L. White and J. F. Roman, *J. Appl. Polym. Sci.*, **20**, 1005 (1976)].

processing because in such operations extrudate swelling occurs under postextrusion conditions that are poorly specified and always nonequilibrium.

The solution of the extrudate swelling problem can, *in principle*, be found by using macroscopic mass and momentum balances over a control volume bound by the capillary exit plane and another at a downstream position where the velocity profile is flat (28). This method has been successfully applied to the solution of extrudate swelling in Newtonian jets (Problem 13.4). The results obtained by such balances in polymers *do not agree* with experiments. A detailed study of extrudate swell analysis by macroscopic balances was done by Bird et al. (29) who distinguish between two regimes: a low Reynolds number regime and a high Reynolds number regime. In the latter regime good analysis can be done using only macroscopic mass and monentum balances, but in the former regime (that includes polymer swelling) the macroscopic mechanical energy balance has to be included in the analysis because of the significant effect of the viscous dissipation term. This renders the analysis more difficult as it requires detailed knowledge of the jet free-surface shape, the distance downstream to fully developed flow, the velocity rearrangement in the die, the Reynolds number, and a new dimensionless group including the primary normal stress difference function $\Psi_1$. Whiple's (29*a*) careful experimental study of the velocity profile in the region before and after capillary exit is an initial step in answering some of these needs. They found that polyer melts "anticipate" the swelling phenomenon in that just before the exit, axial decelerations and radial velocity components are observed. Thus the exit velocity profile is not the same as in the fully developed region and the flow there is not viscometric.*

### Example 13.2

Graessley and co-workers (22) have found that with polystyrene extrudate velocities of 1–3 mm/s in the temperature range of 160°–180°C, about 90%

* The experimental fact that nonviscometric flow prevails at the capillary exit must be taken into account in the discussion of $P_{ex}$ (9) mentioned earlier.

of the final $D/D_0$ value is reached at an axial distance of 0.1 cm past the capillary exit. The rest of the swelling was completed in the next three centimeters. What is the "recoverable strain" at 0.1 cm?

Recoverable strain can also be defined as the tensile strain needed to pull a fully swollen extrudate until its diameter is that of the capillary (22).* Assuming constant density $\pi D^2 L/4 = \pi D_0^2 L_0/4$, or $L_0/L = S_R = (D/D_0)^2$. At 0.1 cm past the capillary exit $S_R = 0.81(D/D_0)^2$, and at 3 cm it is $(D/D_0)^2$. Therefore, at 0.1 cm, 19% of the recoverable tensile strain that the extrudate is capable of undergoing is still present. In other words, if no further swelling were allowed, $0.19(D/D_0)^2$ would be the value of the average "frozen-in" strain in the extrudate.

### *Entrance Pressure Losses*

Earlier in this chapter, in discussing the Bagley correction in capillary viscometry, we pointed out the necessity of eliminating entrance pressure drops to get the correct value of the wall shear stress $\tau_w^*$. As Fig. 13.6 indicates, the level of entrance pressure drops is large for polymer melts and solutions. Figure 13.15 gives specific evidence of the magnitude of the ratio of entrance pressure drop to the shear stress at the wall, $\Delta P_{ent}/\tau_w^*$ is shown for a number of materials (30). According to Eq. 13.1-1, $\Delta P_{ent}/\Delta P_{cap} = (D_0/4L)(\Delta P_{ent}/\tau_w^*)$. Thus for LDPE flowing in an $L/D_0 = 2.4$ capillary, $\Delta P_{ent}/\Delta P_{cap} \cong 1$ at a value of $\Gamma = 2\ s^{-1}$, according to Fig. 13.15. The value of the entrance pressure drop becomes larger than the total capillary pressure drop at higher shear rates. It follows then that in polymer processing, where the length-to-opening ratios are small and shear rates are high, *entrance pressure drops must be included in calculations of the die pressure* in die design equations.

Entrance pressure drops are, of course, observed with all fluids in regions of conduit cross-sectional changes. This is because the conduit shape and the rheological response of the fluid creates extra velocity gradients, that to be sustained, need

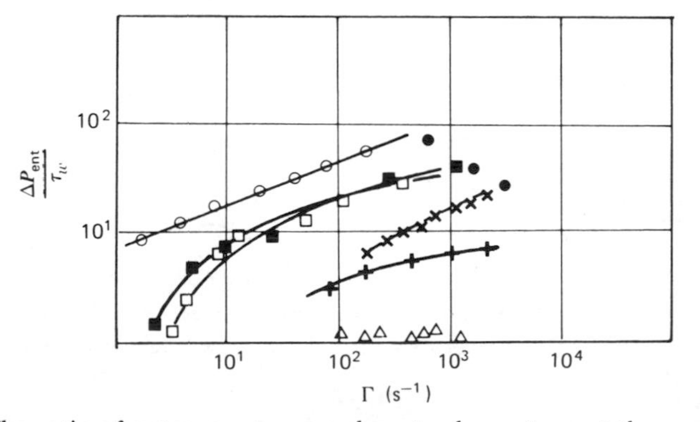


**Fig. 13.15** The ratio of entrance pressure drop to shear stress at the capillary wall versus Newtonian wall shear rate, Γ. ■, PP; □, PS; ○, LDPE; +, HDPE; ●, 2.5% PIB in mineral oil; ×, 10% PIB in decalin; △, NBS-OB oil. [Reprinted with permission from J. L. White, *Appl. Polym. Symp.*, No. 20, 155 (1973)]

* Here we use the "engineering strain", $\varepsilon = l(t)/l_0$, and not the "true strain" used in Section 6.8.

to be "fed" by stress terms, which give rise to extra pressure drops (i.e., $[\nabla \cdot \boldsymbol{\tau}] = -\nabla P$ in the equation of motion). Newtonian fluids in laminar flow in contracting regions exhibit streamlines that radiate in waves in the entire entrance region, as has been shown by Giesekus (31). This simple flow pattern, together with the simple rheological response of Newtonian fluids, gives rise to relatively small entrance pressure losses that have been calculated by Weissberg (32) to be

$$\Delta P_{e,v} = \frac{3\mu Q}{2R^3} \tag{13.2-4}$$

where $Q$ is the volumetric flow rate and $R$ is the capillary radius. Polymer melts and solutions, on the other hand, are rheologically more complex fluids and even under simple radiating flows in the entrance region, would need more stress forms to sustain them, thus larger entrance pressure drops. Additionally, the entrance flow patterns with polymer melts and solutions are typically more complex.* Entrance vortices are observed (Fig. 13.16), with the viscoelastic fluid flowing into the capillary from a "wine glass' region (33). Not all polymers exhibit vortices; HDPE and isotactic PP do not, and all polymer melts and solutions behave like Newtonian fluids at very low shear rates where the viscosity has reached the "Newtonian plateau." As the flow rate is increased, vortices are formed, leading to the conclusion that radiating flow is not compatible with the equation of motion and the constitutive equation describing these fluids. Furthermore, increasing the flow rate results in increasing the vortex size (34). The large entrance pressure losses are a consequence of the large vortices, which effectively increase the capillary length. Large vortices imply small entrance angles $\alpha$, as Fig. 13.16 suggests.† In turn, small

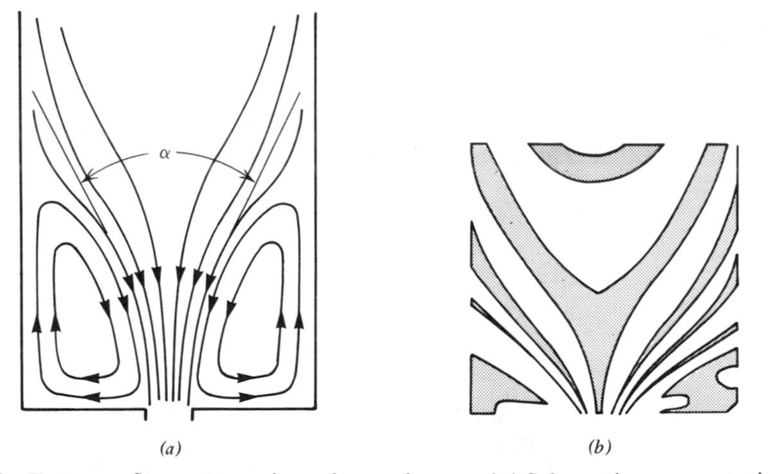


**Fig. 13.16** Entrance flow patterns in molten polymers. (*a*) Schematic representation of the "wine glass" and entrance vortex regions with the entrance angle $\alpha$ (41). (*b*) The observed flow pattern by birefringence of a polystyrene melt. [Reprinted with permission from J. L. White, *Appl. Polym. Symp.*, No. 20, 155 (1973)].

* The viscous contribution to the total entrance pressure loss is very small (C. D. Han, *Am. Inst. Chem. Eng. J.*, **17**, 1480 (1971)].

† It follows then that the capillary entrance angle affects the value of $\Delta P_{ent}$. Han (37) has shown that for HDPE, $\Delta P_{ent}$ decreases with increasing entrance angle, up to 60°, then remains constant from 60 to 180°.

entrance angles give rise to a small elongational extension rate in the region of the "wine glass" stem. This apparently has led Lamb and Cogswell (35) to relate the entrance angle $\alpha$ to the elongational viscosity $\bar{\eta}$, arguing that melts with high elongational viscosity would favor small elongational rates, thus small entrance angles. The relationship proposed is

$$\alpha = \tan^{-1} \frac{2n}{\bar{\eta}} \tag{13.2-5}$$

The ratio of shear to elongational viscosities becomes smaller with increasing deformation rates, giving rise to smaller entrance angles and, consequently, larger entrance vortices, as observed experimentally.

Ballenger and White (34) relate the entrance angle $\alpha$ in degrees to the ratio of the entrance pressure loss to the capillary wall shear stress, $\Delta P_{ent}/\tau_w^*$,

► $$\alpha = 178.5(0.9644)^{\Delta P_{ent}/\tau_w^*} \tag{13.2-6}$$

The relationship is experimental. LaNieve and Bogue (36) have related the entrance pressure losses of polymer solutions to the viscosity and primary normal stress difference coefficient. Thus the works of Ballenger and LaNieve, taken together, seem to imply that the entrance angle (thus the size of the entrance vortices) depends on both the viscosity and the first normal stress difference coefficient. Recently, White and Kondo (38) have shown experimentally that for LDPE and PS

$$\alpha = f\left[\frac{(\tau_{11} - \tau_{22})_w}{(\tau_{12})_w}\right] \tag{13.2-7}$$

Equations 13.2-6 and 13.2-7 seem to imply that the entrance (or "ends") pressure losses are simply related to the first normal stress difference function at the capillary wall. Indeed they find that

$$\frac{\Delta P_e}{(\tau_{11} - \tau_{22})_w} \simeq 2 \tag{13.2-8}$$

A better understanding of the exact origins of the entrance pressure loss in polymer melts requires the experimental determination of the precise flow field in that region. Until such work and the subsequent analysis have been completed, it suffices to state that entrance pressure losses with polymer melts are large, since these fluids are viscoelastic and exhibit large extensional viscosity values. For die design purposes we must have experimentally available data, such as those obtained with zero length capillaries, or with a number of different $L/D_0$ capillaries by extrapolations to $L/D_0 = 0$.

### *Flow Melt Fracture*

In the flow of molten polymers through capillaries and other dies a striking phenomenon is observed at shear stresses of the order of $10^5$ N/m$^2$. As stress is increased, there is a *critical stress* at and above which the emerging polymer stream exhibits irregular distortion. This distortion contributes evidence for some irregularity or resistability in flow.

The author of these remarks is J. P. Tordella (39), who not only investigated the field of unstable polymer melt pressure flow but has written lucidly on it, coining the term *melt fracture* for the above-described phenomenon. The phenomenon was first studied by Spencer and Dillon (40), who found that the critical wall shear stress is independent of the melt temperature but inversely proportional to the weight average molecular weight. These conclusions have remained essentially valid to date. Aside from the review by Tordella (39), two more review papers should be mentioned, one by White (30) on the subject of extrudate distortion, and a more recent general article on polymer processing instabilities by Petrie and Denn (41).

Looking at the melt fracture of specific polymers, we see many similarities and a few differences. Polystyrene extrudates become spiraling from smooth, at $\tau_w^* \simeq 10^5\,N/m^2$, and at higher shear stresses they are grossly distorted. Visual observations show a "wine glass" entrance pattern with vortices that are stable at low stress values and spiral into the capillary and subsequently break down, as $\tau_w^*$ is increased. Clearly melt fracture is an entrance instability phenomenon for this polymer. Similar observations have been made with polypropylene with two qualitative differences. The flow entrance angle is very large, almost 180°, and the observed spiraling of the extrudate is very regular. LDPE extrudates transit from smooth to dull or matte at subcritical values of $\tau_w^*$. With increasing shear stress they become spiraling, over a narrow range of $\tau_w^*$, and subsequently become grossly distorted. Corresponding to smooth and matte extrudates, very small entrance angle but stable patterns are observed at the capillary entrance. This flow pattern spirals in the capillary at high stresses, and at the critical shear stress the "wine glass" stem flow lines are grossly disrupted. With the exception of the appearance of the matte extrudate surface, which is an exit fracture phenomenon as has been demonstrated by Cogswell and Lamb (42) and Vinogradov (43), LDPE behaves in the melt fracture region similarly to PS and PP. In the three polymers just named, two more observations are worth mentioning. First, at the melt fracture onset there is no discontinuity in the flow curve ($\tau_w^*$ vs. $\dot{\gamma}_w^*$). Second, as expected because the entrance is the site of the instability, increasing $L/D_0$ decreases the severity of extrudate distortions.

HDPE exhibits extrudates that are smooth at low $\tau_w^*$ values, exhibit "sharkskin" (a severe form of matte) and regular, helical screw thread surface patterns in the subcritical stress region, followed by grossly distorted shapes, which are accompanied by large pressure fluctuations for constant flow rate flow. That is, in the distorted extrudate region, there is a discontinuity in the flow curve. At higher stresses (flow rates), the extrudate surface becomes smooth again, a fact that is utilized in fast shaping operations such as wire coating and blow molding of HDPE. The distortions either are not affected or become amplified with *increasing* $L/D_0$ (44). The entrance flow pattern at low flow rates corresponding to the matte or smooth extrudates is stable with a very wide entrance angle ($\alpha \sim 180°$). Bagley and Birks (33) have observed only high frequency oscillations into the capillary at the critical shear stress region, whereas White (34), Oyanagi (45), and Bergem (46) have reported spiraling flow patterns into the die, well into the distorted extrudate regions. The sites of the "sharkskin" distortion is again the die exit (Fig. 13.17), and so is the "screw thread" pattern. The site of and the mechanism for the gross extrudate distortion are problems that have no clear answers. The work of White and Ballenger, Oyanagi, den Otter, and Bergem clearly demonstrates that some

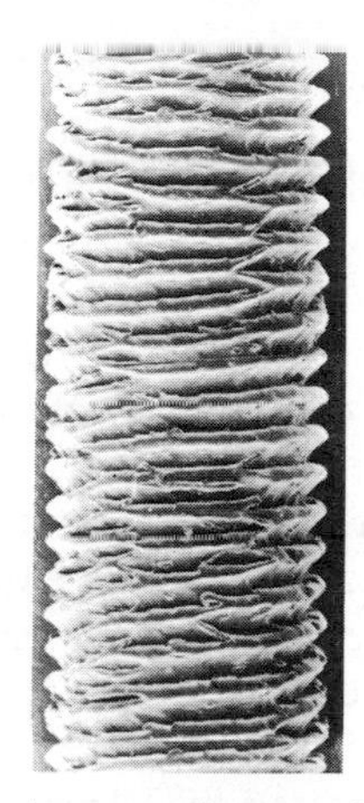

**Fig. 13.17** Scanning electron micrograph of HDPE extruded at a shear rate slightly lower than the oscillation region, showing "sharkskin". [Reprinted with permission from N. Bergem, "Visualization Studies of Polymer Melt Flow Anomalies in Extruders," *Proceedings of the Seventh International Congress on Rheology*, p. 50, Gothenberg (1976).]

instability in the entrance flow patterns is involved in HDPE melt fracture. Clear evidence for this can be found in Fig. 13.18. Slip at the capillary wall, to quote den Otter, "does not appear to be essential for the instability region, although it may occasionally accompany it." The idea of slip at the wall was first proposed by Tordella and later gained popularity because it can be used to explain the discontinuity in the HDPE flow curve (47) and the fact that at shear stresses above the

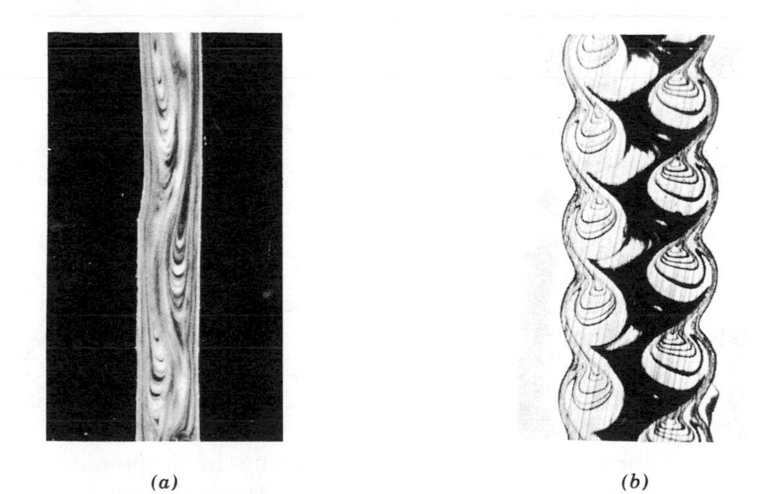

*(a)* *(b)*

**Fig. 13.18** Flow patterns above the oscillating region of HDPE. *(a)* Microtome cut along the cylinder axis of HDPE solidified inside a capillary die, showing that the flow patterns are formed at the die entrance. *(b)* Microtome cut of the HDPE extrudate, resulting under the same conditions as in *(a)*. [Reprinted with permission from N. Bergem, "Visualization Studies of Polymer Melt Flow Anomalies in Extruders," *Proceedings of the Seventh International Congress on Rheology*, p. 50, Gothenberg (1976).]

discontinuity the extrudate becomes smooth again (the melt is continuously slipping at the wall). More recently den Otter (44) has found evidence of slip in the flow of linear elastomers. Thus it is possible that through molecular disentanglement at a certain stress level, a low viscosity layer is formed at the die entrance which, if it is at the periphery, would result in the flow patterns of Fig. 13.18. If disentanglement is involved for the entrance region, there is no reason to exclude it in the capillary where a surface layer of low viscosity film would be formed acting as an effective lubricant.

The magnitude of the melt fracture distortions can be decreased with die entrance streamlining (i.e., decrease of the die entrance angle). Extrudate stretching can eliminate the effects of melt fracture in some polymers, such as polystyrene.

Almost all published data on the phenomenon of melt fracture deal with capillary dies, while in polymer processing dies in shapes of all types are encountered. Slit dies have been investigated by Vlachopoulos and Chan (48) who have found that with polystyrene melts the critical shear stress at the wall is higher in slits than in capillaries. Using the criterion of recoverable shear strain,* which has been found to have a value range of 1–10 at melt fracture, depending on the expression used for the melt compliance (49, 50), he discovered that the ratio of the flow average recoverable shear strain at the slit and capillary walls is 1.4.

Extrudate distortion phenomena are a serious drawback to polymer processing. More industrial and basic research is needed to find ways of processing polymers above the $10^5$ N/m$^2$ shear stress level. An example of such an effort is to be found in the very ingenious method that Tordella used to extrude Teflon, which melt fractures at very low shear rates, at industrially acceptable rates (51). He actually extruded the resin at high pressures as a compacted powder, taking advantage of the very strong dependence of the melting point of PTFE on pressure. Because of this latter property, the compacted powder readily melts upon exiting the die and forms a smooth extrudate.

## 13.3 Fiber Spinning

Spinning is the process of extruding polymer melts or solutions through a metal plate have a number of symmetrically arranged small holes, to form a corresponding number of continuous fluid strands. After the proper postdie treatment, which includes both *melt stretching and cooling* and *cold drawing*, these strands end up as longer fibers of much reduced diameters. The fibers are mechanically very strong, almost totally crystalline, and highly anisotropic. Thus in the total process of fiber formation, not only shaping but also structuring, as discussed in Chapter 3, is taking place, or to quote Mark (52), industrial spinning processes "involve much more than giving shape."

All synthetic fibers are manufactured by spinning. In all spinning methods the liquid to be spun is pressurized in a container that is equipped with screen packs, and is forced out of the spinnerette. If a polymer melt is spun (*melt spinning*), the

* Vlachopoulos et al. (50) have developed for monodisperse PS the melt fracture criterion $S_R \geq 2.65(\bar{M}_z\bar{M}_{z+1}/\bar{M}_w^2)$.

emerging extrudates are stretched and simultaneously cooled with a high velocity cross-current stream. Before the solid fibres are wound on the stretching spools, their cross-sectional area is reduced by a factor of 10–15. This, as discussed in Chapter 3, may result in flow induced crystallization (structuring). Furthermore, before being used in a fabric, the fiber is cold drawn to induce further structural changes, such as those discussed in Section 3.7. This final processing-structuring step significantly enhances fiber strength. Common melt spun fibers are nylon-6 and PET.

Technologically the desired range of viscosity at the spinnerette shear rates is between 100 and 2000 poises. If the polymer melt is too viscous and would require a dangerously high temperature to reach this viscosity range, it is spun in solution form.* Solution spinning is performed industrially in two methods.

*Dry spinning*, where a countercurrent hot air stream is used to evaporate the solvent from the emerging solution stream. The strands are thus transformed from solution, to a gel, to a solid. The solvent evaporation is diffusion controlled and influences both the fiber final diameter and its mechanical properties. Acrylic fibers, for example, are manufactured by this process (a du Pont process).

*Wet spinning*, where the solution extrudates are led into a coagulant bath immediately after they emerge from the spinnerette. The bath contains either a nonsolvent (coagulant) or a liquid that reacts chemically with the dissolved polymer. In both cases the polymer strands are precipitated from solution. Needless to say, the physical or chemical processes occurring in the coagulant bath, in addition to stretching, determine the structure and properties of the fiber.

We discuss only melt spinning in this section, simply because we do not wish to add to the great complexity of the spinning problem the difficult aspect of mass transfer. Such problems were examined by Shih and Middleman (53) and Paul (54). Nevertheless, much of our discussion can be extended to solution spinning, especially as it pertains to die design. Melt spinning is represented schematically in Fig. 13.19. We discuss only the spinnerette flow in this section. The melt strand drawdown region is discussed in Chapter 15 as a secondary forming process.

### The Spinnerette Flow Region

The spinnerette holes are usually short capillaries of $1 < L/D_0 < 5$. They are often streamlined or simply countersunk in an attempt to accommodate the "wine glass" entrance flow region and to minimize extrudate distortion due to possible elastic instabilities. Noncircular spinnerette holes produce "shaped fibers." Because of swelling, the shapes of the hole and the extrudate are different (see Section 13.7). Even in the straight spinnerette dies, the flow cannot be considered viscometric because of the low $L/D_0$ values; capillary entrance phenomena are expected to predominate, or at least to contribute to, the required pressures for spinning. This is illustrated in the example below.

* The "spinnability" of polymer melts or solutions depends not only on their viscosity values, but also on their elastic properties, their ability to undergo large degrees of stretching, and their mass transfer characteristics, in the cases of dry and wet spinning.

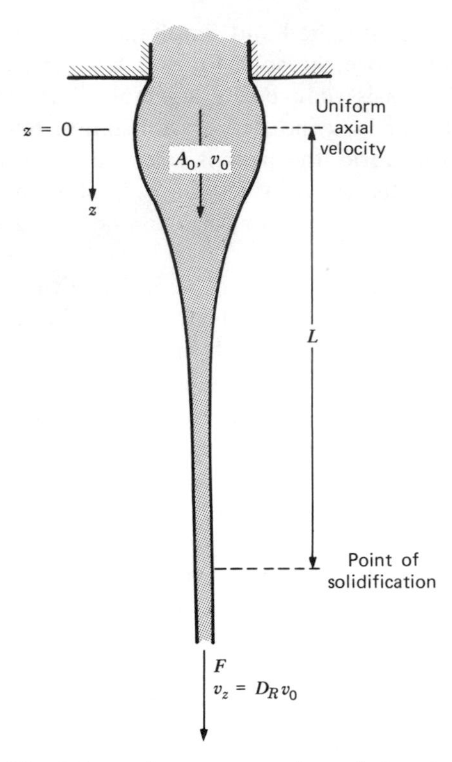


**Fig. 13.19** Schematic representation of the melt spinning process.

**Example 13.3**

Calculate the pressure required to extrude the PP melt using the data shown in Fig. 13.15, at a volumetric flow rate of $4.06 \times 10^{-1}$ cm$^3$/s through a spinnerette that contains 100 identical holes of radius $1.73 \times 10^{-2}$ cm and length $3.46 \times 10^{-2}$ cm.

*Solution*: The volumetric flow rate for each of the spinnerette holes is $4.06 \times 10^{-3}$ cm$^3$/s. Thus the Newtonian shear rate at the capillary wall is

$$\Gamma = \frac{4Q}{\pi R_o^3} = \frac{4.06 \times 10^{-3} \times 4}{3.14(1.73 \times 10^{-2})^3} = 10^3 \text{ s}^{-1}$$

Referring to Fig. 13.15 for the PP melt at $\Gamma = 1000$ s$^{-1}$, we have

$$\frac{\Delta P_e}{\tau_w^*} \simeq 45$$

Since $\tau_w^* = D_0 \Delta P_e / 4L$ and $D_0 = L$, we see that $\Delta P_e = 11.25\ \Delta P_c$. That is, more than 90% of the total pressure is contributed by the entrance region!*

The expected degree of extrudate swelling of *unstretched* melt spun fibers is close to the extrapolated value of $D/D_0$ at $L/D_0 = 0$. For streamlined or counter-

* It is noteworthy that when $L/D_0 = 40$, as would be the case in capillary viscometry, the entrance pressure drop is still 28% of the pressure drop in the capillary $\Delta P_c$.

sunk dies the exit diameter would be appropriate to use for estimating wall shear stress and shear rate values needed for the estimation of $D/D_0$.

In the presence of melt stretching, the value of $D/D_0$ drops rapidly with increasing takeup velocity and takeup stress (27, 55). White and Roman (27) have derived an approximate expression for swelling in the presence of stretching

$$\left[\frac{B}{B(0)}\right]^6 = 1 - \frac{\lambda_{eff}}{\mu}\frac{4F}{\pi D^2}\frac{B^2}{B(0)^6} \tag{13.3-1}$$

where $B = D/D_0$, $B(0)$ the swelling ratio at axial force $F = 0$, and $\lambda_{eff} = \mu/G$ a measure of the relaxation time of the melt. As mentioned earlier, another effect of the stretching of melt spun fibres is the decrease (or even elimination, in some cases at high stretch ratios) of the effects of melt fracture. There appears to be no theoretical work to relate quantitatively these beneficial effects to melt or process variables.

## 13.4 Sheet and Flat Film Forming

Polymer flat film sheets are formed continuously by extruding a polymer through a more or less rectangular "sheeting die," which is quite wide and of a small opening. Because the extruder outlet is by necessity circular and the die rectangular, two fluid particles feeding two arbitrarily chosen die positions will have gone through different flow histories, and this may result in a nonuniform flow rate through the die, dependent on the position along the die width. Thus the design and the choice of the flow passages from the extruder to the die per se are of great importance. A number of sheeting die designs representing different practical as well as theoretical solutions are currently in use.

Upon exiting the die, the sheet extrudate will swell to a level determined by the polymer, the melt temperature, the die length-to-opening ratio, and the shear stress at the die walls. Additionally, flow instabilities will occur at values of the corrected shear stress at the wall of the order of, but *higher* than $10^5$ N/m$^2$, as found by Vlachopoulos and Chan (48), who also concluded that for PS, HDPE, and LDPE the critical $S_R$ in slits is 1.4 times higher than in tubes of circular cross section. Aside from these differences, the information presented in Sections 13.1 and 13.2 applies to slit flow.

Polymer sheets are cooled without stretching by convected cold air (or an inert gas), by immersion into a fluid bath, or by passage over chilled rolls. Flat films are usually stretched and oriented uniaxially and cooled by either of the above-mentioned methods. Films are also cast and cooled on rolls for optimal clarity purposes.

### *Die Design Equations*

The most common mechanical sheet die designs are the center-fed "T" and the "coathanger." In both cases the melt is fed into the center of the manifold, which is of circular or bead-shaped cross section. The manifold distributes the melt into the approach channel through a slit opening running along its entire length. The names "T" and "coathanger" refer to the angle the manifold makes with the flow direction (see Fig. 13.20).

Sheet die design equations were first developed by Carley (56) for T-shaped dies using Newtonian fluids. Pearson (57), whose basic approach we follow below, extended the design equations to power law fluids.

The proper die design delivers a given polymer melt under specified conditions through a constant die opening at a constant rate and temperature (cross machine direction uniformity). We trace the development of a die design equation that has this design objective.* Figure 13.20 presents the geometrical features of the coathanger die, on which the design equation will be developed. The manifold is a tubular, variable radius channel of curved axis $l$. The slit opening $H$ is constant. The

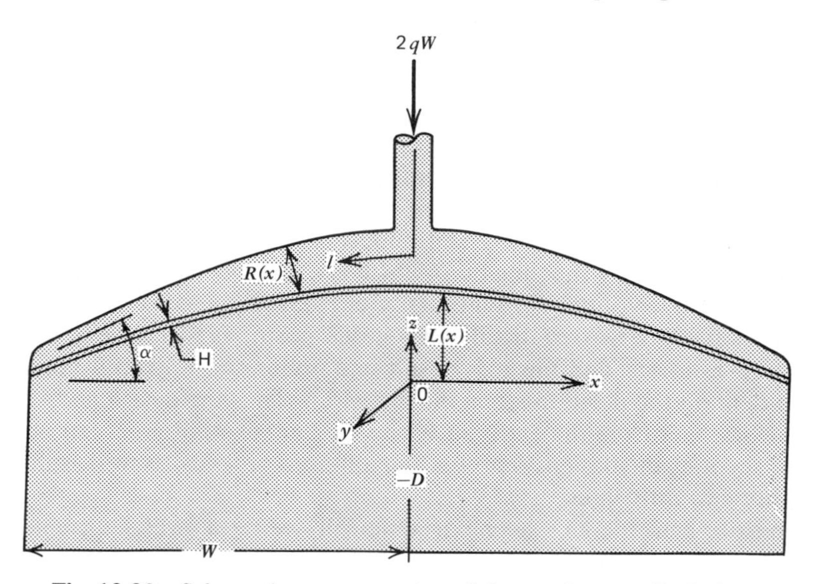


**Fig. 13.20** Schematic representation of the coathanger die design.

only geometric restriction is that the manifold be of small curvature, so that the lubrication approximation can be applied in the manifold region. Also, for the same reason, $dR(x)/dx \ll 1$.

Assuming that the pressure at the manifold entrance is constant and that the problem is isothermal, we have a constant flow rate entering the die (machine direction uniformity). Our objective is to ensure cross-machine direction uniformity, that is, constant flow rate at $z = 0$. But since the slit opening is constant,

$$\frac{dP}{dz} = a = \text{constant} \tag{13.4-1}$$

* In practice, die design does not concern itself simply with the design of the flow passages to control the flow and produce a uniform extrudate—the only aspect we deal with—but also with mechanical design to ensure rigidity of the die under the operating temperatures and pressures, and with chemical and abrasion resistance to the polymers being formed. It also involves the design of an adequate temperature control system, especially in sheeting dies where the surface-to-volume ratio is very large.

This holds everywhere in the parallel plate flow region formed by the slit, where $z$-constant lines are isobars as a consequence. Integrating Eq. 13.4-1 yields

$$[L(0)-L(x)]a = P_0 - P'(l) \tag{13.4-2}$$

Where $P'(l)$ is the manifold pressure at position $l$, $P_0$ is the manifold entrance (delivery) pressure; $x$ and $l$ are related geometrically through the shape of the manifold. Differentiating Eq. 13.4-2 with respect to $l$ results in the following relationship between the pressure gradient in the manifold, its shape, and the pressure gradient in the slit region.

$$\frac{dP'}{dl} = a\frac{dL}{dl} \tag{13.4-3}$$

The relation above is the general design equation satisfying the objective set in the beginning of this section. It holds for any fluid. So, let us give it a usable form by considering a power law fluid flowing in this die. Turning first to the parallel plate region formed by the slit, the $z$-component momentum equation reduces to

$$-\frac{dP}{dy} - \frac{d\tau_{yz}}{dy} = 0 \tag{13.4-4}$$

For a power law fluid,

$$\tau_{yz} = -m\left|\frac{dv_z}{dy}\right|^{n-1}\frac{dv_z}{dy}$$

Using the appropriate boundary conditions in this assumed fully developed flow, we obtain after integration (note that flow is in the negative $z$ direction)

$$v_z(y) = \frac{n}{n+1}\left(\frac{1}{m}\frac{dP}{dz}\right)^{1/n}\left(\frac{H}{2}\right)^{(n+1)/n}\left[\left(\frac{2|y|}{H}\right)^{(n+1)/n} - 1\right] \tag{13.4-5}$$

or

$$v_z(y) = \left(\frac{1}{1+s}\right)\left(\frac{1}{m}\frac{dP}{dz}\right)^{s}\left(\frac{H}{2}\right)^{1+s}[|\xi|^{1+s} - 1] \tag{13.4-6}$$

where $s = 1/n$ and $\xi = 2y/H$. Integrating over the gap opening, we obtain the following relationship between the pressure drop and the flow rate per unit width, $q$, taken to be positive in the negative $z$ direction:

► $$a = \frac{dP}{dz} = 2^{n+1}(2+s)^n m \frac{q^n}{H^{2n+1}} \tag{13.4-7}$$

For the flow inside the manifold of circular cross section* channel, we assume that *locally* we have fully developed tube (capillary) flow. That is, we disregard the channel curvature, the channel tapering off, and the effects of the leak flow into the

* As Fig. 13.1 indicates, the manifold cross section may be bead shaped and not circular. Thus pressure flow in an elliptical cross section channel may be more appropriate for the solution of the manifold flow. Such a problem, for Newtonian incompressible fluids, has been solved analytically (J. G. Knudsen and D. L. Katz, *Fluid Dynamics and Heat Transfer*, McGraw-Hill, New York, 1958). See also Table 13.4 and Fig. 13.31.

slit region. The following relationship was derived previously (see Table 13.2):

$$-\frac{dP'}{dl} = \left(\frac{3+s}{\pi}\right)^n 2m \frac{Q^n(l)}{R(x)^{3n+1}} \tag{13.4-8}$$

From a mass balance point of view, the flow rate in the manifold at any point $Q(l)$ provides for the melt that flows in the slit region from that point on to end of the manifold. Specifically,

$$Q(l) = Q(x) = q(W-x) \tag{13.4-9}$$

Combining Eqs. 13.4-8 and 13.4-9 gives

$$-\frac{dP'}{dl} = \left(\frac{3+s}{\pi}\right)^n 2m \frac{[q(W-x)]^n}{R(x)^{3n+1}} \tag{13.4-10}$$

Inserting Eqs. 13.4-7 and 13.4-10 into the general design equation 13.4-3, we obtain the specific design equation for power law fluids.

► $$\frac{2^n(2+s)^n}{H^{2n+1}}\left(\frac{dL}{dl}\right) + \left(\frac{3+s}{\pi}\right)^n \frac{(W-x)^n}{R(x)^{3n+1}} = 0 \tag{13.4-11}$$

Given the product width $2W$ and the rheological parameter $n$, there are two geometric (die design) parameters available: for a given manifold axis curvature $dL/dl$ or $dL/dx$, there exists a manifold radial taper profile $R(x)$ that results in a uniform pressure at any $z =$ constant line. In particular $P(0) \neq f(x)$. This, together with the fact that $H \neq f(x)$, guarantees the design objectives. Conversely, for a given $R(x)$ there exists an $L(l)$ or $L(x)$ that ensures the die design objectives. Generally, for simplicity of construction, constant $dL/dl$ is used. Note that the expression above provides the value of $R(0)$, which is necessary.

It is worth noting that *not all* $R(x)$ and $L(l)$ or $L(x)$ are acceptable solutions. Any solution that proposes a steeply curving manifold axis or a steeply tapering manifold radius would interefere with the lubrication approximation made during the solution. Furthermore, some solutions may be unacceptable from a machining point of view or because of die strength considerations. Finally, some designs may be preferable over others either because the die design equation applies to them more rigorously (the design is such that the assumptions made are reasonable), or because construction is easier. In the first case more confidence can be placed on the design, and in the second the die can be made more economically.

Figure 13.20 shows that the die does not end at the plane $z = 0$. Because polymer melts are viscoelastic fluids, it extends to $z = -D$, so that a uniform "recent" flow history can be applied on all fluid elements. In deriving the die design equation, we disregarded the viscoelasticity of the melts, taking into account only their shear thinning character.

**Example 13.4**

Specify the coathanger die manifold radius along the entire width of the die, if the manifold axis is straight and makes an angle $\alpha = 5°$ with the $x$-coordinate (see Fig. 13.20). The slit opening is set at $H=0.05$ cm, the half width $W=100$ cm, and the power law index of the polymer melt $n=0.5$ (note: $m$ needs not be specified).

*Solution*: From Eq. 13.4-11 the following expression for $R(x)$ is obtained

$$R(x)^{3n+1}=-\frac{[(3+s)/\pi]^n H^{2n+1}(W-x)^n}{2^n(s+2)^n(dL/dl)}$$

where $dL/dl=-\sin\alpha=-0.0872$. Thus for $n=0.5$ the equation above reduces to

$$R(x)=0.175(W-x)^{0.2}$$

Let us evaluate $R(0)$ and $R(90)$

$$R(0)=(0.175)(2.51)=0.44 \text{ cm}$$

$$R(90)=(0.175)(1.58)=0.277 \text{ cm}$$

The manifold tapered tube is open, to form the slit opening, over an angle $\beta$ such that $\sin[\beta(x)/2]=(H/2)/R(x)$. Thus at $x=0$ $\beta(0)=13°$ and at $x=90$ cm $\beta(90)=21°$. This design results in a very slight taper of the radius of the manifold, about $2\times 10^{-3}$. Also worth noting is that the maximum value of the manifold radius is only about 9 times the slit opening. Finally the taper decreases slightly with decreasing $n$, while the dependence of the manifold radius on $H$ increases with decreasing $n$. The manifold radius becomes infinite at $\alpha=0$ and is very sensitive to $\alpha$ when it has small values.

Other flat film die design equations have been proposed in the literaure. The one advanced by McKelvey and Ito (58) has as a design objective the flow rate uniformity along the die width. This is achieved by varying the final die lip opening. Thus although the resulting flow rate is independent of the width direction, the film or sheet thickness is *not*. Additionally, the wall shear rate for a power law fluid

$$\dot{\gamma}_w=\frac{2(s+2)q}{H^2(x)} \tag{13.4-12}$$

will be width dependent, allowing extrudate swelling to vary.

The die design equation proposed by Pearson (57) utilizes a constant die lip opening, but an approach channel taper that varies with the die width. Thus in the region between the manifold and the die lip opening $P=P(x, z)$ and the flow is two-dimensional. This may affect the flow in the die lip region, since the fluid is viscoelastic and the flow experience recent.

The following criticism applies to all the die design equations for sheet forming that have been proposed so far:

1. The manifold and slit flows are treated independently, disregarding the disturbances in both flow fields as a result of the transition flow from the manifold to the slit, including "entrance" losses. To reduce the latter, tapered, wedge-shaped manifolds are used.

2. The flow is assumed to be isothermal. In any real sheet forming operation some temperature gradients, both in the melt and along the die, will be present (19).
3. The die lip deflection under the pressure applied by the flow is neglected. This could be accounted for, at least approximately, by rather straightforward beam calculations and iterative procedures as suggested by Pearson (57).
4. The die designs developed or mentioned previously are for a specific polymer and specific processing conditions. Nonuniform sheets of another polymer would result if substitutions were made. The same holds true for the same polymer extruded at different temperatures.

For these reasons, die lip opening adjustor bolts are provided for with every sheeting die to make *fine* adjustments. Usually these adjustments are made manually. Because the die flows are often quite fast and manual corrections of sheet thickness nonuniformities result in material waste, feedback systems have been devised to adjust lip opening automatically. Multiple extruders can be used for very wide dies, or a screw may be placed in the manifold of a "T" die.

If "machine direction" thickness uniformity is a problem, it can be best remedied with the use of a gauge detector (beta gauge) that is part of a control system adjusting the speed of the takeup device, to correct for thickness nonuniformities. Small period variations are very difficult to remedy in this fashion.

It can generally be said that the approach to developing die design equations, irrespective of the basic die type, is the following:

1. Simplify the actual flow by assuming that it is a series of well-identified viscometric flows.
2. By applying one or more mass balances, relate the volumetric flow rates in each of the viscometric flows.
3. Allowing for one or more die geometric parameters to be variable, state one or more "extrudate uniformity conditions" that when satisfied (solved for), will determine the geometric variables above as functions of other geometric, process, and material constants.

Obviously this method developing die design equations implies that there is *no unique* die design to achieve product uniformity in the cross-machine direction. Multiple alternative design—thus die design equations—exist because one has an a priori choice of what geometric variables will be allowed to "float."

Two or more polymer melts can be *coextruded* from flat film dies that are fed by more than one extruder. Very thin films (of thickness as small as a fraction of a mil) can be formed in which the placing of each layer serves a specific purpose, such as low diffusivity and chemical inertness (e.g., for food wrapping films) as well as abrasion resistance. The process can be thus looked at as a "coating" operation that achieves very thin coats. Typically the coextrusion die has separate manifolds and restrictor bars (see Fig. 13.1) leading to a common approach channel and die lip opening.

Stratified flow between infinite parallel plates, where the interface is specified, can be easily solved for Newtonian fluids (59) and by trial and error for power law fluids (see Problem 13.6). In reality, stratified flow with polymer melts is more complex in that the interface changes in position and shape with axial distance. Khan and Han (60) have observed that the less viscous component progressively engulfs more of the other melt, wetting more of the die surface and creating a curved interface. This situation becomes worse for longer dies. The problem of interfacial stability is also interesting and important in the fabrication of bicomponent fibers (61–63). Two melt streams extruded side by side into a circular die may end up in a concentric configuration with the less viscous component on the outside. Again, as in the process of mixing of two polymer melts discussed in Chapter 11, the viscosity ratio rather than the melt elasticity seems to be of primary importance (63).

The development of the flat film die design equation follows the same lines as the one for the flat sheet. The obvious differences that must be taken into consideration the the very small die openings and the very high prevailing shear rates, which force one to consider the elastic nature of the fluids and the possibility of nonisothermal flows. Effects of the melt fracture phenomenon do diminish with film stretching.

Upon exiting the die, flat films are stretched axially, and this brings about not only a reduction of the film thickness (film drawdown as in the case of planar extension described in Section 6.8) but also a reduction of the film width ("necking in"). Furthermore, the edges of the flat film are thicker ("bead" formation), which necessitates the elimination of the edges from the final product and the generation of "scrap." At a critical draw ratio of the order of 20, Bergonzoni and DiCresce (64) observed draw resonance in flat film extrusion more than a decade ago. Variations of both the film width and thickness occur: $W_{max}/W_{min}$ can be as high as 10, and $H_{max}/H_{min}$ up to 5. In linearized stability analysis of a two-dimensional stretching of a Newtonian film, a critical draw ratio of 20.2 has been found (65). This subject is discussed in more detail in Section 15.1.

## 13.5 Tube and Tubular (Blown) Film Forming

Plastic tube and tubular films are formed continuously by extruding a polymer through an annular die. The annular flow channel is formed by the outer die body and the die mandrel. A number of annular die designs are currently employed. In the first, the mandrel is supported mechanically onto the outer die body by "spider legs"; Fig. 13.21 illustrates the die. The flow is axisymmetric, and the only serious problem encountered in the cross-machine direction uniformity of the extruded product is that of "weld" lines and streaks caused by the presence of the spider legs, which split the flow. Even though these obstacles are far away from the die lip region, the polymer melt, at normal extrusion speeds, is unable to "heal" completely. That is, the macromolecules comprising the two layers that were split by the spider legs do not establish the entanglement level characteristic of the bulk at the prevailing shear rate and temperature. This is another ramification of incomplete response of polymers because the "experimental time" is smaller than the relaxation time of the system of macromolecules. As expected, weld lines are mechanically weak and have optical properties that differ from those of the bulk, making them visible. Furthermore, they result in film or tube gauge nonuniformities, probably because of the different degree of swelling of the melt in the

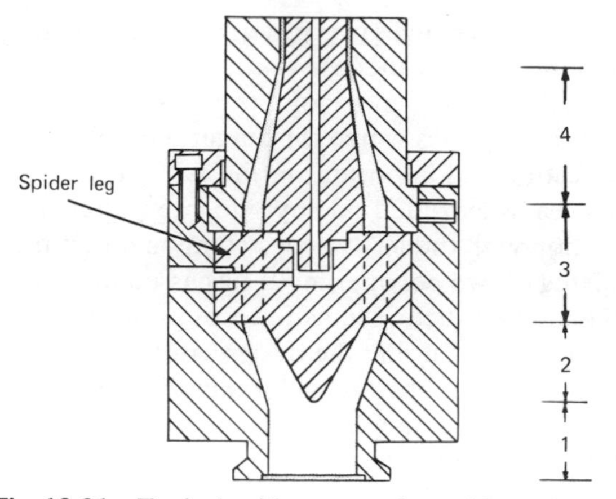


**Fig. 13.21** Typical spider-type tube or blown film die.

neighborhood of the weld line. They also induce cross-machine pressure nonuniformities.

To overcome the above-mentioned problems, basic cross-head die designs (Fig. 13.22) have been devised in which the mandrel is mechanically attached to the die body in such a way that obstacles are not presented to the flow in the annular region.

Unlike the coathanger flat film dies, no simple final film adjustment is possible by lip flexing. Consequently, the order of magnitude of thickness accuracy in tubular dies is ±10% as compared to ±5% in flat sheet and film dies. This larger margin of accuracy in blown film dies is compensated by die rotation, which permits the distribution of the thickness variation across the entire width of the product.

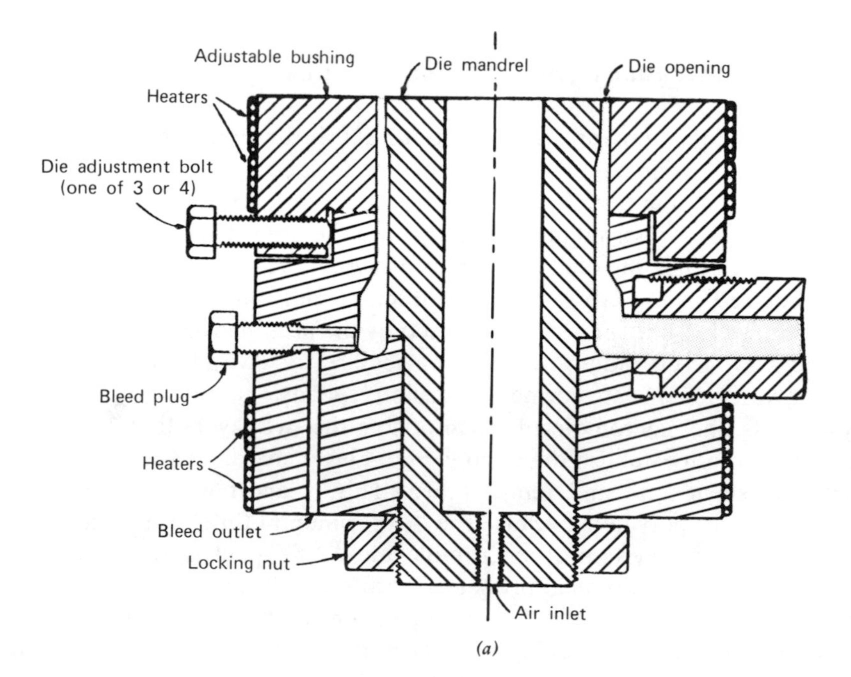


*(a)*

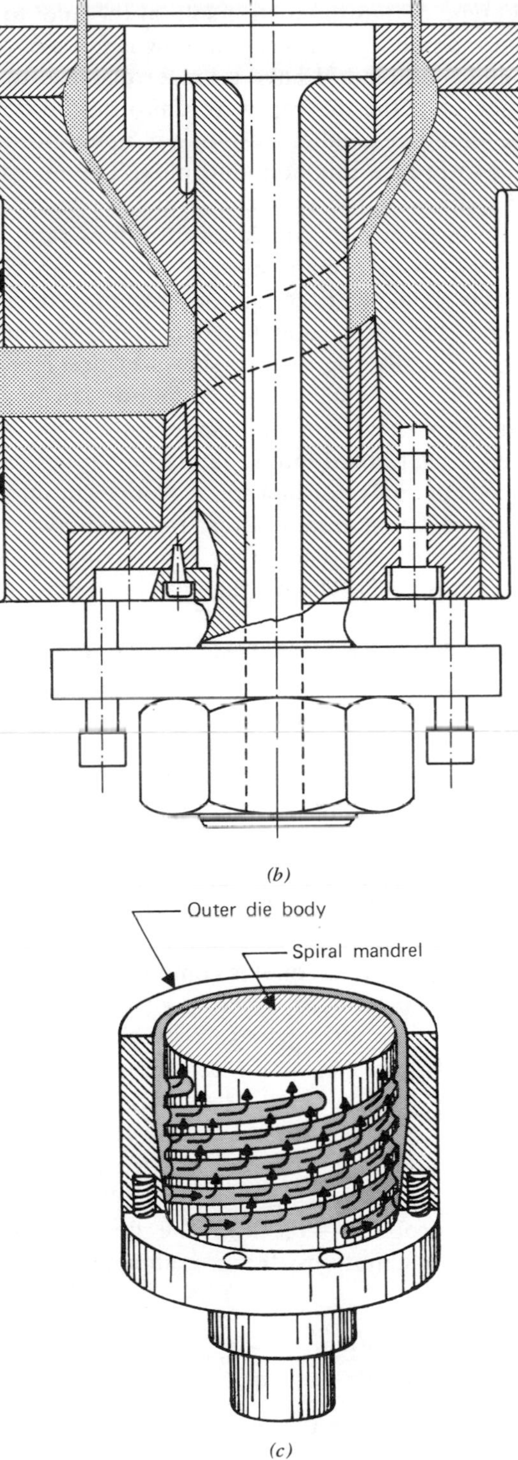


**Fig. 13.22** Tubular dies. (*a*) Side-fed manifold die. (*b*) Blown film die. (*c*) Spiral mandrel die. (Reprinted with permission from J. R. Syms, "Computer Design Dies for Tubular Films," *PIA Course on Accuracy in the Extrusion Process*, October 25, 1973.)

In the cross-head type of dies the melt is split at the inlet to the manifold and recombines 180° from the inlet. Moreover, the flow is not axisymmetric and fluid particles flowing around the mandrel have a longer distance to travel than those that do not. Consequently, if the die gap were uniform, the flow rate 180° away from the die entrance would be smaller, resulting in nonuniform thickness. There are a number of ways to reduce this nonuniformity. The mandrel can be placed eccentrically in the die, allowing for a wider gap at the remote end from the lead port. Such a design can in principle provide uniform flow rate, but the shear rate and temperature histories will remain nonuniform. In another solution (Fig. 13.22*b*), an insert directs the flow at the far end upward, reducing the flow length around the mandrel and eliminating slow flow (stagnant) regions. In addition, the mandrel is also eccentric. Finally, Fig. 13.22*c* shows a spiral mandrel die, currently in common use for film blowing, which allows greater design flexibility in obtaining a uniform flow rate, uniform shear rate and temperature histories, and elimination of weld lines. The feed is distributed into separate flow tubes called "feed ports." Each of these ports feeds the polymer in a spiral groove cut into the mandrel. The spiral decreases in cross-sectional area, whereas the gap between the mandrel and the die increases toward the exit. The result is a mixing or "layering" of melt originating from the various ports. The die flow has been modeled mathematically for design purposes (66).

In the extrusion of tubes, such as rigid PVC or PE pipe, the extrudate passes over a water-cooled mandrel and enters a cold water bath whose length depends on the tube thickness; the tube leaves the bath well below its $T_m$ (if it is crystalline) or $T_g$ (if it is amorphous) and is sectioned to the desired lengths.

Next we discuss the problem of estimating the total pressure drop in tubular dies, and we trace the development of the die design equation.

## *Estimation of Pressure Drop in Tubular Dies*

We know that the tubular die flow channel is composed of a series of more or less annular flow regions, which are straight, tapering, of almost uniform cross section, or interrupted by obstacles. Thus although exact solutions of the flow in them may require numerical methods, we can arrive at a number of useful and simple engineering expressions by examining the steady isothermal flow between two straight concentric cylinders of constant radii $R_o$ and $R_i$.

Consider the annular flow region $R_i \leq r \leq R_o$, $0 \leq z \leq L$, and $0 \leq \theta \leq 2\pi$. An incompressible power law fluid is flowing under steady and isothermal conditions because of the pressure drop $\Delta P = P_0 - P_L$. The flow is assumed to be fully developed. Under these assumptions the $z$-component momentum equation becomes

$$\frac{dP}{dz} = -\frac{1}{r}\frac{d}{dr}(r\tau_{rz}) \tag{13.5-1}$$

When Eq. 13.5-1 is coupled with the power law constitutive equation for this flow

$$\tau_{rz} = -m\left|\frac{dv_z}{dr}\right|^{n-1}\left(\frac{\partial v_z}{\partial r}\right) \tag{13.5-2}$$

the following expression is obtained

$$\left(\frac{r}{m}\right)\frac{dP}{dz}=\frac{d}{dr}\left(r\left|\frac{dv_z}{dr}\right|^{n-1}\frac{dv_z}{dr}\right) \tag{13.5-3}$$

Let $r^*$ be the radial position where the velocity is maximum; that is, $dv_z/dr=0$. Then in region *I*, $R_i \le r \le r^*$, $dv_z^{\mathrm{I}}/dr \ge 0$, and Eq. 13.5-3 becomes

$$\left(\frac{r}{m}\right)\frac{dP}{dz}=\frac{d}{dr}\left[r\left(\frac{dv_z^{\mathrm{I}}}{dr}\right)^n\right] \tag{13.5-4}$$

where $v_z^{\mathrm{I}}(r)$ is the velocity in this region. The accompanying boundary conditions are $v_z^{\mathrm{I}}(R_i)=0$ and $dv_z^{\mathrm{I}}/dr=0$ at $r=r^*$. Similarly in region II, $r^* \le r \le R_o$, $dv_z^{\mathrm{II}}/dr \le 0$, and Eq. 13.5-3 becomes

$$\left(\frac{r}{m}\right)\frac{dP}{dz}=-\frac{d}{dr}\left[r\left(-\frac{dv_z^{\mathrm{II}}}{dr}\right)^n\right] \tag{13.5-5}$$

having the following boundary conditions: $v_z^{\mathrm{II}}(R_o)=0$ and $dv_z^{\mathrm{II}}/dr=0$ at $r=r^*$. Equations 13.5-4 and 13.5-5 can be directly integrated with the foregoing boundary conditions, and the location $r^*$ be obtained by setting $v_z^{\mathrm{I}}(r^*)=v_z^{\mathrm{II}}(r^*)$. This problem was solved by Fredrickson and Bird (67), and the resulting flow rate–pressure drop relationship is

$$Q=\left(\frac{\pi R_o^3}{s+2}\right)\left[\frac{R_o\Delta P}{2mL}\right]^s\left(\frac{\beta-1}{\beta}\right)^{2+s}F(n,\beta) \tag{13.5-6}$$

where $\beta=R_o/R_i$ and $F(n,\beta)$ is the pseudoplasticity and geometry dependent function appearing in Fig. 13.23. For values of $0.4 \le R_i/R_o \le 1.0$, which represent relatively narrow annuli, the function $F$ becomes independent of the degree of pseudoplasticity of the melt. At the limit $\beta \to 1.0$, Table 13.1 can be used to relate the volumetric flow rate to the axial pressure drop, since the geometrical situation corresponds to the flow between parallel plates. In this case

$$Q=\left(\frac{\pi R_o^3}{s+2}\right)\left[\frac{R_o\Delta P}{2mL}\right]\left(\frac{\beta-1}{\beta}\right)^{2+s}\left(\frac{1+\beta}{2\beta}\right) \tag{13.5-7}$$

where in Table 13.1 $W=\pi(R_i+R_o)$. Thus the pressure drop in straight concentric annular sections of tubular dies can be obtained fairly easily using Eq. 13.5-6 with the help of Fig. 13.23, or with Eq. 13.5-7 for very narrow annuli. It should be recalled that in annular pressure flows as in all pressure flows, an error is introduced by using the power law model fluid because of inaccuracies in the low shear rate regions. Moreover, the assumption of isothermicity also introduces errors and can lead to an overestimation of $P_0-P_L$. Cox and Macosko (19) have measured melt surface temperature increases of the order of 10–20°C with LDPE flowing in an $L=0.1$ m, $R_i/R_o=0.5$ annulus at 190°C, and Newtonian wall shear rates of about 200 $s^{-1}$. Such temperature increases would reduce the melt viscosity, especially near the exit of the annulus.

For tapering annular flow channels (regions 2 and 4 of the spider-type die in Fig. 13.21), we can calculate the pressure drop by making the lubrication

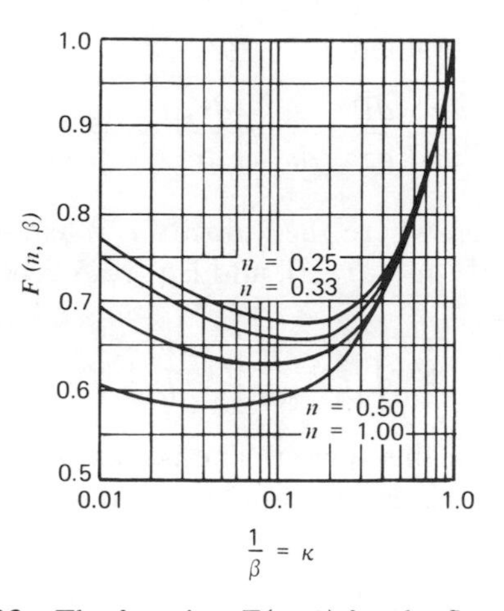


**Fig. 13.23** The function $F(n, \beta)$ for the flow of power law fluids in an annular region. [Reprinted with permission from A. G. Frederickson and R. B. Bird, "Non Newtonian Flow in Annuli," *Ind. Eng. Chem.*, **50**, 347 (1958).]

approximation and using either Eq. 13.5-6 or 13.5-7, depending on the width of the annulus. In both cases $\beta = \beta(z)$ and $H = H(z)$. We can use values of $\beta$ and $H$ that are averaged over the entire length, or we can solve Eq. 13.5-6 or 13.5-7 over small axial length increments $\Delta L_i$, with corresponding $\beta_i$ or $H_i$, then summing up the partial pressure drops. For thin tapered annuli, which can be represented by two almost parallel plates, one can use the Reynolds lubrication equation 5.4-11.

In converging or diverging sections of annular dies, the fluid elements are subjected to axial and radial accelerations. Neglecting the radial spacial accelerations (for small tapers), the $z$-component equation of motion reduces to

$$\rho v_z \frac{\partial v_z}{\partial z} = -\frac{dP}{dz} - \left[\frac{1}{r}\frac{\partial}{\partial r}(r\tau_{rz}) - \frac{\partial \tau_{zz}}{\partial z}\right] \tag{13.5-8}$$

For very viscous fluids the inertial term $\rho v_z(\partial v_z/\partial z)$ is negligible, thus

$$\frac{dP}{dz} = -\frac{1}{r}\frac{\partial}{\partial r}(r\tau_{rz}) - \frac{\partial \tau_{zz}}{\partial z} \tag{13.5-9}$$

For converging channels the first term on the right-hand side of Eq. 13.5-9 increases with increasing axial distance, and because of this $dP/dz$ is not a constant but is $z$-dependent. For a viscous fluid, the value of $\partial\tau_{zz}/\partial z$ is given by the relation

$$\tau_{zz} = -\bar{\eta}\frac{\partial v_z}{\partial z} \tag{13.5-10}$$

A value for $\partial v_z/\partial z$ averaged over the spacing $H$ can be used for approximate calculations. This quantity, since $q = \bar{v}_z(z)H(z)$, is

$$\frac{\partial \bar{v}_z}{\partial z} \cong \frac{\partial}{\partial z}\left(\frac{q}{H(z)}\right) = \frac{\partial}{\partial z}\left(\frac{q}{H_0 - Az}\right) = -\frac{Aq}{(H_0 - Az)^2} \tag{13.5-11}$$

where $A$ is the taper. Considering the result of the equation above as well as Eq. 13.5-10, we conclude that the contribution of the second term of the right-hand side of Eq. 13.5-9 is never zero for tapered channels. For more exact calculations, the dependence of $\partial v_z/\partial z$ on the thickness direction must be taken into account.

Worth and Parnaby (68) have considered the contribution of the elasticity of the polymer melt to the term $\tau_{zz}$. Using a Maxwell-type constitive equation, they find

$$\tau_{zz} - \tau_{rr} \cong \tau_{zz} = \frac{\tau_{rz}^2}{G} \qquad (13.5\text{-}12)$$

Since $\tau_{rz} = f(z)$ for tapered dies, $\partial\tau_{zz}/\partial z$ is nonzero. For reasonable taper values, they find these elastic forces to contribute less than 10% of the viscous pressure drop. They have also calculated the drag and pressure forces on the mandrel.

Turning to the crossfed tubular dies, we note that to develop die design expressions, we must model the two-dimensional flow in the $z$- and $\theta$-directions. This is a task of considerable difficulty. Pearson (69) was the first to model the flow for narrow dies. The flow region was "flattened," and the two-dimensional flow in rectangular coordinates between two plates was considered. The plate separation was allowed to vary in the approach channel so that the resulting output is constant. The final die lip opening is constant, formed by the concentric cylinders. The resulting design equations are complicated, and their solution is computationally demanding. Nevertheless, design expressions for both Newtonian and power law fluids in isothermal flow can be obtained. Gutfinger, Broyer, and Tadmor (70) solved this problem using the flow analysis network (FAN) method discussed in Chapter 16. This approximate but relatively simple numerical method is particularly well suited for two-dimensional slow flow problems in narrow gaps. The results obtained with the FAN method are identical to those of Pearson, but they can be achieved with much smaller computational effort.

As mentioned earlier, there is an additional role that the approach and die lip regions must play. In these regions the polymer melt must be given an opportunity to loose all its "memory" of the crossflow nonuniform strain history. Worth and Parnaby (68) call these regions the "relaxation zone," and by assuming that the melt responds as a simple Voigt fluid (see Section 6.4), they calculate roughly the minimum length for a desired level of relaxation of the strains applied at the entrance.

It is worth noting that although in principle tube and tubular blown film dies are similar, in practice they are quite different in function, size, and complexity. Blown film dies are much longer, have a very small die lip opening, and are subject to more stringent product uniformity criteria because there is no "sizing" equipment downstream. Furthermore, blown film products are almost exclusively LDPE, and occasionally HDPE and PP. On the other hand, HDPE and rigid and plasticized PVC are the common polymers for pipes and tubes.

### Parison Die Design

Another type of tubular die is that used for the formation of the *parison* used in blow molding processes of various types. In a typical reciprocating screw blow molding die head (Fig. 13.24), the choke screw *D* is adjustable to compensate for batch-to-batch or polymer-to-polymer viscosity variations. The choke ring *I* is adjustable to eliminate angular melt pressure variations, and it forms an annular channel that is narrow enough to ensure diminution of the effects of the varying melt flow histories of the incoming melt. The centering screw *J* is used for the final adjustments, which ensure that there is no angular dependence of the parison thickness or diameter. The final die gap is a conical section, slightly tapering,

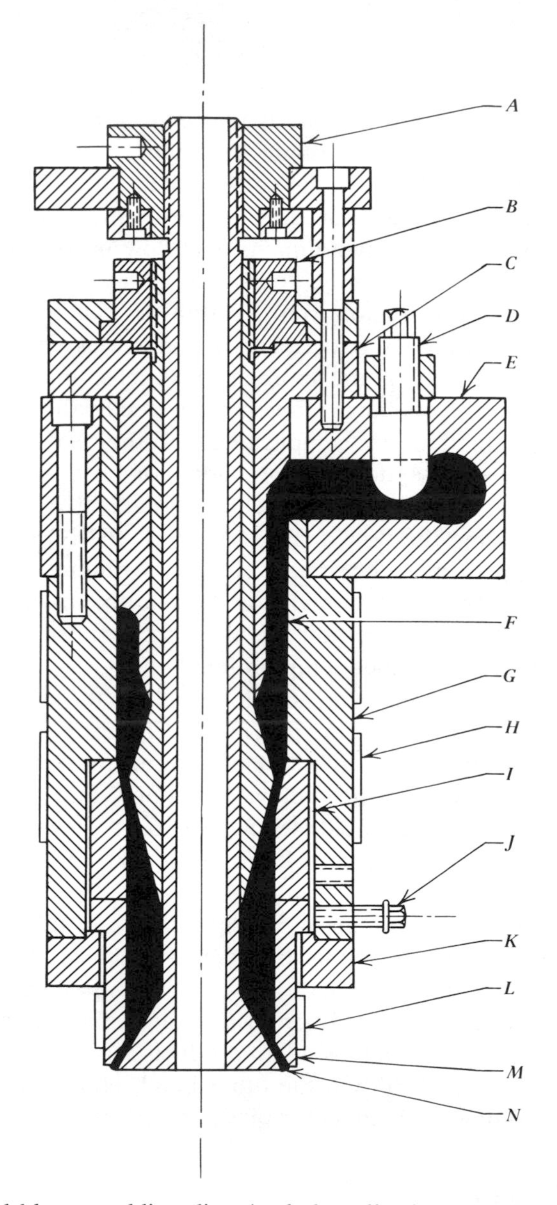


**Fig. 13.24** Typical blow moulding die: *A*, choke adjusting nut; *B*, mandrel adjustment; *C*, feed throat; *D*, choke screw; *E*, die head; *F*, plastic melt; *G*, die barrel; *H*, heater band; *I*, choke ring; *J*, centering screw; *K*, clamp ring; *L*, die heater; *M*, die; *N*, mandrel. [Reprinted with permission from J. D. Frankland, *Trans. Soc. Rheol.*, **19**, 371 (1975).]

annular channel. The cone angle $\Theta$ is appreciable (Fig. 13.25) and is an important die design variable in determining parison diameter, diameter profile, and thickness. Furthermore, an angle $\Theta > 0$ allows die gap adjustment through slight axial mandrel position changes. Such die gap time variations are very desirable to have available, as we shall see below (parison programming).

The large value of $\Theta$ renders the flow in the conical gap nonviscometric, since all velocity components are nonzero and $(v_z/v_r) \simeq \tan \Theta$. It is therefore not totally unexpected that parison formation cannot be well predicted from rheological measurements obtained with steady viscometric flows. In addition, the flow during the parison formation is *transient* in both the ram-accumulator and reciprocating screw processes. The reasons are: the short transient in the ram or screw movement, the compressibility of the melt and, since the times involved are very short, the transient nature of the melt rheological properties. This is another reason for the difficulty in describing the parison formation process completely by material properties obtained during steady rheological measurements.

Finally, at least with HDPE, parison formation flows are above the melt fracture region (see Fig. 15.16). An indication of this is the surface of the upper portion of the parison, formed during the flow rate deceleration part of the cycle, which has a matte texture. Fast parison formation is desirable, and for large parisons essential, because it allows less time for parison creep (sag) under its own weight, and less time for parison diameter reduction or "curtaining" (folding).

From the discussion above it is evident that the exact solution of the flow in the blow molding die is difficult, and only approximate expressions relating flow rate and pressure drop (die design equations) can be obtained. Assuming that the flow is locally straight annular (i.e., making use of the lubrication approximation) and that the fluid can be described by the power law model, Eq. 13.5-6 can be used over sufficiently small channel increments along the conical direction.

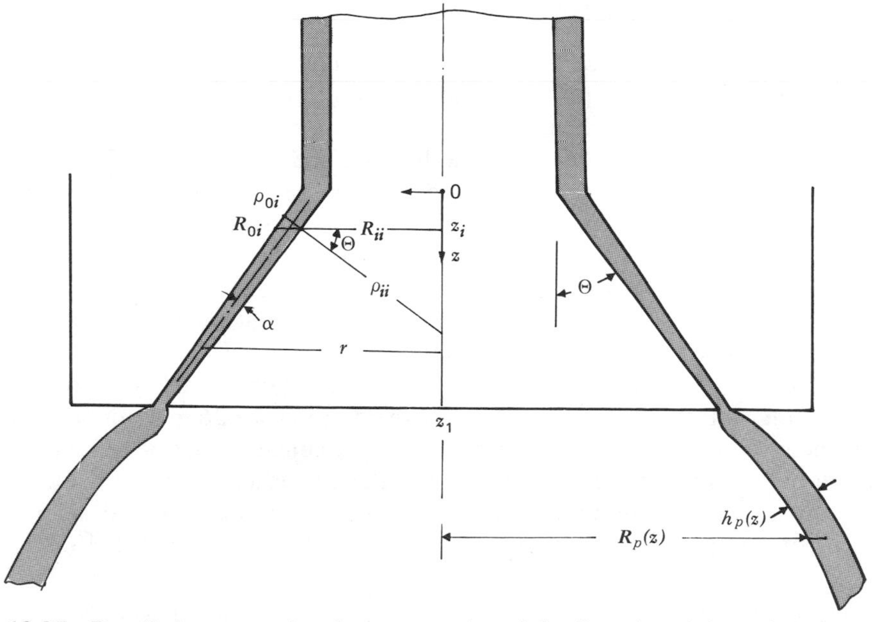


**Fig. 13.25** Detailed cross-sectional representation of the die exit and the parison formed in the blow molding process.

## 13.6 Wire and Cable Coating

Wire and cable coating dies are used in the extrusion process for primary insulation of single conducting wires as well as jacketing or sheathing of a group of wires, already insulated electrically, for mechanical strength and environmental protection purposes.

The bare wire is unwound, sometimes by a controlled tension device, and is preheated to a temperature above the $T_g$ or $T_m$ of the polymer to be extruded; this is done so that the layer next to the bare wire adheres to it, and to drive moisture or oils off the conductor surface. The wire is fed in the back of the cross-head die and into a "guider tube." Upon exiting the guider, it meets the molten plastic, which covers it circumferentially. Since the wire speed, which is controlled by a capstan at the end of the line, is usually higher than the average melt velocity, a certain amount of "drawdown" is imposed on the melt anywhere from a value slightly greater than unity to 4. After exiting the die, the coated wire is exposed to an air or gas flame for the purpose of surface annealing and melt relaxation, which also improves the coating gloss. It then enters a cooling trough, where the cooling medium is usually water. The length of the trough depends on the speed of the wire, the diameter of the wire or cable, and the thickness of insulation; it increases with increasing level of these parameters. As expected, the cooling trough length is longer for crystalline polymers, since the crystallization process is exothermic. For undersea cables cooling troughs not only provide for almost 300 ft of linear travel, but are also divided into several compartments containing water of successively cooler temperatures (typically of a total range 80°–100°C), to avoid fast cooling of the jacket surface, which could cause void formation or thermal stresses.* Upon exiting the cooling trough, the wire passes over the capstan, where its tension is controlled and further cooling can be provided. It then passes through a capacitance measuring device that detects flaws as well as thickness variations. These variations provide information for adjusting the pulling device speeds or the extruder screw speeds. Since defective wire is difficult to reprocess because the product is a composite, the process is closely monitored. In addition, great care is taken to design the coating die properly.

Two types of cross-head die are used for wire and cable coating. The first type is an annular flow or "tubing" die, and Section 13.5 discussed the design of cross-head tubing dies. Here a thin walled tube is extruded; the molten tube is drawn onto the conductor by vacuum after it leaves the die. The vacuum is applied through the clearance between the conductor and the "guide," which is usually of the order of 0.2 mm. Tubing dies are commonly used for jacketing cables or coating very thin wires with polymer melts that are very viscous.

The second type of cross-head die used is the "pressure" type, where the polymer melt contacts the conductor *inside* the die; Fig. 13.26 gives details of such a die. The clearance between the guide and the conductor must be quite small, of the order of 0.05 mm, because at the guide tip the melt is under some pressure. This type of die is quite commonly used for wire coating. From a flow analysis point of view, the pressure die can be broken down into two regions, as in Fig. 13.26.

* Crosslinkable LDPE, used in insulating power cables, requires long residence in steam "curing tubes" 100–300 m long.

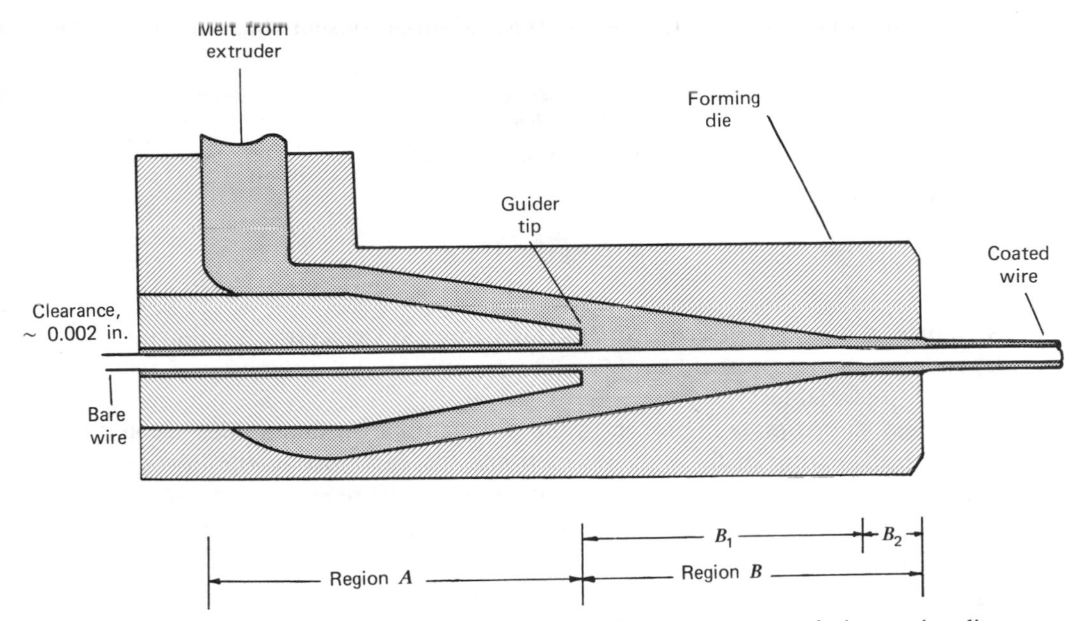


**Fig. 13.26** Detailed schematic representation of the pressure type of wire coating die.

The flow situation in region $A$ is the same as that in the approach channel of side fed tubular dies discussed in the previous section. Thus making the lubrication approximation, the pressure drop in this region can be estimated using Eq. 13.5-6.

In region $B$ one of the containing walls is the conductor wire, which is moving with a velocity $V$. Thus the flow is both drag and pressure induced. In the streamline region $B$, the flow can be treated locally as a combined drag and pressure flow in an annulus of the local thickness (lubrication approximation). Such a flow truly exists in region $B_2$. Thus solving the annular drag and pressure flow will help us in the analysis of the entire region $B$.

### *Combined Isothermal Drag and Pressure Flow in an Annulus*

McKelvey (71) was the first to consider the problem of combined drag and pressure flow in an annulus by examining the steady isothermal flow of an incompressible Newtonian fluid in a die whose annular thickness $H$ is much smaller than the wire diameter $2R_i$. Thus, solving the combined drag and pressure flow between parallel plates in terms of the pressure drop $\Delta P$ gives (see Section 10.2)

$$Q = Q_d + Q_p = \frac{VWH}{2} + \frac{WH^3\,\Delta P}{12\mu L} \tag{13.6-1}$$

where $W$ is the mean circumference of the annular space. The flow rate above must be equal to the rate of polymer wire coating of thickness $h$

$$Q = \pi V[(R_i + h)^2 - R_i^2] = \pi V h(2R_i + h) \tag{13.6-2}$$

Combining Eqs. 13.6-1 and 13.6-2 results in the simple design equation that can be used to estimate the pressure drop $\Delta P$

$$\Delta P = \frac{6\mu VL}{H^2}\left[\frac{4h\pi}{WH}\left(R_i + \frac{h}{2}\right) - 1\right] \tag{13.6-3a}$$

which reduces to

$$\Delta P = \frac{6\mu LV(2h - H)}{H^3} \tag{13.6-3b}$$

by approximating $W = 2\pi(R_i + h/2)$. This expression gives the semiquantitative dependence of the pressure drop on the process $(V, h)$, geometric $(L, H)$, and material $(\mu)$ variables.

A less approximate approach to the problem above, taking into account the pseudoplastic nature of the polymer melt and the curvature of the flow channel, requires numerical solution. We discuss it below. In the region of flow, $R_i \le r \le R_o$, $0 \le z \le L$, $0 \le \theta \le 2\pi$ let there be a radial position $r = r^* = \lambda R_o$, where the velocity is maximum and the shear stress $\tau_{rz}$ zero. For this case only integrating Eq. 13.5-1, which holds for steady, isothermal, and fully developed flow in this region, we have

$$\tau_{rz} = -\left(\frac{r}{2}\right)\left(\frac{dP}{dz}\right) + \frac{C_1}{r} \tag{13.6-4}$$

Invoking the condition $\tau_{rz}(\lambda R_o) = 0$, we arrive at another way of representing the integration constant

$$C_1 = \frac{1}{2}\left(\frac{dP}{dz}\right)(\lambda R_o)^2 \tag{13.6-5}$$

Combining the foregoing results with the power law constitutive equations for this flow, Eq. 13-5.2, we have

$$-\tau_{rz} = \frac{1}{2}\left(\frac{dP}{dz}\right)\left[r - \frac{(\lambda R_o)^2}{r}\right] = m\left|\frac{dv_z}{dr}\right|^{n-1}\frac{dv_z}{dr} \tag{13.6-6}$$

This equation has a different form in the two regions on either side of the maximum velocity.

In region I, $R_i \le r \le \lambda R_o$, $dv_z^{\mathrm{I}}/dr \ge 0$. Thus we have

$$\frac{dv_z^{\mathrm{I}}}{dr} = \left[\frac{1}{2m}\left(\frac{dP}{dz}\right)\left\{r - \frac{(\lambda R_o)^2}{r}\right\}\right]^s \tag{13.6-7}$$

Integrating and noting that $v_z^{\mathrm{I}}(R_i) = V$, we write

$$v_z^{\mathrm{I}}(r) = V + \left(-\frac{1}{2m}\frac{dP}{dz}\right)^s \int_{R_i}^{r}\left[-r + \frac{(\lambda R_o)^2}{r}\right]^s dr \tag{13.6-8}$$

Similarly for region II, $\lambda R_o \le r \le R_o$, $dv_z^{\mathrm{II}}/dr < 0$. Thus after a treatment similar to that in region I and integration, using the boundary condition $v_z(R_o) = 0$, we obtain

$$v_z^{\mathrm{II}}(r) = \left(-\frac{1}{2m}\frac{dP}{dz}\right)^s \int_{r}^{R_o}\left[r - \frac{(\lambda R_o)^2}{r}\right] dr \tag{13.6-9}$$

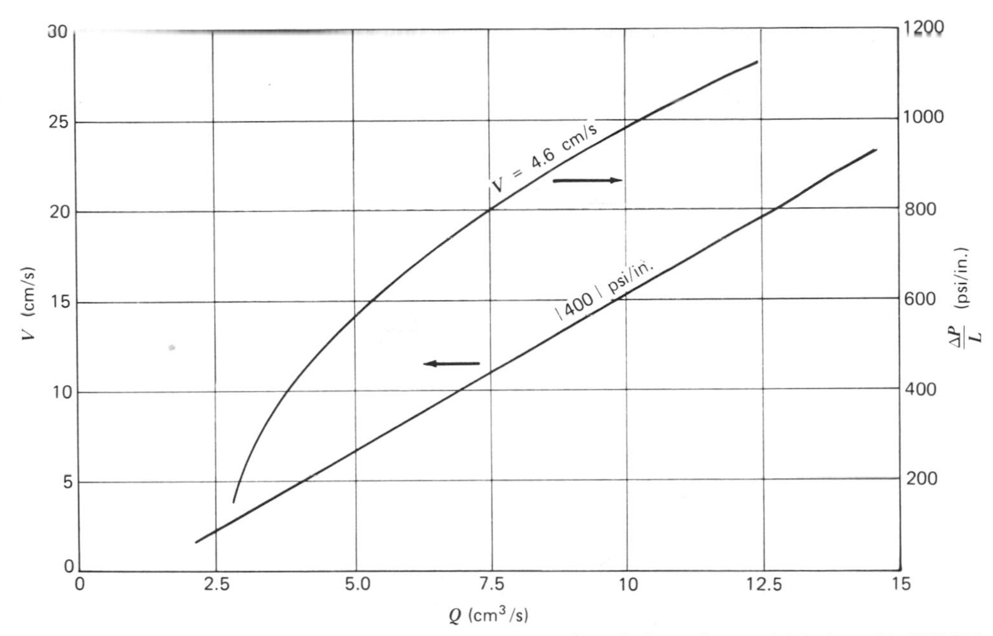


**Fig. 13.27** Numerical results on annular pressure and drag flow of Marlex 50 HDPE flowing at 190°C (72).

The two expressions for $v_z^{\text{I}}$ and $v_z^{\text{II}}$ (i.e., the flow field) cannot be evaluated before a value is found for $\lambda$. The following boundary (no-slip) condition helps in this task

$$v_z^{\text{I}}(\lambda R_o) = v_z^{\text{II}}(\lambda R_o) \tag{13.6-10}$$

Therefore we have

$$V\Big/\left[-\left(\frac{dP}{dz}\right)\Big/2m\right]^s + \int_{R_i}^{\lambda R_o}\left[-r+\frac{(\lambda R_o)^2}{r}\right]dr = \int_{\lambda R_o}^{R_o}\left[r-\frac{(\lambda R_o)^2}{r}\right]dr \tag{13.6-11}$$

and $\lambda$ can be evaluated through the numerical solution of this equation. Once $\lambda$ is evaluated, the flow field $v_z^{\text{I}}$ and $v_z^{\text{II}}$ can be obtained, as well as the flow rate–pressure drop relationship. Numerical (finite difference method) results have been obtained by Cacho-Silvestrini (72). Figure 13.27 shows the resulting relationships between pressure drop, wire velocity, and volumetric flow rate in annular dies with combined drag and pressure flow. Ordinarily such results should also hold for slightly tapered dies, that is, for the die region $B_1$ in Fig. 13.26. But in wire coating dies, the wire speeds are very high and lubrication film type pressure gradients may be generated (see Eq. 10.4-6), which may complicate the resulting flow. Recently Tanner (73) has used the finite element method (FEM) to investigate the flow in the wire coating die in Fig. 13.28*a*. The flow field in the central region of the die, where the lubrication positive gradients are expected, appears in Fig. 13.28*b*. The combined effect of the taper and the wire velocity gives rise to the circulating flow pattern in the upper corner region, which may be significant in that it is similar to die entrance flows, which are related to melt fracture instabilities. In commercial dies the tapers used are very mild and circulating flow are probably avoided.

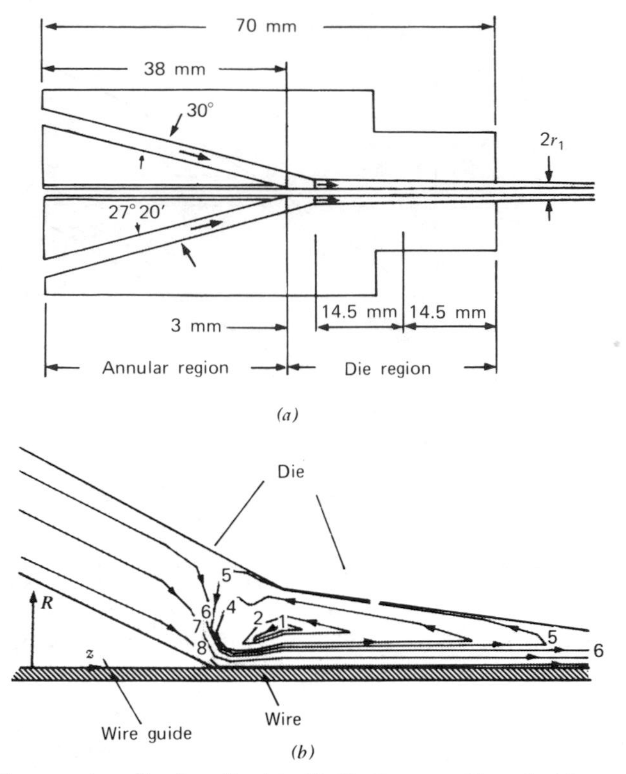


**Fig. 13.28** (*a*) Wire coating die described in R. T. Fenner, *Extruder Screw Design*, Illife, London, 1970. (*b*) Streamlines in the central part of the die using a Newtonian liquid. Wire coating thickness is 0.4 wire diameter. The streamline that becomes the coating surface lies between those labeled 5 and 6 in the figure. (Reprinted with permission from R. I. Tanner, "Some Experiments Using Finite Element Methods in Polymer Processing and Rheology," *Proceedings of the Seventh International Congress on Rheology*, p. 140, Gothenburg, Sweden, 1976.)

Finally, at high wire coating speeds non-isothermal effects may become important. In such a case the energy equation must be solved simultaneously with the equation of motion. The discussion in Section 13.1 is relevant here, in particular with respect to the high local temperature rises. Solution of such a non-isothermal flow problem can be carried out using FDM methods.

Tadmor and Bird (74) have examined the problem of stability in eccentric and noncoaxial wire-die configurations. They conclude, for a CEF fluid, that a negative second normal stress difference would tend to diminish the eccentricity of a wire. Furthermore, they found that shear stresses tend to align the axis of the wire with that of the die (Problem 13.15).

The problem of cooling and solidification of polymer melt coated wires or cables can be solved either by the method of Morrison (75), where the energy balance is stated separately for each phase (for the case of polycrystalline polymers) with a separate statement at the melt interface, or by the method described in Section 9.4, where the first order transition effects are included in the $C_p$ variations with temperature. Specific results on the problem of cooling of polymer coated wires are presented by Biggs and Guenther (76), who solved the problem by assuming constant thermomechanical properties (Problem 9.9).

## 13.7 Profile Extrusion

*Profiles* are all extruded articles having a cross-sectional shape that differs from that of a circle, an annulus, or a very wide and thin rectangle (flat film or sheet). Examples of extruded profiles appear in Fig. 1.1. From these examples it is obvious that the cross-sectional shapes are usually complex, which, in terms of solving the flow problem in profile dies, means complex boundary conditions. Furthermore, profile dies are of nonuniform thickness, raising the possibility of transverse pressure drops and velocity components, and making the prediction of extrudate swelling for viscoelastic fluids very difficult. For these reasons profile dies are built today on a "trial-and-error" basis, and final product shape is achieved with "sizing" devices that act on the extrudate after it leaves the profile die.

The problem of steady, isothermal flow in straight axis channels of noncircular cross section has received considerable theoretical attention. The results of such studies (usually numerical solutions) indicate that for Newtonian fluids, flow involving the axial velocity component alone satisfies the equations of continuity and motion (77–79). The same statement can be made about *inelastic* non-Newtonian fluids, such as the power law fluid, from a mathematical solution point of view. In reality, most non-Newtonian fluids are viscoelastic and exhibit normal stresses. For fluids such as those (i.e., fluids described by constitutive equations that predict normal stresses for viscometric flows), theoretical analyses have shown that secondary flows are created inside channels of nonuniform cross section (80, 81). Specifically it can be shown that a zero second normal stress difference is a necessary (but not sufficient) condition to ensure the absence of secondary flow (81). Of course the analyses of flows in noncircular channels in terms of constitutive equations—which, strictly speaking, hold only for viscometric flows—are expected to yield qualitative results only.

Experimentally low Reynolds number flows in noncircular channels has not been investigated extensively. In particular, only a few studies have been conducted with fluids exhibiting normal stresses (82, 83). Secondary flows, such as vortices in rectangular channels, have been observed using dyes in dilute aqueous solutions of polyacrylamide. Interestingly, these secondary flow vortices (if they exist) seem to have very little effect on the flow rate.

Let us examine more closely some of the problems that arise in designing profile extrusion dies whose origin is to be found in the flow patterns. We consider the square tube flow patterns calculated for a power law fluid of $n = 0.5$ (Fig. 13.29). Although the velocity profiles are symmetric, they still are $\theta$-dependent, $\theta$ being the angle in the cross-sectional plane. Furthermore, it is evident that the velocity gradient $dv_z/dr$, where $r$ is an "effective radius" coordinate, also depends on the angle $\theta$. Therefore in each quadrant, for every value of the angle $\theta$, there is a different velocity gradient variation with $r$. At $\theta = 0$ and $\theta = \pi/2$ the velocity gradients are high, since $r = H$, while at $\theta = \pi/4$, where $r = \sqrt{2}H$, the velocity gradients are small. It follows then that if a polymer melt were flowing in a channel of square cross section, the extrudate would *swell more* at the vicinity of the center of its sides, because of the high prevailing shear rates; the resulting extrudate shape would then show a "bulge" outward at the sides. What is important from a die design point of view is that the cross-sectional shapes of the die and the extrudates are *different*. Simply put, to produce a square cross section extrudate, one needs a

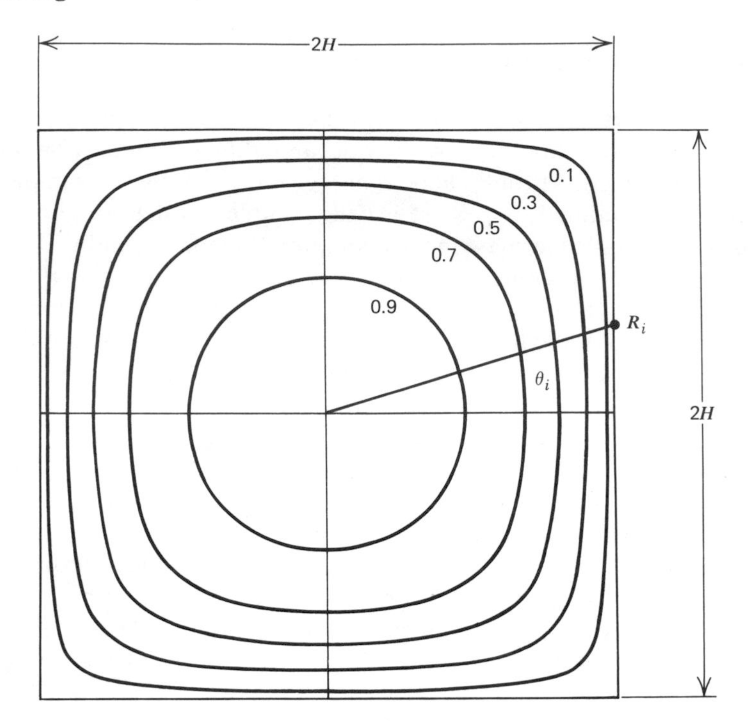


**Fig. 13.29** Isovelocity contours $v/V_0$ of a power law fluid flowing in a square channel, $n = 0.5$. [Reprinted with permission from F. Röthemeyer, *Kunststoffe*, **59**, 333 (1969).]

die looking like a four-cornered star (Fig. 13.30*a*), whose sides are concave in. The curvature of the walls of the die used depends on the variation of extrudate swelling with shear stress for the polymer used (see Problem 13.10). The differences in the shapes and magnitudes of the cross-sectional areas are primarily due to the $\theta$-dependence of the degree of extrudate swelling.

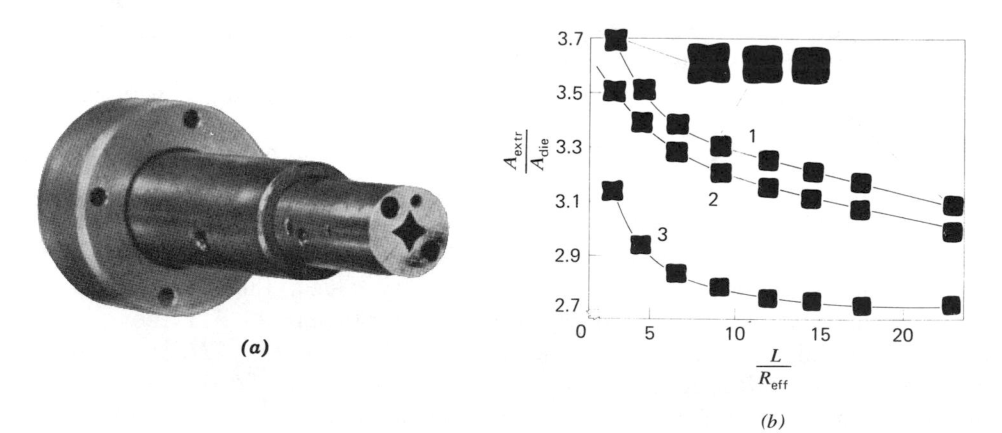


**Fig. 13.30** (*a*) Profile die shape used in obtaining the results in (*b*). (*b*) Plot of $A_{extr}/A_{die}$ versus $L/R_{eff}$ for LDPE, $R_{eff} = (2A_{die}/\text{perimeter of die})$. Curve 1, $T = 196°C$, $\Gamma_{eff} = 40/\pi R_{eff}^3 = 79\ s^{-1}$; curve 2, $T = 189°C$, $\Gamma_{eff} = 56\ s^{-1}$; curve 3, $T = 180°C$, $\Gamma_{eff} = 30\ s^{-1}$. [Reprinted with permission from F. Röthemeyer, *Kunststoffe*, **59**, 333 (1969).]

As Fig. 13.30*b* indicates the degree of swelling and the shape of the extrudate varies with increasing $L/D$ of the die. First, the ratio of areas of the extrudate and the die decreases with increasing $L/R_{\text{eff}}$; this effect was also observed with circular dies. It is again attributed to the loss of memory by the melt of the entrance deformations. The second effect is quite interesting and significant. At very short $L/R_{\text{eff}}$ values, though the degree of swelling is large, the extrudate shape is closer to that of the die than it is at large $L/R_{\text{eff}}$. The reason for this phenomenon is the following: at short capillary lengths the stress field is probably not fully established, at least as far as establishing reasonable shear strains that are independent of the axial distance; thus the effects of the varying velocity gradients and recoverable strains at the various $\theta_i$ are not fully felt. Furthermore, at short $L/R_{\text{eff}}$ values entrance tensile strains predominate and mark the effects of the $\theta$-dependent flow and recoverable shear strain fields. From a die design point of view, though, it seems that short profile dies, in addition to resulting in small pressure drops, also form extrudates of shapes closer to that of the die.

Streamlining of the complex profile dies is as necessary as with any other die shape, but obviously more difficult. For this reason plate dies (84), made up of thin plates inserted in a die housing one behind the other, are common. The channel cross sections in the individual plates differ in such a way as to streamline the polymer melt into the final plate. This construction makes both die modifications and die machining easier. In such complex dies, even approximate design expressions have not been developed yet; in practice, the repeated filing off of metal in the approach plates achieves the desired shape.

Before closing this chapter, we feel that it is useful to list in tabular form some *isothermal pressure flow* relationships commonly used in die flow simulations. Tables 13.1 and 13.2 deal with flow relationships for the parallel plate and circular tube channels using a Newtonian (N), power law (P), and Ellis (E) model fluids. Table 13.3 covers concentric annular channels using Newtonian and power law model fluids. Table 13.4 contains volumetric flow rate–pressure drop (die characteristic) relationships only, which are arrived at by numerical solutions, for Newtonian fluid flow in eccentric annular, elliptical, equilateral, isosceles triangular, semicircular, and circular sector and conical channels. In addition, $Q$ versus $\Delta P$ relationships for rectangular and square channels for Newtonian model fluids are given. Finally Fig. 13.31 presents shape factors for Newtonian fluids flowing in various common shape channels. The shape factor $M_0$ is based on parallel plate pressure flow, namely,

$$Q_{\text{ch}} = Q_{pp} M_0 \tag{13.7-1}$$

where

$$Q_{pp} = \frac{\Delta P B d^3}{12 \mu L} \tag{13.7-2}$$

$Q_{\text{ch}}$ is the volumetric flow rate in any of the channels appearing in the figure and $B$ and $d$ are specified for each channel in the figure.

**Table 13.1 Parallel Plate Pressure Flow:**

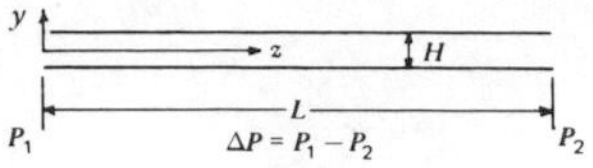


| Ⓝ[a] $\tau_{yz} = -\mu \dfrac{dv_z}{dy}$ | Ⓟ[a] $\tau_{yz} = -m\left\|\dfrac{dv_z}{dy}\right\|^{n-1}\dfrac{dv_z}{dy}$ |
|---|---|
| $\tau_{yz}(y) = \left(\dfrac{\Delta P}{L}\right)y$ | $\tau_{yz}(y) = \left(\dfrac{\Delta P}{L}\right)y$ |
| $\tau_w = \tau_{yz}\left(\dfrac{H}{2}\right) = \dfrac{H\,\Delta P}{2L}$ | $\tau_w = \tau_{yz}\left(\dfrac{H}{2}\right) = \dfrac{H\,\Delta P}{2L}$ |
| $-\dot\gamma_{yz}(y) = \left(\dfrac{\Delta P}{\mu L}\right)y$ | $-\dot\gamma_{yz}(y) = \left(\dfrac{\Delta P}{mL}y\right)^s \quad y \geq 0$ |
| $\dot\gamma_w = -\dot\gamma_{yz}\left(\dfrac{H}{2}\right) = \dfrac{H\,\Delta P}{2\mu L}$ | $\dot\gamma_w = -\dot\gamma_{yz}\left(\dfrac{H}{2}\right) = \left(\dfrac{H\,\Delta P}{2mL}\right)^s$ |
| $v_z(y) = \left(\dfrac{H^2\,\Delta P}{8\mu L}\right)\left[1-\left(\dfrac{2y}{H}\right)^2\right]$ | $v_z(y) = \dfrac{H}{2(s+1)}\left(\dfrac{H\,\Delta P}{2mL}\right)^s\left[1-\left(\dfrac{2y}{H}\right)^{s+1}\right] \quad y \geq 0$ |
| $v_z(0) = v_{\max} = \dfrac{H^2\,\Delta P}{8\mu L}$ | $v_z(0) = v_{\max} = \dfrac{H}{2(s+1)}\left(\dfrac{H\,\Delta P}{2mL}\right)^s$ |
| $\langle v_z\rangle = \dfrac{2}{3}v_{\max}$ | $\langle v_z\rangle = \left(\dfrac{s+1}{s+2}\right)v_{\max}$ |
| $Q = \dfrac{WH^3\,\Delta P}{12\mu L}$ | $Q = \dfrac{WH^2}{2(s+2)}\left(\dfrac{H\,\Delta P}{2mL}\right)^s$ |

Ⓔ[a] $\tau_{yz} = -\eta(\tau)\dfrac{dv_z}{dy} \qquad \eta(\tau) = \dfrac{\eta_0}{1+(\tau/\tau_{1/2})^{\alpha-1}} \qquad \tau = |\tau_{yz}|$

$$\tau_{yz}(y) = \left(\frac{\Delta P}{L}\right)y$$

$$\tau_w = \tau_{yz}\left(\frac{H}{2}\right) = \frac{H\,\Delta P}{2L}$$

$$-\dot\gamma_{yz} = \left(\frac{\Delta P}{\eta_0 L}\right)y\left[1+\left(\frac{\Delta P y}{\tau_{1/2}L}\right)^{\alpha-1}\right]$$

$$\dot\gamma_w = -\dot\gamma_{yz}\left(\frac{H}{2}\right) = \frac{H\,\Delta P}{2\eta_0 L}\left[1+\left(\frac{H\,\Delta P}{2\tau_{1/2}L}\right)^{\alpha-1}\right]$$

$$v_z(y) = \frac{H^2\,\Delta P}{8\eta_0 L}\left\{\left[1-\left(\frac{2y}{H}\right)^2\right]+\left(\frac{2}{1+\alpha}\right)\left(\frac{H\,\Delta P}{2L\tau_{1/2}}\right)^{\alpha-1}\left[1-\left(\frac{2y}{H}\right)^{\alpha+1}\right]\right\}$$

$$v_z(0) = v_{\max} = \frac{H^2\,\Delta P}{8\eta_0 L}\left[1+\left(\frac{2}{1+\alpha}\right)\left(\frac{H\,\Delta P}{2L\tau_{1/2}}\right)^{\alpha-1}\right]$$

$$\langle v_z\rangle = \frac{2}{3}v_{\max}\left[1+\left(\frac{3}{2+\alpha}\right)\left(\frac{H\,\Delta P}{2L\tau_{1/2}}\right)^{\alpha-1}\right]\Big/\left[1+\left(\frac{2}{1+\alpha}\right)\left(\frac{H\,\Delta P}{2L\tau_{1/2}}\right)^{\alpha-1}\right]$$

$$Q = \frac{WH^3\,\Delta P}{12\eta_0 L}\left[1+\left(\frac{3}{2+\alpha}\right)\left(\frac{H\,\Delta P}{2L\tau_{1/2}}\right)^{\alpha-1}\right]$$

[a] Ⓝ Newtonian fluid; Ⓟ Power law model fluid; Ⓔ Ellis fluid.

**Table 13.2 Circular Tube Pressure Flow:**



Ⓝ $\tau_{rz} = -\mu \dfrac{dv_z}{dr}$

$$\tau_{rz}(r) = \left(\frac{\Delta P}{2L}\right) r$$

$$\tau_w = \tau_{rz}(R) = \frac{R\,\Delta P}{2L}$$

$$-\dot{\gamma}_{rz}(r) = \left(\frac{\Delta P}{2\mu L}\right) r$$

$$\dot{\gamma}_w = -\dot{\gamma}_{rz}(R) = \frac{R\,\Delta P}{2\mu L}$$

$$v_z(r) = \frac{R^2\,\Delta P}{4\mu L}\left[1 - \left(\frac{r}{R}\right)^2\right]$$

$$v_z(0) = v_{\max} = \frac{R^2\,\Delta P}{4\mu L}$$

$$\langle v_z \rangle = \tfrac{1}{2} v_{\max}$$

$$Q = \frac{\pi R^4\,\Delta P}{8\mu L}$$

Ⓟ $\tau_{rz} = -m \left|\dfrac{dv_z}{dr}\right|^{n-1} \dfrac{dv_z}{dr}$

$$\tau_{rz}(r) = \left(\frac{\Delta P}{2L}\right) r$$

$$\tau_w = \tau_{rz}(R) = \frac{R\,\Delta P}{2L}$$

$$-\dot{\gamma}_{rz}(r) = \left(\frac{\Delta P r}{2mL}\right)^s$$

$$\dot{\gamma}_w = -\dot{\gamma}_{rz}(R) = \left(\frac{R\,\Delta P}{2mL}\right)^s$$

$$v_z(r) = \frac{R}{1+s}\left(\frac{R\,\Delta P}{2mL}\right)^s \left[1 - \left(\frac{r}{R}\right)^{s+1}\right]$$

$$v_z(0) = v_{\max} = \frac{R}{1+s}\left(\frac{R\,\Delta P}{2mL}\right)^s$$

$$\langle v_z \rangle = \left(\frac{s+1}{s+3}\right) v_{\max}$$

$$Q = \left(\frac{\pi R^3}{s+3}\right)\left(\frac{R\,\Delta P}{2mL}\right)^s$$

Ⓔ $\tau_{rz} = -\eta(\tau)\dfrac{dv_z}{dr}$, where $\eta(\tau) = \dfrac{\eta_0}{1 + (\tau/\tau_{1/2})^{\alpha-1}}$ $\quad \tau = |\tau_{rz}|$

$$\tau_{rz}(r) = \left(\frac{\Delta P}{2L}\right) r$$

$$\tau_w = \tau_{rz}(R) = \frac{R\,\Delta P}{2L}$$

$$-\dot{\gamma}_{rz}(r) = \left(\frac{\Delta P}{2\eta_0 L}\right) r \left[1 + \left(\frac{\Delta P r}{2L\tau_{1/2}}\right)^{\alpha-1}\right]$$

$$\dot{\gamma}_w = -\dot{\gamma}_{rz}(R) = \left(\frac{R\,\Delta P}{2\eta_0 L}\right)\left[1 + \left(\frac{R\,\Delta P}{2L\tau_{1/2}}\right)^{\alpha-1}\right]$$

$$v_z(r) = \frac{R^2\,\Delta P}{4L\eta_0}\left\{\left[1 - \left(\frac{r}{R}\right)^2\right] + \left(\frac{2}{1+\alpha}\right)\left(\frac{R\,\Delta P}{2L\tau_{1/2}}\right)^{\alpha-1}\left[1 - \left(\frac{r}{R}\right)^{\alpha+1}\right]\right\}$$

$$v_z(0) = v_{\max} = \frac{R^2\,\Delta P}{4L\eta_0}\left[1 + \left(\frac{2}{1+\alpha}\right)\left(\frac{R\,\Delta P}{2L\tau_{1/2}}\right)^{\alpha-1}\right]$$

$$\langle v_z \rangle = \tfrac{1}{2} v_{\max}\left[1 + \left(\frac{4}{3+\alpha}\right)\left(\frac{R\,\Delta P}{2L\tau_{1/2}}\right)^{\alpha-1}\right] \Big/ \left[1 + \left(\frac{2}{1+\alpha}\right)\left(\frac{R\,\Delta P}{2L\tau_{1/2}}\right)^{\alpha-1}\right]$$

$$Q = \frac{\pi R^4\,\Delta P}{8\eta_0 L}\left[1 + \left(\frac{4}{3+\alpha}\right)\left(\frac{R\,\Delta P}{2L\tau_{1/2}}\right)^{\alpha-1}\right]$$

**Table 13.3 Concentric Annular Pressure Flow:**

---

$$Ⓝ^{a} \quad \tau_{rz} = -\mu \frac{dv_z}{dr}$$

---

$$\tau_{rz}(r) = \frac{\Delta PR}{2L}\left[\left(\frac{r}{R}\right) - \left(\frac{1-\kappa^2}{2\ln(1/\kappa)}\right)\left(\frac{R}{r}\right)\right]$$

$$-\dot{\gamma}_{rz}(r) = \frac{\Delta PR}{2\mu L}\left[\left(\frac{r}{R}\right) - \left(\frac{1-\kappa^2}{2\ln(1/\kappa)}\right)\frac{R}{r}\right]$$

$$\tau_{w1} = \tau_{rz}(R) = \frac{\Delta PR}{2L}\left[1 - \left(\frac{1-\kappa^2}{2\ln(1/\kappa)}\right)\right]$$

$$\tau_{w2} = \tau_{rz}(\kappa R) = \frac{\Delta PR}{2L}\left[\kappa - \left(\frac{1-\kappa^2}{2\ln(1/\kappa)}\right)\left(\frac{1}{\kappa}\right)\right]$$

$$\dot{\gamma}_{w1} = -\dot{\gamma}_{rz}(R) = \frac{\Delta PR}{2\mu L}\left[1 - \left(\frac{1-\kappa^2}{2\ln(1/\kappa)}\right)\right]$$

$$\dot{\gamma}_{w2} = \dot{\gamma}_{rz}(\kappa R) = \frac{\Delta PR}{2\mu L}\left[\kappa - \left(\frac{1-\kappa^2}{2\ln(1/\kappa)}\right)\left(\frac{1}{\kappa}\right)\right]$$

$$v_z(r) = \frac{\Delta PR^2}{4\mu L}\left[1 - \left(\frac{r}{R}\right)^2 + \left(\frac{1-\kappa^2}{\ln(1/\kappa)}\right)\ln\left(\frac{r}{R}\right)\right]$$

$$v_z(\lambda R) = v_{\max} = \frac{\Delta PR^2}{4\mu L}\left\{1 - \left(\frac{1-\kappa^2}{2\ln(1/\kappa)}\right)\left[1 - \ln\left(\frac{1-\kappa^2}{2\ln(1/\kappa)}\right)\right]\right\} \qquad \lambda^2 = \frac{1-\kappa^2}{2\ln(1/\kappa)}$$

$$\langle v_z \rangle = \frac{\Delta PR^2}{8\mu L}\left[(1+\kappa^2) - \left(\frac{1-\kappa^2}{\ln(1/\kappa)}\right)\right]$$

$$Q = \frac{\pi\,\Delta PR^4}{8\mu L}\left[(1-\kappa^4) - \frac{(1-\kappa^2)^2}{\ln(1/\kappa)}\right]$$

---

$$Ⓟ^{b} \quad \tau_{rz} = -\left|\frac{dv_z}{dr}\right|^{n-1}\left(\frac{dv_z}{dr}\right) \qquad \rho = \frac{r}{R} \qquad s = \frac{1}{n} \qquad \tau_{rz}(\lambda R) = 0$$

---

$$v_z^{\mathrm{I}}(r) = R\left(\frac{\Delta PR}{2mL}\right)^s \int_\kappa^\rho \left(\frac{\lambda^2}{\rho} - \rho\right)^s d\rho \qquad \kappa \le \rho \le \lambda$$

$$v_z^{\mathrm{II}} = R\left(\frac{\Delta PR}{2mL}\right)^s \int_\rho^1 \left(\rho - \frac{\lambda^2}{\rho}\right)^s d\rho \qquad \lambda \le \rho \le 1$$

$\lambda$ is evaluated numerically for the equations above using the boundary condition

$$v_z^{\mathrm{I}}(\lambda R) = v_z^{\mathrm{II}}(\lambda R)$$

$$Q = \frac{\pi R^3}{s+2}\left(\frac{R\,\Delta P}{2mL}\right)^s (1-\kappa)^{s+2} F_1(n, \kappa)$$

$F_1(n, \kappa) = F(n, \beta)$ is given in graphical form in Fig. 13.23, when $\kappa = 1/\beta$

**Very Thin Annuli** ($\kappa \to 1$), $F_1(n, \kappa) \to 1$

---

Ⓔ[c] $\tau_{rz} = -\eta(\tau)\dfrac{dv_z}{dr} \qquad \eta(\tau) = \dfrac{\eta_0}{1+(\tau/\tau_{1/2})^{\alpha-1}} \qquad \tau = |\tau_{rz}|$

---

$$Q = \frac{\tau_{1/2}\pi R^3}{\eta_0}\left\{\frac{\Delta PR}{2\tau_{1/2}}\left[\lambda^4 \ln\frac{1}{\kappa} - \lambda^2(1-\kappa^2) + \tfrac{1}{4}(1-\kappa^4)\right]\right.$$

$$\left. + \left(\frac{\Delta PR}{2\tau_{1/2}}\right)^{\alpha}\left(\sum_{i=0}^{\alpha+1} \varepsilon_{i \atop i\neq(\alpha+3)/2} \lambda^{2i} + F\lambda^{\alpha+3}\right)\right\}$$

$$\varepsilon_i = \binom{\alpha+1}{i}(-1)^i\left(\frac{1+(-1)^{\alpha}\kappa^{\alpha+3-2i}}{\alpha+3-2i}\right)$$

$$F = \binom{\alpha+1}{(\alpha-1)/2}(-1)^{(\alpha-1)/2}\ln\left(\frac{1}{\kappa}\right) \qquad \alpha \text{ odd}$$

$$F = 2\sum_{i=0}^{\alpha+1}\binom{\alpha+1}{i}(-1)^i\frac{1}{2i-\alpha+1} \qquad \alpha \text{ even}$$

---

Ⓔ $\tau_{rz} = -\eta(\tau)\dfrac{dv_z}{dr} \qquad \eta(\tau) = \dfrac{\eta_0}{1+(\tau/\tau_{1/2})^{\alpha-1}} \qquad \tau = |\tau_{rz}|$

---

**Approximately**[d]

$$Q = \frac{\pi R^4 \Delta P \varepsilon^3}{6\eta_0 L}\left[1 + \frac{3}{\alpha+2}\left(\frac{\Delta P\varepsilon R}{2\tau_{1/2}L}\right)^{\alpha-1}\right](1 - \tfrac{1}{2}\varepsilon + \tfrac{1}{60}\varepsilon^2 + \cdots)$$

Where $\varepsilon = 1-\kappa$. This approximate solution is valid for $\kappa > 0.6$.

---

[a] R. B. Bird, W. E. Stewart and E. N. Lightfoot, "Transport Phenomena," Wiley, New York, 1960, p. 51.
[b] A. G. Fredrickson and R. B. Bird, *Ind. Eng. Chem.*, **50**, 347–352 (1958).
[c] D. W. McEachern, *Am. Inst. Chem. Eng. J.*, **12**, 328 (1966).
[d] R. B. Bird, R. C. Armstrong and O. Hassager, "Dynamics of Polymeric Liquids," Vol. I, Wiley, New York, 1977, p. 222.

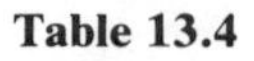

**Table 13.4**

### Ⓝ Eccentric Annulus Pressure Flow[a]

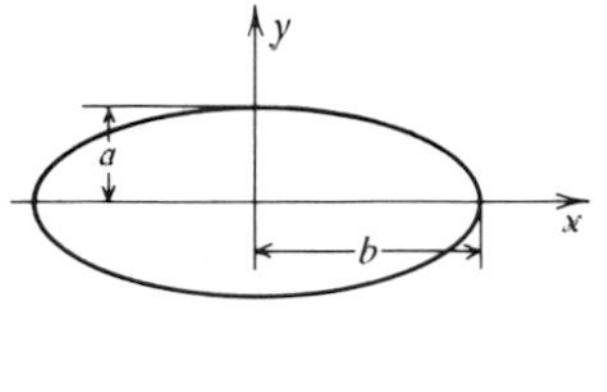


$$Q=\frac{\pi\,\Delta P}{8\mu L}\left\{R^4(1-\kappa^4)-\frac{(R+\kappa R+b)(R+\kappa R-b)(R-\kappa R+b)(R-\kappa R-b)}{\delta-\omega}\right.$$

$$\left.-4b^2\kappa^2R^2\left[1+\frac{\kappa^2R^4}{(R^2-b^2)}+\frac{\kappa^4R^8}{[(R^2-b^2)^2-\kappa^2R^2b^2]^2}+\cdots\right]\right\}$$

where $\omega=\frac{1}{2}\ln\frac{F+M}{F-M}$ $\qquad \delta=\frac{1}{2}\ln\frac{F-b+M}{F-b-M}$

$$F=\frac{R^2-\kappa R^2+b^2}{2b}\qquad M=\sqrt{F^2-R^2}$$

### Ⓝ Elliptical Channel Pressure Flow[a]

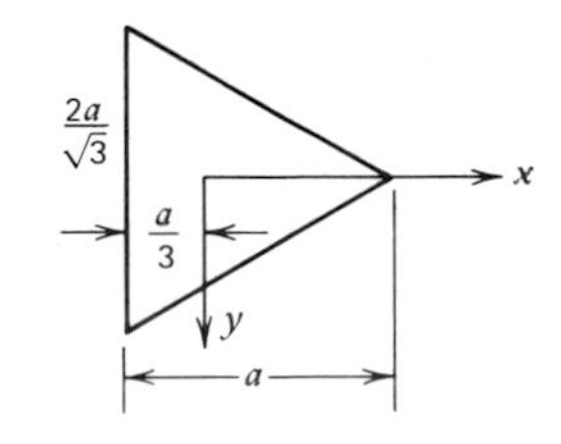


$$v_z(x,y)=\frac{\Delta Pa^2b^2}{2\mu L(a^2+b^2)}\left(1-\frac{x^2}{b^2}-\frac{y^2}{a^2}\right)$$

$$Q=\frac{\pi\,\Delta P}{4\mu L}\frac{a^3b^3}{a^2+b^2}$$

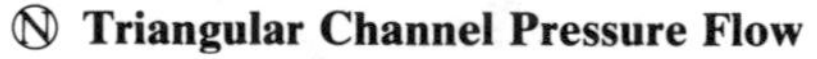

### Ⓝ Triangular Channel Pressure Flow

1. *Equilateral triangle*[a]

$$v_z(x,y)=-\frac{\Delta P}{4a\mu L}[x^3-3xy^2-a(x^2+y^2)+\tfrac{4}{27}a^3+\cdots]$$

$$Q=\frac{\Delta Pa^4}{20\sqrt{3}\mu L}$$

$$\left.\frac{Q_{e\cdot t}}{Q_{\text{tube}}}\right|_{\text{equal area}}=0.72552\qquad \frac{a^2}{\sqrt{3}}=\pi R^2$$

2. *Isosceles triangle (right)*[b]

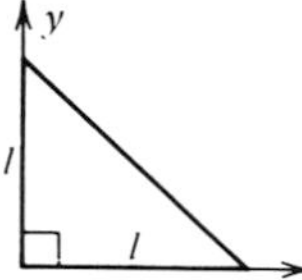


$$v_z(x,y)=\frac{16l^2\,\Delta P}{\pi^4\mu L}\left[\sum_{i=1,3}\sum_{j=2,4}\frac{j\sin(i\pi x/l)\sin(j\pi y/l)}{i(j^2-i^2)(i^2+j^2)^2}+\sum_{i=2,4,\ldots}\sum_{j=1,3,\ldots}\frac{i\sin(i\pi x/l)\sin(j\pi y/l)}{j(i^2-j^2)(i^2+j^2)^2}\right]$$

### Ⓝ Semicircular Channel Pressure Flow[b]

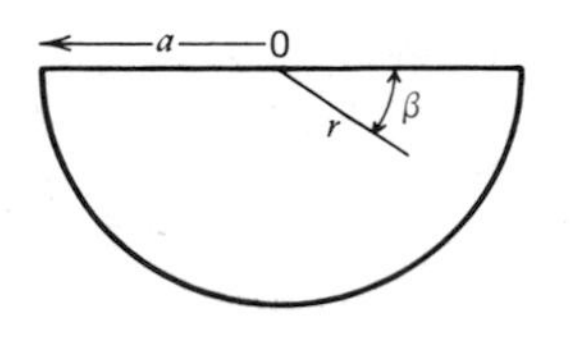


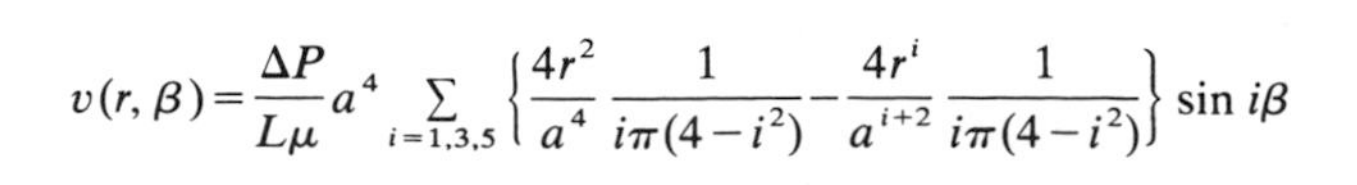

$$v(r,\beta)=\frac{\Delta P}{L\mu}a^4\sum_{i=1,3,5}\left\{\frac{4r^2}{a^4}\frac{1}{i\pi(4-i^2)}-\frac{4r^i}{a^{i+2}}\frac{1}{i\pi(4-i^2)}\right\}\sin i\beta$$

**Ⓝ Circular Section Channel Pressure Flow**[c]

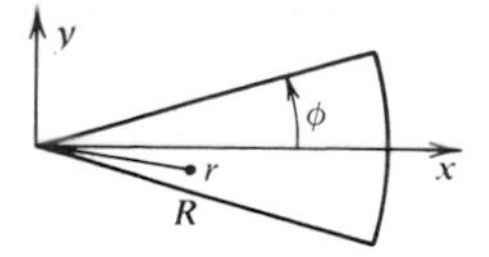


$$v_z(x, y) = \frac{\Delta P}{2\mu L}\left[\frac{x^2 \tan^2 \phi - y^2}{1 - \tan^2 \phi} + \frac{16R^2(2\phi)^2}{\pi^3} \sum_{i=1,3,\ldots}^{\infty} (-1)^{(i+1)/2}\left(\frac{r}{R}\right)^{i\pi/2\phi} \frac{\cos(i\pi\phi)/2\phi}{i[i^2 - (4\phi/\pi)^2]}\right]$$

**Ⓝ Conical Channel Pressure Flow**

L
$R_1$
$R_2$

$$Q = \frac{3\pi \Delta P}{8\mu L}\left[\frac{R_1^3 R_2^3}{R_1^2 + R_1 R_2 + R_2^2}\right]$$

**Ⓟ Conical Channel Pressure Flow**[d] (By the lubrication approximation)

$$Q = \frac{n\pi R_1^3}{3n+1}\left(\frac{R_1 \Delta P a_{13}}{2mL}\right)^s$$

where

$$a_{13} = \frac{3n(R_1/R_2 - 1)}{R_1/R_2([R_1/R_2]^{3n} - 1)}$$

**Ⓝ Rectangular Channel Pressure Flow**[e]

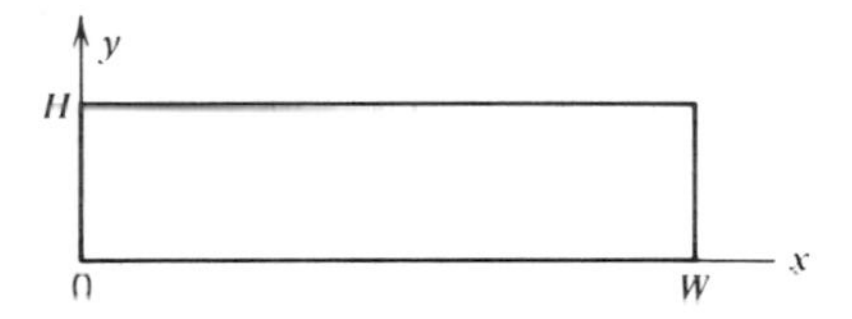


$$v_z(x, y) = \frac{\Delta P}{\mu L}\left\{\frac{y^2}{2} - \frac{yH}{2} + \frac{4H^2}{\pi^3} \sum_{i=1,3,\ldots}^{\infty} \frac{\cosh[(i\pi/2H)(2x - W)]}{i^3 \cosh(i\pi W/H)} \sin\left(\frac{i\pi y}{H}\right)\right\}$$

$$Q = \frac{WH^3}{12\mu}\left(\frac{\Delta P}{L}\right)\left[1 - \frac{192H}{\pi^5 W} \sum_{i=1,3,\ldots}^{\infty} \frac{1}{i^5} \tanh\left(\frac{i\pi W}{2H}\right)\right] = \frac{WH^3}{12\mu} \frac{\Delta P}{L} \cdot F_P$$

$F_p$ is given graphically below

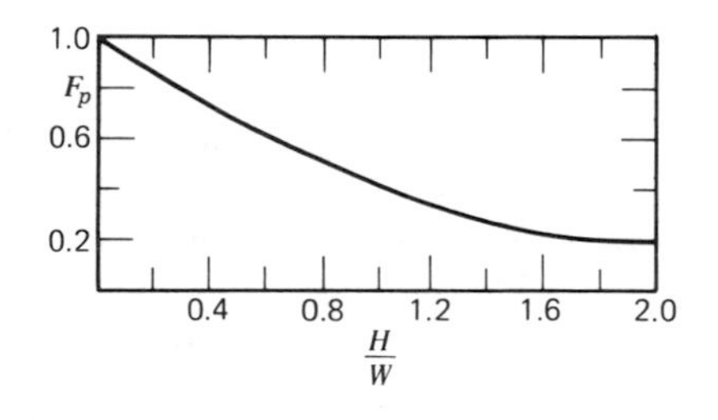


---

[a] J. Happel and H. Brenner, *Low Reynolds Number Hydrodynamics*, Prentice-Hall, Englewood Cliffs, N.J., 1965, Chapter 2.
[b] S. M. Marco and L. S. Han, *Trans. Am. Soc. Mech. Eng.*, **56**, 625 (1955).
[c] E. R. G. Eckert and T. F. Irvine, *Trans. Am. Soc. Mech. Eng.*, **57**, 709 (1956).
[d] J. M. McKelvey, V. Maire, and F. Haupt, *Chem. Eng.*, 95 (September 1976).
[e] M. J. Boussinesq, *J. Math. Pure Appl.* Ser 2, **13**, 377 (1868).

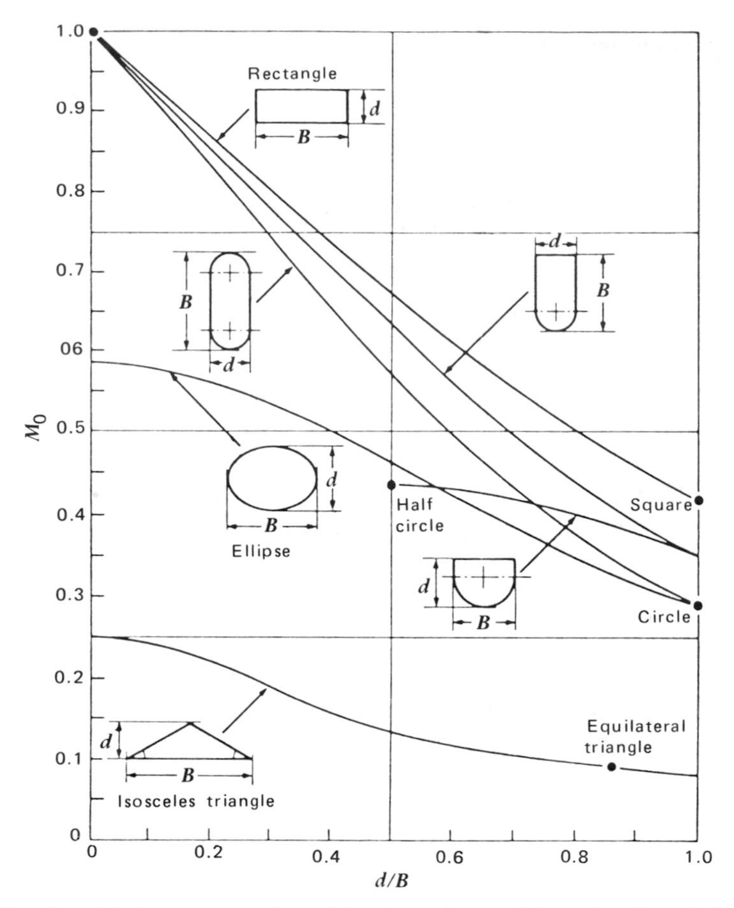


**Fig. 13.31** Graphical representation of the shape factor $M_0$ used in Eq. 13.7-3. [Reprinted with permission from G. P. Lahti, *Soc. Plast. Eng. J.*, (July) 619 (1963).]

## *REFERENCES*


1. E. B. Bagley, "End Corrections in the Capillary Flow of Polyethylene," *J. Appl. Phys.*, **28**, 624 (1957).
2. C. McLuckie and M. Rogers, "Influence of Elastic Effects of Capillary Flow of Molten Polymers," *J. Appl. Polym. Sci.*, **13**, 1049 (1969).
3. N. Hirai and H. Eyring, "Bulk Viscosity of Polymeric Systems," *J. Polym. Sci.*, **37**, 51 (1959).
4. I. J. Duvdevani and I. Klein, "Analysis of Polymer Melt Flow in Capillaries Including Pressure Effects," *Soc. Plast. Eng. J.*, **23**, 41–45 (1967).
5. P. H. Goldblatt and R. S. Porter, "A Comparison of Equations for the Effect of Pressure on the Viscosity of Amorphous Polymers," *J. Appl. Polym. Sci.*, **20**, 1199 (1976).
6. B. C. Sakiades, "Equilibrium Flow of a General Fluid Through a Cylindrical Tube," *Am. Inst. Chem. Eng. J.*, **8**, 317 (1962).
7. C. D. Han, M. Charles, and W. Philippoff, "Measurements of the Axial Pressure Distribution of Molten Polymers in Flow Through a Circular Tube," *Trans. Soc. Rheol.*, **13**, 453 (1969).

8. C. D. Han, "On Slit- and Capillary-Die Rheometry," *Trans. Soc. Rheol.*, **18**, 163 (1974).
9. C. D. Han, *Rheology in Polymer Processing*, Academic Press, New York, 1976, Chapter 5.
10. R. Nahme, *Ing. Arch.*, **11**, 191 (1940).
11. H. A. Brinkman, *Appl. Sci. Res.*, **A2**, 120–124 (1951).
12. R. M. Turian and R. B. Bird, "Viscous Heating in the Cone-and-Plate Viscometer. II. Temperature Dependent Viscosity and Thermal Conductivity," *Chem. Eng. Sci.*, **18**, 689 (1963).
13. J. E. Gerrard, R. E. Steidler, and J. K. Appeldoorn, "Viscous Heating in Capillaries," *Ind. Eng. Chem. Fundam.*, **4**, 332 (1965).
14. R. M. Turian, "Viscous Heating in the Cone-and-Plate Viscometer. III," *Chem. Eng. Sci.*, **20**, 771 (1965).
15. R. A. Morrette and C. G. Gogos, "Viscous Dissipation in Capillary Flow of Rigid PVC and PVC Degradation," *Polym. Eng. Sci.*, **8**, 272 (1968).
16. D. E. Marshall, I. Klein, and R. H. Uhl, "Measurement of Screw and Plastic Temperature Profiles in Extruders," *Soc. Plast. Eng. J.*, **20**, 329 (1964).
17. J. Van Leeuwen, "Stock Temperature Measurement in Plastifying Equipment," *Polym. Eng. Sci.*, **7**, 98–109 (1967).
18. H. T. Kim and E. A. Collins, "Temperature Profiles of Polymer Melts in Tube Flow. Conduction and Shear Heating Corrections," *Polym. Eng. Sci.*, **11**, 83 (1971).
19. H. W. Cox and C. W. Macosko, "Viscous Dissipation in Die Flow," *Am. Inst. Chem. Eng. J.*, **20**, 785 (1974)
20. N. Galili and R. Takserman-Krozer, "Heat Effect in Viscous Flow through a Pipe," *Israel J. Tech.*, **9**, 439 (1971).
21. H. P. Schreiber and E. B. Bagley, "Melt Elasticity in Fractionated HDPE," *Polym. Lett.*, **1**, 365 (1963).
22. W. W. Graessley, S. D. Glasscock, and R. L. Crawley, "Die Swell in Molten Polymers," *Trans. Soc. Rheol.*, **14**, 519 (1970).
23. R. I. Tanner, "A Theory of Die Swell," *J. Polym. Sci.*, **A-28**, 2067 (1970).
24. A. S. Lodge, *Elastic Liquids*, Academic Press, New York, 1964, p. 131.
25. S. I. Abdel-Khalik, O. Hassager, and R. B. Bird, "Prediction of Melt Elasticity from Viscosity Data," *Polym. Eng. Sci.*, **14**, 859 (1974).
26. J. D. Ferry, *Viscoelastic Properties of Polymers*, Wiley, New York, 1971; S. Middleman, *Fundamentals of Polymer Processing*, McGraw-Hill, New York, 1977, p. 472.
27. J. L. White and J. F. Roman, "Extrudate Swell During the Melt Spinning of Fibers—Influence of Rheological Properties and Take-up Force," *J. Appl. Polym. Sci.*, **20**, 1005 (1976).
28. A. B. Metzner, W. T. Houghton, R. A. Sailor, and J. L. White, "A Method for the Measurement of Normal Stresses in Simple Shearing Flow," *Trans. Soc. Rheol.*, **5**, 133 (1961).
29. R. B. Bird, R. K. Prud'homme, and M. Gottlieb, "Extrudate Swell as Analyzed by Macroscopic Balances," The University of Wisconsin, Rheology Research Center Report RRC-35, 1975.
29a. B. Whipple, "Velocity Distributions in Die Swell," Ph.D. dissertation, Washington University, St. Louis, MI, 1974.
30. J. L. White, "Critique on Flow Patterns in Polymer Fluids at the Entrance of a Die and Instabilities Leading to Extrudate Distortion," *Appl. Polym. Symp.*, No. 20, 155 (1973).
31. H. Giesekus, "Verschiedene Phänomene in Strömungen Viskoelastischer Flüssigkeiten durch Düsen," *Rheol. Acta*, **8**, 411 (1969).
32. H. L. Weissberg, "End Corrections for Slow Viscous Flow Through Long Tubes," *Phys. Fluids*, **5**, 1033 (1962).

33. E. B. Bagley and R. M. Birks, "Flow of Polyethylene into a Capillary," *J. Appl. Phys.*, **31**, 556 (1960).
34. T. F. Ballenger and J. L. White, "The Development of the Velocity Field in Polymer Melts into a Reservoir Approaching a Capillary Die," *J. Appl. Polym. Sci.*, **15**, 1849 (1971).
35. P. Lamb and F. N. Cogswell, Paper presented at the International Plastics Congress on Processing Polymer Products, Amsterdam, 1966.
36. H. L. LaNieve, III, and D. C. Bogue, "Correlation of Capillary Entrance Pressure Drops with Normal Stress Data," *J. Appl. Polym. Sci.*, **12**, 353 (1968).
37. C. D. Han, "Influence of the Die Entry Angle in the Entrance Pressure Drop, Recoverable Elastic Energy and Onset of Flow Instability in Polymer Melt Flow," *J. Appl. Polym. Sci.*, **17**, 1403 (1973).
38. J. L. White and A. Kondo, "Rheological Properties of Polymer Melts and Flow Patterns During Extrusion Through a Die Entry Region," *J. Appl. Polym. Sci.*, in press.
39. J. P. Tordella, in *Rheology*, Vol. 4, F. R. Eirich, ed., Academic Press, New York, 1969, Chapter 3.
40. R. S. Spencer and R. D. Dillon, *J. Colloid Inter. Sci.*, **3**, 163 (1940).
41. C. J. S. Petrie and M. M. Denn, "Instabilities in Polymer Processing," *Am. Inst. Chem. Eng. J.*, **22**, 209 (1976).
42. F. N. Cogswell and P. Lamb, *Plast. Today*, 33 (1969).
43. G. V. Vinogradov A., Ya. Malkin, Yu. G. Yanovskii, E. K. Borisenkova, B. V. Yarlykov, and G. V. Berezhnaya, "Viscoelastic Properties and Flow of Narrow Distribution PIB and Polyisoprene," *J. Polym. Sci.*, **A2 10**, 1061 (1972).
44. J. L. den Otter, "Mechanisms of Melt Fracture," *Plast. Polym.*, **38**, 155 (1970).
45. Y. Oyanagi, "A Study of Irregular Flow Behavior of HDPE," *Appl. Polym. Symp.*, No. 20, 123 (1973).
46. N. Bergem, "Visualization Studies of Polymer Melt Flow Anomalies in Extrusion," *Proceedings of the Seventh International Congress on Rheology, Gothenburg, Sweden*, 1976, p. 50.
47. L. L. Blyler, Jr., and A. C. Hart, Jr., "Capillary Flow Instability of Ethylene Polymer Melts," *Polym. Eng. Sci.*, **10**, 193 (1970).
48. J. Vlachopoulos and T. W. Chan, "A Comparison of Melt Fracture Initiation Conditions in Capillaries and Slits," *J. Appl. Polym. Sci.*, in press.
49. W. W. Graessley and L. Segal, "Flow Behavior of Polystyrene Systems in Steady Shearing Flow," *Macromolecule*, **2**, 49 (1969).
50. J. Vlachopoulos, M. Horie, and S. Lidorikis, "An Evaluation of Expressions Predicting Die Swell," *Trans. Soc. Rheol.*, **16**, 669 (1972); J. Vlachopoulos and M. Allam, "Critical Stress and Recoverable Shear for Polymer Melt Fracture," *Polym. Eng. Sci.*, **12**, 184 (1972).
51. J. P. Tordella, "An Unusual Mechanism of Extrusion of PTFE at High Temperature and Pressure," *Trans. Soc. Rheol.*, **7**, 231 (1963); U.S. Patent 2,791,806, 1967.
52. H. F. Mark, in *Rheology*, Vol. 4, F. R. Eirich, ed., Academic Press, New York, 1969, Chapter 7.
53. N. C. Shih and S. Middleman, "Post Extrusion Heat and Solvent Transfer from Polymeric Films," *Polym. Eng. Sci.*, **10**, 4 (1970).
54. D. R. Paul, "Diffusion During the Coagulation Step of Wet-Spinning," *J. Appl. Polym. Sci.*, **12**, 383 (1968).
55. A. Ziabicki, in *Man-Made Fibers*, Vol. 1, H. F. Mark, S. Atlas, and E. Cernia, eds., Wiley, New York, 1967.
56. J. F. Carley, "Flow of Melts in Crosshead-Slit Dies; Criteria for Die Design," *J. Appl. Phys.*, **25**, 1118 (1954).

57. J. R. A. Pearson, "Non-Newtonian Flow and Die Design. Part IV. Flat Film Die Design," *Trans. J. Plast. Inst.*, **32**, 239 (1964).
58. J. M. McKelvey and K. Ito, "Uniformity of Flow from Sheeting Dies," *Polym. Eng. Sci.*, **11**, 258 (1971).
59. R. B. Bird, W. E. Stewart, and E. N. Lightfoot, *Transport Phenomena*, Wiley, New York, 1960, p. 54.
60. A. A. Khan and C. D. Han, "On the Interfacial Deformation in the Stratified Two-phase Flow of Viscoelastic Fluids," *Trans. Soc. Rheol.*, **20**, 595 (1976).
61. J. L. White, R. C. Ufford, K. C. Dharod, and R. L. Price, "Experimental and Theoretical Study of the Extrusion of Two-Phase Molten Polymer Systems," *J. Appl. Polym. Sci.*, **16**, 1313 (1972).
62. J. H. Southern and R. L. Ballman, "Additional Observations on Stratified Bi-Component Flow of Polymer Melts in a Tube," *J. Polym. Sci.*, **A2 13**, 863 (1975).
63. B. L. Lee and J. L. White, "An Experimental Study of Rheological Properties of Polymer Melts in Laminar Shear Flow and of Interfacial Deformation and Its Mechanisms in Two-Phase Stratified Flow," *Trans. Soc. Rheol.*, **18**, 467 (1974).
64. A. Bergonzoni and A. J. DiCresci, "The Phenomenon of Draw Resonance in Polymeric Melts. Parts I and II," *Polym. Eng. Sci.*, **6**, 45–59 (1966).
65. Y. L. Yeow, "On the Stability of Extending Films: A Model for the Film Casting Process," *J. Fluid Mech.*, **66**, 613 (1974).
66. B. Proctor, "Flow Analysis in Extrusion Dies," *Soc. Plast. Eng. J.*, **28** (February 1972); J. R. Syms, "Computer Design Dies for Tubular Films," lecture notes from Plastics Institute of America's course on "Accuracy in the Extrusion Process," October 25, 1973.
67. A. G. Fredrickson and R. B. Bird, "Non-Newtonian Flow in Annuli," *Ind. Eng. Chem.*, **50**, 347 (1958).
68. R. A. Worth and J. Parnaby, "The Design of Dies for Polymer Processing Machinery," *Trans. Inst. Chem. Eng.*, **52**, 368 (1974).
69. J. R. A. Pearson, "Non-Newtonian Flow and Die Design. Part I," *Trans. J. Plast. Inst.*, **30**, 230 (1962).
70. C. Gutfinger, E. Broyer, and Z. Tadmor, "Analysis of a Cross Head Film Blowing Die with the Flow Analysis Network (FAN) Method," *Polym. Eng. Sci.*, **15**, 385–386 (1975).
71. J. M. McKelvey, *Polymer Processing*, Wiley, New York, 1962, p. 111.
72. E. Cacho-Silvestrini, Special Project Report, Department of Chemical Engineering, Stevens Institute of Technology, Hoboken, N.J., 1973.
73. R. I. Tanner, "Some Experiences Using Finite Element Methods in Polymer Processing and Rheology," *Proceedings of the Seventh International Congress on Rheology, Gothenburg, Sweden*, 1975, p. 140.
74. Z. Tadmor and R. B. Bird, "Rheological Analysis of Stabilizing Forces in Wire-Coating Dies," *Polym. Eng. Sci.*, **14**, 124 (1974).
75. M. E. Morrison, "Numerical Evaluation of Temperature Profiles in Filaments Undergoing Solidification," *Am. Inst. Chem. Eng. J.*, **16**, 57 (1970).
76. R. D. Biggs and R. P. Guenther, "Cooling Curves for Extruded PE Wire," *Mod. Plast.*, 126 (May 1963).
77. O. Yandoff, *Acad. Sci., Paris*, **223**, 192 (1946).
78. R. S. Schechter, "On the Steady Flow of a Non-Newtonian Fluid in Cylinder Ducts," *Am. Inst. Chem. Eng. J.*, **7**, 445 (1961).
79. J. A. Wheeler and E. H. Wissler, "Steady Flow of Non-Newtonian Fluids in Square Duct," *Trans. Soc. Rheol.*, **10**, 353 (1966).
80. A. E. Green and R. S. Rivlin, "Steady Flow of Non-Newtonian Fluids Through Tubes," *Quant. Appl. Math.*, **14**, 299 (1956).

81. J. G. Oldroyd, "Some Steady Flows of the General Elastico-viscous Liquid," *Proc. Roy. Soc.*, **A283**, 115 (1965).
82. H. Giesekus, *Rheol. Acta*, **4**, 299 (1965).
83. A. G. Dodson, P. Townsend, and K. Walters, "Non-Newtonian Flow in Pipes of Non-Circular Crossection," submitted to *Comp. Fluids*.
84. S. Matsuhisa, *Japan Plast. Age*, **12**, 25 (1974).

## PROBLEMS

**13.1** ***Bagley Corrections.*** (a) Given the "raw" capillary flow data for two capillaries, plot both $[\Delta P/(4L/D_0)]$ and $\Delta P/4[L/D_0+N(\Gamma)]$ vs $\Gamma$, showing that the flow curves of the former depend on $L/D_0$, while the latter is $L/D_0$ independent.

| Capillary A | | Capillary B | |
|---|---|---|---|
| $L=0.500$ in., $D_0=0.0625$ in. | | $L=2.005$ in., $D_0=0.0501$ in. | |
| $\Delta P$(psi) | $\Gamma(s^{-1})$ | $\Delta P$(psi) | $\Gamma(s^{-1})$ |
| 125.7 | 7.68 | 578.6 | 14.87 |
| 149.2 | 15.36 | 844.5 | 29.73 |
| 227.7 | 38.40 | 1353.1 | 74.34 |
| 394.7 | 76.81 | 1886.4 | 148.67 |
| 610.3 | 153.61 | 2645.8 | 297.34 |
| 972.6 | 384.03 | 3943.6 | 743.35 |
| 1298.7 | 768.07 | 5103.1 | 1486.71 |
| 1688.2 | 1536.13 | 6479.9 | 2973.41 |

(b) Consider the experimental point $L/D_0=50$ and $\Gamma=2000\ s^{-1}$ on Fig. 13.6. What would the magnitude of the error be in evaluating the shear stress at the wall, if the Bagley entrance correction is neglected? Repeat the calculation for $L/D_0=6$, $\Gamma=2000\ s^{-1}$.

**13.2** ***Relative Magnitude of $\Delta P_{ent}$ and $P_{ex}$.*** Han* presents capillary flow data on HDPE at 180°C (Fig. P-13.2) with which a Bagley plot can be constructed for $\Gamma=327.7\ s^{-1}$.

(a) Assuming $P_{ex}=0$ show that for the above shear rate $N=3.3$ and that for $L/D_0=4$ and 8, respectively, $\tau_w^*$ is 20.21 and 20.46 psi.

(b) Taking into account the measured $\Delta P_{ent}$ and $P_{ex}$ values, show that $\tau_w^*=(\Delta P-\Delta P_{ent}-P_{ex})D_0/4L$ and that $\tau_w^*$ values are in this case 21.19 and 21.42 psi for $L/D_0=4$ and 8. That is, including $P_{ex}$ (which is not readily available experimentally) increases the value of $\tau_w^*$ by 5%.

**13.3** ***Estimation of $D/D_0$ from Viscometric Data.*** Use Eqs. 13.2-1 and 13.2-2 and the rheological data on Fig. 6.12 to estimate the extrudate swelling ratio of Tenite 800 LDPE at 160° and 200°C.

* C. D. Han, "On Silt and Capillary-Die Rheometry," *Trans. Soc. Rheol.*, **18**, 163 (1974).

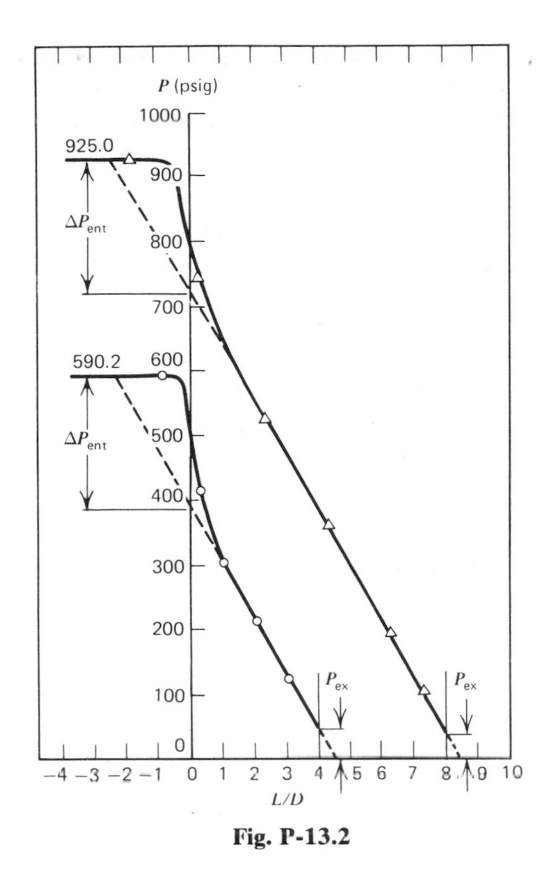


**Fig. P-13.2**

**13.4 *Extrudate Swelling of Newtonian and Power Law Fluids Using Macroscopic Momentum Balance.*** Using a macroscopic momentum balance of the form $\int_0^{D_0/2} \rho v_z(r) 2\pi r\, dr = \rho \bar{V}(\pi D^2/4)$, where $\bar{V}$ is the extrudate velocity after the profile in the capillary has "flattened out," prove that for Newtonian and power law fluids $D/D_0$ is predicted to be, respectively, 0.87 and $[(2n+1)/(3n+1)]^{1/2}$. (See Section 13.2.)

**13.5 *Estimate for Maximum Flow Rate of Production of Flat Sheets.*** Using Eq. 13.4-12, the power law model data appearing in Appendix A and the fact that melt fracture occurs at a value of $\tau_w^*$ of about $10^5$ N/m$^2$, estimate the maximum flow rate (per unit die width) for producing a smooth, flat sheet with a die of 0.05 cm opening for the following polymers: HDPE-Alathon 7040, LDPE-Alathon 1540 at 473°K; PS-Dylene 8, ABS, and HIPS at 483°K.

**13.6 *Coextrusion of Flat Sheets.**** A sheet die is fed by two extruders which deliver two polymer melts at the same temperature but each having power law constants $m_1$, $n_1$ and $m_2$, $n_2$. Fluid 1 is the more viscous of the two. The two streams meet in the approach channel region of the die. That is, the die has separate manifolds and restrictors for each of the melts, but a common

* C. D. Han, "On Silt and Capillary-Die Rheometry," *Trans. Soc. Rheol.*, **18**, 163 (1974).

approach and die lip region (Fig. 13.1). Let the more viscous fluid occupy the region $0 \leq y \leq K$, where $K < H/2$, the half-thickness. Following a solution approach similar to that of Section 13.5 or 13.6 for evaluating the position of the maximum in the velocity profile, obtain expressions for the velocity profile in fluids 1 and 2.

**13.7** ***Pressure Drop Estimation in Spiral Dies.**** Consider a spiral mandrel die similar to the one shown on Fig. 13.22*c*. You are asked to develop an approximate expression for estimating the pressure drop needed to pump a polymer melt of known rheological properties (represented, for example, by power law constants $m$ and $n$). Neglect any "coupling" between the helical flow inside the grooves and axial flow between the flat cylindrical surfaces. Also neglect the taper between the cylindrical surfaces. Express your results in terms of $m$, $n$, $Q$, number and size of the ports and helical grooves, their helical angle, as well as the cylindrical surface spacing and overall length. Make use of Fig. 13.31.

**13.8** ***Flow in the Parison Die Exit Region.*** The flow in the diverging exit region of a parison die, Fig. 13.25, cannot be easily simulated both because of the geometrical complexities involved and because the melt response is not known for such nonviscometric flow. Consider the die exit flow as the *superposition* of annular pressure flow in the $z$-direction and planar extensional flow with stretching in the $\theta$-direction. Set up the appropriate equations to hold for small axial increments $\Delta z$. The annular region over $\Delta z$ is $(\bar{R}_{oi} - \bar{R}_{ii})$ (averaged over $\Delta z$). The planar extension from $z$ to $z + \Delta z$ stretches the polymer melt from $[(R_{oi} + R_{ii})/2]_z$ to $[(R_{oi} + R_{ii})/2]_{z+\Delta z}$ and reduces its thickness from $(R_{oi} - R_{ii})_z$ to $(R_{oi} - R_{ii})_{z+\Delta z}$.

**13.9** ***Wire Coating Pressure Dies—The Lubrication Approximation.*** An example of the pressure wire coating die is shown on Fig. 13.26. Past the guider tip the die contracts with a small taper. Consider a cross section of this region to be formed by two nonparallel plates of spacing $H(z)$, that is, disregard the curvature. The bottom plate is moving with the wire velocity $V$.

(a) Use the lubrication approximation, Eq. 5.4-11, to obtain an expression for $dP/dz$.

(b) Does this result support, qualitatively, the velocity field obtained by Tanner in Fig. 13.28*b*?

**13.10** ***Die for Extruding a Square Extrudate.*** As discussed in Section 13.7, a square die will result in a "bulging" product. In order to produce a square extrudate a four-corner star cross section die has to be constructed (Fig. 13.30*a*) the reason being that the extrudate swells more at higher shear rates and the shear rate increases with decreasing $R_i$ (see Fig. 13.29).

(a) Establish the relationship between the shape of the die and the particular shear rate dependence of extrudate swelling for a given melt.

(b) How does cooling further complicate the final product size problem?

* B. Proctor, "Flow Analysis in Extrusion Dies," *Soc. Plast. Eng. J.*, **28** (February 1972); J. R. Syms, "Computer Design Dies for Tubular Films," lecture notes from Plastic Institute of America's course on "Accuracy in the Extrusion Process," October 25, 1973.

**13.11 *Design Graphs for Dies of Various Shapes But the Same Cross-sectional Area.*** Use Eqs. 13.7-3 and 13.7-4 and Fig. 13.31 to construct $Q$ vs $\Delta P$ graphs for dies which have the same cross-sectional area and the following shapes: circle, ellipse, rectangle, and a rectangle with two rounded off sides in the shape of half circles. Use a Newtonian fluid. How can the equivalent Newtonian fluid concept (see problem 6.14) help you extend the above for non-Newtonian fluids?

**13.12 *Design of Profile Extrusion Dies.**** In designing a die for an extrudate with both thick and thin sections one must achieve uniform extrudate velocities inside the die in both thin and thick cross sections as well as estimate these cross-sectional dimensions in such a way as to give, together with extrudate swelling, the right final dimensions.

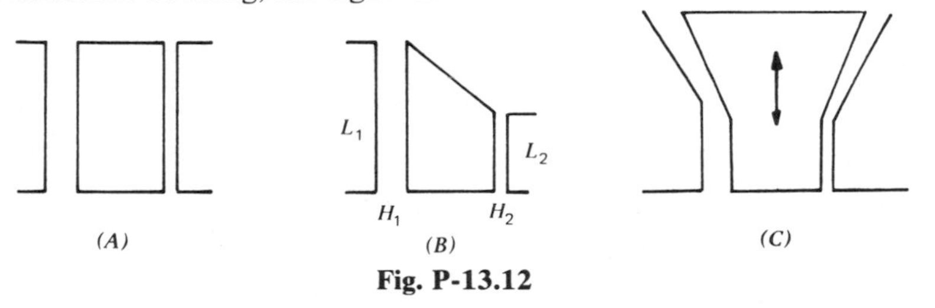


**Fig. P-13.12**

(a) Show that for such a product design $A$ in Fig. P-13.12 is not appropriate.

(b) Calculate the lengths $L_1$ and $L_2$ (for given $H_1$ and $H_2$) which give uniform extrudate velocity (pressure drop) in design $B$. How are $H_1$ and $H_2$ related to the final product thickness? Use a parallel plate flow model (flow rate per unit circumference length). This design results in an awkwardly shaped die.

(c) Analyze design $C$. Prove that if the mandrel can move axially this die can accommodate, in principle, any polymer melt of different rheological properties.

**13.13 *Estimation of Entrance Pressure-Pressure Losses from the Entrance Flow Field.*†** Consider the entrance flow pattern observed with polymer melts and solutions in Fig. 13.16. The flow can be modelled, for small values of $\alpha$, as follows: for $|\theta| \leq \alpha/2$ the fluid is flowing in simple extensional flow (see Section 6.8) and for $\alpha/2 \leq |\theta| \leq \pi/2$ the flow is that between two coaxial cylinders of which the inner is moving with *axial* velocity $V$. The flow in the outer region is a combined drag-pressure flow and, since it is circulatory, the net flow rate is equal to 0. The velocity $V$ can be calculated at any upstream location knowing $\alpha$ and the capillary flow rate.

Use this model for the entrance flow field to get an estimate for the entrance pressure drop.

**13.14 *Coextruded Wire Coating Die Design Equations.*** Two-layer coextruded wire coatings, the inside layer being foamed, have the advantage of

* F. N. Cogswell, *Plastics and Polymers*, "The Scientific Design of Fabrication Processes; Blow Molding," October, 1971.

† A. E. Everage, Jr. and R. L. Ballman, "*An Energy Interpretation of the Flow Patterns in Extrusion through a Die Entry Region,*" 47th Annual Meeting of the Soc. of Rheology, New York, 1977.

increased electrical insulation, while a tough "skin" layer protects the wire mechanically. Following the solution method developed in Section 13.6, derive a design equation for the case where the more viscous fluid 1 occupies the region $R_i \leq r \leq r_1$ while fluid 2 $r_1 \leq r \leq R_o$. $r_1$ is smaller than $r^* = \lambda R_o$, where the maximum in the velocity occurs. No foaming takes place inside the die.

**13.15** ***Stabilizing Forces in Wire Coating Dies.**** Using a CEF equation it can be shown that if the wire in a wire coating die is off centered a lateral stabilizing force arises proportional to the secondary normal stress difference function $\Psi_2$. Use a bipolar coordinate system $\xi$, $\theta$, $\zeta$ (Fig. P-13.15) and the

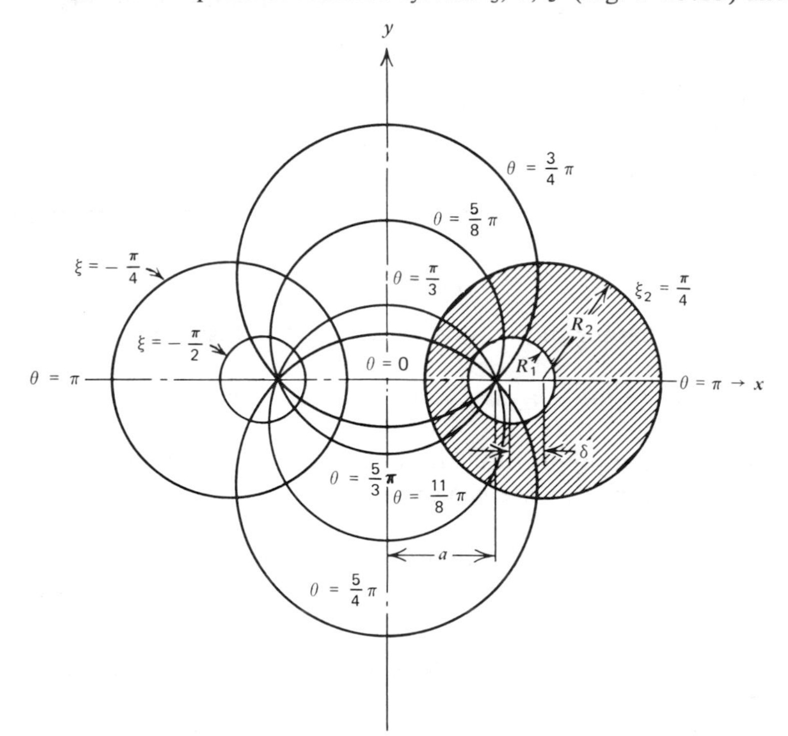


**Fig. P-13.15** Bipolar coordinate system. The shaded area denotes the cross section of the fluid, and the constant $a$ the distance of the pole from the origin. (Reprinted by permission from R. Bird, R. Armstrong, and O. Hassager, *Dynamics of Polymeric Liquids*, Vol. 1, *Fluid Mechanics*, Wiley, New York, 1977.)

components of the equation of continuity and motion in Table P-13.15. Assume that there is no axial pressure gradient and the only nonvanishing velocity component is $v_\zeta(\xi)$, with boundary conditions $v_\zeta(\xi_1, \theta) = V_0$ and $v_\zeta(\xi_2, \theta) = 0$. Further assume the fluid to be incompressible and the flow isothermal.

(a) Show that the velocity profile is given by

$$v_\zeta / V_0 = \frac{\xi - \xi_2}{\xi_1 - \xi_2}$$

* Z. Tadmor and R. B. Bird, "Rheological Analysis of Stabilizing Forces in Wire-Coating Dies," *Polym. Eng. Sci.*, **14**, 124 (1974).

(b) Show that the equation of motion reduces to

$$\frac{\partial P}{\partial \xi} + X\frac{\partial}{\partial \xi}\left(\frac{1}{X}\tau_{\xi\xi}\right) = 0$$

$$\frac{\partial P}{\partial \theta} - \frac{1}{X}\tau_{\xi\xi}\sin\theta = 0$$

where

$$X = \cosh\xi + \cos\theta$$

(c) Show that the lateral force in the wire per unit length $f_x$ is

$$f_x = -\frac{\Psi_2 \pi V_0}{a(\xi_1 - \xi_2)^2}$$

where $a$ is the distance of poles of the bipolar coordinate system from the origin, which is related to the separation of centers of wire and die $\delta$, via

$$\frac{\delta}{R_2} = \sqrt{1 + \left(\frac{a}{R_2}\right)^2} - \sqrt{\left(\frac{R_1}{R_2}\right)^2 + \left(\frac{a}{R_2}\right)^2}$$

Note that

$$\xi_1 = \sinh^{-1}\left(\frac{a}{R_1}\right) \quad \text{and} \quad \xi_2 = \sinh^{-1}\left(\frac{a}{R_2}\right).$$

**Table P-13.15 The Equations of Continuity and Motion in Bipolar Coordinates $(\xi, \theta, \zeta)$[a,b]**

**Continuity**

$$\frac{\partial}{\partial t}\rho + \left(\frac{X}{a}\frac{\partial}{\partial \xi}\rho v_\xi + \frac{X}{a}\frac{\partial}{\partial \theta}\rho v_\theta + \frac{\partial}{\partial \zeta}\rho v_\zeta - \frac{1}{a}\sinh\xi \cdot \rho v_\xi + \frac{1}{a}\sin\theta \cdot \rho v_\theta\right) = 0$$

**Motion**

$$\rho\left[\frac{\partial v_\xi}{\partial t} + v_\xi\left(\frac{X}{a}\frac{\partial}{\partial \xi}v_\xi + \frac{1}{a}v_\theta\sin\theta\right) + v_\theta\left(\frac{X}{a}\frac{\partial}{\partial \theta}v_\xi + \frac{1}{a}v_\theta\sinh\xi\right) + v_\zeta\left(\frac{\partial}{\partial \zeta}v_\xi\right)\right]$$

$$= -\frac{X}{a}\frac{\partial P}{\partial \xi} - \left[\frac{X}{a}\frac{\partial}{\partial \xi}\tau_{\xi\xi} + \frac{X}{a}\frac{\partial}{\partial \theta}\tau_{\theta\xi} + \frac{\partial}{\partial \zeta}\tau_{\zeta\xi} + \frac{1}{a}(\tau_{\theta\theta} - \tau_{\xi\xi})\sinh\xi + \frac{1}{a}(\tau_{\theta\xi} + \tau_{\xi\theta})\sin\theta\right] + \rho g_\xi$$

$$\rho\left[\frac{\partial v_\theta}{\partial t} + v_\xi\left(\frac{X}{a}\frac{\partial}{\partial \xi}v_\theta - \frac{1}{a}v_\xi\sin\theta\right) + v_\theta\left(\frac{X}{a}\frac{\partial}{\partial \theta}v_\theta - \frac{1}{a}v_\xi\sinh\xi\right) + v_\zeta\left(\frac{\partial}{\partial \zeta}v_\theta\right)\right]$$

$$= -\frac{X}{a}\frac{\partial P}{\partial \theta} - \left[\frac{X}{a}\frac{\partial}{\partial \xi}\tau_{\xi\theta} + \frac{X}{a}\frac{\partial}{\partial \theta}\tau_{\theta\theta} + \frac{\partial}{\partial \zeta}\tau_{\zeta\theta} + \frac{1}{a}(\tau_{\theta\theta} - \tau_{\xi\xi})\sin\theta - \frac{1}{a}(\tau_{\theta\xi} + \tau_{\xi\theta})\sinh\xi\right] + \rho g_\theta$$

$$\rho\left[\frac{\partial v_\zeta}{\partial t} + v_\xi\left(\frac{X}{a}\frac{\partial}{\partial \xi}v_\zeta\right) + v_\theta\left(\frac{X}{a}\frac{\partial}{\partial \theta}v_\zeta\right) + v_\zeta\left(\frac{\partial}{\partial \zeta}v_\zeta\right)\right]$$

$$= -\frac{\partial P}{\partial \zeta} - \left[\frac{X}{a}\frac{\partial}{\partial \xi}\tau_{\xi\zeta} + \frac{X}{a}\frac{\partial}{\partial \theta}\tau_{\theta\zeta} + \frac{\partial}{\partial \zeta}\tau_{\zeta\zeta} - \frac{1}{a}\tau_{\xi\zeta}\sinh\xi + \frac{1}{a}\tau_{\theta\zeta}\sin\theta\right] + \rho g_\zeta$$

in which, for Newtonian fluids, $\tau_{ij} = -\mu\{\nabla v) + (\nabla v)^\dagger\}_{ij}$

[a] Z. Tadmor and R. B. Bird, *Polym. Eng. Sci.* **14**, 124 (1974).
[b] For the definition of $X$ and $a$, see Problem 13.15.

# 14

# Molding and Casting



Chapter 1 described the injection molding, compression molding, and casting shaping operations, pointing out that they consist of forcing the polymer into a cavity and duplicating its shape. In the process of casting, the cavity is filled by gravitational flow with a low viscosity liquid (reacting monomer or prepolymer), and following polymerization the liquid solidifies. In compression molding a prepolymer solid mass is heated up or melted and forced to undergo a squeezing flow by hot mold surfaces that close to form a final shape. The prepolymer usually crosslinks and assumes the shape of the closed cavity permanently. In the injection molding process a polymer melt is forced through an orifice (gate) into a closed cold mold, where it solidifies under pressure in the shape of the mold cavity. The polymer is melted, mixed, and injected from the injection unit of the machine. Finally, in the reaction injection molding (RIM) process, low viscosity reacting monomers or prepolymers are mixed just before being injected into a hot cavity, where they react further and solidify. That is, the RIM process is a variation of the casting process; highly reactive liquid systems are injected quickly, rather than being allowed to flow by gravity, into complex shape cavities.

From a process analysis point of view, the basic problems this chapter addresses are (*a*) nonisothermal and transient flow of nonreacting polymer melts, followed by *in situ* cooling and solidification, and (*b*) nonisothermal and transient flow of reacting (polymerizing) liquids followed by *in situ* polymerization and heat transfer. We deal with each of these problems as we discuss the four specific forming processes: injection molding, RIM, compression molding, and casting.

## 14.1 Injection Molding of Polymers

Injection molding involves two distinct processes. The first is melt generation, mixing, and pressurization and flow, which is carried out in the injection unit of the molding machine, and the second is the product shaping, which takes place in the mold cavity. Most injection molding machines are the in-line, reciprocating screw type, as discussed in Chapter 1 and illustrated in Fig. 1.12. The theoretical analysis of the injection unit involves all the facets of the steady continuous plasticating screw extrusion, with the added complication of transient operation due to the periodic screw rotation, on which axial motion is superimposed. In the injection unit the melting step is the dominant one regarding design and operation. Experimental work on melting in injection units (1) has revealed a melting mechanism similar to that in plasticating screw extrusion, which was then used to formulate a mathematical model for the melting process (2). The product of the injection unit is the polymer melt accumulating in the front of the screw. Melt homogeneity affects the filling process and final product quality. However this section covers only the cavity filling process, and we assume that the same quality of well-mixed and uniform temperature melt is produced by the injection unit during each cycle, as well as from one cycle to the next.

To secure the injection of the polymer melt into the mold, the melt must be pressurized. This is achieved by the forward thrust of the screw, which acts as a ram. Hence we have static mechanical pressurization as discussed in Section 10.7, and the resulting flow is a positive displacement type (Section 10.8).

A typical injection mold (Fig. 1.12) is made of at least two parts, one of which is movable so that it can open and close during different parts of the molding cycle (see Fig. 1.13). The entire mold is kept at constant temperature below $T_g$ or $T_m$. The melt exits the nozzle of the injection unit and flows through the mold *sprue*, *runner system*, and *gate* into the mold cavity. Each of these structural elements of the mold performs well-defined functions and affects the molding operation. Thus the sprue forms the overall entrance into the mold. It should not generate large resistance to flow, yet at the same time the melt in it should quickly solidify upon completion of injection and should be extracted from it without difficulty. Finally, the sprue should form a streamlined transition between the nozzle and runner system. All these functions are attainable by a short, diverging conical shape.

The function of the runner system is to transmit the hot melt to the cavities. This should be done with the minimum material and pressure drop "waste." Therefore the runner conduit length must be kept to a minimum level, and the cross section should be *optimally* set for low pressure drop, low material waste, and relatively slow cooling, avoiding premature solidification and "short shots." Generally the runner is about 1.5 times the characteristic thickness of the molded part, and it is of circular cross section to minimize heat loss as well as to facilitate easy machining. Polymer saving and faster cycles occasionally can be achieved by hot runner systems where the polymer in the runners is prevented from solidifying through heating units built around them in the mold. Alternatively, in particular with large parts, it is sufficient to insulate the runner system from the mold. In both cases the sprue can in effect be eliminated from the design.

The gate controls the flow of polymer melt into the die. Its size, shape, and position are affected by a number of considerations. First, a narrow gate is desirable from the standpoint of ease of separation of the molded part from the runner system, as well as solidification after the completion of melt injection, to isolate the cavity from the rest of the system. Of course early solidification must be prevented. Moreover, narrow gates may be detrimental to the finished product because they also bring about large shear rates and stresses (above the melt fracture region), and consequently temperature rises. When the stress level has to be reduced, divergent fan gates are used, spreading the flow. Generally speaking, the gate length is about

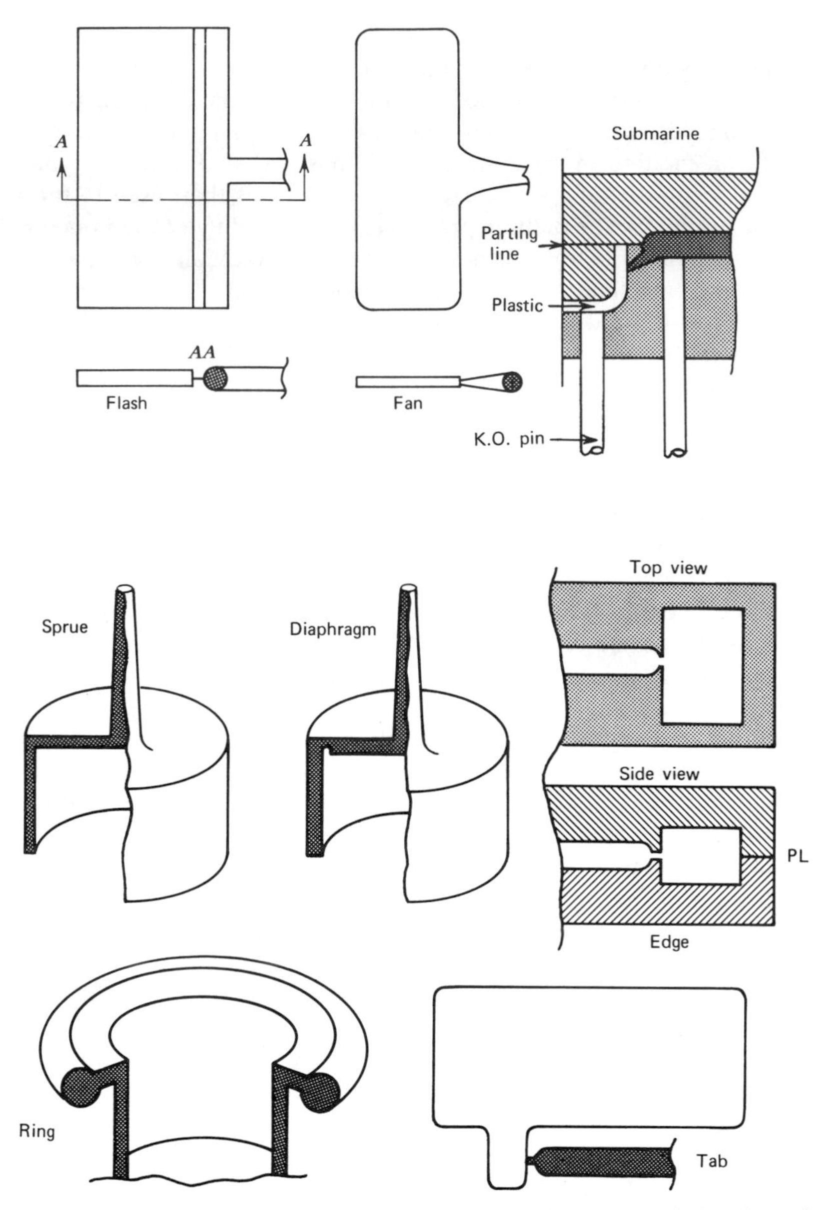


**Fig. 14.1** Typical gate designs and locations. (Reprinted with permission from I. Rubin, *Injection Molding—Theory and Practice*, Wiley, New York, 1972.)

half the characteristic thickness of the section where the gate is attached (usually the heavy sections). It is so positioned that the emerging stream impinges on the opposite wall. Figure 14.1 presents typical gate designs and locations. In multiple cavity molds gates (and runners) also serve the function of balancing flow such that all cavities fill simultaneously. Further information on sprue, runner, and gate design considerations is discussed in the literature (3, 4). The detailed mathematical modeling of the flow of polymer melts through these conduits is not easy, involving most of the complexities of the cavity filling problems, which we discuss below.

**Example 14.1** ***Flow in an Idealized Runner System***

Consider a straight tubular runner of length $L$. A melt following the power law model is injected at constant pressure into the runner. The melt front progresses along the runner until it reaches the gate located at its end. Calculate the melt front position and the instantaneous flow rate as a function of time. Assume an incompressible fluid, isothermal and fully developed flow, and make use of the pseudo-steady state approximation.

The position of the melt front at time $t$ is $Z(t)$ and the instantaneous flow rate $Q(t)$ in terms of the given constant pressure at the inlet to the runner $P_0$ is, from Example 6.3 (or Table 13.2)

$$Q(t)=\frac{\pi R^3}{s+3}\left(\frac{RP_0}{2mZ(t)}\right)^s$$

The position $Z(t)$ is obtained from a mass balance

$$Z(t)=\frac{1}{\pi R^2}\int_0^t Q(t)\,dt$$

Differentiating the equation above we obtain

$$\frac{dZ(t)}{dt}=\frac{Q(t)}{\pi R^2}$$

Finally, substituting the flow rate expression into the preceding equation, followed by integration, gives

$$Z(t)=\left(\frac{1+n}{1+3n}\right)^{n/(1+n)} R\left(\frac{P_0}{2m}\right)^{1/(1+n)} t^{n/(1+n)}$$

and the flow rate is

$$Q(t)=\pi R^3\left(\frac{1+n}{1+3n}\right)^{n/(1+n)}\left(\frac{n}{1+n}\right)\left(\frac{P_0}{2m}\right)^{1/(1+n)}\frac{1}{t^{1/(1+n)}}$$

It is interesting to note that the "penetration depth" $Z(t)$ is proportional to the radius $R$. This implies that the ratio of penetration depths of the same material in two conduits of different radii is given by $Z_1/Z_2=R_1/R_2$, and for constant $P_0$ is dependent only on geometry and not the rheological behavior of the fluid. Assuming a polymer melt with $m=2.18\times10^4\ \mathrm{N\cdot s^n/m^2}$ and $n=0.39$, calculate $Z(t)$ and $Q(t)$ for an applied pressure $P_0=20.6\ \mathrm{MN/m^2}$ in a runner of dimensions $R=2.54$ mm and $L=25.4$ cm.

The expressions for $Z(t)$ and $Q(t)$ for the values just given become

$$Z(t)=0.188\,t^{0.281}$$

$$Q(t)=\frac{1.07\times10^{-6}}{t^{0.719}}$$

Values of $Z(t)$ and $Q(t)$ are listed below for various times:

| $t$(s) | $Z$(m) | $Q(\text{m}^3/\text{s})$ |
|---|---|---|
| 0 | 0 | $\infty$ |
| 0.5 | 0.155 | $1.76\times10^{-6}$ |
| 1 | 0.188 | $1.06\times10^{-6}$ |
| 1.5 | 0.211 | $8\times10^{-7}$ |
| 2 | 0.228 | $6.5\times10^{-7}$ |
| 2.88 | 0.253 | $5\times10^{-7}$ |

These results clearly indicate that we should expect a very high flow rate and quick runner filling initially, followed by a rapid drop in $Q$ and long filling times for the remainder of the long runner. The first half of the runner is filled in 10% of the total runner fill time! From the equation for $Q(t)$ above (under the assumptions made), is it easy to show that for a constant flow rate, an applied pressure linearly increasing with time is required. In practice the initial part of the mold filling cycle is one of increasing applied pressure and almost constant flow rate. If the mold is easy to fill, this situation will persist until mold filling is completed. On the other hand, if the mold flow resistance is high (as for example in Fig. 14.2), the pressure will reach its maximum available value and will remain constant for the rest of the filling time, while the flow rate decreases with time. In the real nonisothermal case, once the flow rate reaches low values the melt has time to cool by conduction to the cold walls, its viscosity increases exponentially, and the flow stops completely, resulting in "short shots."

For comparison purposes, some numerical simulation results of nonisothermal runner filling are included below (5). For the same conditions and using a flow activation energy value of 6 kcal/g · mole, we have

| $t$ (s) | $Z(t)$ | $Q(t)$ | $Z(t)/Z(t)\|_{T=\text{const}}$ | $Q(t)/Q(t)\|_{T=\text{const}}$ |
|---|---|---|---|---|
| 0.5109 | 0.111 | $1.1\times10^{-6}$ | 77% | 63% |
| 0.9703 | 0.140 | $5.10\times10^{-7}$ | 74% | 48% |

It is clear from the tabulation, especially the nonisothermal to isothermal flow ratio, that the polymer is cooling rapidly in the runner and that a "short shot" will result soon after 1 s. Huang's (5) simulation indicates that a "frozen skin" was formed at 0.7 s in the axial region of 2–4 cm from the entrance.

After the cavity has been filled, the injection pressure is still kept on the melt, to "pack" a small amount of melt into the cavity and to compensate for the thermal contraction of the polymer during the cooling and solidification stages. Packing increases the cavity pressure rapidly and appreciably. When the externally applied pressure is removed (by retracting the reciprocating screw or piston of the injection molding machine), backflow out of the cavity takes place, unless the polymer in the gate has solidified or unless such flow is prevented by a one-way valve. At the end of the backflow, if there is any, only cooling of the polymer takes place, together with minute contraction induced local flows. When the polymer has solidified sufficiently to withstand the forces of ejection, the mold is opened and the molded article is removed from the cavity with the aid of the ejection pins. The various

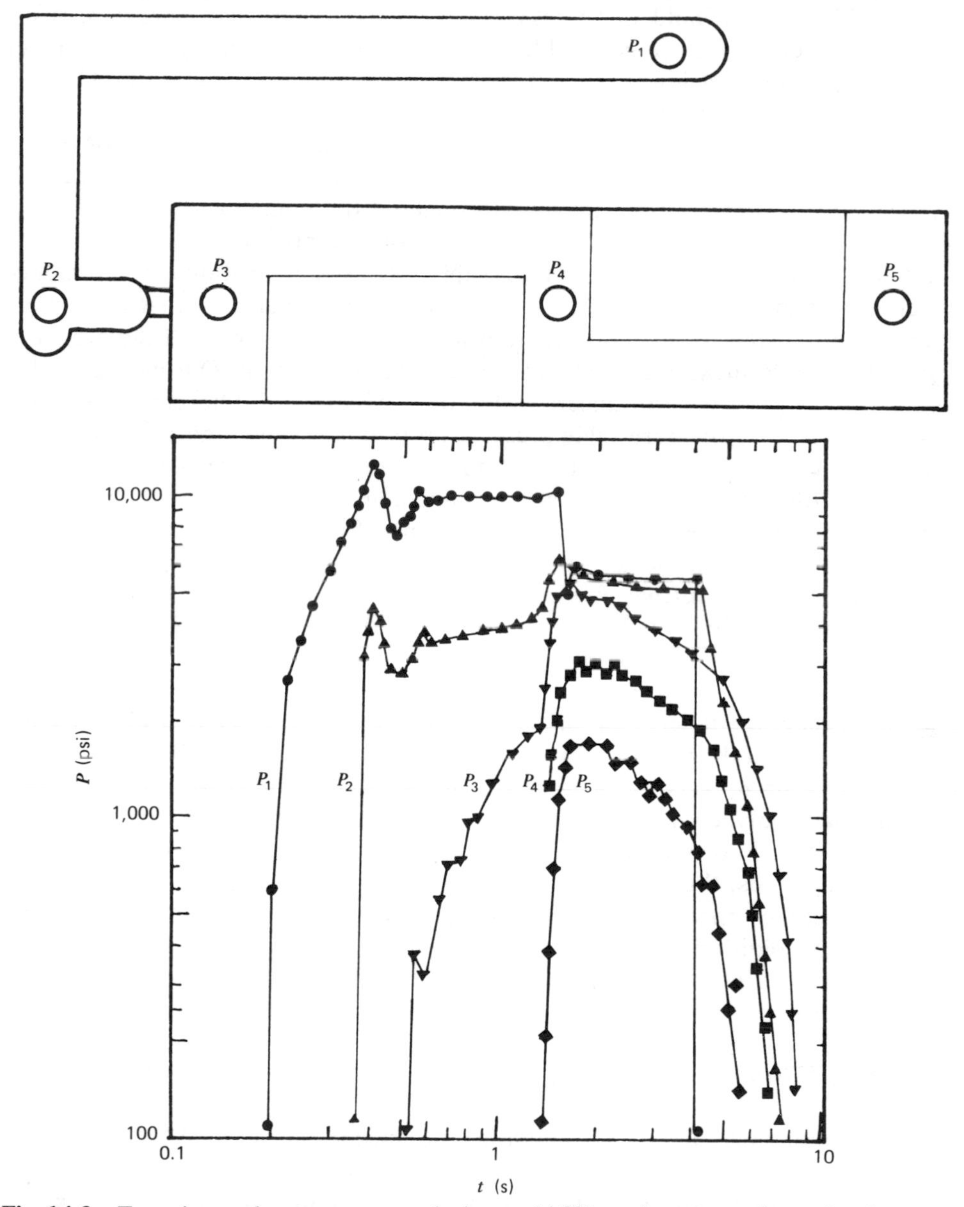


**Fig. 14.2** Experimental pressure traces during mold filling of a rectangular cavity shown on the top with polystyrene at 400°F. $P_1$ is the pressure at the nozzle, $P_2$ the pressure at the end of the runner outside the gate, and $P_3$–$P_5$ are cavity pressures at the locations indicated. Pressure traces $P_4$ and $P_5$ are questionable because of an unintentional preloading of the transducers upon mold closing. (Reprinted with permission from W. L. Krueger and Z. Tadmor, "Injection Molding into Rectangular Cavity with Inserts", Soc. Plastics Eng. 36th Annual Technical Conference, Washington, D.C., 1978, pp. 87–91).

stages of the injection process can be followed by pressure gauges in the mold, as in Fig. 14.2. Here, polystyrene was injected into a shallow rectangular cavity with inserts (top). Pressure transducers were placed in the nozzle, runner, and mold cavity; their output was scanned every 0.02 s and retrieved by a computer (6). The nozzle pressure was set by machine controls to inject at a constant 10,000 psi and upon mold filling, to hold the pressure at 5500 psi. Appreciable pressure overshoots and undershoots are noticeable. The pressure trace at the end of the runner system ($P_2$) follows at lower levels the pattern of the nozzle pressure. The two become equal upon cessation of flow when the mold is full. The difference $P_1–P_2$ indicates the pressure drop over the sprue-runner system. The pressure drop across the gate is approximately given by $P_2–P_3$. We note that just downstream the gate the pressure $P_3$ is increasing with time throughout the filling process (from about 0.4 to 1.3 s). As Example 14.1 pointed out, such a pressure trace approaches conditions of constant filling rate. This is supported by ram position measurements, which were also retrieved at 0.02 s intervals (6). We further note that upon mold filling, when $P_5$ sharply increases, there is also a steep increase in all the pressures except the nozzle pressure, which is then reduced to 5500 psi. During the "hold" time the three pressure transducers within the cavity record different pressures in spite of the absence of appreciable pressure drops due to flow. This is probably the result of skin formation and solidification preventing true core pressure (liquid) recordings. All cavity pressures drop gradually upon solidification, and this gradual pressure drop continues after gate solidification. Of particular practical interest is the residual pressure at the time the mold is opened. If it is near zero, there is the real danger that with further cooling to room temperature, the part will either be smaller than the cavity or will show "sink marks." On the other hand, if the residual pressure is high, the part cannot be easily ejected from the mold and will be "scarred" or deformed during the process.

From the short description of the molding cycle, it is clear that flow, viscous heat generation (filling flow rates are very high), heat transfer, and melt stress relaxation occur to varying degrees of intensity simultaneously. The transport phenomena involved are coupled and, since the solidification times can be comparable to the polymer relaxation times (De ~ 1), molded articles solidify under strained conditions, that is, they contain "frozen-in" strains. Such internal strains greatly affect the properties and morphology of molded articles. Thus we can use the injection molding process to structure polymers. We examine the various stages of injection molding separately.

## *Mold Filling*

It is now clear that there is no simple answer to the question, What should the molding conditions be for the proper molding of a spcecific polymer in a given mold cavity? Figure 14.3, however, illustrates an empirical answer, showing an experimentally determined "molding area" on the melt temperature–injection pressure plane. Within this area the specific polymer is *moldable* in the specific cavity. The area is bounded by four curves. Below the bottom curve the polymer is either a solid or will not flow. Above the top curve the polymer degrades thermally. To the left of the "short shot" curve, the mold cannot be completely filled, and to the right

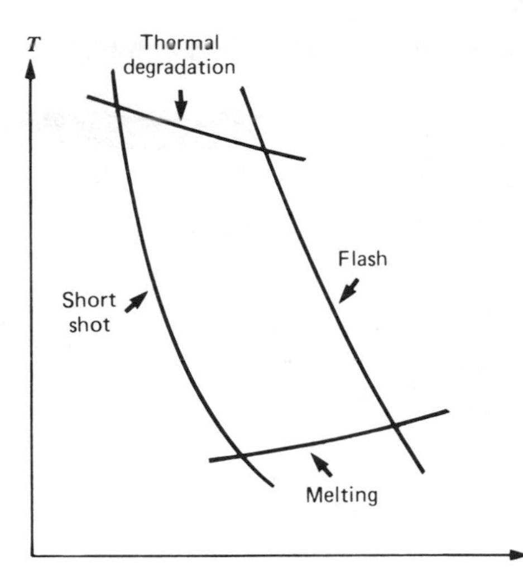


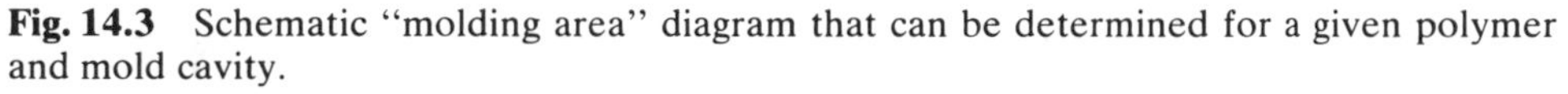

**Fig. 14.3** Schematic "molding area" diagram that can be determined for a given polymer and mold cavity.

of the "flash" curve the melt flows in the gaps formed between the various metal pieces that make up the mold, creating thin webs attached to the molded article at the parting lines. Another practical approach to the question of moldability, especially in comparing one polymer with another, is the use of a standard spiral mold cavity and the measurement, under prescribed molding conditions, of the spiral length filled (7).

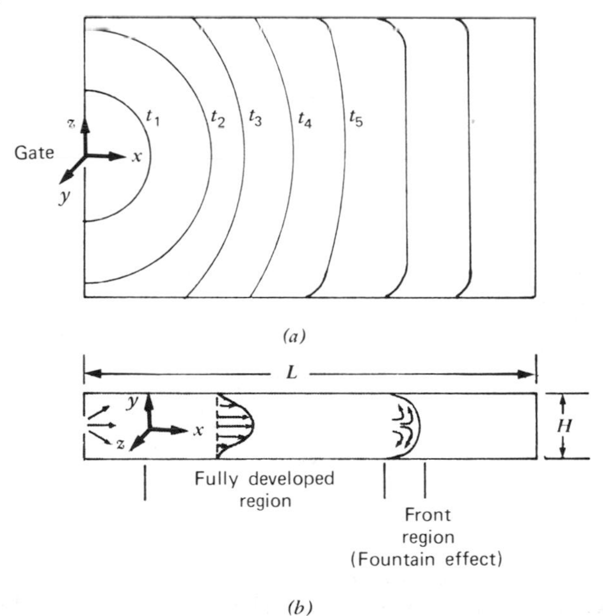


**Fig. 14.4** Schematic representation of the flow patterns during the filling of an end gated rectangular mold whose width is much greater than its thickness. (*a*) Width direction flow fronts at various times. (*b*) Velocity profiles in the fully developed region and schematic representation of the fountain effect in the front region.

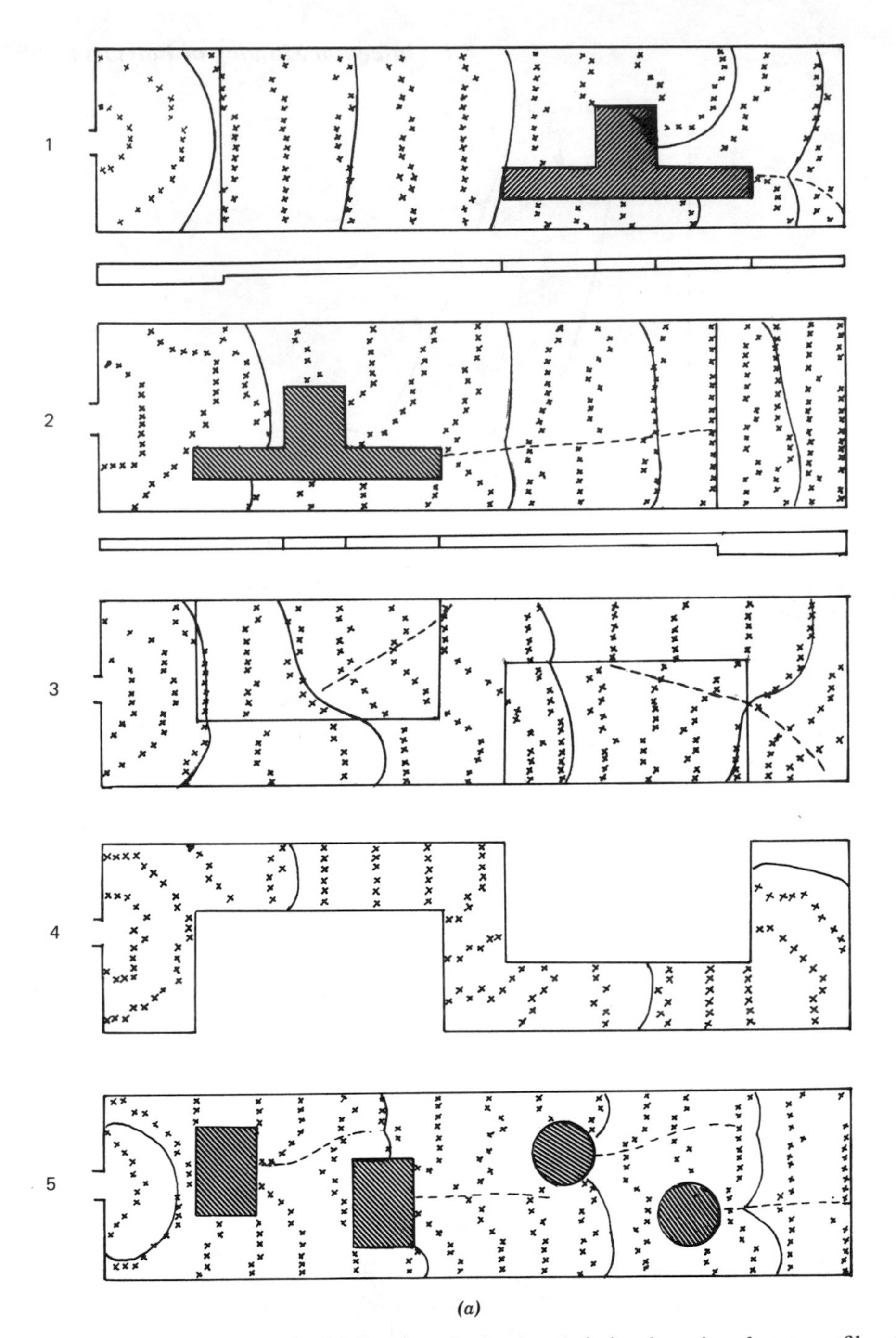


**Fig. 14.5*a*** Experimental (solid lines) and simulated (×) advancing front profiles in a shallow rectangular cavity with inserts of various shapes. Experimental profiles were obtained by short shots (6) as shown in Fig. 14.5b. Cavity dimensions are 3.8 × 15.2 cm (1.5 × 6 in.). Sample (1): A T-shaped insert obstructing completely the flow, and a step reduction in thickness as indicated in the figure. The deep section was 0.335 cm and the thin section varied from 0.168 to 0.180 cm. (In the simulations, the actual thickness distribution was used by measuring the injection molded samples corrected for contraction, but not for mold distortion.) Sample (2) the reverse of sample (1), but the shallow section has a uniform thickness of 0.166 cm. Sample (3) contained two rectangular web inserts, giving rise to a correspondingly shallow section, 0.35 cm thick. Sample (4) had two rectangular

Because mold filling is a complex process, flow visualization studies have been useful and necessary both for the actual mold design and for the mathematical simulation of the process. The first important experimental contributions are due to Gilmore and Spencer (8, 9) whose work forms the basis of a review chapter by Beyer and Spencer (10). Ballman and his co-workers (11–13) conducted mold filling experiments in the early 1960s. A decade later, a time that marks the beginning of serious efforts to solve the injection molding problem, a new wave of mold filling flow visualization studies were reported by Aoba and Odaira (14), Kamal and Kenig (15), White and Dee (16), and Schmidt (17). These studies indicate that the mode of filling at moderate flow rates is an orderly forward flow as shown schematically in Fig. 14.4*a* for a constant depth rectangular cavity. During the early stages of filling, the flow is radial and the melt "front" circular (in this top view). The flow character changes as the melt front advances away from the gate; the predominant velocity component is now $v_x$ and the front shape is either flat for isothermal filling or curved for filling into cold molds (16).

In mold cavities with inserts and nonuniform thickness distribution, the flow pattern is more complicated. This problem was investigated by Krueger and Tadmor (6), using a thin rectangular cavity, $1.5 \times 6$ in. Inserts of various shapes and sizes could be placed in different locations in the mold, creating either obstacles to the flow or regions of different thickness. Polystyrene was injected, and the position and shape of the advancing front could be traced by a series of short shots. Some of the results appear in Fig. 14.5. In Fig. 14.5(1) the outline (shape) of the front is circular in the deep section and becomes somewhat distorted upon entering the thinner region. The flow is split by the "T" insert and reunites past the insert, forming a weld line. The location and the shape of the weld line are determined by the flow profile around the insert. The insert strongly affects the direction of the advancing front, which as we see later, determines the direction of molecular orientation. We would expect, therefore, a highly nonuniform orientation distribution in such a mold. In Fig. 14.5(2) the insert was placed in the thin section close to the gate, completely changing the weld line location and shape, as well as the advancing melt profiles (consequently the orientation distribution). Figure 14.5(3) shows an S-shaped deep section connected by a thin web. We note that the penetration lengths in the deep and shallow sections of the mold, which are being simultaneously filled, follow qualitatively the predictions of Example 14.1, where for $P = \text{constant}$, $Z(t)$ is proportional to the cross-sectional thickness of the channel, and, to a first approximation, is independent of the rheological properties of the melt. This is what is observed (6). There is also formation of two weld lines. The second weld line branches out sideways upon entering the deep section. Figure 14.5(4) shows flow in an S-shaped cavity without weld line formation. Finally, Fig. 14.5(5) shows flow around square and circular inserts, with pronounced weld line formation. These results draw attention to the complex flow patterns occurring even in relatively simple molds. In particular, it is interesting to note the shape of weld lines, which are important not only because they sometimes form visual

---

*Fig. 14.5a cont.*
inserts blocking flow, giving rise to an S-shaped cavity of thickness ranging from 0.165 to 0.173 cm. Sample (5) had two rectangular and two circular inserts obstructing the flow with the rest of the cavity at uniform thickness of 0.323 cm. The broken curves denote visually observable weld lines.

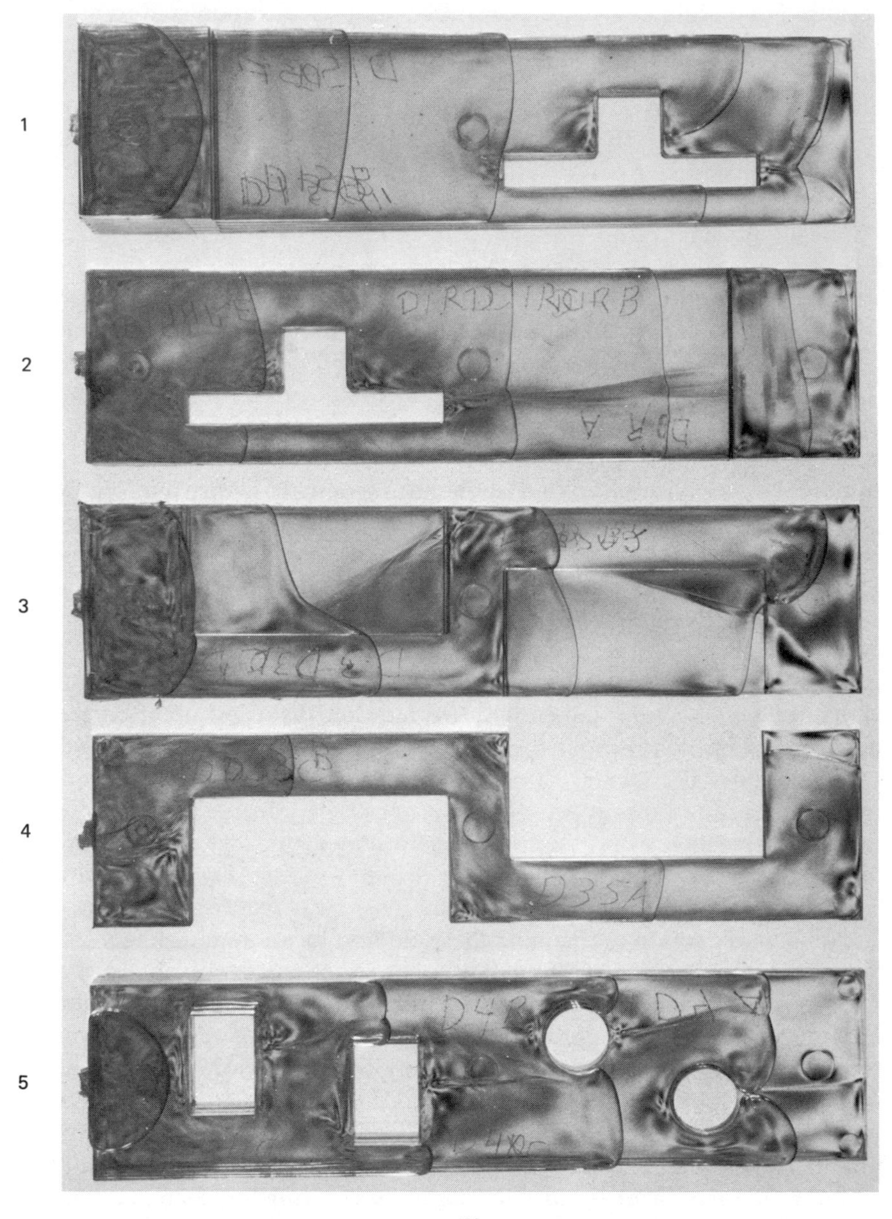


*(b)*

**Fig. 14.5*b*** Photographs of superimposed short shots using polarized light. (Reprinted with permission from W. L. Krueger and Z. Tadmor, "Injection Molding into Rectangular Cavity with Inserts," Society of Plastics Engineers, 36th Annual Technical Conference, Washington D.C., 1978, pp. 87–91.)

defects in the product but also because they generally represent "weak" regions in properties. In general, we can state that weld lines are formed whenever advancing melt fronts "meet," that is, whenever their outward normals are opposite or are converging. The former case occurs in double gate filling and immediately past cavity obstacles, whereas the latter occurs when the front is composed of two segments.

As mentioned in Section 13.5, wèld line interfaces may be characterized by an appreciably different entanglement level as compared to the bulk of the material. The formation and properties of the weld lines can be explained in view of the detailed flow pattern at and near the advancing front, which is different from that in the bulk. In the front region, the melt at the center of the thickness direction, which moves with a high velocity, spills out or *fountains* out to the mold wall to form the surface of the molded article at that location as in Fig. 14.4*b*. This is the only way of filling the region near the wall of the mold, where there is no slip. Thus in the front region the central core decelerates from the maximum velocity at the centerline upstream the front, to the mean velocity at which the front advances. As it decelerates in the direction of flow $x$. it acquires a velocity component in the thickness direction $y$.

The term "fountain effect" was coined and discussed by Rose (18), and it is essentially the reverse of the flow observed near a plunger emptying a fluid out of a channel of the same cross section, as discussed in Section 10.8. The two-dimensional flow in the front region is important in determining the quality and morphology of the surface of the molded article as well as the nature of weld lines. We return to weld line morphology, following the discussion of the flow pattern in the melt front.

When the gate faces a mold wall that is far away and when the filling flow rate is very high, the phenomenon of "jetting" is observed. The melt emerging from the gate forms a jet that rapidly advances until it is stopped by the mold wall opposite to it. Both melt fractured and smooth melt jets have been observed. There are two modes of mold filling under jetting conditions. In the first mode, jetting continues after the jet tip has touched the opposite wall and the jet folds over many times, starting at the impingement surface and continuing toward the gate. When the cavity is almost full of the folded melt jet, regular filling and compression of the folded jet occurs. Thus the filling is in the backward direction. In the second mode of filling, jetting stops after impingement of the jet tip on the opposite wall, and regular forward filling commences. In both cases weld lines may present problems with respect to the optical and mechanical properties of the molded article. It has been experimentally observed that jetting can occur whenever the dimension of the fluid stream is smaller than the smallest dimension in the plane perpendicular to the flow (19). It is thus related both to the gate size and to the degree of extrudate swelling of the melt, rather than to the level of the axial momentum. Filled polymers, which swell less than unfilled melts, exhibit jetting at lower filling rates. Two "cures" for jetting are commonly practiced. First, the gate is positioned so that the emerging melt impinges on a *nearby* wall; second, "fan" gates are used, which increase one of the dimensions of the exiting melt stream, thus decreasing its momentum.

### *Mold Filling Simulations*

A complete mold filling simulation would require the calculation of the detailed velocity and temperature profiles throughout the mold flow region, including the position and shape of the advancing front. This would suffice in principle to determine orientation distribution affecting the article morphology, which evolves upon cooling and solidification. Such a complete model, if available, would be instrumental in assisting the theoretical mold design as well as in optimizing molding conditions for specified property requirements.

The complete problem, of course, is extremely complex even for a relatively simple mold and is hardly soluble for intricate molds. Fortunately, however, a great deal of information and insight can be obtained by simulating (i.e., modeling) selected aspects of the filling problem in isolated "flow regions." Each of these regions requires a unique approach and mathematical tools. Considering the mold filling visualizations in Fig. 14.4*b*, we distinguish among the following regions:

1. The "fully" developed region. During the filling process most of the melt flows in an almost fully developed flow in a narrow gap configuration between cold walls. The nature of this flow determines filling time and part core orientation, as well as the occurrence of short shots. A great deal of insight can be obtained by analyzing one-dimensional flow (either radial, spreading disk, or rectilinear) of hot melt between cold walls. The coupling of the momentum and energy equation eliminates analytical solutions, and finite difference methods can be used.
2. The close neighborhood of the front region. As pointed out earlier, this region determines both surface properties (skin formation) and weld line formations. Hence a detailed analysis of the front region is warranted. This region can be simulated either by approximate analytical or detailed numerical models.
3. The gate region. This region is dominant at the beginning of mold filling. It contributes less as the melt front advances and, because fresh melt is hot in this region, melt memory of the flow experience in this region soon decays.

We discuss some of these regions in some detail below. In addition, we concern ourselves with the *overall* flow pattern during filling. Recall that the manner in which a mold is filled—that is, the location of the advancing melt front—affects the weld line location and the orientation distribution and may be responsible for poor mold filling conditions.

### The Fully Developed Region

A number of mathematical simulations of the flow and heat transfer in the fully developed region have been reported (11, 15, 20–25). We follow here the work of Gogos et al. (23), who simulated the filling of a center gated disk (Fig. 14.6). A frozen surface layer (frozen "skin") can be formed during the filling process, which forces the fluid to flow through a channel of reduced cross section. Assuming constant thermophysical properties, quasi-steady state, $\partial v_r/\partial t = 0$, and neglecting $\tau_{rr}$ and $\tau_{\theta\theta}$ in the $r$-momentum equation, as well as the axial conduction in the energy equation, these balance expressions become

$$\frac{d\tau_{zr}}{\partial z} = -\frac{dP}{dr} \tag{14.1-1}$$

$$\rho C_p\left(\frac{\partial T}{\partial t} + v_r\frac{\partial T}{\partial r}\right) = k\frac{\partial^2 T}{\partial z^2} - \tau_{\theta\theta}\frac{v_r}{r} - \tau_{zr}\frac{\partial v_r}{\partial z} \tag{14.1-2}$$

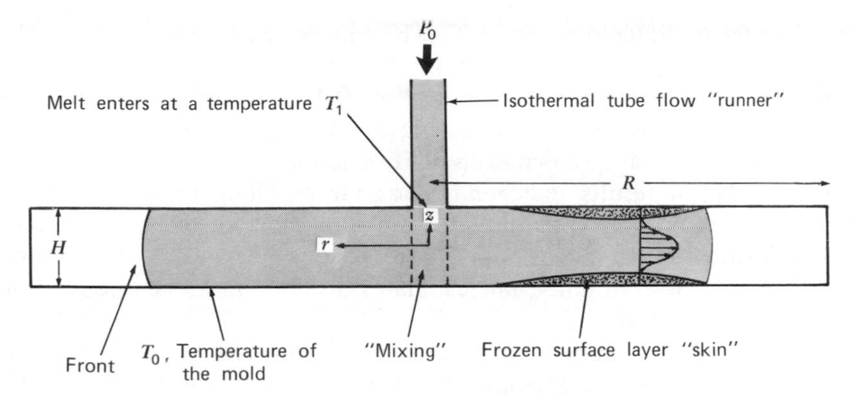


**Fig. 14.6** Cross-sectional view of a center fed disk shaped mold cavity. Indicated schematically are the frozen skin layer that may form during filling as well as the "nipple"-shaped velocity profile.

The term $(\tau_{\theta\theta}(v_r/r))$ is significant only at small values of $r$. Inserting the power law constitutive equation

$$\tau_{zr} = -m\left|\frac{dv_r}{\partial z}\right|^{n-1}\frac{dv_r}{\partial z} \tag{14.1-3}$$

where $m = m_0 \exp(\Delta E/RT)$ and $n$ is a constant. The radial pressure drop is given by (22)

$$\left|\frac{dP}{dr}\right| = \left[\frac{Q(t)}{4\pi r \int_0^{H/2} (z^{1+s}/m^s)\, dz}\right]^{1/s} \tag{14.1-4}$$

The velocity can be obtained by integration of Eq. 14.1-1 and using Eq. 14.1-4

$$v_r(r, z, t) = -\left|\frac{dP}{dr}\right|^s \int_{H/2}^{x} \left(\frac{z}{m}\right)^s dz \tag{14.1-5}$$

For the numerical calculation of the pressure drop as well as the velocity field, one must iterate the pressure at every radial position, until the flow rate in the cavity is the same as that across the tube entrance. In both Eqs. 14.1-4 and 14.1-5 the consistency index $m$ varies with $z$, since the temperature varies in the thickness direction. Two boundary conditions used in the energy equation are of interest. At the advancing front $r = r_{ik}$, the heat transferred to the air in the mold dictates that the term

$$\frac{2r_{ik}h(T_{ik} - T_a)}{(r_{ik}^2 - r_{ik-1}^2)}$$

be included in the right-hand side of Eq. 14.1-2, where $h$ is the heat transfer coefficient to the air. At the mold wall

$$k\left(\frac{\partial T}{\partial z}\right)_{z=H/2} = h\left[T_0 - T\left(r, \frac{H}{2}, t\right)\right] \tag{14.1-6}$$

where $T_0$ is the mold bulk temperature and $h$ is a heat transfer coefficient that is taken to be equal to $k_{\text{mold}}/d$, where $d$ is the distance from the mold surface to the depth where the mold temperature is $T_0$. The energy equation is transformed into a difference equation using an implicit formula and solved by the Crank–Nickolson (26) or O'Brien (27) methods (see section 9.4). The grid size used can be logarithmically decreasing with increasing $z$, so that the details of the rapidly changing temperature and velocity can be revealed.

The simulation results indicate that as far as filling time calculations are concerned, the important variable is the ratio of the rate of heat generated by viscous dissipation to that lost by heat transfer to the cold walls. As a matter of fact, when the ratio is close to unity, fair estimates of filling times can be obtained by assuming isothermal flow.

The flow front is found to advance at an ever decreasing rate, at a constant tube entrance pressure (see Example 14.1). Correspondingly, the filling pressure builds up at an ever increasing rate, as the front advances at constant filling rate. As mentioned earlier a constant filling rate can be assumed if the mold is easy to fill. Realistically the flow rate is constant for the early part of filling, and drops during the latter part. The filling time versus melt temperature at the mold entrance, as well as versus injection pressure is plotted in Fig. 14.7 for an unplasticized PVC

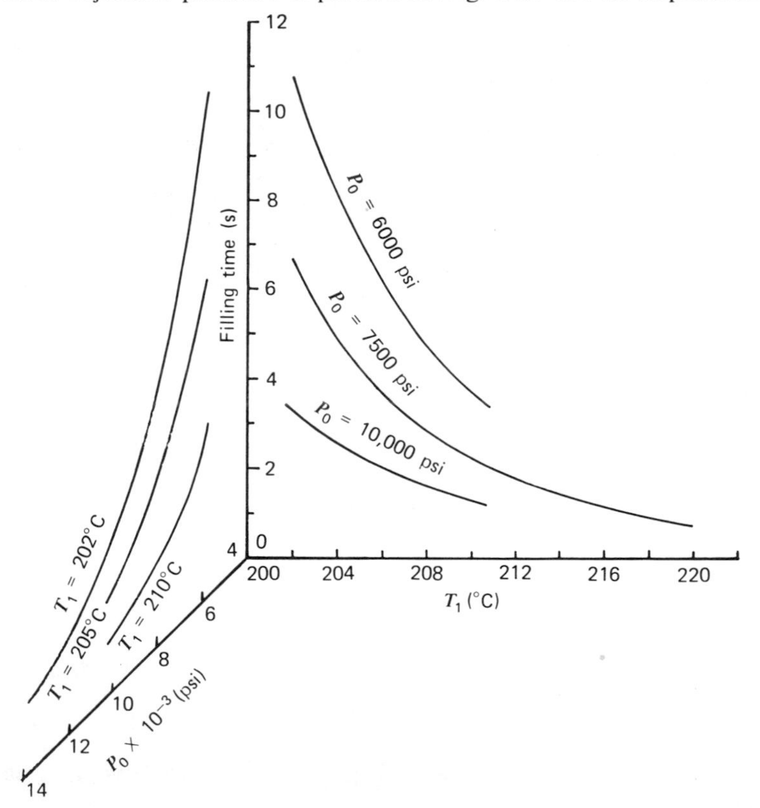


**Fig. 14.7** Filling time versus entrance melt temperature at three constant injection pressures and filling time versus injection pressure at three constants entrance melt temperatures. Mold dimensions are $R = 9$ cm, $H = 0.635$ cm. The polymer is unplasticized PVC of $n = 0.50$, $m$ (202°C) $= 4 \times 10^4$ (N s$^n$/m$^2$), $A = 6.45 \times 10^{-8}$, $\Delta E = 27.8$ (kcal/g · mole), $\rho = 1.3 \times 10^3$ (kg/m$^3$), $C_p = 1.88 \times 10^3$ (J/kg · °K), and $k = 9.6 \times 10^{-2}$ (J/m · s · °K). [Reprinted with permission from P. C. Wu, C. F. Huang, and C. G. Gogos, *Polym. Eng. Sci.*, **14**, 223 (1974)].

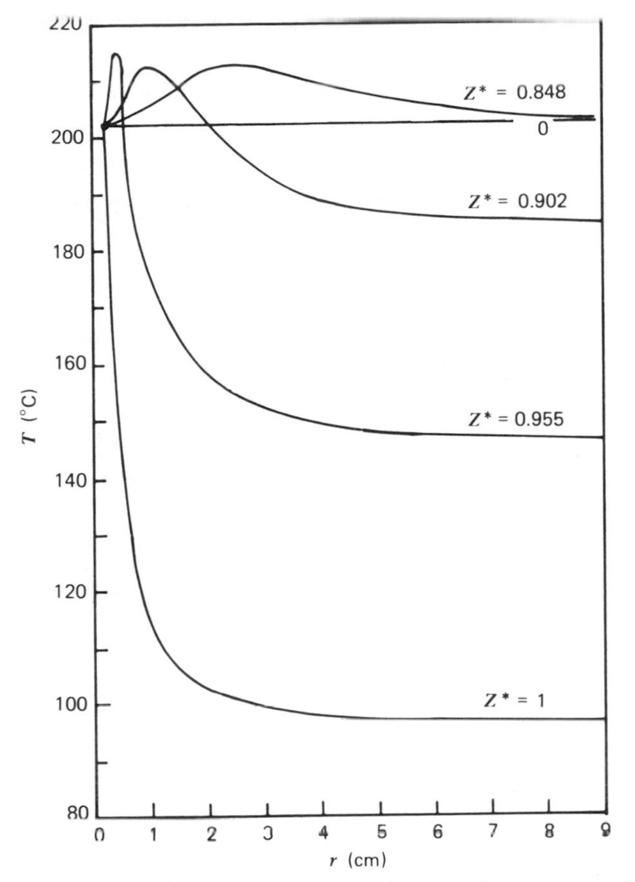


**Fig. 14.8** Temperature distribution at the instant of fill as a function of radial distance and at various values of $Z^* = z/(H/2)$, for PVC at 15,000 psi, $H = 0.635$ cm, $t = 1.45$ s; $R = 9$ cm, $T_1 = 202$°C, $T_0 = 30$°C. [Reprinted with permission from P. C. Wu, C. F. Huang and C. G. Gogos, *Polym. Eng. Sci.*, **14**, 223 (1974)].

resin. The slope of the filling time versus melt temperature curves depends on the activation energy for flow, that is, the temperature sensitivity of the consistency index $m$. On the other hand the slope of the filling time versus injection pressure curves depends on the value of the pseudoplasticity index $n$, increasing with decreasing $n$. The temperature profiles at the instant of complete fill for the same mold and resin entering the mold at 202°C and 15,000 psi are shown in Fig. 14.8. A number of features are interesting. First, almost isothermal conditions prevail in the thickness region from the center of the mold halfway to the wall. This is because the velocity profile is almost flat and heat transfer is negligible. Appreciable and rapid conductive cooling occurs only *very* near the mold cavity wall. If 150°C is taken to be a temperature level, where PVC has practically an infinite viscosity, then an effective thin frozen skin is formed for $r > 2.5$ cm. At lower injection pressures, thicker frozen skin layers are formed. The dependence of the frozen skin profile on molding variables is indicated in Fig. 14.9, where the frozen skin at the moment of fill is plotted under the conditions specified. Its thickness decreases with increasing pressure, melt, and mold temperature and mold thickness. The shape of the frozen skin profile is characterized by a maximum (a "pinch" region). Near the entrance, fresh hot melt keeps the skin to a minimum, and near the front, the melt near the

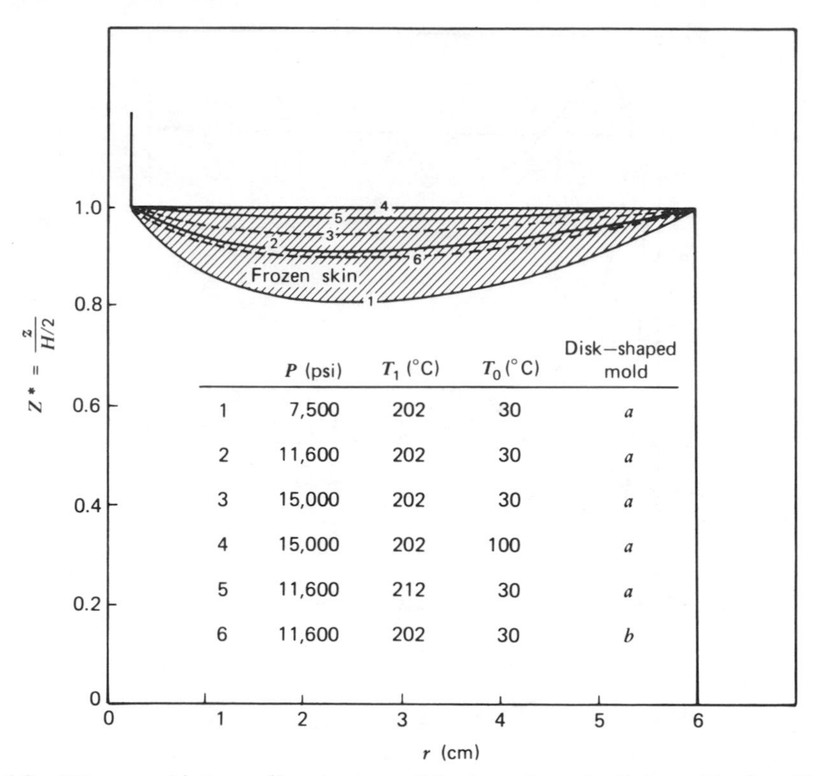


**Fig. 14.9** "Frozen skin" profiles in two disk-shaped center-fed molds ($a$: $R=6$ cm, $H=0.5$ cm; $b$: $R=6$ cm, $H=0.3$ cm). Rigid PVC was considered in the simulation. The frozen skin region is defined by $T<150°C$ (5).

walls is still hot because it originates from the center region. It is the "pinch" region, in which flow would stop first, that would be responsible for short shots at low injection pressures. It is also worth noting that the shape of the frozen skin creates both an axial stretching flow and a $z$-component velocity. This is particularly true in the gate region. The problem has been discussed by Barrie (28). In the disk cavity the stretching flow discussed earlier offsets the assumption of neglecting the term $dv_r/dr$.

### The Front Region

The front region was treated by Tadmor (29) in an attempt to model the experimentally observed molecular orientation distribution in molded articles. Figure 14.10 shows such a distribution, obtained by Wübken and Menges (30), by measuring the shrinkage of microtomed samples at elevated temperatures. Figure 14.10$a$ shows the longitudinal (flow direction) orientation distribution at two injection rates. The characteristic features are the maximum orientation at the surface, a gradual drop in orientation, and a secondary maximum, which is followed by a gradual drop to no orientation at the center. In Fig. 14.10$b$ the maximum longitudinal orientation appears to be a short distance from the wall, whereas the transverse orientation drops continuously from a maximum value at the surface.

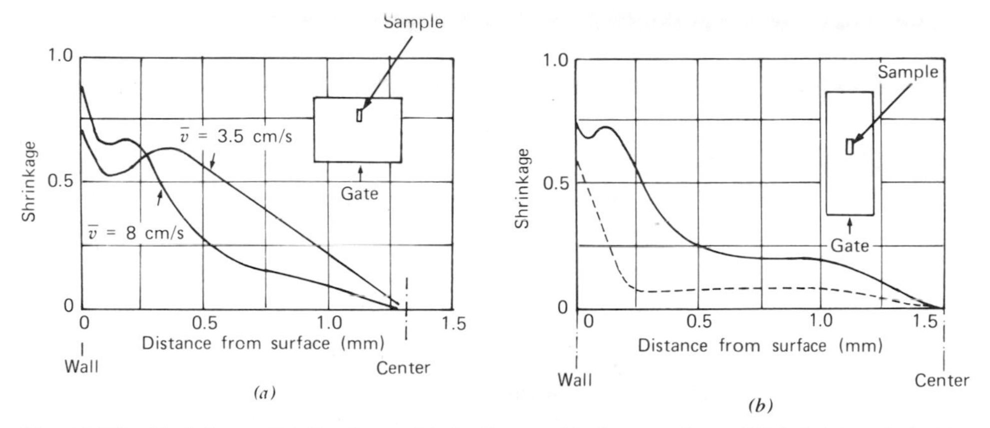


**Fig. 14.10** Shrinkage distributions of injection molded amorphous *PS* (*a*) at two injection rates: longitudinal direction, (*b*) solid curves, longitudinal direction; broken curve, transverse direction. (Reprinted with permission from G. Menges and W. Wübken, "Influence of Processing Conditions on Molecular Orientation in Injection Molds", Soc. Plastics Eng., 31st Annual Technical Conference, Montreal, Canada, 1973).

According to the model proposed by Tadmor (29), both the orientation in the close neighborhood of the wall and the transverse orientation originate from the fountain type of flow in the advancing front region, whereas the source of the rest of the orientation is primarily from the shear flow upstream the front. As Fig. 14.11 indicates, the central fluid element just behind the advancing front, within the fountain flow region, is approximately considered to be undergoing a steady elongational flow. Specifically, for flow in narrow channels, the velocity profile in a coordinate system located on the advancing front and moving with it at the mean velocity $\langle v \rangle$, is $v_x = -\dot{\epsilon}x$, $v_y = \dot{\epsilon}y$, and $v_z = 0$. (This velocity profile also describes incompressible stagnation potential flow.) The molecular orientation is a function of the rate of elongation, which can be estimated by assuming that the maximum velocity $v_{x\,\max}$, upstream the advancing front, drops to the mean front velocity $\langle v \rangle$ within a certain distance. Assuming this distance to be of the order of the gap

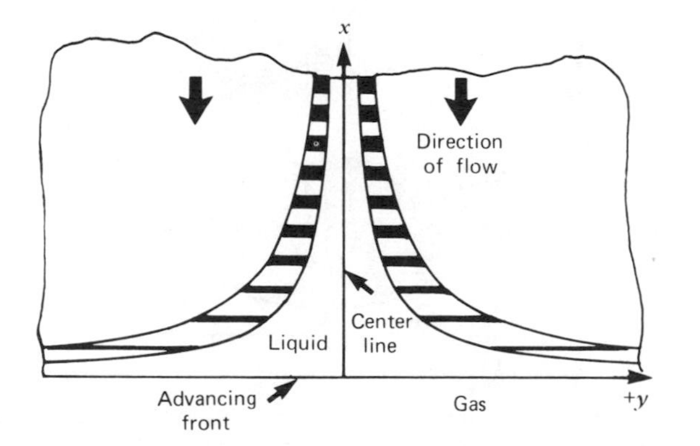


**Fig. 14.11** Schematic representation of the flow pattern in the central portion of the advancing front between two parallel plates. The coordinate system moves in the $x$-direction with the front velocity. Black rectangles denote the stretching deformation the fluid particles experience. [Reprinted with permission from Z. Tadmor, *J. Appl. Polym. Sci.* **18**, 1753 (1974)].

thickness, we obtain the following estimated rate of elongation:

$$\dot{\varepsilon} = \frac{dv_y}{dy} = -\frac{dv_x}{dx} = \frac{\langle v \rangle - v_{\max}}{H} \tag{14.1-7}$$

(Note that $\langle v \rangle < 0$ and $v_{\max} < 0$, since they are in the negative $x$-direction; hence, $\dot{\varepsilon} > 0$.) If $v_{\max}$ is taken from the fully developed flow of a power law model fluid between parallel plates, then

$$-\dot{\varepsilon} = \left(\frac{n}{n+1}\right)\frac{\langle v \rangle}{H} \tag{14.1-8}$$

Clearly the rate of elongation according to this model increases with injection rate, with decreasing gap, and with increasing $n$.

Obviously, the fountain type of flow is not exactly the constant elongational rate flow described by Eq. 14.1-8. Furthermore, the shape of the front is not flat, but bends backward to become tangent to the walls of $y = \pm H/2$. Consequently, the fluid elements that were oriented by the fountain flow in the $y$-direction are deposited on the cold wall with an $x$-direction orientation (Fig. 14.12).

As a result of the fountain type of flow, an oriented polymer layer originating from the central core of the advancing front and experiencing a steady elongation rate given in Eq. 14.1-8 is deposited on the cold wall of the mold. The surface layer

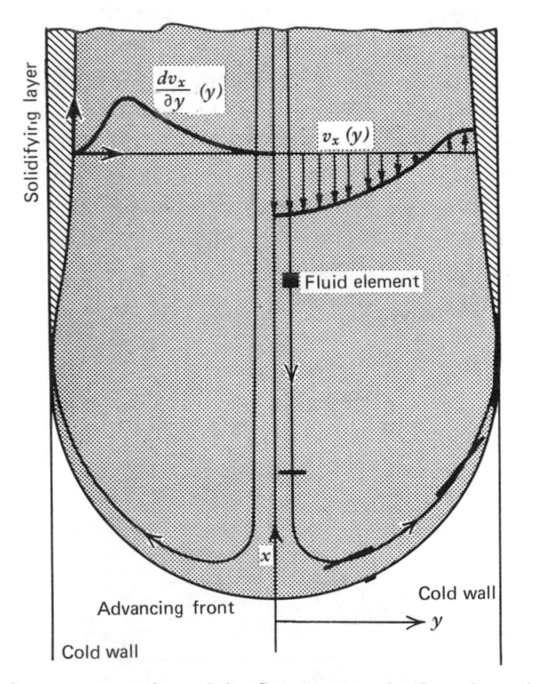


**Fig. 14.12** Schematic representation of the flow pattern in the advancing front between two parallel cold walls. Black rectangles denote the stretching and orientation of a fluid particle approaching the front in the central region of the front. The curved shape of the front causes fluid particles initially oriented in the $y$-direction to end up on the wall, oriented in the $x$-direction. The velocity profile upstream the front is in the $x$-direction. The velocity profile upstream the front is viewed from a coordinate system located *on* the front. [Reprinted with permission from Z. Tadmor *J. Appl. Polym. Sci.*, **18**, 1753 (1974)].

solidifies upon contact with the cold wall, retaining the maximum elongational orientation. Further away from the surface into the skin layer, molecular relaxation occurs, reducing this orientation. The final orientation distribution in the skin layer will be a function of the cooling rate and the spectrum of relaxation times. The fountain-type flow mechanism and the orientation model just described suggest that in narrow gap flow, skin layer orientation is unidirectional in the direction the front advances. If, however, the cross-section of the flow is deep, the fountain-type flow leads to biaxial orientation (i.e., orientation in the $x$-longitudinal and $z$-transverse directions).

Further away from the surface, the fully developed shear flow behind the front is responsible for any molecular orientation that may be present in the final product. Shear orientation is a function of shear rate, which varies over the gap. As discussed earlier, in fully developed flow in molds with cold walls, $\dot{\gamma}$ is almost zero in the immediate vicinity of the wall, exhibits a maximum *near* the wall, and is very low in the central core (Fig. 14.12). Thus the initial shear orientation distribution at any particular location in the mold is approximately determined by the local shear rate distribution at the moment of fill. Shear orientation therefore is unidirectional in the direction of flow. This initial shear orientation relaxes to various degrees, depending on the cooling rate and the relaxation spectrum. A complete orientation distribution can be approximately obtained by superimposing the elongational and shear orientations. The result, depending on the relative magnitude of the two orientation sources, may be complex such as that in Fig. 14.10*a*. Alternatively, if the shear orientation is dominant, a maximum orientation is exhibited at a short distance from the wall where the shear rate was maximum. Clearly, the transverse

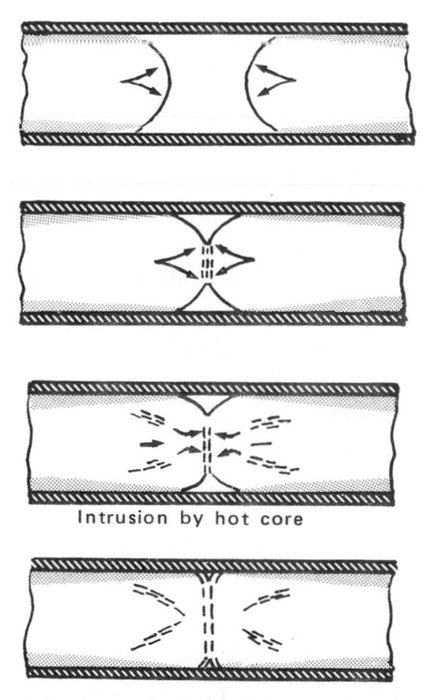


**Fig. 14.13** Schematic representation of the impingement and the subsequent flows in the weld line region: gray areas, cold regions of the melt; dashed lines denote regions undergoing extensional flow. [Reprinted with permission from S. Y. Hobbs, *Polym. Eng. Sci.* **14**, 621 (1974)].

orientation distribution (broken line in Fig. 14.10*b*) exhibits no secondary maximum, lending support to the assumption that it originates from elongational type of flow in the advancing front. We should note that the relative significance of the orientation sources as well as the detailed distribution depend on polymer properties (ability to orient during and relax upon cessation of flow), injection conditions (filling rate, melt and mold temperatures), and mold geometry.

Finally Tadmor (29), assuming that there is a quantitative relationship between shrinkage due to orientation and mean molecular end-to-end distance, following the framework of the above-described model, and assuming a bead-spring molecular model developed by Bird et al. (31), obtained orientation distributions that with a reasonable choice of parameters, agreed semiquantitatively with experimental measurements (30).

Dietz, White, and Clark (32) have recently demonstrated that birefringence measurements during and after filling can be useful in obtaining information about the filling flow kinematics. Birefringence measurements are related through the stress optical law (Eq. 3.9-17) to the stress field. The stress, in turn, is related to the flow kinematics by assuming the appropriate constitutive relations for flow and stress relaxation. Conversely, one can check the validity of assumed velocity fields and constitutive relations by comparing the predicted birefringence to that obtained by experiments. Whether this approach turns out to be quantitatively useful will depend on the feasibility of the analysis after lifting some of the assumptions made in the work mentioned.

Of course a better understanding of the problem involves the solution of the fountain flow in the front region. This is a difficult problem, especially for the nonisothermal flow of viscoelastic fluids. As a free boundary problem it may be amenable to solution with either the marker-and-cell (MAC) approach or the finite element method (FEM). Recently Huang (33) has used the MAC method to investigate the melt front flow. Assuming isothermal Newtonian and power law fluid behavior, he obtained fountain-type flow patterns. Furthermore, his results are in good agreement with the experimental work of Schmidt (17), which we treat later in this section.

Now that the front region fountain-type flow has been discussed, we can turn to the subject of weld line formation, which involves two fronts meeting each other. Figure 14.13 depicts the various stages of flow during weld line formation. When they meet, the two fronts are made of polymer molecules that are aligned with the front shape. Thus they will meet tangentially. Following the first contact, a stagnation-type flow fills the two wedge-shaped regions next to the two mold walls. This flow further stretches the free boundaries of the two fronts and has a tendency to resist the packing flow that is necessary to avoid small sink marks at the weld line.

### *The Overall Flow Pattern*

The most characteristic feature of injection molds is geometrical complexity. In such molds there is a need for the prediction of the overall flow pattern, which provides information on the sequence in which different portions of the mold fill, as well as on short shots, weld line location, and orientation distribution. The more complex a mold, the greater this need is. The irregular complexity of the geometry, which forms the boundary conditions of the flow problem, leads naturally to finite element methods, which are inherently appropriate for handling complex boundary conditions.

By such methods, in principle, the pressure, velocity, and temperature distribution can be determined. However no such attempt appears to have been made. The problem can be greatly simplified by restricting the flow to the narrow gap type of configurations in which locally fully developed flow can be assumed at any instant of time. Many molded articles have a generally thin walled structure where such an analysis is relevant. The FEM formulation of the two-dimensional narrow gap flow problem would closely follow that of anisotropic seepage problems analyzed by Zienkiewicz et al. (34). A simple lattice-type FEM formulation called flow analysis network (FAN) (see Section 16.3) was suggested for this problem by Tadmor et al. (35), assuming isothermal non-Newtonian flow. Krueger and Tadmor (6) simulated with this model the filling of rectangular molds with inserts (Fig. 14.5). A computational grid of $18\times72$ was used. The time dependent gate pressure used in the simulations were based on the experimental pressure traces at location $P_3$ (Fig. 14.2), neglecting the pressure drop from the gate outlet to this location. At the same location the temperature was also measured, and its mean value was used in the simulations. Figure 14.5*a* plots simulation results. Calculated advancing front profiles are marked by $\times$ symbols (sudden jumps in values are the result of the rather coarse grid size) labeled by corresponding filling times. Experimental front profiles obtained by short shots are denoted by solid lines, and experimental (visually observable) weld line locations by dotted lines. The agreement between predicted and measured front profiles is surprisingly good, considering the restrictive isothermal flow assumption in the model and the possible front distortion in short shots. The theoretical model does not account for side wall effects, which clearly appear to restrict flow in the experimental profile. Reasonable agreement was also obtained in predicting mold filling times (6). The reasonably good prediction of the advancing front profiles indicates that no large melt temperature drop occurred during filling under the conditions used; but it also suggests the possibility for the theoretical prediction of orientation distribution and weld line locations.

Finally, simulation studies indicate that the overall flow pattern and the front shapes are only weak functions of the viscous nature of the liquids used; Newtonian liquids exhibit almost the same front patterns as pseudoplastic melts. Such behavior is implied by the results of Example 14.1. This finding was supported experimentally by studies which utilized high speed movie photography during the injection of low viscosity Newtonian fluids in a transparent mold (6). The implications of such findings are twofold: from a modeling point of view it is permissible (to a first approximation) to use the simple Newtonian constitutive equation to predict front positions and shapes; experimentally, it is possible to study mold filling patterns with low viscosity fluids in transparent molds, a convenient system. Of course, filling times and pressures are highly dependent on the viscous properties of the melt used.

### *Cooling of the Molded Part*

Melt cooling takes place from the start of the injection molding cycle because, with the exception of the case of hot runner molding, the entire mold is near room temperature. We saw earlier that during filling, temperature gradients in both the flow and transverse directions appear, and a "frozen skin" develops whose average thickness decreases with increasing entering melt temperature and injection rate. At the end of the filling stage, cooling of the melt is the predominant phenomenon. But because of the resulting specific volume decrease, a small amount of melt must be slowly "packed" into the mold to compensate. Furthermore, if the injection pressure is removed before the gate freezes (or in the absence of a one-way valve) backflow can occur because of the prevailing high pressure in the mold cavity. Finally, during cooling minor secondary flows occur, which may result in appreciable macromolecular orientation. These flows are due to temperature gradients, where melt flows from the hot region to the cold to compensate for the volume contraction. Thus secondary flows are expected to occur in regions of abrupt decreases in the cross-sectional dimension. Whenever such flows are *not* possible, usually due to lack of material, voids are formed in the bulk of the molded article. From an *overall* point of view, the necessary requirement for avoiding void formation is that the mass of the polymer injected is greater than or equal to the product of the atmospheric density times the cavity volume.

In specifying the transport equations that simulate the cooling and solidification of molded parts, we neglect the packing, back, and secondary flows that take place during that stage of molding and contribute convectively to the heat transfer. Kamal and Kuo (36) have simulated cavity packing for the following two cases: fast flows that end in rapid decelerations and negligible thermal contractions during packing, and slow filling flows, where deceleration is negligible but thermal contractions are controlling. In both cases the equation of state of Spencer and Gilmore (Table 5.6) was used.

For an end-fed rectangular mold cavity of a small thickness compared to the other two dimensions, we expect at the end of the filling temperature gradients in the flow and thickness direction, such as those in Fig. 14.7. Thus for a cavity such as that in Fig. 14.4, the energy equation describing mold cooling is

$$\rho C_p \frac{\partial T}{\partial t} = k\left[\frac{\partial^2 T}{\partial y^2} + \frac{\partial^2 T}{\partial x^2}\right] \tag{14.1-9}$$

The temperature field at the end of filling is taken as the initial condition for the equation above. The boundary conditions are (where $l$ is the mold length)

$$\left.\begin{aligned}
&\left(\frac{\partial T}{\partial y}\right)(0, x, t) = 0 \\
&\left(\frac{\partial T}{\partial y}\right)\left(\frac{H}{2}, x, t\right) = -\frac{h}{k}\left[T\left(\frac{H}{2}, x, t\right) - T_0\right] \\
&\left(\frac{\partial T}{\partial x}\right)(y, 0, t) = -\frac{h}{k}[T(y, 0, t) - T_0] \\
&\left(\frac{\partial T}{\partial x}\right)(y, l, t) = -\frac{h}{k}[T(y, l, t) - T_0]
\end{aligned}\right\} \tag{14.1-10}$$

This problem can be solved numerically for the case of constant thermophysical properties. The next section discusses the solution and its results, in covering reactive molding. As Chapter 9 discussed, the case of solidification with crystallization can be treated by stating Eq. 14.1-9 twice, once for the melt and once for the polycrystalline solid, taking into account the heat released during the first order transition by a thermal balance at the interface. Alternatively, Eq. 14.1-9 can be used for both phases if $C_p$, $\rho$, and $k$ are allowed to vary continuously over the entire temperature range. Finally, the conduction in the direction of flow can be neglected, since the gradients in that direction are in general smaller than the transverse gradients. In this case the term $k\partial^2 T/\partial x^2$ is zero, and only the first two boundary conditions of Eq. 14.1-10 have to be used. The numerical solution scheme for this case, but different boundary conditions, is discussed in detail in Section 9.4.

### *Injection Molding Structuring*

There is a broad potential of structuring in injection molding, because the flow field in filling the mold is rapid and complex and can be varied by varying the process variables. Furthermore, the heat transfer can be fast and efficient, at least for the molecules near the surfaces of the mold. In other words, the probability of freezing molecular orientation brought about by flow is high near the surface layers and low near the core, enabling the formation of the "sandwich" type of structure.

From our earlier discussion in this section we expect to have the following macromolecular orientation, starting from the center of the thickness and proceeding outward: (*a*) near the center, no orientation at all because the shear velocity gradients are zero; (*b*) moving away from the center, an increasing amount of shear flow orientation due to the shear gradients that pass through a maximum and the faster cooling as we approach the wall; (*c*) near the wall, we expect only some shear flow orientation (shear gradients are low), in addition to some extensional flow orientation due to the fountain flow; and (*d*) in the wall region, only extensional flow orientation, due to the fountain effect. Although there is no question that shear flow occurs during filling, experimental evidence must be brought forth in support of the fountain flow, since it is so important to the resulting morphology.

We cite two experimental investigations, one by Schmidt (17) and the other by Thamm (37). In Schmidt's work colored tracer particles, which entered the mold at the center of the mold thickness while the mold was partly filled, were found deposited *at* the mold walls at a later time and at an axial distance beyond that of the front at the time of the tracer entrance. (The approximate locations of such particles on the mold wall were predicted by Huang (33) who simulated the melt front flow region problem during filling.) Thamm investigating the morphology of injection molded blends of polypropylene and ethylene-propylene-diene terpolymer (EPDM), found that near the mold surface the EPDM profiles are either elongated for flat narrow molds or disk shaped for more squarelike molds. The EPDM particles in this case act as deformable tracer particles. Thus from Schmidt's work we see that center particles catch up with the surface and flow to the wall, and from Thamm's evidence we can deduce that the flow is either that of simple extension for narrow flat molds, or biaxial extension for square molds. Figure 14.14 demonstrates the extensional nature of the front region flow: the EPDM "tracer" particles are already deformed on either side of the weld line. These observations all but confirm the existence of fountain flow.

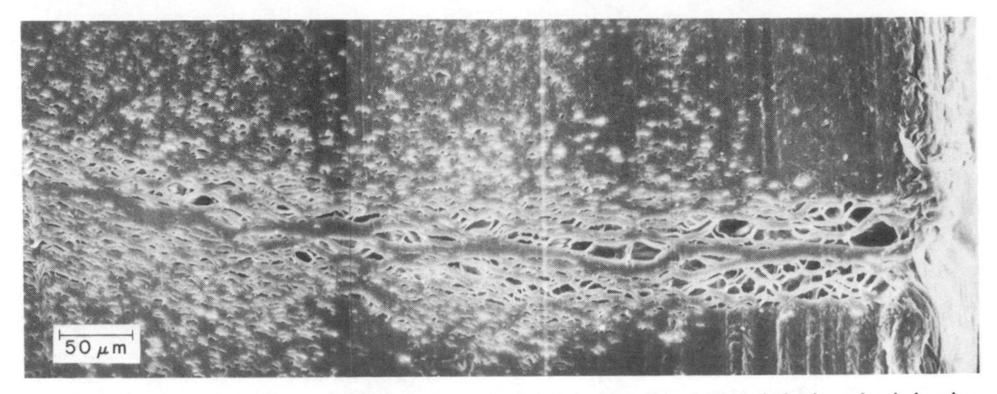


**Fig. 14.14** Scanning electron photomicrograph of a weld line formed during the injection molding of a polypropylene-15% EPDM blend. Surface is hexane-extracted to remove EPDM. [Reprinted with permission from, R. C. Thamm, *Rubber Chem. Technol.*, **50**, 24 (1977)].

In light of this, the crystalline morphology reported by Kantz et al. (38), Clark (39), and others can be understood fairly clearly. At the surface layer row nuclei are formed by chains aligned in the direction of flow (extensional flow) on which lamellae grow in the plane perpendicular to the filling direction. At the layer just below, row nucleation still persists, but there the lamellae are perpendicular to

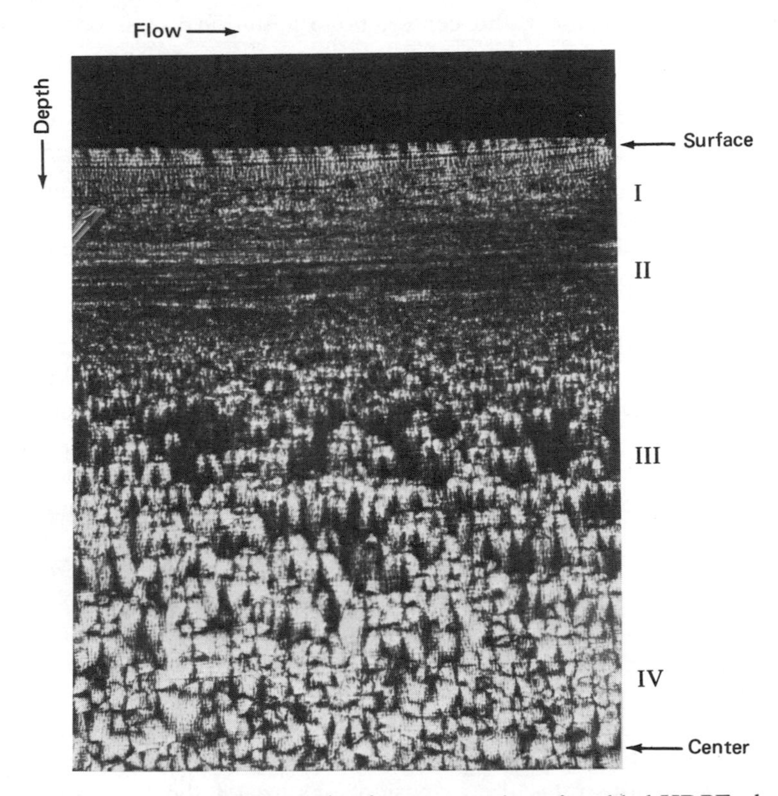


**Fig. 14.15** Birefringence microscopy study of a cross section of molded HDPE, depicting various morphological regions. (Reprinted with permission V. Tan, paper presented at the International Conference on Polymer Processing, MIT, Cambridge, Mass, August 1977.)

the mold surface, but randomly oriented with respect to the filling direction. Shear flow orientation in combination with the prevailing temperature gradients is probably responsible for this morphology. Recall that both shear and elongational flows are capable of producing chain orientation that is intense enough to create row nucleation (see Section 3.6). Spherulitic morphology, indicative of little or no orientation, is observed in the core region. Figure 14.15 shows such variation of crystalline morphology with thickness. Similar morphologies have been reported by Hobbs (40) at and behind weld lines. This is expected in light of our previous discussion.

We note in passing that if nucleating agents were added, the effect of row nuclei would be masked by profuse nucleation on the surfaces of the nucleating agents. Furthermore, the spherulitic core morphology would be grainy. The desirability of adding nucleating agents depends on the mechanical properties that are sought.

For amorphous polymers the skin will be oriented, thus ductile, whereas the core region, being unoriented, will be brittle. Furthermore, the mechanical properties will be anisotropic, since the orientation is predominant in the filling direction. This anisotropy can be overcome for cup-shaped injection molded articles. The male part of the mold can be rotated during filling, giving rise to additional orientation in the $\theta$-direction. This double orientation has been termed "helicoidal" by Cleereman (41), who suggested the process. Figure 14.16 illustrates the resulting impact strength.

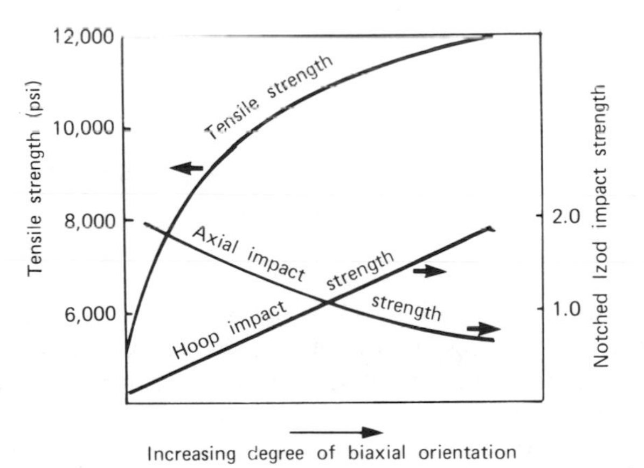


**Fig. 14.16** Mechanical properties of biaxially (helicoidally) oriented PS. [Reprinted with permission from K. J. Cleereman, *Soc. Plastics Eng. J.*, **23**, 43 (October 1967)].

In the last few years a great deal of work has been done in studying injection molding, both from mathematical simulation and structuring-morphology points of view. To be sure, there is a great deal of work left, especially in the area of utilizing the results obtained from the mold filling and cooling simulation in predicting residual stress and orientation or morphology distributions in molded articles. Internal stresses are vitally important to part properties. From the foregoing it is clear that they originate from two sources.

1. The cooling and solidification process, where the part starts solidifying at the surface. This gives rise to thermal (energy elastic) stresses that are

compressive near the surface and tensile in the core section (42). Such stresses can be calculated, neglecting stress relaxation during cooling, by calculating the dimensions of each layer at the moment of solidification which, for the case of a flat part, is dictated by Eq. 14.1-9. Such calculations have been carried out using FEM by Rigdahl (43).

2. In additon to the above-named stresses, entropy elastic stresses induced by flow orientation are also present in molded articles. Presumably one can calculate (assuming a constitutive relationship) the orientation present at the end of filling at each point of the mold and then, with the aid of Eq. 14.1-9, which provides $T(x, y, t)$, calculate the orientation stresses retained for $T < T_g$ or $T_m$. For this task to be carried out, first the exact nature of the fountain flow must be computationally elucidated, since it forms the surface layers that are most important in molded parts. Furthermore, questions about the proper constitutive equation for such complex flows and large deformations must be answered, and the effects of these on the crystallization kinetics and morphology of semicrystalline polymers must be determined. Some indirect and preliminary attempts in this problem have been recently made by Kubat and Rigdahl (44). It is our belief that much progress in this area will come in the next decade; mold design and the specification of the molding variables will be aimed at achieving a *certain* type of *structuring* of the molded part.

## 14.2 Injection Molding of Reacting Polymers

Injection molding has been used to form objects ranging in weight from a fraction of a gram to a few kilograms. As the size of the article to be molded increases, two problems arise in injection molding: (*a*) generating enough homogeneous melt in the injection molding machine and (*b*) maintaining sufficient clamping pressure to keep the mold closed during the filling and packing stages of the injection molding operation. The latter problem becomes serious when the projected area of the molded part is large and requires enormous mold presses.

The reaction injection molding (RIM) process was developed to bypass both problems above. In this process two or more low viscosity liquid streams, which when brought together become reactive, are mixed prior to being injected into large cavities. Some of the polymerization reaction (which may result in either linear, branched, or crosslinked polymers) occurs during the filling stage. The bulk of the reaction, though, takes place after filling and even after removal from the hot mold. Thus the injection pressures needed for filling molds in the RIM process are generally small. Furthermore, homogenization of the reacting fluids is not difficult to achieve since the fluid viscosities are of the order of one poise and accurate metering can be easily achieved; rather simple mixing heads can be used.* As the polymerization proceeds after filling, heat is generated by the reaction which increases the specific volume of the polymer system. On the other hand, a specific volume decrease of the order of 10% accompanies polymerization. Thus packing flows would be necessary, which would require high pressures since the viscosity of

* It should be noted, however, that the mechanical details of such mixing heads are important to product reproducibility.

the reacting system increases with increasing molecular weight or degree of cross-linking. To eliminate the necessity of packing, a *small* amount of a foaming agent is introduced in one of the streams. The resulting foaming action ensures that the RIM article will conform to the shape of the cavity. In this way very large and complex shaped cavities can be formed using a rather small injection pressure, of the order of 1–10 $MN/m^2$, and small clamping presses with inexpensive molds.

Obviously the key to the success of the RIM process has to be economically fast rates of polymerization. Otherwise the process is not competitive to injection molding, but comparable to casting. It follows then that not all polymer systems are good candidates for the RIM process. The most commonly used polymer system, since the commercial inception of the process in the early 1970s, is that of linear and crosslinked polyurethanes, where di- or trialcohols and di- or triisocyanates are the two main reacting species. Fiber filled polyesters have also been used. With the use of proper prepolymers and with the better understanding of the RIM process, which will come with time, it is reasonable to believe that chain addition polymers will be used. Reactive processing is still in its infancy.

From a process simulation point of view there are two main problems: (*a*) nonisothermal and transient flow with chemical reaction, prevalent during the filling stage of the process, and (*b*) conductive heat transfer with heat generation due to the polymerization reaction. We discuss these two problems next, for the case of a linear polyurethene being molded in a long, rectangular, thin mold that is fed by a "gate" occupying the entire $x = 0$ surface in Fig. 14.4.

### *Mold Filling by a Reactive Liquid*

Consider a very long, very wide, and thin mold being fed by a constant temperature mixture of AA, BB molecules. Both types are bifunctional and the feed has a molecular weight $M_0$. The polymerization, assumed to be reversible, proceeds by the reaction of A ends with B ends and follows idealized step polymerization (condensation) kinetics without the generation of a small molecule (45). Specifically, we have

$$\mathrm{AA} + \mathrm{BB} \underset{k_r}{\overset{k_f}{\rightleftharpoons}} \mathrm{AA-BB} \quad [-\Delta H] \tag{14.2-1}$$

Setting $(\mathrm{AA-BB})_x = M_x$, we can write the general reversible condensation reaction

$$\mathrm{M}_x + \mathrm{M}_y \underset{k_r}{\overset{k_f}{\rightleftharpoons}} \mathrm{M}_{x+y} \quad [-\Delta H] \tag{14.2-2}$$

Equation 14.2-1 is second order in A or B ends and first order in A–B links. The rate of change of the concentrations $C_A$, $C_B$, and $C_{AB}$ are

$$\left.\begin{aligned} \frac{dC_A}{dt} &= -k_f C_A C_B + k_r C_{AB} \\ \frac{dC_B}{dt} &= -k_f C_A C_B + k_r C_{AB} \\ \frac{dC_{AB}}{dt} &= k_f C_A C_B - k_r C_{AB} \end{aligned}\right\} \tag{14.2-3}$$

That is,

$$\frac{dC_A}{dt} = \frac{dC_B}{dt} = -\frac{dC_{AB}}{dt}$$

Thus for

$$C_{A_0} = C_{B_0}, \quad \text{and} \quad C_{AB_0} = 0$$

$$C_A = C_B \qquad \text{and} \qquad C_{AB} = C_{A_0} - C_A \tag{14.2-4}$$

and

$$\frac{dC_A}{dt} = -k_f C_A^2 + k_f[C_{A_0} - C_A] \tag{14.2-5}$$

Defining the reaction conversion $\phi$ as

$$C_A = C_{A_0}(1-\phi) \tag{14.2-6}$$

Eq. 14.2-5 becomes

$$\frac{d\phi}{dt} = k_f C_{A_0}(1-\phi)^2 - k_r \phi \tag{14.2-7}$$

For a flowing system with a velocity $v_x = v_x(x, y, t)$, undergoing the chemical reaction above, $d\phi/dt = D\phi/Dt$. Thus Eq. 14.2-7 becomes

$$\frac{\partial \phi}{\partial t} + v_x \frac{\partial \phi}{\partial x} = k_f C_{A_0}(1-\phi)^2 - k_r \phi \tag{14.2-8}$$

The boundary and initial conditions are

$$\begin{aligned} \phi(0, y, t) &= 0 \\ \phi(x, y, 0) &= 0 \end{aligned} \tag{14.2-9}$$

Both reaction rate constants are assumed to obey the Arrhenius temperature dependence

$$\left.\begin{aligned} k_f &= k_{f_0} \exp\left(\frac{-E_{fr}}{RT}\right) \\ k_r &= k_{r_0} \exp\left(\frac{-E_{rr}}{RT}\right) \end{aligned}\right\} \tag{14.2-10}$$

The two reaction activation energies are related to the heat of reaction as follows (46);

$$E_{rr} - E_{fr} = -\Delta H \tag{14.2-11}$$

The reaction has also a characteristic temperature, where $k_f = k_r$. For this reacting system the number and weight average molecular weights are (45)

$$\bar{M}_n = \frac{M_0}{1-\phi} \tag{14.2-12}$$

$$M_w = M_0\left[\frac{1+\phi}{1-\phi}\right] \tag{14.2-13}$$

To solve the filling flow and heat transfer problem with the reacting system above, we need to specify the $x$-direction momentum and energy balances. Following Domine and Gogos (47–48), the $x$-momentum equation during filling is

$$\rho \frac{\partial v_x}{\partial t} = -\frac{\partial P}{\partial x} + \eta \frac{\partial^2 v_x}{\partial y^2} + \frac{\partial \eta}{\partial y}\frac{\partial v_x}{\partial y} \tag{14.2-14}$$

The boundary conditions for Eq. 14.2-14 are

$$\left.\begin{aligned} &\frac{\partial v_x}{\partial y}(x, 0, t) = 0 \\ &v_x\left(x, \frac{H}{2}, t\right) = 0 \\ &\frac{\partial \eta}{\partial y}(x, 0, t) = 0 \\ &\frac{\partial v_x}{\partial t}(0, y, t) = 0 \\ &\frac{\partial P}{\partial x}(0, y, t) = 0 \end{aligned}\right\} \tag{14.2-15}$$

The viscosity is given by the Carreau equation (Eq. 6.5-8), which for melts is

$$\eta = \frac{\eta_0}{[1 + (\lambda \dot{\gamma})^2]^{(1-n)/2}} \tag{14.2-16}$$

where

$$\lambda = \frac{\lambda_0 \eta_0 (\bar{M}_w)^{0.75}}{\rho T} \tag{14.2-17}$$

and $\lambda_0$ is a curve fitting parameter. The zero shear viscosity is primarily a function of the weight average molecular weight

$$\eta_0 = \alpha_i (\bar{M}_w)^{\beta_i} \exp\left[\frac{\Delta E}{RT}\right] \tag{14.2-18}$$

$\alpha_i$ are material parameters, and

$$\Delta E = \delta_i \exp\left(\frac{-\varepsilon_i}{\bar{M}_n}\right) \tag{14.2-19}$$

and $i = 1, 2$, and denote the two regions of no entanglement and entanglement, respectively. Specifically for $i = 1$, $\bar{M}_w < M_e$ (45)

$$\left.\begin{aligned} &\beta_1 = 1 \\ &\delta_1 = \Delta E_0 \exp\left(\frac{\varepsilon_1}{M_0}\right) \\ &\varepsilon_1 = M_0\left(\frac{M_e + M_0}{M_e - M_0}\right) \ln\left(\frac{\Delta E}{\Delta E_0}\right) \end{aligned}\right\} \tag{14.2-20}$$

For $i=2$, $\bar{M}_w > M_e$

$$\left.\begin{aligned} \beta_2 &= 3.4 \\ \delta_2 &= \Delta E \\ \varepsilon_2 &= 0 \end{aligned}\right\} \tag{14.2-21}$$

The energy equation for the filling stage is for constant density and thermal conductivity

$$\rho C_p\left(\frac{\partial T}{\partial t} + v_x \frac{\partial T}{\partial x}\right) = k\frac{\partial^2 T}{\partial y^2} + \eta\left(\frac{\partial v_x}{\partial y}\right)^2 + \frac{\partial \phi}{\partial t}(-\Delta H) \tag{14.2-22}$$

The boundary conditions for Eq. 14.2-22 are

$$\left.\begin{aligned} \frac{\partial T}{\partial y}(x, 0, t) &= 0 \\ \frac{\partial T}{\partial y}\left(x, \frac{H}{2}, t\right) &= -\frac{h}{k}\left[T\left(x, \frac{H}{2}, t\right) - T_w\right] \\ T(0, y, t) &= T_0 \\ \frac{\partial T}{\partial t}(0, y, t) &= 0 \end{aligned}\right\} \tag{14.2-23}$$

where $T_w$ is the mold temperature and $T_0$ is the temperature of the feed, both time independent.

Section 14.1 indicated that the fountain flow in the front region could be neglected in the simulation of the mold filling process and calculations of the time dependent front position and filling pressure. In RIM, unfortunately, this is no longer true, because the fluid viscosity depends on the molecular weight, and to know the molecular weight of any fluid element at any time, we must know where the fluid has been since entering the mold. Domine (47) uses a "demon" that computationally moves material from the central region of the front toward the front mold wall in a manner similar to that of the fountain flow.

The species balance relation (Eq. 14.2-8) is transformed to a difference equation using the forward difference on the time derivative and the backward difference on the space derivative. The finite difference form of the $x$-momentum equation 14.2-14 is obtained by using the forward difference on all derivatives and is solved by the Crank–Nickolson method. The same is true for the energy equation 14.2-22.

The results of the simulation show that for reaction rates that are common in the conventional RIM process, the chemical reaction cannot be ignored during the filling step. In other words, RIM *is not casting.* Appreciable conversion and nonisothermicity can be obtained during filling. Figure 14.17 gives the conversion and temperature fields at the moment of fill. Both temperature and conversion increase with increasing flow direction distance; this is simply the result of larger residence times. It is worth noting that the fountain flow taking the reactive fluid from the center and depositing it on the wall makes both profiles flatter. As the feed reaction constant and the fill time increases, there is more chance for chemical reaction

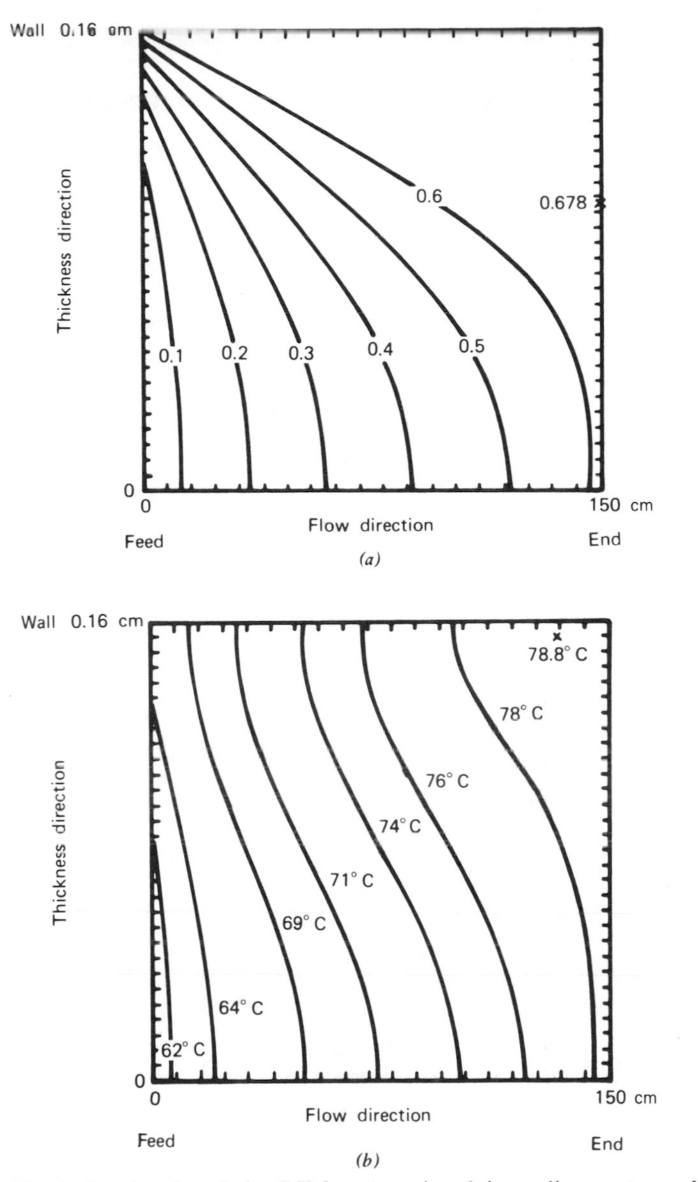


**Fig. 14.17** Simulation results of the RIM process involving a linear step polymerization: $T_0 = T_w = 60°C$, $k_f = 0.5$ l/mole · s, $t_{fill} = 2.4$ s. (*a*) Conversion contours at the time of fill. (*b*) Temperature contours at the time of fill. [Reprinted with permission from J. D. Domine and C. G. Gogos, "Computer Simulations of Injection Molding of a Reactive Linear Condensation Polymer", Soc. Plastics Eng., 34th Annual Technical Conference, Atlantic City, N.J., 1976].

during filling. Figure 14.18 presents the results of such increases, for the case where the product of $(k_f \times t_{fill})$ is 4 times that of the preceding case. Conversion levels exceeding 90% are obtained, giving rise to a rather thick "reacted skin," as Fig. 14.18*b* indicates by the line for velocity of 10 cm/s. Furthermore, there is a "pinching" effect on the flow midway in the flow direction, which results in a nipplelike velocity profile with very high velocity gradients and VED levels. As a matter of fact, for the system studied there appears to be an upper limit in feed

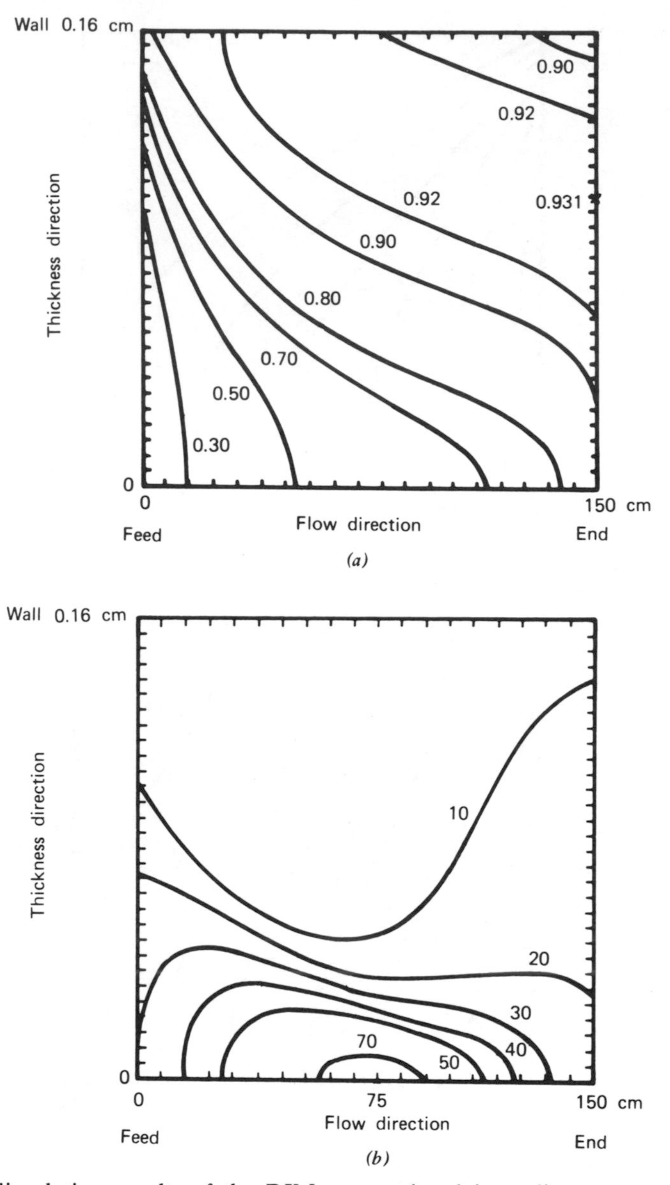


**Fig. 14.18** Simulation results of the RIM process involving a linear step polymerization. $T_0 = T_w = 60°C$, $k_f = 1.0$ l/mole · s, $t_{fill} = 4.8$ s (*a*) Conversion contours at the time of fill. (*b*) Temperature contours at the time of fill. (Reprinted with permission from J. D. Domine and C. G. Gogos, "Computer Simulations of Injection Molding of Reactive Linear Condensation Polymer", Soc. Plastics Eng., 34th Annual Technical Conference, Atlantic City, N.J., 1976).

condition reaction rate constant because of local thermal problems that arise by way of high chemical and viscous heat generation terms. This problem may be reduced by increasing the thermal conductivity of the reacting system (by incorporating conductive additives) and by making the system more pseudoplastic (perhaps by the addition of dissolved elastomer). At any rate the effects of both the material and process variables must be studied for the understanding of the filling step of the RIM process. This has been done by Domine (47, 48).

At the end of the filling stage, only heat transfer with chemical reaction occurs, which can be described by the following species and energy balance equations:

$$\frac{\partial \phi}{\partial t} = k_f C_{A_0}(1-\phi)^2 - k_r \phi \tag{14.2-24}$$

$$\rho C_p \frac{\partial T}{\partial t} = k\frac{\partial^2 T}{\partial y^2} + k\frac{\partial^2 T}{\partial x^2} + \frac{\partial \phi}{\partial t}(-\Delta H) \tag{14.2-25}$$

The axial conduction, which was neglected in Eq. 14.2-22 as being smaller than axial convection, is included now. The two equations above hold for $t \geq t_{\text{fill}}$. The temperature boundary conditions for an adiabatic feed surface are

$$\left.\begin{aligned} &\frac{\partial T}{\partial x}(0, y, t) = 0 \\ &\frac{\partial T}{\partial x}(l, y, t) = -\frac{h}{k}[T(l, y, t) - T_w] \\ &\frac{\partial T}{\partial y}\left(x, \frac{H}{2}, t\right) = -\frac{h}{k}\left[T\left(x, \frac{H}{2}, t\right) - T_w\right] \\ &\frac{\partial T}{\partial y}(x, 0, t) = 0 \end{aligned}\right\} \tag{14.2-26}$$

Since the energy balance involves second order derivatives in both the $x$- and $y$-directions, the alternating direction implicit (ADI) method is used (49). This method requires three time levels of temperature and involves the solution of the equation twice, once in each direction.

The "postfill cure" and heat transfer continues until the thickness–average tensile modulus is high enough at every $x$ position for the part to be removed. The tensile modulus is dependent on the number average molecular weight (47). With this procedure the *demold time* is obtained. The demold time for the case corresponding to Fig. 14.17 is 62.4 s, compared to 12 s for the case corresponding to Fig. 14.18. The low demold time in the latter case is the result of the thick reacted skin formed during the filling process, as mentioned previously.

Analyses such as the foregoing are necessary to understand the interrelations among the chemical, process, and rheological variables in RIM.

### Thermoplastic Foam Injection Molding

In thermoplastic foam injection molding, which is in principle a RIM-like process, a gas is introduced into the molten polymer in the injection molding machine (50), or a gas producing compound (usually in fine powder form) is mixed with the polymer pellets or powder prior to processing (51). In either case, upon injection into the mold, the gas can come out of solution because of the prevailing low pressures, especially as the advancing front is approached. The product formed can be one of a very dense "skin" and a foamed core of density 20–50% that of the unfoamed polymer. The surface contains only a few voids because of the phenomenon of skin formation (i.e., fast cooling; Fig. 14.9). Nevertheless the surface is not void free because of the low pressures during the fountain flow phenomenon. Typical density profiles of thermoplastic structural foam articles indicate that the solid surface skin is usually about one-quarter of the half-thickness. The density decreases *rapidly* to a constant low value in the core region.

These facts justify the statement made previously that this process is similar to RIM, since physicochemical reactions occur concurrently with mold filling. On the other hand, in contrast to RIM, gas generation can occur *well before* the melt reaches the mold, necessitating the consideration of the chemical reaction during the melting, melt storage, and pumping steps of the process (i.e., inside the injection molding machine). Throne and Griskey (52, 53) have written comprehensive reviews on the subject. In the brief discussion, below, we follow their work.

The rate of gas generation from gas producing additives can often be expressed in terms of first order kinetics (e.g, azodicarbonamide). Since the kinetic constant is temperature dependent, the amount of gas generated depends on the *entire* particle temperature history. On the other hand, the gas generated can be dissolved into the molten polymer under high pressures. For relatively low gas concentrations, Durrill and Griskey (54) have found that Henry's law applies for a number of polymer melt–gas pairs. Henry's law constant increases exponentially with temperature.

Bubbles can grow from dissolved gases in melts when the pressure is lowered. The solubility of the gas in the melt, its diffusivity and the pressure, of course, govern the rate of growth of such bubbles. The process is complicated and is not well understood. Gent and Tomkins (55) have done extensive work with elastomers, as have Street et al., (56, 57). Recently Prud'homme and Bird (58) have considered the process of dilation of a gas bubble in a polymer melt, finding that the dilatational viscosity plays an important role in the rate of bubble growth. Experimentally it is usually learned that an effective incubation period precedes bubble growth (57). This incubation period decreases with increasing size of the initial bubble, which is controlled in practice by the size of the nucleating agents, such as mineral fillers, which are usually added at 1–5% concentration. It follows that the dispersive mixing of the nucleating and foaming powders in the polymer particulate solids, as well as the dispersive and distributive mixing in the molten mixture are important to the final size and size distribution of the cells. Finely dispersed nucleating and foaming agents result in a reduced level of surface roughness; the nucleating site is small and the incubation period large, thus the surface that is formed by fountain-type flow contains smaller cells. It also follows that the longer the distance from the gate, the coarser the surface and the thinner the solid skin. In a process patented by the Farrel company the filling occurs very fast and at high pressure (to avoid foaming altogether), then for a very short period the mold is opened to allow for a pressure drop, which initiates foaming; this is an attempt to improve surface quality.

The process above has not been simulated as a coupled mass, momentum, and heat transfer problem. The simulation is feasible and is similar to that in RIM, but the solution would be hindered by ill-studied and concentration dependent transport and thermodynamic properties, and by the fact that the coupled transport processes occur in almost the entire injection molding machine.

## 14.3 Compression Molding

In the compression molding process a thermoplastic or partially polymerized thermosetting polymer, usually in a preheated, preformed shape vaguely corresponding to that of the cavity, is placed in the heated cavity; the mold begins to close, and pressure is applied to the preform, forcing it to further heat up close to the mold temperature and flow to fill the mold cavity. While under pressure, the polymer undergoes complete polymerization (crosslinking), aided by the heat transferred from the mold. After that the mold is opened, the part is ejected, and the cycle starts again. Only occasionally is the process used for thermoplastics forming (PVC phonograph records), its primary use being either with elastomeric or hard thermosetting polymers. This process wastes very little material (no runners and sprue) and can produce large parts. It is difficult to produce parts with very close tolerances, however, because the final size of the compression molded article—that is, the degree to which the mold closes—depends on the *exact* amount of the preform. Furthermore, the process does not easily lend itself to molding of intricate parts with deep undercuts.

Figure 14.19 represents the various stages of the compression molding cycle from the point of view of the plunger force needed to close the mold at a constant

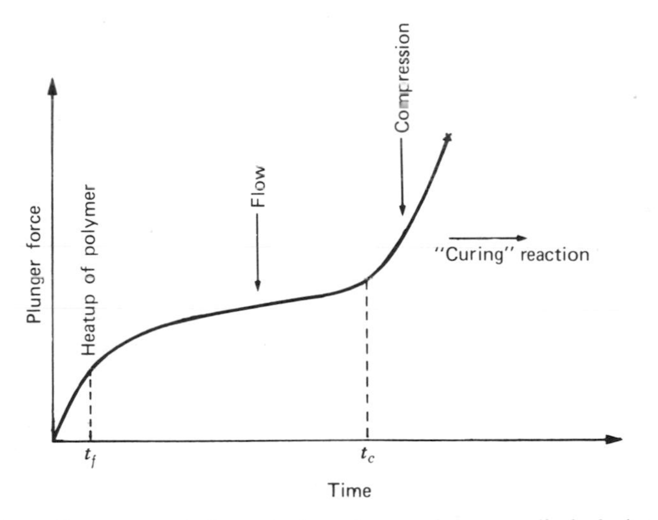


**Fig. 14.19** Schematic representation of the plunger force applied during compression molding to move the plunger at a constant speed. The various stages of the process are depicted.

rate. In the first region $t \le t_f$ the force increases rapidly as the preform is squeezed and heated. At $t_f$ the polymer is presumably in the molten state and, as such, is forced to flow in the cavity and fill it. Filling terminates at $t_c$ when compression of the polymer melt takes place, to compensate for the volume contraction that results from the polymerization reaction. The bulk of the chemical reaction occurs after $t_c$. Below we comment on each of the steps of the compression molding process.

During the preform heating part of the cycle, the main problems to be considered are heat transfer and flow (or elastic deformation) of the compressed

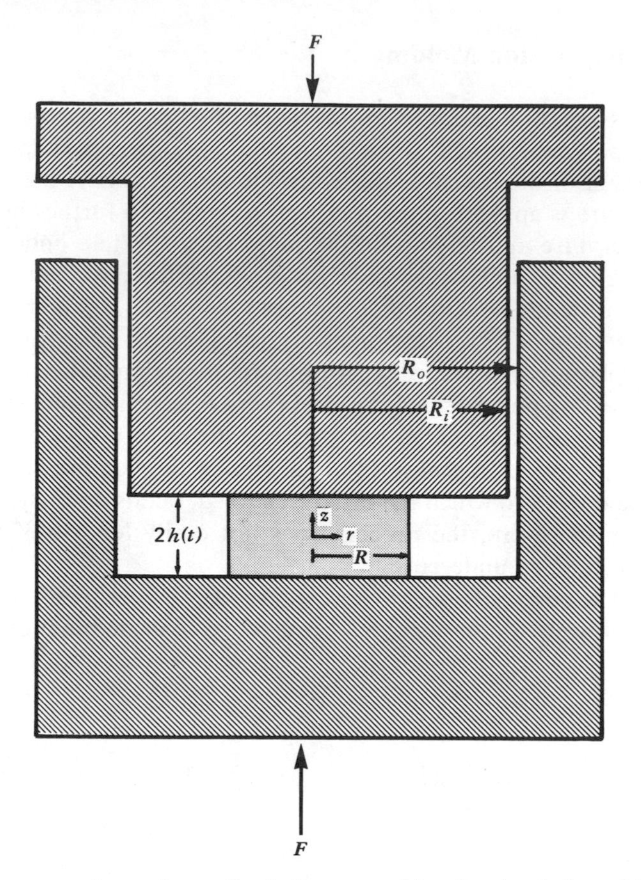


**Fig. 14.20** The geometry and coordinate frame used for the simulation of the compression molding process of a cup-shaped cavity.

particulate matter. Referring to Fig. 14.20, the heat transfer problem can be described with the following form of the energy equation

$$\rho C_p \frac{\partial T}{\partial t} = k \frac{\partial^2 T}{\partial z} \tag{14.3-1}$$

where the following assumptions have been made: the convective and dissipative effects associated with $v_r$ are negligible, as is the radial conductive heat transfer, and the thermophysical properties are taken to be constant. The temperature of the preform is constant at the start of the process, and the boundary conditions are

$$\left. \begin{aligned} \frac{\partial T}{\partial z}(0, t) &= 0 \\ \frac{\partial T}{\partial z}(h(t), t) &= -U[T(h(t), t) - T_w] \end{aligned} \right\} \tag{14.3-2}$$

where $U$ is the ratio of the heat transfer coefficient to the thermal conductivity. Furthermore, because of the plunger motion

$$h(t) = h_0 - \dot{h}t \tag{14.3-3}$$

The heat transfer problem above can be solved in a fashion similar to the one used in Section 9.4, to yield $T(z, t)$. In principle, once the temperature field is known in the preform at any time before $t_f$, the plunger force can be calculated. The preform can be taken as a solid that slips at the mold surface and has a temperature dependent compressive modulus. At any time $t < t_f$ each layer of the preform will deform by an amount such that (a) the force on *every* layer of thickness $\Delta z$ is the same (and equal to the unknown quantity), and (b) the sum of the compressive deformations of all the layers equals the deformation imposed on the preform by the plunger at the given time. The force can also be estimated by assuming that the preform is a viscous liquid with a temperature dependent viscosity undergoing an overall constant rate squeezing flow. The problem can then be solved in a manner similar to that in Section 10.9. An average temperature value can be used, or each layer can be considered to flow at a rate such that (a) the force on every layer is independent of $z$, and (b) the sum of all the squeezing rates is equal to the one applied by the plunger $\dot{h}$.

Assuming that $T(z, t_f) = T_w$, we can deal with the flow problem alone for $t_f \le t \le t_c$. Referring to Fig. 14.20 again, as long as the preform radius is less than $R_o$, we can treat the problem as an isothermal radial flow of an incompressible fluid flowing between two disks that approach each other at a constant rate $\dot{h}$ (see Section 10.9). Summarizing the results for the velocity field, pressure distribution, and plunger force needed to squeeze an isothermal power law fluid at a constant slow or moderate squeeze rate, we have

$$v_r(z, r, t) = \frac{h^{1+s}}{1+s}\left(-\frac{1}{m}\frac{\partial P}{\partial r}\right)^s\left[1-\left(\frac{z}{h}\right)^{1+s}\right] \tag{14.3-4}$$

where the pressure gradient can be obtained from

$$P = P_a + \frac{m(2+s)^n}{2^n(n+1)}\frac{(-\dot{h})R^{n+1}}{h^{2n+1}}\left[1-\left(\frac{r}{R}\right)^{n+1}\right] \tag{14.3-5}$$

The plunger force $F_N$ can be calculated from the foregoing.

$$F_N = \frac{m\pi(2+s)^n}{2^n(n+3)}\frac{(-\dot{h})^n R^{n+3}}{h^{2n+1}} \tag{14.3-6}$$

Since the fluid is assumed to be incompressible and nonreacting at this stage, its volume is constant, implying that

$$h(t)R^2(t) = C_1 \tag{14.3-7}$$

Thus the Scott equation 14.3-6 becomes

$$F_N = \frac{m\pi(2+s)^n}{2^n(n+3)}\frac{(-\dot{h})^n}{C_1^{2n+1}}R^{5(n+1)} \tag{14.3-8}$$

When the radius of the flowing preform reaches the value of $R_o$ (Fig. 14.20), the fluid is forced to flow in the annular space $R_o - R_i$. For a constant squeeze rate, the rate of increase of the axial annular distance occupied by the melt is

$$-\pi R_i^2\dot{h}(t) = \pi(R_o^2 - R_i^2)\dot{l} \tag{14.3-9}$$

For small annular spacings $\Delta R \ll \bar{R}$, Eq. 14.3-9 becomes

$$\dot{l} = \dot{h}\left[\frac{\bar{R}}{2\Delta R}\right] \tag{14.3-10}$$

Once annular flow occurs, there is an additional force term acting on the plunger. The pressure at $r = R_i$ is not atmospheric but that which is needed to support the flow in the annulus. To calculate this pressure, we first turn to the volumetric flow rate in the annular region, which is

$$Q = \pi(R_o^2 - R_i^2)\dot{l} \tag{14.3-11}$$

For a thin annulus, $\Delta R \ll \bar{R}$, the plunger travel rate $\dot{h}$ is very small compared to $\dot{l}$ (according to Eq. 14.3-10), and the annular flow can be considered to be a pressure, not a combined pressure and drag flow. For a thin annulus Eq. 14.3-11 reduces to

$$Q = 2\pi\bar{R}\,\Delta R\dot{l} \tag{14.3-12}$$

For isothermal annular pressure flow of a incompressible power law fluid, Frederickson and Bird (59) have calculated the following relationship between $Q$ and $\Delta P$

$$Q = \frac{\pi R_0^3}{s+2}\left(\frac{R_0\Delta P}{2ml}\right)^s (1-\kappa)^{s+2} F_1(n, \kappa) \tag{14.3-13}$$

where $1/\beta = \kappa = R_i/R_o$. Figure 13.23 plots $F(n, \kappa)$. For a thin annulus, Eq. 14.3-13 reduces (see Table 13.3) to

$$Q = \frac{\pi R_0^3}{s+2}\left(\frac{R_0\Delta P}{2ml}\right)^s (1-\kappa)^{s+2} \tag{14.3-14}$$

Therefore, to calculate the added pressure at $r = R_i$, we use for a wide annulus Eqs. 14.3-11 and 14.3-13, or for a thin annulus Eqs. 14.3-12 and 14.3-14. For the latter case, we write

$$P(R) - P_{atm} = \frac{2ml}{R_0^{3n+1}} \frac{[2(s+2)\bar{R}\Delta R\dot{l}]^n}{(1-\kappa)^{1+2n}} \tag{14.3-15}$$

When this is multiplied by the plunger area $\pi R^2$ and added to the right-hand side of Eq. 14.3-6, the plunger force is obtained for the case where annular flow is present in a cup-shaped cavity compression molding. Similar expressions can be obtained for the entire flow stage $t_f < t < t_c$, during the compression molding of other shapes making use of the quasi-steady state and, when there is a need, the lubrication approximation.

The reaction stage of the compression molding process can be described by Eqs. 14.2-24 and 14.2-25, employed in the simulation of the postfilling reaction stage in the RIM process. Of course Eq. 14.2-24 is applicable only to a linear and reversible step polymerization. Furthermore, we have assumed that the melt is at a uniform temperature at the beginning of reaction. Therefore in Eq. 14.2-25 conduction should occur only in the thickness direction. Broyer and Macosko (60) have solved numerically the problem of heat transfer with a crosslinking polymerization reaction that is more representative of the compression molding process. For a thin rectangular mold of half-thickness $h$ and temperature $T_w$, as well as $n$th order kinetics and constant thermophysical properties, the governing equations in dimensionless form are

$$-\frac{dc_A^*}{dt^*} = k^* C_A^{*n} \exp B\left[\frac{\Delta T'_{ad} T^*}{\Delta T'_{ad} T^* + 1}\right] \tag{14.3-16}$$

and

$$\frac{\partial T^*}{\partial t^*} = \frac{\partial^2 T^*}{\partial y^{*2}} + k^* C_A^{*n} \exp B\left[\frac{\Delta T'_{ad} T^*}{\Delta T'_{ad} T^* + 1}\right] \tag{14.3-17}$$

where $C_A^* = C_A/C_{A_0}$, $t^* = \alpha t/h^2$, $y^* = y/h$, $h$ is the half-thickness of the polymerizing slab, $k^* = C_{A_0}^{n-1} h^2 A\, e^{-B}/\alpha$, $B = E/RT_0$, $E$ is the reaction activation energy and $A$ is the frequency factor, $T^* = (T \quad T_0)/(T_{ad} - T_0)$, and the adiabatic temperature rise is

$$\Delta T'_{ad} = \frac{T_{ad} - T_0}{T_0} = \frac{\Delta H C_{A_0}}{\rho C_p T_0}.$$

$\Delta H$ is the heat of reaction. $C_A$ is the concentration of the A functional group in the trifunctional group–bifunctional group step reaction

$$A_3 + B_2 \rightarrow \text{Crosslinked polymer}$$

The system of Eqs. 14.3-16 and 14.3-17 can be solved for the adiabatic, isothermal, or constant wall flux cases, using the Crank–Nickolson method. The thermomechanical and reaction data for such systems were evaluated by Lipshitz, Macosko, and Mussatti (61) at 45°C for a polyester triol and a chain extended 1,6-hexamethylene diisocyanate (HDI) with dibutyltin as a catalyst. Figure 14.21 gives the temperature profiles for the isothermal wall case. Because of the high heat of polyurethane formation and the low conductivity of the system, the center of the slab shows nearly an adiabatic temperature rise. The peaks come closer to the adiabatic temperature also when $k^*$ is increased.

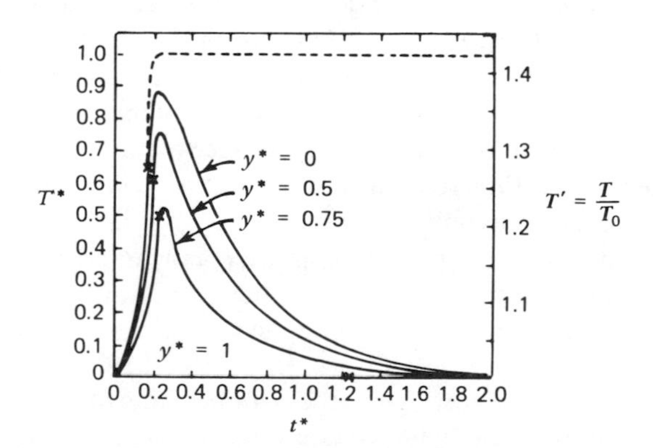


**Fig. 14.21** Temperature distribution in a reacting polyurethane slab; isothermal wall simulations. Dotted line denotes the adiabatic temperature rise and x indicate gel points. $k^* = 1$, $n = 1$, $\Delta T'_{ad} = 0.423$, and $B = 18.7$, $\Phi_{gel} = 0.707$. [Reprinted with permission from E. Broyer and C. W. Macosko, *Am. Inst. Chem. Eng. J.* **22**, 268 (1976).]

The extent of reaction $\phi$ can be related to the weight–average molecular weight for the triol-HDI-polyurethane system under equal stoichiometry (62) as follows:

$$\bar{M}_w = \frac{\frac{2}{3}(1+\phi^2)M^2{}_{A_3} + (1+2\phi^2)M^2_{B_2} + 4\phi M_{A_3}M_{B_2}}{(\frac{2}{3}M_{A_3} + M_{B_2})(1-2\phi^2)} \qquad (14.3\text{-}18)$$

The "gel point" is defined to be the condition when $\bar{M}_w$ goes to infinity, which occurs at $\phi = (\frac{1}{2})^{1/2}$. It follows from Fig. 14.21 that the center of the slab will gel faster than the skin. A more uniform extent of reaction can be achieved by maintaining the mold walls at a higher temperature, assuming that this does not result in excessive surface reaction and interference with the filling and the compression processes. Knowing the temperature, the conversion, and the molecular weight distribution as a function of thickness and reaction time is essential in determining the required compression mold cycle or the time and temperature in the postcuring step, if it is necessary.

## 14.4 Casting

Casting is a forming process that has been used in metal forming for centuries. It is also a process that was introduced very early in shaping polymers. In it a liquid prepolymer system, which either crosslinks or further polymerizes to form larger linear or branched chains, is poured into a mold. The fluid fills the mold under the influence of gravity. The initial viscosity of the system must, therefore, be small. The polymer is removed after the reaction has taken place in the mold cavity. Large parts with thick cross sections can be produced economically by casting, without tieing up expensive molds. The process, on the other hand, is limited to forming relatively simple shapes, and it is a low production volume operation. In the field of thermoplastics it is widely used in forming acrylic (PMMA) flat sheets. In this process a viscous prepolymer solution fills completely the space formed by a highly polished metal sheet and a large "window frame" flexible gasket; it is covered by another metal sheet whose top surface serves as the bottom surface for another casting cavity. In this fashion a vertical battery of cavities is formed. The flexible gasket is used so that the mold can *shrink* during the polymerization process, which results in a volume contraction. In this way voids can be eliminated. In the ordinary casting process void formation can be a major problem unless one of the surfaces is allowed to contract. Commonly the top surface, exposed to the atmosphere, does not polymerize rapidly, thus remains flexible to "cave in" and avoids the creation of negative pressures in the bulk of the reacting system, which would cause void formation. The deformed top surface is then machined away. This problem, of course, exists because casting is an atmospheric pressure process and not one, like compression molding, where the reacting system is in a compressed state while a prepolymer.

The simulation of the casting process can be achieved by the following general system of species and energy balance equations for a mold which is "infinite" in the $z$-direction is, $H$ thick and $W$ wide

$$-\frac{\partial C_A}{\partial t} = k_f C_A^n \qquad (14.4\text{-}1)$$

where $C_A = C_A(x, y, t)$, due to the temperature dependence of $k_{fs}$ and

$$\rho C_p \frac{\partial T}{\partial t} = k \frac{\partial^2 T}{\partial x^2} + k \frac{\partial^2 T}{\partial y^2} - \frac{\partial C_A}{\partial t}(-\Delta H) \tag{14.4-2}$$

The boundary and initial conditions for $T(x, y, t)$ are as follows:

$$\left.\begin{aligned} &\left(\frac{\partial T}{\partial x}\right)(0, y, t) = 0 \\ &\left(\frac{\partial T}{\partial y}\right)(x, 0, t) = 0 \\ &\frac{\partial T}{\partial x}\left(\frac{W}{2}, y, t\right) = -\frac{h}{k}\left[T\left(\frac{W}{2}, y, t\right) - T_w\right] \\ &\frac{\partial T}{\partial y}\left(x, \frac{H}{2}, t\right) = \frac{h}{k}\left[T\left(x, \frac{H}{2}, t\right) - T_w\right] \\ &T(x, y, 0) = T_0 \end{aligned}\right\} \tag{14.4-3}$$

Equation 14.4-2 has two conduction terms in the two cross-sectional directions because usually cast parts are not narrow. Equations 14.4-1 to 14.4-3 can yield the conversion and temperature fields at any time during the reaction, once the reaction kinetics and thermodynamics have been specified. In this way estimates of the casting time required for parts with given properties can be made. As mentioned in the previous section, the degree of conversion can be related to a molecular weight average that in turn, together with the temperature field, leads to estimates of part properties such as the tensile modulus and hardness (47). From the brief discussion above it is clear that casting can be considered to be a special case of the RIM or compression molding processes and can be described in the same fashion as the postfill reaction step in either of these processes.

## REFERENCES


1. R. C. Donovan, D. E. Thomas, and L. D. Leversen, "An Experimental Study of Plasticating in a Reciprocating-Screw Injection Molding Machine," *Polym. Eng. Sci.*, **11**, 353 (1971).
2. S. D. Lipshitz, R. Lavie, and Z. Tadmor, "A Melting Model for Reciprocating Screw Injection Molding Machines," *Polym. Eng. Sci.*, **14**, 553 (1974).
3. G. B. Thayer, J. W. Mighton, R. B. Dahl and C. E. Beyer, "Injection Molding," in *Processing of Thermoplastic Materials*, E. C. Bernhardt, ed., Reinhold, New York, 1959, Chapter 5.
4. I. I. Rubin, *Injection Molding—Theory and Practice*, Wiley-Interscience, New York, 1972.
5. C. F. Huang, "Numerical Simulations of the Filling Process with a Cup-Shaped Mold," M.S. thesis, Department of Chemical Engineering, Stevens Institute of Technology, Hoboken, N.J., 1974.
6. W. L. Krueger and Z. Tadmor, "Injection Molding Into Rectangular Cavity Containing

Inserts," Society of Plastics Engineers, 36th Annual Technical Conference, Washington, D.C., 1978, pp. 87–91.
7. F. C. Caras, "Spiral Mold for Thermosets," *Mod. Plast.*, **41**, 140 (1963).
8. G. D. Gilmore and R. S. Spencer, "Role of Pressure, Temperature and Time in the Injection Molding Process," *Mod. Plast.*, 143 (April 1950); "Residual Strains in Injection Molded Polystyrene," *Mod. Plast.*, 97 (December 1950).
9. R. S. Spencer and G. D. Gilmore, "Some Flow Phenomena in the Injection Molding of PS," *J. Colloid Sci.*, **6**, 118 (1951).
10. C. E. Beyer and R. S. Spencer, "Rheology in Molding," in *Rheology*, Vol. 3, F. R. Eirich, ed., Academic Press, New York, 1960.
11. R. L. Ballman, T. Shusman, and H. L. Toor, "Injection Molding—Flow into a Cold Cavity," *Ind. Eng. Chem.*, **51**, 847 (1959).
12. R. L. Ballman, T. Shusman, and H. L. Toor, "Injection Molding: A Rheological Interpretation," *Mod. Plast.*, **37**, 105 (1959); *ibid.*, 115.
13. G. B. Jackson and R. L. Ballman, "The Effect of Orientation on the Physical Properties of Injection Moldings," *Soc. Plast. Eng. J.*, **16**, 1147–1152 (1960).
14. T. Aoba and H. Odaira, *Proceedings of the 14th Japanese Congress on Materials Research, Kyoto*, Japan, 1970, p. 124.
15. M. R. Kamal and S. Kenig, "The Injection Molding of Thermoplastics. Part I. Theoretical Model," *Polym. Eng. Sci.*, **12**, 294 (1972).
16. J. L. White and H. B. Dee, "Flow Visualization of Injection Molding Polyethylene and Polystyrene Melts and Sandwich Molding," *Polym. Eng. Sci.*, **14**, 212 (1974).
17. L. R. Schmidt, "A special Mold and Tracer Technique for Studying Shear and Extensional Flows in a Mold Cavity During Injection Molding," *Polym. Eng. Sci.*, **14**, 797 (1974).
18. W. Rose, "Fluid–Fluid Interfaces in Steady Motion," *Nature*, **191**, 242–243 (1961).
19. K. Oda, J. L. White, and E. S. Clark, "Jetting Phenomena in Injection Mold Filling," *Polym. Eng. Sci.*, **16**, 585 (1976).
20. J. R. A. Pearson, *Mechanical Principles of Polymer Melt Processing*, Pergamon Press, Oxford, 1966, p. 128.
21. D. H. Harry and R. G. Parrott, "Numerical Simulation of Injection Mold Filling," *Polym. Eng. Sci.*, **10**, 209 (1970).
22. J. L. Berger and C. G. Gogos, *Technical Papers*, 29th Annual Technical Conference of the Society of Plastics Engineers, Washington, D.C., 1971, Vol. 17, p. 1; *Polym. Eng. Sci.*, **13**, 102 (1973).
23. P. C. Wu, C. F. Huang, and C. G. Gogos, "Simulation of the Mold Filling Process," *Polym. Eng. Sci.*, **14**, 223 (1974).
24. B. R. Laurencena and M. C. Williams, "Radial Flow of Non-Newtonian Fluids Between Parallel Plates," *Trans. Soc. Rheol.*, **18**, 331 (1974).
25. G. Williams and H. A. Lord, "Mold Filling Studies for the Injection Molding of Thermoplastics," *Polym. Eng. Sci.*, **15**, 553 (1975).
26. J. Crank and P. Nickolson, *Proc. Camb. Phil. Soc.*, **43**, 50 (1947).
27. G. G. O'Brien, M. A. Hyman, and S. Kaplan, *J. Math. Phys.*, **29**, 223 (1951).
28. I. T. Barrie, *Soc. Plast. Eng. J.*, **27** (8), 64 (1971).
29. Z. Tadmor, "Molecular Orientation in Injection Molding," *J. Appl. Polym. Sci.*, **18**, 1753–1772 (1974).
30. G. Menges and G. Wübken, "Influence of Processing Conditions on Molecular Orientation in Injection Molds", Soc. Plastics Eng., 31st Annual Technical Conference, Montreal, Canada, 1973; p. 519.
31. R. B. Bird, H. R. Warner, Jr., and D. C. Evans, *Advan. Polym. Sci.*, **8**, 1 (1971).
32. W. Dietz, J. L. White, and E. S. Clark, "Orientation Development and Relaxation in Molding of Amorphous Polymers," *Polym. Eng. Sci.*, **18**, 273–281 (1978).

33. C. F. Huang, Ph.D. dissertation, Department of Chemical Engineering, Stevens Institute of Technology, Hoboken, N.J., 1978.
34. O. C. Zienkiewicz, P. Mayer, and Y. K. Cheung, "Solution of Anisotropic Seepage Problems by Finite Elements," *Proc. Am. Soc. Civ. Eng.*, **92**, EM1, 111–120 (1964).
35. E. Broyer, C. Gutfinger, and Z. Tadmor, "A Theoretical Model for the Cavity Filling Process in Injection Molding," *Trans. Soc. Rheol.*, **19**, 423 (1975).
36. Y. Kuo and M. R. Kamal, "Flows of Thermoplastics in the Filling and Packing Stages of Injection Molding," Paper presented at the International Conference on Polymer Processing, MIT, Cambridge, MA. August 16, 1977.
37. R. C. Thamm, "Phase Morphology of High-Impact-Strength Blends of EPDM and Polypropylene. Knit-Line Behavior", *Rubber Chem. Technol.*, **50**, 24–34 (1977).
38. M. R. Kantz, H. H. Newman, Jr., and F. H. Stigale, "The Skin-Core Morphology and Structure-Properties Relationships in Injection Molded PP," *J. Appl. Polym. Sci.*, **16**, 1249 (1972).
39. E. S. Clark, *Appl. Polym. Symp.*, **24**, 45 (1974).
40. S. Y. Hobbs, "Some Observations in the Morphology and Fracture Characteristics of Knit Lines," *Polym. Eng. Sci.*, **14**, 621 (1974).
41. K. J. Cleereman, "Injection Molding of Shapes of Rotational Symmetry with Multiaxial Orientation," *Soc. Plast. Eng. J.*, **23**, 43 (October 1967).
42. W. Knappe, "Die Festigheit Thermoplastischer Kunststoffe in Abhängigkeit von der Verarbeitungsbendingungen," *Kunststoffe*, **51**, 562 (1961).
43. M. Rigdahl, "Calculations of Residual Thermal Stresses in Injection Molded Amorphous Polymers by the Finite Element Method." *Int. J. Polym. Mater.*, in press.
44. J. Kubat and M. Rigdahl, "A Simple Model for Stress Relaxation in Injection Molded Plastics with an Internal Stress Distribution," *Mater. Sci. Eng.*, **21**, 63 (1975).
45. G. Odian, *Principles of Polymerization*, McGraw-Hill, New York, 1970; T. G. Fox and V. R. Allen, *J. Chem. Phys.*, **41**, 344 (1964).
46. W. J. Moore, *Physical Chemistry*, Prentice-Hall, Englewood Cliffs, N. J., 1962, p. 274.
47. J. D. Domine, "Computer Simulation of the Injection Molding of a Liquid Undergoing a Linear Step Polymerization," Ph.D. thesis, Department of Chemical Engineering, Stevens Institute of Technology, Hoboken, N.J., 1976.
48. J. D. Domine and C. G. Gogos, "Computer Simulations of Injection Molding of a Reactive Linear Condensation Polymer", Society of Plastics Engineers, 34th Annual Technical Conference, Atlantic City, N. J., 1976.
49. D. W. Peaceman and H. H. Rachford, Jr., *J. SIAM*, **3**, 28 (1955).
50. R. G. Angell, Jr. (to Union Carbide), U.S. Patent 3,268,636 (Aug. 23, 1966).
51. Uniroyal Chemicals Div., Uniroyal, Inc., Bulletin ASP-1533.
52. J. L. Throne, *J. Cellular Plast.*, **10**, 208 (1972).
53. J. L. Throne and R. G. Griskey, "Structural Thermoplastic Foam—A Low Energy Processed Material," *Polym. Eng. Sci.*, **15**, 747 (1975).
54. P. L. Durrill and R. G. Griskey, "Diffusion and Solution of Gases in Thermally Softened or Molten Polymers. Part I," *Am. Inst. Chem. Eng. J.*, **12**, 1147 (1966).
55. A. N. Gent and D. A. Tomkins, "Nucleation and Growth of Gas Bubbles in Elastomers," *J. Appl. Phys.*, **40**, 2520 (1969).
56. J. R. Street, A. L. Fricke, and L. P. Reiss, *Ind. Eng. Chem. Fundam.*, **10**, 54 (1971).
57. J. L. McCormic and J. R. Street, "The Dynamics of Phase Growth in Viscous non-Newtonian Fluids. 2. Growth in Fluids of Finite Extent" Unpublished report, Shell Development Co., Emeriville, Calif., 1973.
58. R. K. Prud'homme and R. B. Bird, "The Dilatational Viscosity of Suspensions of Gas Bubbles in Newtonian and non-Newtonian Fluids," Rheology Research Center, University of Wisconsin, August 1976.

59. A. G. Fredrickson and R. B. Bird, "Non-Newtonian Flow in Annuli," *Ind. Eng. Chem.*, **50**, 347 (1958).
60. E. Broyer and C. W. Macosko, "Heat Transfer and Curing in Polymer Reaction Molding," *Am. Inst. Chem. Eng. J.*, **22**, 268 (1976).
61. S. D. Lipshitz, C. W. Macosko, and F. G. Mussatti, Technical Papers, 33rd Annual Technical Conference of the Society of Plastics Engineers, Montreal, Canada, 1975; vol. 21, p. 239.
62. C. W. Macosko and D. R. Miller, "A New Derivation of Post Gel Properties of Network Polymers," *Macromolecules*, **9**, 206 (1976).

## *PROBLEMS*

**14.1** ***Injection Mold Runner Filling at a Constant Flow Rate.*** As mentioned in Section 14.1, when the "flow resistance" of the runner is not very high or the injection pressure is high, then the runner (or even the runner-mold cavity system) can fill at a constant flow rate. Using the viscosity data and runner dimensions used in Example 14.1 calculate the required injection pressure to fill the entire runner at a constant rate of $1.2 \times 10^{-6}\ \text{m}^3/\text{s}$.

**14.2** ***Filling of an Injection Mold Runner of Noncircular Cross Section.*** Consider the filling of a runner, the cross section of which is formed by three rectangle sides and a semicircle. The filling takes place at a constant applied injection pressure of $20\ \text{MN/m}^2$. The dimensions of the runner cross section (see Fig. 13.31) are $d/B = 0.8$ and $B = 5.0$ mm, while its length is 25 cm.

(a) Assuming that the polymer melt has a viscosity which is shear rate independent and equal to $6 \times 10^3\ \text{N} \cdot \text{s/m}^2$ and making the pseudo-steady state assumption calculate $Q(t)$ and $Z(t)$. Use the results in Section 13.7 and Fig. 13.31.

(b) How would the concept of the "equivalent Newtonian" viscosity (Problem 6.14) be helpful in handling a pseudoplastic polymer melt?

**14.3** ***Relative Pressure Drops in the Runner and Gate in Injection Molding.*** Consider the pressure traces on Fig. 14.2 as well as the location of the pressure transducers. Assuming that the filling process is isothermal, neglecting the "elbows" in the runner and the distance of $P_3$ from the gate, calculate $P_1$–$P_2$ (runner pressure drop) and $P_2$–$P_3$ (gate pressure drop) at 0.7 s. The runner distance from $P_1$ to $P_2$ is 8.0 in., its width is 0.43 in. and the thickness 0.317 in. The gate dimensions are 0.25 in. wide, 0.07 in. long, and 0.089 in. deep. The rheological properties of the PS melt used are $\ln \eta = A_0 + A_1 \ln \dot{\gamma} + A_{11}(\ln \dot{\gamma})^2 + A_2 T + A_{22} T^2 + A_{12} T \ln \dot{\gamma}$ where $\dot{\gamma}$ is in $\text{s}^{-1}$ and, $T$ in °F and $\eta$ in $\text{lb}_f \cdot \text{s/in.}^2$ The coefficients are: $A_0 = 0.14070 \times 10^2$, $A_1 = -0.80596 \times 10^0$, $A_{11} = -0.22504 \times 10^{-1}$, $A_2 = -0.44972 \times 10^{-1}$, $A_{22} = 0.38399 \times 10^{-4}$, $A_{12} = 0.99782 \times 10^{-3}$. Compare your answers to the $P_1$–$P_2$ and $P_2$–$P_3$ transducer values appearing on Fig. 14.2. Use Fig. 13.31 using an equivalent Newtonian viscosity. Do you need to consider entrance pressure losses?

**14.4 "Packing" Flow During Injection Molding.** In Fig. 14.2, assuming that during the period $1.5<t<3$ s no appreciable skin has been formed at the positions of the transducer $P_1$, $P_2$, and $P_3$, obtain an estimate of the "packing flow" rate either from $P_1$–$P_2$ or $P_2$–$P_3$. Use the runner and gate dimensions given in Problem 14.3. The fluid can be assumed to be Newtonian at these slow flow rates with a viscosity evaluated from the rheological data in Problem 14.3. Check your answer by calculating the corresponding thermal contraction of the melt in the mold in the period of 1 s. The thermal expansion coefficient of the polystyrene melt is $6\times10^{-4}\,°K^{-1}$, the entering melt temperature is 400°F and the mold temperature is 75°F.

**14.5 Design of a Multiple Cavity Runner System.** Multicavity molds must have center-fed runner systems that fill the various cavities as symmetrically as possible to avoid building pressure in one part of the mold and creating excessive flash there. Furthermore, it is desirable for uniform properties of the molded parts from each cavity to start and end the filling process simultaneously in all cavities. Consider the runner-cavity system shown on Fig.

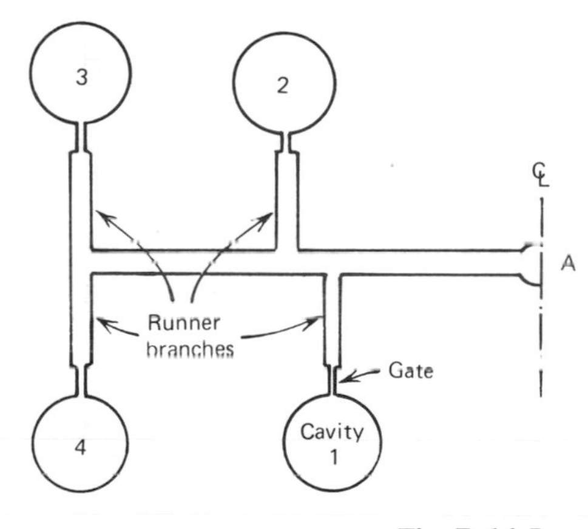


**Fig. P-14.5**

P-14.5 (only half plane is shown). Assuming that the injection pressure is constant at $A$ and that filling is isothermal:

(a) Design the runner branches for each of the four cavities in order to get simultaneous filling; the gates are identical.

(b) Design the gates so that there is simultaneous filling in each of the cavities; the runner branches are identical. Consider the polymer melt to be Newtonian.

(c) How would you solve (a) or (b) for a pseudoplastic melt?

**14.6 The "Molding Area" in Injection Molding.** (a) Discuss the dependence of each of the curves making up the molding area in Fig. 14.3 on polymer parameters such as $T_g$, $T_m$, $m(T)$, $n$, $k$, $m(P)$, and $T_m(P)$ and thermal degradation (Fig. 9.1).

(b) Apply the above ideas to three polymers PVC, nylon and HDPE, whose properties appear in Appendix A. To simulate "flashing," consider parallel plate flow of 0.001 in spacing, with maximum allowable flow length of 0.05 cm during a 1 s period.

**14.7** ***The Assumption of Isothermal Cavity Filling.*** As stated in Section 14.1 (The Overall Flow Pattern) good estimates of filling rates can be obtained by assuming that the cavity filling flow is isothermal. The success of this assumption is illustrated in Fig. 14.4 where the predicted positions of the short shots are compared with the experimental.* In an attempt to investigate when is the isothermicity assumption good, Gogos et al.† compared isothermal and nonisothermal calculated filling times for two molds (Fig. P-14.7). The material parameters for the PVC used in the simulations are shown on Fig. 14.7.

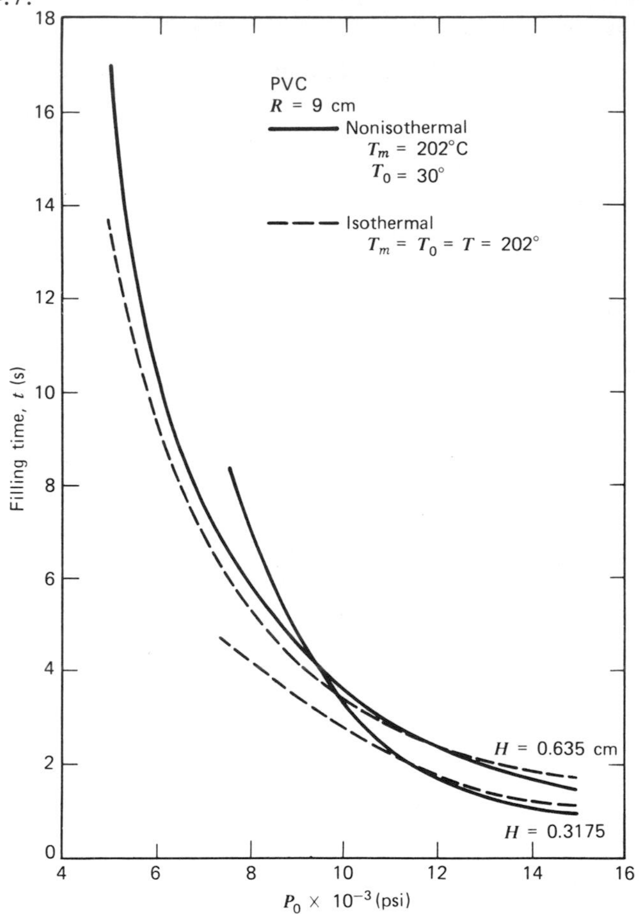


**Fig. P-14.7**

(a) Discuss the results in terms of the "balance" between heat generated and heat lost during filling.

(b) How do the isothermal and nonisothermal velocity profiles look and what physical conditions interrelate them? What can you conclude from this condition about the temperature field?

(c) How would the results be for LDPE?

* W. L. Krueger and Z. Tadmor, "Injection Molding Into Rectangular Cavity Containing Inserts," Society of Plastics Engineers, 36th Annual Technical Conference, Washington, D.C., 1978, p. 87.

† P. C. Wu, C. F. Huang, and C. G. Gogos, "Simulation of the Mold Filling Process," *Polym. Eng. Sci.*, **14**, 223 (1974).

**14.8** ***The Assumption of Constant Frozen Skin Thickness.*** Barrie* considering the filling flow of large area articles suggests that they be treated as isothermal flows between two plates not of the actual separation $h$, but $(h-2\Delta x)$, where $\Delta x$ is the frozen skin thickness. The $\Delta x$ is taken to be independent of the flow distance. Evidence supporting this assumption is brought from structural foam molding, where the solid skin thickness does not vary much. Empirical estimates of $\Delta x$ indicate $\Delta x \propto \tau^{1/3}$, where $\tau = Ah/Q$, $A$ is the area covered during filling and $Q$ is the filling rate. On the other hand, from heat transfer calculations, $\Delta x \propto \tau^{1/2}$. Prove the second relationship.

**14.9** ***Sandwich Injection Molding.*** In the ICI sandwich molding process, two injection machines are used to fill a mold. First, a melt fills a fraction of the mold (say $\frac{1}{5}$-$\frac{1}{10}$) and immediately following, the second injection machine injects a melt with a foaming agent. It is observed that the first melt forms the surface area of the *entire* mold. Explain the flow mechanism—sketching it out at its various stages—that makes this process possible. (A similar process has been used to mold articles of "virgin" polymer skin and recycled core†).

**14.10** ***Nonisothermal Flow During Compression Molding.*** In Section 14.3 we assumed that the squeezing flow in compression molding which takes place at $t<t_f$ is isothermal. An alternative assumption is that at $t_f$ a nonuniform temperature field exists in the preform, which, as a function of the dimensionless variable $z^* = z/h$, remains constant throughout the flow stage of the process (this stage is very fast)

(a) Assuming a constant squeeze rate $\dot{h}$, solve the problem of nonisothermal squeeze flow between two parallel disks, to obtain a modified Scott equation.

(b) From the Scott equation 14.3-8 and force measurements in constant squeeze rate compression molding processes, can you determine if the flow is isothermal?

(c) If polymerization takes place at the same time as flow, how should it be taken into account in the solution of the problem?

* I. T. Barrie, *Soc. Plast. Eng. J.*, **27** (8), 64 (1971): "An Application of Rheology to the Injection Molding of Large-Area Articles." *Plast. Polym.* (February) 47–51 (1970).

† G. Williams and H. A. Lord, "Mold Filling Studies for the Injection Molding of Thermoplastics," *Polym. Eng. Sci.*, **15**, 553 (1975).

# 15

# Secondary Shaping



This chapter discusses four "secondary" shaping operations: melt spun fiber stretching, tubular film ("bubble") forming, blow molding, and thermoforming. The first three occur immediately after die forming. Thermoforming involves polymer sheets or films that have been extruded and solidified in a separate operation. Sections 13.3–13.5 cover the die design aspects of the die forming operations associated with the four secondary shaping methods named.

## 15.1 Melt Spun Fiber Stretching

There is no clear point of demarcation where postextrudate swelling ends and melt stretching begins. The phenomena occur simultaneously, especially near the die exit, where the rapid rate of swelling occurs ordinarily. Experimental data from actual melt spinning runs (1) indicate that the melt strand cross-sectional area decreases hyperbolically from the spinnerette exit to the takeup rolls. Figure 15.1 gives typical melt strand area and radius axial profiles. The melt drawdown region extends to about 200 cm from the spinnerette exit. There is no specific indication of where the melt strand begins to solidify ("frost line").

From $A(z)$ and $R(z)$ in Fig. 15.1, it is obvious that the flow field in this region has the general form $v_z = v_z(r, z)$, $v_\theta = 0$, $v_r = v_r(r, z)$. To determine the flow, therefore, we must solve simultaneously the $r$ and $z$ components of the equation of motion and the thermal energy balance, the continuity equation and the appropriate boundary conditions. This task is very difficult, especially in light of the nonlinear constitutive rheological equations that must be employed. Thus neither the detailed flow field in the region where the strand radius decreases with the axial distance rapidly, nor the exact rate of radius decrease, can be predicted at the present time.

Several attempts have been made to obtain expressions for the velocity, the fiber radius, and the temperature as functions of the axial distance. Kase and Matsuo (2) were the first to consider nonisothermal fiber stretching. Han in a

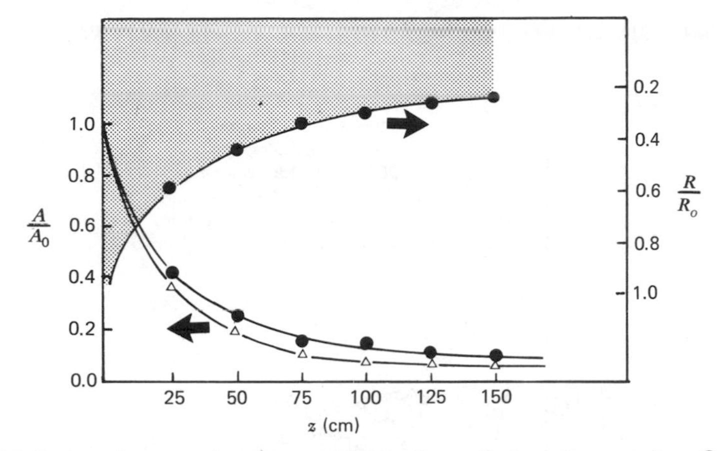


**Fig. 15.1** Melt strand area and radius profiles in the melt drawdown region: ●, nylon 6 at 265°C and takeup velocity of 300 m/min; △, PP at 262°C and takeup velocity of 350 m/min. (Reprinted with permission from H. F. Mark, in *Rheology*, Vol. 4, F. R. Eirich, ed., Academic Press, New York, 1969.)

recent book (3a) has generalized their approach and has arrived at the following two equations, assuming steady state, that the only velocity component is $v_z(z)$, and $T = T(z)$ (see Problem 15.1)

$$v_z \frac{dv_z}{dz} = g_z - v_z \frac{d}{dz}\left(\frac{\tau_{zz}}{\rho v_z}\right) - 2\left(\frac{\pi v_z}{\rho G}\right)^{1/2} F_D \tag{15.1-1}$$

$$\frac{dT}{dz} = -\frac{2}{C_p}\left(\frac{\pi}{\rho G v_z}\right)^{1/2} [h(T - T_a) + \sigma\varepsilon(T^4 - T_a^4)] \tag{15.1-2}$$

where $\varepsilon$ is the emissivity and $G = \rho\pi R^2 v_z$, $F_D$ is the air drag force per unit area equal to

$$F_D = \left(\frac{0.843}{\pi R^2}\right)\left(\frac{\rho_a}{\rho}\right) G v_z \left[\frac{\pi\rho\mu_a}{\rho_a}\frac{(L-z)}{G}\right]^{0.915} \tag{15.1-3}$$

$h$ is the heat transfer coefficient given by

$$\frac{hR}{k_a} = 0.21(1+K)\left(\frac{2R\rho_a v_z}{\mu_a}\right)^{0.334} \tag{15.1-4}$$

$K$ is an adjustible parameter, and the subscript $a$ refers to ambient air. Han coupled these two transport equations with a "power law in tension" constitutive relationship containing a temperature dependent viscosity term

$$\tau_{zz} = -3\alpha e^{\beta/T}\left[k_1 + k_2\left(\frac{dv_z}{dz}\right)^{n-1}\right]\frac{dv_z}{dz} \tag{15.1-5}$$

where

$$\alpha = \eta_0\, e^{-\Delta E/RT_0} = \eta_0\, e^{-\beta/T_0} \tag{15.1-6}$$

This system of equations is solved numerically. The results obtained are physically reasonable up to the axial position where crystallization commences, when the rate of cooling slows down because of the exothermic solidification phenomenon. This is indicated in Fig. 15.2, where the fiber surface temperature is

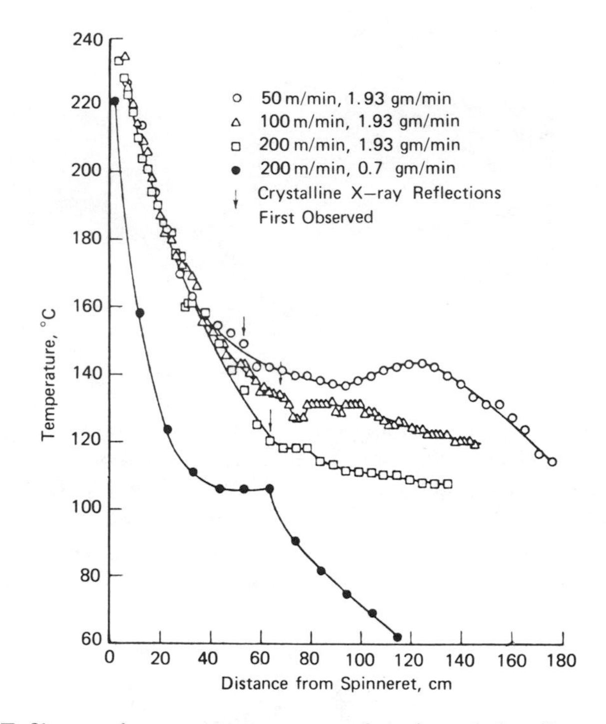


**Fig. 15.2** HDPE fiber surface temperature as a function of the distance from the spinnerete. [Reprinted with permission from J. R. Dees and J. E. Spruiell, *J. Appl. Polym. Sci.*, **18**, 1053 (1974).]

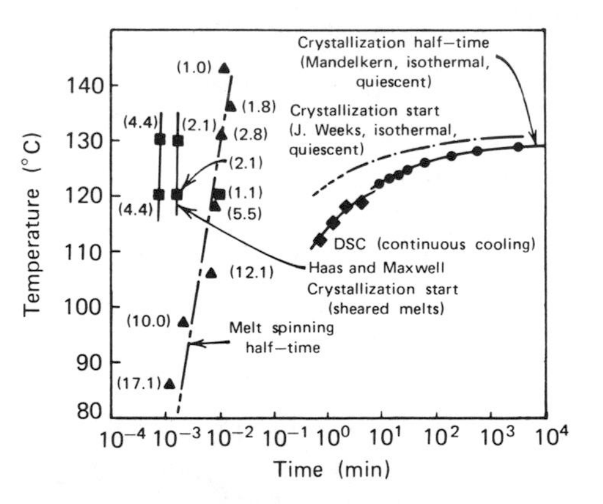


Fig. 15.3 Continuous cooling transformation curves and isothermal crystallization results for HDPE crystallized in various ways. Numbers in parentheses are the stresses in (dynes/cm$^2 \times 10^{-6}$). The works of the investigators in parenthesis are cited in Ref. 5. [Reprinted with permission from J. E. Spruiell and J. L. White, *Polym. Eng. Sci.*, **15**, 660 (1975).]

measured during spinning as a function of $z$. It is worth noting that fiber surface temperature can increase with $z$, as a result of inner layer crystallization.

The flow and heat transfer problems during fiber stretching and solidification are by no means solved. As the solution to Problem 15.2 indicates, radial temperature gradients must be taken into account. This has been done by Morrison (4) Furthermore, in the solution of the cooling problem the *actual* crystallization kinetics must be considered. In melt spun fibers we observe the phenomenon of flow induced crystallization (see Section 3.6). In Fig. 15.3, Spruiell and White (5) compare the crystallization kinetics of HDPE under quiescent, continuous shear, and spinning conditions. It is clear that the onset of crystallization occurs in melt spun fibers at times that are at least 2 orders of magnitude shorter, than in the quiescent state.

The structuring effects of melt spinning were briefly discussed in Section 3.6. We note here the work of Dees and Spruiell (6) with HDPE. They report that the observed orientation function behavior during melt spinning (see Fig. 3.28) can be explained with a morphological model assuming that at low spinline stresses or takeup velocities, spherulitic structures are obtained. Increasing the takeup velocity results in row nucleated twisted and, at even higher speeds, in row nucleated untwisted lamellae. This morphological model can lead to the morphology of the hard elastic structures, discussed in Section 3.6, which are obtained when high

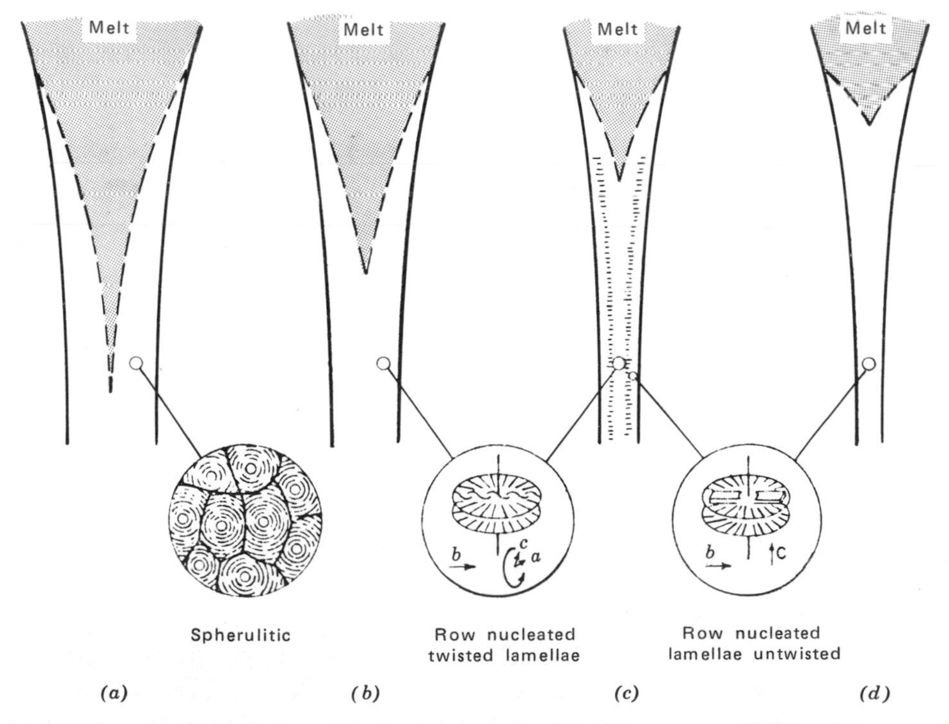


**Fig. 15.4** Morphological model of structures developed in as-spun HDPE. Takeup velocities are (a) very low; (b) low; (c) medium; and (d) high. [Reprinted with permission from J. E. Spruiell and J. L. White, *Polym. Eng. Sci.*, **15**, 660 (1975.]

stresses are applied in either fiber or film forming. Figure 15.4 represents the model of Dees and Spruiell.

*Draw resonance* and *spinnability*, two more problems related to the stability of the melt spinning process, have received considerable attention recently. In the first case the fiber strand exhibits regular and sustained periodic variations of its diameter. Spinnability involves the ability of polymer melts to be drawn without breaking because of either "necking" or cohesive fracture. Petrie and Denn (7) have recently reviewed the work in these two areas. Physically the occurrence of draw resonance can be viewed as follows. In the region between the spinnerette exit and the takeup rolls there can be a time variation of the total extrudate mass: although the rate of mass entering this region is constant, the rate of leaving is not controlled, since only the takeup speed is regulated, not the fiber diameter. Thus if the strand thins out near the takeup rolls, the diameter of the strand above it will increase, creating (from the spinnerette exit) a thick-thin strand. But the thick portion soon reaches the takeup rolls. Mass leaves the region at a high rate and the strands thins out upstream, creating a thin-thick strand. The process can repeat itself. This may explain the reports that if solidification occurs before the takeup rolls, no resonance is observed (8, 9), as well as the observations of increased resonance period with increased residence time in the spinline (10).

Isothermal draw resonance is found to be independent of the flow rate. It occurs at a critical value of draw ratio (i.e., the ratio of the strand speed at the takeup rolls to that at the spinnerette exit). For fluids that are almost Newtonian, such as PET and polysiloxane, the critical draw ratio is about 20. For polymer melts such as HDPE, LDPE, PS, and PP, which are all pseudoplastic and viscoelastic, the critical draw ratio value can be as low as 3 (11). The maximum to minimum diameter ratio decreases with decreasing draw ratio and decreasing drawdown length.

Linear stability analyses of the isothermal draw resonance phenomenon have been carried out in the last decade. Thus Pearson and Shah (12) have examined the behavior of inelastic fluids. For Newtonian fluids a critical draw ratio of 20.2 is found. Flow thinning and flow thickening fluids exhibit critical drawn ratios that are smaller or larger, respectively, than 20.2. Fisher and Denn (13) have carried out both infinitesimal (linearized) and finite amplitude analyses of the draw resonance problem. They find that Newtonian fluids are stable to finite amplitude disturbances for draw ratios of less than 20.2. Linearized stability analysis revealed that for fluids that obey a White–Metzner type equation, the critical draw ratio depends on the power law index $n$ and the viscoelastic dimensionless number $N$

$$N = 3^{(1-s/2)}\left(\frac{m}{G}\right)^s\left(\frac{V_0}{L}\right) \qquad (15.1\text{-}7)$$

where $s = 1/n$, $L$ is the spinlike length, $G$ is the tensile modulus, and $V_0$ is the spinnerette velocity. The results appear in Fig. 15.5. Of interest is the "nose" region of the curves, which indicates that one could eliminate the draw resonance phenomenon by an *increase* in the draw ratio. Of interest is also the work of Han (3b), who finds experimentally that as the temperature level is decreased in isothermal spinning, draw resonance occurs at lower draw ratios. This seems reasonable from the figure. In the "nose" region, decreasing the temperature increases $G$ and decreases $m$, which in turn decreases $N$, bringing about lower draw ratio values.

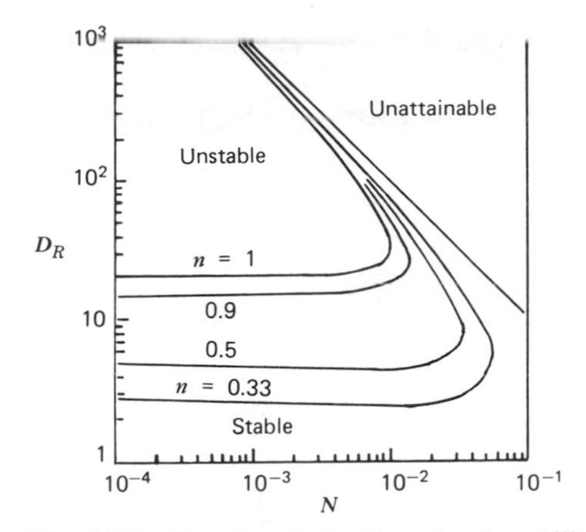


**Fig. 15.5** Results of the linearized stability analysis for a White–Metzner type fluid, indicating the dependence of the critical draw ratio on $n$ and $N$. [Reprinted with permission from R. J. Fisher and M. M. Denn, *A.I.Ch.E.I.*, **22**, 236 (1976).]

Chen et al. (14) and White and Ide (10) have presented experimental and theoretical (isothermal linear stability analysis) results that indicate the following: first, that polymer melts respond similarly to uniform elongational flow and to melt spinning; second, that polymers whose elongational viscosity $\bar{\eta}^+(t, \dot{\varepsilon})$ increases with time or strain (see Fig. 6.16) result in a stable spinline, do not exhibit draw resonance, and undergo cohesive failure at high draw ratios. A prime example of such behavior is LDPE. On the other hand, polymer melts with a decreasing $\bar{\eta}^+(t, \dot{\varepsilon})$ exhibit draw resonance at low draw ratios and break in a ductile fashion (after "necking") at high draw ratios. Typical polymers in this category are HDPE and PP.

In nonisothermal melt spinning the process becomes increasingly stable to draw resonance with increasing Stanton number $St = h/C_p m$, where $m$ is the mass flux (15). Linearized stability analysis supports the stabilizing effect of increasing Stanton number.

The problems of draw resonance and spinnability are *not fully* understood. In particular, the effects of the viscoelasticity of the melts, their molecular structure, and the nonisothermal nature of the process must be systematically studied.

## 15.2 The Film Blowing Process

As Chapters 1 and 13 mentioned, tubular dies are commonly used for the manufacture of blown (slightly biaxially oriented) film. Figure 15.6 schematizes the film blowing process. The tubular film extrudate thins out axially because of the internal pressure applied by the central air stream and by an axial tension that is deliberately imposed by the rotation of the nip rolls. This is why the orientation is biaxial to degrees that can be varied by varying the internal pressure and the axial

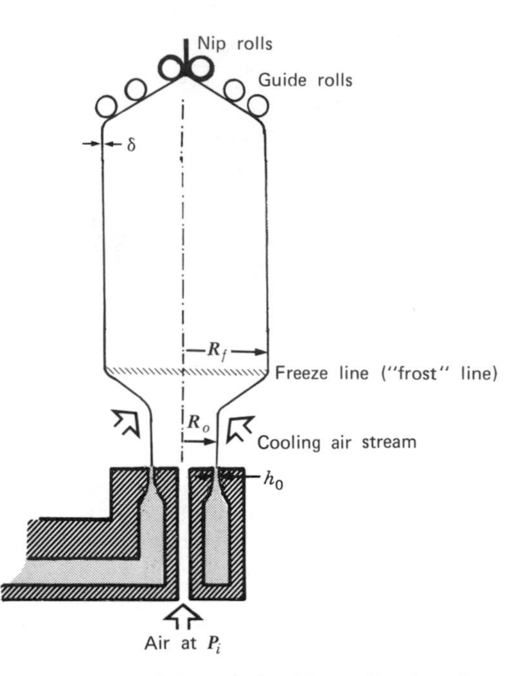


**Fig. 15.6** Schematic representation of the tubular blown film forming operation.

tension. The nip rolls also act as a seal for the film bubble. The bubble diameter increases and reaches a constant value $R_f$ in the neighborhood of the freezing line, where $T_{\text{film}} = T_m$. The axial distance between the die exit and the freezing line is controlled by the rate of cooling supplied by cold air jets from an air ring. After the nip rolls, the flattened bubble is cut at the creases and wound onto two separate rolls. In a number of processes further axial orientation of the cold film is employed to increase the strength in that direction. The film thicknesses produced are of the order of 10 to 100 microns, and the linear film production rates are very high.

The tubular film stretching process has been analyzed by a number of investigators, all of whom assume that $\delta/R \ll 1$, thus avoiding shear stresses and moments in the film. The object of the analyses is to predict the bubble radius and film thickness as a function of the axial distance from the die exit. Pearson and Petrie (17) have examined the process by assuming that it is isothermal and that the melt is inelastic. Petrie (18) has simulated the process using either a Newtonian fluid or an elastic solid; in the Newtonian case he inserted the temperature profile obtained by Ast (19) to specify the temperature dependent viscosity. Finally, Han and Park (20–22) have used coupled force and thermal energy balances to take care of the nonisothermal nature of the process. They also used a power law type of viscosity relationship. Experimental results indicate that the material is viscoelastic during this process. Petrie's isothermal elastic and nonisothermal Newtonian solutions *bracket* the experimental results for $R(z)$ (18). Furthermore, $R(z)$ is sensitive to values of the overall heat transfer coefficient and the level of radiation heat losses (23). This is supported by the simulation results of Han and Park. In practice, the cooling of the bubble is the limiting factor to high production rates.

The kinematics of the unequal biaxial extension can be obtained in a straightforward manner (see Problem 15.3). For an incompressible fluid

$$\dot{\boldsymbol{\gamma}} = \frac{Q\cos\theta}{\pi R\delta}\begin{pmatrix} -\left(\frac{1}{\delta}\frac{d\delta}{dz}+\frac{1}{R}\frac{dR}{dz}\right) & 0 & 0 \\ 0 & \frac{1}{\delta}\frac{d\delta}{dz} & 0 \\ 0 & 0 & \frac{1}{R}\frac{dR}{dz} \end{pmatrix} \tag{15.2-1}$$

where $Q$ is the volumetric flow rate. The geometric variables appear in Fig. P.15.3.

If the melt is assumed to be Newtonian then

$$\boldsymbol{\pi} = P\boldsymbol{\delta} - \mu\dot{\boldsymbol{\gamma}} \tag{15.2-2}$$

Setting $\pi_{22}=0$, we obtain the hydrostatic pressure

$$P = \frac{Q\mu\cos\theta}{\pi R\delta^2}\frac{d\delta}{dz} \tag{15.2-3}$$

From the foregoing, the other two total normal stress components can be evaluated

$$\pi_{11} = \frac{Q\mu\cos\theta}{\pi R\delta}\left(\frac{2}{\delta}\frac{d\delta}{dz}+\frac{1}{R}\frac{dR}{dz}\right) \tag{15.2-4}$$

$$\pi_{33} = \frac{Q\mu\cos\theta}{\pi R\delta}\left(\frac{1}{\delta}\frac{d\delta}{dz}-\frac{1}{R}\frac{dR}{dz}\right) \tag{15.2-5}$$

To solve for $\delta(z)$ and $R(z)$, one needs to state the force balance equations for the blown film. The simplest form of these equations, disregarding inertial and gravity forces, are the classical thin film equations. In terms of the principal radii of curvature $R_C = R\sqrt{1+\dot{R}^2} = R/\cos\theta$ and $R_L = -(1+\dot{R}^2)^{3/2}/\ddot{R} = -\sec^3\theta/\ddot{R}$ where $\dot{R} = dR/dz$ and $\ddot{R} = d\dot{R}/dz$, and the maximum bubble radius $R_f$, they are

$$\Delta P = -\delta\left(\frac{\pi_{11}}{R_L}+\frac{\pi_{33}}{R_C}\right) \tag{15.2-6}$$

$$F_z = -2\pi R\cos\theta\,\delta\pi_{11} + \pi\,\Delta P(R_f^2 - R^2) \tag{15.2-7}$$

where $F_z$ is the force in the axial direction acting on the bubble for $z \geqslant Z_f$. Equation 15.2-6 is a generalization of the free surface boundary condition 5.3-1, and Eq. 15.2-7 is the $z$-direction force balance on a portion of the bubble bound by two planes, one at $z$ and the other at the freezing line $z = Z_f$. Substituting Eqs. 15.2-4 and 15.2-5 into Eqs. 15.2-6 and 15.2-7, we obtain two differential equations, one for the radius and the other for the thickness. In terms of the dimensionless parameters $r = R/R_o$, $w = \delta/R_o$, and $\zeta = z/R_o$, these equations are

$$2r^2(A + r^2B)\ddot{r} = 6\dot{r} + r(1+\dot{r}^2)(A - 3r^2B) \tag{15.2-8}$$

where $\dot{r} = dr/d\zeta$ and $\ddot{r} = d\dot{r}/d\zeta$ and subject to

$$r(0) = 1$$

$$\dot{r}\left(\frac{Z_f}{R_0}\right) = 0$$

and

$$\dot{w} = -w\left[\frac{\dot{r}}{2r} + \frac{(1+\dot{r}^2)(A + r^2 B)}{4}\right] \tag{15.2-9}$$

where $\dot{w} = dw/d\zeta$ and subject to

$$w(0) = \delta_0/R_0$$

where

$$A = \frac{R_0 F_z}{\mu Q} - B\left(\frac{R_f}{R_o}\right)^2 \tag{15.2-10}$$

and

$$B = \frac{\pi R_o^3 \Delta P}{\mu Q} \tag{15.2-11}$$

Note that to solve these equations, the position of the freeze line $Z_f$ and the value of $\delta_0$, the die exit thickness, must be known. Neither quantity can be specified a priori.

Han and Park (20), in place of Eq. 15.2-2, use a power law type of temperature dependent viscosity with $\bar{\eta}_0$ being the viscosity at melt temperature $T_0$ at $z = 0$ of the form

$$\bar{\eta}_B(\text{II}_{\dot{\gamma}}, T) = \bar{\eta}_0 \exp\left[\frac{\Delta E}{R}\left(\frac{1}{T} - \frac{1}{T_0}\right)\right] \sqrt{\left(\frac{\text{II}_{\dot{\gamma}}}{2}\right)}^{\,n-1} \tag{15.2-12}$$

Inserting the above into the force balances as before results for the nonisothermal case in the following differential equations, corresponding to Eqs. 15.2-8 and 15.2-9

► $$2r^2(A_n + r^2 B)\ddot{r} = 6\dot{r}\frac{\bar{\eta}_B}{\bar{\eta}_0} + r(1+\dot{r}^2)(A_n - 3r^2 B) \tag{15.2-13}$$

► $$\dot{w} = -w\left[\frac{\dot{r}}{2r} + \frac{\bar{\eta}_0}{\bar{\eta}_B}\frac{(1+\dot{r}^2)(A_n + r^2 B)}{4}\right] \tag{15.2-14}$$

where

$$A_n = \frac{R_0 F_z}{\bar{\eta}_0 Q} - B\left(\frac{R_f}{R_0}\right)^2 \tag{15.2-15}$$

$$B = \frac{\pi R_0^3 \Delta P}{\bar{\eta}_0 Q} \tag{15.2-16}$$

As in the case of fiber spinning, the temperature is assumed to be $z$-dependent only, and the following differential equation results from the thermal energy balance with no heat transfer at the inner surface and convective and radiative heat transfer at the outer surface:

$$-\rho C_v\left(\frac{Q \cos \theta}{2\pi R}\right)\frac{dT}{dz} = h(T - T_a) + \sigma\varepsilon(T^4 - T_a^4) \qquad (15.2\text{-}17)$$

The differential equations above can be solved numerically.

The same investigators (3) report experimental and theoretical results including gravitational effects as in Fig. 15.7. The agreement is good in this case and fair in general.

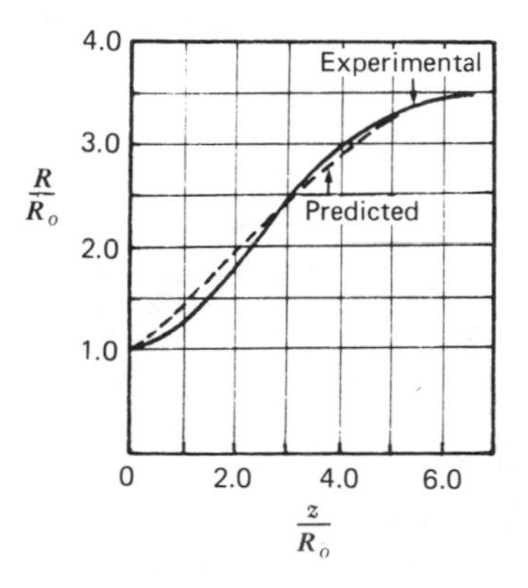


**Fig. 15.7** Experimental and predicted $R/R_o$ for LDPE during the blown film process at $T = 180°C$, $Q = 25.11\ cm^3/s$, $\Delta P = 3.95 \times 10^{-3}$ psi and a stretch ratio of 12.4. [Reprinted with permission from C. D. Han, *Rheology in Polymer Processing*, Academic Press, New York, 1976, Fig. 9.14.]

In industrial scale film blowing, the weight of the film is not negligible. Including gravitational forces, the equilibrium equations become

$$\Delta P = -\delta\left(\frac{\pi_{11}}{R_L} + \frac{\pi_{33}}{R_C}\right) - \rho g \delta \sin \theta \qquad (15.2\text{-}18)$$

$$F_z = -2\pi R \cos \theta h \pi_{11} + \pi \Delta P(R_f^2 - R^2) + 2\pi\rho g \int_z^{Z_f} Rh \sec \theta \, dz \qquad (15.2\text{-}19)$$

These equations, together with Eqs. 15.2-4 and 15.2-5, yield expressions similar to the differential equations 15.2-8 and 15.2-9, subject to the same boundary conditions.

Finally Han and Park (22) investigated the instability of the bubble forming process and have found that under uniaxial stretching ($P = P_a$) a more or less regularly varying bubble diameter with axial distance can be generated. This draw resonance-type phenomenon starts at a critical stretch ratio. The amplitude and frequency of variations of the diameter increase with increasing stretch ratio. A second type of bubble "instability" is observed under biaxial extension, when a step change is imposed in the takeup speed. The bubble changes shape irregularly with time, becoming stable if the size changes are not large. Furthermore, they report that increasing the melt temperature makes the bubble forming process more unstable.

## 15.3 Analysis of the Free Sheet Blowing (Thermoforming) Process: Elastic Behavior

Free bubble blowing is a commonly practiced method of thermoforming, in particular with acrylic sheets. Schmidt and Carley (24) analyzed this process in detail both experimentally and theoretically with a variety of polymers in an attempt to develop a thermoformability test method. Bubble blowing is an operation within a group classified by Alfrey (25) as involving *membranes and rotational symmetry.* Most secondary shaping operations, as well as film blowing, fall within this group. The theoretical aspects of the response of a viscoelastic liquid to sudden extension were discussed in section 6.8.

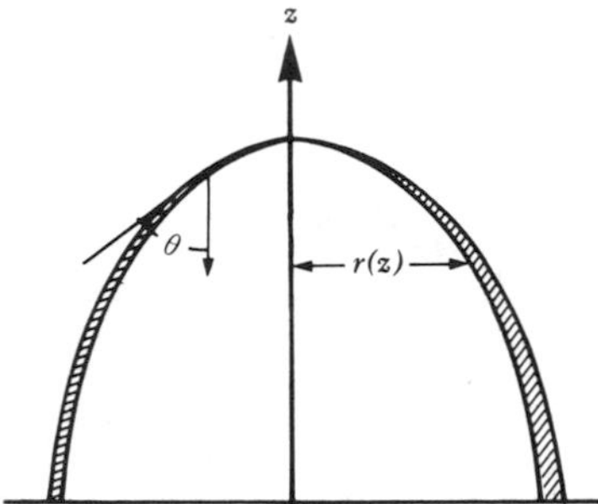


**Fig. 15.8** Schematic representation of a free bubble with rotational symmetry.

Consider a general rotationally symmetric membrane (Fig. 15.8). The membrane is specified by the meridian curve $r(z)$ and the thickness distribution $\delta(z)$. The two principal radii of curvature of the surface $R_L$ and $R_C$ in the meridian (longitudinal) and circumferential directions can be related to $r(z)$ as follows:

$$R_C = r\sqrt{1+(dr/dz)^2} \tag{15.3-1}$$

and

$$R_L = -\frac{[1+(dr/dz)^2]^{3/2}}{d^2r/dz^2} \tag{15.3-2}$$

By symmetry, the two principal directions of stress and strain are in the meridian $\pi_{11}$ and circumferential "hoop" $\pi_{33}$ directions. The third principal stress is normal to the surface and is taken to be zero.

In these processes, we are generally dealing with large deformations of *incompressible* materials, and deformation can be conveniently defined in terms of the principal extension ratios $\lambda_L$ and $\lambda_C$, where the extension ratio in the normal direction is given by

$$\lambda_N = \frac{1}{\lambda_C \lambda_L} \tag{15.3-3}$$

During a bubble blowing operation, there is a pressure difference exerting a net normal force per unit area $\Delta P$. In addition, we consider the possibility of a tangential (e.g., frictional) force acting per unit area of the membrane surface $F_T$; in free bubble blowing $F_T = 0$. Neglecting body and acceleration forces, and for membranes of small thickness gradients, the following equations of equilibrium result by appropriate force balances (24), the same force balances that yielded Eqs.

15.2-6 and 15.2-7.

$$\Delta P = -\delta\left(\frac{\pi_{11}}{R_L} + \frac{\pi_{33}}{R_C}\right) \tag{15.3-4}$$

$$-\frac{d}{dz}(r\delta\pi_{11}\cos\theta) = r\left(\frac{dr}{dz}\Delta P - F_T\right) \tag{15.3-5}$$

where $\theta$ is the angle in Fig. 15.8. To solve the problem at hand, in addition to initial and boundary conditions, a constitutive equation is needed to relate stresses to deformations. The equilibrium, small strain elastic behavior of an incompressible rubbery polymer can be specified by a single modulus, which can be related to molecular structure. The problem is more complex for large elastic straining, since the stress-strain relation is nonlinear. An appropriate constitutive equation for purely elastic and isotropic materials is that first proposed by Finger (26)

$$\boldsymbol{\tau} = -\nu_1 \mathbf{C}^{-1} + \nu_2 \mathbf{C} \tag{15.3-6}$$

where $\nu_1$ and $\nu_2$ are nonlinear elastic moduli and $\mathbf{C}$ is the Finger strain tensor

$$\mathbf{C} = \boldsymbol{\delta} - \boldsymbol{\gamma}_{[0]} \tag{15.3-7}$$

Since all the mechanical energy during deformation is stored, from the second law of thermodynamics one can obtain for incompressible materials (27)

$$\boldsymbol{\tau} = -2\frac{\partial W}{\partial I_{\mathbf{C}}}\mathbf{C}^{-1} + 2\frac{\partial W}{\partial II_{\mathbf{C}}}\mathbf{C} \tag{15.3-8}$$

where $W = U/\rho$ is the strain energy function (internal energy per unit volume for an adiabatic deformation), and $I_{\mathbf{C}}$ and $II_{\mathbf{C}}$ are the first and second invariants of $\mathbf{C}$. Thus we have

$$\frac{\partial \nu_1}{\partial II_{\mathbf{C}}} = \frac{\partial \nu_2}{\partial I_{\mathbf{C}}} \tag{15.3-9}$$

Schmidt and Carley (24) have proposed the following relationships, which give a good empirical fit with the observed thickness distribution during their free bubble blowing process

$$\frac{\partial W}{\partial I_{\mathbf{C}}} = a_1 \tag{15.3-10}$$

$$\frac{\partial W}{\partial II_{\mathbf{C}}} = a_2(II_{\mathbf{C}} - 3) \tag{15.3-11}$$

where $a_1$ and $a_2$ are the empirical constants.*

Solution of the equilibrium equations in conjunction with the strain energy function equation results in the description of the bubble blowing operation (i.e.,

* Alfrey (28) points out that a multiparameter constitutive equation such as the foregoing calls for the identification of more than one *structural* parameter, which affects the response of the material. Thus the "effective crosslink density" that provides the molecular-structural link to the kinetic theory of rubber elasticity, is no longer sufficient, and additional structural features (chain flexibility, cohesive energy density, etc.) are required.

**Fig. 15.9** High impact PS bubbles. *Top:* outside surfaces; *bottom:* the interiors and thickness profiles. [Reprinted with permission from L. R. Schmidt and J. F. Carley, *Polym. Eng. Sci.*, **15**, 51 (1975).]

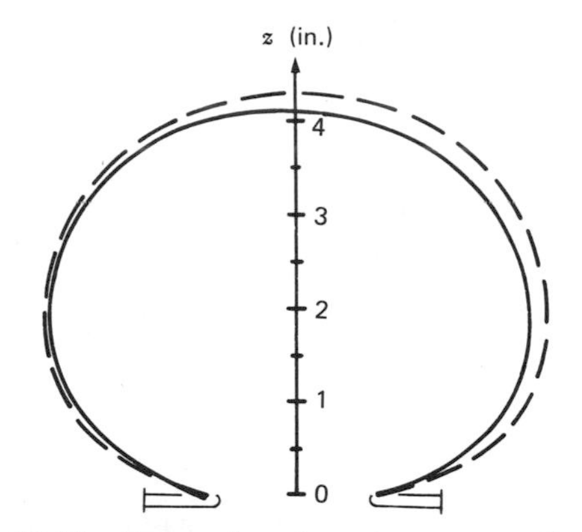


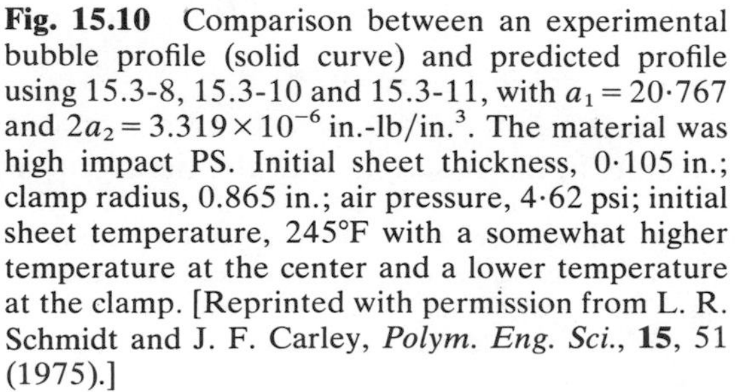

**Fig. 15.10** Comparison between an experimental bubble profile (solid curve) and predicted profile using 15.3-8, 15.3-10 and 15.3-11, with $a_1 = 20{\cdot}767$ and $2a_2 = 3.319 \times 10^{-6}$ in.-lb/in.$^3$. The material was high impact PS. Initial sheet thickness, 0·105 in.; clamp radius, 0.865 in.; air pressure, 4·62 psi; initial sheet temperature, 245°F with a somewhat higher temperature at the center and a lower temperature at the clamp. [Reprinted with permission from L. R. Schmidt and J. F. Carley, *Polym. Eng. Sci.*, **15**, 51 (1975).]

the prediction of both the bubble shape and the thickness distribution). Figures 15.9–15.11 give some experimental data and compare predictions using the procedure of Schmidt and Carley (24), as outlined previously. The materials used covered a broad range of polymers including PS, high impact PS, cellulose acetate butyrate, which could be blown to shapes from hemispheres to large spheroidal bubbles, and rigid PVC, acrylic modified PVC, cast PMMA, and PC, which could not be deformed beyond the hemispherical shape without rupture. The bubbles in Fig. 15.9 were blown from a 2.66 mm thick, high impact PS sheet. The samples were preprinted with a set of concentric circles to follow the deformation and to calculate

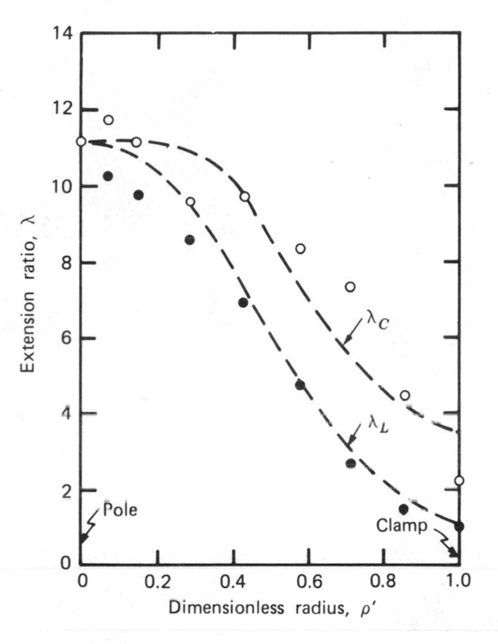


**Fig. 15.11** Comparison of experimental extension ratios and predicted values for the experiment of Fig. 15.10: $\lambda_C$ is in the extension ratio in the meridian direction, and $\lambda_L$ is in the longitudinal direction; $\rho'$ is a reduced radius. [Reprinted with permission from J. R. Schmidt and J. F. Carley, *Polym. Eng. Sci.*, **15**, 51 (1975).]

the extension ratios. We note a significant variation in shell thicknesses from the bottom to the top of the bubble. Figure 15.10 compares the experimental and predicted bubble shape, and Fig. 15.11 shows experimental and theoretical data on extensional ratios. The agreement between prediction and experiments is very good, supporting the assumption that the inflation of heat softened polymeric sheets can be viewed as an elastic process (cf. pp. 187–189). The inflation process lasted about 8 s, with about 90% of the area formed in the last 1.5 s. With the aid of high speed movie photography, Schmidt and Carley also observed that when rupture of

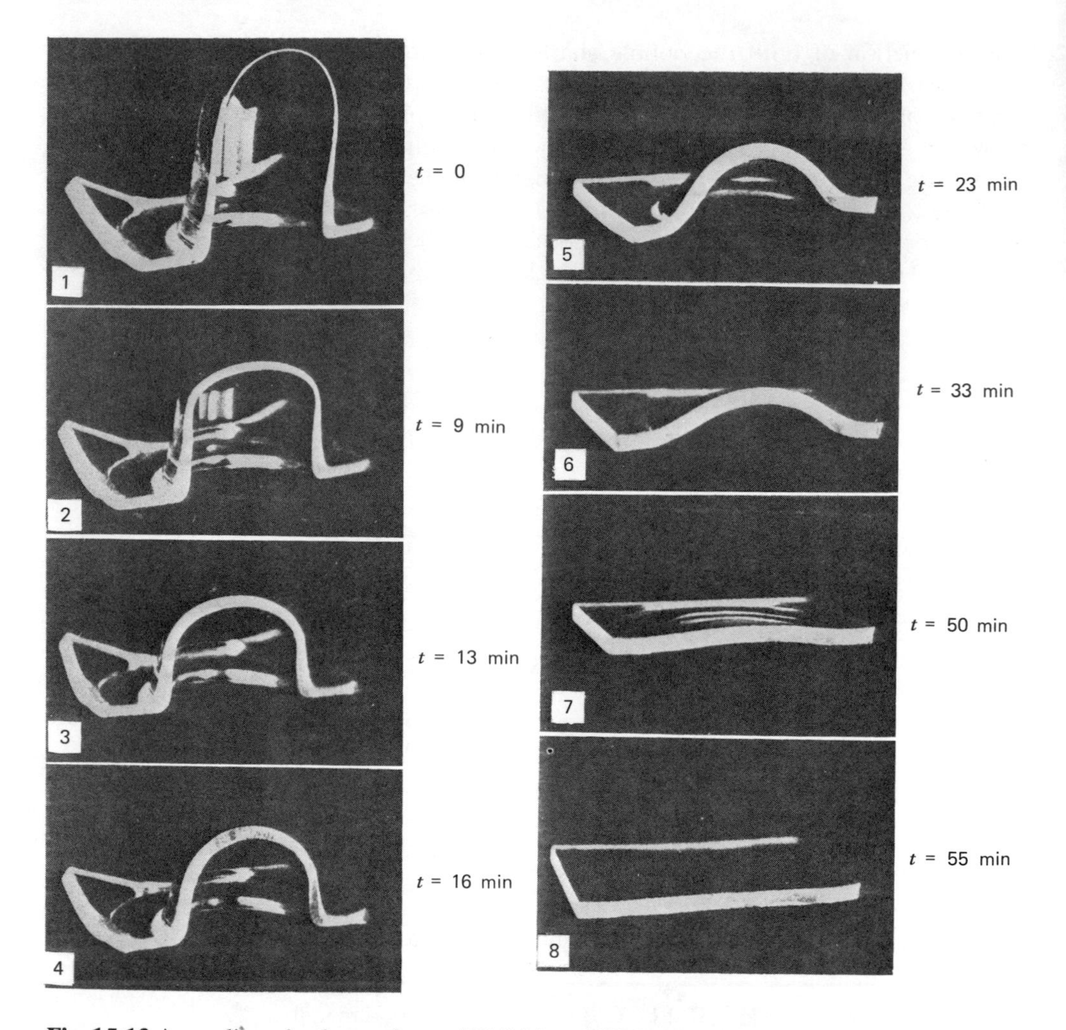


**Fig. 15.12** Annealing of a thermoformed PMMA at 160°C. After 55 min, the original shape is regained. (Reprinted with permission from N. Rosenzweig, *Process Simulation of Thermoplastic Sheet Forming*, M.S. thesis, Dep. Chem. Eng., Technion-Israel Institute of Technology, Haifa, 1975.)

the bubble occurred, the spheroidal shape retracted to the original disk-like shape in less than 1/700 s. A similar total recovery of thermoformed PMMA deep cup upon annealing at 160°C for 55 min (Fig. 15.12) was also observed by Rosenzweig (29). From Fig. 15.11 and Eq. 15.3-3, we can calculate $\lambda_N = \delta/\delta_0$, where $\delta$ is the sheet thickness at any position of the inflated sheet and for any given stage of the process. At the top of the bubble $\lambda_N = 8.26 \times 10^{-3}$, and near the clamp, $\lambda_N = 2.2 \times 10^{-1}$; that is, the thickness varies along the bubble meridian by a factor of 26. This is important to note from the practical thermoforming point of view, but also in connection with the biaxial extension rheological experiments, which use the bubble blowing method and assume a uniform biaxial extensional flow (Eq. 6.8-9). Only near the top of the bubble is the thickness almost constant.

## 15.4 The Thermoforming Process

### *Forming Temperature and the Heating (Melting) Step*

Chapter 1 demonstrated that there is an imaginative variety of thermoforming processes, showing some of them schematically in Fig. 1.22. The nature of the actual shaping (deformation) process, however, is similar in all these processes. Section 15.3 discussed the deformation aspects of this shaping process. Another common operational aspect to all thermoforming processes is the heating step, dealt with in detail in Chapter 9—in particular, in Sections 9.3 and 9.4. Calculation procedures outlined there are directly applicable to thermoforming. Two aspects of the "melting" step should be emphasized. First, the processing temperature is an important operational variable. The lowest practical forming temperature is that at which a square box with fairly sharp corners can be drawn from the sheet without "whitening" or other visible damage. The highest forming temperature is that which brings about no excessive sagging of the sheet in the clamp and does not damage the polymer. Sagging is the combined result of thermal expansion and gravitational deformation. It is noteworthy that polymers commonly used in thermoforming (ABS, high impact PS) exhibit high "yield" values in the melt stress-strain relationships, which help them avoid sagging. The actual thermoforming temperature selected within the operating range involves a number of considerations. Low temperatures are advantageous in reducing both the heating and cooling segments of the cycle. Furthermore, a higher degree of the imposed biaxial orientation will be retained, which will increase the impact strength of the thermoform. Higher temperatures are advantageous because high extensions and exact mold reproduction are achieved. The sheets for thermoforming are generally extruded. Die forming imparts to the sheets considerable nonisotropic molecular orientation. Schmidt and Carley (24), for example, have observed a 31% shrinkage of 1.52 mm thick, high impact PS sheets in the extrusion direction with only slight shrinkage in the perpendicular direction. The heating step brings about annealing of some of this previous strain history. High forming temperatures facilitate this annealing process.

The other aspect of the melting step is sheet temperature distribution. One common requirement is stringent temperature uniformity throughout the sheet. Local temperature variations may cause undesirable local thickness variations. But even if the sheet is thermally balanced, the common vacuum forming method induces thickness variations. The deforming sheet comes in contact with the cold mold at different times. Hence the parts of the mold that are formed last (e.g., sharp corners) are the thinnest. This nonuniformity, inherent in the straight vacuum forming, is superimposed on the nonuniformity due to the deformation process itself, discussed in the preceding section. A simple model for simulating the thickness distribution is discussed below (Example 15.1). There are a number of ways to reduce thickness nonuniformity. Some of the process modifications in Fig. 1.22 are aimed to achieve this objective; a different, efficient means, however, involves "pattern heating" or "programmable zone heating" techniques. By this method a preset temperature distribution, $T(x, y)$ (where $x$–$y$ is the sheet plane) is imparted to the sheet, such that sheet temperatures in regions of high deformation (thin sections) are lower. For the quantitative analysis of the melting step under such boundary conditions, we must turn to numerical methods, either finite difference or finite element techniques. These were discussed in sections 9.4 and 16.3 respectively. Moreover, the theoretical analysis of the deformation process of such a nonisothermal sheet would probably be best achieved by finite element methods.

### Thickness Distribution in "Straight" Thermoforming

By "straight" thermoforming we mean a process such as that in Fig. 1.22*a*, where the preheated sheet is vacuum formed or pressure formed directly onto the walls of a cavity, without an auxiliary mechanical aid. The sheet undergoes a biaxial extension ("free blowing") until it touches the cold mold surfaces. Further bubble expansion and contact with the mold surface, in a rolling fashion, is determined by the shape of the mold. The sheet cooling process commences upon contact with the cold mold surface. The surface cools quickly, and this process, together with the frictional forces existing between the sheet and the mold, reduces the chances of any further thickness reduction at any position of the sheet in contact with the mold. Thus only the remaining (free) portion of the bubble continues to deform and decrease in thickness to a degree that is larger than a comparable "free blowing" process. The process is terminated when all the free surface of the expanding sheet contacts the mold surface.

Clearly, the part thickness distribution is technologically important. Such work has been carried out by Sheryshev et al. (30) and by Rosenzweig (29). The salient points of such investigations are illustrated in the example below.

#### Example 15.1 *Wall Thickness Distribution in a Conical Mold*

Consider a straight thermoforming process into a conical mold as in Fig. 15.13. We want to derive a theoretical expression for the thickness distribution of the final conical-shaped product. The sheet has an initial uniform thickness of $h_0$ and it is isothermal.

The following assumptions are made: the polymer is incompressible; the polymer deforms as an elastic solid*; the free bubble is uniform in thickness and has a spherical shape; the free bubble remains isothermal, but the sheet solidifies upon contact with the mold walls; there is no slip on the walls, and

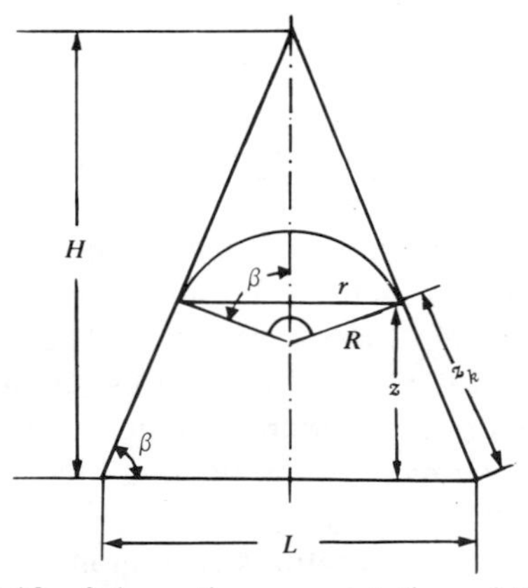


**Fig. 15.13** Schematic representation of thermoforming a polymer sheet into a conical mold.

* See sections 15.3 and 6.8.

the bubble thickness is very small compared to its size. With regard to the assumption that the free bubble is constant in thickness, we recall that Schmidt and Carley (24) observed broad thickness distributions in sheets that were *rapidly* stretched biaxially only *beyond* the hemispherical shape. Moreover, Denson and Gallo (31) observed a very narrow thickness distribution during *slow* deformation rates, of the order of $10^{-3}\ s^{-1}$, and for sheets inflated to shapes less than hemispherical. Thus the present analysis holds for thermoforming processes when the bubble is less than hemispherical.

In Fig. 15.13 we note the after a certain time the free bubble contacts the mold at height $z$ and has a spherical shape of radius $R$. The radius is determined by mold geometry and bubble position and is given by the following expression:

$$R = \frac{H - z_k \sin\beta}{\sin\beta \tan\beta} \tag{15.4-1}$$

where $H$ is the cone altitude, $z_k$ is the slant height to the point of contact, and $\beta$ is the angle indicated in Fig. 15.13. The surface area $A$ of the spherical bubble is

$$A = 2\pi R^2(1 - \cos\beta) \tag{15.4-2}$$

The thickness distribution of the wall is obtained by making a differential mass (volume) balance.

$$2\pi R^2(1-\cos\beta)h|_{z_k} - 2\pi R^2(1-\cos\beta)h|_{z_k+\Delta z_k} = 2\pi rh\ \Delta z_k \tag{15.4-3}$$

where $h$ is the thickness at $z_k$. The resulting differential equation for $h$, after substituting $r = R\sin\beta$, followed by some rearrangement, is

$$-\frac{d}{dz_k}(R^2 h) = \frac{Rh\sin\beta}{1-\cos\beta} \tag{15.4-4}$$

From Eq. 15.4-1 we get

$$\frac{dR}{dz_k} = -\frac{1}{\tan\beta} \tag{15.4-5}$$

and Eq. 15.4-4, with Eqs. 15.4-5 and 15.4-1, can be rearranged to

$$\frac{dh}{h} = \left(2 - \frac{\tan\beta\sin\beta}{1-\cos\beta}\right)\sin\beta\frac{dz_k}{H - z_k\sin\beta} \tag{15.4-6}$$

Integrating Eq. 15.4-6 with the initial condition $h(0) = h_1$, where $h_1$ is the initial thickness of the bubble tangent to the cone at $z_k = 0$ results in

$$\frac{h}{h_1} = \left(1 - \frac{z_k}{H}\sin\beta\right)^{(\sec\beta - 1)} \tag{15.4-7}$$

Finally, the initial thickness $h_1$ can be related to the original sheet thickness $h_0$ with Eq. 15.4-2 where $R = L/2\sin\beta$

$$\frac{\pi L^2 h_0}{4} = \frac{\pi L^2(1-\cos\beta)}{2\sin^2\beta}h_1 \tag{15.4-8}$$

The thickness distribution, therefore, is given by

$$\frac{h}{h_0}=\frac{1+\cos\beta}{2}\left(1-\frac{z_k}{H}\sin\beta\right)^{(\sec\beta-1)} \tag{15.4-9}$$

By a similar analysis thickness distributions can be derived in other relatively simple but more realistic molds, such as truncated cones. In such cases the model above holds until the bubble comes in contact with the bottom of the mold at its center. Following this point, new balance equations must be derived, providing the thickness distribution on the mold bottom and mold sides. A comparison of calculated thickness distributions with experimental data measured by Neitzert (32) indicated 10–45% lower values in thickness (29), with better agreement in shallow truncated cones than in deep ones. One source of disagreement is probably the substantial drawing of the polymer from under the clamped portion of the sheet. Otherwise, the model does generally predict the correct shape of the distribution.

## 15.5 The Blow Molding Process

Chapter 1 described in some detail the salient features of the different types of blow molding, and Section 13.5 dealt with the parison die design. This section covers the following aspects of the process: the problems associated with parison shape and

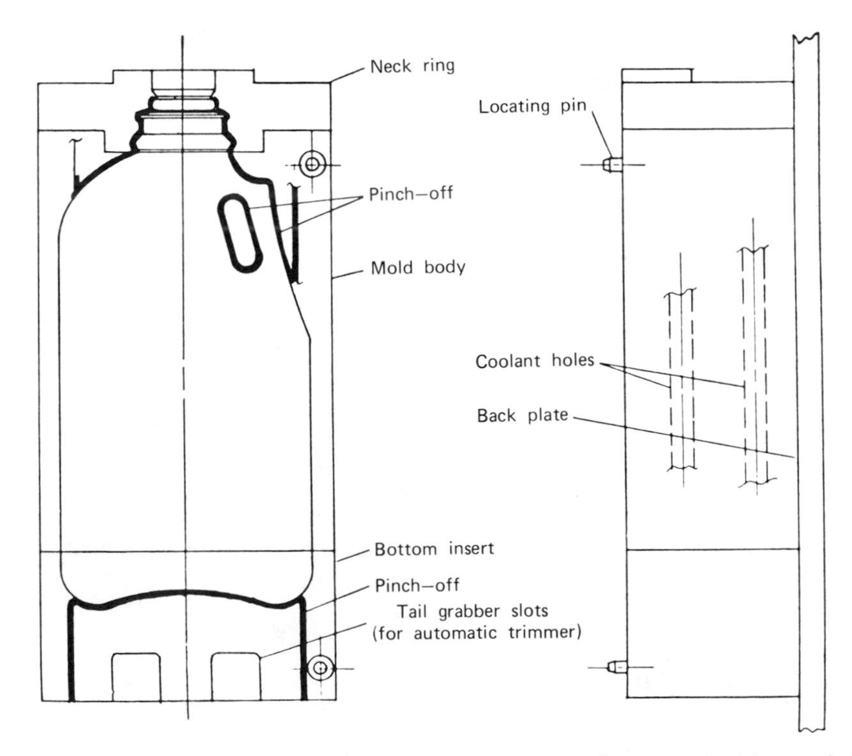


**Fig. 15.14** A blow molding mold for a container with a handle. [Reprinted with permission from J. D. Frankland, *Trans. Soc. Rheol.*, **19**, 371 (1975).]

thickness uniformity, the parison stretching and cooling processes, and finally, the question of part "structuring." These matters are discussed in connection with the intermittent extrusion blow molding process, including the ram-accumulator, which is used for large parisons, where difficulties with uniformity and parison stability are encountered. The continuous blow molding process, of course, is a special case of intermittent extrusion. Figure 15.14 illustrates the type of mold and blown article to be discussed. Its large size is one of the advantages of intermittent blow molding. Blow molding and thermoforming are similar only during the stretching process of the molten tube (parison) and the sheet, respectively. But even in this process there exist differences. First in the blow molding process the stretching can be described by the planar extension kinematics (Eq. 6.8-7), and in thermoforming by biaxial extension (Eq. 6.8-9). Second, the temperature level at which stretching occurs is much higher for the blow molding process, since the parison was just produced by a reciprocating screw device and had to flow rapidly in narrow channels. Thus during stretching the parison is expected to be "less elastic" than the corresponding sheet that is being thermoformed. It follows then from the viscoelastic fluid analysis of Section 6.8 that parison stretching is more a viscous process occurring near or below the critical strain rate $\dot{\varepsilon}_0 = (2\lambda_{max})^{-1}$, whereas the stretching of thermoformed films can be best described as an elastic process, as in the work of Schmidt and Carley (24). Our analysis of the parison inflation process reflects this point.

### The Parison Formation

From both the product design and the economic point of view, the shape and thickness of the parison throughout its entire length, and before mold closing, must be controllable. If the degree of inflation is independent of the axial distance, the parison thickness must be uniform. If it varies, it should be thicker where the degree of inflation is larger. Only under such conditions can acceptable product strength levels be reached with the *minimum* product weight. The parison diameter must also be controlled, not only because it contributes to the product thickness uniformity but for two additional reasons. First, its "lay flat" width ($\pi D_{parison}/2$) should not exceed the mold diameter, since this will interfere with mold closing. Second, if the blown product has a handle, then at the axial height of the handle, half the "lay flat" must be larger than the radial distance of the outer handle edge. This is because handle "pinch off" occurs before any appreciable radial stretching of the parison.

From the discussion above, we can conclude that because of the flow rate variation during parison formation and the presence of gravitational forces, the parison diameter and thickness are not uniform along the axial direction. Furthermore, these parameters are not easily controllable and certainly not predictable from basic rheological measurements. Therefore practical solutions must be found and approximate analyses proposed.

Referring to Fig. 13.25, the parison thickness $h_p(z)$ and radius $R_p(z)$ are related to each other and to the volumetric flow rate. For a parison die head extrusion angle $\Theta = 0$, we can, in principle, estimate the parison thickness from extrudate swelling experiments with capillaries, at the same wall shear stresses. In this case, the following problems must be taken into consideration. First, the flow rate (thus shear stresses) varies with time, and second, only the leading portion of the parison is expected to achieve its full extrudate swelling value. Gravitational forces acting on the rest of the parison impose a constant tensile stress on it, which restricts swelling and causes axial deformation. This deformation at short times is viscoelastic. Commercial HDPE used in blow molding has very high molecular weights to avoid sagging of the parison (i.e., to make the deformation more elastic). Cogswell *et al.* (33) found that for such resins, even for large parisons, 70% of the "sag" is elastic after 8 s. Furthermore, he offers a very approximate method of evaluating the cross-sectional area and thickness reduction, and the accompanying parison length increase, due to gravity. Finally, the problem of "bounce," caused by the parison deceleration, is discussed in terms of the large inertial forces that are present, especially in large parisons. The problem of the parison bounce is important in practice, because the length of the parison at the moment of mold closing depends on it.

When $\Theta > 0$ (Fig. 13.25), the flow in the conical section annular channel is not viscometric; thus capillary flow extrudate swelling results do not correlate with the observed $h_p(z)$. The parison radius $R_p(z)$ is an even more difficult quantity to predict, because it depends not only on the melt flow experience inside the die but also on the forces acting on the parison, its elastic "strength," and probably its elongational viscosity. Miller (34) attempted to correlate the ratio of the final radius of the leading edge of the parison to that at the exit, $R_f/R_1$, to the structure and conventional rheological properties of a number of HDPE resins. No correlation was found.

Schaul, Hannon and Wissbrun (35) discuss another problem related to parison formation, parison "pleating" or "curtaining." The parison upon exiting the die is smooth and fully blown, but at a given downstream distance (or a certain time) it may form pleats (Fig. 15.15). Pleating involves buckling of the parison under its own weight. Thus the occurrence of pleating should be a function of the parison wall thickness and should depend on the angle of extrusion (larger gravity component and a larger moment arm). From a melt property viewpoint, increased melt strength* and higher extrusion rates (which make the melt more elastic and allow less time for creep) alleviate pleating. The statements above are borne out experimentally (35).

In the absence of the ability to quantitatively analyze the parison formation problem and to relate it to fundamental rheological properties, Wissbrun et al. (35) have approached it in an empirical way: four parison "properties" were evaluated at various levels of peak shot pressure and die gap, the two available process variables in reciprocating screw blow molding. Response surfaces were experimentally obtained for the final parison diameter, product (bottle) weight, severity of melt fracture, and pleating. By specifying minimum acceptable levels for each property (plotting the four "acceptable level" curves on one graph), they were able

* *Melt strength* or drawdown tension is the force measured during the drawdown of a melt strand exiting a constant flow rate viscometer die. Both the isothermal (36) and nonisothermal (37) responses have been studied. In the latter case, where the strands are air cooled, melt strength is found to be nearly independent of the takeup speed, but dependent the flow activation energy. Wissbrun (38) suggests that it might be related to the steady elongational but not solely dependent on it. Frequently it is found to correlate with processing behavior such as parison diameter (35).

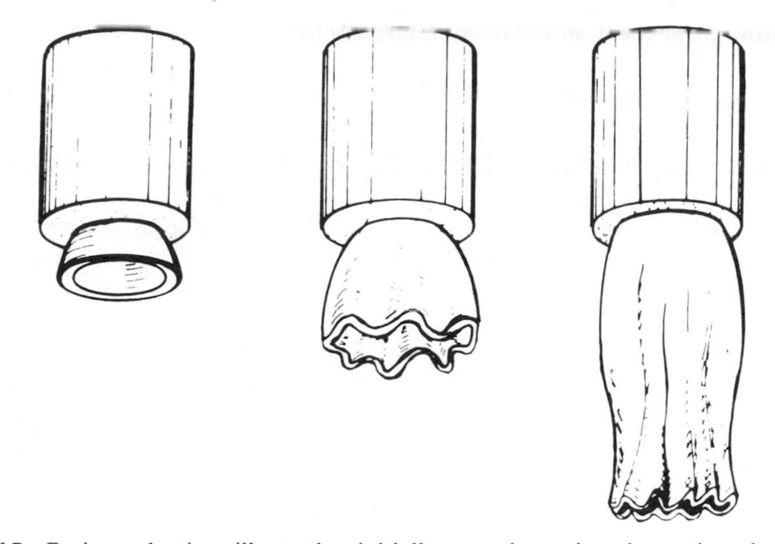

**Fig. 15.15** Parison pleating, illustrating initially smooth partison becoming pleated with increased length. [Reprinted with permission from J. S. Schaul, M. J. Hannon and K. F. Wissbrun, *Trans. Soc. Rheol.*, **19**, 351 (1975).]

to obtain "operating lines," given in Fig. 15.16. It is important to emphasize that these results are *specific* to the polymer and parison forming system used. The heavy line represents the acceptable range of shot pressure and die gap values to produce the specified product. It is worth mentioning again that the acceptable region of the shot pressure–die gap space is in the range *beyond* the melt fracture related flow curve discontinuity (see Section 13.2).

Turning to parison "programming" or "profiling"—that is, the intentional variation of the parison thickness by varying the die gap during forming—we refer to Fig. 13.25. An axial movement $\Delta z$ of the core of the die downward reduces the

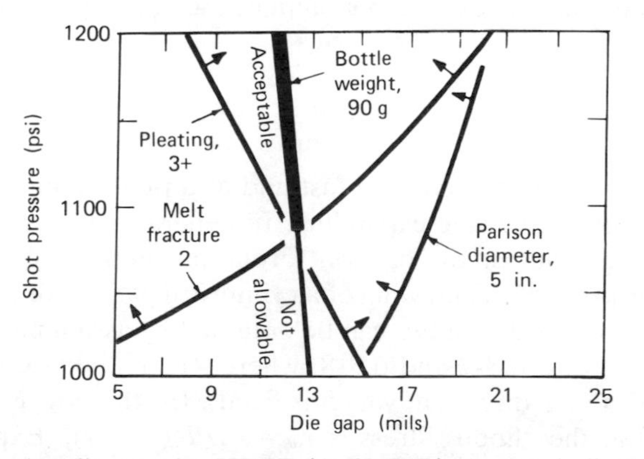


**Fig. 15.16** Operating diagram for HDPE (A-60-70R). Arrows indicate regions of acceptable operation; heavy line is the allowable operating line for this combination of polymer, product, and machine conditions. [Reprinted with permission from J. S. Schaul, M. J. Hannon and K. F. Wissbrun, *Trans. Soc. Rheol.*, **19**, 351 (1975).]

inside radius $\rho_{ii}$ at location $z_i$ (assuming $\alpha = 0$) by

$$\Delta\rho_{ii} = \Delta z \sin \Theta \tag{15.5-1}$$

This will affect the parison die design equation because the radius ratio $\beta$ will change, and so will $F(n, \beta_i)$, the shape factor. Assuming, for simplicity, that the flow in the conical annular channel can be approximated by that of pressure flow between parallel plates, we can use Table 13.1 (for a power law model fluid) where $H = (\rho_{0i} - \rho_{ii})$ and $q = Q/\pi(R_{0i} + R_{ii})$

$$q = \frac{H^2}{2(s+2)}\left[\frac{H \cos\theta}{2m}\left(-\frac{dP}{dz}\right)\right]^s \tag{15.5-2}$$

where the pressure gradient in the direction of flow is $-\cos\theta\, dP/dz$ was substituted for the pressure drop. The shear rate at the wall from Table 13.1 is

$$\dot{\gamma}_w = \left[\frac{H \cos\theta}{2m}\left(-\frac{dP}{dz}\right)\right]^s \tag{15.5-3}$$

and combining Eqs. 15.5-2 and 15.5-3, $\dot{\gamma}_w$ can be expressed in terms of $q$ and $H$ as

$$\dot{\gamma}_w = \frac{2(s+2)q}{H^2} \tag{15.5-4}$$

Let us now examine the effects of a downward die core movement on the parison thickness, assuming a constant shot pressure. First of all, the parison will widen by $\Delta H = \Delta\rho_{ii} = \Delta z \sin\theta$, as much as the die opens, and the flow rate will increase according to Eq. 15.5-2. In addition, since the melt undergoes swelling, there will be a change in swelling. Referring to Eq. 15.5-3, the wall shear rate will increase with increasing $H$, and since swelling increases with increasing $\dot{\gamma}_w$, there will be an increased amount of swelling compared to that before the core movement.*

This analysis of parison programming is approximate. In practice, the core movement programming $\Delta z(t)$ must be specified, or at least "fine tuned" empirically, using the resin, die, and extrusion system of the actual process.

### Parison Inflation

In practice, the parison is inflated very fast and at a predetermined rate, which is such that it does not burst while expanding. In general, parison inflation is a less critical problem than that of the parison formation. Following Denson (39), an approximate description of the blowing of a cylindrical parison of uniform radius $R_i$ and thickness $h_i$ to that of $R_o$ and $h_o$ can be obtained by assuming that (*a*) the flow is planar extension (Eqs. 6.8-7 and 6.8-8, where "1" is the $\Theta$-direction, "2" the thickness, and "3" the $z$-direction, which is fixed), (*b*) the flow is isothermal, and (*c*) $h/R \ll 1$, so that the "hoop" stress is $\tau_{\theta\theta} = -[PR(t)/h(t)]$. Experimentally the

* No quantitative predictions are possible from capillary extrudate swelling, since the conical section annular flow is complex and nonviscometric.

planar extension viscosity, at *very low* strain rates, can be expressed as

$$\bar{\eta}_{pl} = K(\dot{\varepsilon}_{pl})^{n-1} \tag{15.5-5}$$

as shown for polyisobutylene in Fig. 15.17. Thus

$$\tau_{\theta\theta} = -\frac{PR(t)}{h(t)} = -K(\dot{\varepsilon}_{pl})^{n-1}\dot{\varepsilon}_{pl} \tag{15.5-6}$$

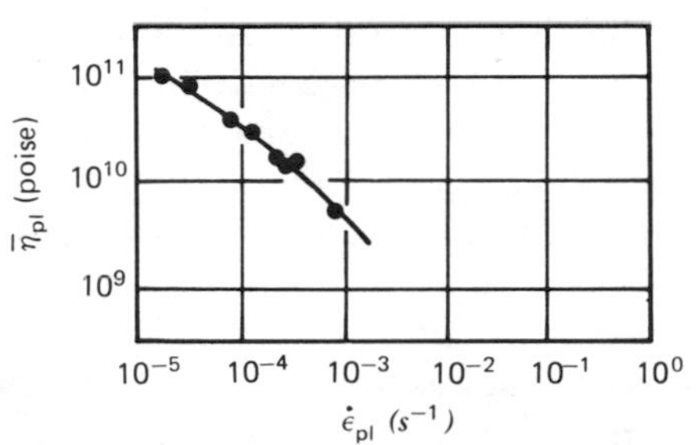


**Fig. 15.17** Planar extension viscosity as a function of strain rate for a high molecular weight PIB. [Reprinted with permission from C. D. Denson, *Polym. Eng. Sci.*, **13**, 125 (1973).]

Since

$$\dot{\varepsilon}_{pl} = \frac{1}{R(t)}\frac{dR(t)}{dt} \tag{15.5-7}$$

and

$$V = 2\pi R(t)h(t)L = \text{constant} \tag{15.5-8}$$

we obtain the expression

$$\frac{dR}{dt} = CP^{s}R^{2s+1} \tag{15.5-9}$$

where

$$C = \left(\frac{2\pi L}{VK}\right)^{s} \tag{15.5-10}$$

Equation 15.5-9 can be solved for any time dependent or constant pressure to give the radial value at any time. For example, if $P$ is constant, parison inflation time is

► $$t = \frac{1}{2sCP^{s}}\left[\left(\frac{1}{R_i}\right)^{2s} - \left(\frac{1}{R_0}\right)^{2s}\right] \tag{15.5-11}$$

The analysis above is based on the assumption that the behavior of the cylindrical parison responds according to Eqs. 15.5-5 and 15.5-6. As discussed in Section 6.8, viscous flow behavior is expected to occur below the "critical" strain rate $\dot{\varepsilon} = (2\lambda_{max})^{-1}$. Since the stretching rates in the blow molding process are fast, the analysis of Denson (39) may hold only for high melt temperatures, where the maximum relaxation time is small. Yet a good deal of orientation in the $\theta$-direction is retained in blow molded containers, suggesting that the elastic (deformation) nature of the stretching process may have to be taken into consideration.

## Parison Cooling

In general, parison cooling presents problems (in increasing the overall blow molding cycle) only when the final parison thickness is large. In thin blown articles the mold is opened when the "pinched-off" parts have solidified so that they can be easily stripped off; thus they are the rate controlling elements as far as cooling is concerned. Of course it is important to note that for fast blow molding of even very thin articles, the *rate of crystallization* must be fast. For this reason HDPE, which crystallizes fast is ideally suited, as are amorphous polymers, which do not crystallize.

For thick blown articles the process of cooling and solidification takes an appreciable time. For this reason, it is important to so conduct the parison inflation step, and to apply sufficient pressure, that the entire cooling part is in good contact with the cold wall of the mold, thus maintaining a high heat transfer coefficient. Additionally, cooling can be supplied through the inflating gas. Assuming that the inflated parison can be represented by a flat sheet that is infinite in width and length, the temperature field can be obtained by the methods developed in Section 9.4.

From a structuring point of view, thin cylindrical blow molded articles solidify while the molecules are preferentially oriented in the $\theta$-direction. If the orientation is extensive, one should expect row nucleation in the $z$-direction. In thick crystallizable articles only the skin is expected to retain any of the applied orientation during inflation.

A recent successful attempt to incorporate biaxial orientation into blow molded cylindrical containers, such as bottles, is the stretch blow molding process briefly discussed in Chapter 1. In this process the parison is first stretched and immediately afterward blown radially. Axial temperature uniformity is essential, otherwise bursting may occur. Furthermore, the temperature (averaged across the parison thickness) must be only a few degrees above $T_g$, for amorphous polymers, which are commonly used in this process. Thus the relaxation times of the melt at such a low temperature level are larger than the time required to cool the part, which results in extensive orientation and structuring of the product. Lightweight, impact resistant bottles can be produced in this fashion, attesting to the potential of structuring in polymer processing.

## *REFERENCES*

1. H. F. Mark, in *Rheology*, Vol. 4, F. R. Eirich, ed., Academic Press, New York, 1969, Chapter 7.
2. S. Kase and T. Matsuo, "Studies on Melt Spinning. I. Fundamental Equations on the Dynamics of Melt Spinning," *J. Polym. Sci.*, Part A, **3**, 2541 (1965).

3. C. D. Han, *Rheology in Polymer Processing*, Academic Press, New York, 1976. (a) Chapter 8; (b) Section 12.3.1.
4. M. E. Morrison, "Numerical Evaluation of Temperature Profiles in Filaments Undergoing Solidification," *Am. Inst. Chem. Eng. J.*, **16**, 57 (1970).
5. J. E. Spruiell and J. L. White, "Structure Development During Polymer Processing: Studies of Melt Spinning of PE and PP," *Polym. Eng. Sci.*, **15**, 660 (1975).
6. J. R. Dees and J. E. Spruiell, "Structure Development During Melt Spinning of Linear Polyethylene Fibers," *J. Appl. Polym. Sci.*, **18**, 1053 (1974).
7. C. J. S. Petrie and M. M. Denn, "Instabilities in Polymer Processing," *Am. Inst. Chem. Eng. J.*, **22**, 209 (1976).
8. J. C. Miller, *Soc. Plast. Eng. Trans.*, **3**, 134 (1963).
9. S. Kase, "Studies on Melt Spinning. IV. On the Stability of Melt Spinning," *J. Appl. Polym. Sci.*, **18**, 3279 (1974).
10. J. L. White and Y. Ide, "Instabilities and Failure in Elongational Flow and Melt Spinning of Fibers," *J. Appl. Polym. Sci.*, in press.
11. G. F. Cruz-Saenz, G. J. Donnelly, and C. B. Weinberger, "Onset of Draw Resonance During Isothermal Melt Spinning," *Am. Inst. Chem., Eng. J.*, **22**, 441 (1976).
12. J. R. A. Pearson and Y. T. Shah, "Stability Analysis of the Fiber Spinning Process," *Trans. Soc. Rheol.*, **16**, 519 (1972).
13. R. J. Fisher and M. M. Denn, "Finite Amplitude Stability and Draw Resonance in Isothermal Melt Spinning," *Chem. Eng. Sci.*, **30**, 1129 (1975).
14. I. J. Chen, G. E. Hagler, L. E. Abbott, D. C. Bogue, and J. L. White, "Interpretation of Tensile and Melt Spinning Experiments in LDPE and HDPE," *Trans. Soc. Rheol.*, **16**, 473 (1972).
15. G. Vassilatos, "On the Stability of Drawdown of Polymer Melts," Paper presented at the 68th Annual Meeting of the American Institute of Chemical Engineers, Los Angeles, 1975.
16. J. R. A. Pearson and Y. T. Shah, "On the Stability of Isothermal and Nonisothermal Fiber Spinning of Power Law Fluids," *Ind. Eng. Chem. Fundam.*, **13**, 134 (1974).
17. J. R. A. Pearson and C. J. S. Petrie, "The Flow of a Tubular Film. Part I. Formal Mathematical Representation," *J. Fluid Mech.*, **40**, 1 (1970).
18. C. J. S. Petrie, "A Comparison of Theoretical Predictions with Published Experimental Measurements on the Blown Film Process," *Am. Inst. Chem. Eng. J.*, **21**, 275 (1975).
19. W. Ast, "Der Abkählvorgang bein Herstellen von Blasfolien aus Polyäthylen Niedriger Dichte," *Kunststoffe*, **63**, 427 (1973).
20. C. D. Han and J. Y. Park, "Studies on Blown Film Extrusion. II. Analysis of the Deformation and Heat Transfer Processes," *J. Appl. Polym. Sci.*, **19**, 3277 (1975).
21. C. D. Han and J. Y. Park, "Studies on Blown Film Extrusion. I. Experimental Determination of Elongational Viscosity," *J. Appl. Polym. Sci.*, **19**, 3257 (1975).
22. C. D. Han and J. Y. Park, "Studies on Blown Film Extrusion. III. Bubble Instability," *J. Appl. Polym. Sci.*, **19**, 3291 (1975).
23. G. Menges and W. O. Predöhl, "Certain Aspects of Film Blowing of LDPE," *Polym. Eng. Sci.*, **15**, 394 (1975).
24. L. R. Schmidt and J. F. Carley, "Biaxial Stretching of Heat Softened Plastic Sheets: Experiments and Results," *Polym. Eng. Sci.*, **15**, 51–62 (1975).
25. T. Alfrey, Jr., "Fabrication of Thermoplastic Polymers," *Appl. Polym. Symp.* No. 17, 3–24 (1971).
26. J. Finger, *Akad. Wiss. Wein. Sitzber.*, **103**, 1073 (1894).
27. R. S. Rivlin, *Phil. Trans. Roy. Soc.*, **A240**, 459 (1948).
28. T. Alfrey, Jr., in *Applied Polymer Science*, J. K. Craver and R. W. Tess, ed., American Chemical Society, Washington, D.C., 1975, Chapter 5 "Structure—Property Relations in Polymers," p. 51.

29. N. Rosenzweig, "Process Simulation of Thermoplastic Sheet Forming," M.S. thesis, Department of Chemical Engineering, Technion—Israel Institute of Technology, Haifa, 1975. See also N. Rosenzweig, M. Narkis and Z. Tadmor, "Wall Thickness Distribution in Thermoforming", Submitted to *Polym. Eng. Sci.* 1978.
30. M. A. Sheryshev et al., *Sov. Plast.*, No. 11, 33 (1970).
31. C. D. Denson and R. J. Gallo, "Measurements on the Biaxial Extension Viscosity of Bulk Polymers: The Inflation of a Thin Sheet," *Polym. Eng. Sci.*, **11**, 174 (1971).
32. W. A. Nietzert, *Plasterarbeiter*, **18**, 316 (1967).
33. F. N. Cogswell, P. C. Webb, J. C. Weeks, S. G. Maskell, and P. D. R. Rice, "The Scientific Design of Fabrication Processes: Blow Molding," *Plast. Polym.*, **39**, 340 (1971).
34. J. C. Miller, "A Rheological Product Problem in Blow Molding," *Trans. Soc. Rheol.*, **19**, 341 (1975).
35. J. S. Schaul, M. J. Hannon, and K. F. Wissbrun, "Analysis of Factors Determining Parison Properties in High Shear Rate Blow Molding," *Trans. Soc. Rheol.*, **19**, 351 (1975).
36. G. M. Fehn, "Steady-State Drawing of Polymer Melts," *J. Polymer Sci.*, **6**, 247 (1968).
37. A. Bergonzoni and A. J. DiCresce, *Polym. Eng. Sci.*, **6**, 50 (1966).
38. K. F. Wissbrun, "Interpretation of the Melt Strength Test," *Polym. Eng. Sci.*, **13**, 342 (1973).
39. C. D. Denson, "Implications of Extensional Flows in Polymer Fabrication Processes," *Polym. Eng. Sci.*, **13**, 125 (1973).

## *PROBLEMS*

**15.1 *Transport Equations for Nonisothermal Fiber Stretching.*** Starting from the momentum and thermal energy balances derive Eqs. 15.1-1 and 15.1-2 which are used for the description of the nonisothermal stretching of molten polymer fibers.

(a) Clearly discuss the assumptions made, and specifically discuss: the nature of $F_D$ and its inclusion in the momentum balance (e.g., "Where did it come from?"); the relative magnitude of gravity (use real $\bar{\eta}$ and $\dot{\varepsilon}$ data); the form of the entire Eq. 15.1-2 and the absence of radial temperature gradients.

(b) What additional complexities to the solution of the problem would the inclusion of $dT/dr$ bring?

**15.2 *Estimation of $T$ ($r = 0, t$) in Melt-Spun Fibers.*** From $T(x_1, \alpha t)$ values (Fig. P-15.2) calculate the centerline temperature of a HDPE melt strand exiting the spinnerette at 240°C at a takeup speed of 50 m/min and a mass flow rate of 1.93 g/min, at a distance 50 cm below the spinnerette plate (Fig. 15.2).

(a) Assume that no change in phase occurs and that the heat transfer coefficient is $10^{-3}$ cal/cm$^2 \cdot$ s $\cdot$ °K, $k = 8 \times 10^{-4}$ cal/cm $\cdot$ s $\cdot$ °K, $\rho = 0.75$ g/cm$^3$, $C_p = 0.5$ cal/g and the cooling medium temperature is 25°C.

(b) What can you conclude about the magnitude of $dT/dr$ relative to $dT/dz$? Repeat for a shorter spinnline distance of, say, 10 cm.

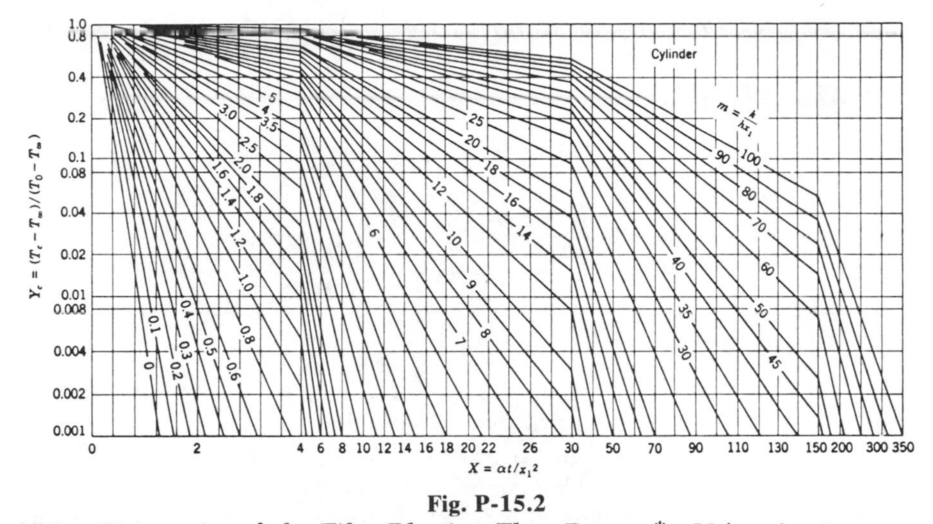


**Fig. P-15.2**

**15.3** ***Kinematics of the Film Blowing Flow Process.**** Using the $\xi_i$ coordinates which are embedded in the *inner* surface of the bubble and considering that $v_2 = d\delta/dt$ at $\xi_2 = \delta$, $v_3 = 2\pi(dR/dt)$ calculate the nonzero components of $\dot{\gamma}$ for this flow (Eq. 15.2-1). The coordinate system is shown on Fig. P-15.3. Assume that $\delta/R \ll 1$.

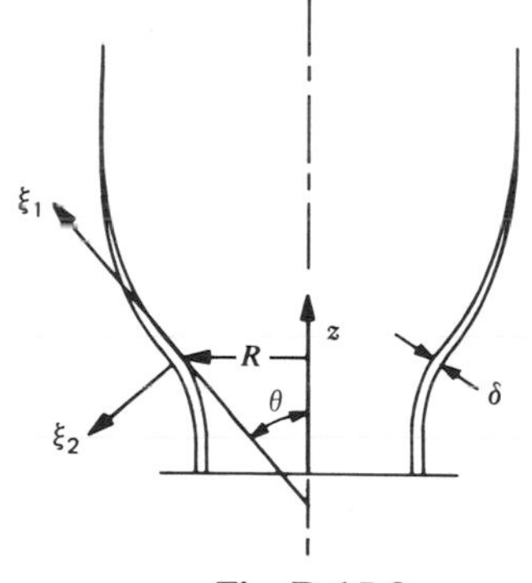


**Fig. P-15.3**

**15.4** ***Wall Thickness—Thermoforming a Cup.*** Example 15.1 treated the thickness distribution of a flat sheet which is thermoformed into a conical mold. Consider a cup 6 cm in diameter throughout its height and 10 cm high, which has to be made out of a 1.5 mm thick molten HIPS sheet. Following a procedure similar to that in Example 15.1, derive an expression for the thickness distribution assuming that: the free bubble is of spherical shape until its top reaches the mold bottom; once the bubble touches the mold, no further deformation occurs; the thickness of the free bubble at any stage of its deformation is spatially uniform. Once the melt touches the bottom of the mold, the deformation which fills the corners can be represented by spherical sections of successively smaller radii and centers which move diametrically to the corners.

* S. Middleman, *Fundamentals of Polymer Processing*, McGraw-Hill, New York, 1977, Chapter 10.

**15.5 *Blown Film Equations of Equilibrium.*** Derive the blown film force balance (equilibrium) equations 15.2-6 and 15.2-7. The two principal radii of curvature are given by:

$$R_C = R\left[1+\left(\frac{dR}{dz}\right)^2\right]^{1/2} = \frac{R}{\cos\theta}$$

$$R_L = -\frac{\left[1+\left(\frac{dR}{dz}\right)^2\right]^{3/2}}{d^2R/dz^2} = -\frac{\sec^3\theta}{d^2R/dz^2}$$

where $R$ and $\theta$ are shown on Fig. P.15-3.

**15.6 *Blown Film Deformation During Folding by the "Tent" and the Nip Rolls.**** * In the film blowing process the solidified round film bubble comes into contact with a series of guides (slats or rollers) that form the "tent," and gradually collapses into a folded configuration between the nip rolls. Folding the bubble and feeding it into the nip rolls impose a deformation on the film, which is a function of the angle $\theta$ in the plane of the cross section of the bubble. The deformation results from the fact that each fiber of the film, that is, at each $\theta$, travels a distance $L(\theta)$ which is different from the last round cross section to the nip rolls. Derive an expression for this deformation. What effect does it have on the wraping step following the slitting?

* William Arruda, private communication.

# 16

# Calendering



## 16.1 The Calendering Process

Section 1.1 briefly described the calendering process. The number of rolls of a calender is determined by the nature of material processed and the product. Rubber can be calendered on a two roll calender, with four roll calenders generally used for double coating of substrates (Fig. 16.1*a*). However, the surface quality requirements of calendered thermoplastic polymer requires four roll calenders (Figs. 16.1*b*, 16.1*c*). In the latter case the polymer passes three nip regions. The first pass is the "feed" pass, the second the "metering" pass, and the third "sheet formation", "gauging and finishing" pass (1). Calenders with the five rolls in various arrangements are also used. Transfer from one roll to the next is accomplished by some combination of differentials in roll speed, temperature, and surface finish (2). The width of the sheet (when the speed of both rolls is equal) changes at each nip in inverse proportion to the decrease in thickness.

The production rate of a calendering line, when not limited by the mixing and melting capacity upstream, is determined primarily by the size and surface requirements of the product and the properties of the polymer (1). Thus heavy sheets of 0.25 mm and up can be produced at 60 m/min without difficulties. Even higher speeds are possible if the sheet is posttreated (e.g., embossed, top coated). But certain rigid, glossy, roller polished sheets are produced at much lower rates of 10–35 m/min. Thin flexible films can be produced at 100 m/min at the roll and 125 m/min at the winder. The higher speed at the winder is due to a drawdown process that helps in producing thin films (0.04 mm and below); films of such thickness would be hard to separate from the roll. Calender sizes range up to 90 cm (36 in.) in diameter and 250 cm (97 in.) wide, with polymer throughputs up to 4000 kg/hr.

Surface temperature of the rolls is carefully controlled by using drilled rolls—that is, axially drilled holes all around the periphery—in which a temperature controlling liquid is circulated.

The calendering process is commonly used for shaping high melt viscosity thermoplastic sheets. The calendering process is particularly suitable for polymers susceptible to thermal degradation or containing substantial amounts of solid

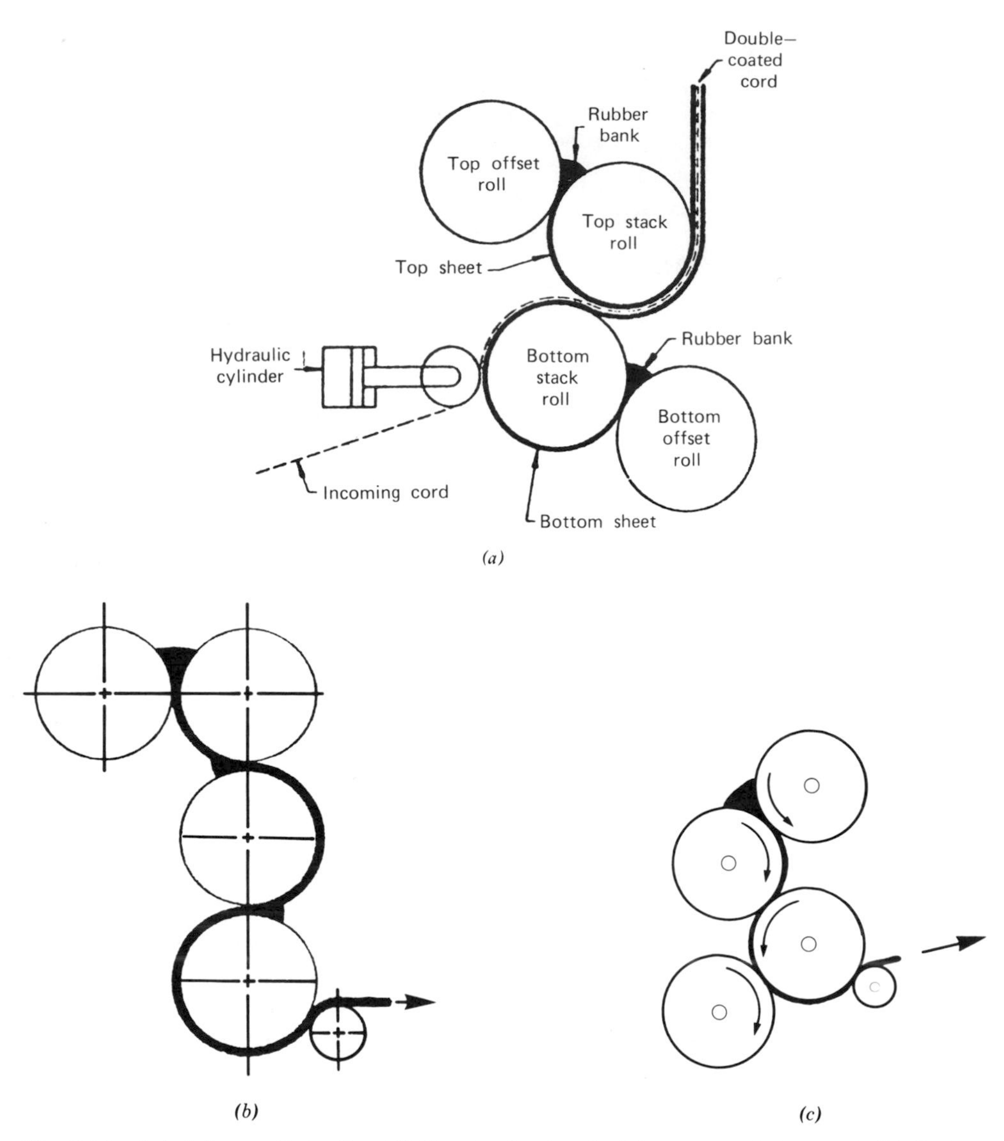


**Fig. 16.1** (*a*) A four roll, inclined "Z" calender for double casting of tire cord. (*b*) A four roll, inverted "L" calender. (*c*) A four roll "Z" calender.

additives. This is because the calender can convey large rates of melt with a small mechanical energy input (compared to an extruder). The thickness of the calendered product must be uniform in both the machine and cross-machine directions. Any variation in gap size due to roll dimensions, setting, thermal effects, and roll distortion due to high pressures developing in the gap, will result in product nonuniformity in the cross-machine direction. Eccentricity of the roll with respect to the roll shaft, as well as roll vibration and feed uniformity, must be tightly controlled to avoid nonuniformity in the machine direction. A uniform empty gap size will be distorted in operation because of hydrodynamic forces, developed in the nip, which deflect the rolls. The resulting product will be thick in the middle and thin at the edges (Fig. 16.2). Three common methods are employed to compensate for this deformation, which are commonly referred to as "roll crown," "roll-crossing," and "roll-bending." "Roll crown" indicates that the roll diameter at the

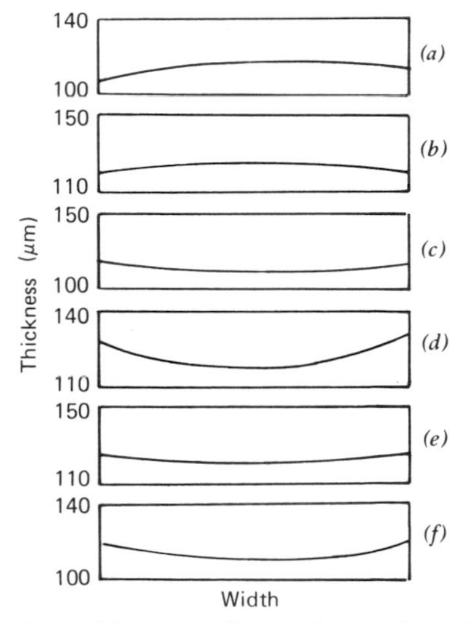


**Fig. 16.2** Effect of roll-crossing and bending on web uniformity, in a calender 1.8 m wide. (*a*) No crossing or bending. (*b*) Crossing, 4 mm; no bending. (*c*) Crossing, 8 mm; no bending. (*d*) Crossing, 12 mm; no bending. (*e*) No crossing, bending of 10 Mp. (*f*) No crossing, bending of 16 Mp. (D. Katz, "An Inquiry on the Behavior of Bingham Materials in Calender Processing," M.S. thesis, Department of Mechanics, Technion—Israel Institute of Technology, Haifa, 1973.)

center is slightly greater than at the edges. In principle, by an appropriate roll diameter, profile roll deflection can be exactly compensated for *given* operating conditions. Roll-crossing and roll-bending provide means for continuous adjustment of gap size distribution. Roll-crossing results in wider gap at the edges and can easily be visualized giving the same effect as increasing the "roll crown." In roll-bending a bending moment is applied on both ends of the roll by two additional bearings, which can increase or decrease the bending due to hydrodynamic forces. Figure 16.2 shows the effect of roll-crossing and roll-bending on product uniformity. An exact knowledge of the hydrodynamic pressure distribution in the nip is therefore necessary for predicting by structural analysis the exact gap thickness distribution, as well as the load on the bearings.

Accurate gap thickness control and the stringent roll temperature uniformity requirements are indicative of the sensitivity of the product quality to minor variations in conditions. It is not surprising, therefore, that a calender line takes a long time, sometimes hours, to "stabilize"—that is, to reach steady state. Consequently calender lines are best utilized in long production runs. The ruggedness and basic simplicity of the machine elements involved are fully compatible with such long runs.

## 16.2 Mathematical Modeling of Calendering

A comprehensive mathematical model of the calendering process should consist of a hydrodynamic and roll structural analysis (coupled because of roll deflection), heat transfer in the deforming polymer and the rolls, and product response to drawdown. By taking into account the rheological properties of the material, feed conditions, and operating conditions, such as roll speeds and temperatures, gap separation, roll crossing and bending, the following matters can be ascertained: the exact nature of the flow in the nip, the width variation from nip to nip, the thickness and temperature distribution, as well as the effect of these conditions on the transfer of the material from roll to roll, and the onset of instabilities.

Such a model would assist the calender designer to select roll size, gap separation, roll "crown," "crossing," and "bending" requirements, and operating conditions for given production rates and quality requirements.

The first step in developing such a model (cf. Section 5.2) is to gain a clear *qualitative* picture on the exact nature of the flow mechanism. A viscoelastic melt is fed into the first nip in strips. The melt accumulates in the center zone of the nip area and simultaneously undergoes flow into the nip and sideways; the drag induced flow leads to pressure buildup, which inevitably produces pressure gradients in the machine and cross-machine directions, resulting in the flow above. Experimental evidence of such a pressure distribution is given by Unkrüer (3), who reports on detailed calendering studies of PVC and PS. Figure 16.3 gives pressure profiles at three cross-machine locations. Thus a complex, three-dimensional flow field is set up with an a priori unspecified free boundary. Axial flow (cross-machine direction) continues throughout the nip zone all the way to the exit, but the rate

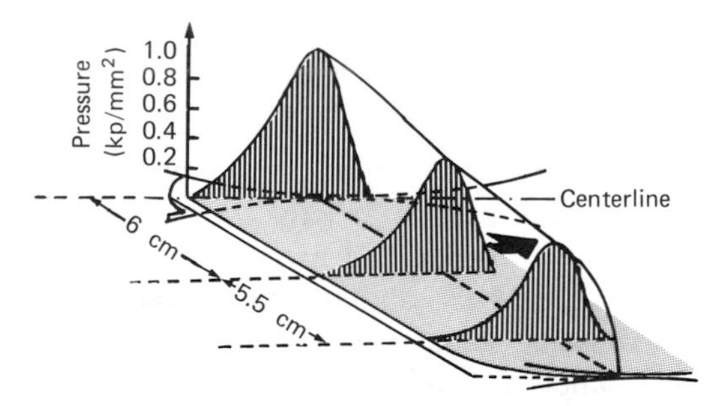


**Fig. 16.3** Pressure profiles in the calender gap at various axial positions, with rigid PVC (Vestolit Z 1877) at equal roll speeds of 5 cm/s and roll temperature of 185°C: Minimum gap, 0.6 mm; roll diameter, 30 cm; width, 50 cm. Note the drop in pressure in the cross-machine direction with distance from the centerline, which drops to zero at the end of the rolled web. (Reprinted with permission from W. Unkrüer, Doctoral thesis, IKV, Technischen Hochschule, Aachen, 1970.)

varies because of the varying gap size. That is, in the narrow region of the nip drag flow in the direction of rotation is predominant as compared to cross-machine pressure flow. According to Marshall (2), it can be assumed that the increase in width is virtually limited to the entrance zone up to where the peak pressure is obtained. The actual flow in the nip area is further complicated because the gap clearance varies axially as a result of built-in roll-crowns, hydrodynamic flexing, and bending of the rolls. All these factors should bring about such a flow distribution in the nip area that results in uniform flow rate per unit width. Minor variations of pressure profiles in the direction of rotation will cause variations in detachment locations, hence thickness variations.

In light of this qualitative picture we can interpret recent developments of the calendering process, whereby an extruder equipped with a simple and short, sheet forming die, feeds the nips uniformly throughout the nip width of only one pair of calender rolls. The sheet forming die, therefore, performs the functions of the first nip—namely, spreading out the material and feeding it at a more or less uniform rate to the second nip.

The function of the second and third nips are a further reduction of the thickness and of the flow rate nonuniformities in both machine and cross-machine directions. The rolling banks in all nips act as reservoirs that can accommodate and attenuate flow rate fluctuations. Thus no sharp qualitative distinction should be made between the functions of the three nips; it is the relative significance of the various functions that changes from nip to nip. All nips "meter" flow rate, reduce thickness, and "wash out" variations in thickness and flow rate to various degrees (just as in plasticating screw extruders, flow rate is determined by the whole length of the screw, not merely by the "metering" section).

Clearly there is no simple solution to this three-dimensional flow, in a complex geometry (variable gap thickness in two directions) with rheologically complex fluids under nonisothermal conditions. To our knowledge no such comprehensive solution has been proposed or attempted.

Most models proposed in the literature are based on Gaskell's (4) model, which was discussed in detail in Sections 10.5 and 11.8. This is a one-dimensional, rather restrictive model. Recall that to use the model, we must know the location $X_1$ where the sheet detaches from one of the rolls ($X_1$ is uniquely related to $X_2$, the upstream location where the rolls come in contact with the polymer). This is tantamount to an a priori knowledge of the exiting sheet thickness $2H_1$. The latter, however, for a given flow rate $Q$, depends on the exiting sheet width $W_1$

$$Q = 2H_1 W_1 U \tag{16.2-1}$$

where $U$ is the velocity of the roll surface. But $W_1$ cannot be predicted from a one-dimensional model (which implicitly assumes infinitely wide rolls); hence as McKelvey (5) pointed out, $X_1$ (or $H_1$) must be considered to be an experimentally determined parameter of the model. This, of course, restricts the predictive capability of the model. To overcome this problem, the previously discussed cross-machine flow must be incorporated into the model.

This, however, is not the only limitation of the Gaskell model. As discussed in Section 10.5, it fails to predict the experimentally observed flow patterns in the inlet region because it neglects the effect of the incoming melt stream on the flow in the bank, as well as the non-Newtonian elastic effects. Consequently the model does not predict satisfactorily the observed pressure profiles as shown by Bergen and Scott (6), Unkrüer (3), and others (see Section 10.5).

Following Gaskell's work a great deal of effort was invested by numerous workers in the field to improve on his model. Most of this effort, however, concentrated on solving basically the Gaskell model with more realistic constitutive equations and attempts to account for nonisothermal effects. In the original Gaskell model a purely viscous (nonelastic and time independent) fluid model is assumed, with specific solutions for Newtonian and Bingham plastic fluids, and a brief treatment of nonsymmetric calenders. McKelvey (5) and Brazinsky et al. (7) extended the model to power law fluids (as discussed in Section 10.5), and Alston and Astill (8) investigated fluids whose shear rate dependent viscosity can be represented by a hyperbolic tangent function. Flow of viscoelastic fluids in the roll geometry was considered by Paslay (9), Tokita and White (10), and Chong (11); Paslay's analysis is essentially based on a three constant Oldroyd model.* He analyzes the interrelations of the parameters of the constitutive equation with flow kinematics, but neglects the normal stresses in the equation of motion. Tokita and White (10) relate experimental observations on milling of elastomers to rheological parameters of a second order Rivlin–Ericksen asymptotic expansion fluid and point out the significance of the Deborah and Weisenberg† numbers in milling and calendering. Following their analysis, Example 16.1 briefly explores the significance of normal stresses in calendering. However velocity and pressure profiles were not obtained by them. Chong (11) analyzed a power law model fluid, a three constant Oldroyd fluid, and a modified second order Rivlin–Ericksen equation. He incorrectly stated that the shear rate and the shear stress attain maximum values at minimum clearance location, and in integrating for the velocity profile with the power law model fluids did not properly account for the sign of the pressure gradient. The velocity profile for the Oldroyd fluid cannot be obtained analytically. Therefore Chong obtained an approximate pressure distribution by assuming Newtonian flow kinematics, and he analyzed the flow pattern with the Rivlin–Ericksen flow equation in terms of dimensionless groups only. He also measured the separating force at various calendering conditions of cellulose acetate. Like Tokita and White (10) he found, upon analyzing experimental data of calendering cellulose acetate, that the Weissenberg number is an important number in determining the onset of a nonuniform internal strain pattern, called "nerve" in calendering. Calendering defects with PVC were also studied by Agassant et al. (12), who also measured separating force, torque power, and reservoir height-to-gap ratio ($H_2/H_0$) as a function of calendering conditions.

With regard to constitutive equations, White (13) notes that in view of the short residence time of the polymer in the nip region (of the order of magnitude of seconds), it would be far more realistic to use a constitutive equation that includes viscoelastic transient effects such as stress overshoot, a situation comparable to that of squeezing flows discussed in Section 10.9.

* The three constant Oldroyd model is a non-linear constitutive equation of the differential corrotational type, such as the ZFD fluid Eq. 6.3-10. For details, see R. B. Bird, R. C. Armstrong and O. Hassager, *Dynamics of Polymeric Liquids*, Vol. 1, Wiley, New York, 1977, § 8.1.

† The ratio of the primary normal stress difference to shear stress is the Weissenberg number; for single relaxation convected coordinate constitutive equations, such as the ZFD and White–Metzner fluids, this ratio equals $\dot{\gamma}\lambda$. $\dot{\gamma}$ can be approximated by $V/L$, where $V$ is a characteristic velocity and $L$ a characteristic length.

**Example 16.1** ***The Significance of Normal Stresses*** (10, 13)

Consider the calender geometry of Fig. 10.23. We make the same simplifying assumptions as in Section 10.5, but instead of a Newtonian or power law model fluid, we assume a CEF model that exhibits normal stresses in viscometric flows. By accepting the lubrication approximation, we assume that locally we have a fully developed viscometric flow because there is only one velocity component $v_x$, which is a function of only one spatial variable $y$.

An analysis similar to that carried out in Example 6.4 leads to the following nonvanishing stress components

$$\tau_{xx} = -(\Psi_1 + \Psi_2)\dot{\gamma}^2 \tag{16.2-2}$$

$$\tau_{yy} = -\Psi_2\dot{\gamma}^2 \tag{16.2-3}$$

where $\dot{\gamma} = |\dot{\gamma}_{yx}|$ is the shear rate and

$$\dot{\gamma}_{yx} = \frac{dv_x}{dy} \tag{16.2-5}$$

which is assumed to be independent of $x$ locally.

The equation of motion then reduces to

$$\frac{\partial P}{\partial x} = -\frac{\partial \tau_{yx}}{\partial y} \tag{16.2-6}$$

$$\frac{\partial P}{\partial y} = -\frac{\partial \tau_{yy}}{\partial y} \tag{16.2-7}$$

and

$$\frac{\partial P}{\partial z} = 0 \tag{16.2-8}$$

Comparing to the solution in Section 10.5, we observe that instead of a single differential equation for the velocity profile (16.2-6), two coupled (through $\dot{\gamma}$) differential equations are obtained. However the kinematics can be well approximated by assuming $\partial P/\partial y = 0$, which then will lead to the same velocity profile given in Section 10.5. Moreover, we note that the pressure at the roll surface will differ from that of the simple model by a term $-\Psi_2\dot{\gamma}^2$. Since $\Psi_2$ is found to be negative, this normal stress contribution adds to the pressure at the roll surface. Hence it is the secondary normal difference function that plays a role in calculating the forces on the calender roll. This can probably be assumed to be small.

The present analysis was based on the lubrication approximation; that is, we neglected changes in the $x$-direction. If this assumption is lifted, we are faced with a flow field in which two nonvanishing velocities exist which are functions of two spatial coordinates, $v_x(x, y)$, $v_y(x, y)$. This is clearly a non-viscometric flow situation, and the CEF equation is not applicable. White (13) made an order of magnitude evaluation of normal stress effects for this more realistic flow situation. In this case the equation of motion reduces to

$$-\frac{\partial P}{\partial x}=\frac{\partial \tau_{yx}}{\partial y}+\frac{\partial \tau_{xx}}{\partial x} \tag{16.2-9}$$

and

$$-\frac{\partial P}{\partial y}=\frac{\partial \tau_{yx}}{\partial x}+\frac{\partial \tau_{yy}}{\partial y} \tag{16.2-10}$$

which can be combined by respective differentiation into one equation

$$\frac{\partial^2 \tau_{yx}}{\partial y^2}-\frac{\partial^2 \tau_{yx}}{\partial x^2}+\frac{\partial^2}{\partial x\,\partial y}(\tau_{xx}-\tau_{yy})=0 \tag{16.2-11}$$

Expressing the various terms in Eq. 16.2-11 at the roll surface as orders of magnitude, we get

$$\frac{\tau_w}{H_0^2};\ \frac{\tau_w}{R^2};\ \frac{\tau_{xx}-\tau_{yy}}{H_0 R} \tag{16.2-12}$$

where $H_0$ and $R$ are gap clearance and roll radius, respectively. If $R \gg H_0$, the second term is negligible. The third term, which reflects the primary normal stress difference, is also negligible, provided

$$\frac{R}{H_0}\tau_w \gg \tau_{xx}-\tau_{yy} \tag{16.2-13}$$

The condition above is met at low shear rates, but it begins to break down with increasing shear rate when $\tau_{xx}-\tau_{yy}$ increases rapidly, as indicated in Fig. 6.12.

In the Gaskell model the flow geometry is simplified to facilitate the solution (see Eq. 10.5-9). This geometrical simplification can be avoided either by using bipolar coordinates or finite element methods. Both provide a convenient way to treat calenders of unequal rolls and unequal speeds. The former approach was taken by Finston (14). Takserman-Krozer et al. (15), and Bekin et al. (16), whereas the latter was chosen by Vlachopoulos et al. (17).

Finally, the Gaskell model is isothermal, whereas in calendering significant nonisothermal effects arise because of viscous dissipation and heat conduction to the temperature controlled rolls. Finston (14) was the first to deal with viscous heating of Newtonian fluids. Torner (18) reported on an experimental study by Petrusanskii et al. (19) on the calendering of SBR on a $12\times32$ cm calender. Figure 16.4 is a schematic view of the reported temperature profiles. Characteristic to the temperature profiles is the existence of two maxima in the vicinity of the rolls. This is the combined effect of a shear rate profile with a maximum value at the roll surface and heat conduction to the temperature controlled roll surface. The temperature profile has a minimum at the center plane. The temperature profiles change in the machine direction, with a gradual temperature rise at the center plane and more complex behavior in the vicinity of the rolls. It should be noted that these temperature profiles do not refer to recirculating regions in the entry to the calender gap. Temperature effects were also studied by Bekin et al. (16), using

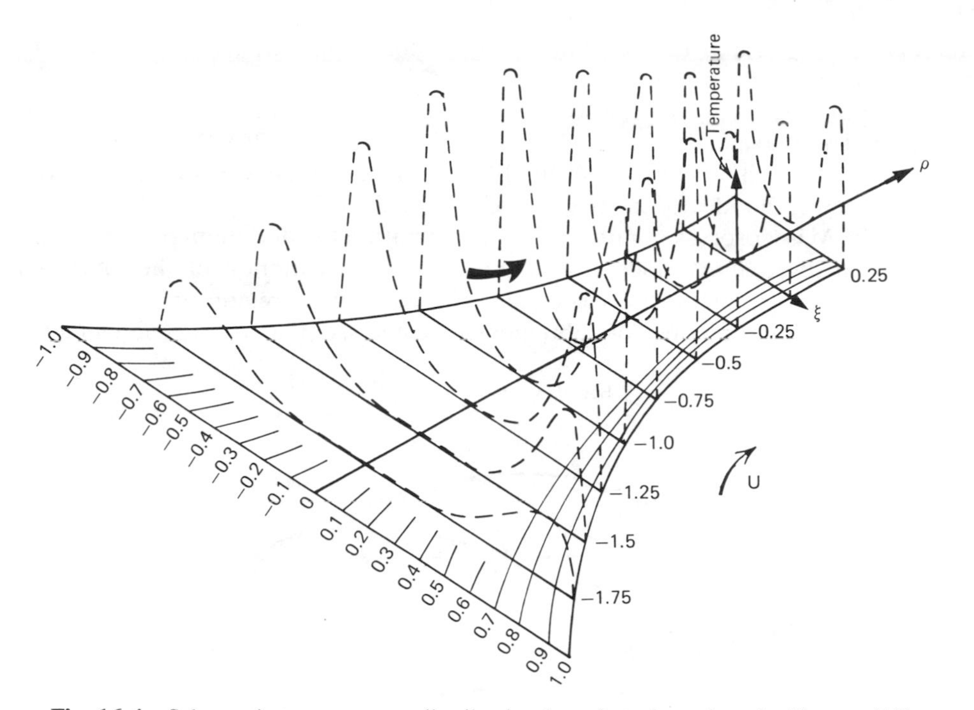


**Fig. 16.4** Schematic temperature distribution based on data given by Torner (18).

bipolar coordinates and temperature dependent fluid viscosity, and by Vlachopoulos et al. (20, 21), who combined the FEM with an implicit finite difference method. The FEM provides the velocity profiles, which then enable the solution of the energy equation for the temperature profiles using the finite difference method. Some of the computed results obtained by this method are discussed later (Section 16.4).

It transpires from the discussion in this section that a rigorous theoretical analysis incorporating all the practically significant effects is quite difficult. Among the various approaches presented, it appears that the FEM provides most flexibility in dealing with the real geometry, realistic constitutive equations, and thermal effects. This method, moreover, can in principle intermesh with an FEM structural analysis for roll deflection calculations as well as with finite difference methods for heat transfer calculation or free boundary analysis (e.g., with the MAC method). The next section, therefore, discusses the principles of the FEM. Devoting this attention to the FEM is further justified by the increasing role it plays in the analysis of other shaping methods (22), for example, in die forming and molding.

## 16.3 The Finite Element Method (FEM)

The FEM has been used for many years in the structural analysis of solids. Its application to fluid flow is a relatively recent development of growing importance. This is particularly true in the field of polymer processing, where the nature of the problems—namely, flow of non-Newtonian fluids in complex geometries—makes the FEM a most appropriate mathematical tool. Applications of the FEM to polymer processing were reviewed by Tanner (22).

It is not possible within the scope of this text to present a thorough review of the FEM. This is available in a number of excellent texts (23–26). We describe very briefly some basic characteristics of the FEM, to give readers who are not familiar with this method an idea of the basic concepts involved.

The FEM is a process of numerical approximation to continuum problems that provides an *approximate*, piecewise, continuous representation of the unknown field variables (e.g., pressures, velocities). The continuous region or body is subdivided into a *finite* number of subregions or *elements* (Fig. 16.5). The elements

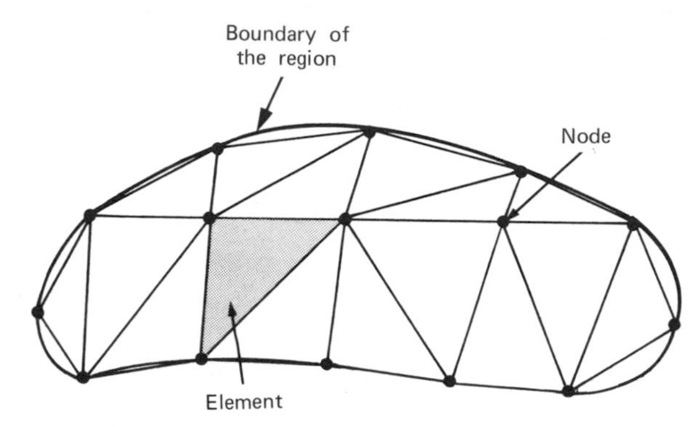


**Fig. 16.5** Two-dimensional region represented as an assemblage of triangular elements.

may be of variable size and shape, and they are so chosen that closely fit the body. This is in sharp contrast to finite difference methods, which are characterized by a regular size mesh, describable by the coordinates that describe the boundaries of the body. The crossing of two curves bounding adjacent elements forms *nodes. The values of the field variables at the nodes form the desired solution.* Common shapes of finite elements are triangular (Fig. 16.5), rectangular, and quadrilateral in two-dimensional problems, and rectangular prismatic and tetrahedral in three-dimensional problems. Within each element an interpolation function for the variable is *assumed.* These assumed functions, called *trial functions* or *field variable models*, are relatively simple functions such as truncated polynomials. The number of terms (coefficients) in the polynomial selected to represent the unknown function must at least equal the degrees of freedom associated with the element. For example, in a simple one-dimensional case (Fig. 16.6a), we have two degrees of freedom $P_i$ and $P_j$ for a field variable $P(x)$ in element $e$. Additional conditions are needed for more terms (e.g., derivatives at nodes $i$ and $j$ or additional internal nodes). The chosen function must satisfy certain additional requirements. Not only must it be continuous throughout the element but also compatible across element interfaces. In the simple case (Fig. 16.6*b*) this means $P_a = P_b$ at the node $m$ common to elements $a$ and $b$. Thus the coefficients of a selected trial function can be expressed in terms of the (unknown) values of the field variables at the nodes. For a two-dimensional case we can write for field variable $u$

$$u^{(m)}(x, y) = \sum_{i=1}^{r} N_i(x, y) u_i^{(m)} \qquad (16.3\text{-}1)$$

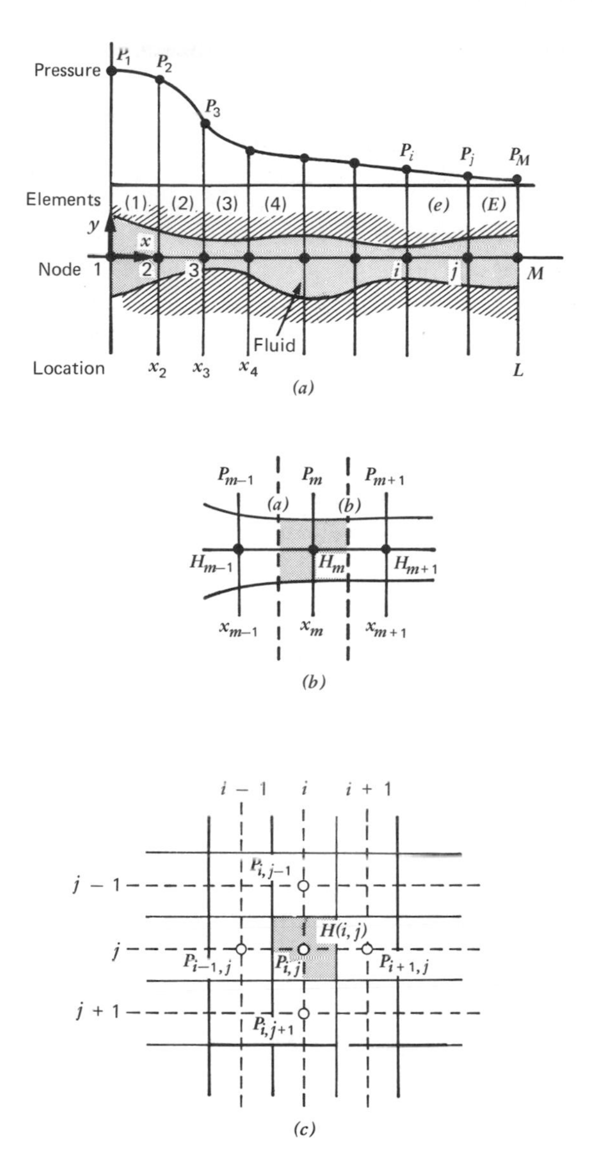


**Fig. 16.6** (*a*) Side view of two infinite plates with a variable gap in the *x*-direction. The region of interest $0<x<L$ is broken up into $E$ elements forming $M$ nodes. A schematic pressure profile for one-dimensional pressure flow is plotted on the top. (*b*) Details of two neighboring elements *a* and *b* with the common node *m*. (*c*) Two-dimensional square elements representing flow analysis network model.

where the superscript denotes that the field is for the *m*th element, $r$ is the number of nodes associated with this element, $u_i^{(m)}$ are the nodal values of the variable $u$, and $N_i(x, y)$ is the *shape function*. The function $N_i(x, y)$ is determined by the shape of the element, the location of the nodes, and the number of terms in the polynomial. Again, the objective is the numerical evaluation of $u_i$. The common approach to obtaining this is to set up the finite element equations either by "*variational*" *methods* or by "*residual*" methods such as the *Galerkin method.*

These, together with appropriate boundary conditions, result in a set of linear or nonlinear algebraic equations with the nodal variables $u_i$ as unknowns.

Structural analysis, initially developed on an intuitive basis, became later identified with variational calculus in which the Ritz procedure is used to minimize a *functional* derived mathematically or arrived at directly from physical principles. By substituting the final solutions into the variational statement of the problem and minimizing the latter, the FEM equations are obtained. Example 16.2 gives a very simple demonstration of this procedure.

In many cases (general non-Newtonian flow problems being among them), however, a variational principle either does not exist or its existence is not obvious. Nevertheless these problems can often be defined by a set of differential equations (e.g., the equations of continuity and motion with a constitutive equation), together with their boundary conditions. In such cases weighted residual methods such as the collocation and the Galerkin methods produce a simpler and direct formulation of the FEM equations (27). In the Galerkin method the approximate interpolation function is substituted into the governing differential equations, multiplied by the weighting function, which is the relevant shape function, and integrated over the body. The resulting expressions are set to zero, leading to a set of algebraic equations.

Inherent in the FEM is the flexibility in dealing with complex geometries as well as mixed boundary conditions (e.g., stress and velocity boundary conditions as in a die-swell problem). Moreover, computationally the FEM is not difficult to carry out. Not only can a continuous domain of complex boundary be easily broken down into well-fitting finite elements, but the inherent possibility exists of using elements of various sizes and shapes. This permits a refined solution in critical regions (corners, sudden changes in geometry, etc.) without the penalty of excessive computation in the rest of the regions, as would be the case with the more limited finite difference methods. Finally, it is noted by Zienkiewicz (28) that by certain function choices, the standard finite difference processes can be included in the general finite elements concept.

To elucidate some of the basic concepts of the FEM, we shall solve by this method in the following example, a one-dimensional, very simple problem.

**Example 16.2** ***FEM Formulation of Isothermal Steady Pressure Flow in Narrow but Variable Thickness Gap of a Newtonian Fluid.***

The governing differential equation is the Reynolds equation given for a two-dimensional flow in Eq. 5.4-11. To demonstrate the FEM formulation, we consider the one-dimensional flow case, for which Eq. 5.4-11 reduces to

$$\frac{d}{dx}\left(\frac{H(x)^3}{\mu}\frac{dP}{dx}\right)=0 \qquad (16.3\text{-}2)$$

We have retained the viscosity because we want to treat approximately the non-Newtonian fluids case later. If the function $H(x)$ is known, the differential equation above can be solved analytically or numerically for $P(x)$ in a straightforward method without turning to FEM. Our purpose here, however, is to demonstrate the FEM method and following Myers (26) we do so in step-by-step fashion.

The flow configuration appears in Fig. 16.6*a*. The one-dimensional conduit of length $L$ is broken down into $E$ elements, bounded by $M$ nodes. Our objective is to set up the FEM equations that will give the pressure values $P_i$.

The first step is to derive the *variational statement* of the problem. This can be done with the aid of the Lagrange–Euler equation

$$\frac{\partial F}{\partial P} - \frac{d}{dx}\left(\frac{\partial F}{\partial \dot{P}}\right) = 0 \tag{16.3-3}$$

which must be satisfied for the following *functional* $I$

$$I = \int_0^L F(x, P, \dot{P})\, dx \tag{16.3-4}$$

to be an *extremum*. Comparison of Eqs. 16.3-2 and 16.3-3 gives the following expressions for $F$:

$$\frac{\partial F}{\partial P} = 0 \tag{16.3-5}$$

and

$$\frac{\partial F}{\partial \dot{P}} = \frac{H^3}{\mu} \dot{P} \tag{16.3-6}$$

Integration of Eqs. 16.3-5 and 16.3-6, respectively, gives

$$F = K_0 + f(\dot{P}) \tag{16.3-7}$$

and

$$F = \frac{H^3}{2\mu}(\dot{P})^2 + g(P) \tag{16.3-8}$$

Comparing Eqs. 16.3-7 nnd 16.3-8, we note $g(P) = K_0$ and $f(\dot{P}) = (H^3/2\mu)(\dot{P})^2$; thus $F$ can be written as

$$F = K_0 + \frac{1}{2}\frac{H^3}{\mu}(\dot{P})^2 \tag{16.3-9}$$

Hence the variational statement of this problem reduces to obtaining the extremum of the functional

$$I = \int_0^L \left[K_0 + \frac{1}{2}\frac{H^3}{\mu}(\dot{P})^2\right] dx \tag{16.3-10}$$

that is, we are searching for the unknown *function* $P(x)$ that when substituted into Eq. 16.3-10, gives an extremum for $I$. To evaluate $I$, we break it down into $E$ subintegrals corresponding to the $E$ elements

$$I = I^{(1)} + I^{(2)} + \cdots + I^{(E)} = \sum_{e=1}^{E} I^{(e)} \tag{16.3-11}$$

The integral $I^{(e)}$ over a typical finite element is

$$I^{(e)} = \int_{x_i}^{x_j} \left[ K_0 + \frac{1}{2}\frac{H^3}{\mu}(\dot{P})^2 \right] dx \tag{16.3-12}$$

We now *assume* a linear trial function for the variation of the pressure within each element

$$P^{(e)} = c_1^{(e)} + c_2^{(e)} x \tag{16.3-13}$$

We thus have two coefficients, and since we have two degrees of freedom—the (unknown) nodal values of the pressures—we can express the former in terms of the latter, and Eq. 16.3-13 can be written as

$$P^{(e)} = \left(\frac{x_j - x}{x_j - x_i}\right) P_i + \left(\frac{x - x_i}{x_j - x_i}\right) P_j \tag{16.3-14}$$

Note that Eq. 16.3-14 is of the same form as Eq. 16.3-1. Next we take the derivative of $P^{(e)}$ given in Eq. 16.3-14 with respect to $x$

$$\dot{P} = \frac{dP^{(e)}}{\partial x} = \frac{P_j - P_i}{x_j - x_i} \tag{16.3-15}$$

and substitute it into Eq. 16.3-12 which, after integration, gives

$$I^{(e)} = K_0(x_j - x_i) + \frac{1}{2}\frac{H^3}{\mu}\left(\frac{P_j - P_i}{x_j - x_i}\right)^2 \tag{16.3-16}$$

We have assumed in the foregoing integration that $\mu$ is constant within the element and equal to the average value in it.

Next we differentiate $I^{(e)}$ with respect to the nodal pressures $P_i$ and $P_j$

$$\frac{\partial I^{(e)}}{\partial P_i} = -\frac{H^3}{\mu}\frac{P_j - P_i}{x_j - x_i} \tag{16.3-17}$$

and

$$\frac{\partial I^{(e)}}{\partial P_j} = \frac{H^3}{\mu}\frac{P_j - P_i}{x_j - x_i} \tag{16.3-18}$$

$I^{(e)}$ is a function of $P_i$ and $P_j$ only, whereas $I$ (in Eq. 16.3-11) is a function of $P_1, P_2, \ldots, P_M$. To find the extremum of $I$, we must differentiate $I$ with respect to all $P_i$ and set the results equal to zero, obtaining $M$ equations. Thus differentiating Eq. 16.3-11 with respect to a typical nodal pressure $P_m$, we get

$$\frac{\partial I}{\partial P_m} = \frac{\partial I^{(1)}}{\partial P_m} + \frac{\partial I^{(2)}}{\partial P_m} + \cdots + \frac{\partial I^{(E)}}{\partial P_m} = 0 \tag{16.3-19}$$

But the pressure $P_m$ appears only in two neighboring elements, as Fig. 16.6*b* shows. For element *a* we set $i = m - 1$ and $j = m$ in Eqs. 16.3-17 and 16.3-18, and for element *b* we set $i = m$ and $j = m + 1$ in the two equations above, resulting in

$$\frac{\partial I^{(a)}}{\partial P_m} = \left(\frac{H^3}{\mu}\right)^{(a)} \frac{P_m - P_{m-1}}{x_m - x_{m-1}} \tag{16.3-20}$$

and

$$\frac{\partial I^{(b)}}{\partial P_m} = -\left(\frac{H^3}{\mu}\right)^{(b)} \frac{P_{m+1} - P_m}{x_{m+1} - x_m} \tag{16.3-21}$$

where the superscripts $a$ and $b$ on $H^3/\mu$ indicate that mean local values are used. Adding Eqs. 16.3-20 and 16.3-21 and equating the sum to zero, we get

$$\left(\frac{H^3}{\mu}\right)^{(a)} \frac{P_m - P_{m-1}}{x_m - x_{m-1}} + \left(\frac{H^3}{\mu}\right)^{(b)} \frac{P_m - P_{m+1}}{x_{m+1} - x_m} = 0 \tag{16.3-22}$$

Since $m$ is any interior nodal point, Eq. 16.3-22 is in a set of $M-2$ algebraic equations, the solution of which provides the required pressure field (profile).

As a numerical example, consider a linearly decreasing gap broken down into 4 equal length elements. The gaps at the entrance and at the exit are 1 and 0.5 cm, respectively. Thus $H_1 = 1$, $H_2 = 0.875$, $H_3 = 0.75$, $H_4 = 0.625$, and $H_5 = 0.5$. The inlet pressure is 1 atm, and the exit pressure is zero. The resulting equations from Eq. 16.3-22, with constant viscosity, are

$$-1.53618 + 2.53618P_2 - P_3 = 0$$

$$-1.6506P_2 + 2.6506P_3 - P_4 = 0$$

$$-1.8257P_3 + 2.8257P_4 = 0$$

which upon solution give $P_2 = 0.897$, $P_3 = 0.738$, and $P_4 = 0.477$. The exact analytical solutions obtained by integrating Eq. 16.3-2 are $P_2 = 0.8980$, $P_3 = 0.7407$, and $P_4 = 0.480$, which agree well with the FEM solution using only four elements.

Equation 16.3-22 can also be derived by a "controlled volume" approach. Consider the $a$ element confining node $m$ in Fig. 16.6$b$ (shaded area). For an incompressible fluid and under the same assumptions as above, we can make the following flow rate balance

$$\underbrace{\frac{1}{12}\left(\frac{H^3}{\mu}\right)^{(a)} \frac{P_{m-1} - P_m}{x_m - x_{m-1}}}_{\text{Rate of flow into element}} = \underbrace{\frac{1}{12}\left(\frac{H^3}{\mu}\right)^{(b)} \frac{P_m - P_{m+1}}{x_{m+1} - x_m}}_{\text{Rate of flow out of element}} \tag{16.3-23}$$

Clearly Eq. 16.3-23 is identical to Eq. 16.3-22. This is the basis for the flow analysis network (FAN) method developed by Tadmor et al. (29) to solve two-dimensional steady or quasi-steady state flow problems in injection molds and extrusion dies. In two-dimensional flows the pressure distribution is obtained by dividing the flow region into an equal-sized mesh of square elements (Fig. 16.6$c$). At the center of each element there is a node. The nodes of adjacent elements are interconnected by links. Thus the total flow field is represented by a network of nodes and links. The fluid flows out of each node through the links and into the adjacent nodes of the network. The local gap separation determines the "resistance" to flow between nodes. Making the quasi-steady state approximation, a mass (or volume) flow rate balance can be made about each node (as done earlier for one-dimensional flow), to give the following set of algebraic equations

$$X_{i,j}(P_{i,j}-P_{i+1,j})+X_{i-1,j}(P_{i,j}-P_{i-1,j})+Z_{i,j}(P_{i,j}-P_{i,j+1})+Z_{i,j-1}(P_{i,j}-P_{i,j-1})=0 \tag{16.3-24}$$

where $X_{i,j}$ and $Z_{i,j}$ are "flow conductances" in the $x$- and $z$-directions, respectively

$$X_{i,j}=\frac{1}{12\mu}\left(\frac{H_{i,j}+H_{i+1,j}}{2}\right)^3 \tag{16.3-25}$$

and

$$Z_{i,j}=\frac{1}{12\mu}\left(\frac{H_{i,j}+H_{i,j+1}}{2}\right)^3 \tag{16.3-26}$$

This two-dimensional formulation of the flow problem is identical in concept to the "discrete element method" or "lattice models" of classical structural analysis. Physically, the FEM concept differs from the lattice analogy in that the elements themselves are two- or three-dimensional bodies (30). The FAN method, however, is a straightforward simple method, which was extended to deal with non-Newtonian fluids by replacing the Newtonian viscosity with an "equivalent Newtonian viscosity" (31). The latter is uniquely related to the local shear stress at the wall, which in turn depends on the local pressure gradient. Both can be converged upon by repeated solutions of the set of algebraic equations for $P_{ij}$, while in each iteration the viscosities are recalculated. This method was applied to both cross-head die flow and mold filling problems. Both are two-dimensional, the first being steady while the second is assumed to be quasi-steady.

The FEM formulation of two-dimensional problems is not different in principle from the simple one-dimensional case just described. For two-dimensional problems, however, the algebra becomes involved and matrix notation is required to keep it manageable.

## 16.4 Analysis of Calendering with the FEM

As pointed out earlier, the calendering process was analyzed with FEM by Vlachopoulos et al. (17, 20, 21) using the Galerkin weighted residual method. The flow region was broken down into both triangular and quadrilateral elements (Figs. 16.7 and 16.8). The two cases lead to virtually the same results. They used the general framework of the Gaskell model, invoking the lubrication approximation, and considered only liquids of the generalized Newtonian fluid type. Hence the governing differential equations are the equation of motion, which reduces to

$$\frac{\partial P}{\partial x}=\frac{\partial \tau_{yx}}{\partial y} \tag{16.4-1}$$

and the equation of continuity

$$\frac{\partial (hv_x)}{\partial x}=0 \tag{16.4-2}$$

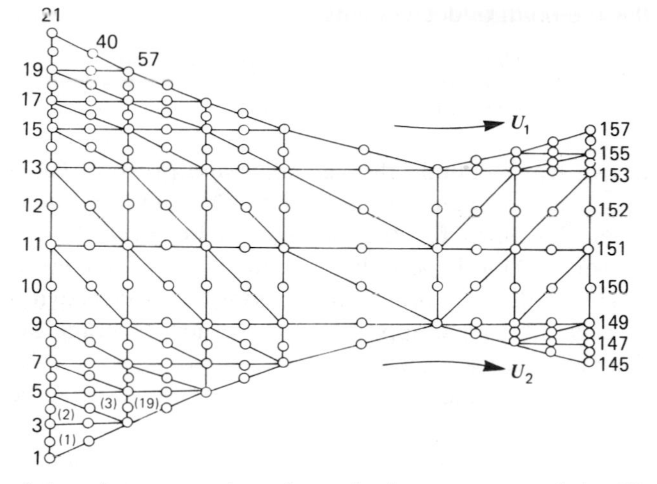


**Fig. 16.7** The finite element mesh and numbering system used by Kiparissides and Vlachopoulos (17) with triangular orthogonal elements.

where $2h$ is the gap between the rolls. The boundary conditions for the coordinate system shown in Fig. 10.23 are

$$P(X_2)=0 \tag{16.4-3a}$$

where $X_2<0$ is the point of contact.

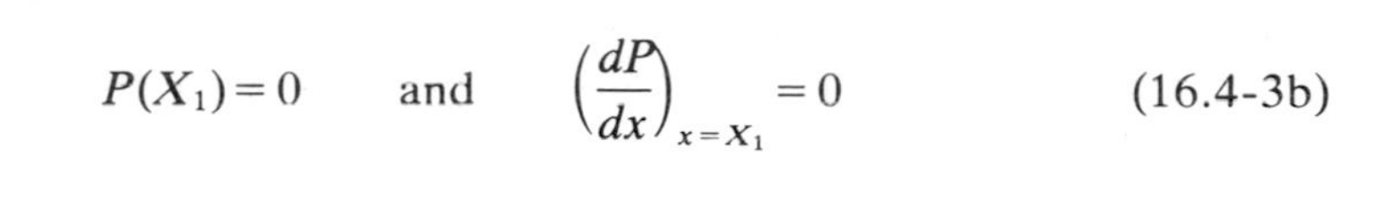

$$P(X_1)=0 \qquad \text{and} \qquad \left(\frac{dP}{dx}\right)_{x=X_1}=0 \tag{16.4-3b}$$

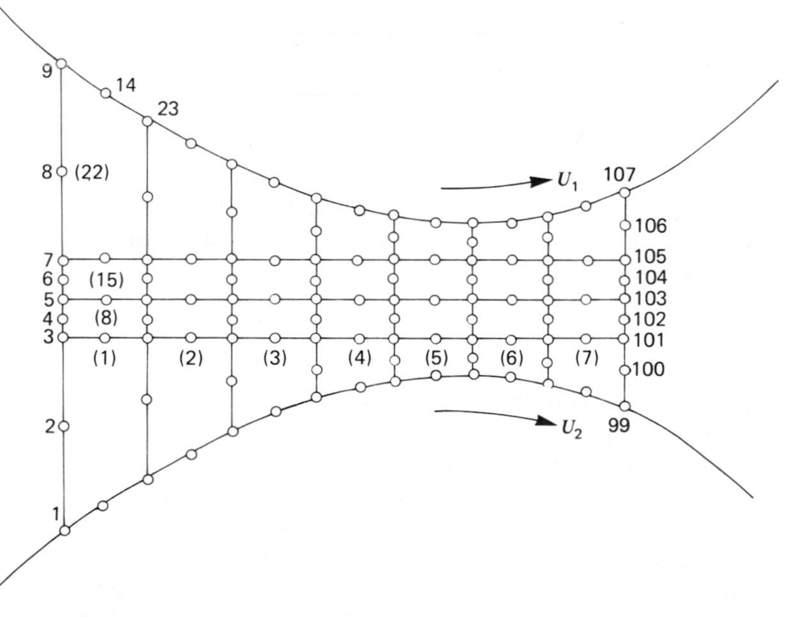


**Fig. 16.8** The finite element mesh and numbering system used by Kiparissides and Vlachopoulos (17) with quadrilateral elements.

where $X_1 > 0$ is the point of detachment.

$$v_x(h) = U_1 \quad \text{at the ``upper'' roll surface} \tag{16.4-3c}$$

$$v_x(-h) = U_2 \quad \text{at the ``lower'' roll surface} \tag{16.4-3d}$$

The objective was to obtain the velocities and pressure values at the nodes. Parabolic shape functions were assumed within the elements for the velocity field and linear shape functions for the pressure field. Both Newtonian and non-Newtonian (power law and hyperbolic tangent) fluids were considered. The latter case was solved by an iterative procedure, making an initial Newtonian approximation. Then, in each successive approximation, a new (average over the element) non-Newtonian viscosity was used.

Results obtained by the FEM were in general good agreement with the analytical solutions for both Newtonian and non-Newtonian cases, with differences attributable to the coarseness of the mesh size. The advantage of the FEM, however, becomes apparent in analyzing cases where analytical solutions are not readily available—for example, asymmetric calendering. We can distinguish two cases of asymmetric calendering depending on whether roll speeds or roll diameters are unequal. The Newtonian model in the former case indicates that the expected pressure profile is identical to that of an equal roll speed calender turning at $U = (U_1 + U_2)/2$. Similarly in the latter case the expected pressure profile is identical to that of hypothetical calender of equal size rolls of $R = (R_1 + R_2)/2$. The non-Newtonian case, however, is different. Specifically, roll speed differences drastically alter the velocity distribution and the shear rate distribution between the rolls; thus we would expect a different response from a shear rate dependent fluid. Figure 16.9 demonstrates this effect for a power law model fluid with $n = 0.25$. A speed ratio of $U_1/U_2 = 20/40$ generates only about 33% of the maximum pressure at $U_1 = U_2 = 40$, 38% of the maximum pressure at $U_1 = U_2 = 30$ (instead of the Newtonian 100%), and 44% of the maximum pressure at $U_1 = U_2 = 20$. Different roll sizes at equal tangential speeds (not frequencies of rotation) show a less drastic effect. For the same fluid with $\lambda = 0.3$, $U = 40$ cm/s, and $H_0 = 0.01$ cm, the maximum pressure with equal sized 15 cm roll radii is $3.3 \times 10^6$ dynes/cm$^2$, whereas for a 10–20 cm radii roll pair, the maximum pressure is only $2.9 \times 10^6$ dynes/cm$^2$.

In considering the nonisothermal problem (20, 21), the energy equation assuming constant thermophysical properties and within the framework of the lubrication approximation reduces to

$$\rho C_p v_x \frac{\partial T}{\partial x} = k \frac{\partial^2 T}{\partial x^2} + \tau_{yx} \frac{dv_x}{dy} \tag{16.4-4}$$

and must be solved in conjunction with Eqs. 16.4-1 and 16.4-2.

The boundary conditions are constant inlet temperature, which equals the roll surface temperatures. This equation was solved numerically by using an implicit *finite difference* method with $v_x$ and $dv_x/dy$ values obtained from the FEM. According to the authors, the combination of FEM and FDM helps reduce computer storage and computer time requirements.

Computed temperature profiles at various $\rho$ positions confirm the type of the temperature field in Fig. 16.4, that is, near the entry the temperature rise is confined in the vicinity of the roll surfaces. Downstream, the temperature of the

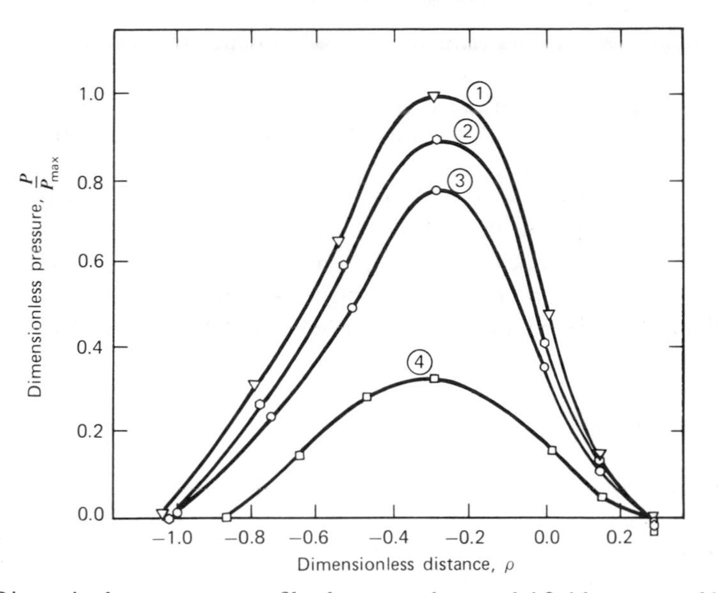


**Fig. 16.9** Dimensionless pressure profiles for power law model fluids computed by FEM for calendering with equal sized rolls moving at various speeds: $R = 10$ cm, $2H_0 = 0.01$ cm, $n = 0.25$, $m = 10^4$ dyne s$^n$/cm$^2$, $P_{max} = 3.2 \times 10^6$ dyne/cm$^2$. Curve 1, $U_1 = U_2 = 40$ cm/s; curve 2, $U_1 = U_2 = 30$ cm/s; curve 3, $U_1 = U_2 = 20$ cm/s; curve 4, $U_1 = 20$ cm/s, $U_2 = 40$ cm/s. [Reprinted with permission from C. Kiparissides and J. Vlachopoulos, "Finite Element Analysis of Calendering," *Polym. Eng. Sci.*, **16**, 712 (1976).]

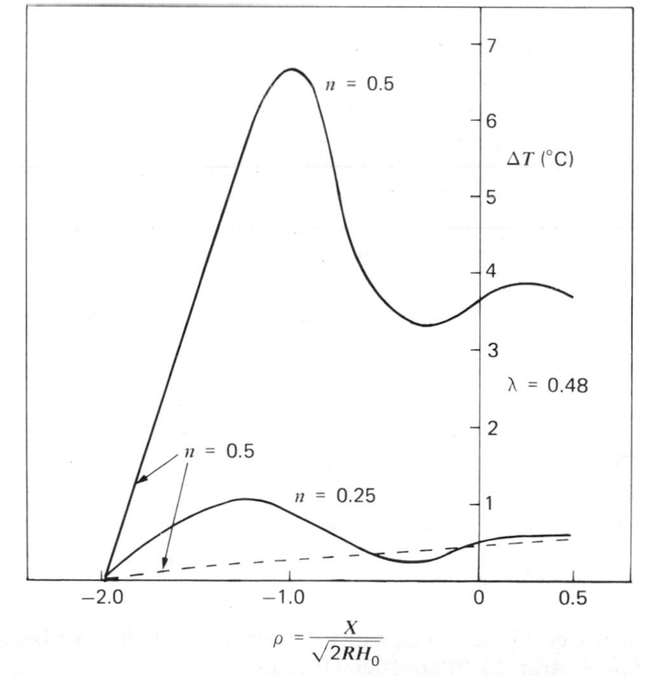


**Fig. 16.10** Downstream profile of the maximum (solid curves) and centerplane (broken curve) temperature increases with $n$ as a parameter. $m = 2.5 \times 10^4$ N · s$^n$/m$^2$, $R = 15$ cm, $H_0 = 0.025$ cm, $U = 40$ cm/s, $\lambda = 0.48$, $\rho = 1$ g/cm$^3$, $C_p = 2.1$ J/g K, $k = 1.7 \times 10^{-3}$ J/cm K. [Reprinted with permission from C. Kiparissides and J. Vlachopoulos, *Polym. Eng. Sci.*, **18**, 210 (1978).]

whole flow field rises, yet the temperature differences diminish. Figure 16.10 gives the maximum temperature as a function of the $\rho$ position, with $n$ as a parameter. It is interesting to note that the downstream profile of the maximum temperature exhibits two maxima. This is the result of the interaction of the viscous dissipation distribution (determined by the shear rate distribution), with the heat conduction to the rolls and heat convection. Significantly, the exiting temperature does not reflect the higher temperature the fluid experienced in the gap, and this temperature is very sensitive to $n$. Vlachopoulos et al. (21) suggest that the magnitude and location of the temperature maxima might explain the occasional appearance of blisters in calendered sheets. These have been observed (14) to originate at a short distance below the sheet surface.

## *REFERENCES*


1. G. W. Eighmy, Jr., "Calendering," in *Modern Plastics Encyclopedia*, McGraw-Hill, New York, 1977, p. 234.
2. D. I. Marshall, "Calendering," in *Processing of Thermoplastics Materials*, E. C. Bernhardt, ed., Reinhold, New York, 1959, Chapter 6.
3. W. Unkrüer, "Beitrag zur Ermittlung des Druckverlaufes und der Fliessvorgänge im Walzsplat bei der Kalanderverarbeitung von PVC Hart zu Folien," Doctoral thesis, IKV, Technischen Hochschule, Aachen, 1970.
4. R. E. Gaskell, "The Calendering of Plastic Materials," *J. Appl. Mech.*, **17**, 334–336 (1950).
5. J. M. McKelvey, *Polymer Processing*, Wiley, New York, 1962, Chapter 9.
6. J. T. Bergen and G. W. Scott, "Pressure Distribution in Calendering of Plastic Materials," *J. Appl. Mech.*, **18**, 101–106 (1951).
7. I. Brazinsky, H. F. Cosway, C. F. Valle, Jr., R. Jones, R. Clark, and V. Story, "A Theoretical Study of Liquid Film Spread Heights in the Calendering of Newtonian and Power Law Fluids," *J. Appl. Polym. Sci.*, **14**, 2771 (1970).
8. W. W. Alston and K. N. Astill, "An Analysis for the Calendering of Non-Newtonian Fluids," *J. Appl. Polym. Sci.*, **17**, 3157 (1973).
9. P. R. Paslay, "Calendering of Viscoelastic Materials," *J. Appl. Mech.*, **24**, 602 (1957).
10. N. Tokita and J. L. White, "Milling Behavior of Gum Elastomers," *J. Appl. Polym. Sci.*, **10**, 1011 (1966).
11. J. S. Chong, "Calendering Thermoplastic Materials," *J. Appl. Polym. Sci.*, **12**, 191–212 (1968).
12. J. F. Agassant and P. Avenas, "Calendering of PVC—Forecase of Stresses and Torques," Paper presented at the Second International Symposium on PVC, Lyon, France, 1976; J. L. Bourgeois and J. F. Agassant, "Calendering of PVC—Defects in Calendered PVC Films and Sheets," Paper presented at the Second International Symposium on PVC, Lyon, France, 1976.
13. J. L. White, "Elastomer Rheology and Processing," *Rubber Chem. Technol.*, **42**, 257–338 (1969).
14. M. Finston, "Thermal Effects in Calendering of Plastic Materials," *J. Appl. Mech.*, **18**, 12 (1951).
15. R. Takserman-Kozer, G. Schenkel, and G. Ehrmann, "Fluid Flow Between Rotating Cylinders," *Rheol. Acta*, **14**, 1066–1076 (1975).
16. N. G. Bekin, V. V. Litvinov, and V. Yu. Petrusanskii, "Method of Calculation of the Energy and Hydrodynamic Characteristics of the Calendering of Polymeric Materials," *Kauc. Rezina*, **8**, 32 (1975); English trans., *Int. Polym. Sci. Technol.*, **3**, T 55–T 58 (1976).

17. C. Kiparissides and J. Vlachopoulos, "Finite Element Analysis of Calendering," *Polym. Eng. Sci.*, **16**, 712 (1976).
18. R. V. Torner, "Grundprozesse der Verarbeitung von Polymerer," VEB Deutscher Verlag für Grundstoffindustrie, Leipzig, 1974 (translated from Russian).
19. V. Yu. Petrusanskii and A. I. Stachaev, *Uch. Zap. Jaroslavsk. Technol. Inst.*, t. 23 (1971).
20. J. Vlachopoulos and C. Kiparissides, "An Analysis of Thermoplastics in Calendering," Paper presented at the 26th Canadian Chemical Engineering Conference, Toronto, 1976.
21. C. Kiparissides and J. Vlachopoulos, "A Study of Viscous Dissipation in the Calendering of Power Law Fluids," *Polym. Eng. Sci.*, **18**, 210–213 (1978).
22. R. I. Tanner, "Some Experiences Using Finite Element Methods in Polymer Processing and Rheology," *Proceedings of the Seventh International Congress on Rheology, Gothenburg, Sweden, 1976*, pp. 140–145.
23. O. C. Zienkiewicz, *The Finite Element Method in Engineering Science*, McGraw-Hill, London, 1971.
24. H. C. Martin and G. F. Carey, *Introduction to Finite Element Analysis—Theory and Application*, McGraw-Hill, New York, 1973.
25. K. H. Huebner, *The Finite Element Method for Engineers*, Wiley, New York, 1975.
26. G. E. Myers, *Analytical Methods in Conduction Heat Transfer*, McGraw-Hill, New York, 1971, Chapter 9, "Finite Elements."
27. O. C. Zienkiewicz and C. Taylor, "Weighted Residual Processes in Finite Elements with Particular Reference to Some Transient and Coupled Problems," in *Lectures on Finite Element Methods in Continuum Mechanics*, J. T. Oden and E. R. A. Oliveria, eds., U. A. H. Press, Huntsville, Ala., 1973.
28. J. T. Oden, O. C. Zienkiewicz, R. H. Gallagher, and C. Taylor, eds., *Finite Elements in Flow Problems*, U. A. H. Press, Huntsville, Ala., 1974, p. 4.
29. Z. Tadmor, E. Broyer, and C. Gutfinger, "Flow Analysis Network—A Method for Solving Flow Problems in Polymer Processing," *Polym. Eng. Sci.*, **14**, 660–665 (1974).
30. C. S. Desai and J. F. Abel, *Introduction to the Finite Element Method—A Numerical Method for Engineering Analysis*, Van Nostrand Reinhold, New York, 1972, p. 68.
31. E. Broyer, C. Gutfinger, and Z. Tadmor, "Evaluating Flows of a Non-Newtonian Fluid by the Method of Equivalent Newtonian Viscosity," *Am. Inst. Chem. Eng. J.*, **21**, 198–200 (1975).

## PROBLEMS

**16.1** ***Calendering of Polymers—The Newtonian Gaskell Model.*** A calender with 200 cm diameter, 100 cm wide equal sized rolls operates at a speed of 50 cm/s. At a gap separation of 0.02 cm it produces a 0.022 cm thick film. Assuming a Newtonian viscosity of $10^4$ poise, calculate in the last nip (a) the maximum pressure, (b) the separating force, and (c) estimate the mean temperature rise.

**16.2** ***Separating Force Between Rolls in an Experimental Calender.*** A cellulose acetate based polymeric compound is calendered on a laboratory inverted L-shaped calender with 16 in. wide rolls of 8 in. diameter. The minimum gap between the rolls is 15 mils. The sheet width is 15 in. Calculate the separation force and the maximum pressure between a pair of rolls as a function of exiting film thickness, assuming that film thickness equals the gap separation at the point of detachment. Both rolls turn at 10 rpm. The polymer at the calendered temperature of 90°C follows a power law model with $m = 3 \times 10^6$ dyne. $s^n/cm^2$ and $n = 0.4$. [Data based partly on J. S. Chong, "Calendering Thermoplastic Materials," *J. Appl. Polym. Sci.*, **12**, 191–212 (1968).]

**16.3** ***Design Considerations of a Calender.*** It is desired to manufacture a 2 m wide, 0.1 mm thick PVC film at a rate of 1200 Kg/hr with an inverted L-calender. Suggest a design procedure to select roll sizes, gap separations and operating conditions.

**16.4** ***Dissipated Work in Calendering.*** Calculate the dissipated mechanical work ($\boldsymbol{\tau}:\dot{\boldsymbol{\gamma}}$) during the forming of the sheet by calendering as described in problem 16.1. How much work would be dissipated if the sheet were extruded at the same rate through a sheet die of opening 0.02 cm and die lip length 10 cm?

# 17

# A Guide to the Analysis of Polymer Processing in Terms of Elementary Steps and Shaping Methods

The main goals of this text are to define the scope of the field of polymer processing and to formulate a logical and useful framework for its analysis. It is evident that the scope of the field extends from the fundamentals of polymer science to those of engineering. This is perhaps not surprising, since the field of polymer processing forms the meeting ground of these activities. The concept of "structuring" is the "practical" thread that intertwines the two. This book suggests that the breakdown of polymer processing into shaping operations and a set of clearly defined elementary steps provides a logical and useful framework for its analysis. A schematic view of the breakdown appears in Fig. 1.24, and Section 1.2 discusses the reasoning behind it. This approach claims that what is happening to the polymer in a certain type of machine is *not unique*; polymers go through similar experiences in other machines, and these experiences can be described by a set of *elementary steps* that provide the *unified* view of the field. On the other hand, it is emphasized that the unique features of each machine are the *particular mechanisms* utilized in each of the elementary steps, *and* the *particular design solutions*. The mechanisms emerge systematically from the elementary step approach. They are hinged on some basic geometrical configurations, *which are the building blocks for the engineering conception of machines.* Based on these building blocks, a multitude of practical design solutions can be conceived. We recall, for example, that the parallel plate geometry is one of the basic configurations or building blocks of the drag induced pressurization mechanism. Chapter 10 demonstrated how the single screw extruder can be created from this geometry and indicated that a "flat spiral extruder" or a "rotating shaft extruder," with the spiral channel cut in the barrel, could form other (less useful) design solutions. Similarly, we found that the counterrotating intermeshing twin screw geometry, the gear pump, and the piston-cylinder pump, are alternative design solutions based on a static pressurization mechanism, and originating from a building block consisting of a surface moving normal to its plane and generating positive displacement flow. In designing new machinery, however, it is usually necessary to consider more than one elementary step. *The design process could in effect be viewed as a synthesis of a machine from elementary steps.*

The elementary step approach, however, is useful not only in design and synthesis but also in *analysis*. This was demonstrated foremost in Chapter 12, where the single screw, plasticating screw extrusion process was analyzed in terms of elementary steps, and also in certain shaping operations where elementary steps concur with the shaping operation. Calendering and "mold coating" can serve as examples to illustrate the latter point. By identifying the mechanism of pressurization in calendering (drag induced pressurization in non-parallel plate geometry), the shaping operation itself occurring over several nips can be better understood. Similarly, by identifying that dip coating, slush molding, electrostatic coating, rotational molding, and so on, involve essentially conduction melting without melt removal, we are permitted to take a unified view of these shaping methods, and arrive at the definition of the "mold coating" shaping method.

In spite of the occasional simultaneous occurrence of shaping and elementary steps, this text divorces the two (in concept) and extracts the shaping operation itself from the process. This separation leads not only to a more systematic classification of the shaping methods on the basis of the principal mechanism involved, but also may reduce the frequently arising restriction on imaginative thinking when a certain shaping method is conceptually tied to a given process. From this point of view, blow molding, for example, was classified as a secondary shaping method where the dominant mechanism is the elongational deformation of a simple shape (parison). The latter can be obtained by die forming (the common blow molding process), by injection molding (the injection blow molding process), or we could conceive of a dip coating process on a porous core followed by blowing, or a rotational molded parison followed by blowing. Similarly, the top forming process (Section 1.1) is a thermoforming shaping method whereby the preformed "sheet" is prepared in a sequence of injection and compression molding steps. Clearly, by understanding the basic shaping methods and dealing with them individually, the principles of many other useful process combinations could be conceived. Paralleling the elementary step approach, we could view these basic shaping methods as "elementary shaping steps." Finally, by separating the treatment of the shaping method from that of the process, attention can be focused on detailed design equations as well as the structuring aspects of the process.

The analysis of polymer processing in terms of elementary steps and shaping methods can also be regarded as a process of disassembly of a complex mechanism into basic components. This should facilitate the understanding of the field and should provide a unified view of it. But, as it inevitably happens in using such an approach, the basic components relevant to a given *particular process* are to be found under different headings in various chapters. Therefore, to aid the casual reader and to provide a handy reference, we present a schematic guide (Figs. 17.1–17.5) to the most important processes, including processes that have not been treated individually in the text. These figures suggest where the reader can find the material necessary to understand and analyze any portion of the process. The following information is contained in these figures:

1. Overall schematic representation of the entire process, including both preprocessing and postprocessing steps.
2. All the stages of the entire thermal and mechanical "experience" of the polymer. Each element of this total experience is identified separately in connection with specific parts of the process and associated machinery.
3. The relevant elementary steps, elementary step mechanisms, and shaping methods associated with the polymer experience in the process, together with the corresponding section in the text that treats each one.

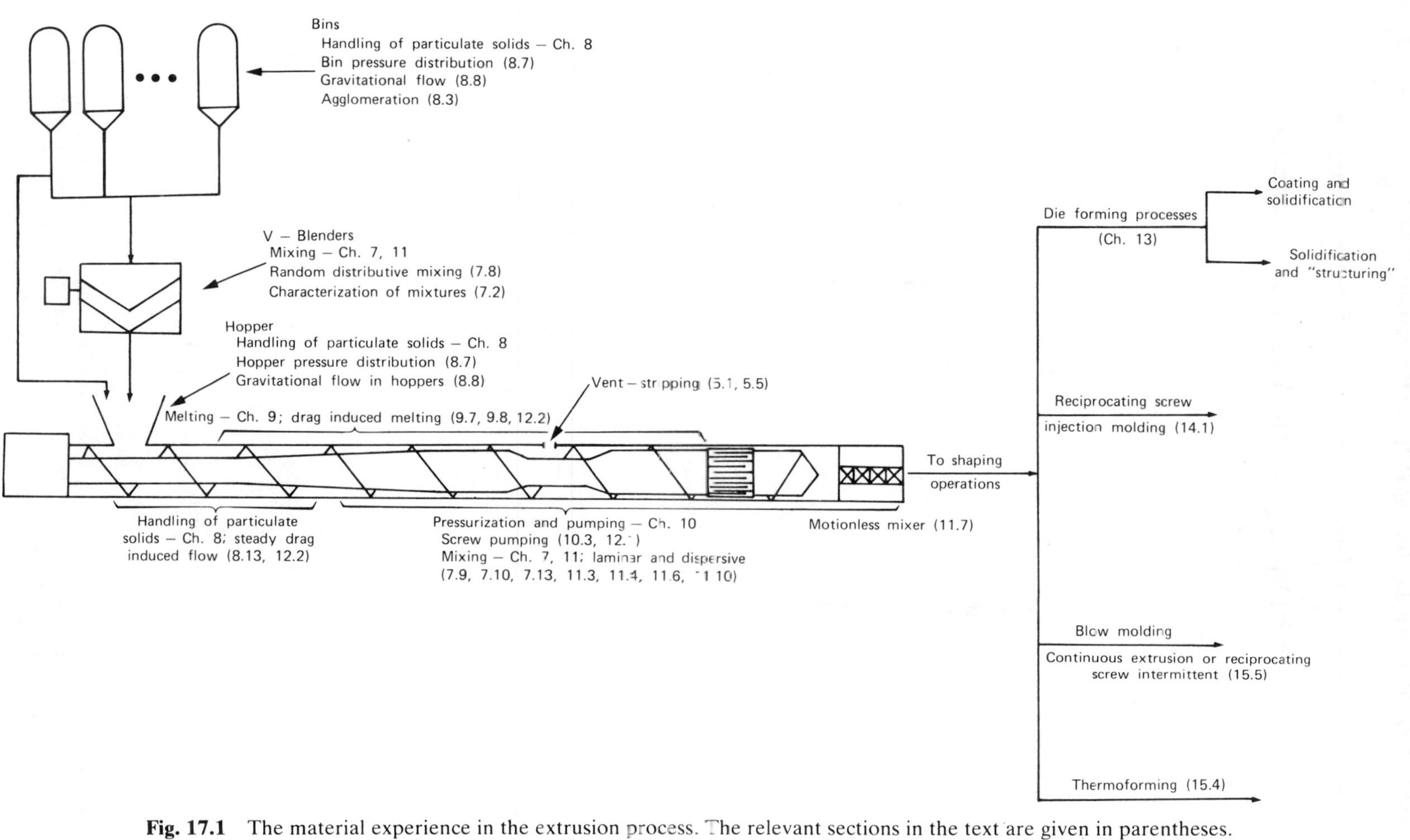


**Fig. 17.1** The material experience in the extrusion process. The relevant sections in the text are given in parentheses.

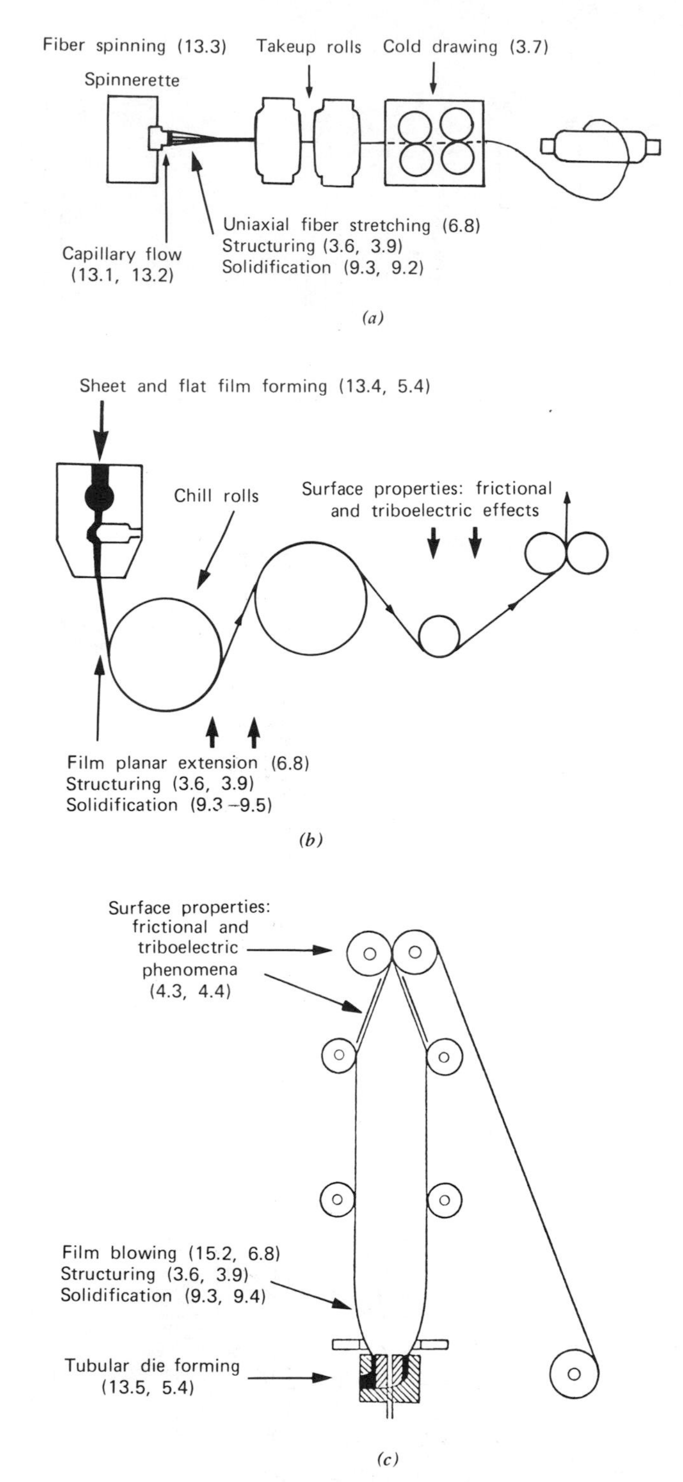


**Fig. 17.2** The material experiences in the die forming methods of (*a*) fiber spinning, (*b*) flat film forming, and (*c*) tubular film blowing. The relevant text sections are given in parentheses. See in particular Chapter 13.

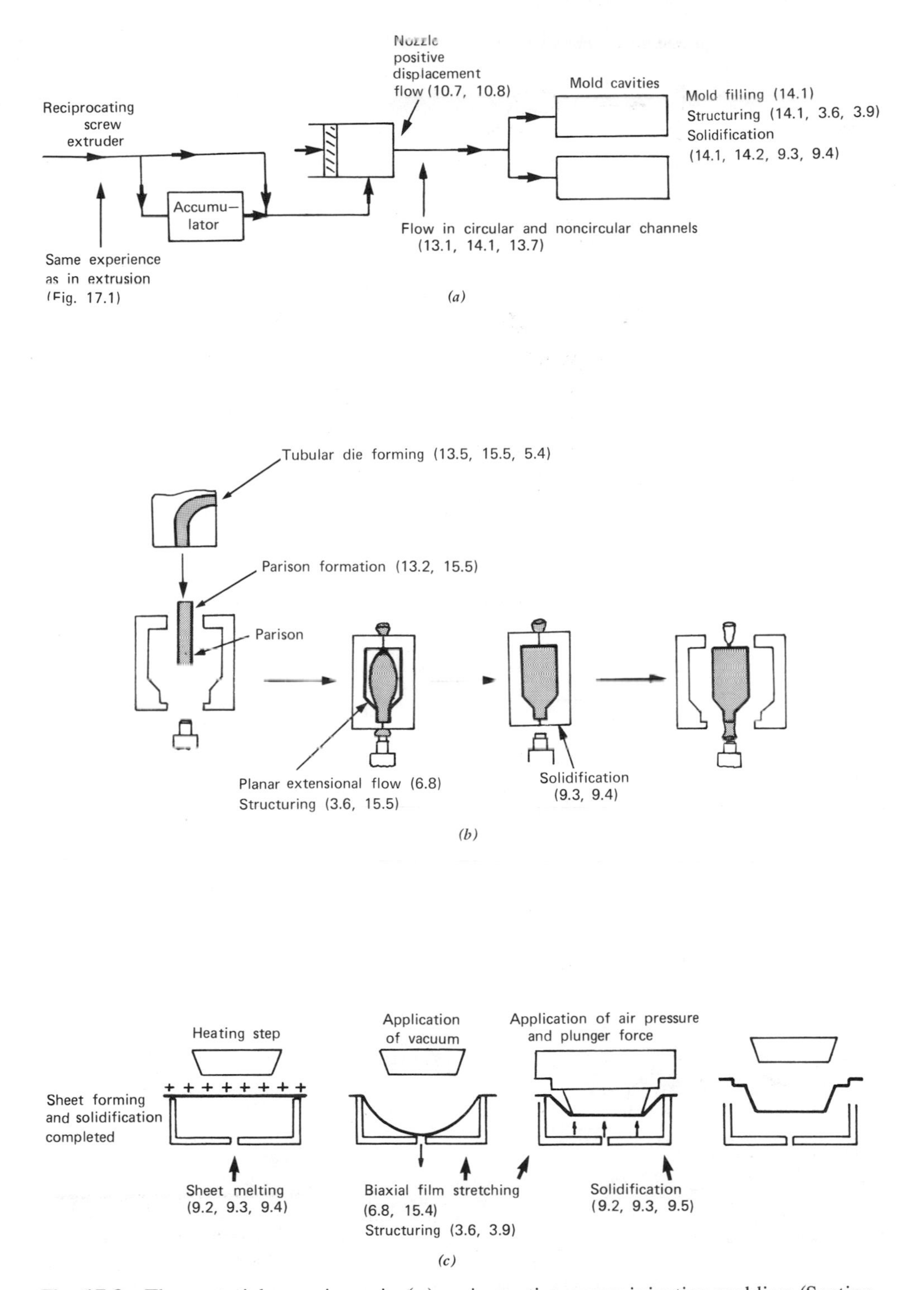


**Fig. 17.3** The material experience in (*a*) reciprocating screw injection molding (Section 14.1), (*b*) blow molding (Section 15.5), and (*c*) thermoforming (Section 15.4). Other relevant sections in the text are given in the figure.

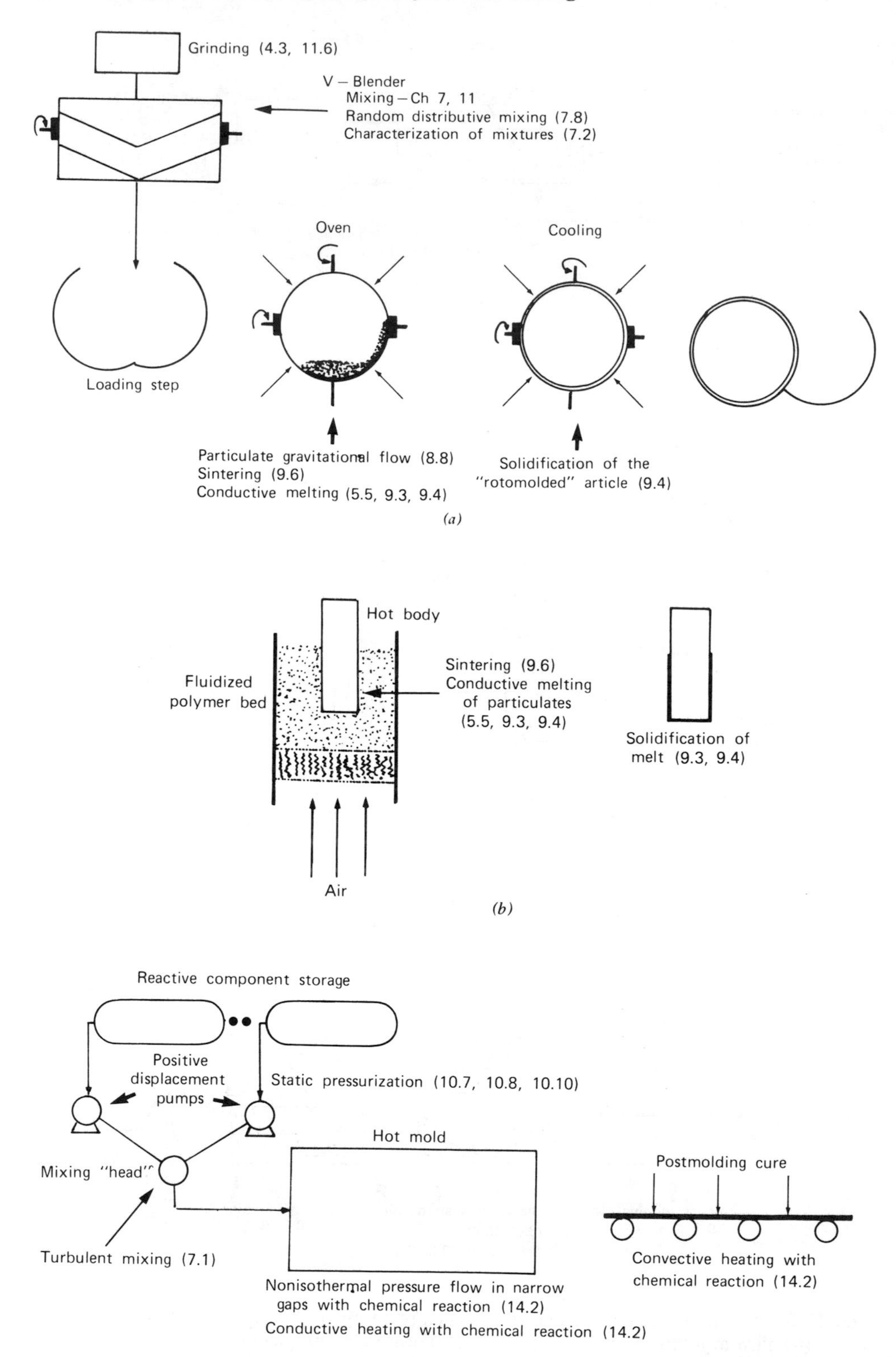

Grinding (4.3, 11.6)
V – Blender
Mixing – Ch 7, 11
Random distributive mixing (7.8)
Characterization of mixtures (7.2)
Oven
Cooling
Loading step
Particulate gravitational flow (8.8)
Sintering (9.6)
Conductive melting (5.5, 9.3, 9.4)
Solidification of the "rotomolded" article (9.4)
(a)
Hot body
Fluidized polymer bed
Sintering (9.6)
Conductive melting of particulates (5.5, 9.3, 9.4)
Solidification of melt (9.3, 9.4)
Air
(b)
Reactive component storage
Positive displacement pumps
Static pressurization (10.7, 10.8, 10.10)
Hot mold
Mixing "head"
Postmolding cure
Turbulent mixing (7.1)
Nonisothermal pressure flow in narrow gaps with chemical reaction (14.2)
Conductive heating with chemical reaction (14.2)
Convective heating with chemical reaction (14.2)
(c)

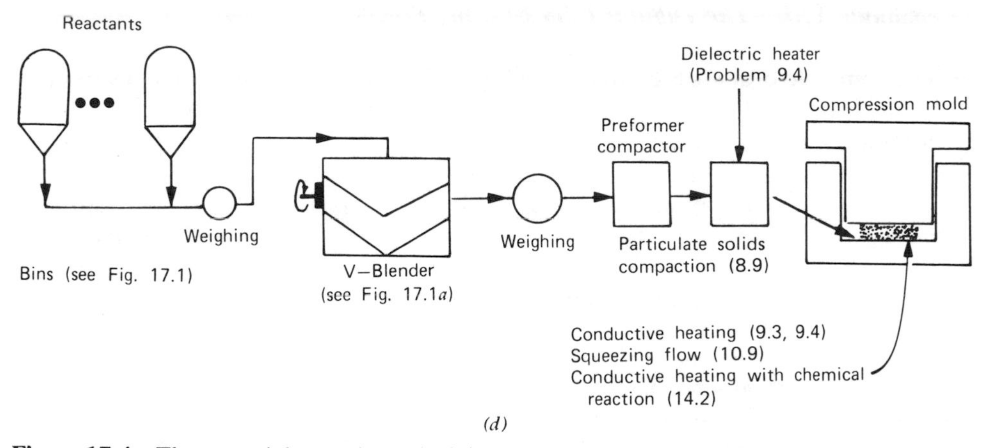


**Figure 17.4** The material experience in (*a*) rotational molding (Problem 9.3), (*b*) fluidized bed coating (Problem 9.6), (*c*) reactive injection molding (Section 14.2), and (*d*) compression molding (Section 14.3). Other relevant sections in the text are given in the figure.

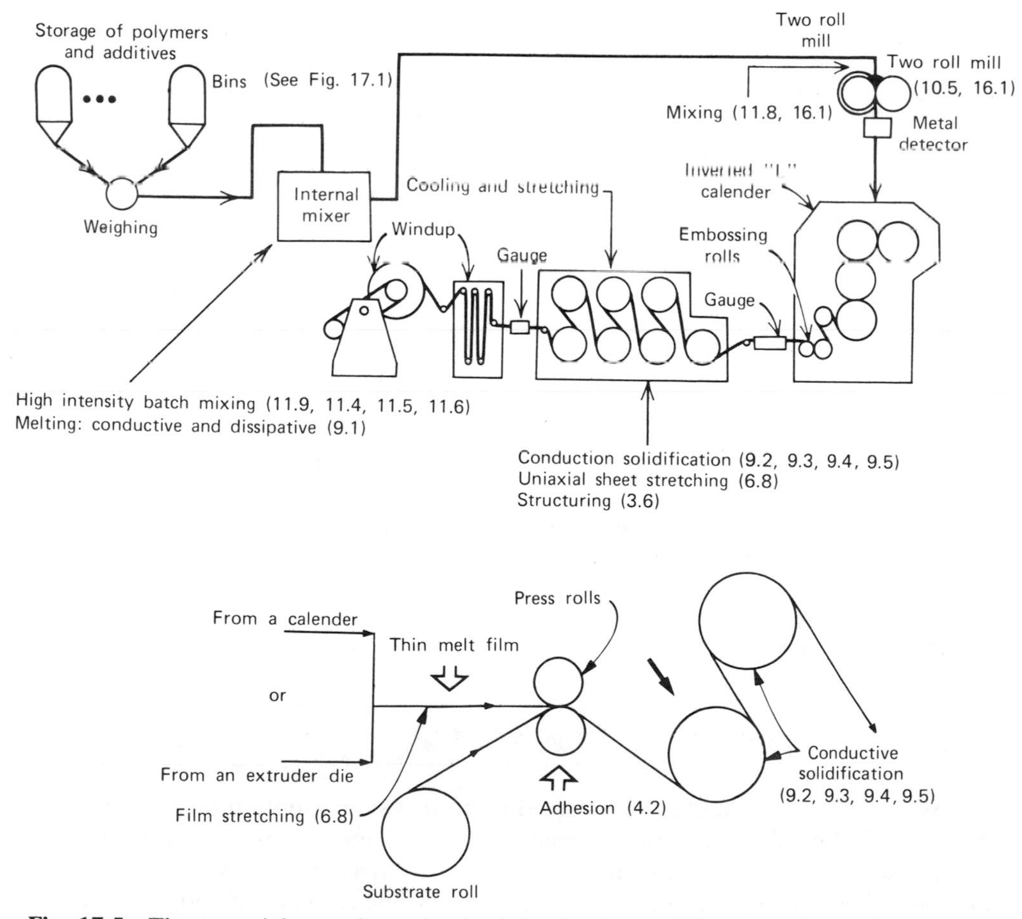


**Fig. 17.5** The material experience in the (*a*) calendering (Chapter 16) and (*b*) coating processes. Relevant sections in the text are given in parentheses.

**Example 17.1** ***The Tubular Film Blowing Process***

The complete process is contained in Figs. 17.1 and 17.2*c*. The mechanisms of interest, the relevant sections in the text, and some of the important relevant equations are summarized below:

| Equipment | Mechanisms/Phenomena of Interest | Relevant Sections | Relevant Equations |
|---|---|---|---|
| Storage bins | Pressure distribution in bins | 8.7 | 8.7-4 to 8.7-7 |
| | Agglomeration | 8.3 | |
| | Gravitational flow in bins | 8.8 | |
| | Background material | 4.3, 4.4, 8.1, 8.2 | |
| V-Blender | Random distributive mixing | 7.8 | |
| | Characterization of mixtures | 7.2, 7.3 | 7.3-1 to 7.3-7 |
| | Background material | 7.1, 11.1, 11.2 | |
| Hopper | Pressure distribution in hoppers | 8.7 | 8.7-4 to 8.7-13 |
| | Gravitational flow in hoppers | 8.8 | |
| | Background material: same as for storage bins | | |
| Screw | Steady drag induced flow of particulate solids | 8.13, 12.2 | 8.13-7, 12.2-5 to 12.2-10 |
| | Compaction of solids | 8.9 | 8.9-1, 8.9-2 |
| | Background material | 4.3 | |
| | Drag induced melting | 9.7, 9.8, 12.2 | 9.8-22 to 9.8-24, 9.8-36, 9.8-53, 12.2-21, 12.2-25, 12.2-32 to 12.2-37 |
| | Dissipative mix-melting | 9.10 | |
| | Pumping and pressurization | 10.3, 12.1 | 10.3-25, 10.3-32, 12.1-3, 12.1-9 |
| | Laminar and dispersive mixing | 7.9, 7.10, 7.13, 11.3, 11.4, 11.6, 11.10 | 11.10-26 |
| Mixing device | Depends on the particular type used | 7.8, 11.7 | 11.7-1, 11.7-2 |
| Tubular film die | Die forming–tube and tubular film forming | 13.5, 5.4 | 13.5-6 |
| | Background material | 13.1, 13.2 | |
| Blowing step | Biaxial film stretching | 6.8, 15.2 | 15.2-3 to 15.2-17 |
| | Structuring | 3.6, 3.9 | |
| | Solidification | 9.3, 9.4, 9.5 | |
| Takeup rolls | Friction and tribolectric phenomena | 4.3, 4.4 | |

We conclude this text, as we started it, by focusing attention on the polymer itself. After discussing the various elements of processing and the complete process as well, several questions arise naturally. Which polymers or polymer grades are suitable to use in any particular process. Why is one polymer or polymer grade more "processable" than another? And finally, can quantitative criteria, in terms of independently measured properties, be given to the illusive concept of "processability"?

"Processability" can be loosely defined as the ability to process and shape a polymer uniformly, predictably, and economically. This definition leads to a number of general conclusions:

1. Generally we can say that for a polymer to be processable it must be thermally stable at the processing temperature for a period of time that is a small multiple of the mean residence time in the processing equipment. Furthermore, it should have a molecular weight such that at these temperatures, it flows easily.
2. Since there is such a great variety of polymer shaping methods and the real possibility exists that a given polymer is processable by one method and not by others, it might be more appropriate to talk about a *certain type* of "processability" *relative to a particular forming method*, such as "spinnability," "moldability," or "extrudability." We have already touched on the "spinnability" of polymers in Section 13.3, relative to the acceptable level of melt viscosity for economically acceptable production rates, and in Section 15.1 relative to the phenomena of fiber cohesive failure (at desirable fiber draw ratios) and draw resonance. Reaching the critical wall shear stress at economically acceptable shear rates (flow rates) is an "extrudability" problem, discussed in connection with PTFE in Section 13.2. "Moldability" is usually tested in an empirical way. The polymer in question is forced under standard conditions into a spiral runner-type mold. The degree of filling of the spiral mold by the polymer is taken as a measure of its moldability. Clearly the main properties "tested" here are the steady shear viscosity, the flow activation energy, the degree of "stress overshoot" (Fig. 6.5), and the thermal properties of the polymer. LDPE can be readily formed into blown film because its elongational viscosity $\bar{\eta}^+(\dot{\varepsilon})$ increases with increasing tensile strains, as discussed in Section 6.8. In thermoforming, the polymer melt must exhibit a relatively high "yield stress" to avoid "sagging" (see Section 15.4). Thus one can generalize from the statements above to the effect that if a polymer is not processable because of the *shaping step*, the properties responsible are its flow properties, that is $\eta$, $\eta(t)$, $\bar{\eta}^+(\dot{\varepsilon})$, $\Delta E_f$, $\Psi_1$ and $\Psi_2$ and $H(\ln \lambda)$.
3. Some polymer processing methods require that the product be *structured* by the process itself. That is, special orientation and morphological features must be introduced to the final product. Examples of such operations are fiber spinning and stretch blow molding. In the first, the fiber must be capable of being "cold drawn" following the spinning step, to attain the morphological features that impart strength in the axial direction (see Section 3.7). In the stretch blow molding process the polymer must have relaxation times at the forming temperature large enough that prior to being vitrified, much of the imposed biaxial orientation is retained. Amorphous polymers just above $T_g$ are of this type. We can refer to the foregoing as "*structurability*," which depends on the flow properties of the polymer melt and its mechanical properties at $T_g < T < T_m$.
4. The inability of a polymer to be formed (processed) by a certain method does not necessarily hinge on the final forming step alone; *all the elementary processing* steps preceding it must be implemented effectively. That is, whether a polymer is "extrudable" may not depend only (or primarily) on its inability to flow easily through the required extruder die; rather, it may be because the polymer is not able to perform well in any of the elementary steps of solids conveying, melting, or pumping and homogenization.

(Inability to perform in the last step would imply inability to flow well in the die.) An example can be found in LDPE recycle "fluff," which is extruded only with difficulty, primarily because it cannot feed properly. Its melting and melt flow behavior is the same as "virgin" LDPE in pellet form. Other examples are polymers with low melt viscosity and high melting point that melt at lower rates in extruders and reciprocating screw injection molding machines, because of the small amount of the viscous energy dissipation. PP and nylons are such polymers. Asbestos filled rigid PVC, used for floor tiles, could be formed by pressure flow through a sheeting die, but if this die were fed by an extruder, two problems would arise: first the PVC compound would be susceptible to thermal degradation because of the vigorous shearing imposed by the extruder and because PVC is very viscous; second, excessive screw and barrel wear would result because of the abrasive nature of this PVC compound and the by-product of the potential thermal degradation, hydrogen chloride, a corrosive material that promotes wear. Shaping the floor tile sheets by the calendering process, which imparts a much smaller amount of mechanical work to the polymer (much smaller shear rates, because *both* rolls are moving), is a more practical processing method for this polymer.

Many other examples can be given pointing to the following conclusion: the lack of processability of a polymer is often due not to its flow performance in the final shaping step, but to its incapacity in one or more of the elementary steps preceding forming, that is, an inability of the polymer to respond to one or more aspects of the *total* thermal and flow (deformation) experience during processing. Properties responsible for the lack of processability in these cases are very low bulk density, low friction coefficient, low viscosity, sensitivity to thermal and oxidative degradation, and the "mechanical properties" of the melt, such as cohesive fracture at low elongations, which contribute to poor dispersion of additives to polymers in mixing devices such as two roll mills*.

5. In either of the cases above—that is, lack of processability because of inappropriate flow properties in shaping or because of other properties in any of the elementary processing steps—we can state that there cannot be a simple relation between a critical level of the responsible property and the lack of processability. The reason is simple: in processing, even in "simple" die flows, the deformation, flow, and temperature histories are much more complex than those imposed during testing for flow, mechanical, and thermal properties. For example, most processing flows are nonviscometric and nonisothermal, whereas viscometry is carried out in viscometric isothermal flows. The same comments can be made about "external" elongational flows, which are neither homogeneous nor isothermal. This point, concerning our inability to relate specific levels or specific kinds of "tested" properties to the lack of processability, was discussed in Sections 6.8 and 15.5 for the blown film and blow molding processes, respectively.

It is clear that to better understand polymer processing in its totality, and processability in particular, we need more information than is presently available on polymer properties evaluated *under conditions similar to those encountered in processing.* This knowledge would also help us understand in more quantitative terms the structuring effects of polymer processing.

* J. L. White, "Processibility of Rubbers and Rheological Behavior," *Rubber Chem. Technol.*, **50**, 163–185 (1977).

# Appendix A: Rheological and Thermodynamic Properties

## A.1 Melt Viscosity

1. Oscillatory shear flow experiments, in connection with the Cox–Merz rule,* were used to determine the low shear rate melt viscosity, as well as the zero shear rate viscosity $\eta_0$. Dr. K. F. Wissbrun conducted the measurements at Celanese Research Co., Summit, N.J., using a Mechanical Spectrometer.

2. Capillary flow experiments were conducted for the determination of melt viscosity at high shear rates by Mr. W. Rahim of Stevens Institute of Technology using the Instron Capillary Rheometer. Rabinowitch and Bagley corrections were applied on the data.

3. Figures A.1, A.2, and A.3 plot the shear rate viscosity for 13 commercial polymers as a function of shear rate at one temperature.

4. Table A.1 lists the power law model (Eq. 6.5-2), the Ellis model (Eq. 6.5-6), and the Carreau model (Eq. 6.5-8) constant, which were evaluated from $\eta(\dot{\gamma})$ as in Figs. A.1, A.2, and A.3 at three different temperatures, as indicated in the Table.

## A.2 Heat Capacity, Particulate Solids and Melt Density

1. Heat capacity measurements were conducted with the Differential Screening Calorimeter, Perkin–Elmer Model 1B, by Mr. W. Rahim.

2. Particulate solids at room temperature and melt densities at 150°C† were measured by Mr. W. Rahim using the Instron Capillary Rheometer. The exit to the rheometer barrel (reservoir) was closed with a brass plug and the polymer was compressed, recording the volume occupied by the polymer sample of known weight, as well as the pressure level.

Figures A.4–A.16 plot the above data for the commercial polymers listed in Table A.1.

* The *Cox–Merz rule* is an empirical relationship predicting that the magnitude of the complex viscosity $|\eta^*(\omega)|$ should be compared with the viscosity at equal values of frequency and shear rate: $|\eta^*(\omega)| = \sqrt{\eta'(\omega)^2 + \eta''(\omega)^2} = \eta(\dot{\gamma})_{\dot{\gamma}=\omega}$ [W. P. Cox and E. H. Merz, *J. Polym. Sci.*, **28**, 619–622 (1958)].

† Note from the $C_p$ curves that for some of the polymers, 150°C is below either $T_m$ or $T_g$, in which case the polymer is a soft solid.

**Table A.1**

| Polymer | Temperature | Power law model | | | Carreau model | | | | Ellis model | | | |
|---|---|---|---|---|---|---|---|---|---|---|---|---|
| | Temperature (K) | Shear rate range ($s^{-1}$) | $m$ $\left(\frac{N \cdot s^n}{m^2}\right)$ | $n$ | $\eta_0$ $\left(\frac{N \cdot s}{m^2}\right)$ | Shear rate range ($s^{-1}$) | $n$ | $\lambda$ (s) | $\eta_0$ $\left(\frac{N \cdot s}{m^2}\right)$ | Shear stress range $\left(\frac{N}{m^2}\right) \times 10^{-5}$ | $\alpha$ | $\tau_{1/2}$ $\left(\frac{N}{m^2}\right)$ |
| High impact | 443 | 100–7000 | $7.58 \times 10^4$ | 0.20 | $2.1 \times 10^5$ | 50–1000 | 0.26 | 6.77 | $2.1 \times 10^5$ | 2.0–4.0 | 4.66 | $5.09 \times 10^4$ |
| polystyrene (HIPS) | 463 | 100–7000 | $4.57 \times 10^4$ | 0.21 | $1.48 \times 10^5$ | 50–1000 | 0.22 | 5.31 | $1.48 \times 10^5$ | 0.9–1.6 | 4.80 | $3.29 \times 10^4$ |
| LX-2400[a] | 483 | 100–7000 | $3.61 \times 10^4$ | 0.19 | $1.05 \times 10^5$ | 100–3000 | 0.16 | 2.57 | $1.05 \times 10^5$ | 0.5–1.9 | 4.74 | $2.22 \times 10^4$ |
| Polystyrene (PS) | 463 | 100–4500 | $4.47 \times 10^4$ | 0.22 | $1.4 \times 10^4$ | 30–6000 | 0.24 | 0.27 | $1.4 \times 10^5$ | 1.0–2.7 | 4.51 | $6.04 \times 10^4$ |
| Dylene™ 8[b] | 483 | 100–4000 | $2.38 \times 10^4$ | 0.25 | $9.2 \times 10^3$ | 30–4000 | 0.27 | 0.32 | $9.2 \times 10^3$ | 0.6–1.8 | 3.85 | $3.17 \times 10^4$ |
| | 498 | 100–5000 | $1.56 \times 10^4$ | 0.28 | $6.6 \times 10^3$ | 30–4000 | 0.30 | 0.35 | $6.6 \times 10^3$ | 0.5–1.7 | 3.67 | $2.27 \times 10^4$ |
| Styrene acrylonitrile | 463 | 100–9000 | $9.0 \times 10^4$ | 0.21 | $2.2 \times 10^4$ | 100–8000 | 0.21 | 0.17 | $2.2 \times 10^4$ | 2.3–5.6 | 4.40 | $1.26 \times 10^5$ |
| (SAN) Lustran™ | 493 | 100–8000 | $3.22 \times 10^4$ | 0.27 | $9.0 \times 10^3$ | 80–8000 | 0.28 | 0.18 | $9.0 \times 10^3$ | 1.0–3.2 | 3.59 | $5.03 \times 10^4$ |
| 31-1000[a] | 523 | 100–8000 | $1.11 \times 10^4$ | 0.35 | $4.2 \times 10^3$ | 100–5000 | 0.36 | 0.25 | $4.2 \times 10^3$ | 0.5–2.0 | 2.76 | $1.80 \times 10^4$ |
| Thermoplastic | 473 | 100–5000 | $2.75 \times 10^4$ | 0.27 | $3.6 \times 10^4$ | 40–5000 | 0.28 | 1.62 | $3.6 \times 10^4$ | 1.3–2.7 | 3.69 | $2.41 \times 10^4$ |
| olefin (TPO) | 493 | 100–4000 | $1.83 \times 10^4$ | 0.30 | $2.15 \times 10^4$ | 70–2000 | 0.31 | 1.42 | $2.15 \times 10^4$ | 0.7–1.7 | 3.32 | $1.68 \times 10^4$ |
| Vistaflex™ 905B[c] | 513 | 100–3000 | $1.99 \times 10^4$ | 0.28 | $1.35 \times 10^4$ | 70–2000 | 0.31 | 0.72 | $1.35 \times 10^4$ | 0.6–1.6 | 3.36 | $1.99 \times 10^4$ |
| Acrylonitrile buta- | 443 | 100–5500 | $1.19 \times 10^5$ | 0.25 | $7.95 \times 10^4$ | 100–2000 | 0.29 | 0.82 | $7.95 \times 10^4$ | 3.0–6.0 | 3.31 | $8.98 \times 10^4$ |
| diene styrene (ABS) | 463 | 100–6000 | $6.29 \times 10^4$ | 0.25 | $4.4 \times 10^4$ | 100–3000 | 0.26 | 0.73 | $4.4 \times 10^4$ | 1.8–5.0 | 3.85 | $4.79 \times 10^4$ |
| AM-1000[d] | 483 | 100–7000 | $3.93 \times 10^4$ | 0.25 | $2.6 \times 10^4$ | 40–4000 | 0.25 | 0.57 | $2.6 \times 10^4$ | 1.0–2.5 | 3.33 | $3.18 \times 10^4$ |
| Polypropylene (PP) | 453 | 100–4000 | $6.79 \times 10^3$ | 0.37 | $4.21 \times 10^3$ | 70–4000 | 0.38 | 0.49 | $4.21 \times 10^3$ | 0.3–1.4 | 2.72 | $9.57 \times 10^3$ |
| CD 460[c] | 463 | 100–3500 | $4.89 \times 10^3$ | 0.41 | $3.2 \times 10^3$ | 70–4000 | 0.41 | 0.51 | $3.2 \times 10^3$ | 0.2–1.4 | 2.50 | $7.19 \times 10^3$ |
| | 473 | 100–4000 | $4.35 \times 10^3$ | 0.41 | $2.5 \times 10^3$ | 50–3000 | 0.41 | 0.40 | $2.5 \times 10^3$ | 0.2–1.3 | 2.49 | $7.17 \times 10^3$ |

| | | | | | | | | | | | | |
|---|---|---|---|---|---|---|---|---|---|---|---|---|
| Ethylene ethyl | 443 | 100–6000 | $1.21\times10^4$ | 0.38 | $5.4\times10^3$ | 80–6000 | 0.42 | 0.40 | $5.4\times10^3$ | 0.6–3.0 | 2.71 | $2.24\times10^4$ |
| acrylate | 463 | 100–4000 | $6.91\times10^3$ | 0.43 | $3.5\times10^3$ | 10–3000 | 0.58 | 1.08 | $3.5\times10^3$ | 0.3–2.5 | 2.39 | $1.40\times10^4$ |
| DPDA-6169[e] | 483 | 100–6000 | $3.77\times10^3$ | 0.48 | $2.3\times10^3$ | 40–1000 | 0.50 | 0.51 | $2.3\times10^3$ | 0.2–2.5 | 2.13 | $7.02\times10^3$ |
| High density poly- | 453 | 100–1000 | $6.19\times10^3$ | 0.56 | $2.1\times10^3$ | 100–1200 | 0.54 | 0.07 | $2.1\times10^3$ | 0.8–2.8 | 2.57 | $7.50\times10^4$ |
| ethylene (HDPE) | 473 | 100–1000 | $4.68\times10^3$ | 0.59 | $1.52\times10^3$ | 100–1200 | 0.50 | 0.08 | $1.52\times10^3$ | 0.6–2.5 | 2.51 | $7.49\times10^4$ |
| Alathon™ 7040[f] | 493 | 100–1000 | $3.73\times10^3$ | 0.61 | $1.17\times10^3$ | 180–1400 | 0.58 | 0.05 | $1.17\times10^3$ | 0.5–2.0 | 2.49 | $7.67\times10^4$ |
| Low density | 433 | 100–4000 | $9.36\times10^3$ | 0.41 | $6.3\times10^3$ | 80–1000 | 0.42 | 0.59 | $6.3\times10^3$ | 0.6–3.0 | 2.56 | $1.52\times10^4$ |
| polyethylene (LDPE) | 453 | 100–6500 | $5.21\times10^3$ | 0.46 | $3.2\times10^3$ | 100–7000 | 0.47 | 0.47 | $3.2\times10^3$ | 0.4–2.0 | 2.22 | $9.06\times10^3$ |
| Alathon™ 1540[f] | 473 | 100–6000 | $4.31\times10^3$ | 0.47 | $1.7\times10^3$ | 100–1000 | 0.48 | 0.21 | $1.7\times10^3$ | 0.3–1.8 | 2.23 | $1.2\times10^4$ |
| Nylon | 498 | 100–2500 | $2.62\times10^3$ | 0.63 | $1.6\times10^3$ | 100–2000 | 0.63 | 0.27 | $1.6\times10^3$ | 0.4–3.0 | 1.64 | $1.06\times10^4$ |
| Capron™ 8200[g] | 503 | 100–2000 | $1.95\times10^3$ | 0.66 | $1.3\times10^3$ | 100–2000 | 0.65 | 0.32 | $1.3\times10^3$ | 0.3–3.0 | 1.70 | $1.3\times10^4$ |
| | 508 | 100–2300 | $1.81\times10^3$ | 0.66 | $1.1\times10^3$ | 100–2000 | 0.68 | 0.36 | $1.1\times10^3$ | 0.3–3.0 | 1.61 | $1.04\times10^4$ |
| Polypropylene (PP) | 483 | 100–3000 | $3.21\times10^4$ | 0.25 | $3.5\times10^4$ | 100–3500 | 0.24 | 1.05 | $3.5\times10^4$ | 0.001–0.3 | 4.19 | $3.35\times10^4$ |
| E 612[c] | 513 | 50–3000 | $2.24\times10^4$ | 0.28 | $1.98\times10^4$ | 50–3500 | 0.27 | 0.81 | $1.98\times10^4$ | 0.0005–0.3 | 3.76 | $2.52\times10^4$ |
| Polymethylmetha- | 493 | 100–6000 | $8.83\times10^4$ | 0.19 | $1.3\times10^4$ | 100–4000 | 0.19 | 0.09 | $1.3\times10^4$ | 2.0–4.5 | 5.23 | $1.44\times10^5$ |
| crylate (PMMA) | 513 | 100–6000 | $4.27\times10^4$ | 0.25 | $6.0\times10^3$ | 100–4000 | 0.25 | 0.07 | $6.0\times10^3$ | 1.2–3.3 | 3.68 | $7.36\times10^4$ |
| Lucite™ 147[f] | 533 | 100–7000 | $2.62\times10^4$ | 0.27 | $2.9\times10^3$ | 100–4000 | 0.36 | 0.11 | $2.9\times10^3$ | 0.9–2.4 | 4.06 | $8.56\times10^4$ |
| Polycarbonate (PC) | 553 | 100–1000 | $8.39\times10^3$ | 0.64 | $1.52\times10^3$ | 0.01–3000 | 0.26 | 0.002 | $1.52\times10^3$ | 1.2–8.0 | 2.23 | $9.46\times10^5$ |
| Lexan™[g] | 573 | 100–1000 | $4.31\times10^3$ | 0.67 | $8\times10^2$ | 0.01–2000 | 0.26 | 0.002 | $8\times10^2$ | 0.8–5.8 | 2.76 | $6.38\times10^5$ |
| | 593 | 100–1000 | $1.08\times10^3$ | 0.80 | $4.2\times10^2$ | 0.01–2500 | 0.53 | 0.002 | $4.2\times10^2$ | 0.4–3.0 | 2.06 | $7.34\times10^5$ |

[a] Monsanto Co.
[b] ARCO.
[c] Exxon Chemical Co., USA.
[d] Borg-Warner Corp.
[e] Union Carbide Corp.
[f] E. I. Dupont de Nemours and Co., Inc.
[g] Allied Chemical Corp.
[h] General Electric Co. Plastics Div.

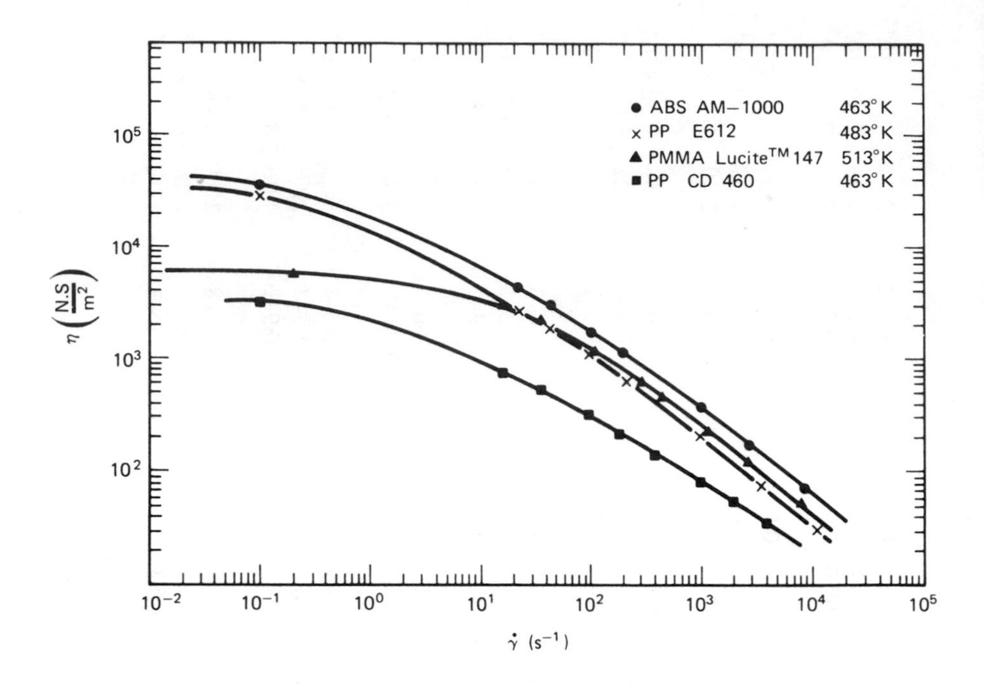

● ABS AM–1000 463°K
× PP E612 483°K
▲ PMMA Lucite™ 147 513°K
■ PP CD 460 463°K
$\eta\left(\frac{N.S}{m^2}\right)$
$10^5$
$10^4$
$10^3$
$10^2$
$10^{-2}$ $10^{-1}$ $10^0$ $10^1$ $10^2$ $10^3$ $10^4$ $10^5$
$\dot{\gamma}$ $(s^{-1})$


**Fig. A.1**

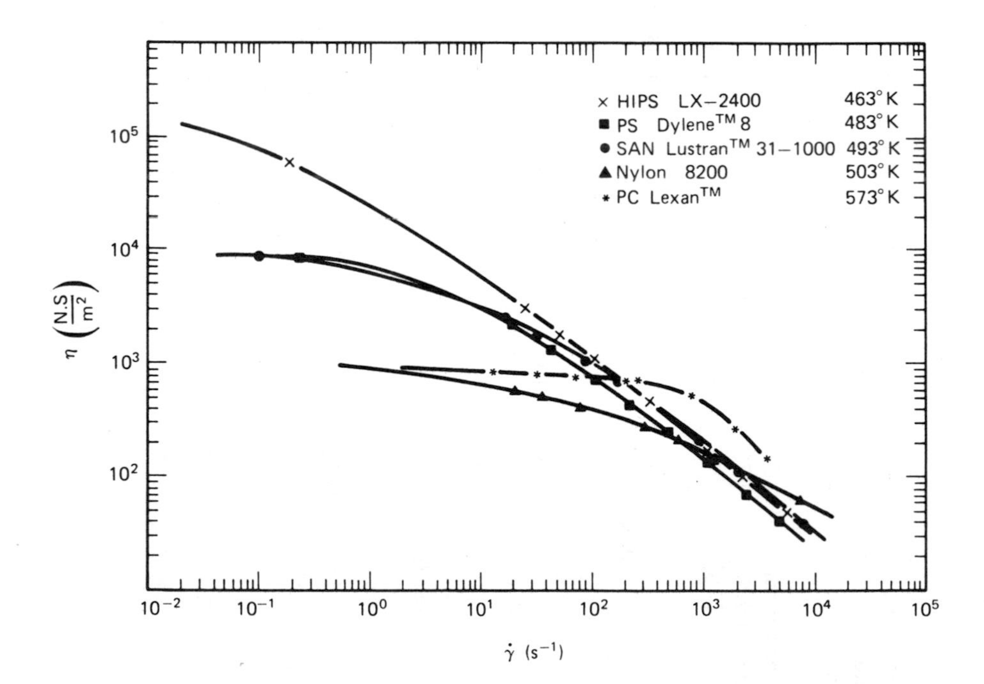

× HIPS LX–2400 463°K
■ PS Dylene™ 8 483°K
● SAN Lustran™ 31–1000 493°K
▲ Nylon 8200 503°K
* PC Lexan™ 573°K
$\eta\left(\frac{N.S}{m^2}\right)$
$10^5$
$10^4$
$10^3$
$10^2$
$10^{-2}$ $10^{-1}$ $10^0$ $10^1$ $10^2$ $10^3$ $10^4$ $10^5$
$\dot{\gamma}$ $(s^{-1})$


**Fig. A.2**

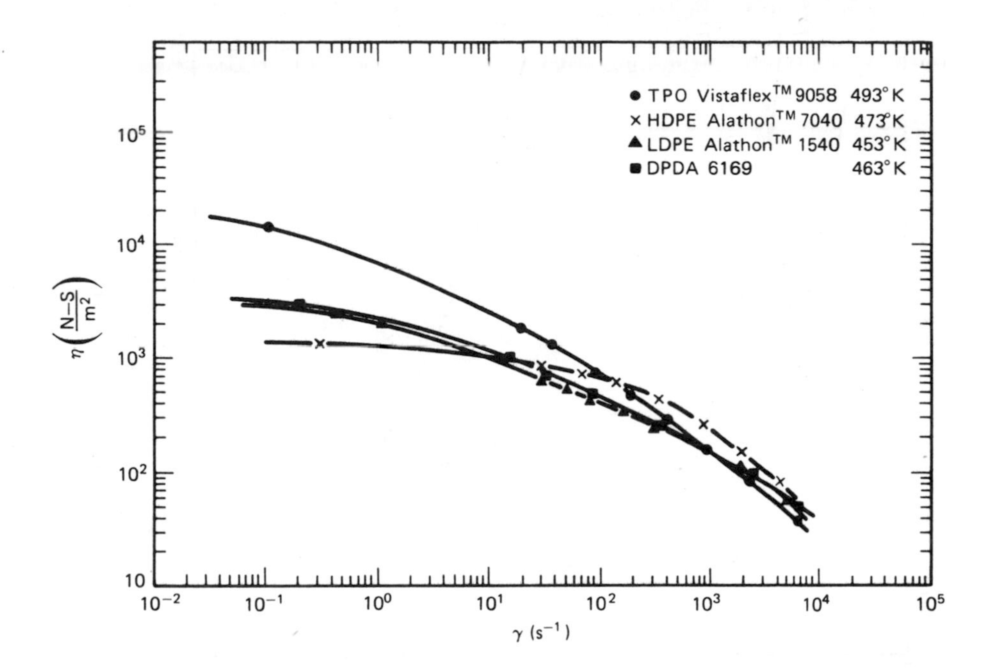


**Fig. A.3**

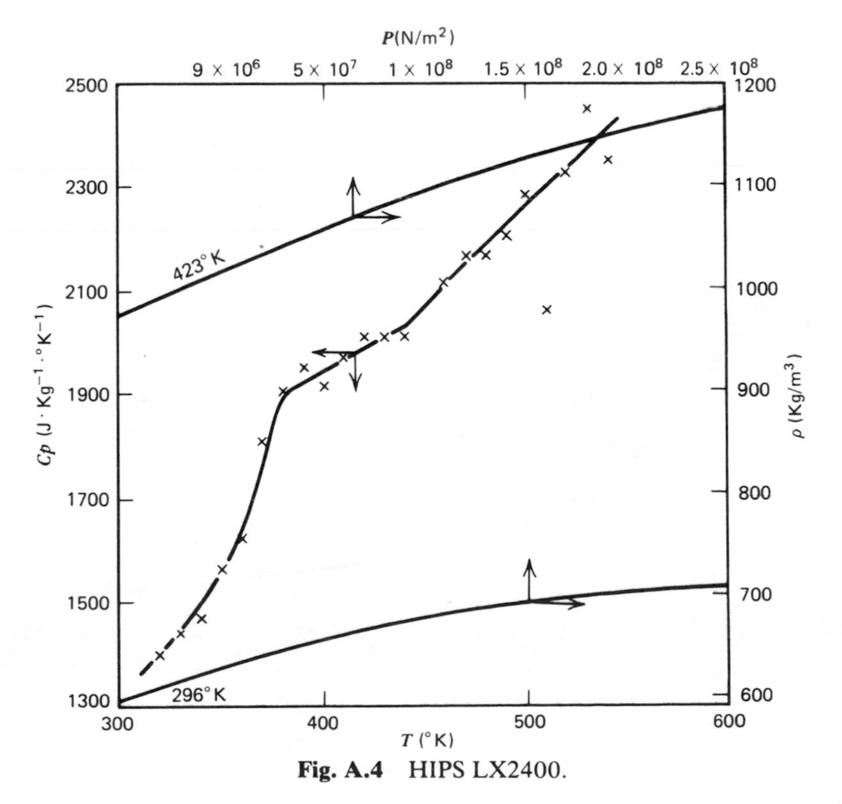


**Fig. A.4** HIPS LX2400.

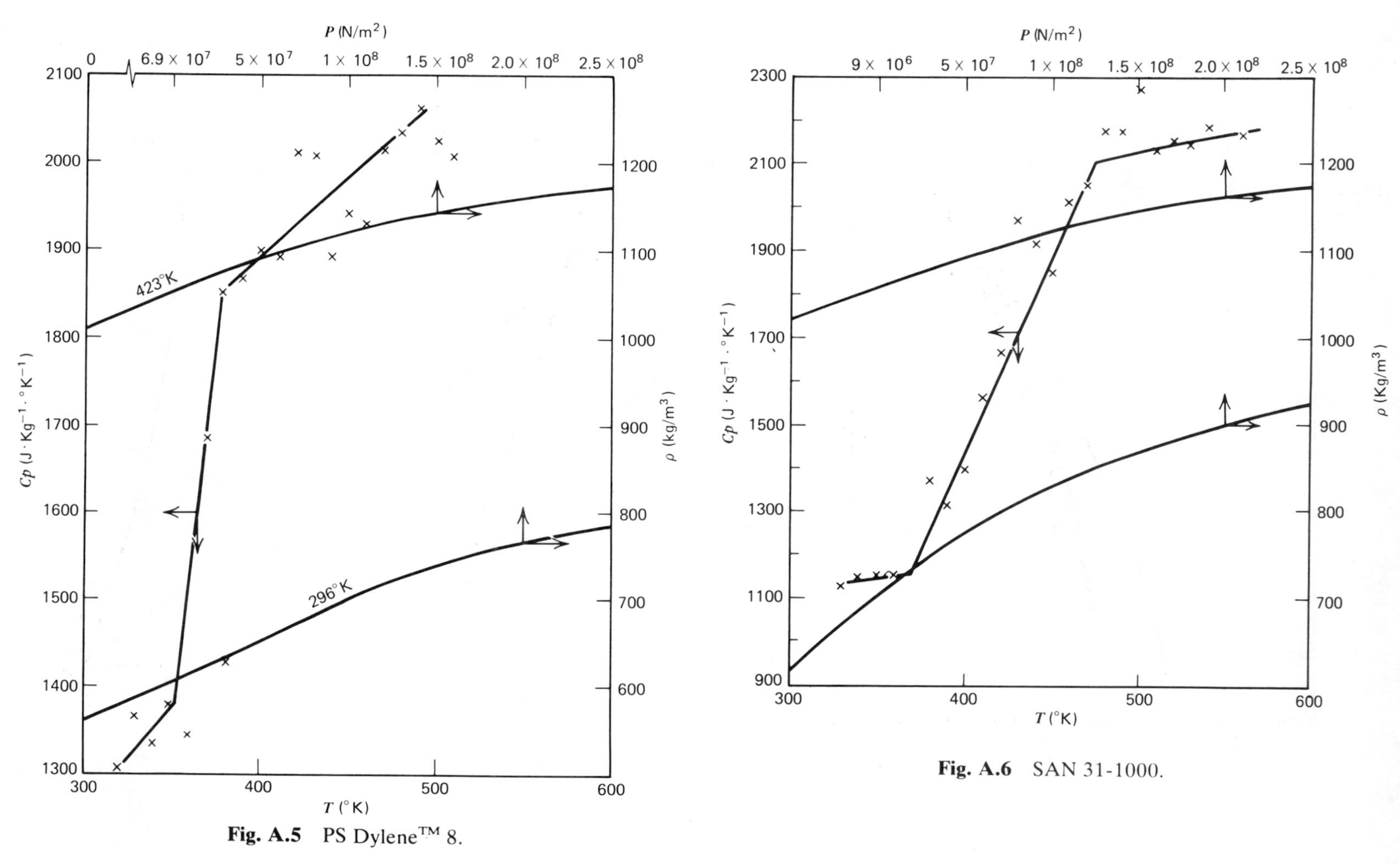


**Fig. A.5** PS Dylene™ 8.



**Fig. A.6** SAN 31-1000.

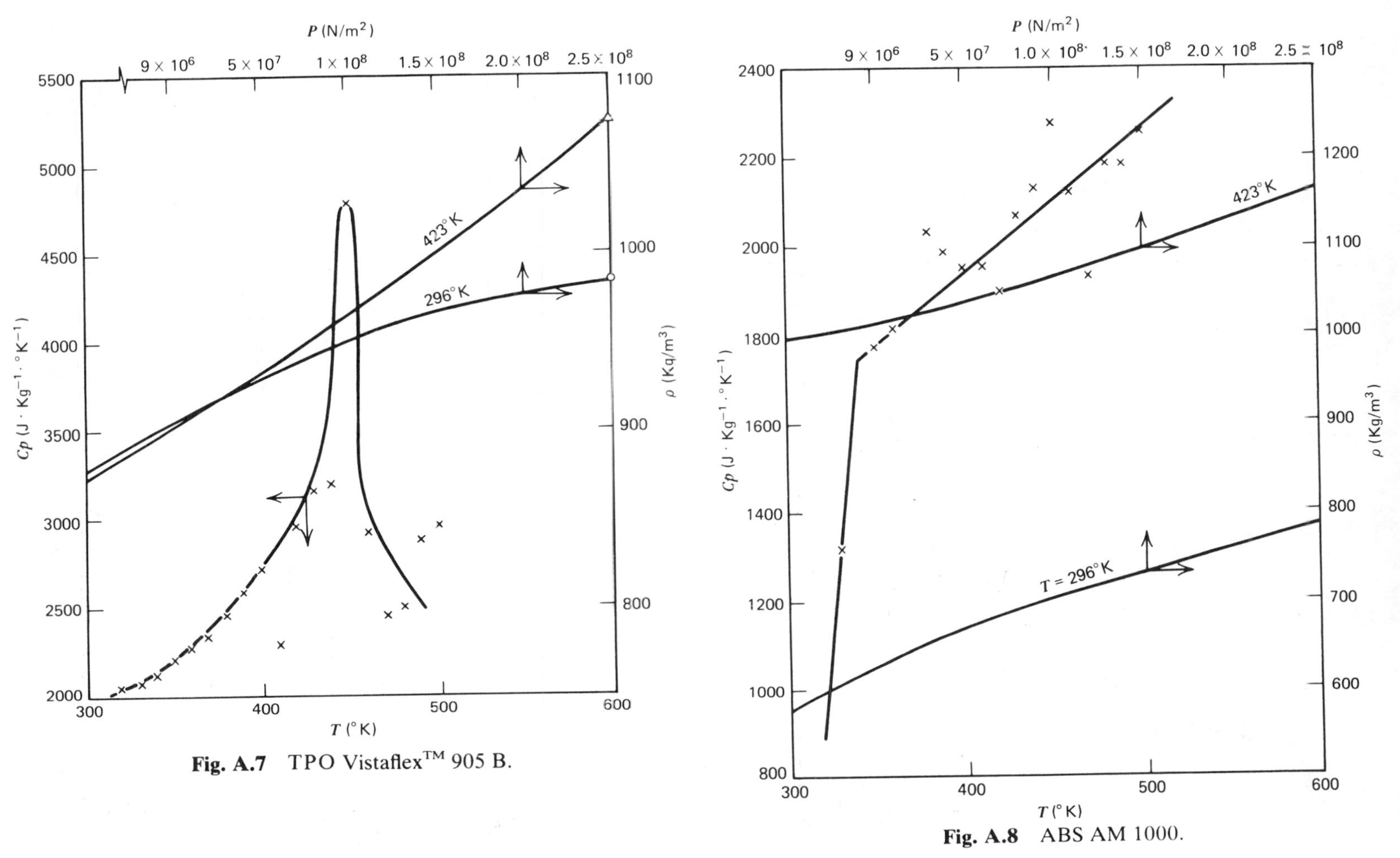


**Fig. A.7** TPO Vistaflex™ 905 B.

**Fig. A.8** ABS AM 1000.

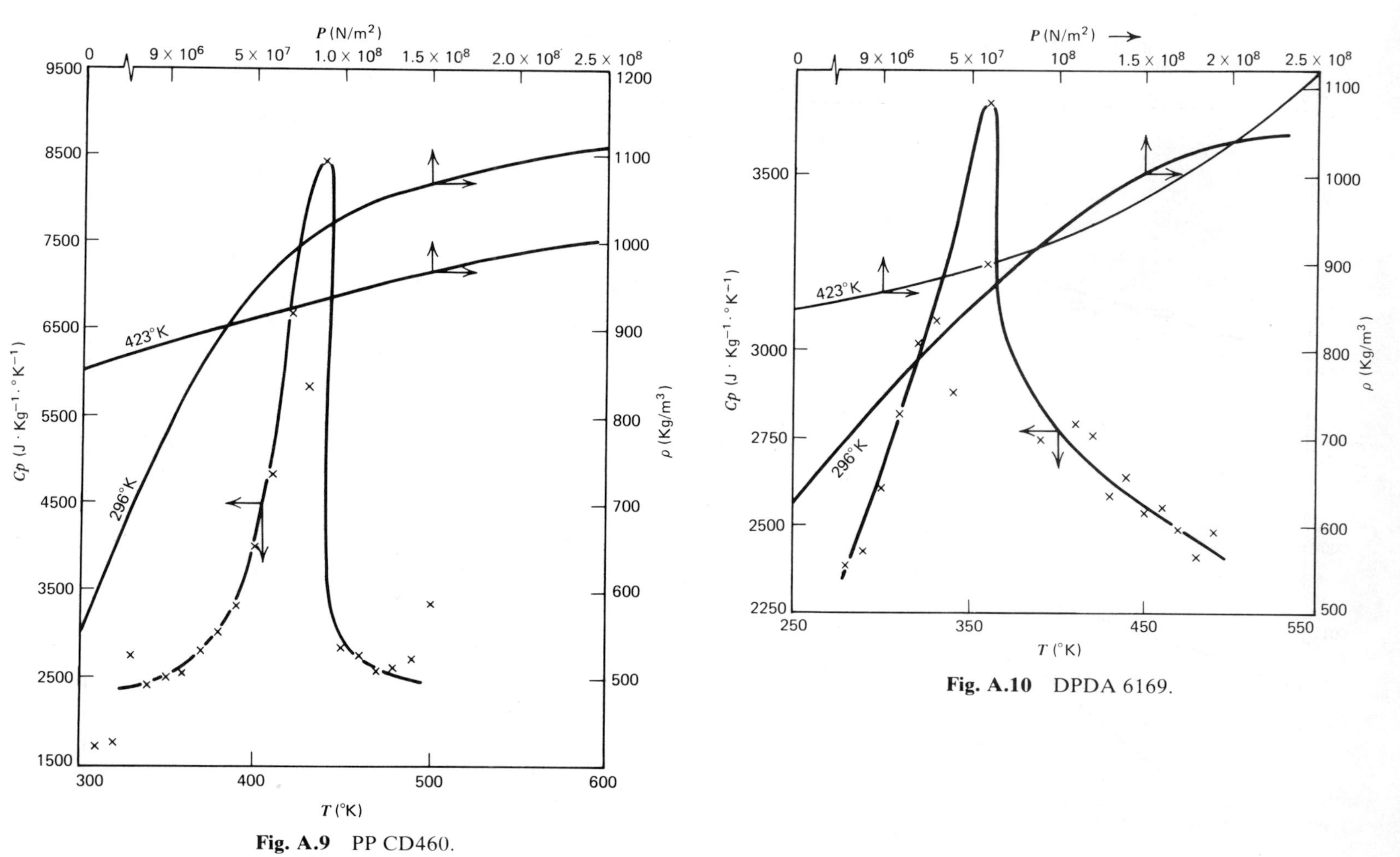


**Fig. A.9** PP CD460.

**Fig. A.10** DPDA 6169.

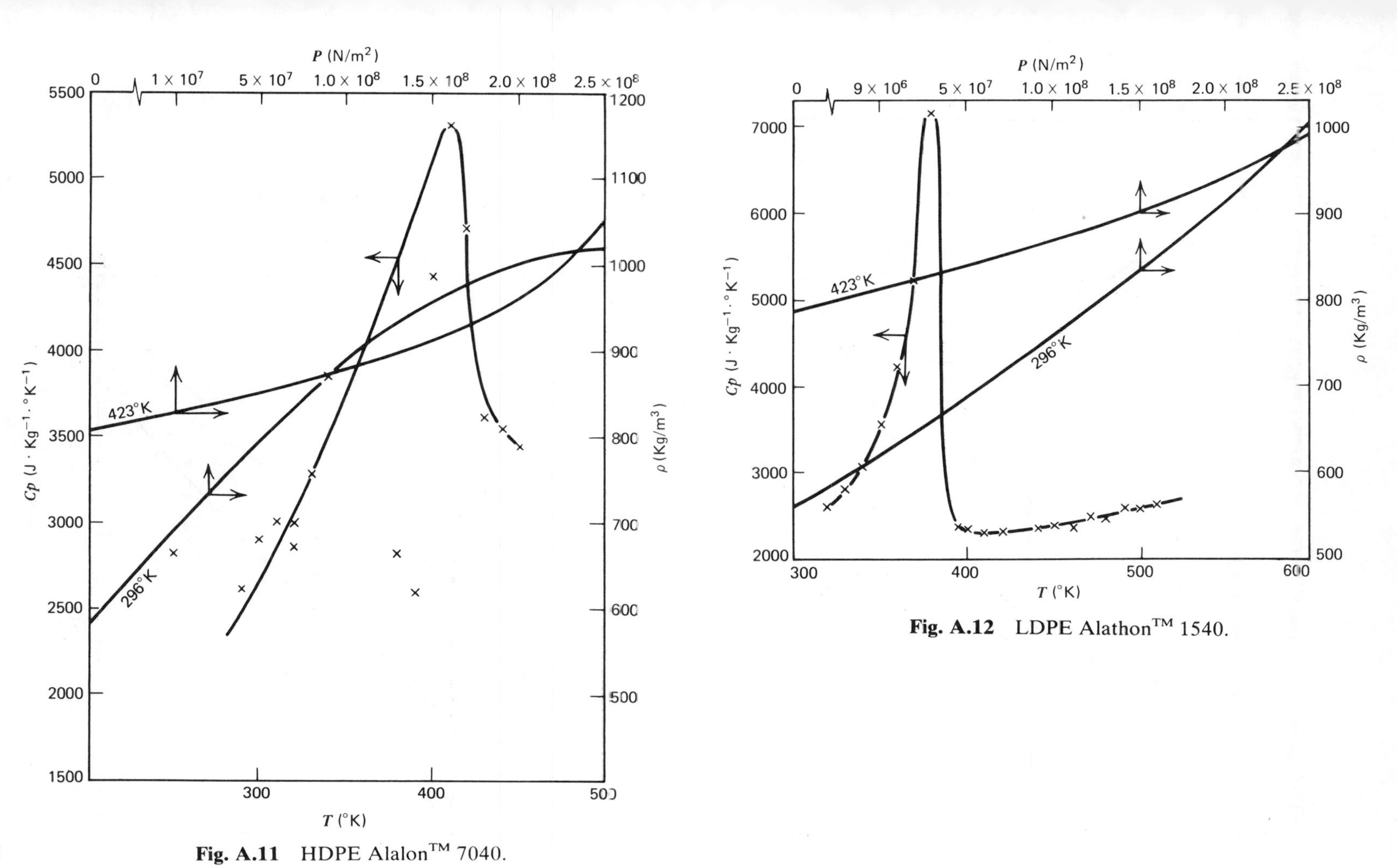


**Fig. A.11** HDPE Alalon™ 7040.



**Fig. A.12** LDPE Alathon™ 1540.

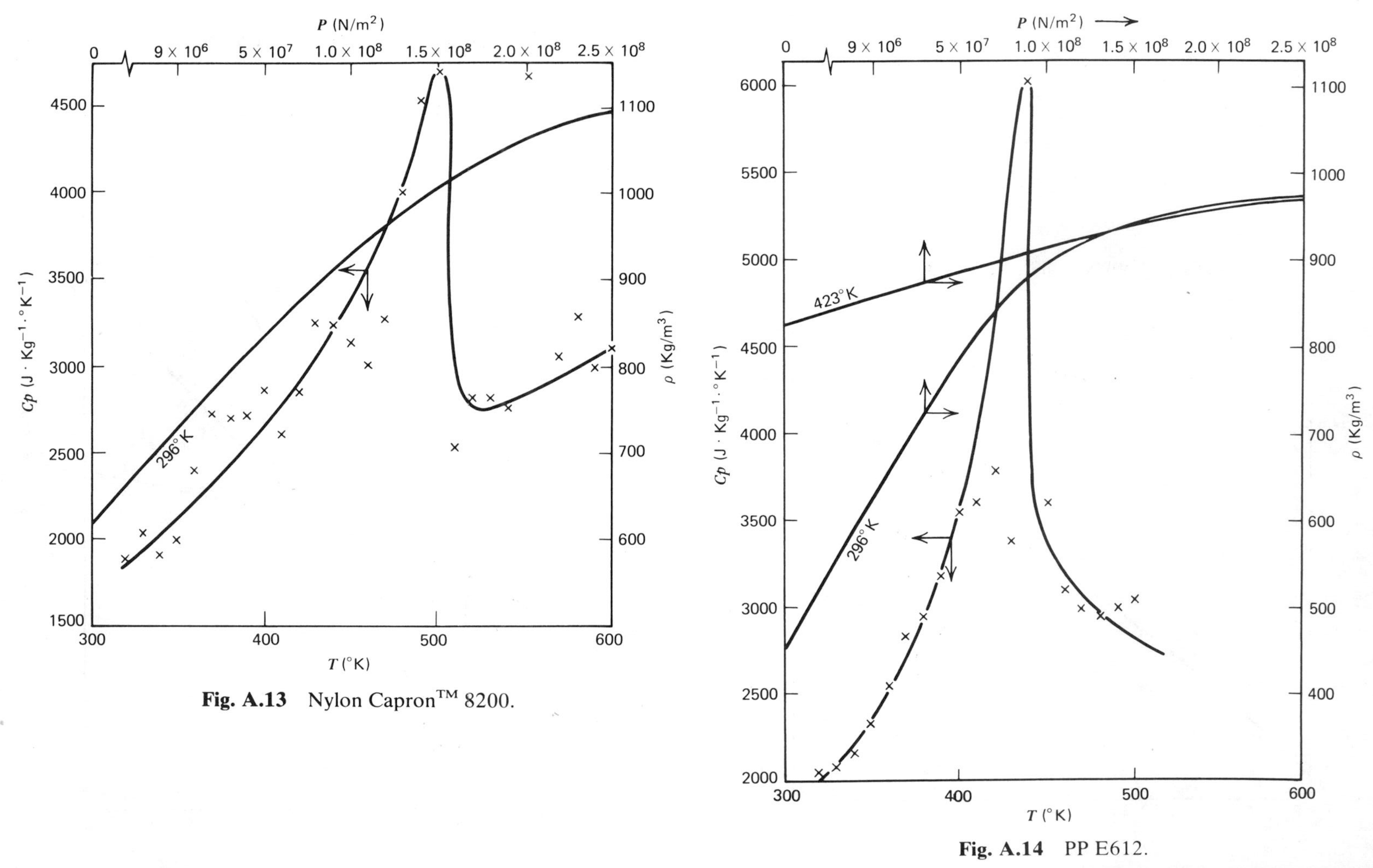


**Fig. A.13** Nylon Capron™ 8200.

**Fig. A.14** PP E612.

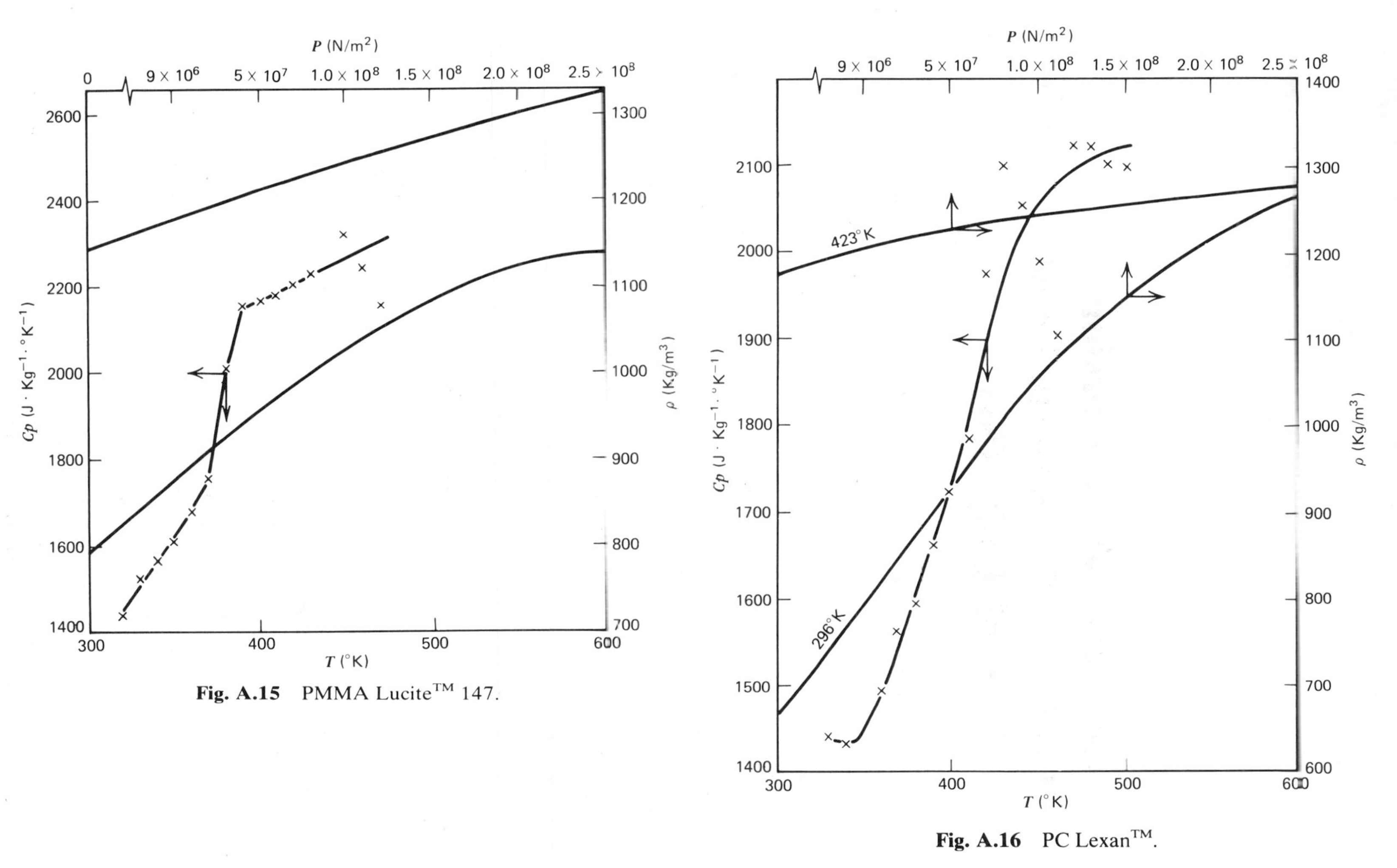


**Fig. A.15** PMMA Lucite™ 147.



**Fig. A.16** PC Lexan™.

# Appendix B: Conversion Tables to the International System of Units (SI)

*The International System of Units (SI) and Conversion Tables**

| Quantity | Unit | SI Symbol | |
|---|---|---|---|
| **Basic units** | | | |
| Length | meter | m | |
| Mass | kilogram | kg | |
| Time | second | s | |
| Electric current | ampere | A | |
| Thermodynamic temperature | kelvin | K | |
| Luminous intensity | candela | cd | |
| *Supplementary units* | | | |
| Plane angle | radian | rad | |
| Solid angle | steradian | sr | |
| **Derived units** | | | |
| Acceleration | meter per second squared | — | $m/s^2$ |
| Activity (of a radioactive source) | disintegration per second | — | (disintegration/s) |
| Angular acceleration | radian per second squared | — | $rad/s^2$ |
| Angular velocity | radian per second | — | rad/s |
| Area | square meter | — | $m^2$ |
| Density | kilogram per cubic meter | — | $kg/m^3$ |
| Electric capacitance | farad | F | A·s/V |
| Electric field strength | volt per meter | — | V/m |
| Electric inductance | henry | H | V·s/A |
| Electric potential difference | volt | V | W/A |
| Electric resistance | ohm | Ω | V/A |
| Electromotive force | volt | V | W/A |
| Energy | joule | J | N·m |
| Entropy | joule per kelvin | — | J/K |
| Force | newton | N | $kg{\cdot}m/s^2$ |
| Frequency | hertz | Hz | — |
| Magnetomotive force | ampere | A | — |
| Power | watt | W | J/s |

*E. A. Mechtly "The International System of Units," NASA SP-7012, Washington, D.C. 1969; *Am. Inst. Chem. Eng. J.*, **17**, 511 (1971).

| Quantity | Unit | SI Symbol | |
|---|---|---|---|
| Pressure | newton per square meter | — | N/m |
| Quantity of electricity | coulomb | C | A·s |
| Quantity of heat | joule | J | N·m |
| Radiant intensity | watt per steradian | — | W/sr |
| Specific heat | joule per kilogram-kelvin | — | J/kg·K |
| Stress | newton per square meter | — | $N/m^2$ |
| Thermal conductivity | watt per meter-kelvin | — | W/m·K |
| Velocity | meter per second | — | m/s |
| Viscosity, dynamic | Newton-second per square meter | — | $N·s/m^2$ |
| Viscosity, kinematic | square meter per second | — | $m^2/s$ |
| Voltage | volt | V | W/A |
| Volume | cubic meter | — | $m^3$ |
| Wavenumber | reciprocal meter | — | (wave)/m |
| Work | joule | J | N·m |

*SI Prefixes*

| Factor | Prefix | Symbol | Factor | Prefix | Symbol |
|---|---|---|---|---|---|
| $10^{12}$ | tera | T | $10^{-1}$ | deci | d |
| $10^{9}$ | giga | G | $10^{-2}$ | centi | c |
| $10^{6}$ | mega | M | $10^{-3}$ | milli | m |
| $10^{3}$ | kilo | k | $10^{-6}$ | micro | $\mu$ |
| $10^{2}$ | hecto | h | $10^{-9}$ | nano | n |
| $10^{1}$ | deka | da | $10^{-12}$ | pico | p |

*Physical Constants*

| | Unit | Value | |
|---|---|---|---|
| Avogadro constant | $k\ mole^{-1}$ | 6.0222 | E+26* |
| Gas law constant | J/kmole·K | 8.3143 | E+3 |
| Boltzmann constant | J/K | 1.3806 | E−23 |
| Stefan–Boltzmann constant | $W/m^2K^4$ | 5.66916 | E−8 |
| Planck constant | J·s | 6.6262 | E−34 |
| Gravitational acceleration | $m/s^2$ | 9.80665 | E+00 |

* E+26 denotes $10^{26}$.

*Conversion Table to SI Units*

| To Convert from | To | Multiply by |
|---|---|---|
| angstrom | meter (m) | 1.000 000* E−10 |
| atmosphere (normal) | newton/meter² ($N/m^2$) | 1.013 250* E+05 |
| barrel (for petroleum, 42 gal) | meter³ ($m^3$) | 1.589 873 E−01 |
| Bar | newton/meter² ($N/m^2$) | 1.000 000* E+05 |

*Conversion Table to SI Units—contd.*

| To Convert from | To | Multiply by | |
|---|---|---|---|
| British thermal unit (International Table) | joule (J) | 1.055 04 | E+03 |
| Btu/lbm-°F (heat capacity) | joule/kilogram-kelvin (J/kg·K) | 4.186 800* | E+03 |
| Btu/second | watt (W) | 1.054 350 | E+03 |
| Btu/ft$^2$-hr-°F (heat transfer coefficient) | joule/meter$^2$-second-kelvin (J/m$^2$·s·K) | 5.678 264 | E+00 |
| Btu/ft$^2$-hr (heat flux) | joule/meter$^2$-second (J/m$^2$·s) | 3.154 591 | E+00 |
| Btu/ft-hr-deg F (thermal conductivity) | joule/meter-second-kelvin (J/m·s·K) | 1.730 735 | E+00 |
| Calorie (International Table) | joule (J) | 4.186 800* | E+00 |
| cal/g-°C | joule/kilogram·kelvin (J/kg·K) | 4.186 800* | E+03 |
| cal/sec-cm-K | joule/meter-second-kelvin (J/m·s·K) | 4.186 800* | E+02 |
| centimeter | meter (m) | 1.000 000* | E−02 |
| centimeter$^2$/second | meter$^2$/second (m$^2$/s) | 1.000 000* | E−04 |
| centimeter of mercury (0°C) | newton/meter$^2$ (N/m$^2$) | 1.333 22 | E+03 |
| centimeter of water (4°C) | newton/meter$^2$ (N/m$^2$) | 9.806 38 | E+01 |
| centipoise | newton-second/meter$^2$ (N·s/m$^2$) | 1.000 000* | E−03 |
| centistoke | meter$^2$/second (m$^2$/s) | 1.000 000* | E−06 |
| degree Celsius | kelvin (K) | $t_K = t_C + 273.15$ | |
| degree Fahrenheit | kelvin (K) | $t_K = (t_F + 459.67)/1.8$ | |
| degree Rankine | kelvin (K) | $t_K = t_R/1.8$ | |
| dyne | newton (N) | 1.000 000* | E−05 |
| dynes/centimeter$^2$ | newton/meter$^2$ (N/m$^2$) | 1.000 000* | E−01 |
| erg | joule (J) | 1.000 000* | E−07 |
| fluid ounce (U.S.) | meter$^3$ (m$^3$) | 2.957 353 | E−05 |
| foot | meter (m) | 3.048 000* | E−01 |
| foot (U.S. survey) | meter (m) | 3.048 006 | E−01 |
| foot of water (39.2°F) | newton/meter$^2$ (N/m$^2$) | 2.988 98 | E+03 |
| foot$^2$ | meter$^2$ (m$^2$) | 9.290 304* | E−02 |
| foot/second$^2$ | meter/second$^2$ (m/s$^2$) | 3.048 000* | E−01 |
| foot$^2$/hour | meter$^2$/second (m$^2$/s) | 2.580 640* | E−05 |
| foot-pound-force | joule (J) | 1.355 818 | E+00 |
| foot$^2$/second | meter$^2$/second (m$^2$/s) | 9.290 304* | E−02 |
| foot$^3$ | meter$^3$ (m$^3$) | 2.831 685 | E−02 |
| gallon (U.S. liquid) | meter$^3$ (m$^3$) | 3.785 412 | E−03 |
| gram | kilogram (kg) | 1.000 000* | E−03 |
| gram/centimeter$^3$ | kilogram/meter$^3$ (kg/m$^3$) | 1.000 000* | E+03 |
| horsepower (550 ft·lb$_f$/s) | watt (W) | 7.456 999 | E+02 |
| horsepower-hour | joule (J) | 2.6845 | E+06 |
| hour (mean solar) | second (s) | 3.600 000 | E+03 |
| inch | meter (m) | 2.540 000* | E−02 |
| inch of mercury (60°F) | newton/meter$^2$ (N/m$^2$) | 3.376 85 | E+03 |

*Conversion Table to SI Units—contd.*

| To Convert from | To | Multiply by | |
|---|---|---|---|
| inch of water (60°F) | newton/meter$^2$ (N/m$^2$) | 2.488 4 | E+02 |
| inch$^2$ | meter$^2$ (m$^2$) | 6.451 600* | E−04 |
| inch$^3$ | meter$^3$ (m$^3$) | 1.638 706 | E−05 |
| kilocalorie | joule (J) | 4.186 800* | E+03 |
| kilogram-force (kgf) | newton (N) | 9.806 650* | E+00 |
| kilowatt-hour | joules (J) | 3.600 000 | E+06 |
| liter | meter$^3$ (m$^3$) | 1.000 000* | E−03 |
| micron | meter (m) | 1.000 000* | E−06 |
| mil | meter (m) | 2.540 000* | E−05 |
| mile (U.S. statute) | meter (m) | 1.609 344* | E+03 |
| mile/hour | meter/second (m/s) | 4.470 400* | E−01 |
| millimeter of mercury (0°C) (torr) | newton/meter$^2$ (N/m$^2$) | 1.333 224 | E+02 |
| minute (angle) | radian (rad) | 2.908 882 | E−04 |
| minute (mean solar) | second (s) | 6.000 000* | E+01 |
| ohm (international of 1948) | ohm (Ω) | 1.000 495 | E+00 |
| ounce-mass (avoirdupois) | kilogram (kg) | 2.834 952 | E−02 |
| ounce (U.S. fluid) | meter$^3$ (m$^3$) | 2.957 353 | E−05 |
| pint (U.S. liquid) | meter$^3$ (m$^3$) | 4.731 765 | E−04 |
| poise (absolute viscosity) | newton-second/meter$^2$ (N · s/m$^2$) | 1.000 000* | E−01 |
| poundal | newton (N) | 1.382 550 | E−01 |
| pound-force (lbf avoirdupois) | newton (N) | 4.448 222 | E+00 |
| pound-force-second/foot$^2$ | newton-second/meter$^2$ (N · s/m$^2$) | 4.788 025 | E+01 |
| pound-force-second/inch$^2$ | newton-second/meter$^2$ (N · s/m$^2$) | 6.894 757 | E+03 |
| pound-mass (lbm avoirdupois) | kilogram (kg) | 4.535 924 | E−01 |
| pound-mass/foot$^3$ | kilogram/meter$^3$ (kg/m$^3$) | 1.601 846 | E+01 |
| pound-mass/foot-second | newton-second/meter$^2$ (N·s/m$^2$) | 1.488 164 | E+00 |
| pound-mass/foot$^2$-second (mass transfer coefficient) | kilogram/meter$^2$-second (kg/m$^2$·s) | 4.88243 | E+00 |
| psi (pounds per inch$^2$) | newton/meter$^2$ (N/m$^2$) | 6.894 757 | E+03 |
| quart (U.S. liquid) | meter$^3$ (m$^3$) | 9.463 529 | E−04 |
| second (angle) | radian (rad) | 4.848 137 | E−06 |
| slug | kilogram (kg) | 1.459 390 | E+01 |
| stoke (kinematic viscosity) | meter$^2$/second (m$^2$/s) | 1.000 000* | E−04 |
| ton (long, 2240 lbm) | kilogram (kg) | 1.016 047 | E+03 |
| ton (short, 2000 lbm) | kilogram (kg) | 9.071 847 | E+02 |
| torr (mm Hg, 0°C) | newton/meter$^2$ (N/m$^2$) | 1.333 22 | E+02 |
| volt (international of 1948) | volt (absolute) (V) | 1.000 330 | E+00 |
| watt (international of 1948) | watt (W) | 1.000 165 | E+00 |
| watt-hour | joule (J) | 3.600 000* | E+03 |
| watt/centimeter$^2$-°K | joule/meter$^2$-second-°K (J/m$^2$·s·K) | 1.000 000* | E+04 |

* An asterisk after the sixth decimal place indicates the conversion factor is exact and all subsequent digits are zero.

# Appendix C: Notation

Only symbols used frequently in the text are included here. Reference is made to the equation, section, or example where the symbol is first used.

$a$ — Temperature dependence coefficient of the viscosity parameter in the power law model (6.5-5)
$a_i$ — Velocity gradients in elongational flows (6.8-1)
$a_T$ — Time-temperature superposition shift factor (6.4-10)
$A$ — Taper of the screw channel (12.2-23)
$A$ — Interfacial area element (7.9)

$b$ — $= -a(T_1 - T_0)$: dimensionless dependence coefficient of the viscosity parameter in the power law model (10.2-43)
$b'$ — $= b/n$ (10.2-45)
Br — Brinkman number (9.8-26)

$c$ — Coefficient of cohesion (8.6-2)
$c_w$ — Coefficient of cohesion at the wall (8.6-8)
$C_i$ — Molar concentrations of species $i$.
$C_p$ — Specific heat at constant pressure (5.1)
$C_v$ — Specific heat at constant volume (5.1)
$C_s$ — Specific heat at constant pressure of a solid polymer (9.8)
$C_m$ — Specific heat at constant pressure of a polymer melt (9.8)

$D$ — Diameter
$D_s$ — Diameter of the root of a screw (12.2-1)
$D_b$ — Inside diameter of the barrel of a screw extruder (12.2-1)
$D_f$ — Diameter of a screw at the tip of the flight (10.11-6)
$\mathscr{D}_{AB}$, $\mathscr{D}$ — Mass diffusivities (5.1-23, 5.5-4)
De — Deborah number (2.1-5)

$e$ — Flight width of screws (10.3-6)
$e_v$ — Rate of viscous heat dissipation per unit volume (13.1-5a)
$E$ — Young's modulus in tension
$E$ — Number of elements of static mixers in series (11.7-1)
$E_v$ — Rate of conversion of mechanical energy into thermal energy (11.3-19)

$f$ — Kinematic coefficient of friction (4.3)
$f'$ — Static coefficient of friction (4.3-4)
$f_w, f'_w$ — $f$ and $f'$ at the container wall (8.6-8)
$f$ — Orientation function (3.9-3)
$ff$ — Flow factor (8.8-1)
$f(t)\,dt$ — Residence time distribution function (7.11)
$f(\gamma)\,d\gamma$ — Strain distribution function (7.10)
$f_L$ — Leakage flow correction for the pressure flow term in screw extruders (10.3-33)

| | |
|---|---|
| $\mathbf{F}$ | Force |
| $F_N$ | Normal force |
| $F_d$ | Shape factor for drag flow in screw channel (10.3-26) |
| $F_p$ | Shape factor for pressure flow in screw channel (10.3-27) |
| $F(t)$ | Cumulative residence time distribution function (7.11-2) |
| $F(\gamma)$ | Cumulative strain distribution function (7.10-11) |
| $F(n, \beta)$ | Shape factor for power law model fluids in annular flow (13.5-6) |
| $\mathcal{F}$ | Combined configuration-emissivity factor (9.2-2) |
| $\mathbf{g}$ | Gravitational acceleration |
| $g(t)\,dt$ | Internal residence time distribution function (7.11) |
| $g(\gamma)\,d\gamma$ | Internal strain distribution function (7.10) |
| $G$ | Dimensionless pressure gradient (10.2-19) |
| $G$ | Mass flow rate (12.2-1) |
| $G'(\omega)$ | In-phase dynamic modulus (Ex. 6.2) |
| $G''(\omega)$ | Out-of-phase dynamic modulus (Ex. 6.2) |
| $G(t)$ | Cumulative internal residence time distribution function (7.11-1) |
| $G(\gamma)$ | Cumulative internal strain distribution function (7.10-1) |
| $G(t-t')$ | Relaxation modulus in the Goddard–Miller equation; also relaxation modulus of a linear viscoelastic body (6.3-7) |
| $h$ | Half-separation between two rolls, two parallel plates, or two approaching disks (10.5) |
| $h$ | $= H/W$: channel aspect ratio in screws (10.3-23) |
| $h$ | Heat transfer coefficient (9.2-1) |
| $H$ | Enthalpy (3.6-1) |
| $H$ | Channel depth |
| $H$ | Separation between parallel plates (10.2-6) |
| $H_0$ | Half-minimum gap between rolls (10.5) |
| $H_1$ | Half-thickness of the sheet leaving the roll mill also half separation between rolls at location $X_1$ (10.5) |
| $H_2$ | Half-separation between rolls at location $X_2$ (10.5) |
| $H(\ln \lambda)$ | Relaxation spectrum function (6.4-7) |
| $I$ | Intensity of segregation (7.6-1) |
| $j_{A_i}$ | Mass flux components for species $A$ in the $i$-direction relative to the mass average velocity (5.1-23) |
| $J(t)$ | Time dependent creep compliance (6.4-4) |
| $J_n$ | $n$th order Bessel function (Table 9.1) |
| $k$ | Thermal conductivity |
| $k_s$, $k_m$ | Thermal conductivity of solid and molten polymers (9.8) |
| $k$ | Boltzmann constant |
| $k_f$, $k_r$ | Forward and reverse chemical reaction rate constants (14.2-1) |
| $K$ | Constant in the Stefan–Neumann equation (9.3-34) |
| $K$ | Ratio of compressive stresses in the horizontal and vertical directions in bins (8.7-2) |
| $K$ | Die (flow conductance) constant (12.1-4) |
| $l$ | Screw axial length (10.3-7) |
| $L$ | Characteristic length of flow channels |
| $L_s$ | Lead of a screw (10.3-5) |
| $L^*$ | Effective capillary length to account for end effects (13.1-1) |
| $m$ | Parameter of the power law model (consistency index) (6.5-3) |

| | |
|---|---|
| $m_0$ | $= m(T_0)$, where $T_0$ is a reference temperature (9.8-1) |
| $m'_1, m'_2$ | Parameters similar to $m$ in the power law model representation of $\Psi_1$ and $\Psi_2$ (6.7) |
| $M$ | Molecular weight |
| $M_0$ | Mer molecular weight |
| $\bar{M}_n$ | Number average molecular weight (2.1-1) |
| $\bar{M}_w$ | Weight average molecular weight (2.1-2) |
| $\bar{M}_z$ | $= M_0 \sum x^3 C_x / \sum x^2 C_x$ $z$ average molecular weight $C_x$ is the molar concentration of chains with $x$ mer units |
| $\bar{M}_{z+1}$ | $= M_0 \sum x^4 C_x / \sum x^3 C_x$ $z+1$ average molecular weight |
| $M$ | Dimensionless ratio of latent to sensible heats (9.8-25) |
| $M_1, M_2$ | Memory functions of the BKZ model (6.3-18) |
| $M(t-t')$ | Memory functions in linear viscoelasticity and in codeformational single integral models (6.3-14) |
| | |
| $\mathbf{n}$ | Unit outward normal vector |
| $n$ | Parameter in the power law model; also parameter in the Carreau model (6.5-3, 6.5-8) |
| $n'_1, n'_2$ | Parameters similar to $n$ in the power law model representation of $\Psi_1$ and $\Psi_2$ (6.7) |
| $N$ | Frequency of screw rotation (10.3-1) |
| $N_{ent}$ | Entrance capillary length correction factor (13.1-1a) |
| $N_{ex}$ | Exit capillary length correction factor (13.1-1a) |
| $N(\Gamma)$ | Bagley capillary entrance correction factor (13.1-1) |
| Nu | Nusselt number (13.1-10) |
| $N_i(x, y)$ | Shape function in the finite element method (FEM) (16.3-1) |
| | |
| $p$ | Volume fraction of the minor component (7.3-1) |
| $P$ | Pressure |
| $P_a$ | Atmospheric pressure |
| $P_w$ | Power (10.5-22) |
| $\mathscr{P}$ | $= P + \rho g z$ (12.1-2) |
| | |
| $\mathbf{q}$ | Heat flux vector (5.1-22) |
| $q$ | Volumetric flow rate per unit width (10.2-7) |
| $q_d$ | Drag volumetric flow rate per unit width (10.2-8) |
| $q_p$ | Pressure volumetric flow rate per unit width (10.2-9) |
| $Q$ | Volumetric flow rate (7.11-5) |
| $Q_d$ | Drag volumetric flow rate (10.3) |
| $Q_p$ | Pressure volumetric flow rate (10.3) |
| $Q_h$ | Total heat rate added to a mixer (11.3-20) |
| $Q_D$ | Volumetric flow rate through an extruder die (12.1-4) |
| $Q_s$ | Volumetric flow rate through an extruder (12.1-3) |
| | |
| $r$ | Radial coordinate in cylindrical and spherical coordinates |
| $r$ | Striation thickness (7.8-1) |
| $R$ | Radius |
| $R_i$ | Inside annular radius (13.5) |
| $R_0$ | Outside annular radius (13.5) |
| $R$ | Gas constant |
| $R_f$ | Blown film bubble radius at the freeze line (15.2) |
| $R_L, R_C$ | Longitudinal and circumferential radii of curvature (15.3-1, 15.3-2) |

| | |
|---|---|
| $R(r)$ | Coefficient of correlation (7.5-2) |
| $R(\tau)$ | Autocorrelation coefficient (7.13-11) |
| $s$ | $=1/n$: reciprocal of the power law model parameter (10.2) |
| sign | Sign function (10.2-21) |
| $\mathscr{s}$ | Scale of segregation (7.5-1) |
| $S^2$ | Variance of test samples (7.3-6) |
| $S$ | Entropy |
| $S_R$ | Recoverable shear strain (13.2-2) |
| $t$ | Time |
| $t_0$ | Minimum residence time (7.11-2) |
| $\bar{t}$ | Mean residence time (7.11-4) |
| $t_{1/2}$ | Half time in squeezing flow (10.9-13) |
| $t_f$ | Fraction of time a fluid particle spends in the upper portion of the screw channel (11.10-8) |
| $T$ | Temperature |
| $T_m$ | Melting point |
| $T_g$ | Glass transition temperature (2.1) |
| $T_c$ | Crystallization temperature |
| $T_s$ | Temperature of a polymer solid (9.3) |
| $T'$ | Excess temperature (9.5-3) |
| $T_w$ | Wall temperature |
| $\mathscr{T}$ | Torque (6.7-16) |
| $u_i$ | Dimensionless velocity components (10.2-18) |
| $U$ | Internal energy (5.1) |
| $U$ | Tangential velocity of calender rolls (10.5-3) |
| $U_1$ | Viscous dissipation dimensionless term in polymer melting (9.8-51) |
| $U_2$ | Dimensionless factor for temperature effects in drag flow (9.8-49) |
| $\boldsymbol{v}$ | Velocity vector |
| $v_i$ | Velocity components |
| $V$ | Volume |
| $V_0$ | Plate velocity in parallel plate flow (10.2) |
| $V_b$ | Velocity of the extruder barrel relative to the screw (10.3-8) |
| $\hat{V}$, $\hat{V}_c$, $\hat{V}_a$ | Specific volume of a polymer and its crystalline and amorphous portions (3.5-1) |
| $w_A$ | Rate of melting per unit area (9.3-39) |
| $w_L$ | Rate of melting per unit length (9.8-35) |
| $w_T$ | Total rate of melting (9.9-1) |
| $W$ | Width of flow channels |
| $x, y, z$ | Cartesian coordinates |
| $X$ | Solid bed width (12.2-13) |
| $X_1, X_2$ | Axial locations where the polymer detaches and contacts the rolls in calendering (10.5) |
| $y_i$ | Mole fraction (2.1-1) |
| $Z_s$ | Helical length of one turn at the root of the twin extruder screw (10.11-3) |
| $Z_T$ | Total helical length required for melting (12.2-29) |

### *Greek Letters*

| | |
|---|---|
| $\alpha$ | Thermal diffusivity (5.5) |
| $\alpha_s, \alpha_l$ | Thermal diffusivity of solid and molten polymers (9.3) |
| $\alpha$ | Dimensionless parameter in the Ellis model (6.5-6) |
| $\alpha_x, \alpha_y, \alpha_z$ | Angles of directional cosines (7.9-2) |
| $\alpha$ | Entrance angle formed in the "wine glass" capillary entrance flow region (13.2-5) |
| $\alpha_m$ | Orientation of the principal stress plane in particulate solids (8.4-3) |
| $\beta$ | $= R_o/R_i$: ratio of radii in concentric cylinders (11.3-5) |
| $\beta, \beta_w$ | Angle of internal friction inside particulate solids and at the container wall (8.6-1, 8.6-8) |
| $\beta$ | Parameter for the pressure dependence of viscosity (6.1-1) |
| $\gamma$ | Shear strain, total strain (6.4-1, 7.9) |
| $\bar{\gamma}$ | Mean strain (7.10-2) |
| $\dot{\boldsymbol{\gamma}}$ | Rate of strain tensor (5.1-26) |
| $\dot{\gamma}$ | Magnitude of $\dot{\boldsymbol{\gamma}}$ (6.5-1) |
| $\dot{\gamma}_w$ | Shear rate at the wall (6.7-11) |
| $\Gamma_w$ | $=4Q/\pi R^3$: Newtonian shear rate at the capillary wall (6.7-10) |
| $\Gamma$ | Surface tension (4.1) |
| $\delta$ | Effective angle of friction (8.6-6) |
| $\delta$ | Melt film thickness (9.8) |
| $\boldsymbol{\delta}$ | Unit tensor |
| $\boldsymbol{\delta}_i$ | Unit vectors (5.1-7) |
| $\delta_f$ | Radial flight clearance in screw extruders (10.3) |
| $\Delta E$ | Activation energy for flow (6.1-1) |
| $\Delta P$ | Pressure difference over a finite channel length or a flow region |
| $\varepsilon$ | Tensile strain |
| $\dot{\varepsilon}$ | Uniaxial elongational strain rate (6.8-5) |
| $\dot{\varepsilon}_{\text{pl}}$ | Planar elongational strain rate (6.8-7) |
| $\dot{\varepsilon}_{\text{bi}}$ | Biaxial elongational strain rate (6.8-9) |
| $\zeta$ | $= H/H_1$: dimensionless height between nonparallel plates (10.4-5) |
| $\eta$ | Non-Newtonian shear rate dependent viscosity (6.3-5) |
| $\eta_0$ | Zero shear rate viscosity |
| $\bar{\eta}$ | Elongational viscosity (6.8-17) |
| $\bar{\eta}^+$ | Elongational stress growth function (viscosity) (6.8-24) |
| $\eta$ | $= y/\delta$: dimensionless coordinate (9.8-16) |
| $\theta$ | Angle used in cylindrical and spherical coordinates |
| $\theta$ | Helix angle in screw extruders (10.3) |
| $\theta_b, \theta_s$ | Helix angle evaluated at the barrel and screw root (12.1-3) |
| $\Theta$ | Dimensionless temperature (9.3-14) |
| $\lambda$ | Relaxation time (2.1) |
| $\lambda$ | Parameter of the Carreau fluid model (6.5-8) |
| $\lambda$ | Heat of fusion (9.3-29) |
| $\lambda$ | Dimensionless axial location in roll-mill flow (10.5-12) |
| $\mu$ | Viscosity of a Newtonian fluid (5.1-21) |
| $\mu_f$ | Viscosity in the region of the screw flight clearance (10.3-33) |

| | |
|---|---|
| $\zeta$ | Dimensionless $x$, $y$, or $z$ coordinate |
| $\boldsymbol{\pi}$ | Total stress tensor $(P\boldsymbol{\delta}+\boldsymbol{\tau})$ (5.1-16) |
| $\boldsymbol{\pi}'$ | $= -\boldsymbol{\pi}^\dagger$ |
| $\rho$ | Density |
| $\rho_m$, $\rho_s$ | Density of molten and solid polymers (9.8) |
| $\rho$ | $= x\sqrt{2RH_0}$: transformation variable (10.5-10) |
| $\rho$ | Dimensionless radial coordinate (11.3-5) |
| $\boldsymbol{\rho}_i$ | Position vectors |
| $\sigma$, $\sigma_w$ | Normal stress (pressure), normal stress at the wall (8.2-1, 8.6-8) |
| $\sigma$ | Stefan–Boltzmann radiation constant (9.2-2) |
| $\sigma^2$ | Variance |
| $\sigma_{min}$, $\sigma_{max}$ | Principal stresses in particulate solids (8.4-4, 8.4-5) |
| $\tau$ | Shear stress (8.2-1) |
| $\boldsymbol{\tau}$ | Deviatoric stress tensor (5.1-16) |
| $\tau$ | Dimensionless time $(\alpha t/b^2)$ (9.4-6) |
| $\tau_w$, $\tau_w^*$ | Shear stress at the wall, corrected shear stress at the wall (6.7-3, 13.1-1) |
| $\tau_m$, $\tau_{max}$ | Maximum shear stress along streamlines in roll-mills, overall maximum $\tau_m$ (11.8-4) |
| $\phi$ | Spherical coordinate |
| $\phi$ | Solids conveying angle (8.13-1) |
| $\phi$ | Fraction of gap occupied by the minor component (11.4-2) |
| $\Phi$ | Dimensional group in melting with drag removal (12.2-20) |
| $\chi$ | Dimensionless $x$ coordinate (10.3-23) |
| $\psi$ | Dimensionless rate of melting in screw extruders (12.2-22) |
| $\Psi_1$, $\Psi_2$ | Primary (first) and secondary (second) normal stress coefficients (6.3-5) |
| $\boldsymbol{\omega}$ | Vorticity tensor (5.1-27) |
| $\Omega$ | Angular velocity |

# Author Index

# Subject Index